T0177197

Quantum Computation and Quantum Information

10th Anniversary Edition

One of the most cited books in physics of all time, *Quantum Computation and Quantum Information* remains the best textbook in this exciting field of science. This 10th Anniversary Edition includes a new Introduction and Afterword from the authors setting the work in context.

This comprehensive textbook describes such remarkable effects as fast quantum algorithms, quantum teleportation, quantum cryptography, and quantum error-correction. Quantum mechanics and computer science are introduced, before moving on to describe what a quantum computer is, how it can be used to solve problems faster than "classical" computers, and its real-world implementation. It concludes with an in-depth treatment of quantum information.

Containing a wealth of figures and exercises, this well-known textbook is ideal for courses on the subject, and will interest beginning graduate students and researchers in physics, computer science, mathematics, and electrical engineering.

MICHAEL NIELSEN was educated at the University of Queensland, and as a Fulbright Scholar at the University of New Mexico. He worked at Los Alamos National Laboratory, as the Richard Chace Tolman Fellow at Caltech, was Foundation Professor of Quantum Information Science and a Federation Fellow at the University of Queensland, and a Senior Faculty Member at the Perimeter Institute for Theoretical Physics. He left Perimeter Institute to write a book about open science and now lives in Toronto.

ISAAC CHUANG is a Professor at the Massachusetts Institute of Technology, jointly appointed in Electrical Engineering & Computer Science, and in Physics. He leads the quanta research group at the Center for Ultracold Atoms, in the MIT Research Laboratory of Electronics, which seeks to understand and create information technology and intelligence from the fundamental building blocks of physical systems, atoms, and molecules.

In praise of the book 10 years after publication

Ten years after its initial publication, "Mike and Ike" (as it's affectionately called) remains the quantum computing textbook to which all others are compared. No other book in the field matches its scope: from experimental implementation to complexity classes, from the philosophical justifications for the Church-Turing Thesis to the nitty-gritty of bra/ket manipulation. A dog-eared copy sits on my desk; the section on trace distance and fidelity alone has been worth many times the price of the book to me.

Scott Aaronson, Massachusetts Institute of Technology

Quantum information processing has become a huge interdisciplinary field at the intersection of both, theoretical and experimental quantum physics, computer science, mathematics, quantum engineering and, more recently, even quantum metrology. The book by Michael Nielsen and Isaac Chuang was seminal in many ways: it paved the way for a broader, yet deep understanding of the underlying science, it introduced a common language now widely used by a growing community and it became the standard book in the field for a whole decade. In spite of the fast progress in the field, even after 10 years the book provides the basic introduction into the field for students and scholars alike and the 10th anniversary edition will remain a bestseller for a long time to come. The foundations of quantum computation and quantum information processing are excellently laid out in this book and it also provides an overview over some experimental techniques that have become the testing ground for quantum information processing during the last decade. In view of the rapid progress of the field the book will continue to be extremely valuable for all entering this highly interdisciplinary research area and it will always provide the reference for those who grew up with it. This is an excellent book, well written, highly commendable, and in fact imperative for everybody in the field.

Rainer Blatt, Universität Innsbruck

My well-perused copy of Nielsen and Chuang is, as always, close at hand as I write this. It appears that the material that Mike and Ike chose to cover, which was a lot, has turned out to be a large portion of what will become the eternal verities of this still-young field. When another researcher asks me to give her a clear explanation of some important point of quantum information science, I breathe a sigh of relief when I recall that it is in this book – my job is easy, I just send her there.

David DiVincenzo, IBM T. J. Watson Research Center

If there is anything you want to know, or remind yourself, about quantum information science, then look no further than this comprehensive compendium by Ike and Mike. Whether you are an expert, a student or a casual reader, tap into this treasure chest of useful and well presented information.

Artur Ekert, Mathematical Institute, University of Oxford

Nearly every child who has read Harry Potter believes that if you just say the right thing or do the right thing, you can coerce matter to do something fantastic. But what adult would believe it? Until quantum computation and quantum information came along in the early 1990s, nearly none. The quantum computer is the Philosopher's Stone of our century, and Nielsen and Chuang is our basic book of incantations. Ten years have passed since its publication, and it is as basic to the field as it ever was. Matter will do wonderful things if asked to, but we must first understand its language. No book written since (there was no before) does the job of teaching the language of quantum theory's possibilities like Nielsen and Chuang's.

Chris Fuchs, Perimeter Institute for Theoretical Physics

Nielsen and Chuang is the bible of the quantum information field. It appeared 10 years ago, yet even though the field has changed enormously in these 10 years - the book still covers most of the important concepts of the field.

Lov Grover, Bell Labs

Quantum Computation and Quantum Information, commonly referred to as "Mike and Ike," continues to be a most valuable resource for background information on quantum information processing. As a mathematically-impaired experimentalist, I particularly appreciate the fact that armed with a modest background in quantum mechanics, it is possible to pick up at any point in the book and readily grasp the basic ideas being discussed. To me, it is still "the" book on the subject.

David Wineland, National Institute of Standards and Technology, Boulder, Colorado

Endorsements for the original publication

Chuang and Nielsen have produced the first comprehensive study of quantum computation. To develop a robust understanding of this subject one must integrate many ideas whose origins are variously within physics, computer science, or mathematics. Until this text, putting together the essential material, much less mastering it, has been a challenge. Our Universe has intrinsic capabilities and limitations on the processing of information. What these are will ultimately determine the course of technology and shape our efforts to find a fundamental physical theory. This book is an excellent way for any scientist or graduate student – in any of the related fields – to enter the discussion.

Michael Freedman, Fields Medalist, Microsoft

Nielsen and Chuang's new text is remarkably thorough and up-to-date, covering many aspects of this rapidly evolving field from a physics perspective, complementing the computer science perspective of Gruska's 1999 text. The authors have succeeded in producing a self-contained book accessible to anyone with a good undergraduate grounding in math, computer science or physical sciences. An independent student could spend an enjoyable year reading this book and emerge ready to tackle the current literature and do serious research. To streamline the exposition, footnotes have been gathered into short but lively History and Further Reading sections at the end of each chapter.

Charles H Bennett, IBM

This is an excellent book. The field is already too big to cover completely in one book, but Nielsen and Chuang have made a good selection of topics, and explain the topics they have chosen very well.

Peter Shor, Massachusetts Institute of Technology

Quantum Computation and Quantum Information

10th Anniversary Edition

Michael A. Nielsen & Isaac L. Chuang

CAMBRIDGE
UNIVERSITY PRESS

Shaftesbury Road, Cambridge CB2 8EA, United Kingdom

One Liberty Plaza, 20th Floor, New York, NY 10006, USA

477 Williamstown Road, Port Melbourne, VIC 3207, Australia

314–321, 3rd Floor, Plot 3, Splendor Forum, Jasola District Centre, New Delhi – 110025, India

103 Penang Road, #05–06/07, Visioncrest Commercial, Singapore 238467

Cambridge University Press is part of Cambridge University Press & Assessment, a department of the University of Cambridge.

We share the University's mission to contribute to society through the pursuit of education, learning and research at the highest international levels of excellence.

www.cambridge.org
Information on this title: www.cambridge.org/9781107002173

First published 2000
10th Anniversary edition published 2010 (version 13, October 2023)

Printed in the United Kingdom by TJ Books Limited, Padstow Cornwall

A catalogue record for this publication is available from the British Library

ISBN 978-1-107-00217-3 Hardback

To our parents,
and our teachers

Contents

Introduction to the Tenth Anniversary Edition

Quantum mechanics has the curious distinction of being simultaneously the most successful and the most mysterious of our scientific theories. It was developed in fits and starts over a remarkable period from 1900 to the 1920s, maturing into its current form in the late 1920s. In the decades following the 1920s, physicists had great success applying quantum mechanics to understand the fundamental particles and forces of nature, culminating in the development of the standard model of particle physics. Over the same period, physicists had equally great success in applying quantum mechanics to understand an astonishing range of phenomena in our world, from polymers to semiconductors, from superfluids to superconductors. But, while these developments profoundly advanced our understanding of the natural world, they did only a little to improve our understanding of quantum mechanics.

This began to change in the 1970s and 1980s, when a few pioneers were inspired to ask whether some of the fundamental questions of computer science and information theory could be applied to the study of quantum systems. Instead of looking at quantum systems purely as phenomena to be explained as they are found in nature, they looked at them as systems that can be *designed*. This seems a small change in perspective, but the implications are profound. No longer is the quantum world taken merely as presented, but instead it can be created. The result was a new perspective that inspired both a resurgence of interest in the fundamentals of quantum mechanics, and also many new questions combining physics, computer science, and information theory. These include questions such as: what are the fundamental physical limitations on the space and time required to construct a quantum state? How much time and space are required for a given dynamical operation? What makes quantum systems difficult to understand and simulate by conventional classical means?

Writing this book in the late 1990s, we were fortunate to be writing at a time when these and other fundamental questions had just crystallized out. Ten years later it is clear such questions offer a sustained force encouraging a broad research program at the foundations of physics and computer science. Quantum information science is here to stay. Although the theoretical foundations of the field remain similar to what we discussed 10 years ago, detailed knowledge in many areas has greatly progressed. Originally, this book served as a comprehensive overview of the field, bringing readers near to the forefront of research. Today, the book provides a basic foundation for understanding the field, appropriate either for someone who desires a broad perspective on quantum information science, or an entryway for further investigation of the latest research literature. Of course, many fundamental challenges remain, and meeting those challenges promises to stimulate exciting and unexpected links among many disparate parts of physics, computer science, and information theory. We look forward to the decades ahead!

– Michael A. Nielsen and Isaac L. Chuang, March, 2010.

Afterword to the Tenth Anniversary Edition

An enormous amount has happened in quantum information science in the 10 years since the first edition of this book, and in this afterword we cannot summarize even a tiny fraction of that work. But a few especially striking developments merit comment, and may perhaps whet your appetite for more.

Perhaps the most impressive progress has been in the area of experimental implementation. While we are still many years from building large-scale quantum computers, much progress has been made. Superconducting circuits have been used to implement simple two-qubit quantum algorithms, and three-qubit systems are nearly within reach. Qubits based on nuclear spins and single photons have been used, respectively, to demonstrate proof-of-principle for simple forms of quantum error correction and quantum simulation. But the most impressive progress of all has been made with trapped ion systems, which have been used to implement many two- and three-qubit algorithms and algorithmic building blocks, including the quantum search algorithm and the quantum Fourier transform. Trapped ions have also been used to demonstrate basic quantum communication primitives, including quantum error correction and quantum teleportation.

A second area of progress has been in understanding what physical resources are required to quantum compute. Perhaps the most intriguing breakthrough here has been the discovery that quantum computation can be done via measurement alone. For many years, the conventional wisdom was that coherent superposition-preserving unitary dynamics was an essential part of the power of quantum computers. This conventional wisdom was blown away by the realization that quantum computation can be done without any unitary dynamics at all. Instead, in some new models of quantum computation, quantum measurements alone can be used to do arbitrary quantum computations. The only coherent resource in these models is quantum memory, i.e., the ability to store quantum information. An especially interesting example of these models is the one-way quantum computer, or cluster-state computer. To quantum compute in the cluster-state model requires only that the experimenter have possession of a fixed universal state known as the cluster state. With a cluster state in hand, quantum computation can be implemented simply by doing a sequence of single-qubit measurements, with the particular computation done being determined by which qubits are measured, when they are measured, and how they are measured. This is remarkable: you're given a fixed quantum state, and then quantum compute by "looking" at the individual qubits in appropriate ways.

A third area of progress has been in *classically* simulating quantum systems. Feynman's pioneering 1982 paper on quantum computing was motivated in part by the observation that quantum systems often seem hard to simulate on conventional classical computers. Of course, at the time there was only a limited understanding of how difficult it is to simulate different quantum systems on ordinary classical computers. But in the 1990s and, especially, in the 2000s, we have learned much about which quantum systems are easy to simulate, and which are hard. Ingenious algorithms have been developed to classically simulate many quantum systems that were formerly thought to be hard to simulate, in particular, many quantum systems in one spatial dimension, and certain two-dimensional quantum systems. These classical algorithms have been made possible by the development of insightful classical descriptions that capture in a compact way much or all of the essential physics of the system in question. At the same time, we have learned that some systems that formerly seemed simple are surprisingly complex. For example, it has long been known that quantum systems based on a certain type of optical component – what are called linear optical systems – are easily simulated classically. So it was surprising when it was discovered that adding two seemingly innocuous components – single-photon sources and photodetectors – gave linear optics the full power of quantum computation. These and similar investigations have deepened our understanding of which quantum systems are easy to simulate, which quantum systems are hard to simulate, and why.

A fourth area of progress has been a greatly deepened understanding of quantum communication channels. A beautiful and complete theory has been developed of how entangled quantum states can assist classical communication over quantum channels. A plethora of different quantum protocols for communication have been organized into a comprehensive family (headed by "mother" and "father" protocols), unifying much of our understanding of the different types of communication possible with quantum information. A sign of the progress is the disproof of one of the key unsolved conjectures reported in this book (p. 554), namely, that the communication capacity of a quantum channel with product states is equal to the unconstrained capacity (i.e., the capacity with any entangled state allowed as input). But, despite the progress, much remains beyond our understanding. Only very recently, for example, it was discovered, to considerable surprise, that two quantum channels, each with zero quantum capacity, can have a positive quantum capacity when used together; the analogous result, with classical capacities over classical channels, is known to be impossible.

One of the main motivations for work in quantum information science is the prospect of fast quantum algorithms to solve important computational problems. Here, the progress over the past decade has been mixed. Despite great ingenuity and effort, the chief algorithmic insights stand as they were 10 years ago. There has been considerable technical progress, but we do not yet understand what exactly it is that makes quantum computers powerful, or on what class of problems they can be expected to outperform classical computers.

What is exciting, though, is that ideas from quantum computation have been used to prove a variety of theorems about classical computation. These have included, for example, results about the difficulty of finding certain hidden vectors in a discrete lattice of points. The striking feature is that these proofs, utilizing ideas of quantum computation, are sometimes considerably simpler and more elegant than prior, classical proofs. Thus, an awareness has grown that quantum computation may be a more natural model of computation than the classical model, and perhaps fundamental results may be more easily revealed through the ideas of quantum computation.

Preface

This book provides an introduction to the main ideas and techniques of the field of quantum computation and quantum information. The rapid rate of progress in this field and its cross-disciplinary nature have made it difficult for newcomers to obtain a broad overview of the most important techniques and results of the field.

Our purpose in this book is therefore twofold. First, we introduce the background material in computer science, mathematics and physics necessary to understand quantum computation and quantum information. This is done at a level comprehensible to readers with a background at least the equal of a beginning graduate student in one or more of these three disciplines; the most important requirements are a certain level of mathematical maturity, and the desire to learn about quantum computation and quantum information. The second purpose of the book is to develop in detail the central results of quantum computation and quantum information. With thorough study the reader should develop a working understanding of the fundamental tools and results of this exciting field, either as part of their general education, or as a prelude to independent research in quantum computation and quantum information.

Structure of the book

The basic structure of the book is depicted in Figure 1. The book is divided into three parts. The general strategy is to proceed from the concrete to the more abstract whenever possible. Thus we study quantum computation before quantum information; specific quantum error-correcting codes before the more general results of quantum information theory; and throughout the book try to introduce examples before developing general theory.

Part I provides a broad overview of the main ideas and results of the field of quantum computation and quantum information, and develops the background material in computer science, mathematics and physics necessary to understand quantum computation and quantum information in depth. Chapter 1 is an introductory chapter which outlines the historical development and fundamental concepts of the field, highlighting some important open problems along the way. The material has been structured so as to be accessible even without a background in computer science or physics. The background material needed for a more detailed understanding is developed in Chapters 2 and 3, which treat in depth the fundamental notions of quantum mechanics and computer science, respectively. You may elect to concentrate more or less heavily on different chapters of Part I, depending upon your background, returning later as necessary to fill any gaps in your knowledge of the fundamentals of quantum mechanics and computer science.

Part II describes quantum computation in detail. Chapter 4 describes the fundamen-

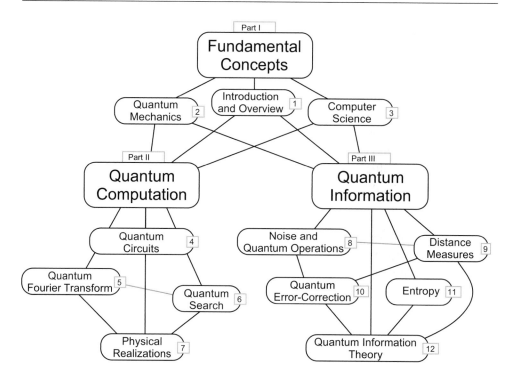

Figure 1. Structure of the book.

tal elements needed to perform quantum computation, and presents many elementary operations which may be used to develop more sophisticated applications of quantum computation. Chapters 5 and 6 describe the quantum Fourier transform and the quantum search algorithm, the two fundamental quantum algorithms presently known. Chapter 5 also explains how the quantum Fourier transform may be used to solve the factoring and discrete logarithm problems, and the importance of these results to cryptography. Chapter 7 describes general design principles and criteria for good physical implementations of quantum computers, using as examples several realizations which have been successfully demonstrated in the laboratory.

Part III is about quantum information: what it is, how information is represented and communicated using quantum states, and how to describe and deal with the corruption of quantum and classical information. Chapter 8 describes the properties of *quantum noise* which are needed to understand real-world quantum information processing, and the *quantum operations formalism*, a powerful mathematical tool for understanding quantum noise. Chapter 9 describes *distance measures* for quantum information which allow us to make quantitatively precise what it means to say that two items of quantum information are similar. Chapter 10 explains quantum error-correcting codes, which may be used to protect quantum computations against the effect of noise. An important result in this chapter is the *threshold theorem*, which shows that for realistic noise models, noise is *in principle* not a serious impediment to quantum computation. Chapter 11 introduces the fundamental information-theoretic concept of *entropy*, explaining many properties of entropy in both classical and quantum information theory. Finally, Chapter 12 discusses the information carrying properties of quantum states and quantum communication channels, detailing many of the strange and interesting properties such systems can have for the transmission of information both classical and quantum, and for the transmission of secret information.

A large number of exercises and problems appear throughout the book. Exercises are intended to solidify understanding of basic material and appear within the main body of the text. With few exceptions these should be easily solved with a few minutes work. Problems appear at the end of each chapter, and are intended to introduce you to new and interesting material for which there was not enough space in the main text. Often the problems are in multiple parts, intended to develop a particular line of thought in some depth. A few of the problems were unsolved as the book went to press. When this is the case it is noted in the statement of the problem. Each chapter concludes with a summary of the main results of the chapter, and with a 'History and further reading' section that charts the development of the main ideas in the chapter, giving citations and references for the whole chapter, as well as providing recommendations for further reading.

The front matter of the book contains a detailed Table of Contents, which we encourage you to browse. There is also a guide to nomenclature and notation to assist you as you read.

The end matter of the book contains six appendices, a bibliography, and an index.

Appendix 1 reviews some basic definitions, notations, and results in elementary probability theory. This material is assumed to be familiar to readers, and is included for ease of reference. Similarly, Apendix 2 reviews some elementary concepts from group theory, and is included mainly for convenience. Appendix 3 contains a proof of the Solovay–Kitaev theorem, an important result for quantum computation, which shows that a finite set of quantum gates can be used to quickly approximate an arbitrary quantum gate. Appendix 4 reviews the elementary material on number theory needed to understand the quantum algorithms for factoring and discrete logarithm, and the RSA cryptosystem, which is itself reviewed in Appendix 5. Appendix 6 contains a proof of Lieb's theorem, one of the most important results in quantum computation and quantum information, and a precursor to important entropy inequalities such as the celebrated strong subadditivity inequality. The proofs of the Solovay–Kitaev theorem and Lieb's theorem are lengthy enough that we felt they justified a treatment apart from the main text.

The bibliography contains a listing of all reference materials cited in the text of the book. Our apologies to any researcher whose work we have inadvertently omitted from citation.

The field of quantum computation and quantum information has grown so rapidly in recent years that we have not been able to cover all topics in as much depth as we would have liked. Three topics deserve special mention. The first is the subject of *entanglement measures*. As we explain in the book, entanglement is a key element in effects such as quantum teleportation, fast quantum algorithms, and quantum error-correction. It is, in short, a resource of great utility in quantum computation and quantum information. There is a thriving research community currently fleshing out the notion of entanglement as a new type of physical resource, finding principles which govern its manipulation and utilization. We felt that these investigations, while enormously promising, are not yet complete enough to warrant the more extensive coverage we have given to other subjects in this book, and we restrict ourselves to a brief taste in Chapter 12. Similarly, the subject of distributed quantum computation (sometimes known as quantum communication complexity) is an enormously promising subject under such active development that we have not given it a treatment for fear of being obsolete before publication of the book. The implementation of quantum information processing machines has also developed into a fascinating and rich area, and we limit ourselves to but a single chapter on this subject. Clearly, much more can be said about physical implementations, but this would begin to involve many more areas of physics, chemistry, and engineering, which we do not have room for here.

How to use this book

This book may be used in a wide variety of ways. It can be used as the basis for a variety of courses, from short lecture courses on a specific topic in quantum computation and quantum information, through to full-year classes covering the entire field. It can be used for independent study by people who would like to learn just a little about quantum computation and quantum information, or by people who would like to be brought up to the research frontier. It is also intended to act as a reference work for current researchers in the field. We hope that it will be found especially valuable as an introduction for researchers new to the field.

Note to the independent reader

The book is designed to be accessible to the independent reader. A large number of exercises are peppered throughout the text, which can be used as self-tests for understanding of the material in the main text. The Table of Contents and end of chapter summaries should enable you to quickly determine which chapters you wish to study in most depth. The dependency diagram, Figure 1, will help you determine in what order material in the book may be covered.

Note to the teacher

This book covers a diverse range of topics, and can therefore be used as the basis for a wide variety of courses.

A one-semester course on quantum computation could be based upon a selection of material from Chapters 1 through 3, depending on the background of the class, followed by Chapter 4 on quantum circuits, Chapters 5 and 6 on quantum algorithms, and a selection from Chapter 7 on physical implementations, and Chapters 8 through 10 to understand quantum error-correction, with an especial focus on Chapter 10.

A one-semester course on quantum information could be based upon a selection of material from Chapters 1 through 3, depending on the background of the class. Following that, Chapters 8 through 10 on quantum error-correction, followed by Chapters 11 and 12 on quantum entropy and quantum information theory, respectively.

A full year class could cover all material in the book, with time for additional readings selected from the 'History and further reading' section of several chapters. Quantum computation and quantum information also lend themselves ideally to independent research projects for students.

Aside from classes on quantum computation and quantum information, there is another way we hope the book will be used, which is as the text for an introductory class in quantum mechanics for physics students. Conventional introductions to quantum mechanics rely heavily on the mathematical machinery of partial differential equations. We believe this often obscures the fundamental ideas. Quantum computation and quantum information offers an excellent conceptual laboratory for understanding the basic concepts and unique aspects of quantum mechanics, without the use of heavy mathematical machinery. Such a class would focus on the introduction to quantum mechanics in Chapter 2, basic material on quantum circuits in Chapter 4, a selection of material on quantum algorithms from Chapters 5 and 6, Chapter 7 on physical implementations of quantum computation, and then almost any selection of material from Part III of the book, depending upon taste.

Note to the student

We have written the book to be as self-contained as possible. The main exception is that occasionally we have omitted arguments that one really needs to work through oneself to believe; these are usually given as exercises. Let us suggest that you should at least attempt all the exercises as you work through the book. With few exceptions the exercises can be worked out in a few minutes. If you are having a lot of difficulty with many of the exercises it may be a sign that you need to go back and pick up one or more key concepts.

Further reading

As already noted, each chapter concludes with a 'History and further reading' section. There are also a few broad-ranging references that might be of interest to readers. Preskill's[Pre98b] superb lecture notes approach quantum computation and quantum information from a somewhat different point of view than this book. Good overview articles on specific subjects include (in order of their appearance in this book): Aharonov's review of quantum computation[Aha99b], Kitaev's review of algorithms and error-correction[Kit97b], Mosca's thesis on quantum algorithms[Mos99], Fuchs' thesis[Fuc96] on distinguishability and distance measures in quantum information, Gottesman's thesis on quantum error-correction[Got97], Preskill's review of quantum error-correction[Pre97], Nielsen's thesis on quantum information theory[Nie98], and the reviews of quantum information theory by Bennett and Shor[BS98] and by Bennett and DiVincenzo[BD00]. Other useful references include Gruska's book[Gru99], and the collection of review articles edited by Lo, Spiller, and Popescu[LSP98].

Errors

Any lengthy document contains errors and omissions, and this book is surely no exception to the rule. If you find any errors or have other comments to make about the book, please email them to: qci@squint.org. As errata are found, we will add them to a list maintained at the book web site: http://www.squint.org/qci/.

Acknowledgements

A few people have decisively influenced how we think about quantum computation and quantum information. For many enjoyable discussions which have helped us shape and refine our views, MAN thanks Carl Caves, Chris Fuchs, Gerard Milburn, John Preskill and Ben Schumacher, and ILC thanks Tom Cover, Umesh Vazirani, Yoshi Yamamoto, and Bernie Yurke.

An enormous number of people have helped in the construction of this book, both directly and indirectly. A partial list includes Dorit Aharonov, Andris Ambainis, Nabil Amer, Howard Barnum, Dave Beckman, Harry Buhrman, the Caltech Quantum Optics Foosballers, Andrew Childs, Fred Chong, Richard Cleve, John Conway, John Cortese, Michael DeShazo, Ronald de Wolf, David DiVincenzo, Steven van Enk, Henry Everitt, Ron Fagin, Mike Freedman, Michael Gagen, Neil Gershenfeld, Daniel Gottesman, Jim Harris, Alexander Holevo, Andrew Huibers, Julia Kempe, Alesha Kitaev, Manny Knill, Shing Kong, Raymond Laflamme, Andrew Landahl, Ron Legere, Debbie Leung, Daniel Lidar, Elliott Lieb, Theresa Lynn, Hideo Mabuchi, Yu Manin, Mike Mosca, Alex Pines, Sridhar Rajagopalan, Bill Risk, Beth Ruskai, Sara Schneider, Robert Schrader, Peter Shor, Sheri Stoll, Volker Strassen, Armin Uhlmann, Lieven Vandersypen, Anne Verhulst, Debby Wallach, Mike Westmoreland, Dave Wineland, Howard Wiseman, John Yard, Xinlan Zhou, and Wojtek Zurek.

Thanks to the folks at Cambridge University Press for their help turning this book from an idea into reality. Our especial thanks go to our thoughtful and enthusiastic editor Simon Capelin, who shepherded this project along for more than three years, and to Margaret Patterson, for her timely and thorough copy-editing of the manuscript.

Parts of this book were completed while MAN was a Tolman Prize Fellow at the California Institute of Technology, a member of the T-6 Theoretical Astrophysics Group at the Los Alamos National Laboratory, and a member of the University of New Mexico Center for Advanced Studies, and while ILC was a Research Staff Member at the IBM Almaden Research Center, a consulting Assistant Professor of Electrical Engineering at Stanford University, a visiting researcher at the University of California Berkeley Department of Computer Science, a member of the Los Alamos National Laboratory T-6 Theoretical Astrophysics Group, and a visiting researcher at the University of California Santa Barbara Institute for Theoretical Physics. We also appreciate the warmth and hospitality of the Aspen Center for Physics, where the final page proofs of this book were finished.

MAN and ILC gratefully acknowledge support from DARPA under the NMRQC research initiative and the QUIC Institute administered by the Army Research Office. We also thank the National Science Foundation, the National Security Agency, the Office of Naval Research, and IBM for their generous support.

Nomenclature and notation

There are several items of nomenclature and notation which have two or more meanings in common use in the field of quantum computation and quantum information. To prevent confusion from arising, this section collects many of the more frequently used of these items, together with the conventions that will be adhered to in this book.

Linear algebra and quantum mechanics
All vector spaces are assumed to be finite dimensional, unless otherwise noted. In many instances this restriction is unnecessary, or can be removed with some additional technical work, but making the restriction globally makes the presentation more easily comprehensible, and doesn't detract much from many of the intended applications of the results.

A *positive* operator A is one for which $\langle \psi | A | \psi \rangle \geq 0$ for all $|\psi\rangle$. A *positive definite* operator A is one for which $\langle \psi | A | \psi \rangle > 0$ for all $|\psi\rangle \neq 0$. The *support* of an operator is defined to be the vector space orthogonal to its kernel. For a Hermitian operator, this means the vector space spanned by eigenvectors of the operator with non-zero eigenvalues.

The notation U (and often but not always V) will generically be used to denote a unitary operator or matrix. H is usually used to denote a quantum logic gate, the *Hadamard gate*, and sometimes to denote the *Hamiltonian* for a quantum system, with the meaning clear from context.

Vectors will sometimes be written in column format, as for example,

$$\begin{bmatrix} 1 \\ 2 \end{bmatrix}, \tag{0.1}$$

and sometimes for readability in the format $(1, 2)$. The latter should be understood as shorthand for a column vector. For two-level quantum systems used as qubits, we shall usually identify the state $|0\rangle$ with the vector $(1, 0)$, and similarly $|1\rangle$ with $(0, 1)$. We also define the Pauli sigma matrices in the conventional way – see 'Frequently used quantum gates and circuit symbols', below. Most significantly, the convention for the Pauli sigma z matrix is that $\sigma_z |0\rangle = |0\rangle$ and $\sigma_z |1\rangle = -|1\rangle$, which is reverse of what some physicists (but usually not computer scientists or mathematicians) intuitively expect. The origin of this dissonance is that the $+1$ eigenstate of σ_z is often identified by physicists with a so-called 'excited state', and it seems natural to many to identify this with $|1\rangle$, rather than with $|0\rangle$ as is done in this book. Our choice is made in order to be consistent with the usual indexing of matrix elements in linear algebra, which makes it natural to identify the first column of σ_z with the action of σ_z on $|0\rangle$, and the second column with the action on $|1\rangle$. This choice is also in use throughout the quantum computation and quantum information community. In addition to the conventional notations σ_x, σ_y and σ_z for the Pauli sigma matrices, it will also be convenient to use the notations $\sigma_1, \sigma_2, \sigma_3$ for these three matrices, and to define σ_0 as the $2{\times}2$ identity matrix. Most often, however, we use the notations I, X, Y and Z for $\sigma_0, \sigma_1, \sigma_2$ and σ_3, respectively.

Information theory and probability

As befits good information theorists, logarithms are *always* taken to base two, unless otherwise noted. We use $\log(x)$ to denote logarithms to base 2, and $\ln(x)$ on those rare occasions when we wish to take a natural logarithm. The term *probability distribution* is used to refer to a finite set of real numbers, p_x, such that $p_x \geq 0$ and $\sum_x p_x = 1$. The *relative entropy* of a positive operator A with respect to a positive operator B is defined by $S(A\|B) \equiv \mathrm{tr}(A \log A) - \mathrm{tr}(A \log B)$.

Miscellanea

$\oplus$ denotes modulo two addition. Throughout this book 'z' is pronounced 'zed'.

Frequently used quantum gates and circuit symbols

Certain schematic symbols are often used to denote unitary transforms which are useful in the design of quantum circuits. For the reader's convenience, many of these are gathered together below. The rows and columns of the unitary transforms are labeled from left to right and top to bottom as $00\ldots0, 00\ldots1$ to $11\ldots1$ with the bottom-most wire being the least significant bit. Note that $e^{i\pi/4}$ is the square root of i, so that the $\pi/8$ gate is the square root of the phase gate, which itself is the square root of the Pauli-Z gate.

$$
\text{Hadamard} \quad -\boxed{H}- \quad \frac{1}{\sqrt{2}}\begin{bmatrix} 1 & 1 \\ 1 & -1 \end{bmatrix}
$$

$$
\text{Pauli-}X \quad -\boxed{X}- \quad \begin{bmatrix} 0 & 1 \\ 1 & 0 \end{bmatrix}
$$

$$
\text{Pauli-}Y \quad -\boxed{Y}- \quad \begin{bmatrix} 0 & -i \\ i & 0 \end{bmatrix}
$$

$$
\text{Pauli-}Z \quad -\boxed{Z}- \quad \begin{bmatrix} 1 & 0 \\ 0 & -1 \end{bmatrix}
$$

$$
\text{Phase} \quad -\boxed{S}- \quad \begin{bmatrix} 1 & 0 \\ 0 & i \end{bmatrix}
$$

$$
\pi/8 \quad -\boxed{T}- \quad \begin{bmatrix} 1 & 0 \\ 0 & e^{i\pi/4} \end{bmatrix}
$$

controlled-NOT		$\begin{bmatrix} 1 & 0 & 0 & 0 \\ 0 & 1 & 0 & 0 \\ 0 & 0 & 0 & 1 \\ 0 & 0 & 1 & 0 \end{bmatrix}$
swap		$\begin{bmatrix} 1 & 0 & 0 & 0 \\ 0 & 0 & 1 & 0 \\ 0 & 1 & 0 & 0 \\ 0 & 0 & 0 & 1 \end{bmatrix}$
controlled-Z	$=$	$\begin{bmatrix} 1 & 0 & 0 & 0 \\ 0 & 1 & 0 & 0 \\ 0 & 0 & 1 & 0 \\ 0 & 0 & 0 & -1 \end{bmatrix}$
controlled-phase	S	$\begin{bmatrix} 1 & 0 & 0 & 0 \\ 0 & 1 & 0 & 0 \\ 0 & 0 & 1 & 0 \\ 0 & 0 & 0 & i \end{bmatrix}$
Toffoli		$\begin{bmatrix} 1 & 0 & 0 & 0 & 0 & 0 & 0 & 0 \\ 0 & 1 & 0 & 0 & 0 & 0 & 0 & 0 \\ 0 & 0 & 1 & 0 & 0 & 0 & 0 & 0 \\ 0 & 0 & 0 & 1 & 0 & 0 & 0 & 0 \\ 0 & 0 & 0 & 0 & 1 & 0 & 0 & 0 \\ 0 & 0 & 0 & 0 & 0 & 1 & 0 & 0 \\ 0 & 0 & 0 & 0 & 0 & 0 & 0 & 1 \\ 0 & 0 & 0 & 0 & 0 & 0 & 1 & 0 \end{bmatrix}$
Fredkin (controlled-swap)		$\begin{bmatrix} 1 & 0 & 0 & 0 & 0 & 0 & 0 & 0 \\ 0 & 1 & 0 & 0 & 0 & 0 & 0 & 0 \\ 0 & 0 & 1 & 0 & 0 & 0 & 0 & 0 \\ 0 & 0 & 0 & 1 & 0 & 0 & 0 & 0 \\ 0 & 0 & 0 & 0 & 1 & 0 & 0 & 0 \\ 0 & 0 & 0 & 0 & 0 & 0 & 1 & 0 \\ 0 & 0 & 0 & 0 & 0 & 1 & 0 & 0 \\ 0 & 0 & 0 & 0 & 0 & 0 & 0 & 1 \end{bmatrix}$
measurement		Projection onto $|0\rangle$ and $|1\rangle$
qubit		wire carrying a single qubit (time goes left to right)
classical bit		wire carrying a single classical bit
n qubits		wire carrying n qubits

I Fundamental concepts

1 Introduction and overview

Science offers the boldest metaphysics of the age. It is a thoroughly human construct, driven by the faith that if we dream, press to discover, explain, and dream again, thereby plunging repeatedly into new terrain, the world will somehow come clearer and we will grasp the true strangeness of the universe. And the strangeness will all prove to be connected, and make sense.
– Edward O. Wilson

Information is physical.
– Rolf Landauer

What are the fundamental concepts of quantum computation and quantum information? How did these concepts develop? To what uses may they be put? How will they be presented in this book? The purpose of this introductory chapter is to answer these questions by developing in broad brushstrokes a picture of the field of quantum computation and quantum information. The intent is to communicate a basic understanding of the central concepts of the field, perspective on how they have been developed, and to help you decide how to approach the rest of the book.

Our story begins in Section 1.1 with an account of the historical context in which quantum computation and quantum information has developed. Each remaining section in the chapter gives a brief introduction to one or more fundamental concepts from the field: quantum bits (Section 1.2), quantum computers, quantum gates and quantum circuits (Section 1.3), quantum algorithms (Section 1.4), experimental quantum information processing (Section 1.5), and quantum information and communication (Section 1.6).

Along the way, illustrative and easily accessible developments such as quantum teleportation and some simple quantum algorithms are given, using the basic mathematics taught in this chapter. The presentation is self-contained, and designed to be accessible even without a background in computer science or physics. As we move along, we give pointers to more in-depth discussions in later chapters, where references and suggestions for further reading may also be found.

If as you read you're finding the going rough, skip on to a spot where you feel more comfortable. At points we haven't been able to avoid using a little technical lingo which won't be completely explained until later in the book. Simply accept it for now, and come back later when you understand all the terminology in more detail. The emphasis in this first chapter is on the big picture, with the details to be filled in later.

1.1 Global perspectives

Quantum computation and quantum information is the study of the information processing tasks that can be accomplished using quantum mechanical systems. Sounds pretty simple and obvious, doesn't it? Like many simple but profound ideas it was a long time before anybody thought of doing information processing using quantum mechanical systems. To see why this is the case, we must go back in time and look in turn at each of the fields which have contributed fundamental ideas to quantum computation and quantum information – quantum mechanics, computer science, information theory, and cryptography. As we take our short historical tour of these fields, think of yourself first as a physicist, then as a computer scientist, then as an information theorist, and finally as a cryptographer, in order to get some feel for the disparate perspectives which have come together in quantum computation and quantum information.

1.1.1 History of quantum computation and quantum information

Our story begins at the turn of the twentieth century when an unheralded revolution was underway in science. A series of crises had arisen in physics. The problem was that the theories of physics at that time (now dubbed *classical physics*) were predicting absurdities such as the existence of an 'ultraviolet catastrophe' involving infinite energies, or electrons spiraling inexorably into the atomic nucleus. At first such problems were resolved with the addition of *ad hoc* hypotheses to classical physics, but as a better understanding of atoms and radiation was gained these attempted explanations became more and more convoluted. The crisis came to a head in the early 1920s after a quarter century of turmoil, and resulted in the creation of the modern theory of *quantum mechanics*. Quantum mechanics has been an indispensable part of science ever since, and has been applied with enormous success to everything under and inside the Sun, including the structure of the atom, nuclear fusion in stars, superconductors, the structure of DNA, and the elementary particles of Nature.

What is quantum mechanics? Quantum mechanics is a mathematical framework or set of rules for the construction of physical theories. For example, there is a physical theory known as *quantum electrodynamics* which describes with fantastic accuracy the interaction of atoms and light. Quantum electrodynamics is built up within the framework of quantum mechanics, but it contains specific rules not determined by quantum mechanics. The relationship of quantum mechanics to specific physical theories like quantum electrodynamics is rather like the relationship of a computer's operating system to specific applications software – the operating system sets certain basic parameters and modes of operation, but leaves open how specific tasks are accomplished by the applications.

The rules of quantum mechanics are simple but even experts find them counter-intuitive, and the earliest antecedents of quantum computation and quantum information may be found in the long-standing desire of physicists to better understand quantum mechanics. The best known critic of quantum mechanics, Albert Einstein, went to his grave unreconciled with the theory he helped invent. Generations of physicists since have wrestled with quantum mechanics in an effort to make its predictions more palatable. One of the goals of quantum computation and quantum information is to develop tools which sharpen our intuition about quantum mechanics, and make its predictions more transparent to human minds.

For example, in the early 1980s, interest arose in whether it might be possible to use quantum effects to signal faster than light – a big no-no according to Einstein's theory of relativity. The resolution of this problem turns out to hinge on whether it is possible to *clone* an unknown quantum state, that is, construct a copy of a quantum state. If cloning were possible, then it would be possible to signal faster than light using quantum effects.

However, cloning – so easy to accomplish with classical information (consider the words in front of you, and where they came from!) – turns out not to be possible in general in quantum mechanics. This *no-cloning theorem*, discovered in the early 1980s, is one of the earliest results of quantum computation and quantum information. Many refinements of the no–cloning theorem have since been developed, and we now have conceptual tools which allow us to understand how well a (necessarily imperfect) quantum cloning device might work. These tools, in turn, have been applied to understand other aspects of quantum mechanics.

A related historical strand contributing to the development of quantum computation and quantum information is the interest, dating to the 1970s, of obtaining *complete control over single quantum systems*. Applications of quantum mechanics prior to the 1970s typically involved a gross level of control over a bulk sample containing an enormous number of quantum mechanical systems, none of them directly accessible. For example, superconductivity has a superb quantum mechanical explanation. However, because a superconductor involves a huge (compared to the atomic scale) sample of conducting metal, we can only probe a few aspects of its quantum mechanical nature, with the individual quantum systems constituting the superconductor remaining inaccessible. Systems such as particle accelerators do allow limited access to individual quantum systems, but again provide little control over the constituent systems.

Since the 1970s many techniques for controlling single quantum systems have been developed. For example, methods have been developed for trapping a single atom in an 'atom trap', isolating it from the rest of the world and allowing us to probe many different aspects of its behavior with incredible precision. The scanning tunneling microscope has been used to move single atoms around, creating designer arrays of atoms at will. Electronic devices whose operation involves the transfer of only single electrons have been demonstrated.

Why all this effort to attain complete control over single quantum systems? Setting aside the many technological reasons and concentrating on pure science, the principal answer is that researchers have done this on a hunch. Often the most profound insights in science come when we develop a method for probing a new regime of Nature. For example, the invention of radio astronomy in the 1930s and 1940s led to a spectacular sequence of discoveries, including the galactic core of the Milky Way galaxy, pulsars, and quasars. Low temperature physics has achieved its amazing successes by finding ways to lower the temperatures of different systems. In a similar way, by obtaining complete control over single quantum systems, we are exploring untouched regimes of Nature in the hope of discovering new and unexpected phenomena. We are just now taking our first steps along these lines, and already a few interesting surprises have been discovered in this regime. What else shall we discover as we obtain more complete control over single quantum systems, and extend it to more complex systems?

Quantum computation and quantum information fit naturally into this program. They provide a useful series of challenges at varied levels of difficulty for people devising methods to better manipulate single quantum systems, and stimulate the development of new experimental techniques and provide guidance as to the most interesting directions in which to take experiment. Conversely, the ability to control single quantum systems is essential if we are to harness the power of quantum mechanics for applications to quantum computation and quantum information.

Despite this intense interest, efforts to build quantum information processing systems have resulted in modest success to date. Small quantum computers, capable of doing dozens of operations on a few quantum bits (or *qubits*) represent the state of the art in quantum computation. Experimental prototypes for doing *quantum cryptography* – a way of communicating in secret across long distances – have been demonstrated, and are even at the level where they may be useful for some real-world applications. However, it remains a great challenge to physicists and engineers of the future to develop techniques for making large-scale quantum information processing a reality.

Let us turn our attention from quantum mechanics to another of the great intellectual triumphs of the twentieth century, computer science. The origins of computer science are lost in the depths of history. For example, cuneiform tablets indicate that by the time of Hammurabi (circa 1750 B.C.) the Babylonians had developed some fairly sophisticated algorithmic ideas, and it is likely that many of those ideas date to even earlier times.

The modern incarnation of computer science was announced by the great mathematician Alan Turing in a remarkable 1936 paper. Turing developed in detail an abstract notion of what we would now call a programmable computer, a model for computation now known as the *Turing machine*, in his honor. Turing showed that there is a *Universal Turing Machine* that can be used to simulate any other Turing machine. Furthermore, he claimed that the Universal Turing Machine *completely captures* what it means to perform a task by algorithmic means. That is, if an algorithm can be performed on *any* piece of hardware (say, a modern personal computer), then there is an equivalent algorithm for a Universal Turing Machine which performs exactly the same task as the algorithm running on the personal computer. This assertion, known as the *Church–Turing thesis* in honor of Turing and another pioneer of computer science, Alonzo Church, asserts the equivalence between the physical concept of what class of algorithms can be performed on *some physical device* with the rigorous mathematical concept of a Universal Turing Machine. The broad acceptance of this thesis laid the foundation for the development of a rich theory of computer science.

Not long after Turing's paper, the first computers constructed from electronic components were developed. John von Neumann developed a simple theoretical model for how to put together in a practical fashion all the components necessary for a computer to be fully as capable as a Universal Turing Machine. Hardware development truly took off, though, in 1947, when John Bardeen, Walter Brattain, and Will Shockley developed the transistor. Computer hardware has grown in power at an amazing pace ever since, so much so that the growth was codified by Gordon Moore in 1965 in what has come to be known as *Moore's law*, which states that computer power will double for constant cost roughly once every two years.

Amazingly enough, Moore's law has approximately held true in the decades since the 1960s. Nevertheless, most observers expect that this dream run will end some time during the first two decades of the twenty-first century. Conventional approaches to the fabrication of computer technology are beginning to run up against fundamental difficulties of size. Quantum effects are beginning to interfere in the functioning of electronic devices as they are made smaller and smaller.

One possible solution to the problem posed by the eventual failure of Moore's law is to move to a different computing paradigm. One such paradigm is provided by the theory of quantum computation, which is based on the idea of using quantum mechanics to perform computations, instead of classical physics. It turns out that while an ordinary computer can be used to simulate a quantum computer, it appears to be impossible to perform the simulation in an *efficient* fashion. Thus quantum computers offer an essential speed advantage over classical computers. This speed advantage is so significant that many researchers believe that *no* conceivable amount of progress in classical computation would be able to overcome the gap between the power of a classical computer and the power of a quantum computer.

What do we mean by 'efficient' versus 'inefficient' simulations of a quantum computer? Many of the key notions needed to answer this question were actually invented before the notion of a quantum computer had even arisen. In particular, the idea of *efficient* and *inefficient* algorithms was made mathematically precise by the field of *computational complexity*. Roughly speaking, an efficient algorithm is one which runs in time polynomial in the size of the problem solved. In contrast, an inefficient algorithm requires super-polynomial (typically exponential) time. What was noticed in the late 1960s and early 1970s was that it seemed as though the Turing machine model of computation was at least as powerful as any other model of computation, in the sense that a problem which could be solved efficiently in some model of computation could also be solved efficiently in the Turing machine model, by using the Turing machine to simulate the other model of computation. This observation was codified into a strengthened version of the Church–Turing thesis:

Any algorithmic process can be simulated efficiently using a Turing machine.

The key strengthening in the strong Church–Turing thesis is the word *efficiently*. If the strong Church–Turing thesis is correct, then it implies that no matter what type of machine we use to perform our algorithms, that machine can be simulated efficiently using a standard Turing machine. This is an important strengthening, as it implies that for the purposes of analyzing whether a given computational task can be accomplished efficiently, we may restrict ourselves to the analysis of the Turing machine model of computation.

One class of challenges to the strong Church–Turing thesis comes from the field of *analog computation*. In the years since Turing, many different teams of researchers have noticed that certain types of analog computers can efficiently solve problems believed to have no efficient solution on a Turing machine. At first glance these analog computers appear to violate the strong form of the Church–Turing thesis. Unfortunately for analog computation, it turns out that when realistic assumptions about the presence of noise in analog computers are made, their power disappears in all known instances; they cannot efficiently solve problems which are not efficiently solvable on a Turing machine. This lesson – that the effects of realistic noise must be taken into account in evaluating the efficiency of a computational model – was one of the great early challenges of quantum computation and quantum information, a challenge successfully met by the development of a theory of *quantum error-correcting codes* and *fault-tolerant quantum computation*. Thus, unlike analog computation, quantum computation can in principle tolerate a finite amount of noise and still retain its computational advantages.

The first major challenge to the strong Church–Turing thesis arose in the mid 1970s, when Robert Solovay and Volker Strassen showed that it is possible to test whether an integer is prime or composite using a *randomized algorithm*. That is, the Solovay–Strassen test for primality used randomness as an *essential* part of the algorithm. The algorithm did not determine whether a given integer was prime or composite with certainty. Instead, the algorithm could determine that a number was *probably* prime or else composite *with certainty*. By repeating the Solovay–Strassen test a few times it is possible to determine with near certainty whether a number is prime or composite. The Solovay-Strassen test was of especial significance at the time it was proposed as no deterministic test for primality was then known, nor is one known at the time of this writing. Thus, it seemed as though computers with access to a random number generator would be able to efficiently perform computational tasks with no efficient solution on a conventional deterministic Turing machine. This discovery inspired a search for other randomized algorithms which has paid off handsomely, with the field blossoming into a thriving area of research.

Randomized algorithms pose a challenge to the strong Church–Turing thesis, suggesting that there are efficiently soluble problems which, nevertheless, cannot be efficiently solved on a deterministic Turing machine. This challenge appears to be easily resolved by a simple modification of the strong Church–Turing thesis:

> *Any algorithmic process can be simulated efficiently using a probabilistic Turing machine.*

This *ad hoc* modification of the strong Church–Turing thesis should leave you feeling rather queasy. Might it not turn out at some later date that yet another model of computation allows one to efficiently solve problems that are not efficiently soluble within Turing's model of computation? Is there any way we can find a single model of computation which is guaranteed to be able to efficiently simulate any other model of computation?

Motivated by this question, in 1985 David Deutsch asked whether the laws of physics could be use to *derive* an even stronger version of the Church–Turing thesis. Instead of adopting *ad hoc* hypotheses, Deutsch looked to physical theory to provide a foundation for the Church–Turing thesis that would be as secure as the status of that physical theory. In particular, Deutsch attempted to define a computational device that would be capable of efficiently simulating an *arbitrary* physical system. Because the laws of physics are ultimately quantum mechanical, Deutsch was naturally led to consider computing devices based upon the principles of quantum mechanics. These devices, quantum analogues of the machines defined forty-nine years earlier by Turing, led ultimately to the modern conception of a quantum computer used in this book.

At the time of writing it is not clear whether Deutsch's notion of a Universal Quantum Computer is sufficient to efficiently simulate an arbitrary physical system. Proving or refuting this conjecture is one of the great open problems of the field of quantum computation and quantum information. It is possible, for example, that some effect of quantum field theory or an even more esoteric effect based in string theory, quantum gravity or some other physical theory may take us beyond Deutsch's Universal Quantum Computer, giving us a still more powerful model for computation. At this stage, we simply don't know.

What Deutsch's model of a quantum computer did enable was a challenge to the strong form of the Church–Turing thesis. Deutsch asked whether it is possible for a quantum computer to efficiently solve computational problems which have no efficient solution on a classical computer, even a probabilistic Turing machine. He then constructed a simple example suggesting that, indeed, quantum computers might have computational powers exceeding those of classical computers.

This remarkable first step taken by Deutsch was improved in the subsequent decade by many people, culminating in Peter Shor's 1994 demonstration that two enormously important problems – the problem of finding the prime factors of an integer, and the so-called 'discrete logarithm' problem – could be solved efficiently on a quantum computer. This attracted widespread interest because these two problems were and still are widely believed to have no efficient solution on a classical computer. Shor's results are a powerful indication that quantum computers are more powerful than Turing machines, even probabilistic Turing machines. Further evidence for the power of quantum computers came in 1995 when Lov Grover showed that another important problem – the problem of conducting a search through some unstructured search space – could also be sped up on a quantum computer. While Grover's algorithm did not provide as spectacular a speed-up as Shor's algorithms, the widespread applicability of search-based methodologies has excited considerable interest in Grover's algorithm.

At about the same time as Shor's and Grover's algorithms were discovered, many people were developing an idea Richard Feynman had suggested in 1982. Feynman had pointed out that there seemed to be essential difficulties in simulating quantum mechanical systems on classical computers, and suggested that building computers based on the principles of quantum mechanics would allow us to avoid those difficulties. In the 1990s several teams of researchers began fleshing this idea out, showing that it is indeed possible to use quantum computers to efficiently simulate systems that have no known efficient simulation on a classical computer. It is likely that one of the major applications of quantum computers in the future will be performing simulations of quantum mechanical systems too difficult to simulate on a classical computer, a problem with profound scientific and technological implications.

What other problems can quantum computers solve more quickly than classical computers? The short answer is that we don't know. Coming up with good quantum algorithms seems to be *hard*. A pessimist might think that's because there's nothing quantum computers are good for other than the applications already discovered! We take a different view. Algorithm design for quantum computers is hard because designers face two difficult problems not faced in the construction of algorithms for classical computers. First, our human intuition is rooted in the classical world. If we use that intuition as an aid to the construction of algorithms, then the algorithmic ideas we come up with will be classical ideas. To design good quantum algorithms one must 'turn off' one's classical intuition for at least part of the design process, using truly quantum effects to achieve the desired algorithmic end. Second, to be truly interesting it is not enough to design an algorithm that is merely quantum mechanical. The algorithm must be *better* than any existing classical algorithm! Thus, it is possible that one may find an algorithm which makes use of truly quantum aspects of quantum mechanics, that is nevertheless not of widespread interest because classical algorithms with comparable performance characteristics exist. The combination of these two problems makes the construction of new quantum algorithms a challenging problem for the future.

Even more broadly, we can ask if there are any generalizations we can make about the power of quantum computers versus classical computers. What is it that makes quantum computers more powerful than classical computers – assuming that this is indeed the case? What class of problems can be solved efficiently on a quantum computer, and how does that class compare to the class of problems that can be solved efficiently on a classical computer? One of the most exciting things about quantum computation and quantum information is how *little* is known about the answers to these questions! It is a great challenge for the future to understand these questions better.

Having come up to the frontier of quantum computation, let's switch to the history of another strand of thought contributing to quantum computation and quantum information: information theory. At the same time computer science was exploding in the 1940s, another revolution was taking place in our understanding of *communication*. In 1948 Claude Shannon published a remarkable pair of papers laying the foundations for the modern theory of information and communication.

Perhaps the key step taken by Shannon was *to mathematically define the concept of information*. In many mathematical sciences there is considerable flexibility in the choice of fundamental definitions. Try thinking naively for a few minutes about the following question: how would you go about mathematically defining the notion of an information source? Several *different* answers to this problem have found widespread use; however, the definition Shannon came up with seems to be far and away the most fruitful in terms of increased understanding, leading to a plethora of deep results and a theory with a rich structure which seems to accurately reflect many (though not all) real-world communications problems.

Shannon was interested in two key questions related to the communication of information over a communications channel. First, what resources are required to send information over a communications channel? For example, telephone companies need to know how much information they can reliably transmit over a given telephone cable. Second, can information be transmitted in such a way that it is protected against noise in the communications channel?

Shannon answered these two questions by proving the two fundamental theorems of information theory. The first, Shannon's *noiseless channel coding theorem*, quantifies the physical resources required to store the output from an information source. Shannon's second fundamental theorem, the *noisy channel coding theorem*, quantifies how much information it is possible to reliably transmit through a noisy communications channel. To achieve reliable transmission in the presence of noise, Shannon showed that *error-correcting codes* could be used to protect the information being sent. Shannon's noisy channel coding theorem gives an upper limit on the protection afforded by error-correcting codes. Unfortunately, Shannon's theorem does not explicitly give a practically useful set of error-correcting codes to achieve that limit. From the time of Shannon's papers until today, researchers have constructed more and better classes of error-correcting codes in their attempts to come closer to the limit set by Shannon's theorem. A sophisticated theory of error-correcting codes now exists offering the user a plethora of choices in their quest to design a good error-correcting code. Such codes are used in a multitude of places including, for example, compact disc players, computer modems, and satellite communications systems.

Quantum information theory has followed with similar developments. In 1995, Ben Schumacher provided an analogue to Shannon's noiseless coding theorem, and in the process defined the 'quantum bit' or 'qubit' as a tangible physical resource. However, no analogue to Shannon's noisy channel coding theorem is yet known for quantum information. Nevertheless, in analogy to their classical counterparts, a theory of quantum error-correction has been developed which, as already mentioned, allows quantum computers to compute effectively in the presence of noise, and also allows communication over noisy *quantum* channels to take place reliably.

Indeed, classical ideas of error-correction have proved to be enormously important in developing and understanding quantum error-correcting codes. In 1996, two groups working independently, Robert Calderbank and Peter Shor, and Andrew Steane, discovered an important class of quantum codes now known as CSS codes after their initials. This work has since been subsumed by the stabilizer codes, independently discovered by Robert Calderbank, Eric Rains, Peter Shor and Neil Sloane, and by Daniel Gottesman. By building upon the basic ideas of classical linear coding theory, these discoveries greatly facilitated a rapid understanding of quantum error-correcting codes and their application to quantum computation and quantum information.

The theory of quantum error-correcting codes was developed to protect quantum states against noise. What about transmitting ordinary *classical* information using a quantum channel? How efficiently can this be done? A few surprises have been discovered in this arena. In 1992 Charles Bennett and Stephen Wiesner explained how to transmit *two* classical bits of information, while only transmitting *one* quantum bit from sender to receiver, a result dubbed *superdense coding*.

Even more interesting are the results in *distributed quantum computation*. Imagine you have two computers networked, trying to solve a particular problem. How much communication is required to solve the problem? Recently it has been shown that quantum computers can require *exponentially less* communication to solve certain problems than would be required if the networked computers were classical! Unfortunately, as yet these problems are not especially important in a practical setting, and suffer from some undesirable technical restrictions. A major challenge for the future of quantum computation and quantum information is to find problems of real-world importance for which distributed quantum computation offers a substantial advantage over distributed classical computation.

Let's return to information theory proper. The study of information theory begins with the properties of a single communications channel. In applications we often do not deal with a single communications channel, but rather with networks of many channels. The subject of *networked information theory* deals with the information carrying properties of such networks of communications channels, and has been developed into a rich and intricate subject.

By contrast, the study of networked quantum information theory is very much in its infancy. Even for very basic questions we know little about the information carrying abilities of networks of quantum channels. Several rather striking preliminary results have been found in the past few years; however, no unifying theory of networked information theory exists for quantum channels. One example of networked quantum information theory should suffice to convince you of the value such a general theory would have. Imagine that we are attempting to send quantum information from Alice to Bob through a noisy quantum channel. If that channel has zero capacity for quantum information, then it is impossible to reliably send *any* information from Alice to Bob. Imagine instead that we consider two copies of the channel, operating in synchrony. Intuitively it is clear (and can be rigorously justified) that such a channel also has zero capacity to send quantum information. However, if we instead *reverse* the direction of one of the channels, as illustrated in Figure 1.1, it turns out that sometimes we can obtain a non-zero capacity for the transmission of information from Alice to Bob! Counter-intuitive properties like this illustrate the strange nature of quantum information. Better understanding the information carrying properties of networks of quantum channels is a major open problem of quantum computation and quantum information.

Let's switch fields one last time, moving to the venerable old art and science of *cryptography*. Broadly speaking, cryptography is the problem of doing *communication* or

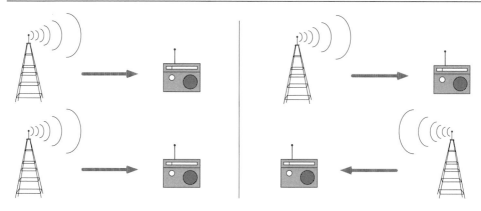

Figure 1.1. Classically, if we have two very noisy channels of zero capacity running side by side, then the combined channel has zero capacity to send information. Not surprisingly, if we reverse the direction of one of the channels, we still have zero capacity to send information. Quantum mechanically, reversing one of the zero capacity channels can actually allow us to send information!

computation involving two or more parties *who may not trust one another*. The best known cryptographic problem is the transmission of secret messages. Suppose two parties wish to communicate in secret. For example, you may wish to give your credit card number to a merchant in exchange for goods, hopefully without any malevolent third party intercepting your credit card number. The way this is done is to use a *cryptographic protocol*. We'll describe in detail how cryptographic protocols work later in the book, but for now it will suffice to make a few simple distinctions. The most important distinction is between *private key cryptosystems* and *public key cryptosystems*.

The way a private key cryptosystem works is that two parties, 'Alice' and 'Bob', wish to communicate by sharing a *private key*, which only they know. The exact form of the key doesn't matter at this point – think of a string of zeroes and ones. The point is that this key is used by Alice to *encrypt* the information she wishes to send to Bob. After Alice encrypts she sends the encrypted information to Bob, who must now recover the original information. Exactly how Alice encrypts the message *depends upon the private key*, so that to recover the original message Bob needs to know the private key, in order to undo the transformation Alice applied.

Unfortunately, private key cryptosystems have some severe problems in many contexts. The most basic problem is how to distribute the keys? In many ways, the key distribution problem is just as difficult as the original problem of communicating in private – a malevolent third party may be eavesdropping on the key distribution, and then use the intercepted key to decrypt some of the message transmission.

One of the earliest discoveries in quantum computation and quantum information was that quantum mechanics can be used to do key distribution in such a way that Alice and Bob's security can not be compromised. This procedure is known as *quantum cryptography* or *quantum key distribution*. The basic idea is to exploit the quantum mechanical principle that observation in general disturbs the system being observed. Thus, if there is an eavesdropper listening in as Alice and Bob attempt to transmit their key, the presence of the eavesdropper will be visible as a disturbance of the communications channel Alice and Bob are using to establish the key. Alice and Bob can then throw out the key bits established while the eavesdropper was listening in, and start over. The first quantum cryptographic ideas were proposed by Stephen Wiesner in the late 1960s, but unfortunately were not accepted for publication! In 1984 Charles Bennett and Gilles Brassard, building on Wiesner's earlier work, proposed a protocol using quantum mechanics to distribute keys between Alice and Bob, without any possibility of a compromise. Since then numerous quantum cryptographic protocols have been proposed, and experimental prototypes developed. At the time of this writing, the experimental prototypes are nearing the stage where they may be useful in limited-scale real-world applications.

The second major type of cryptosystem is the *public key cryptosystem*. Public key cryptosystems don't rely on Alice and Bob sharing a secret key in advance. Instead, Bob simply publishes a 'public key', *which is made available to the general public*. Alice can make use of this public key to encrypt a message which she sends to Bob. What is interesting is that a third party *cannot* use Bob's public key to decrypt the message! Strictly speaking, we shouldn't say *cannot*. Rather, the encryption transformation is chosen in a very clever and non-trivial way so that it is *extremely difficult* (though not impossible) to invert, given only knowledge of the public key. To make inversion easy, Bob has a *secret key* matched to his public key, which together enable him to *easily* perform the decryption. This secret key is not known to anybody other than Bob, who can therefore be confident that only he can read the contents of Alice's transmission, to the extent that it is unlikely that anybody else has the computational power to invert the encryption, given only the public key. Public key cryptosystems solve the key distribution problem by making it unnecessary for Alice and Bob to share a private key before communicating.

Rather remarkably, public key cryptography did not achieve widespread use until the mid-1970s, when it was proposed independently by Whitfield Diffie and Martin Hellman, and by Ralph Merkle, revolutionizing the field of cryptography. A little later, Ronald Rivest, Adi Shamir, and Leonard Adleman developed the *RSA cryptosystem*, which at the time of writing is the most widely deployed public key cryptosystem, believed to offer a fine balance of security and practical usability. In 1997 it was disclosed that these ideas – public key cryptography, the Diffie–Hellman and RSA cryptosystems – were actually invented in the late 1960s and early 1970s by researchers working at the British intelligence agency GCHQ.

The key to the security of public key cryptosystems is that it should be difficult to invert the encryption stage if only the public key is available. For example, it turns out that inverting the encryption stage of RSA is a problem closely related to factoring. Much of the presumed security of RSA comes from the belief that factoring is a problem hard to solve on a classical computer. However, Shor's fast algorithm for factoring on a quantum computer could be used to break RSA! Similarly, there are other public key cryptosystems which can be broken if a fast algorithm for solving the discrete logarithm problem – like Shor's quantum algorithm for discrete logarithm – were known. This practical application of quantum computers to the breaking of cryptographic codes has excited much of the interest in quantum computation and quantum information.

We have been looking at the historical antecedents for quantum computation and quantum information. Of course, as the field has grown and matured, it has sprouted its own subfields of research, whose antecedents lie mainly within quantum computation and quantum information.

Perhaps the most striking of these is the study of *quantum entanglement*. Entanglement is a uniquely quantum mechanical *resource* that plays a key role in many of the most interesting applications of quantum computation and quantum information; entanglement is iron to the classical world's bronze age. In recent years there has been a tremendous effort trying to better understand the properties of entanglement considered as a fundamental resource of Nature, of comparable importance to energy, information, entropy, or any other fundamental resource. Although there is as yet no complete theory of entanglement, some progress has been made in understanding this strange property of quantum mechanics. It is hoped by many researchers that further study of the properties of entanglement will yield insights that facilitate the development of new applications in quantum computation and quantum information.

1.1.2 Future directions

We've looked at some of the history and present status of quantum computation and quantum information. What of the future? What can quantum computation and quantum information offer to science, to technology, and to humanity? What benefits does quantum computation and quantum information confer upon its parent fields of computer science, information theory, and physics? What are the key open problems of quantum computation and quantum information? We will make a few very brief remarks about these overarching questions before moving onto more detailed investigations.

Quantum computation and quantum information has taught us to *think physically about computation*, and we have discovered that this approach yields many new and exciting capabilities for information processing and communication. Computer scientists and information theorists have been gifted with a new and rich paradigm for exploration. Indeed, in the broadest terms we have learned that *any physical theory*, not just quantum mechanics, may be used as the basis for a theory of information processing and communication. The fruits of these explorations may one day result in information processing devices with capabilities far beyond today's computing and communications systems, with concomitant benefits and drawbacks for society as a whole.

Quantum computation and quantum information certainly offer challenges aplenty to physicists, but it is perhaps a little subtle what quantum computation and quantum information offers to physics in the long term. We believe that just as we have learned to think physically about computation, we can also learn to *think computationally about physics*. Whereas physics has traditionally been a discipline focused on understanding 'elementary' objects and simple systems, many interesting aspects of Nature arise only when things become larger and more complicated. Chemistry and engineering deal with such complexity to some extent, but most often in a rather *ad hoc* fashion. One of the messages of quantum computation and information is that new tools are available for traversing the gulf between the small and the relatively complex: computation and algorithms provide systematic means for constructing and understanding such systems. Applying ideas from these fields is already beginning to yield new insights into physics. It is our hope that this perspective will blossom in years to come into a fruitful way of understanding all aspects of physics.

We've briefly examined some of the key motivations and ideas underlying quantum computation and quantum information. Over the remaining sections of this chapter we give a more technical but still accessible introduction to these motivations and ideas, with the hope of giving you a bird's-eye view of the field as it is presently poised.

1.2 Quantum bits

The *bit* is the fundamental concept of classical computation and classical information. Quantum computation and quantum information are built upon an analogous concept, the *quantum bit*, or *qubit* for short. In this section we introduce the properties of single and multiple qubits, comparing and contrasting their properties to those of classical bits.

What is a qubit? We're going to describe qubits as *mathematical objects* with certain specific properties. 'But hang on', you say, 'I thought qubits were physical objects.' It's true that qubits, like bits, are realized as actual physical systems, and in Section 1.5 and Chapter 7 we describe in detail how this connection between the abstract mathematical point of view and real systems is made. However, for the most part we treat qubits as abstract mathematical objects. The beauty of treating qubits as abstract entities is that it gives us the freedom to construct a general theory of quantum computation and quantum information which does not depend upon a specific system for its realization.

What then is a qubit? Just as a classical bit has a *state* – either 0 or 1 – a qubit also has a state. Two possible states for a qubit are the states $|0\rangle$ and $|1\rangle$, which as you might guess correspond to the states 0 and 1 for a classical bit. Notation like '$|\ \rangle$' is called the *Dirac notation*, and we'll be seeing it often, as it's the standard notation for states in quantum mechanics. The difference between bits and qubits is that a qubit can be in a state *other* than $|0\rangle$ or $|1\rangle$. It is also possible to form *linear combinations* of states, often called *superpositions*:

$$|\psi\rangle = \alpha|0\rangle + \beta|1\rangle. \tag{1.1}$$

The numbers α and β are complex numbers, although for many purposes not much is lost by thinking of them as real numbers. Put another way, the state of a qubit is a vector in a two-dimensional complex vector space. The special states $|0\rangle$ and $|1\rangle$ are known as *computational basis states*, and form an orthonormal basis for this vector space.

We can examine a bit to determine whether it is in the state 0 or 1. For example, computers do this all the time when they retrieve the contents of their memory. Rather remarkably, we cannot examine a qubit to determine its quantum state, that is, the values of α and β. Instead, quantum mechanics tells us that we can only acquire much more restricted information about the quantum state. When we measure a qubit we get either the result 0, with probability $|\alpha|^2$, or the result 1, with probability $|\beta|^2$. Naturally, $|\alpha|^2 + |\beta|^2 = 1$, since the probabilities must sum to one. Geometrically, we can interpret this as the condition that the qubit's state be normalized to length 1. Thus, in general a qubit's state is a unit vector in a two-dimensional complex vector space.

This dichotomy between the unobservable state of a qubit and the observations we can make lies at the heart of quantum computation and quantum information. In most of our abstract models of the world, there is a direct correspondence between elements of the abstraction and the real world, just as an architect's plans for a building are in correspondence with the final building. The lack of this direct correspondence in quantum mechanics makes it difficult to intuit the behavior of quantum systems; however, there is an indirect correspondence, for qubit states can be manipulated and transformed in ways which lead to measurement outcomes which depend distinctly on the different properties of the state. Thus, these quantum states have real, experimentally verifiable consequences, which we shall see are essential to the power of quantum computation and quantum information.

The ability of a qubit to be in a superposition state runs counter to our 'common sense' understanding of the physical world around us. A classical bit is like a coin: either heads or tails up. For imperfect coins, there may be intermediate states like having it balanced on an edge, but those can be disregarded in the ideal case. By contrast, a qubit can exist in a *continuum* of states between $|0\rangle$ and $|1\rangle$ – until it is observed. Let us emphasize again that when a qubit is measured, it only ever gives '0' or '1' as the measurement result – probabilistically. For example, a qubit can be in the state

$$\frac{1}{\sqrt{2}}|0\rangle + \frac{1}{\sqrt{2}}|1\rangle, \tag{1.2}$$

which, when measured, gives the result 0 fifty percent ($|1/\sqrt{2}|^2$) of the time, and the result 1 fifty percent of the time. We will return often to this state, which is sometimes denoted $|+\rangle$.

Despite this strangeness, qubits are decidedly real, their existence and behavior extensively validated by experiments (discussed in Section 1.5 and Chapter 7), and many different physical systems can be used to realize qubits. To get a concrete feel for how a qubit can be realized it may be helpful to list some of the ways this realization may occur: as the two different polarizations of a photon; as the alignment of a nuclear spin in a uniform magnetic field; as two states of an electron orbiting a single atom such as shown in Figure 1.2. In the atom model, the electron can exist in either the so-called 'ground' or 'excited' states, which we'll call $|0\rangle$ and $|1\rangle$, respectively. By shining light on the atom, with appropriate energy and for an appropriate length of time, it is possible to move the electron from the $|0\rangle$ state to the $|1\rangle$ state and vice versa. But more interestingly, by reducing the time we shine the light, an electron initially in the state $|0\rangle$ can be moved 'halfway' between $|0\rangle$ and $|1\rangle$, into the $|+\rangle$ state.

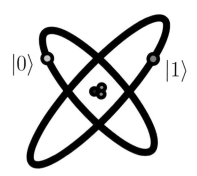

Figure 1.2. Qubit represented by two electronic levels in an atom.

Naturally, a great deal of attention has been given to the 'meaning' or 'interpretation' that might be attached to superposition states, and of the inherently probabilistic nature of observations on quantum systems. However, by and large, we shall not concern ourselves with such discussions in this book. Instead, our intent will be to develop mathematical and conceptual pictures which are predictive.

One picture useful in thinking about qubits is the following geometric representation.

Because $|\alpha|^2 + |\beta|^2 = 1$, we may rewrite Equation (1.1) as

$$|\psi\rangle = e^{i\gamma} \left(\cos \frac{\theta}{2} |0\rangle + e^{i\varphi} \sin \frac{\theta}{2} |1\rangle \right), \tag{1.3}$$

where θ, φ and γ are real numbers. In Chapter 2 we will see that we can *ignore* the factor of $e^{i\gamma}$ out the front, because it has *no observable effects*, and for that reason we can effectively write

$$|\psi\rangle = \cos \frac{\theta}{2} |0\rangle + e^{i\varphi} \sin \frac{\theta}{2} |1\rangle. \tag{1.4}$$

The numbers θ and φ define a point on the unit three-dimensional sphere, as shown in Figure 1.3. This sphere is often called the *Bloch sphere*; it provides a useful means of visualizing the state of a single qubit, and often serves as an excellent testbed for ideas about quantum computation and quantum information. Many of the operations on single qubits which we describe later in this chapter are neatly described within the Bloch sphere picture. However, it must be kept in mind that this intuition is limited because there is no simple generalization of the Bloch sphere known for multiple qubits.

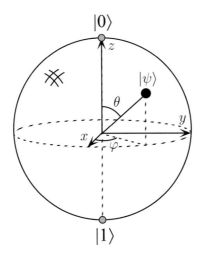

Figure 1.3. Bloch sphere representation of a qubit.

How much information is represented by a qubit? Paradoxically, there are an infinite number of points on the unit sphere, so that in principle one could store an entire text of Shakespeare in the infinite binary expansion of θ. However, this conclusion turns out to be misleading, because of the behavior of a qubit when observed. Recall that measurement of a qubit will give *only* either 0 or 1. Furthermore, measurement *changes* the state of a qubit, collapsing it from its superposition of $|0\rangle$ and $|1\rangle$ to the specific state consistent with the measurement result. For example, if measurement of $|+\rangle$ gives 0, then the post-measurement state of the qubit will be $|0\rangle$. Why does this type of collapse occur? Nobody knows. As discussed in Chapter 2, this behavior is simply one of the *fundamental postulates* of quantum mechanics. What is relevant for our purposes is that from a single measurement one obtains only a single bit of information about the state of the qubit, thus resolving the apparent paradox. It turns out that only if infinitely many identically prepared qubits were measured would one be able to determine α and β for a qubit in the state given in Equation (1.1).

But an even more interesting question to ask might be: how much information is represented by a qubit *if we do not measure it?* This is a trick question, because how can one quantify information if it cannot be measured? Nevertheless, there is something conceptually important here, because when Nature evolves a closed quantum system of qubits, not performing any 'measurements', she apparently does keep track of all the continuous variables describing the state, like α and β. In a sense, in the state of a qubit, Nature conceals a great deal of 'hidden information'. And even more interestingly, we will see shortly that the potential amount of this extra 'information' grows exponentially with the number of qubits. Understanding this hidden *quantum information* is a question that we grapple with for much of this book, and which lies at the heart of what makes quantum mechanics a powerful tool for information processing.

1.2.1 Multiple qubits

Hilbert space is a big place.
– Carlton Caves

Suppose we have two qubits. If these were two classical bits, then there would be four possible states, 00, 01, 10, and 11. Correspondingly, a two qubit system has four *computational basis states* denoted $|00\rangle, |01\rangle, |10\rangle, |11\rangle$. A pair of qubits can also exist in superpositions of these four states, so the quantum state of two qubits involves associating a complex coefficient – sometimes called an *amplitude* – with each computational basis state, such that the state vector describing the two qubits is

$$|\psi\rangle = \alpha_{00}|00\rangle + \alpha_{01}|01\rangle + \alpha_{10}|10\rangle + \alpha_{11}|11\rangle. \tag{1.5}$$

Similar to the case for a single qubit, the measurement result x (= 00, 01, 10 or 11) occurs with probability $|\alpha_x|^2$, with the state of the qubits after the measurement being $|x\rangle$. The condition that probabilities sum to one is therefore expressed by the *normalization condition* that $\sum_{x\in\{0,1\}^2} |\alpha_x|^2 = 1$, where the notation '$\{0,1\}^2$' means 'the set of strings of length two with each letter being either zero or one'. For a two qubit system, we could measure just a subset of the qubits, say the first qubit, and you can probably guess how this works: measuring the first qubit alone gives 0 with probability $|\alpha_{00}|^2 + |\alpha_{01}|^2$, leaving the post-measurement state

$$|\psi'\rangle = \frac{\alpha_{00}|00\rangle + \alpha_{01}|01\rangle}{\sqrt{|\alpha_{00}|^2 + |\alpha_{01}|^2}}. \tag{1.6}$$

Note how the post-measurement state is *re-normalized* by the factor $\sqrt{|\alpha_{00}|^2 + |\alpha_{01}|^2}$ so that it still satisfies the normalization condition, just as we expect for a legitimate quantum state.

An important two qubit state is the *Bell state* or *EPR pair*,

$$\frac{|00\rangle + |11\rangle}{\sqrt{2}}. \tag{1.7}$$

This innocuous-looking state is responsible for many surprises in quantum computation and quantum information. It is the key ingredient in quantum teleportation and super-dense coding, which we'll come to in Section 1.3.7 and Section 2.3, respectively, and the prototype for many other interesting quantum states. The Bell state has the property that upon measuring the first qubit, one obtains two possible results: 0 with probability $1/2$, leaving the post-measurement state $|\varphi'\rangle = |00\rangle$, and 1 with probability $1/2$, leaving $|\varphi'\rangle = |11\rangle$. As a result, a measurement of the second qubit always gives the same result as the measurement of the first qubit. That is, the measurement outcomes are *correlated*. Indeed, it turns out that other types of measurements can be performed on the Bell state, by first applying some operations to the first or second qubit, and that interesting correlations still exist between the result of a measurement on the first and second qubit. These correlations have been the subject of intense interest ever since a famous paper by Einstein, Podolsky and Rosen, in which they first pointed out the strange properties of states like the Bell state. EPR's insights were taken up and greatly improved by John Bell, who proved an amazing result: the measurement correlations in the Bell state are *stronger than could ever exist between classical systems*. These results, described in detail in Section 2.6, were the first intimation that quantum mechanics allows information processing beyond what is possible in the classical world.

More generally, we may consider a system of n qubits. The computational basis states of this system are of the form $|x_1 x_2 \ldots x_n\rangle$, and so a quantum state of such a system is specified by 2^n amplitudes. For $n = 500$ this number is larger than the estimated number of atoms in the Universe! Trying to store all these complex numbers would not be possible on any conceivable classical computer. Hilbert space is indeed a big place. In principle, however, Nature manipulates such enormous quantities of data, even for systems containing only a few hundred atoms. It is as if Nature were keeping 2^{500} hidden pieces of scratch paper on the side, on which she performs her calculations as the system evolves. This enormous potential computational power is something we would very much like to take advantage of. But how can we think of quantum mechanics as computation?

1.3 Quantum computation

Changes occurring to a quantum state can be described using the language of *quantum computation*. Analogous to the way a classical computer is built from an electrical circuit containing wires and logic gates, a quantum computer is built from a *quantum circuit* containing wires and elementary *quantum gates* to carry around and manipulate the quantum information. In this section we describe some simple quantum gates, and present several example circuits illustrating their application, including a circuit which teleports qubits!

1.3.1 Single qubit gates

Classical computer circuits consist of *wires* and *logic gates*. The wires are used to carry information around the circuit, while the logic gates perform manipulations of the information, converting it from one form to another. Consider, for example, classical single bit logic gates. The only non-trivial member of this class is the NOT gate, whose operation is defined by its *truth table*, in which $0 \rightarrow 1$ and $1 \rightarrow 0$, that is, the 0 and 1 states are interchanged.

Can an analogous quantum NOT gate for qubits be defined? Imagine that we had some process which took the state $|0\rangle$ to the state $|1\rangle$, and vice versa. Such a process would obviously be a good candidate for a quantum analogue to the NOT gate. However, specifying the action of the gate on the states $|0\rangle$ and $|1\rangle$ does not tell us what happens to superpositions of the states $|0\rangle$ and $|1\rangle$, without further knowledge about the properties of quantum gates. In fact, the quantum NOT gate acts *linearly*, that is, it takes the state

$$\alpha|0\rangle + \beta|1\rangle \tag{1.8}$$

to the corresponding state in which the role of $|0\rangle$ and $|1\rangle$ have been interchanged,

$$\alpha|1\rangle + \beta|0\rangle. \tag{1.9}$$

Why the quantum NOT gate acts linearly and not in some nonlinear fashion is a very interesting question, and the answer is not at all obvious. It turns out that this linear behavior is a general property of quantum mechanics, and very well motivated empirically; moreover, nonlinear behavior can lead to apparent paradoxes such as time travel, faster-than-light communication, and violations of the second laws of thermodynamics. We'll explore this point in more depth in later chapters, but for now we'll just take it as given.

There is a convenient way of representing the quantum NOT gate in matrix form, which follows directly from the linearity of quantum gates. Suppose we define a matrix X to represent the quantum NOT gate as follows:

$$X \equiv \begin{bmatrix} 0 & 1 \\ 1 & 0 \end{bmatrix}. \tag{1.10}$$

(The notation X for the quantum NOT is used for historical reasons.) If the quantum state $\alpha|0\rangle + \beta|1\rangle$ is written in a vector notation as

$$\begin{bmatrix} \alpha \\ \beta \end{bmatrix}, \tag{1.11}$$

with the top entry corresponding to the amplitude for $|0\rangle$ and the bottom entry the amplitude for $|1\rangle$, then the corresponding output from the quantum NOT gate is

$$X \begin{bmatrix} \alpha \\ \beta \end{bmatrix} = \begin{bmatrix} \beta \\ \alpha \end{bmatrix}. \tag{1.12}$$

Notice that the action of the NOT gate is to take the state $|0\rangle$ and replace it by the state corresponding to the first column of the matrix X. Similarly, the state $|1\rangle$ is replaced by the state corresponding to the second column of the matrix X.

So quantum gates on a single qubit can be described by two by two matrices. Are there any constraints on what matrices may be used as quantum gates? It turns out that there are. Recall that the normalization condition requires $|\alpha|^2 + |\beta|^2 = 1$ for a quantum state $\alpha|0\rangle + \beta|1\rangle$. This must also be true of the quantum state $|\psi'\rangle = \alpha'|0\rangle + \beta'|1\rangle$ after the gate has acted. It turns out that the appropriate condition on the matrix representing the gate is that the matrix U describing the single qubit gate be *unitary*, that is $U^\dagger U = I$, where $U^\dagger$ is the *adjoint* of U (obtained by transposing and then complex conjugating U), and I is the two by two identity matrix. For example, for the NOT gate it is easy to verify that $X^\dagger X = I$.

Amazingly, this *unitarity* constraint is the *only* constraint on quantum gates. Any unitary matrix specifies a valid quantum gate! The interesting implication is that in contrast to the classical case, where only one non-trivial single bit gate exists – the NOT

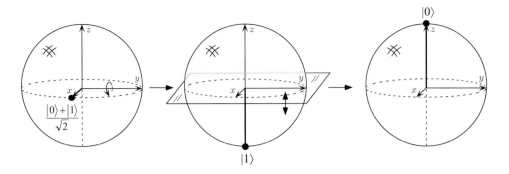

Figure 1.4. Visualization of the Hadamard gate on the Bloch sphere, acting on the input state $(|0\rangle + |1\rangle)/\sqrt{2}$.

gate – there are many non-trivial single qubit gates. Two important ones which we shall use later are the Z gate:

$$Z \equiv \begin{bmatrix} 1 & 0 \\ 0 & -1 \end{bmatrix}, \tag{1.13}$$

which leaves $|0\rangle$ unchanged, and flips the sign of $|1\rangle$ to give $-|1\rangle$, and the *Hadamard* gate,

$$H \equiv \frac{1}{\sqrt{2}} \begin{bmatrix} 1 & 1 \\ 1 & -1 \end{bmatrix}. \tag{1.14}$$

This gate is sometimes described as being like a 'square-root of NOT' gate, in that it turns a $|0\rangle$ into $(|0\rangle + |1\rangle)/\sqrt{2}$ (first column of H), 'halfway' between $|0\rangle$ and $|1\rangle$, and turns $|1\rangle$ into $(|0\rangle - |1\rangle)/\sqrt{2}$ (second column of H), which is also 'halfway' between $|0\rangle$ and $|1\rangle$. Note, however, that H^2 is not a NOT gate, as simple algebra shows that $H^2 = I$, and thus applying H twice to a state does nothing to it.

The Hadamard gate is one of the most useful quantum gates, and it is worth trying to visualize its operation by considering the Bloch sphere picture. In this picture, it turns out that single qubit gates correspond to rotations and reflections of the sphere. The Hadamard operation is just a rotation of the sphere about the $\hat{y}$ axis by $90°$, followed by a rotation about the $\hat{x}$ axis by $180°$, as illustrated in Figure 1.4. Some important single qubit gates are shown in Figure 1.5, and contrasted with the classical case.

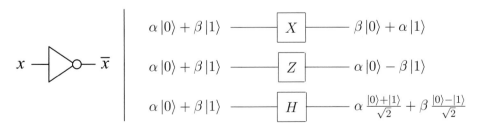

Figure 1.5. Single bit (left) and qubit (right) logic gates.

There are infinitely many two by two unitary matrices, and thus infinitely many single qubit gates. However, it turns out that the properties of the complete set can be understood from the properties of a much smaller set. For example, as explained in Box 1.1, an arbitrary single qubit unitary gate can be decomposed as a product of rotations

$$\begin{bmatrix} \cos\frac{\gamma}{2} & -\sin\frac{\gamma}{2} \\ \sin\frac{\gamma}{2} & \cos\frac{\gamma}{2} \end{bmatrix},\tag{1.15}$$

and a gate which we'll later understand as being a rotation about the $\hat{z}$ axis,

$$\begin{bmatrix} e^{-i\beta/2} & 0 \\ 0 & e^{i\beta/2} \end{bmatrix},\tag{1.16}$$

together with a *(global) phase shift* – a constant multiplier of the form $e^{i\alpha}$. These gates can be broken down further – we don't need to be able to do these gates for arbitrary α, β and γ, but can build arbitrarily good approximations to such gates using only certain special *fixed* values of α, β and γ. In this way it is possible to build up an arbitrary single qubit gate using a *finite* set of quantum gates. More generally, an arbitrary quantum computation on any number of qubits can be generated by a finite set of gates that is said to be *universal* for quantum computation. To obtain such a universal set we first need to introduce some quantum gates involving multiple qubits.

Box 1.1: Decomposing single qubit operations

In Section 4.2 starting on page 174 we prove that an arbitrary 2×2 unitary matrix may be decomposed as

$$U = e^{i\alpha}\begin{bmatrix} e^{-i\beta/2} & 0 \\ 0 & e^{i\beta/2} \end{bmatrix}\begin{bmatrix} \cos\frac{\gamma}{2} & -\sin\frac{\gamma}{2} \\ \sin\frac{\gamma}{2} & \cos\frac{\gamma}{2} \end{bmatrix}\begin{bmatrix} e^{-i\delta/2} & 0 \\ 0 & e^{i\delta/2} \end{bmatrix},\tag{1.17}$$

where α, β, γ, and δ are real-valued. Notice that the second matrix is just an ordinary rotation. It turns out that the first and last matrices can also be understood as rotations in a different plane. This decomposition can be used to give an exact prescription for performing an *arbitrary* single qubit quantum logic gate.

1.3.2 Multiple qubit gates

Now let us generalize from one to multiple qubits. Figure 1.6 shows five notable multiple bit classical gates, the AND, OR, XOR (exclusive-OR), NAND and NOR gates. An important theoretical result is that any function on bits can be computed from the composition of NAND gates alone, which is thus known as a *universal* gate. By contrast, the XOR alone or even together with NOT is not universal. One way of seeing this is to note that applying an XOR gate does not change the total parity of the bits. As a result, any circuit involving only NOT and XOR gates will, if two inputs x and y have the same parity, give outputs with the same parity, restricting the class of functions which may be computed, and thus precluding universality.

The prototypical multi-qubit quantum logic gate is the *controlled*-NOT or CNOT gate. This gate has two input qubits, known as the *control* qubit and the *target* qubit, respectively. The circuit representation for the CNOT is shown in the top right of Figure 1.6; the top line represents the control qubit, while the bottom line represents the target

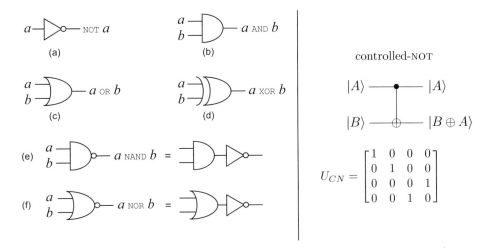

Figure 1.6. On the left are some standard single and multiple bit gates, while on the right is the prototypical multiple qubit gate, the controlled-NOT. The matrix representation of the controlled-NOT, U_{CN}, is written with respect to the amplitudes for $|00\rangle$, $|01\rangle$, $|10\rangle$, and $|11\rangle$, in that order.

qubit. The action of the gate may be described as follows. If the control qubit is set to 0, then the target qubit is left alone. If the control qubit is set to 1, then the target qubit is flipped. In equations:

$$|00\rangle \rightarrow |00\rangle; \ |01\rangle \rightarrow |01\rangle; \ |10\rangle \rightarrow |11\rangle; \ |11\rangle \rightarrow |10\rangle. \tag{1.18}$$

Another way of describing the CNOT is as a generalization of the classical XOR gate, since the action of the gate may be summarized as $|A, B\rangle \rightarrow |A, B \oplus A\rangle$, where $\oplus$ is addition modulo two, which is exactly what the XOR gate does. That is, the control qubit and the target qubit are XORed and stored in the target qubit.

Yet another way of describing the action of the CNOT is to give a matrix representation, as shown in the bottom right of Figure 1.6. You can easily verify that the first column of U_{CN} describes the transformation that occurs to $|00\rangle$, and similarly for the other computational basis states, $|01\rangle$, $|10\rangle$, and $|11\rangle$. As for the single qubit case, the requirement that probability be conserved is expressed in the fact that U_{CN} is a *unitary matrix*, that is, $U_{CN}^{\dagger} U_{CN} = I$.

We noticed that the CNOT can be regarded as a type of generalized-XOR gate. Can other classical gates such as the NAND or the regular XOR gate be understood as unitary gates in a sense similar to the way the quantum NOT gate represents the classical NOT gate? It turns out that this is not possible. The reason is because the XOR and NAND gates are essentially *irreversible* or *non-invertible*. For example, given the output $A \oplus B$ from an XOR gate, it is not possible to determine what the inputs A and B were; there is an irretrievable *loss of information* associated with the irreversible action of the XOR gate. On the other hand, unitary quantum gates are *always* invertible, since the inverse of a unitary matrix is also a unitary matrix, and thus a quantum gate can always be inverted by another quantum gate. Understanding how to do classical logic in this *reversible* or *invertible* sense will be a crucial step in understanding how to harness the power of quantum mechanics for computation. We'll explain the basic idea of how to do reversible computation in Section 1.4.1.

Of course, there are many interesting quantum gates other than the controlled-NOT. However, in a sense the controlled-NOT and single qubit gates are the prototypes for *all* other gates because of the following remarkable *universality* result: *Any multiple qubit logic gate may be composed from* CNOT *and single qubit gates*. The proof is given in Section 4.5, and is the quantum parallel of the universality of the NAND gate.

1.3.3 Measurements in bases other than the computational basis

We've described quantum measurements of a single qubit in the state $\alpha|0\rangle + \beta|1\rangle$ as yielding the result 0 or 1 and leaving the qubit in the corresponding state $|0\rangle$ or $|1\rangle$, with respective probabilities $|\alpha|^2$ and $|\beta|^2$. In fact, quantum mechanics allows somewhat more versatility in the class of measurements that may be performed, although certainly nowhere near enough to recover α and β from a single measurement!

Note that the states $|0\rangle$ and $|1\rangle$ represent just one of many possible choices of basis states for a qubit. Another possible choice is the set $|+\rangle \equiv (|0\rangle + |1\rangle)/\sqrt{2}$ and $|-\rangle \equiv (|0\rangle - |1\rangle)/\sqrt{2}$. An arbitrary state $|\psi\rangle = \alpha|0\rangle + \beta|1\rangle$ can be re-expressed in terms of the states $|+\rangle$ and $|-\rangle$:

$$|\psi\rangle = \alpha|0\rangle + \beta|1\rangle = \alpha\frac{|+\rangle + |-\rangle}{\sqrt{2}} + \beta\frac{|+\rangle - |-\rangle}{\sqrt{2}} = \frac{\alpha + \beta}{\sqrt{2}}|+\rangle + \frac{\alpha - \beta}{\sqrt{2}}|-\rangle. \quad (1.19)$$

It turns out that it is possible to treat the $|+\rangle$ and $|-\rangle$ states as though they were the computational basis states, and measure with respect to this new basis. Naturally, measuring with respect to the $|+\rangle, |-\rangle$ basis results in the result '+' with probability $|\alpha + \beta|^2/2$ and the result '−' with probability $|\alpha - \beta|^2/2$, with corresponding post-measurement states $|+\rangle$ and $|-\rangle$, respectively.

More generally, given any basis states $|a\rangle$ and $|b\rangle$ for a qubit, it is possible to express an arbitrary state as a linear combination $\alpha|a\rangle + \beta|b\rangle$ of those states. Furthermore, provided the states are *orthonormal*, it is possible to *perform a measurement with respect to the $|a\rangle, |b\rangle$ basis*, giving the result a with probability $|\alpha|^2$ and b with probability $|\beta|^2$. The orthonormality constraint is necessary in order that $|\alpha|^2 + |\beta|^2 = 1$ as we expect for probabilities. In an analogous way it is possible in principle to measure a quantum system of many qubits with respect to an arbitrary orthonormal basis. However, just because it is possible in principle does not mean that such a measurement can be done easily, and we return later to the question of how efficiently a measurement in an arbitrary basis can be performed.

There are many reasons for using this extended formalism for quantum measurements, but ultimately the best one is this: the formalism allows us to describe observed experimental results, as we will see in our discussion of the Stern–Gerlach experiment in Section 1.5.1. An even more sophisticated and convenient (but essentially equivalent) formalism for describing quantum measurements is described in the next chapter, in Section 2.2.3.

1.3.4 Quantum circuits

We've already met a few simple quantum circuits. Let's look in a little more detail at the elements of a quantum circuit. A simple quantum circuit containing three quantum gates is shown in Figure 1.7. The circuit is to be read from left-to-right. Each line in the circuit represents a *wire* in the quantum circuit. This wire does not necessarily correspond to a physical wire; it may correspond instead to the passage of time, or perhaps to a physical particle such as a photon – a particle of light – moving from one location to another through space. It is conventional to assume that the state input to the circuit is a computational basis state, usually the state consisting of all $|0\rangle$s. This rule is broken frequently in the literature on quantum computation and quantum information, but it is considered polite to inform the reader when this is the case.

The circuit in Figure 1.7 accomplishes a simple but useful task – it swaps the states of the two qubits. To see that this circuit accomplishes the swap operation, note that the sequence of gates has the following sequence of effects on a computational basis state $|a, b\rangle$,

$$
\begin{aligned}
|a, b\rangle &\longrightarrow |a, a \oplus b\rangle \\
&\longrightarrow |a \oplus (a \oplus b), a \oplus b\rangle = |b, a \oplus b\rangle \\
&\longrightarrow |b, (a \oplus b) \oplus b\rangle = |b, a\rangle ,
\end{aligned} \tag{1.20}
$$

where all additions are done modulo 2. The effect of the circuit, therefore, is to interchange the state of the two qubits.

Figure 1.7. Circuit swapping two qubits, and an equivalent schematic symbol notation for this common and useful circuit.

There are a few features allowed in classical circuits that are not usually present in quantum circuits. First of all, we don't allow 'loops', that is, feedback from one part of the quantum circuit to another; we say the circuit is *acyclic*. Second, classical circuits allow wires to be 'joined' together, an operation known as FANIN, with the resulting single wire containing the bitwise OR of the inputs. Obviously this operation is not reversible and therefore not unitary, so we don't allow FANIN in our quantum circuits. Third, the inverse operation, FANOUT, whereby several copies of a bit are produced is also not allowed in quantum circuits. In fact, it turns out that quantum mechanics forbids the copying of a qubit, making the FANOUT operation impossible! We'll see an example of this in the next section when we attempt to design a circuit to copy a qubit.

As we proceed we'll introduce new quantum gates as needed. It's convenient to introduce another convention about quantum circuits at this point. This convention is illustrated in Figure 1.8. Suppose U is *any* unitary matrix acting on some number n of qubits, so U can be regarded as a quantum gate on those qubits. Then we can define a *controlled-U* gate which is a natural extension of the controlled-NOT gate. Such a gate has a single *control qubit*, indicated by the line with the black dot, and n *target qubits*, indicated by the boxed U. If the control qubit is set to 0 then nothing happens to the target qubits. If the control qubit is set to 1 then the gate U is applied to the target qubits. The prototypical example of the controlled-U gate is the controlled-NOT gate, which is a controlled-U gate with $U = X$, as illustrated in Figure 1.9.

Another important operation is measurement, which we represent by a 'meter' symbol,

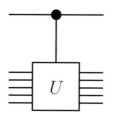

Figure 1.8. Controlled-U gate.

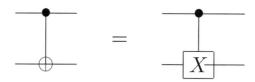

Figure 1.9. Two different representations for the controlled-NOT.

as shown in Figure 1.10. As previously described, this operation converts a single qubit state $|\psi\rangle = \alpha|0\rangle + \beta|1\rangle$ into a probabilistic classical bit M (distinguished from a qubit by drawing it as a double-line wire), which is 0 with probability $|\alpha|^2$, or 1 with probability $|\beta|^2$.

Figure 1.10. Quantum circuit symbol for measurement.

We shall find quantum circuits useful as models of all quantum processes, including but not limited to computation, communication, and even quantum noise. Several simple examples illustrate this below.

1.3.5 Qubit copying circuit?

The CNOT gate is useful for demonstrating one particularly fundamental property of quantum information. Consider the task of copying a classical bit. This may be done using a classical CNOT gate, which takes in the bit to copy (in some unknown state x) and a 'scratchpad' bit initialized to zero, as illustrated in Figure 1.11. The output is two bits, both of which are in the same state x.

Suppose we try to copy a qubit in the unknown state $|\psi\rangle = a|0\rangle + b|1\rangle$ in the same manner by using a CNOT gate. The input state of the two qubits may be written as

$$\left[a|0\rangle + b|1\rangle\right]|0\rangle = a|00\rangle + b|10\rangle, \tag{1.21}$$

The function of CNOT is to negate the second qubit when the first qubit is 1, and thus the output is simply $a|00\rangle + b|11\rangle$. Have we successfully copied $|\psi\rangle$? That is, have we created the state $|\psi\rangle|\psi\rangle$? In the case where $|\psi\rangle = |0\rangle$ or $|\psi\rangle = |1\rangle$ that is indeed what this circuit does; it is possible to use quantum circuits to copy classical information encoded as a $|0\rangle$ or a $|1\rangle$. However, for a general state $|\psi\rangle$ we see that

$$|\psi\rangle|\psi\rangle = a^2|00\rangle + ab|01\rangle + ab|10\rangle + b^2|11\rangle. \tag{1.22}$$

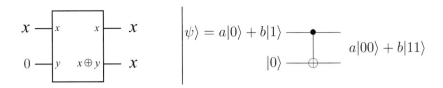

Figure 1.11. Classical and quantum circuits to 'copy' an unknown bit or qubit.

Comparing with $a|00\rangle + b|11\rangle$, we see that unless $ab = 0$ the 'copying circuit' above does *not* copy the quantum state input. In fact, it turns out to be *impossible* to make a copy of an unknown quantum state. This property, that qubits cannot be copied, is known as the *no-cloning* theorem, and it is one of the chief differences between quantum and classical information. The no-cloning theorem is discussed at more length in Box 12.1 on page 532; the proof is very simple, and we encourage you to skip ahead and read the proof now.

There is another way of looking at the failure of the circuit in Figure 1.11, based on the intuition that a qubit somehow contains 'hidden' information not directly accessible to measurement. Consider what happens when we measure one of the qubits of the state $a|00\rangle + b|11\rangle$. As previously described, we obtain either 0 or 1 with probabilities $|a|^2$ and $|b|^2$. However, once one qubit is measured, the state of the other one is completely determined, and no additional information can be gained about a and b. In this sense, the extra hidden information carried in the original qubit $|\psi\rangle$ was lost in the first measurement, and cannot be regained. If, however, the qubit had been copied, then the state of the other qubit should still contain some of that hidden information. Therefore, a copy cannot have been created.

1.3.6 Example: Bell states

Let's consider a slightly more complicated circuit, shown in Figure 1.12, which has a Hadamard gate followed by a CNOT, and transforms the four computational basis states according to the table given. As an explicit example, the Hadamard gate takes the input $|00\rangle$ to $(|0\rangle + |1\rangle)|0\rangle/\sqrt{2}$, and then the CNOT gives the output state $(|00\rangle + |11\rangle)/\sqrt{2}$. Note how this works: first, the Hadamard transform puts the top qubit in a superposition; this then acts as a control input to the CNOT, and the target gets inverted only when the control is 1. The output states

$$|\beta_{00}\rangle = \frac{|00\rangle + |11\rangle}{\sqrt{2}};\tag{1.23}$$

$$|\beta_{01}\rangle = \frac{|01\rangle + |10\rangle}{\sqrt{2}};\tag{1.24}$$

$$|\beta_{10}\rangle = \frac{|00\rangle - |11\rangle}{\sqrt{2}};\ \text{and}\tag{1.25}$$

$$|\beta_{11}\rangle = \frac{|01\rangle - |10\rangle}{\sqrt{2}},\tag{1.26}$$

are known as the *Bell states*, or sometimes the *EPR states* or *EPR pairs*, after some of the people – Bell, and Einstein, Podolsky, and Rosen – who first pointed out the strange properties of states like these. The mnemonic notation $|\beta_{00}\rangle, |\beta_{01}\rangle, |\beta_{10}\rangle, |\beta_{11}\rangle$ may be understood via the equations

$$|\beta_{xy}\rangle \equiv \frac{|0,y\rangle + (-1)^x |1,\bar{y}\rangle}{\sqrt{2}},\tag{1.27}$$

where $\bar{y}$ is the negation of y.

In	Out
$|00\rangle$	$(|00\rangle + |11\rangle)/\sqrt{2} \equiv |\beta_{00}\rangle$
$|01\rangle$	$(|01\rangle + |10\rangle)/\sqrt{2} \equiv |\beta_{01}\rangle$
$|10\rangle$	$(|00\rangle - |11\rangle)/\sqrt{2} \equiv |\beta_{10}\rangle$
$|11\rangle$	$(|01\rangle - |10\rangle)/\sqrt{2} \equiv |\beta_{11}\rangle$

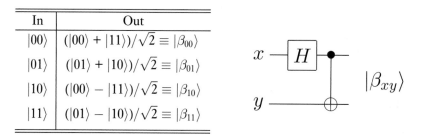

Figure 1.12. Quantum circuit to create Bell states, and its input–ouput quantum 'truth table'.

1.3.7 Example: quantum teleportation

We will now apply the techniques of the last few pages to understand something non-trivial, surprising, and a lot of fun – quantum teleportation! Quantum teleportation is a technique for moving quantum states around, even in the absence of a quantum communications channel linking the sender of the quantum state to the recipient.

Here's how quantum teleportation works. Alice and Bob met long ago but now live far apart. While together they generated an EPR pair, each taking one qubit of the EPR pair when they separated. Many years later, Bob is in hiding, and Alice's mission, should she choose to accept it, is to deliver a qubit $|\psi\rangle$ to Bob. She does not know the state of the qubit, and moreover can only send *classical* information to Bob. Should Alice accept the mission?

Intuitively, things look pretty bad for Alice. She doesn't know the state $|\psi\rangle$ of the qubit she has to send to Bob, and the laws of quantum mechanics prevent her from determining the state when she only has a single copy of $|\psi\rangle$ in her possession. What's worse, even if she did know the state $|\psi\rangle$, describing it precisely takes an infinite amount of classical information since $|\psi\rangle$ takes values in a *continuous* space. So even if she did know $|\psi\rangle$, it would take forever for Alice to describe the state to Bob. It's not looking good for Alice. Fortunately for Alice, quantum teleportation is a way of utilizing the entangled EPR pair in order to send $|\psi\rangle$ to Bob, with only a small overhead of classical communication.

In outline, the steps of the solution are as follows: Alice interacts the qubit $|\psi\rangle$ with her half of the EPR pair, and then measures the two qubits in her possession, obtaining one of four possible classical results, 00, 01, 10, and 11. She sends this information to Bob. Depending on Alice's classical message, Bob performs one of four operations on his half of the EPR pair. Amazingly, by doing this he can recover the original state $|\psi\rangle$!

The quantum circuit shown in Figure 1.13 gives a more precise description of quantum teleportation. The state to be teleported is $|\psi\rangle = \alpha|0\rangle + \beta|1\rangle$, where α and β are unknown amplitudes. The state input into the circuit $|\psi_0\rangle$ is

$$|\psi_0\rangle = |\psi\rangle|\beta_{00}\rangle\tag{1.28}$$

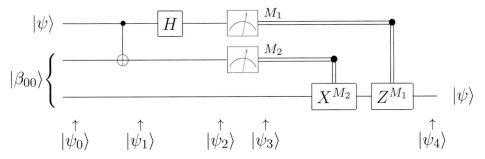

Figure 1.13. Quantum circuit for teleporting a qubit. The two top lines represent Alice's system, while the bottom line is Bob's system. The meters represent measurement, and the double lines coming out of them carry classical bits (recall that single lines denote qubits).

$$= \frac{1}{\sqrt{2}} \left[\alpha|0\rangle(|00\rangle + |11\rangle) + \beta|1\rangle(|00\rangle + |11\rangle) \right], \tag{1.29}$$

where we use the convention that the first two qubits (on the left) belong to Alice, and the third qubit to Bob. As we explained previously, Alice's second qubit and Bob's qubit start out in an EPR state. Alice sends her qubits through a CNOT gate, obtaining

$$|\psi_1\rangle = \frac{1}{\sqrt{2}} \left[\alpha|0\rangle(|00\rangle + |11\rangle) + \beta|1\rangle(|10\rangle + |01\rangle) \right]. \tag{1.30}$$

She then sends the first qubit through a Hadamard gate, obtaining

$$|\psi_2\rangle = \frac{1}{2} \left[\alpha(|0\rangle + |1\rangle)(|00\rangle + |11\rangle) + \beta(|0\rangle - |1\rangle)(|10\rangle + |01\rangle) \right]. \tag{1.31}$$

This state may be re-written in the following way, simply by regrouping terms:

$$|\psi_2\rangle = \frac{1}{2} \left[|00\rangle \left(\alpha|0\rangle + \beta|1\rangle \right) + |01\rangle \left(\alpha|1\rangle + \beta|0\rangle \right) \right.$$
$$\left. + |10\rangle \left(\alpha|0\rangle - \beta|1\rangle \right) + |11\rangle \left(\alpha|1\rangle - \beta|0\rangle \right) \right]. \tag{1.32}$$

This expression naturally breaks down into four terms. The first term has Alice's qubits in the state $|00\rangle$, and Bob's qubit in the state $\alpha|0\rangle + \beta|1\rangle$ – which is the original state $|\psi\rangle$. If Alice performs a measurement and obtains the result 00 then Bob's system will be in the state $|\psi\rangle$. Similarly, from the previous expression we can read off Bob's post-measurement state, given the result of Alice's measurement:

$$00 \longmapsto |\psi_3(00)\rangle \equiv \left[\alpha|0\rangle + \beta|1\rangle \right] \tag{1.33}$$

$$01 \longmapsto |\psi_3(01)\rangle \equiv \left[\alpha|1\rangle + \beta|0\rangle \right] \tag{1.34}$$

$$10 \longmapsto |\psi_3(10)\rangle \equiv \left[\alpha|0\rangle - \beta|1\rangle \right] \tag{1.35}$$

$$11 \longmapsto |\psi_3(11)\rangle \equiv \left[\alpha|1\rangle - \beta|0\rangle \right]. \tag{1.36}$$

Depending on Alice's measurement outcome, Bob's qubit will end up in one of these four possible states. Of course, to know which state it is in, Bob must be told the result of Alice's measurement – we will show later that it is this fact which prevents teleportation from being used to transmit information faster than light. Once Bob has learned the measurement outcome, Bob can 'fix up' his state, recovering $|\psi\rangle$, by applying the appropriate quantum gate. For example, in the case where the measurement yields 00, Bob doesn't need to do anything. If the measurement is 01 then Bob can fix up his state by applying the X gate. If the measurement is 10 then Bob can fix up his state by applying the Z gate. If the measurement is 11 then Bob can fix up his state by applying first an X and then a Z gate. Summing up, Bob needs to apply the transformation $Z^{M_1} X^{M_2}$ (note how time goes from left to right in circuit diagrams, but in matrix products terms on the *right* happen *first*) to his qubit, and he will recover the state $|\psi\rangle$.

There are many interesting features of teleportation, some of which we shall return to later in the book. For now we content ourselves with commenting on a couple of aspects. First, doesn't teleportation allow one to transmit quantum states faster than light? This would be rather peculiar, because the theory of relativity implies that faster than light information transfer could be used to send information backwards in time. Fortunately, quantum teleportation does not enable faster than light communication, because to complete the teleportation Alice must transmit her measurement result to Bob over a classical communications channel. We will show in Section 2.4.3 that without this classical communication, teleportation does not convey *any* information at all. The classical channel is limited by the speed of light, so it follows that quantum teleportation cannot be accomplished faster than the speed of light, resolving the apparent paradox.

A second puzzle about teleportation is that it appears to create a copy of the quantum state being teleported, in apparent violation of the no-cloning theorem discussed in Section 1.3.5. This violation is only illusory since after the teleportation process only the target qubit is left in the state $|\psi\rangle$, and the original data qubit ends up in one of the computational basis states $|0\rangle$ or $|1\rangle$, depending upon the measurement result on the first qubit.

What can we learn from quantum teleportation? Quite a lot! It's much more than just a neat trick one can do with quantum states. Quantum teleportation emphasizes the interchangeability of *different* resources in quantum mechanics, showing that one shared EPR pair together with two classical bits of communication is a resource at least the equal of one qubit of communication. Quantum computation and quantum information has revealed a plethora of methods for interchanging resources, many built upon quantum teleportation. In particular, in Chapter 10 we explain how teleportation can be used to build quantum gates which are resistant to the effects of noise, and in Chapter 12 we show that teleportation is intimately connected with the properties of quantum error-correcting codes. Despite these connections with other subjects, it is fair to say that we are only beginning to understand *why* it is that quantum teleportation is possible in quantum mechanics; in later chapters we endeavor to explain some of the insights that make such an understanding possible.

1.4 Quantum algorithms

What class of computations can be performed using quantum circuits? How does that class compare with the computations which can be performed using classical logical circuits? Can we find a task which a quantum computer may perform better than a classical computer? In this section we investigate these questions, explaining how to perform classical computations on quantum computers, giving some examples of problems for which quantum computers offer an advantage over classical computers, and summarizing the known quantum algorithms.

1.4.1 Classical computations on a quantum computer

Can we simulate a classical logic circuit using a quantum circuit? Not surprisingly, the answer to this question turns out to be yes. It would be very surprising if this were not the case, as physicists believe that all aspects of the world around us, including classical logic circuits, can ultimately be explained using quantum mechanics. As pointed out earlier, the reason quantum circuits cannot be used to directly simulate classical circuits is because unitary quantum logic gates are inherently *reversible*, whereas many classical logic gates such as the NAND gate are inherently irreversible.

Any classical circuit can be replaced by an equivalent circuit containing only *reversible* elements, by making use of a reversible gate known as the *Toffoli gate*. The Toffoli gate has three input bits and three output bits, as illustrated in Figure 1.14. Two of the bits are *control bits* that are unaffected by the action of the Toffoli gate. The third bit is a *target bit* that is flipped if both control bits are set to 1, and otherwise is left alone. Note that applying the Toffoli gate twice to a set of bits has the effect $(a, b, c) \rightarrow (a, b, c \oplus ab) \rightarrow (a, b, c)$, and thus the Toffoli gate is a reversible gate, since it has an inverse – itself.

Inputs			Outputs		
a	b	c	a'	b'	c'
0	0	0	0	0	0
0	0	1	0	0	1
0	1	0	0	1	0
0	1	1	0	1	1
1	0	0	1	0	0
1	0	1	1	0	1
1	1	0	1	1	1
1	1	1	1	1	0

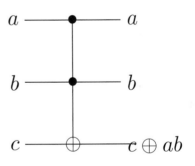

Figure 1.14. Truth table for the Toffoli gate, and its circuit representation.

The Toffoli gate can be used to simulate NAND gates, as shown in Figure 1.15, and can also be used to do FANOUT, as shown in Figure 1.16. With these two operations it becomes possible to simulate all other elements in a classical circuit, and thus an arbitrary classical circuit can be simulated by an equivalent reversible circuit.

The Toffoli gate has been described as a classical gate, but it can also be implemented as a quantum logic gate. By definition, the quantum logic implementation of the Toffoli gate simply permutes computational basis states in the same way as the classical Toffoli gate. For example, the quantum Toffoli gate acting on the state $|110\rangle$ flips the third qubit because the first two are set, resulting in the state $|111\rangle$. It is tedious but not difficult to write this transformation out as an 8 by 8 matrix, U, and verify explicitly that U is a unitary matrix, and thus the Toffoli gate is a legitimate quantum gate. The quantum Toffoli gate can be used to simulate irreversible classical logic gates, just as the classical

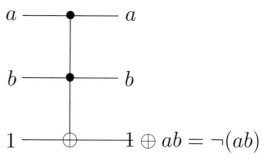

Figure 1.15. Classical circuit implementing a NAND gate using a Toffoli gate. The top two bits represent the input to the NAND, while the third bit is prepared in the standard state 1, sometimes known as an *ancilla* state. The output from the NAND is on the third bit.

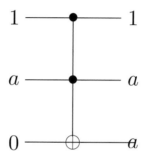

Figure 1.16. FANOUT with the Toffoli gate, with the second bit being the input to the FANOUT (and the other two bits standard ancilla states), and the output from FANOUT appearing on the second and third bits.

Toffoli gate was, and ensures that quantum computers are capable of performing any computation which a classical (deterministic) computer may do.

What if the classical computer is non-deterministic, that is, has the ability to generate random bits to be used in the computation? Not surprisingly, it is easy for a quantum computer to simulate this. To perform such a simulation it turns out to be sufficient to produce random fair coin tosses, which can be done by preparing a qubit in the state $|0\rangle$, sending it through a Hadamard gate to produce $(|0\rangle + |1\rangle)/\sqrt{2}$, and then measuring the state. The result will be $|0\rangle$ or $|1\rangle$ with 50/50 probability. This provides a quantum computer with the ability to efficiently simulate a non-deterministic classical computer.

Of course, if the ability to simulate classical computers were the only feature of quantum computers there would be little point in going to all the trouble of exploiting quantum effects! The advantage of quantum computing is that much more powerful functions may be computed using qubits and quantum gates. In the next few sections we explain how to do this, culminating in the Deutsch–Jozsa algorithm, our first example of a quantum algorithm able to solve a problem faster than any classical algorithm.

1.4.2 Quantum parallelism

Quantum parallelism is a fundamental feature of many quantum algorithms. Heuristically, and at the risk of over-simplifying, quantum parallelism allows quantum computers to evaluate a function $f(x)$ for many *different* values of x simultaneously. In this section we explain how quantum parallelism works, and some of its limitations.

Suppose $f(x) : \{0, 1\} \rightarrow \{0, 1\}$ is a function with a one-bit domain and range. A convenient way of computing this function on a quantum computer is to consider a two qubit quantum computer which starts in the state $|x, y\rangle$. With an appropriate sequence of logic gates it is possible to transform this state into $|x, y \oplus f(x)\rangle$, where $\oplus$ indicates addition modulo 2; the first register is called the 'data' register, and the second register the 'target' register. We give the transformation defined by the map $|x, y\rangle \rightarrow |x, y \oplus f(x)\rangle$ a name, U_f, and note that it is easily shown to be unitary. If $y = 0$, then the final state of the second qubit is just the value $f(x)$. (In Section 3.2.5 we show that given a classical circuit for computing f there is a quantum circuit of comparable efficiency which computes the transformation U_f on a quantum computer. For our purposes it can be considered to be a black box.)

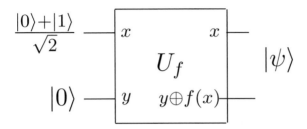

Figure 1.17. Quantum circuit for evaluating $f(0)$ and $f(1)$ simultaneously. U_f is the quantum circuit which takes inputs like $|x, y\rangle$ to $|x, y \oplus f(x)\rangle$.

Consider the circuit shown in Figure 1.17, which applies U_f to an input not in the computational basis. Instead, the data register is prepared in the superposition $(|0\rangle + |1\rangle)/\sqrt{2}$, which can be created with a Hadamard gate acting on $|0\rangle$. Then we apply U_f, resulting in the state:

$$\frac{|0, f(0)\rangle + |1, f(1)\rangle}{\sqrt{2}}. \tag{1.37}$$

This is a remarkable state! The different terms contain information about both $f(0)$ and $f(1)$; it is almost as if we have evaluated $f(x)$ for two values of x simultaneously, a feature known as 'quantum parallelism'. Unlike classical parallelism, where multiple circuits each built to compute $f(x)$ are executed simultaneously, here a *single* $f(x)$ circuit is employed to evaluate the function for multiple values of x simultaneously, by exploiting the ability of a quantum computer to be in superpositions of different states.

This procedure can easily be generalized to functions on an arbitrary number of bits, by using a general operation known as the *Hadamard transform*, or sometimes the *Walsh–Hadamard transform*. This operation is just n Hadamard gates acting in parallel on n qubits. For example, shown in Figure 1.18 is the case $n = 2$ with qubits initially prepared as $|0\rangle$, which gives

$$\left(\frac{|0\rangle + |1\rangle}{\sqrt{2}}\right)\left(\frac{|0\rangle + |1\rangle}{\sqrt{2}}\right) = \frac{|00\rangle + |01\rangle + |10\rangle + |11\rangle}{2} \tag{1.38}$$

as output. We write $H^{\otimes 2}$ to denote the parallel action of two Hadamard gates, and read '$\otimes$' as 'tensor'. More generally, the result of performing the Hadamard transform on n

qubits initially in the all $|0\rangle$ state is

$$\frac{1}{\sqrt{2^n}} \sum_x |x\rangle, \tag{1.39}$$

where the sum is over all possible values of x, and we write $H^{\otimes n}$ to denote this action. That is, the Hadamard transform produces an equal superposition of all computational basis states. Moreover, it does this extremely efficiently, producing a superposition of 2^n states using just n gates.

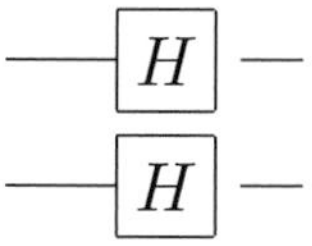

Figure 1.18. The Hadamard transform $H^{\otimes 2}$ on two qubits.

Quantum parallel evaluation of a function with an n bit input x and 1 bit output, $f(x)$, can thus be performed in the following manner. Prepare the $n + 1$ qubit state $|0\rangle^{\otimes n}|0\rangle$, then apply the Hadamard transform to the first n qubits, followed by the quantum circuit implementing U_f. This produces the state

$$\frac{1}{\sqrt{2^n}} \sum_x |x\rangle|f(x)\rangle. \tag{1.40}$$

In some sense, quantum parallelism enables all possible values of the function f to be evaluated simultaneously, even though we apparently only evaluated f once. However, this parallelism is *not* immediately useful. In our single qubit example, measurement of the state gives only *either* $|0, f(0)\rangle$ or $|1, f(1)\rangle$! Similarly, in the general case, measurement of the state $\sum_x |x, f(x)\rangle$ would give only $f(x)$ for a single value of x. Of course, a classical computer can do this easily! Quantum computation requires something more than just quantum parallelism to be useful; it requires the ability to *extract* information about more than one value of $f(x)$ from superposition states like $\sum_x |x, f(x)\rangle$. Over the next two sections we investigate examples of how this may be done.

1.4.3 Deutsch's algorithm

A simple modification of the circuit in Figure 1.17 demonstrates how quantum circuits can outperform classical ones by implementing *Deutsch's algorithm* (we actually present a simplified and improved version of the original algorithm; see 'History and further reading' at the end of the chapter). Deutsch's algorithm combines quantum parallelism with a property of quantum mechanics known as *interference*. As before, let us use the Hadamard gate to prepare the first qubit as the superposition $(|0\rangle + |1\rangle)/\sqrt{2}$, but now let us prepare the second qubit y as the superposition $(|0\rangle - |1\rangle)/\sqrt{2}$, using a Hadamard gate applied to the state $|1\rangle$. Let us follow the states along to see what happens in this circuit, shown in Figure 1.19.

The input state

$$|\psi_0\rangle = |01\rangle \tag{1.41}$$

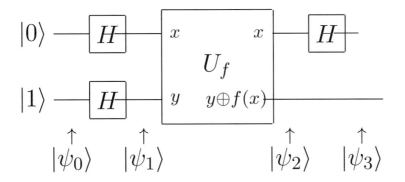

Figure 1.19. Quantum circuit implementing Deutsch's algorithm.

is sent through two Hadamard gates to give

$$|\psi_1\rangle = \left[\frac{|0\rangle + |1\rangle}{\sqrt{2}}\right]\left[\frac{|0\rangle - |1\rangle}{\sqrt{2}}\right]. \tag{1.42}$$

A little thought shows that if we apply U_f to the state $|x\rangle(|0\rangle - |1\rangle)/\sqrt{2}$ then we obtain the state $(-1)^{f(x)}|x\rangle(|0\rangle - |1\rangle)/\sqrt{2}$. Applying U_f to $|\psi_1\rangle$ therefore leaves us with one of two possibilities:

$$|\psi_2\rangle = \begin{cases} \pm\left[\dfrac{|0\rangle + |1\rangle}{\sqrt{2}}\right]\left[\dfrac{|0\rangle - |1\rangle}{\sqrt{2}}\right] & \text{if } f(0) = f(1) \\[2ex] \pm\left[\dfrac{|0\rangle - |1\rangle}{\sqrt{2}}\right]\left[\dfrac{|0\rangle - |1\rangle}{\sqrt{2}}\right] & \text{if } f(0) \neq f(1). \end{cases} \tag{1.43}$$

The final Hadamard gate on the first qubit thus gives us

$$|\psi_3\rangle = \begin{cases} \pm|0\rangle\left[\dfrac{|0\rangle - |1\rangle}{\sqrt{2}}\right] & \text{if } f(0) = f(1) \\[2ex] \pm|1\rangle\left[\dfrac{|0\rangle - |1\rangle}{\sqrt{2}}\right] & \text{if } f(0) \neq f(1). \end{cases} \tag{1.44}$$

Realizing that $f(0) \oplus f(1)$ is 0 if $f(0) = f(1)$ and 1 otherwise, we can rewrite this result concisely as

$$|\psi_3\rangle = \pm|f(0) \oplus f(1)\rangle\left[\frac{|0\rangle - |1\rangle}{\sqrt{2}}\right], \tag{1.45}$$

so by measuring the first qubit we may determine $f(0) \oplus f(1)$. This is very interesting indeed: the quantum circuit has given us the ability to determine a *global property* of $f(x)$, namely $f(0) \oplus f(1)$, using only *one* evaluation of $f(x)$! This is faster than is possible with a classical apparatus, which would require at least two evaluations.

This example highlights the difference between quantum parallelism and classical randomized algorithms. Naively, one might think that the state $|0\rangle|f(0)\rangle + |1\rangle|f(1)\rangle$ corresponds rather closely to a probabilistic classical computer that evaluates $f(0)$ with probability one-half, or $f(1)$ with probability one-half. The difference is that in a classical computer these two alternatives forever exclude one another; in a quantum computer it is possible for the two alternatives to *interfere* with one another to yield some global property of the function f, by using something like the Hadamard gate to recombine the different alternatives, as was done in Deutsch's algorithm. The essence of the design of many quantum algorithms is that a clever choice of function and final transformation allows efficient determination of useful global information about the function – information which cannot be attained quickly on a classical computer.

1.4.4 The Deutsch–Jozsa algorithm

Deutsch's algorithm is a simple case of a more general quantum algorithm, which we shall refer to as the Deutsch–Jozsa algorithm. The application, known as *Deutsch's problem*, may be described as the following game. Alice, in Amsterdam, selects a number x from 0 to $2^n - 1$, and mails it in a letter to Bob, in Boston. Bob calculates some function $f(x)$ and replies with the result, which is either 0 or 1. Now, Bob has promised to use a function f which is of one of two kinds; either $f(x)$ is *constant* for all values of x, or else $f(x)$ is *balanced*, that is, equal to 1 for exactly half of all the possible x, and 0 for the other half. Alice's goal is to determine with certainty whether Bob has chosen a constant or a balanced function, corresponding with him as little as possible. How fast can she succeed?

In the classical case, Alice may only send Bob one value of x in each letter. At worst, Alice will need to query Bob at least $2^n/2 + 1$ times, since she may receive $2^n/2$ 0s before finally getting a 1, telling her that Bob's function is balanced. The best deterministic classical algorithm she can use therefore requires $2^n/2 + 1$ queries. Note that in each letter, Alice sends Bob n bits of information. Furthermore, in this example, physical distance is being used to artificially elevate the cost of calculating $f(x)$, but this is not needed in the general problem, where $f(x)$ may be inherently difficult to calculate.

If Bob and Alice were able to exchange qubits, instead of just classical bits, and if Bob agreed to calculate $f(x)$ using a unitary transform U_f, then Alice could achieve her goal in just *one* correspondence with Bob, using the following algorithm.

Analogously to Deutsch's algorithm, Alice has an n qubit register to store her query in, and a single qubit register which she will give to Bob, to store the answer in. She begins by preparing both her query and answer registers in a superposition state. Bob will evaluate $f(x)$ using quantum parallelism and leave the result in the answer register. Alice then interferes states in the superposition using a Hadamard transform on the query register, and finishes by performing a suitable measurement to determine whether f was constant or balanced.

The specific steps of the algorithm are depicted in Figure 1.20. Let us follow the states through this circuit. The input state

$$|\psi_0\rangle = |0\rangle^{\otimes n}|1\rangle \tag{1.46}$$

is similar to that of Equation (1.41), but here the query register describes the state of n qubits all prepared in the $|0\rangle$ state. After the Hadamard transform on the query register and the Hadamard gate on the answer register we have

$$|\psi_1\rangle = \sum_{x\in\{0,1\}^n} \frac{|x\rangle}{\sqrt{2^n}} \left[\frac{|0\rangle - |1\rangle}{\sqrt{2}}\right]. \tag{1.47}$$

The query register is now a superposition of all values, and the answer register is in an

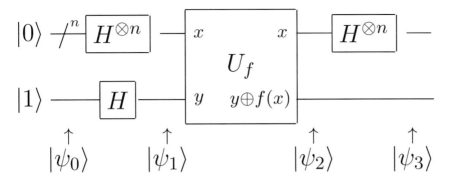

Figure 1.20. Quantum circuit implementing the general Deutsch–Jozsa algorithm. The wire with a '/' through it represents a set of n qubits, similar to the common engineering notation.

evenly weighted superposition of 0 and 1. Next, the function f is evaluated (by Bob) using $U_f : |x, y\rangle \rightarrow |x, y \oplus f(x)\rangle$, giving

$$|\psi_2\rangle = \sum_x \frac{(-1)^{f(x)}|x\rangle}{\sqrt{2^n}} \left[\frac{|0\rangle - |1\rangle}{\sqrt{2}} \right]. \tag{1.48}$$

Alice now has a set of qubits in which the result of Bob's function evaluation is stored in the amplitude of the qubit superposition state. She now interferes terms in the super-position using a Hadamard transform on the query register. To determine the result of the Hadamard transform it helps to first calculate the effect of the Hadamard transform on a state $|x\rangle$. By checking the cases $x = 0$ and $x = 1$ separately we see that for a single qubit $H|x\rangle = \sum_z (-1)^{xz}|z\rangle/\sqrt{2}$. Thus

$$H^{\otimes n}|x_1, \ldots, x_n\rangle = \frac{\sum_{z_1, \ldots, z_n}(-1)^{x_1 z_1 + \cdots + x_n z_n}|z_1, \ldots, z_n\rangle}{\sqrt{2^n}}. \tag{1.49}$$

This can be summarized more succinctly in the very useful equation

$$H^{\otimes n}|x\rangle = \frac{\sum_z (-1)^{x \cdot z}|z\rangle}{\sqrt{2^n}}, \tag{1.50}$$

where $x \cdot z$ is the bitwise inner product of x and z, modulo 2. Using this equation and (1.48) we can now evaluate $|\psi_3\rangle$,

$$|\psi_3\rangle = \sum_z \sum_x \frac{(-1)^{x \cdot z + f(x)}|z\rangle}{2^n} \left[\frac{|0\rangle - |1\rangle}{\sqrt{2}} \right]. \tag{1.51}$$

Alice now observes the query register. Note that the amplitude for the state $|0\rangle^{\otimes n}$ is $\sum_x(-1)^{f(x)}/2^n$. Let's look at the two possible cases – f constant and f balanced – to discern what happens. In the case where f is constant the amplitude for $|0\rangle^{\otimes n}$ is $+1$ or -1, depending on the constant value $f(x)$ takes. Because $|\psi_3\rangle$ is of unit length it follows that all the other amplitudes must be zero, and an observation will yield 0s for all qubits in the query register. If f is balanced then the positive and negative contributions to the amplitude for $|0\rangle^{\otimes n}$ cancel, leaving an amplitude of zero, and a measurement must yield a result other than 0 on at least one qubit in the query register. Summarizing, if Alice measures all 0s then the function is constant; otherwise the function is balanced. The Deutsch–Jozsa algorithm is summarized below.

Algorithm: Deutsch–Jozsa

Inputs: (1) A black box U_f which performs the transformation $|x\rangle|y\rangle \rightarrow |x\rangle|y \oplus f(x)\rangle$, for $x \in \{0, \dots, 2^n - 1\}$ and $f(x) \in \{0, 1\}$. It is promised that $f(x)$ is either *constant* for all values of x, or else $f(x)$ is *balanced*, that is, equal to 1 for exactly half of all the possible x, and 0 for the other half.

Outputs: 0 if and only if f is constant.

Runtime: One evaluation of U_f. Always succeeds.

Procedure:

1. $|0\rangle^{\otimes n}|1\rangle$ initialize state

2. $\rightarrow \dfrac{1}{\sqrt{2^n}} \displaystyle\sum_{x=0}^{2^n-1} |x\rangle \left[\dfrac{|0\rangle - |1\rangle}{\sqrt{2}} \right]$ create superposition using Hadamard gates

3. $\rightarrow \displaystyle\sum_x (-1)^{f(x)}|x\rangle \left[\dfrac{|0\rangle - |1\rangle}{\sqrt{2}} \right]$ calculate function f using U_f

4. $\rightarrow \displaystyle\sum_z \sum_x \dfrac{(-1)^{x \cdot z + f(x)}|z\rangle}{\sqrt{2^n}} \left[\dfrac{|0\rangle - |1\rangle}{\sqrt{2}} \right]$ perform Hadamard transform

5. $\rightarrow z$ measure to obtain final output z

We've shown that a quantum computer can solve Deutsch's problem with one evaluation of the function f compared to the classical requirement for $2^n/2 + 1$ evaluations. This appears impressive, but there are several important caveats. First, Deutsch's problem is not an especially important problem; it has no known applications. Second, the comparison between classical and quantum algorithms is in some ways an apples and oranges comparison, as the method for evaluating the function is quite different in the two cases. Third, if Alice is allowed to use a probabilistic classical computer, then by asking Bob to evaluate $f(x)$ for a few randomly chosen x she can very quickly determine with high probability whether f is constant or balanced. This probabilistic scenario is perhaps more realistic than the deterministic scenario we have been considering. Despite these caveats, the Deutsch–Jozsa algorithm contains the seeds for more impressive quantum algorithms, and it is enlightening to attempt to understand the principles behind its operation.

Exercise 1.1: (Probabilistic classical algorithm) Suppose that the problem is not to distinguish between the constant and balanced functions *with certainty*, but rather, with some probability of error $\epsilon < 1/2$. What is the performance of the best classical algorithm for this problem?

1.4.5 Quantum algorithms summarized

The Deutsch–Jozsa algorithm suggests that quantum computers may be capable of solving some computational problems much more efficiently than classical computers. Unfortunately, the problem it solves is of little practical interest. Are there more interesting problems whose solution may be obtained more efficiently using quantum algorithms? What are the principles underlying such algorithms? What are the ultimate limits of a quantum computer's computational power?

Broadly speaking, there are three classes of quantum algorithms which provide an advantage over known classical algorithms. First, there is the class of algorithms based upon quantum versions of the Fourier transform, a tool which is also widely used in classical algorithms. The Deutsch–Jozsa algorithm is an example of this type of algorithm, as are Shor's algorithms for factoring and discrete logarithm. The second class of algorithms is quantum search algorithms. The third class of algorithms is quantum simulation, whereby a quantum computer is used to simulate a quantum system. We now briefly describe each of these classes of algorithms, and then summarize what is known or suspected about the computational power of quantum computers.

Quantum algorithms based upon the Fourier transform

The discrete Fourier transform is usually described as transforming a set $x_0, \ldots, x_{N-1}$ of N complex numbers into a set of complex numbers $y_0, \ldots, y_{N-1}$ defined by

$$y_k \equiv \frac{1}{\sqrt{N}} \sum_{j=0}^{N-1} e^{2\pi i j k / N} x_j . \tag{1.52}$$

Of course, this transformation has an enormous number of applications in many branches of science; the Fourier transformed version of a problem is often easier than the original problem, enabling a solution.

The Fourier transform has proved so useful that a beautiful generalized theory of Fourier transforms has been developed which goes beyond the definition (1.52). This general theory involves some technical ideas from the character theory of finite groups, and we will not attempt to describe it here. What is important is that the Hadamard transform used in the Deutsch–Jozsa algorithm is an example of this generalized class of Fourier transforms. Moreover, many of the other important quantum algorithms also involve some type of Fourier transform.

The most important quantum algorithms known, Shor's fast algorithms for factoring and discrete logarithm, are two examples of algorithms based upon the Fourier transform defined in Equation (1.52). The Equation (1.52) does not appear terribly quantum mechanical in the form we have written it. Imagine, however, that we define a linear transformation U on n qubits by its action on computational basis states $|j\rangle$, where $0 \le j \le 2^n - 1$,

$$|j\rangle \longrightarrow \frac{1}{\sqrt{2^n}} \sum_{k=0}^{2^n-1} e^{2\pi i j k / 2^n} |k\rangle . \tag{1.53}$$

It can be checked that this transformation is unitary, and in fact can be realized as a quantum circuit. Moreover, if we write out its action on superpositions,

$$\sum_{j=0}^{2^n-1} x_j |j\rangle \longrightarrow \frac{1}{\sqrt{2^n}} \sum_{k=0}^{2^n-1} \left[\sum_{j=0}^{2^n-1} e^{2\pi i j k / 2^n} x_j \right] |k\rangle = \sum_{k=0}^{2^n-1} y_k |k\rangle , \tag{1.54}$$

we see that it corresponds to a vector notation for the Fourier transform (1.52) for the case $N = 2^n$.

How quickly can we perform the Fourier transform? Classically, the fast Fourier transform takes roughly $N \log(N) = n2^n$ steps to Fourier transform $N = 2^n$ numbers. On a quantum computer, the Fourier transform can be accomplished using about $\log^2(N) = n^2$ steps, an exponential saving! The quantum circuit to do this is explained in Chapter 5.

This result seems to indicate that quantum computers can be used to very quickly compute the Fourier transform of a vector of 2^n complex numbers, which would be fantastically useful in a wide range of applications. However, that is *not* exactly the case; the Fourier transform is being performed on the information 'hidden' in the amplitudes of the quantum state. This information is not directly accessible to measurement. The catch, of course, is that if the output state is measured, it will collapse each qubit into the state $|0\rangle$ or $|1\rangle$, preventing us from learning the transform result y_k directly. This example speaks to the heart of the conundrum of devising a quantum algorithm. On the one hand, we can perform certain calculations on the 2^n amplitudes associated with n qubits far more efficiently than would be possible on a classical computer. But on the other hand, the results of such a calculation are not available to us if we go about it in a straightforward manner. More cleverness is required in order to harness the power of quantum computation.

Fortunately, it does turn out to be possible to utilize the quantum Fourier transform to efficiently solve several problems that are believed to have no efficient solution on a classical computer. These problems include Deutsch's problem, and Shor's algorithms for discrete logarithm and factoring. This line of thought culminated in Kitaev's discovery of a method to solve the *Abelian stabilizer problem*, and the generalization to the *hidden subgroup problem*,

> Let f be a function from a finitely generated group G to a finite set X such that f is constant on the cosets of a subgroup K, and distinct on each coset. Given a quantum black box for performing the unitary transform $U|g\rangle|h\rangle = |g\rangle|h \oplus f(g)\rangle$, for $g \in G$, $h \in X$, and $\oplus$ an appropriately chosen binary operation on X, find a generating set for K.

The Deutsch–Jozsa algorithm, Shor's algorithms, and related 'exponentially fast' quantum algorithms can all be viewed as special cases of this algorithm. The quantum Fourier transform and its applications are described in Chapter 5.

Quantum search algorithms

A completely different class of algorithms is represented by the quantum search algorithm, whose basic principles were discovered by Grover. The quantum search algorithm solves the following problem: Given a search space of size N, and no prior knowledge about the structure of the information in it, we want to find an element of that search space satisfying a known property. How long does it take to find an element satisfying that property? Classically, this problem requires approximately N operations, but the quantum search algorithm allows it to be solved using approximately $\sqrt{N}$ operations.

The quantum search algorithm offers only a quadratic speedup, as opposed to the more impressive exponential speedup offered by algorithms based on the quantum Fourier transform. However, the quantum search algorithm is still of great interest, since searching heuristics have a wider range of application than the problems solved using the quantum Fourier transform, and adaptations of the quantum search algorithm may have utility for a very wide range of problems. The quantum search algorithm and its applications are described in Chapter 6.

Quantum simulation

Simulating naturally occurring quantum mechanical systems is an obvious candidate for a task at which quantum computers may excel, yet which is believed to be difficult on a classical computer. Classical computers have difficulty simulating general quantum systems for much the same reasons they have difficulty simulating quantum computers – the number of complex numbers needed to describe a quantum system generally grows *exponentially* with the size of the system, rather than linearly, as occurs in classical systems. In general, storing the quantum state of a system with n distinct components takes something like c^n bits of memory on a classical computer, where c is a constant which depends upon details of the system being simulated, and the desired accuracy of the simulation.

By contrast, a quantum computer can perform the simulation using kn qubits, where k is again a constant which depends upon the details of the system being simulated. This allows quantum computers to efficiently perform simulations of quantum mechanical systems that are believed not to be efficiently simulatable on a classical computer. A significant caveat is that even though a quantum computer can simulate many quantum systems far more efficiently than a classical computer, this does not mean that the fast simulation will allow the desired information about the quantum system to be obtained. When measured, a kn qubit simulation will collapse into a definite state, giving only kn bits of information; the c^n bits of 'hidden information' in the wavefunction is not entirely accessible. Thus, a crucial step in making quantum simulations useful is development of systematic means by which desired answers can be efficiently extracted; how to do this is only partially understood.

Despite this caveat, quantum simulation is likely to be an important application of quantum computers. The simulation of quantum systems is an important problem in many fields, notably quantum chemistry, where the computational constraints imposed by classical computers make it difficult to accurately simulate the behavior of even moderately sized molecules, much less the very large molecules that occur in many important biological systems. Obtaining faster and more accurate simulations of such systems may therefore have the welcome effect of enabling advances in other fields in which quantum phenomena are important.

In the future we may discover a physical phenomenon in Nature which cannot be efficiently simulated on a quantum computer. Far from being bad news, this would be wonderful! At the least, it will stimulate us to extend our models of computation to encompass the new phenomenon, and increase the power of our computational models beyond the existing quantum computing model. It also seems likely that very interesting new physical effects will be associated with any such phenomenon!

Another application for quantum simulation is as a general method to obtain insight into other quantum algorithms; for example, in Section 6.2 we explain how the quantum search algorithm can be viewed as the solution to a problem of quantum simulation. By approaching the problem in this fashion it becomes much easier to understand the origin of the quantum search algorithm.

Finally, quantum simulation also gives rise to an interesting and optimistic 'quantum corollary' to Moore's law. Recall that Moore's law states that the power of classical computers will double once every two years or so, for constant cost. However, suppose we are simulating a quantum system on a classical computer, and want to add a single qubit (or a larger system) to the system being simulated. This doubles or more the memory requirements needed for a classical computer to store a description of the state of the quantum system, with a similar or greater cost in the time needed to simulate the dynamics. The quantum corollary to Moore's law follows from this observation, stating that quantum computers are keeping pace with classical computers provided a *single qubit* is added to the quantum computer every two years. This corollary should not be taken too seriously, as the exact nature of the gain, if any, of quantum computation over classical is not yet clear. Nevertheless, this heuristic statement helps convey why we should be interested in quantum computers, and hopeful that they will one day be able to outperform the most powerful classical computers, at least for some applications.

The power of quantum computation

How powerful are quantum computers? What gives them their power? Nobody yet knows the answers to these questions, despite the suspicions fostered by examples such as factoring, which strongly suggest that quantum computers are more powerful than classical computers. It is still possible that quantum computers are no more powerful than classical computers, in the sense that any problem which can be efficiently solved on a quantum computer can also be efficiently solved on a classical computer. On the other hand, it may eventually be proved that quantum computers are much more powerful than classical computers. We now take a brief look at what is known about the power of quantum computation.

Computational complexity theory is the subject of classifying the difficulty of various computational problems, both classical and quantum, and to understand the power of quantum computers we will first examine some general ideas from computational complexity. The most basic idea is that of a *complexity class*. A complexity class can be thought of as a collection of computational problems, all of which share some common feature with respect to the computational resources needed to solve those problems.

Two of the most important complexity classes go by the names **P** and **NP**. Roughly speaking, **P** is the class of computational problems that can be solved quickly on a classical computer. **NP** is the class of problems which have *solutions* which can be quickly checked on a classical computer. To understand the distinction between **P** and **NP**, consider the problem of finding the prime factors of an integer, n. So far as is known there is no fast way of solving this problem on a classical computer, which suggests that the problem is not in **P**. On the other hand, if somebody tells you that some number p is a factor of n, then we can quickly check that this is correct by dividing p into n, so factoring is a problem in **NP**.

It is clear that **P** is a subset of **NP**, since the ability to solve a problem implies the ability to check potential solutions. What is not so clear is whether or not there are problems in **NP** that are not in **P**. Perhaps the most important unsolved problem in theoretical computer science is to determine whether these two classes are different:

$$\mathbf{P} \overset{?}{\neq} \mathbf{NP}. \tag{1.55}$$

Most researchers believe that **NP** contains problems that are not in **P**. In particular, there is an important subclass of the **NP** problems, the **NP**-complete problems, that are of especial importance for two reasons. First, there are thousands of problems, many highly important, that are known to be **NP**-complete. Second, any given **NP**-complete problem is in some sense 'at least as hard' as all other problems in **NP**. More precisely, an algorithm to solve a specific **NP**-complete problem can be adapted to solve any other problem in **NP**, with a small overhead. In particular, if **P** ≠ **NP**, then it will follow that no **NP**-complete problem can be efficiently solved on a classical computer.

It is not known whether quantum computers can be used to quickly solve all the problems in **NP**, despite the fact that they can be used to solve some problems – like factoring – which are believed by many people to be in **NP** but not in **P**. (Note that factoring is not known to be **NP**-complete, otherwise we would already know how to efficiently solve all problems in **NP** using quantum computers.) It would certainly be very exciting if it were possible to solve all the problems in **NP** efficiently on a quantum computer. There is a very interesting negative result known in this direction which rules out using a simple variant of quantum parallelism to solve all the problems in **NP**. Specifically, one approach to the problem of solving problems in **NP** on a quantum computer is to try to use some form of quantum parallelism to search in parallel through all the possible solutions to the problem. In Section 6.6 we will show that no approach based upon such a search-based methodology can yield an efficient solution to all the problems in **NP**. While it is disappointing that this approach fails, it does not rule out that some deeper structure exists in the problems in **NP** that will allow them all to be solved quickly using a quantum computer.

P and **NP** are just two of a plethora of complexity classes that have been defined. Another important complexity class is **PSPACE**. Roughly speaking, **PSPACE** consists of those problems which can be solved using resources which are few in spatial size (that is, the computer is 'small'), but not necessarily in time ('long' computations are fine). **PSPACE** is believed to be strictly larger than both **P** and **NP** although, again, this has never been proved. Finally, the complexity class **BPP** is the class of problems that can be solved using randomized algorithms in polynomial time, if a bounded probability of error (say $1/4$) is allowed in the solution to the problem. **BPP** is widely regarded as being, even more so than **P**, the class of problems which should be considered efficiently soluble on a classical computer. We have elected to concentrate here on **P** rather than **BPP** because **P** has been studied in more depth, however many similar ideas and conclusions arise in connection with **BPP**.

What of quantum complexity classes? We can define **BQP** to be the class of all computational problems which can be solved efficiently on a quantum computer, where a bounded probability of error is allowed. (Strictly speaking this makes **BQP** more analogous to the classical complexity class **BPP** than to **P**, however we will ignore this subtlety for the purposes of the present discussion, and treat it as the analogue of **P**.) Exactly where **BQP** fits with respect to **P**, **NP** and **PSPACE** is as yet unknown. What is known is that quantum computers can solve all the problems in **P** efficiently, but that there are no problems outside of **PSPACE** which they can solve efficiently. Therefore, **BQP** lies somewhere between **P** and **PSPACE**, as illustrated in Figure 1.21. An important implication is that if it is proved that quantum computers are strictly more powerful than classical computers, then it will follow that **P** is not equal to **PSPACE**. Proving this latter result has been attempted without success by many computer scientists, suggesting that it may be non-trivial to prove that quantum computers are more powerful than classical computers, despite much evidence in favor of this proposition.

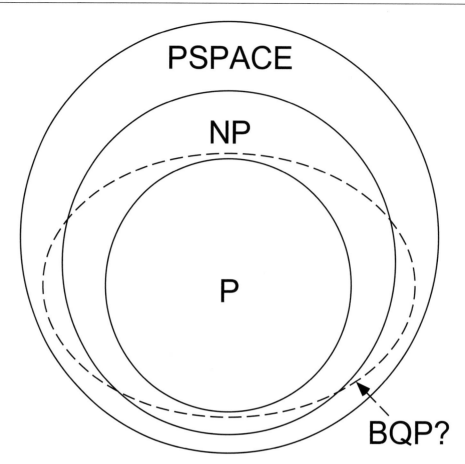

Figure 1.21. The relationship between classical and quantum complexity classes. Quantum computers can quickly solve any problem in **P**, and it is known that they can't solve problems outside of **PSPACE** quickly. Where quantum computers fit between **P** and **PSPACE** is not known, in part because we don't even know whether **PSPACE** is bigger than **P**!

We won't speculate further on the ultimate power of quantum computation now, preferring to wait until after we have better understood the principles on which fast quantum algorithms are based, a topic which occupies us for most of Part II of this book. What is already clear is that the *theory* of quantum computation poses interesting and significant challenges to the traditional notions of computation. What makes this an important challenge is that the theoretical model of quantum computation is believed to be *experimentally* realizable, because – to the best of our knowledge – this theory is consistent with the way Nature works. If this were not so then quantum computation would be just another mathematical curiosity.

1.5 Experimental quantum information processing

Quantum computation and quantum information is a wonderful theoretical discovery, but its central concepts, such as superpositions and entanglement, run counter to the intuition we garner from the everyday world around us. What evidence do we have that these ideas truly describe how Nature operates? Will the realization of large-scale quantum computers be experimentally feasible? Or might there be some principle of physics which fundamentally prohibits their eventual scaling? In the next two sections we address these questions. We begin with a review of the famous 'Stern–Gerlach' experiment, which provides evidence for the existence of qubits in Nature. We then widen our scope, addressing the broader problem of how to build practical quantum information processing systems.

1.5.1 The Stern–Gerlach experiment

The qubit is a fundamental element for quantum computation and quantum information. How do we know that systems with the properties of qubits exist in Nature? At the time of writing there is an enormous amount of evidence that this is so, but in the early days of quantum mechanics the qubit structure was not at all obvious, and people struggled with phenomena that we may now understand in terms of qubits, that is, in terms of two level quantum systems.

A decisive (and very famous) early experiment indicating the qubit structure was conceived by Stern in 1921 and performed with Gerlach in 1922 in Frankfurt. In the original Stern–Gerlach experiment, hot atoms were 'beamed' from an oven through a magnetic field which caused the atoms to be deflected, and then the position of each atom was recorded, as illustrated in Figure 1.22. The original experiment was done with silver atoms, which have a complicated structure that obscures the effects we are discussing. What we describe below actually follows a 1927 experiment done using hydrogen atoms. The same basic effect is observed, but with hydrogen atoms the discussion is easier to follow. Keep in mind, though, that this privilege wasn't available to people in the early 1920s, and they had to be very ingenious to think up explanations for the more complicated effects they observed.

Hydrogen atoms contain a proton and an orbiting electron. You can think of this electron as a little 'electric current' around the proton. This electric current causes the atom to have a magnetic field; each atom has what physicists call a 'magnetic dipole moment'. As a result each atom behaves like a little bar magnet with an axis corresponding to the axis the electron is spinning around. Throwing little bar magnets through a magnetic field causes the magnets to be deflected by the field, and we expect to see a similar deflection of atoms in the Stern–Gerlach experiment.

How the atom is deflected depends upon both the atom's magnetic dipole moment – the axis the electron is spinning around – and the magnetic field generated by the Stern–Gerlach device. We won't go through the details, but suffice to say that by constructing the Stern–Gerlach device appropriately, we can cause the atom to be deflected by an amount that depends upon the $\hat{z}$ component of the atom's magnetic dipole moment, where $\hat{z}$ is some fixed external axis.

Two major surprises emerge when this experiment is performed. First, since the hot atoms exiting the oven would naturally be expected to have their dipoles oriented randomly in every direction, it would follow that there would be a continuous distribution of atoms seen at all angles exiting from the Stern–Gerlach device. Instead, what is seen is atoms emerging from a *discrete* set of angles. Physicists were able to explain this by assuming that the magnetic dipole moment of the atoms is *quantized*, that is, comes in discrete multiples of some fundamental amount.

This observation of quantization in the Stern–Gerlach experiment was surprising to physicists of the 1920s, but not completely astonishing because evidence for quantization effects in other systems was becoming widespread at that time. What was truly surprising was the *number* of peaks seen in the experiment. The hydrogen atoms being used were such that they should have had *zero* magnetic dipole moment. Classically, this is surprising in itself, since it corresponds to no orbital motion of the electron, but based on what was known of quantum mechanics at that time this was an acceptable notion. Since the hydrogen atoms would therefore have zero magnetic moment, it was expected that only one beam of atoms would be seen, and this beam would not be deflected by the magnetic field. Instead, two beams were seen, one deflected up by the magnetic field, and the other deflected down!

This puzzling doubling was explained after considerable effort by positing that the electron in the hydrogen atom has associated with it a quantity called *spin*. This spin is not in any way associated to the usual rotational motion of the electron around the proton; it is an entirely new quantity to be associated with an electron. The great physicist Heisenberg labeled the idea 'brave' at the time it was suggested, and it is a brave idea, since it introduces an essentially new physical quantity into Nature. The spin of the electron is posited to make an *extra* contribution to the magnetic dipole moment of a hydrogen atom, in addition to the contribution due to the rotational motion of the electron.

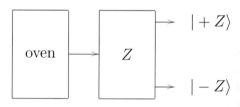

Figure 1.22. Abstract schematic of the Stern–Gerlach experiment. Hot hydrogen atoms are beamed from an oven through a magnetic field, causing a deflection either up ($|+Z\rangle$) or down ($|-Z\rangle$).

What is the proper description of the spin of the electron? As a first guess, we might hypothesize that the spin is specified by a single bit, telling the hydrogen atom to go up or down. Additional experimental results provide further useful information to determine if this guess needs refinement or replacement. Let's represent the original Stern–Gerlach apparatus as shown in Figure 1.22. Its outputs are two beams of atoms, which we shall call $|+Z\rangle$ and $|-Z\rangle$. (We're using suggestive notation which looks quantum mechanical, but of course you're free to use whatever notation you prefer.) Now suppose we cascade two Stern–Gerlach apparatus together, as shown in Figure 1.23. We arrange it so that the second apparatus is *tipped sideways*, so the magnetic field deflects atoms along the $\hat{x}$ axis. In our thought-experiment we'll block off the $|-Z\rangle$ output from the first Stern–Gerlach apparatus, while the $|+Z\rangle$ output is sent through a second apparatus oriented along the $\hat{x}$ axis. A detector is placed at the final output to measure the distribution of atoms along the $\hat{x}$ axis.

A classical magnetic dipole pointed in the $+\hat{z}$ direction has no net magnetic moment in the $\hat{x}$ direction, so we might expect that the final output would have one central peak. However, experimentally it is observed that there are two peaks of equal intensity! So perhaps these atoms are peculiar, and have definite magnetic moments along each axis, independently. That is, maybe each atom passing through the second apparatus can be

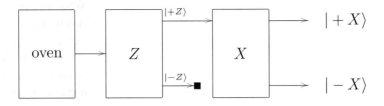

Figure 1.23. Cascaded Stern–Gerlach measurements.

described as being in a state we might write as $|+Z\rangle|+X\rangle$ or $|+Z\rangle|-X\rangle$, to indicate the two values for spin that might be observed.

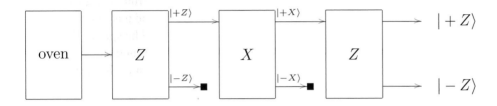

Figure 1.24. Three stage cascaded Stern–Gerlach measurements.

Another experiment, shown in Figure 1.24, can test this hypothesis by sending one beam of the previous output through a second $\hat{z}$ oriented Stern–Gerlach apparatus. If the atoms had retained their $|+Z\rangle$ orientation, then the output would be expected to have only one peak, at the $|+Z\rangle$ output. However, again *two* beams are observed at the final output, of equal intensity. Thus, the conclusion would seem to be that contrary to classical expectations, a $|+Z\rangle$ state consists of equal portions of $|+X\rangle$ and $|-X\rangle$ states, and a $|+X\rangle$ state consists of equal portions of $|+Z\rangle$ and $|-Z\rangle$ states. Similar conclusions can be reached if the Stern–Gerlach apparatus is aligned along some other axis, like the $\hat{y}$ axis.

The qubit model provides a simple explanation of this experimentally observed behavior. Let $|0\rangle$ and $|1\rangle$ be the states of a qubit, and make the assignments

$$|+Z\rangle \leftarrow |0\rangle \tag{1.56}$$
$$|-Z\rangle \leftarrow |1\rangle \tag{1.57}$$
$$|+X\rangle \leftarrow (|0\rangle + |1\rangle)/\sqrt{2}. \tag{1.58}$$
$$|-X\rangle \leftarrow (|0\rangle - |1\rangle)/\sqrt{2} \tag{1.59}$$

Then the results of the cascaded Stern–Gerlach experiment can be explained by assuming that the $\hat{z}$ Stern–Gerlach apparatus measures the spin (that is, the qubit) in the computational basis $|0\rangle, |1\rangle$, and the $\hat{x}$ Stern–Gerlach apparatus measures the spin with respect to the basis $(|0\rangle + |1\rangle)/\sqrt{2}, (|0\rangle - |1\rangle)/\sqrt{2}$. For example, in the cascaded $\hat{z}$-$\hat{x}$-$\hat{z}$ experiment, if we assume that the spins are in the state $|+Z\rangle = |0\rangle = (|+X\rangle + |-X\rangle)/\sqrt{2}$ after exiting the first Stern–Gerlach experiment, then the probability for obtaining $|+X\rangle$ out of the second apparatus is $1/2$, and the probability for $|-X\rangle$ is $1/2$. Similarly, the probability for obtaining $|+Z\rangle$ out of the third apparatus is $1/2$. A qubit model thus properly predicts results from this type of cascaded Stern–Gerlach experiment.

This example demonstrates how qubits could be a believable way of modeling systems in Nature. Of course it doesn't establish beyond all doubt that the qubit model is the correct way of understanding electron spin – far more experimental corroboration is required. Nevertheless, because of many experiments like these, we now believe that electron spin is best described by the qubit model. What is more, we believe that the qubit model (and generalizations of it to higher dimensions; quantum mechanics, in other words) is capable of describing *every* physical system. We now turn to the question of what systems are especially well adapted to quantum information processing.

1.5.2 Prospects for practical quantum information processing

Building quantum information processing devices is a great challenge for scientists and engineers of the third millennium. Will we rise to meet this challenge? Is it possible at all? Is it worth attempting? If so, how might the feat be accomplished? These are difficult and important questions, to which we essay brief answers in this section, to be expanded upon throughout the book.

The most fundamental question is whether there is any point of principle that prohibits us from doing one or more forms of quantum information processing? Two possible obstructions suggest themselves: that noise may place a fundamental barrier to useful quantum information processing; or that quantum mechanics may fail to be correct.

Noise is without a doubt a significant obstruction to the development of practical quantum information processing devices. Is it a *fundamentally irremovable* obstruction that will forever prevent the development of large-scale quantum information processing devices? The theory of quantum error-correcting codes strongly suggests that while quantum noise is a practical problem that needs to be addressed, it does not present a fundamental problem of *principle*. In particular, there is a *threshold theorem* for quantum computation, which states, roughly speaking, that provided the level of noise in a quantum computer can be reduced below a certain constant 'threshold' value, quantum error-correcting codes can be used to push it down even further, essentially *ad infinitum*, for a small overhead in the complexity of the computation. The threshold theorem makes some broad assumptions about the nature and magnitude of the noise occurring in a quantum computer, and the architecture available for performing quantum computation; however, provided those assumptions are satisfied, the effects of noise can be made essentially negligible for quantum information processing. Chapters 8, 10 and 12 discuss quantum noise, quantum error-correction and the threshold theorem in detail.

A second possibility that may preclude quantum information processing is if quantum mechanics is incorrect. Indeed, probing the validity of quantum mechanics (both relativistic and non-relativistic) is one reason for being interested in building quantum information processing devices. Never before have we explored a regime of Nature in which complete control has been obtained over large-scale quantum systems, and perhaps Nature may reveal some new surprises in this regime which are not adequately explained by quantum mechanics. If this occurs, it will be a momentous discovery in the history of science, and can be expected to have considerable consequences in other areas of science and technology, as did the discovery of quantum mechanics. Such a discovery might also impact quantum computation and quantum information; however, whether the impact would enhance, detract or not affect the power of quantum information processing cannot be predicted in advance. Until and unless such effects are found we have no way of knowing how they might affect information processing, so for the remainder of this book we go with all the evidence to date and assume that quantum mechanics is a complete and correct description of the world.

Given that there is no fundamental obstacle to building quantum information processing devices, why should we invest enormous amounts of time and money in the attempt to do so? We have already discussed several reasons for wanting to do so: practical applications such as quantum cryptography and the factoring of large composite numbers; and the desire to obtain fundamental insights into Nature and into information processing.

These are good reasons, and justify a considerable investment of time and money in the effort to build quantum information processing devices. However, it is fair to say that a clearer picture of the relative power of quantum and classical information processing is needed in order to assess their relative merits. To obtain such a picture requires further theoretical work on the foundations of quantum computation and quantum information. Of particular interest is a decisive answer to the question 'Are quantum computers more powerful than classical computers?' Even if the answer to such a question eludes us for the time being, it would be useful to have a clear path of interesting applications at varying levels of complexity to aid researchers aiming to experimentally realize quantum information processing. Historically, the advance of technology is often hastened by the use of short- to medium-term incentives as a stepping-stone to long-term goals. Consider that microprocessors were initially used as controllers for elevators and other simple devices, before graduating to be the fundamental component in personal computers (and then on to who-knows-what). Below we sketch out a path of short- to medium-term goals for people interested in achieving the long-term goal of large-scale quantum information processing.

Surprisingly many small-scale applications of quantum computation and quantum information are known. Not all are as flashy as cousins like the quantum factoring algorithm, but the relative ease of implementing small-scale applications makes them extremely important as medium-term goals in themselves.

Quantum state tomography and quantum process tomography are two elementary processes whose perfection is of great importance to quantum computation and quantum information, as well as being of independent interest in their own right. Quantum state tomography is a method for determining the quantum state of a system. To do this, it has to overcome the 'hidden' nature of the quantum state – remember, the state can't be directly determined by a measurement – by performing repeated preparations of the same quantum state, which is then measured in different ways in order to build up a complete description of the quantum state. Quantum process tomography is a more ambitious (but closely related) procedure to completely characterize the *dynamics* of a quantum system. Quantum process tomography can, for example, be used to characterize the performance of an alleged quantum gate or quantum communications channel, or to determine the types and magnitudes of different noise processes in a system. Beside obvious applications to quantum computation and quantum information, quantum process tomography can be expected to have significant applications as a diagnostic tool to aid in the evaluation and improvement of primitive operations in any field of science and technology where quantum effects are important. Quantum state tomography and quantum process tomography are described in more detail in Chapter 8.

Various small-scale communications primitives are also of great interest. We have already mentioned quantum cryptography and quantum teleportation. The former is likely to be useful in practical applications involving the distribution of a small amount of key material that needs to be highly secure. The uses of quantum teleportation are perhaps more open to question. We will see in Chapter 12 that teleportation may be an extremely useful primitive for transmitting quantum states between distant nodes in a network, in the presence of noise. The idea is to focus one's efforts on distributing EPR pairs between the nodes that wish to communicate. The EPR pairs may be corrupted during communication, but special 'entanglement distillation' protocols can then be used to 'clean up' the EPR pairs, enabling them to be used to teleport quantum states from one location to another. In fact, procotols based upon entanglement distillation and teleportation offer performance superior to more conventional quantum error-correction techniques in enabling noise free communication of qubits.

What of the medium-scale? A promising medium-scale application of quantum information processing is to the simulation of quantum systems. To simulate a quantum system containing even a few dozen 'qubits' (or the equivalent in terms of some other basic system) strains the resources of even the largest supercomputers. A simple calculation is instructive. Suppose we have a system containing 50 qubits. To describe the state of such a system requires $2^{50} \approx 10^{15}$ complex amplitudes. If the amplitudes are stored to 128 bits of precision, then it requires 256 bits or 32 bytes in order to store each amplitude, for a total of 32×10^{15} bytes of information, or about 32 thousand terabytes of information, well beyond the capacity of existing computers, and corresponding to about the storage capacity that might be expected to appear in supercomputers during the second decade of the twenty-first century, presuming that Moore's law continues on schedule. 90 qubits at the same level of precision requires 32×10^{27} bytes, which, even if implemented using single atoms to represent bits, would require kilograms (or more) of matter.

How useful will quantum simulations be? It seems likely that conventional methods will still be used to determine elementary properties of materials, such as bond strengths and basic spectroscopic properties. However, once the basic properties are well understood, it seems likely that quantum simulation will be of great utility as a laboratory for the design and testing of properties of novel molecules. In a conventional laboratory setup, many different types of 'hardware' – chemicals, detectors, and so on – may be required to test a wide variety of possible designs for a molecule. On a quantum computer, these different types of hardware can all be simulated in software, which is likely to be much less expensive and much faster. Of course, final design and testing must be performed with real physical systems; however, quantum computers may enable a much larger range of potential designs to be explored and evaluated *en route* to a better final design. It is interesting to note that such *ab initio* calculations to aid in the design of new molecules have been attempted on classical computers; however, they have met with limited success due to the enormous computational resources needed to simulate quantum mechanics on a classical computer. Quantum computers should be able to do much better in the relatively near future.

What of large-scale applications? Aside from scaling up applications like quantum simulation and quantum cryptography, relatively few large-scale applications are known: the factoring of large numbers, taking discrete logarithms, and quantum searching. Interest in the first two of these derives mainly from the *negative* effect they would have of limiting the viability of existing public key cryptographic systems. (They might also be of substantial practical interest to mathematicians interested in these problems simply for their own sake.) So it does not seem likely that factoring and discrete logarithm will be all that important as applications for the long run. Quantum searching may be of tremendous use because of the wide utility of the search heuristic, and we discuss some possible applications in Chapter 6. What would really be superb are many more large-scale applications of quantum information processing. This is a great goal for the future!

Given a path of potential applications for quantum information processing, how can it be achieved in real physical systems? At the small scale of a few qubits there are already several working proposals for quantum information processing devices. Perhaps the easiest to realize are based upon *optical* techniques, that is, electromagnetic radiation. Simple devices like mirrors and beamsplitters can be used to do elementary manipulations of photons. Interestingly, a major difficulty has been producing single photons on demand; experimentalists have instead opted to use schemes which produce single photons 'every now and then', at random, and wait for such an event to occur. Quantum cryptography, superdense coding, and quantum teleportation have all been realized using such optical techniques. A major advantage of the optical techniques is that photons tend to be highly stable carriers of quantum mechanical information. A major disadvantage is that photons don't directly interact with one another. Instead, the interaction has to be mediated by something else, like an atom, which introduces additional noise and complications into the experiment. An *effective* interaction between two photons is set up, which essentially works in two steps: photon number one interacts with the atom, which in turn interacts with the second photon, causing an overall interaction between the two photons.

An alternative scheme is based upon methods for trapping different types of atom: there is the *ion trap*, in which a small number of charged atoms are trapped in a confined space; and *neutral atom traps*, for trapping uncharged atoms in a confined space. Quantum information processing schemes based upon atom traps use the atoms to store qubits. Electromagnetic radiation also shows up in these schemes, but in a rather different way than in what we referred to as the 'optical' approach to quantum information processing. In these schemes, photons are used to manipulate the information stored in the atoms themselves, rather than as the place the information is stored. Single qubit quantum gates can be performed by applying appropriate pulses of electromagnetic radiation to individual atoms. Neighboring atoms can interact with one another via (for example) dipole forces that enable quantum gates to be accomplished. Moreover, the exact nature of the interaction between neighboring atoms can be modified by applying appropriate pulses of electromagnetic radiation to the atoms, giving the experimentalist control over what gates are performed in the system. Finally, quantum measurement can be accomplished in these systems using the long established *quantum jumps* technique, which implements with superb accuracy the measurements in the computational basis used for quantum computation.

Another class of quantum information processing schemes is based upon *Nuclear Magnetic Resonance*, often known by its initials, NMR. These schemes store quantum information in the *nuclear spin* of atoms in a molecule, and manipulate that information using electromagnetic radiation. Such schemes pose special difficulties, because in NMR it is not possible to directly access individual nuclei. Instead, a huge number (typically around 10^{15}) of essentially identical molecules are stored in solution. Electromagnetic pulses are applied to the sample, causing each molecule to respond in roughly the same way. You should think of each molecule as being an independent computer, and the sample as containing a huge number of computers all running in parallel (classically).

NMR quantum information processing faces three special difficulties that make it rather different from other quantum information processing schemes. First, the molecules are typically prepared by letting them equilibrate at room temperature, which is so much higher than typical spin flip energies that the spins become nearly completely randomly oriented. This fact makes the initial state rather more 'noisy' than is desirable for quantum information processing. How this noise may be overcome is an interesting story that we tell in Chapter 7. A second problem is that the class of measurements that may be performed in NMR falls well short of the most general measurements we would like to perform in quantum information processing. Nevertheless, for many instances of quantum information processing the class of measurements allowed in NMR is sufficient. Third, because molecules cannot be individually addressed in NMR you might ask how it is that individual qubits can be manipulated in an appropriate way. Fortunately, different nuclei in the molecule can have different properties that allow them to be individually addressed – or at least addressed at a sufficiently fine-grained scale to allow the operations essential for quantum computation.

Many of the elements required to perform large-scale quantum information processing can be found in existing proposals: superb state preparation and quantum measurements can be performed on a small number of qubits in the ion trap; superb dynamics can be performed in small molecules using NMR; fabrication technology in solid state systems allows designs to be scaled up tremendously. A single system having all these elements would be a long way down the road to a dream quantum computer. Unfortunately, all these systems are very different, and we are many, many years from having large-scale quantum computers. However, we believe that the existence of all these properties in existing (albeit different) systems does bode well for the long-term existence of large-scale quantum information processors. Furthermore, it suggests that there is a great deal of merit to pursuing *hybrid* designs which attempt to marry the best features of two or more existing technologies. For example, there is much work being done on trapping atoms inside *electromagnetic cavities*. This enables flexible manipulation of the atom inside the cavity via optical techniques, and makes possible real-time feedback control of single atoms in ways unavailable in conventional atom traps.

To conclude, note that it is important not to assess quantum information processing as though it were just another technology for information processing. For example, it is tempting to dismiss quantum computation as yet another technological fad in the evolution of the computer that will pass in time, much as other fads have passed – for example, the 'bubble memories' widely touted as the next big thing in memory during the early 1980s. This is a mistake, since quantum computation is an *abstract paradigm* for information processing that may have many *different* implementations in technology. One can compare two different proposals for quantum computing as regards their technological merits – it makes sense to compare a 'good' proposal to a 'bad' proposal – however even a very poor proposal for a quantum computer is of a different qualitative nature from a superb design for a classical computer.

1.6 Quantum information

The term 'quantum information' is used in two distinct ways in the field of quantum computation and quantum information. The first usage is as a broad catch-all for all manner of operations that might be interpreted as related to information processing using quantum mechanics. This use encompasses subjects such as quantum computation, quantum teleportation, the no–cloning theorem, and virtually all other topics in this book.

The second use of 'quantum information' is much more specialized: it refers to the study of *elementary* quantum information processing tasks. It does not typically include, for example, quantum algorithm design, since the details of specific quantum algorithms are beyond the scope of 'elementary'. To avoid confusion we will use the term 'quantum information theory' to refer to this more specialized field, in parallel with the widely used term '(classical) information theory' to describe the corresponding classical field. Of course, the term 'quantum information theory' has a drawback of its own – it might be seen as implying that theoretical considerations are all that matter! Of course, this is not the case, and experimental demonstration of the elementary processes studied by quantum information theory is of great interest.

The purpose of this section is to introduce the basic ideas of quantum information theory. Even with the restriction to elementary quantum information processing tasks, quantum information theory may look like a disordered zoo to the beginner, with many apparently unrelated subjects falling under the 'quantum information theory' rubric. In part, that's because the subject is still under development, and it's not yet clear how all the pieces fit together. However, we can identify a few fundamental goals uniting work on quantum information theory:

(1) **Identify elementary classes of static resources in quantum mechanics**. An example is the qubit. Another example is the *bit*; classical physics arises as a special case of quantum physics, so it should not be surprising that elementary static resources appearing in classical information theory should also be of great relevance in quantum information theory. Yet another example of an elementary class of static resources is a Bell state shared between two distant parties.

(2) **Identify elementary classes of dynamical processes in quantum mechanics**. A simple example is *memory*, the ability to store a quantum state over some period of time. Less trivial processes are quantum information transmission between two parties, Alice and Bob; copying (or trying to copy) a quantum state, and the process of protecting quantum information processing against the effects of noise.

(3) **Quantify resource tradeoffs incurred performing elementary dynamical processes**. For example, what are the minimal resources required to reliably transfer quantum information between two parties using a noisy communications channel?

Similar goals define classical information theory; however, quantum information theory is broader in scope than classical information theory, for quantum information theory includes all the static and dynamic elements of classical information theory, as well as *additional* static and dynamic elements.

The remainder of this section describes some examples of questions studied by quantum information theory, in each case emphasizing the fundamental static and dynamic elements under consideration, and the resource tradeoffs being considered. We begin with an example that will appear quite familiar to classical information theorists: the problem of sending classical information through a quantum channel. We then begin to branch out and explore some of the new static and dynamic processes present in quantum mechanics, such as quantum error-correction, the problem of distinguishing quantum states, and entanglement transformation. The chapter concludes with some reflections on how the tools of quantum information theory can be applied elsewhere in quantum computation and quantum information.

1.6.1 Quantum information theory: example problems

Classical information through quantum channels

The fundamental results of classical information theory are Shannon's *noiseless channel coding theorem* and Shannon's *noisy channel coding theorem*. The noiseless channel coding theorem quantifies how many bits are required to store information being emitted by a source of information, while the noisy channel coding theorem quantifies how much information can be reliably transmitted through a noisy communications channel.

What do we mean by an *information source*? Defining this notion is a fundamental problem of classical and quantum information theory, one we'll re-examine several times. For now, let's go with a provisional definition: a classical information source is described by a set of probabilities p_j, $j = 1, 2, \ldots, d$. Each use of the source results in the 'letter' j being emitted, chosen at random with probability p_j, independently for each use of the source. For instance, if the source were of English text, then the numbers j might correspond to letters of the alphabet and punctuation, with the probabilities p_j giving the relative frequencies with which the different letters appear in regular English text. Although it is not true that the letters in English appear in an independent fashion, for our purposes it will be a good enough approximation.

Regular English text includes a considerable amount of redundancy, and it is possible to exploit that redundancy to *compress* the text. For example, the letter 'e' occurs much more frequently in regular English text than does the letter 'z'. A good scheme for compressing English text will therefore represent the letter 'e' using fewer bits of information than it uses to represent 'z'. Shannon's noiseless channel coding theorem quantifies exactly how well such a compression scheme can be made to work. More precisely, the noiseless channel coding theorem tells us that a classical source described by probabilities p_j can be compressed so that on average each use of the source can be represented using $H(p_j)$ bits of information, where $H(p_j) \equiv -\sum_j p_j \log(p_j)$ is a function of the source probability distribution known as the *Shannon entropy*. Moreover, the noiseless channel coding theorem tells us that to attempt to represent the source using fewer bits than this will result in a high probability of error when the information is decompressed. (Shannon's noiseless channel coding theorem is discussed in much greater detail in Chapter 12.)

Shannon's noiseless coding theorem provides a good example where the goals of information theory listed earlier are all met. Two static resources are identified (goal number 1): the bit and the information source. A two-stage dynamic process is identified (goal 2), compressing an information source, and then decompressing to recover the information source. Finally a quantitative criterion for determining the resources consumed (goal 3) by an optimal data compression scheme is found.

Shannon's second major result, the noisy channel coding theorem, quantifies the amount of information that can be reliably transmitted through a noisy channel. In particular, suppose we wish to transfer the information being produced by some information source to another location through a noisy channel. That location may be at another point in space, or at another point in time – the latter is the problem of storing information in the presence of noise. The idea in both instances is to encode the information being produced using error-correcting codes, so that any noise introduced by the channel can be corrected at the other end of the channel. The way error-correcting codes achieve this is by introducing enough redundancy into the information sent through the channel so that even after some of the information has been corrupted it is still possible to recover the original message. For example, suppose the noisy channel is for the transmission of single bits, and the noise in the channel is such that to achieve reliable transmission each bit produced by the source must be encoded using two bits before being sent through the channel. We say that such a channel has a *capacity* of half a bit, since each use of the channel can be used to reliably convey roughly half a bit of information. Shannon's noisy channel coding theorem provides a general procedure for calculating the capacity of an arbitrary noisy channel.

Shannon's noisy channel coding theorem also achieves the three goals of information theory we stated earlier. Two types of static resources are involved (goal 1), the information source, and the bits being sent through the channel. Three dynamical processes are involved (goal 2). The primary process is the noise in the channel. To combat this noise we perform the dual processes of encoding and decoding the state in an error-correcting code. For a fixed noise model, Shannon's theorem tells us how much redundancy must be introduced by an optimal error-correction scheme if reliable information transmission is to be achieved (goal 3).

For both the noiseless and noisy channel coding theorems Shannon restricted himself to storing the output from an information source in classical systems – bits and the like. A natural question for quantum information theory is what happens if the storage medium is changed so that classical information is transmitted using quantum states as the medium. For example, it may be that Alice wishes to compress some classical information produced by an information source, transmitting the compressed information to Bob, who then decompresses it. If the medium used to store the compressed information is a quantum state, then Shannon's noiseless channel coding theorem cannot be used to determine the optimal compression and decompression scheme. One might wonder, for example, if using qubits allows a better compression rate than is possible classically. We'll study this question in Chapter 12, and prove that, in fact, qubits do not allow any significant saving in the amount of communication required to transmit information over a noiseless channel.

Naturally, the next step is to investigate the problem of transmitting classical information through a *noisy* quantum channel. Ideally, what we'd like is a result that quantifies the *capacity* of such a channel for the transmission of information. Evaluating the capacity is a very tricky job for several reasons. Quantum mechanics gives us a huge variety of noise models, since it takes place in a continuous space, and it is not at all obvious how to adapt classical error-correction techniques to combat the noise. Might it be advantageous, for example, to encode the classical information using *entangled* states, which are then transmitted one piece at a time through the noisy channel? Or perhaps it will be advantageous to decode using entangled measurements? In Chapter 12 we'll prove the *HSW (Holevo–Schumacher–Westmoreland) theorem*, which provides a lower bound on the capacity of such a channel. Indeed, it is widely believed that the HSW theorem provides an exact evaluation of the capacity, although a complete proof of this is not yet known! What remains at issue is whether or not encoding using entangled states can be used to raise the capacity beyond the lower bound provided by the HSW theorem. All evidence to date suggests that this doesn't help raise the capacity, but it is still a fascinating open problem of quantum information theory to determine the truth or falsity of this conjecture.

Quantum information through quantum channels

Classical information is, of course, not the only static resource available in quantum mechanics. Quantum states themselves are a natural static resource, even more natural than classical information. Let's look at a *different* quantum analogue of Shannon's coding theorems, this time involving the compression and decompression of quantum states.

To begin, we need to define some quantum notion of an information source, analogous to the classical definition of an information source. As in the classical case, there are several different ways of doing this, but for the sake of definiteness let's make the provisional definition that a quantum source is described by a set of probabilities p_j and corresponding quantum states $|\psi_j\rangle$. Each use of the source produces a state $|\psi_j\rangle$ with probability p_j, with different uses of the source being independent of one another.

Is it possible to compress the output from such a quantum mechanical source? Consider the case of a qubit source which outputs the state $|0\rangle$ with probability p and the state $|1\rangle$ with probability $1 - p$. This is essentially the same as a classical source emitting single bits, either 0 with probability p, or 1 with probability $1 - p$, so it is not surprising that similar techniques can be used to compress the source so that only $H(p, 1 - p)$ qubits are required to store the compressed source, where $H(\cdot)$ is again the Shannon entropy function.

What if the source had instead been producing the state $|0\rangle$ with probability p, and the state $(|0\rangle + |1\rangle)/\sqrt{2}$ with probability $1 - p$? The standard techniques of classical data compression no longer apply, since in general it is not possible for us to distinguish the states $|0\rangle$ and $(|0\rangle + |1\rangle)/\sqrt{2}$. Might it still be possible to perform some type of compression operation?

It turns out that a type of compression is still possible, even in this instance. What is interesting is that the compression may no longer be *error-free*, in the sense that the quantum states being produced by the source may be slightly distorted by the compression–decompression procedure. Nevertheless, we require that this distortion ought to become very small and ultimately negligible in the limit of large blocks of source output being compressed. To quantify the distortion we introduce a *fidelity* measure for the compression scheme, which measures the average distortion introduced by the compression scheme. The idea of quantum data compression is that the compressed data should be recovered with very good fidelity. Think of the fidelity as being analogous to the probability of doing the decompression correctly – in the limit of large block lengths, it should tend towards the no error limit of 1.

Schumacher's noiseless channel coding theorem quantifies the resources required to do quantum data compression, with the restriction that it be possible to recover the source with fidelity close to 1. In the case of a source producing orthogonal quantum states $|\psi_j\rangle$ with probabilities p_j Schumacher's theorem reduces to telling us that the source may be compressed down to but not beyond the classical limit $H(p_j)$. However, in the more general case of non-orthogonal states being produced by the source, Schumacher's theorem tells us how much a quantum source may be compressed, and the answer is *not* the Shannon entropy $H(p_j)$! Instead, a new entropic quantity, the *von Neumann* entropy, turns out to be the correct answer. In general, the von Neumann entropy agrees with the Shannon entropy if and only if the states $|\psi_j\rangle$ are orthogonal. Otherwise, the von Neumann entropy for the source $p_j, |\psi_j\rangle$ is in general strictly *smaller* than the Shannon entropy $H(p_j)$. Thus, for example, a source producing the state $|0\rangle$ with probability p

and $(|0\rangle + |1\rangle)/\sqrt{2}$ with probability $1 - p$ can be reliably compressed using fewer than $H(p, 1 - p)$ qubits per use of the source!

The basic intuition for this decrease in resources required can be understood quite easily. Suppose the source emitting states $|0\rangle$ with probability p and $(|0\rangle + |1\rangle)/\sqrt{2}$ with probability $1 - p$ is used a large number n times. Then by the law of large numbers, with high probability the source emits about np copies of $|0\rangle$ and $n(1 - p)$ copies of $(|0\rangle + |1\rangle)/\sqrt{2}$. That is, it has the form

$$|0\rangle^{\otimes np} \left(\frac{|0\rangle + |1\rangle}{\sqrt{2}} \right)^{\otimes n(1-p)}, \tag{1.60}$$

up to re-ordering of the systems involved. Suppose we expand the product of $|0\rangle + |1\rangle$ terms on the right hand side. Since $n(1 - p)$ is large, we can again use the law of large numbers to deduce that the terms in the product will be roughly one-half $|0\rangle$s and one-half $|1\rangle$s. That is, the $|0\rangle + |1\rangle$ product can be well approximated by a superposition of states of the form

$$|0\rangle^{\otimes n(1-p)/2} |1\rangle^{\otimes n(1-p)/2}. \tag{1.61}$$

Thus the state emitted by the source can be approximated as a superposition of terms of the form

$$|0\rangle^{\otimes n(1+p)/2} |1\rangle^{\otimes n(1-p)/2}. \tag{1.62}$$

How many states of this form are there? Roughly n choose $n(1 + p)/2$, which by Stirling's approximation is equal to $N \equiv 2^{nH[(1+p)/2,(1-p)/2]}$. A simple compression method then is to label all states of the form (1.62) $|c_1\rangle$ through $|c_N\rangle$. It is possible to perform a unitary transform on the n qubits emitted from the source that takes $|c_j\rangle$ to $|j\rangle|0\rangle^{\otimes n - nH[(1+p)/2,(1-p)/2]}$, since j is an $nH[(1 + p)/2, (1 - p)/2]$ bit number. The compression operation is to perform this unitary transformation, and then drop the final $n - nH[(1+p)/2, (1-p)/2]$ qubits, leaving a compressed state of $nH[(1+p)/2, (1-p)/2]$ qubits. To decompress we append the state $|0\rangle^{\otimes n - nH[(1+p)/2,(1-p)/2]}$ to the compressed state, and perform the inverse unitary transformation.

This procedure for quantum data compression and decompression results in a storage requirement of $H[(1 + p)/2, (1 - p)/2]$ qubits per use of the source, which whenever $p \geq 1/3$ is an improvement over the $H(p, 1 - p)$ qubits we might naively have expected from Shannon's noiseless channel coding theorem. In fact, Schumacher's noiseless channel coding theorem allows us to do somewhat better even than this, as we will see in Chapter 12; however, the essential reason in that construction is the same as the reason we were able to compress here: we exploited the fact that $|0\rangle$ and $(|0\rangle + |1\rangle)/\sqrt{2}$ are not orthogonal. Intuitively, the states contain some redundancy since both have a component in the $|0\rangle$ direction, which results in more physical similarity than would be obtained from orthogonal states. It is this redundancy that we have exploited in the coding scheme just described, and which is used in the full proof of Schumacher's noiseless channel coding theorem. Note that the restriction $p \geq 1/3$ arises because when $p < 1/3$ this particular scheme doesn't exploit the redundancy in the states: we end up effectively *increasing* the redundancy present in the problem! Of course, this is an artifact of the particular scheme we have chosen, and the general solution exploits the redundancy in a much more sensible way to achieve data compression.

Schumacher's noiseless channel coding theorem is an analogue of Shannon's noiseless channel coding theorem for the compression and decompression of quantum states. Can we find an analogue of Shannon's noisy channel coding theorem? Considerable progress on this important question has been made, using the theory of quantum error-correcting codes; however, a fully satisfactory analogue has not yet been found. We review some of what is known about the quantum channel capacity in Chapter 12.

Quantum distinguishability

Thus far all the dynamical processes we have considered – compression, decompression, noise, encoding and decoding error-correcting codes – arise in both classical and quantum information theory. However, the introduction of new types of information, such as quantum states, enlarges the class of dynamical processes beyond those considered in classical information theory. A good example is the problem of distinguishing quantum states. Classically, we are used to being able to distinguish different items of information, at least in principle. In practice, of course, a smudged letter 'a' written on a page may be very difficult to distinguish from a letter 'o', but in principle it is possible to distinguish between the two possibilities with perfect certainty.

On the other hand, quantum mechanically it is *not* always possible to distinguish between arbitrary states. For example, there is no process allowed by quantum mechanics that will reliably distinguish between the states $|0\rangle$ and $(|0\rangle + |1\rangle)/\sqrt{2}$. Proving this rigorously requires tools we don't presently have available (it is done in Chapter 2), but by considering examples it's pretty easy to convince oneself that it is not possible. Suppose, for example, that we try to distinguish the two states by measuring in the computational basis. Then, if we have been given the state $|0\rangle$, the measurement will yield 0 with probability 1. However, when we measure $(|0\rangle + |1\rangle)/\sqrt{2}$ the measurement yields 0 with probability 1/2 and 1 with probability 1/2. Thus, while a measurement result of 1 implies that state must have been $(|0\rangle + |1\rangle)/\sqrt{2}$, since it couldn't have been $|0\rangle$, we can't infer anything about the identity of the quantum state from a measurement result of 0.

This indistinguishability of non-orthogonal quantum states is at the heart of quantum computation and quantum information. It is the essence of our assertion that a quantum state contains hidden information that is not accessible to measurement, and thus plays a key role in quantum algorithms and quantum cryptography. One of the central problems of quantum information theory is to develop measures quantifying how well non-orthogonal quantum states may be distinguished, and much of Chapters 9 and 12 is concerned with this goal. In this introduction we'll limit ourselves to pointing out two interesting aspects of indistinguishability – a connection with the possibility of faster-than-light communication, and an application to 'quantum money.'

Imagine for a moment that we could distinguish between arbitrary quantum states. We'll show that this implies the ability to communicate faster than light, using entanglement. Suppose Alice and Bob share an entangled pair of qubits in the state $(|00\rangle + |11\rangle)/\sqrt{2}$. Then, if Alice measures in the computational basis, the post-measurement states will be $|00\rangle$ with probability 1/2, and $|11\rangle$ with probability 1/2. Thus Bob's system is either in the state $|0\rangle$, with probability 1/2, or in the state $|1\rangle$, with probability 1/2. Suppose, however, that Alice had instead measured in the $|+\rangle, |-\rangle$ basis. Recall that $|0\rangle = (|+\rangle + |-\rangle)/\sqrt{2}$ and $|1\rangle = (|+\rangle - |-\rangle)/\sqrt{2}$. A little algebra shows that the initial state of Alice and Bob's system may be rewritten as $(|++\rangle + |--\rangle)/\sqrt{2}$. Therefore, if Alice measures in the $|+\rangle, |-\rangle$ basis, the state of Bob's system after the measurement will be $|+\rangle$ or $|-\rangle$ with probability $1/2$ each. So far, this is all basic quantum mechanics. But if Bob had access to a device that could distinguish the four states $|0\rangle, |1\rangle, |+\rangle, |-\rangle$ from one another, then he could tell whether Alice had measured in the computational basis, or in the $|+\rangle, |-\rangle$ basis. Moreover, he could get that information *instantaneously*, as soon as Alice had made the measurement, providing a means by which Alice and Bob could achieve faster-than-light communication! Of course, we know that it is not possible to distinguish non-orthogonal quantum states; this example shows that this restriction is also intimately tied to other physical properties which we expect the world to obey.

The indistinguishability of non-orthogonal quantum states need not always be a handicap. Sometimes it can be a boon. Imagine that a bank produces banknotes imprinted with a (classical) serial number, and a sequence of qubits each in either the state $|0\rangle$ or $(|0\rangle + |1\rangle)/\sqrt{2}$. Nobody but the bank knows what sequence of these two states is embedded in the note, and the bank maintains a list matching serial numbers to embedded states. The note is impossible to counterfeit exactly, because it is impossible for a would-be counterfeiter to determine with certainty the state of the qubits in the original note, without destroying them. When presented with the banknote a merchant (of certifiable repute) can verify that it is not a counterfeit by calling the bank, telling them the serial number, and then asking what sequence of states were embedded in the note. They can then check that the note is genuine by measuring the qubits in the $|0\rangle, |1\rangle$ or $(|0\rangle + |1\rangle)/\sqrt{2}, (|0\rangle - |1\rangle)/\sqrt{2}$ basis, as directed by the bank. With probability which increases exponentially to one with the number of qubits checked, any would-be counterfeiter will be detected at this stage! This idea is the basis for numerous other quantum cryptographic protocols, and demonstrates the utility of the indistinguishability of non-orthogonal quantum states.

Exercise 1.2: Explain how a device which, upon input of one of two non-orthogonal quantum states $|\psi\rangle$ or $|\varphi\rangle$ correctly identified the state, could be used to build a device which cloned the states $|\psi\rangle$ and $|\varphi\rangle$, in violation of the no-cloning theorem. Conversely, explain how a device for cloning could be used to distinguish non-orthogonal quantum states.

Creation and transformation of entanglement

Entanglement is another elementary static resource of quantum mechanics. Its properties are amazingly different from those of the resources most familiar from classical information theory, and they are not yet well understood; we have at best an incomplete collage of results related to entanglement. We don't yet have all the language needed to understand the solutions, but let's at least look at two information-theoretic problems related to entanglement.

Creating entanglement is a simple dynamical process of interest in quantum information theory. How many qubits must two parties exchange if they are to create a particular entangled state shared between them, given that they share no prior entanglement? A second dynamical process of interest is *transforming entanglement* from one form into another. Suppose, for example, that Alice and Bob share between them a Bell state, and wish to transform it into some other type of entangled state. What resources do they need to accomplish this task? Can they do it without communicating? With classical communication only? If quantum communication is required then how much quantum communication is required?

Answering these and more complex questions about the creation and transformation of entanglement forms a fascinating area of study in its own right, and also promises to give insight into tasks such as quantum computation. For example, a distributed quantum computation may be viewed as simply a method for generating entanglement between two or more parties; lower bounds on the amount of communication that must be done to perform such a distributed quantum computation then follow from lower bounds on the amount of communication that must be performed to create appropriate entangled states.

1.6.2 Quantum information in a wider context

We have given but the barest glimpse of quantum information theory. Part III of this book discusses quantum information theory in much greater detail, especially Chapter 11, which deals with fundamental properties of entropy in quantum and classical information theory, and Chapter 12, which focuses on pure quantum information theory.

Quantum information theory is the most abstract part of quantum computation and quantum information, yet in some sense it is also the most fundamental. The question driving quantum information theory, and ultimately all of quantum computation and quantum information, is *what makes quantum information processing tick?* What is it that separates the quantum and the classical world? What resources, unavailable in a classical world, are being utilized in a quantum computation? Existing answers to these questions are foggy and incomplete; it is our hope that the fog may yet lift in the years to come, and we will obtain a clear appreciation for the possibilities and limitations of quantum information processing.

Problem 1.1: (Feynman–Gates conversation)   Construct a friendly imaginary
    discussion of about 2000 words between Bill Gates and Richard Feynman, set in
    the present, on the future of computation. (*Comment*: You might like to try
    waiting until you've read the rest of the book before attempting this question.
    See the 'History and further reading' below for pointers to one possible answer
    for this question.)

Problem 1.2:   What is the most significant discovery yet made in quantum
    computation and quantum information? Write an essay of about 2000 words for
    an educated lay audience about the discovery. (*Comment*: As for the previous
    problem, you might like to try waiting until you've read the rest of the book
    before attempting this question.)

History and further reading

Most of the material in this chapter is revisited in more depth in later chapters. Therefore the historical references and further reading below are limited to material which does not recur in later chapters.

Piecing together the historical context in which quantum computation and quantum information have developed requires a broad overview of the history of many fields. We have tried to tie this history together in this chapter, but inevitably much background material was omitted due to limited space and expertise. The following recommendations attempt to redress this omission.

The history of quantum mechanics has been told in many places. We recommend especially the outstanding works of Pais[Pai82, Pai86, Pai91]. Of these three, [Pai86] is most directly concerned with the development of quantum mechanics; however, Pais' biographies of Einstein[Pai82] and of Bohr[Pai91] also contain much material of interest, at a less intense level. The rise of technologies based upon quantum mechanics has been described by Milburn[Mil97, Mil98]. Turing's marvelous paper on the foundations of computer science[Tur36] is well worth reading. It can be found in the valuable historical collection of Davis[Dav65]. Hofstadter[Hof79] and Penrose[Pen89] contain entertaining and informative discussions of the foundations of computer science. Shasha and Lazere's biography of fifteen leading computer scientists[SL98] gives considerable insight into many different facets of the history of computer science. Finally, Knuth's awesome series of books[Knu97, Knu98a, Knu98b] contain an amazing amount of historical information. Shannon's brilliant papers founding information theory make excellent reading[Sha48] (also reprinted in [SW49]). MacWilliams and Sloane[MS77] is not only an excellent text on error-correcting codes, but also contains an enormous amount of useful historical information. Similarly, Cover and Thomas[CT91] is an excellent text on information theory, with extensive historical information. Shannon's collected works, together with many useful historical items have been collected in a large volume[SW93] edited by Sloane and Wyner. Slepian has also collected a useful set of reprints on information theory[Sle74]. Cryptography is an ancient art with an intricate and often interesting history. Kahn[Kah96] is a huge history of cryptography containing a wealth of information. For more recent developments we recommend the books by Menezes, van Oorschot, and Vanstone[MvOV96], Schneier[Sch96a], and by Diffie and Landau[DL98].

Quantum teleportation was discovered by Bennett, Brassard, Crépeau, Jozsa, Peres, and Wootters[BBC+93], and later experimentally realized in various different forms by Boschi, Branca, De Martini, Hardy and Popescu[BBM+98] using optical techniques, by Bouwmeester, Pan, Mattle, Eibl, Weinfurter, and Zeilinger[BPM+97] using photon polarization, by Furusawa, Sørensen, Braunstein, Fuchs, Kimble, and Polzik using 'squeezed' states of light[FSB+98], and by Nielsen, Knill, and Laflamme using NMR[NKL98].

Deutsch's problem was posed by Deutsch[Deu85], and a one-bit solution was given in the same paper. The extension to the general n-bit case was given by Deutsch and Jozsa[DJ92]. The algorithms in these early papers have been substantially improved subsequently by Cleve, Ekert, Macchiavello, and Mosca[CEMM98], and independently in unpublished work by Tapp. In this chapter we have given the improved version of the algorithm, which fits very nicely into the hidden subgroup problem framework that will later be discussed in Chapter 5. The original algorithm of Deutsch only worked probabilistically; Deutsch and Jozsa improved this to obtain a deterministic algorithm, but their method required two function evaluations, in contrast to the improved algorithms presented in this chapter. Nevertheless, it is still conventional to refer to these algorithms as Deutsch's algorithm and the Deutsch–Jozsa algorithm in honor of two huge leaps forward: the concrete demonstration by Deutsch that a quantum computer could do something faster than a classical computer; and the extension by Deutsch and Jozsa which demonstrated for the first time a similar gap for the scaling of the time required to solve a problem.

Excellent discussions of the Stern–Gerlach experiment can be found in standard quantum mechanics textbooks such as the texts by Sakurai[Sak95], Volume III of Feynman, Leighton and Sands[FLS65a], and Cohen-Tannoudji, Diu and Laloë[CTDL77a, CTDL77b].

Problem 1.1 was suggested by the lovely article of Rahim[Rah99].

2 Introduction to quantum mechanics

I ain't no physicist but I know what matters.
– Popeye the Sailor

Quantum mechanics: Real Black Magic Calculus
– Albert Einstein

Quantum mechanics is the most accurate and complete description of the world known. It is also the basis for an understanding of quantum computation and quantum information. This chapter provides all the necessary background knowledge of quantum mechanics needed for a thorough grasp of quantum computation and quantum information. No prior knowledge of quantum mechanics is assumed.

Quantum mechanics is easy to learn, despite its reputation as a difficult subject. The reputation comes from the difficulty of some *applications*, like understanding the structure of complicated molecules, which aren't fundamental to a grasp of the subject; we won't be discussing such applications. The only prerequisite for understanding is some familiarity with elementary linear algebra. Provided you have this background you can begin working out simple problems in a few hours, even with no prior knowledge of the subject.

Readers already familiar with quantum mechanics can quickly skim through this chapter, to become familiar with our (mostly standard) notational conventions, and to assure themselves of familiarity with all the material. Readers with little or no prior knowledge should work through the chapter in detail, pausing to attempt the exercises. If you have difficulty with an exercise, move on, and return later to make another attempt.

The chapter begins with a review of some material from linear algebra in Section 2.1. This section assumes familiarity with elementary linear algebra, but introduces the notation used by physicists to describe quantum mechanics, which is different to that used in most introductions to linear algebra. Section 2.2 describes the basic postulates of quantum mechanics. Upon completion of the section, you will have understood *all* of the fundamental principles of quantum mechanics. This section contains numerous simple exercises designed to help consolidate your grasp of this material. The remaining sections of the chapter, and of this book, elucidate upon this material, without introducing fundamentally new physical principles. Section 2.3 explains *superdense coding*, a surprising and illuminating example of quantum information processing which combines many of the postulates of quantum mechanics in a simple setting. Sections 2.4 and 2.5 develop powerful mathematical tools – the *density operator, purifications*, and the *Schmidt decomposition* – which are especially useful in the study of quantum computation and quantum information. Understanding these tools will also help you consolidate your understanding of elementary quantum mechanics. Finally, Section 2.6 examines the question of how quantum mechanics goes beyond the usual 'classical' understanding of the way the world works.

2.1 Linear algebra

This book is written as much to disturb and annoy as to instruct.
– The first line of *About Vectors*, by Banesh Hoffmann.

Life is complex – it has both real and imaginary parts.
– Anonymous

Linear algebra is the study of vector spaces and of linear operations on those vector spaces. A good understanding of quantum mechanics is based upon a solid grasp of elementary linear algebra. In this section we review some basic concepts from linear algebra, and describe the standard notations which are used for these concepts in the study of quantum mechanics. These notations are summarized in Figure 2.1 on page 62, with the quantum notation in the left column, and the linear-algebraic description in the right column. You may like to glance at the table, and see how many of the concepts in the right column you recognize.

In our opinion the chief obstacle to assimilation of the postulates of quantum mechanics is not the postulates themselves, but rather the large body of linear algebraic notions required to understand them. Coupled with the unusual Dirac notation adopted by physicists for quantum mechanics, it can appear (falsely) quite fearsome. For these reasons, we advise the reader not familiar with quantum mechanics to quickly read through the material which follows, pausing mainly to concentrate on understanding the absolute basics of the notation being used. Then proceed to a careful study of the main topic of the chapter – the postulates of quantum mechanics – returning to study the necessary linear algebraic notions and notations in more depth, as required.

The basic objects of linear algebra are *vector spaces*. The vector space of most interest to us is $\mathbf{C}^n$, the space of all n-tuples of complex numbers, $(z_1, \ldots, z_n)$. The elements of a vector space are called *vectors*, and we will sometimes use the column matrix notation

$$\begin{bmatrix} z_1 \\ \vdots \\ z_n \end{bmatrix} \tag{2.1}$$

to indicate a vector. There is an *addition* operation defined which takes pairs of vectors to other vectors. In $\mathbf{C}^n$ the addition operation for vectors is defined by

$$\begin{bmatrix} z_1 \\ \vdots \\ z_n \end{bmatrix} + \begin{bmatrix} z_1' \\ \vdots \\ z_n' \end{bmatrix} \equiv \begin{bmatrix} z_1 + z_1' \\ \vdots \\ z_n + z_n' \end{bmatrix}, \tag{2.2}$$

where the addition operations on the right are just ordinary additions of complex numbers. Furthermore, in a vector space there is a *multiplication by a scalar* operation. In $\mathbf{C}^n$ this operation is defined by

$$z \begin{bmatrix} z_1 \\ \vdots \\ z_n \end{bmatrix} \equiv \begin{bmatrix} zz_1 \\ \vdots \\ zz_n \end{bmatrix}, \tag{2.3}$$

where z is a *scalar*, that is, a complex number, and the multiplications on the right are ordinary multiplication of complex numbers. Physicists sometimes refer to complex numbers as *c-numbers*.

Quantum mechanics is our main motivation for studying linear algebra, so we will use the standard notation of quantum mechanics for linear algebraic concepts. The standard quantum mechanical notation for a vector in a vector space is the following:

$$|\psi\rangle. \tag{2.4}$$

ψ is a label for the vector (any label is valid, although we prefer to use simple labels like ψ and φ). The $|\cdot\rangle$ notation is used to indicate that the object is a vector. The entire object $|\psi\rangle$ is sometimes called a *ket*, although we won't use that terminology often.

A vector space also contains a special *zero vector*, which we denote by 0. It satisfies the property that for any other vector $|v\rangle$, $|v\rangle + 0 = |v\rangle$. Note that we do not use the ket notation for the zero vector – it is the only exception we shall make. The reason for making the exception is because it is conventional to use the 'obvious' notation for the zero vector, $|0\rangle$, to mean something else entirely. The scalar multiplication operation is such that $z0 = 0$ for any complex number z. For convenience, we use the notation $(z_1, \ldots, z_n)$ to denote a column matrix with entries $z_1, \ldots, z_n$. In $\mathbf{C}^n$ the zero element is $(0, 0, \ldots, 0)$. A *vector subspace* of a vector space V is a subset W of V such that W is also a vector space, that is, W must be closed under scalar multiplication and addition.

Notation	Description
z^*	Complex conjugate of the complex number z.
	$(1 + i)^* = 1 - i$
$\|\psi\rangle$	Vector. Also known as a *ket*.
$\langle\psi\|$	Vector dual to $\|\psi\rangle$. Also known as a *bra*.
$\langle\varphi\|\psi\rangle$	Inner product between the vectors $\|\varphi\rangle$ and $\|\psi\rangle$.
$\|\varphi\rangle \otimes \|\psi\rangle$	Tensor product of $\|\varphi\rangle$ and $\|\psi\rangle$.
$\|\varphi\rangle\|\psi\rangle$	Abbreviated notation for tensor product of $\|\varphi\rangle$ and $\|\psi\rangle$.
A^*	Complex conjugate of the A matrix.
A^T	Transpose of the A matrix.
$A^\dagger$	Hermitian conjugate or adjoint of the A matrix, $A^\dagger = (A^T)^*$.
	$\begin{bmatrix} a & b \\ c & d \end{bmatrix}^\dagger = \begin{bmatrix} a^* & c^* \\ b^* & d^* \end{bmatrix}.$
$\langle\varphi\|A\|\psi\rangle$	Inner product between $\|\varphi\rangle$ and $A\|\psi\rangle$.
	Equivalently, inner product between $A^\dagger\|\varphi\rangle$ and $\|\psi\rangle$.

Figure 2.1. Summary of some standard quantum mechanical notation for notions from linear algebra. This style of notation is known as the *Dirac* notation.

2.1.1 Bases and linear independence

A *spanning set* for a vector space is a set of vectors $|v_1\rangle, \ldots, |v_n\rangle$ such that any vector $|v\rangle$ in the vector space can be written as a linear combination $|v\rangle = \sum_i a_i |v_i\rangle$ of vectors in that set. For example, a spanning set for the vector space $\mathbf{C}^2$ is the set

$$|v_1\rangle \equiv \begin{bmatrix} 1 \\ 0 \end{bmatrix} ; \quad |v_2\rangle \equiv \begin{bmatrix} 0 \\ 1 \end{bmatrix} , \tag{2.5}$$

since any vector

$$|v\rangle = \begin{bmatrix} a_1 \\ a_2 \end{bmatrix} \tag{2.6}$$

in $\mathbf{C}^2$ can be written as a linear combination $|v\rangle = a_1|v_1\rangle + a_2|v_2\rangle$ of the vectors $|v_1\rangle$ and $|v_2\rangle$. We say that the vectors $|v_1\rangle$ and $|v_2\rangle$ *span* the vector space $\mathbf{C}^2$.

Generally, a vector space may have many different spanning sets. A second spanning set for the vector space $\mathbf{C}^2$ is the set

$$|v_1\rangle \equiv \frac{1}{\sqrt{2}} \begin{bmatrix} 1 \\ 1 \end{bmatrix} ; \quad |v_2\rangle \equiv \frac{1}{\sqrt{2}} \begin{bmatrix} 1 \\ -1 \end{bmatrix} , \tag{2.7}$$

since an arbitrary vector $|v\rangle = (a_1, a_2)$ can be written as a linear combination of $|v_1\rangle$ and $|v_2\rangle$,

$$|v\rangle = \frac{a_1 + a_2}{\sqrt{2}}|v_1\rangle + \frac{a_1 - a_2}{\sqrt{2}}|v_2\rangle. \tag{2.8}$$

A set of non-zero vectors $|v_1\rangle, \ldots, |v_n\rangle$ are *linearly dependent* if there exists a set of complex numbers $a_1, \ldots, a_n$ with $a_i \neq 0$ for at least one value of i, such that

$$a_1|v_1\rangle + a_2|v_2\rangle + \cdots + a_n|v_n\rangle = 0. \tag{2.9}$$

A set of vectors is *linearly independent* if it is not linearly dependent. It can be shown that any two sets of linearly independent vectors which span a vector space V contain the same number of elements. We call such a set a *basis* for V. Furthermore, such a basis set always exists. The number of elements in the basis is defined to be the *dimension* of V. In this book we will only be interested in *finite dimensional* vector spaces. There are many interesting and often difficult questions associated with infinite dimensional vector spaces. We won't need to worry about these questions.

Exercise 2.1: (Linear dependence: example) Show that $(1, -1), (1, 2)$ and $(2, 1)$ are linearly dependent.

2.1.2 Linear operators and matrices
A *linear operator* between vector spaces V and W is defined to be any function $A : V \to W$ which is linear in its inputs,

$$A \left(\sum_i a_i|v_i\rangle \right) = \sum_i a_i A \left(|v_i\rangle \right) . \tag{2.10}$$

Usually we just write $A|v\rangle$ to denote $A(|v\rangle)$. When we say that a linear operator A is defined *on* a vector space, V, we mean that A is a linear operator from V to V. An important linear operator on any vector space V is the *identity operator*, I_V, defined by the equation $I_V|v\rangle \equiv |v\rangle$ for all vectors $|v\rangle$. Where no chance of confusion arises we drop the subscript V and just write I to denote the identity operator. Another important linear operator is the *zero operator*, which we denote 0. The zero operator maps all vectors to the zero vector, $0|v\rangle \equiv 0$. It is clear from (2.10) that once the action of a linear operator A on a basis is specified, the action of A is completely determined on all inputs.

Suppose V, W, and X are vector spaces, and $A : V \rightarrow W$ and $B : W \rightarrow X$ are linear operators. Then we use the notation BA to denote the *composition* of B with A, defined by $(BA)(|v\rangle) \equiv B(A(|v\rangle))$. Once again, we write $BA|v\rangle$ as an abbreviation for $(BA)(|v\rangle)$.

The most convenient way to understand linear operators is in terms of their *matrix representations*. In fact, the linear operator and matrix viewpoints turn out to be completely equivalent. The matrix viewpoint may be more familiar to you, however. To see the connection, it helps to first understand that an m by n complex matrix A with entries A_{ij} is in fact a linear operator sending vectors in the vector space $\mathbf{C}^n$ to the vector space $\mathbf{C}^m$, under matrix multiplication of the matrix A by a vector in $\mathbf{C}^n$. More precisely, the claim that the matrix A is a linear operator just means that

$$A\left(\sum_i a_i|v_i\rangle\right) = \sum_i a_i A|v_i\rangle \tag{2.11}$$

is true as an equation where the operation is matrix multiplication of A by column vectors. Clearly, this is true!

We've seen that matrices can be regarded as linear operators. Can linear operators be given a matrix representation? In fact they can, as we now explain. This equivalence between the two viewpoints justifies our interchanging terms from matrix theory and operator theory throughout the book. Suppose $A : V \rightarrow W$ is a linear operator between vector spaces V and W. Suppose $|v_1\rangle, \ldots, |v_m\rangle$ is a basis for V and $|w_1\rangle, \ldots, |w_n\rangle$ is a basis for W. Then for each j in the range $1, \ldots, m$, there exist complex numbers A_{1j} through A_{nj} such that

$$A|v_j\rangle = \sum_i A_{ij}|w_i\rangle. \tag{2.12}$$

The matrix whose entries are the values A_{ij} is said to form a *matrix representation* of the operator A. This matrix representation of A is completely equivalent to the operator A, and we will use the matrix representation and abstract operator viewpoints interchangeably. Note that to make the connection between matrices and linear operators we must specify a set of input and output basis states for the input and output vector spaces of the linear operator.

Exercise 2.2: (Matrix representations: example) Suppose V is a vector space with basis vectors $|0\rangle$ and $|1\rangle$, and A is a linear operator from V to V such that $A|0\rangle = |1\rangle$ and $A|1\rangle = |0\rangle$. Give a matrix representation for A, with respect to the input basis $|0\rangle, |1\rangle$, and the output basis $|0\rangle, |1\rangle$. Find input and output bases which give rise to a different matrix representation of A.

Exercise 2.3: (Matrix representation for operator products) Suppose A is a linear operator from vector space V to vector space W, and B is a linear operator from vector space W to vector space X. Let $|v_i\rangle, |w_j\rangle$, and $|x_k\rangle$ be bases for the vector spaces V, W, and X, respectively. Show that the matrix representation for the linear transformation BA is the matrix product of the matrix representations for B and A, with respect to the appropriate bases.

Exercise 2.4: (Matrix representation for identity) Show that the identity operator on a vector space V has a matrix representation which is one along the diagonal and zero everywhere else, if the matrix representation is taken with respect to the same input and output bases. This matrix is known as the *identity matrix*.

2.1.3 The Pauli matrices

Four extremely useful matrices which we shall often have occasion to use are the *Pauli matrices*. These are 2 by 2 matrices, which go by a variety of notations. The matrices, and their corresponding notations, are depicted in Figure 2.2. The Pauli matrices are so useful in the study of quantum computation and quantum information that we encourage you to memorize them by working through in detail the many examples and exercises based upon them in subsequent sections.

$$\sigma_0 \equiv I \equiv \begin{bmatrix} 1 & 0 \\ 0 & 1 \end{bmatrix} \qquad \sigma_1 \equiv \sigma_x \equiv X \equiv \begin{bmatrix} 0 & 1 \\ 1 & 0 \end{bmatrix}$$

$$\sigma_2 \equiv \sigma_y \equiv Y \equiv \begin{bmatrix} 0 & -i \\ i & 0 \end{bmatrix} \qquad \sigma_3 \equiv \sigma_z \equiv Z \equiv \begin{bmatrix} 1 & 0 \\ 0 & -1 \end{bmatrix}$$

Figure 2.2. The Pauli matrices. Sometimes I is omitted from the list with just X, Y and Z known as the Pauli matrices.

2.1.4 Inner products

An *inner product* is a function which takes as input two vectors $|v\rangle$ and $|w\rangle$ from a vector space and produces a complex number as output. For the time being, it will be convenient to write the inner product of $|v\rangle$ and $|w\rangle$ as $(|v\rangle, |w\rangle)$. This is not the standard quantum mechanical notation; for pedagogical clarity the $(\cdot, \cdot)$ notation will be useful occasionally in this chapter. The standard quantum mechanical notation for the inner product $(|v\rangle, |w\rangle)$ is $\langle v|w\rangle$, where $|v\rangle$ and $|w\rangle$ are vectors in the inner product space, and the notation $\langle v|$ is used for the *dual vector* to the vector $|v\rangle$; the dual is a linear operator from the inner product space V to the complex numbers $\mathbf{C}$, defined by $\langle v|(|w\rangle) \equiv \langle v|w\rangle \equiv (|v\rangle, |w\rangle)$. We will see shortly that the matrix representation of dual vectors is just a row vector.

A function $(\cdot, \cdot)$ from $V \times V$ to $\mathbf{C}$ is an inner product if it satisfies the requirements that:

(1) $(\cdot, \cdot)$ is linear in the second argument,

$$\left(|v\rangle, \sum_i \lambda_i |w_i\rangle \right) = \sum_i \lambda_i \left(|v\rangle, |w_i\rangle \right). \tag{2.13}$$

(2) $(|v\rangle, |w\rangle) = (|w\rangle, |v\rangle)^*$.

(3) $(|v\rangle, |v\rangle) \geq 0$ with equality if and only if $|v\rangle = 0$.

For example, $\mathbf{C}^n$ has an inner product defined by

$$((y_1, \dots, y_n), (z_1, \dots, z_n)) \equiv \sum_i y_i^* z_i = \begin{bmatrix} y_1^* \cdots y_n^* \end{bmatrix} \begin{bmatrix} z_1 \\ \vdots \\ z_n \end{bmatrix}. \tag{2.14}$$

We call a vector space equipped with an inner product an *inner product space*.

Exercise 2.5: Verify that $(\cdot, \cdot)$ just defined is an inner product on $\mathbf{C}^n$.

Exercise 2.6: Show that any inner product $(\cdot, \cdot)$ is conjugate-linear in the first argument,

$$\left(\sum_i \lambda_i |w_i\rangle, |v\rangle\right) = \sum_i \lambda_i^* (|w_i\rangle, |v\rangle). \tag{2.15}$$

Discussions of quantum mechanics often refer to *Hilbert space*. In the finite dimensional complex vector spaces that come up in quantum computation and quantum information, a Hilbert space is *exactly the same thing* as an inner product space. From now on we use the two terms interchangeably, preferring the term Hilbert space. In infinite dimensions Hilbert spaces satisfy additional technical restrictions above and beyond inner product spaces, which we will not need to worry about.

Vectors $|w\rangle$ and $|v\rangle$ are *orthogonal* if their inner product is zero. For example, $|w\rangle \equiv (1, 0)$ and $|v\rangle \equiv (0, 1)$ are orthogonal with respect to the inner product defined by (2.14). We define the *norm* of a vector $|v\rangle$ by

$$\||v\rangle\| \equiv \sqrt{\langle v|v\rangle}. \tag{2.16}$$

A *unit vector* is a vector $|v\rangle$ such that $\||v\rangle\| = 1$. We also say that $|v\rangle$ is *normalized* if $\||v\rangle\| = 1$. It is convenient to talk of *normalizing* a vector by dividing by its norm; thus $|v\rangle / \||v\rangle\|$ is the *normalized* form of $|v\rangle$, for any non-zero vector $|v\rangle$. A set $|i\rangle$ of vectors with index i is *orthonormal* if each vector is a unit vector, and distinct vectors in the set are orthogonal, that is, $\langle i|j\rangle = \delta_{ij}$, where i and j are both chosen from the index set.

Exercise 2.7: Verify that $|w\rangle \equiv (1, 1)$ and $|v\rangle \equiv (1, -1)$ are orthogonal. What are the normalized forms of these vectors?

Suppose $|w_1\rangle, \ldots, |w_d\rangle$ is a basis set for some vector space V with an inner product. There is a useful method, the *Gram–Schmidt* procedure, which can be used to produce an orthonormal basis set $|v_1\rangle, \ldots, |v_d\rangle$ for the vector space V. Define $|v_1\rangle \equiv |w_1\rangle / \||w_1\rangle\|$, and for $1 \le k \le d - 1$ define $|v_{k+1}\rangle$ inductively by

$$|v_{k+1}\rangle \equiv \frac{|w_{k+1}\rangle - \sum_{i=1}^{k} \langle v_i|w_{k+1}\rangle |v_i\rangle}{\||w_{k+1}\rangle - \sum_{i=1}^{k} \langle v_i|w_{k+1}\rangle |v_i\rangle\|}. \tag{2.17}$$

It is not difficult to verify that the vectors $|v_1\rangle, \ldots, |v_d\rangle$ form an orthonormal set which is also a basis for V. Thus, any finite dimensional vector space of dimension d has an orthonormal basis, $|v_1\rangle, \ldots, |v_d\rangle$.

Exercise 2.8: Prove that the Gram–Schmidt procedure produces an orthonormal basis for V.

From now on, when we speak of a matrix representation for a linear operator, we mean a matrix representation with respect to orthonormal input and output bases. We also use the convention that if the input and output spaces for a linear operator are the same, then the input and output bases are the same, unless noted otherwise.

With these conventions, the inner product on a Hilbert space can be given a convenient matrix representation. Let $|w\rangle = \sum_i w_i |i\rangle$ and $|v\rangle = \sum_j v_j |j\rangle$ be representations of vectors $|w\rangle$ and $|v\rangle$ with respect to some orthonormal basis $|i\rangle$. Then, since $\langle i|j\rangle = \delta_{ij}$,

$$\langle v|w\rangle = \left(\sum_i v_i |i\rangle, \sum_j w_j |j\rangle \right) = \sum_{ij} v_i^* w_j \delta_{ij} = \sum_i v_i^* w_i \tag{2.18}$$

$$= [v_1^* \ldots v_n^*] \begin{bmatrix} w_1 \\ \vdots \\ w_n \end{bmatrix} . \tag{2.19}$$

That is, the inner product of two vectors is equal to the vector inner product between two matrix representations of those vectors, provided the representations are written with respect to the same orthonormal basis. We also see that the dual vector $\langle v|$ has a nice interpretation as the row vector whose components are complex conjugates of the corresponding components of the column vector representation of $|v\rangle$.

There is a useful way of representing linear operators which makes use of the inner product, known as the *outer product* representation. Suppose $|v\rangle$ is a vector in an inner product space V, and $|w\rangle$ is a vector in an inner product space W. Define $|w\rangle\langle v|$ to be the linear operator from V to W whose action is defined by

$$(|w\rangle\langle v|) \, (|v'\rangle) \equiv |w\rangle \, \langle v|v'\rangle = \langle v|v'\rangle|w\rangle. \tag{2.20}$$

This equation fits beautifully into our notational conventions, according to which the expression $|w\rangle\langle v|v'\rangle$ could potentially have one of two meanings: we will use it to denote the result when the *operator* $|w\rangle\langle v|$ acts on $|v'\rangle$, and it has an existing interpretation as the result of multiplying $|w\rangle$ by the complex number $\langle v|v'\rangle$. Our definitions are chosen so that these two potential meanings coincide. Indeed, we *define* the former in terms of the latter!

We can take linear combinations of outer product operators $|w\rangle\langle v|$ in the obvious way. By definition $\sum_i a_i |w_i\rangle\langle v_i|$ is the linear operator which, when acting on $|v'\rangle$, produces $\sum_i a_i |w_i\rangle\langle v_i|v'\rangle$ as output.

The usefulness of the outer product notation can be discerned from an important result known as the *completeness relation* for orthonormal vectors. Let $|i\rangle$ be any orthonormal basis for the vector space V, so an arbitrary vector $|v\rangle$ can be written $|v\rangle = \sum_i v_i |i\rangle$ for some set of complex numbers v_i. Note that $\langle i|v\rangle = v_i$ and therefore

$$\left(\sum_i |i\rangle\langle i| \right) |v\rangle = \sum_i |i\rangle\langle i|v\rangle = \sum_i v_i |i\rangle = |v\rangle. \tag{2.21}$$

Since the last equation is true for all $|v\rangle$ it follows that

$$\sum_i |i\rangle\langle i| = I. \tag{2.22}$$

This equation is known as the *completeness relation*. One application of the completeness relation is to give a means for representing any operator in the outer product notation. Suppose $A : V \rightarrow W$ is a linear operator, $|v_i\rangle$ is an orthonormal basis for V, and $|w_j\rangle$ an orthonormal basis for W. Using the completeness relation twice we obtain

$$A = I_W A I_V \tag{2.23}$$

$$= \sum_{ij} |w_j\rangle\langle w_j|A|v_i\rangle\langle v_i| \tag{2.24}$$

$$= \sum_{ij} \langle w_j|A|v_i\rangle|w_j\rangle\langle v_i|, \tag{2.25}$$

which is the outer product representation for A. We also see from this equation that A has matrix element $\langle w_j|A|v_i\rangle$ in the ith column and jth row, with respect to the input basis $|v_i\rangle$ and output basis $|w_j\rangle$.

A second application illustrating the usefulness of the completeness relation is the *Cauchy–Schwarz inequality*. This important result is discussed in Box 2.1, on this page.

Exercise 2.9: (Pauli operators and the outer product) The Pauli matrices (Figure 2.2 on page 65) can be considered as operators with respect to an orthonormal basis $|0\rangle, |1\rangle$ for a two-dimensional Hilbert space. Express each of the Pauli operators in the outer product notation.

Exercise 2.10: Suppose $|v_i\rangle$ is an orthonormal basis for an inner product space V. What is the matrix representation for the operator $|v_j\rangle\langle v_k|$, with respect to the $|v_i\rangle$ basis?

Box 2.1: The Cauchy-Schwarz inequality

The *Cauchy–Schwarz inequality* is an important geometric fact about Hilbert spaces. It states that for any two vectors $|v\rangle$ and $|w\rangle$, $|\langle v|w\rangle|^2 \le \langle v|v\rangle\langle w|w\rangle$. To see this, use the Gram–Schmidt procedure to construct an orthonormal basis $|i\rangle$ for the vector space such that the first member of the basis $|i\rangle$ is $|w\rangle/\sqrt{\langle w|w\rangle}$. Using the completeness relation $\sum_i |i\rangle\langle i| = I$, and dropping some non-negative terms gives

$$\langle v|v\rangle\langle w|w\rangle = \sum_i \langle v|i\rangle\langle i|v\rangle\langle w|w\rangle \tag{2.26}$$

$$\ge \frac{\langle v|w\rangle\langle w|v\rangle}{\langle w|w\rangle}\langle w|w\rangle \tag{2.27}$$

$$= \langle v|w\rangle\langle w|v\rangle = |\langle v|w\rangle|^2, \tag{2.28}$$

as required. A little thought shows that equality occurs if and only if $|v\rangle$ and $|w\rangle$ are linearly related, $|v\rangle = z|w\rangle$ or $|w\rangle = z|v\rangle$, for some scalar z.

2.1.5 Eigenvectors and eigenvalues

An *eigenvector* of a linear operator A on a vector space is a non-zero vector $|v\rangle$ such that $A|v\rangle = v|v\rangle$, where v is a complex number known as the *eigenvalue* of A corresponding to $|v\rangle$. It will often be convenient to use the notation v both as a label for the eigenvector, and to represent the eigenvalue. We assume that you are familiar with the elementary properties of eigenvalues and eigenvectors – in particular, how to find them, via the characteristic equation. The *characteristic function* is defined to be $c(\lambda) \equiv \det|A - \lambda I|$, where det is the *determinant* function for matrices; it can be shown that the characteristic function depends only upon the operator A, and not on the specific matrix representation used for A. The solutions of the *characteristic equation* $c(\lambda) = 0$ are the eigenvalues of the operator A. By the fundamental theorem of algebra, every polynomial has at least one complex root, so every operator A has at least one eigenvalue, and a corresponding eigenvector. The *eigenspace* corresponding to an eigenvalue v is the set of vectors which have eigenvalue v. It is a vector subspace of the vector space on which A acts.

A *diagonal representation* for an operator A on a vector space V is a representation $A = \sum_i \lambda_i |i\rangle\langle i|$, where the vectors $|i\rangle$ form an orthonormal set of eigenvectors for A, with corresponding eigenvalues λ_i. An operator is said to be *diagonalizable* if it has a diagonal representation. In the next section we will find a simple set of necessary and sufficient conditions for an operator on a Hilbert space to be diagonalizable. As an example of a diagonal representation, note that the Pauli Z matrix may be written

$$ Z = \begin{bmatrix} 1 & 0 \\ 0 & -1 \end{bmatrix} = |0\rangle\langle 0| - |1\rangle\langle 1|, \tag{2.29} $$

where the matrix representation is with respect to orthonormal vectors $|0\rangle$ and $|1\rangle$, respectively. Diagonal representations are sometimes also known as *orthonormal decompositions*.

When an eigenspace is more than one dimensional we say that it is *degenerate*. For example, the matrix A defined by

$$ A \equiv \begin{bmatrix} 2 & 0 & 0 \\ 0 & 2 & 0 \\ 0 & 0 & 0 \end{bmatrix} \tag{2.30} $$

has a two-dimensional eigenspace corresponding to the eigenvalue 2. The eigenvectors $(1, 0, 0)$ and $(0, 1, 0)$ are said to be *degenerate* because they are linearly independent eigenvectors of A with the same eigenvalue.

Exercise 2.11: (Eigendecomposition of the Pauli matrices)   Find the eigenvectors, eigenvalues, and diagonal representations of the Pauli matrices $X, Y,$ and Z.

Exercise 2.12: Prove that the matrix

$$ \begin{bmatrix} 1 & 0 \\ 1 & 1 \end{bmatrix} \tag{2.31} $$

is not diagonalizable.

2.1.6 Adjoints and Hermitian operators

Suppose A is any linear operator on a Hilbert space, V. It turns out that there exists a unique linear operator $A^\dagger$ on V such that for all vectors $|v\rangle, |w\rangle \in V$,

$$ (|v\rangle, A|w\rangle) = (A^\dagger|v\rangle, |w\rangle). \tag{2.32} $$

This linear operator is known as the *adjoint* or *Hermitian conjugate* of the operator A. From the definition it is easy to see that $(AB)^\dagger = B^\dagger A^\dagger$. By convention, if $|v\rangle$ is a vector, then we define $|v\rangle^\dagger \equiv \langle v|$. With this definition it is not difficult to see that $(A|v\rangle)^\dagger = \langle v|A^\dagger$.

Exercise 2.13: If $|w\rangle$ and $|v\rangle$ are any two vectors, show that $(|w\rangle\langle v|)^\dagger = |v\rangle\langle w|$.

Exercise 2.14: (Anti-linearity of the adjoint) Show that the adjoint operation is anti-linear,

$$\left(\sum_i a_i A_i\right)^\dagger = \sum_i a_i^* A_i^\dagger. \tag{2.33}$$

Exercise 2.15: Show that $(A^\dagger)^\dagger = A$.

In a matrix representation of an operator A, the action of the Hermitian conjugation operation is to take the matrix of A to the conjugate-transpose matrix, $A^\dagger \equiv (A^*)^T$, where the $*$ indicates complex conjugation, and T indicates the transpose operation. For example, we have

$$\begin{bmatrix} 1+3i & 2i \\ 1+i & 1-4i \end{bmatrix}^\dagger = \begin{bmatrix} 1-3i & 1-i \\ -2i & 1+4i \end{bmatrix}. \tag{2.34}$$

An operator A whose adjoint is A is known as a *Hermitian* or *self-adjoint* operator. An important class of Hermitian operators is the *projectors*. Suppose W is a k-dimensional vector subspace of the d-dimensional vector space V. Using the Gram–Schmidt procedure it is possible to construct an orthonormal basis $|1\rangle, \ldots, |d\rangle$ for V such that $|1\rangle, \ldots, |k\rangle$ is an orthonormal basis for W. By definition,

$$P \equiv \sum_{i=1}^{k} |i\rangle\langle i| \tag{2.35}$$

is the *projector* onto the subspace W. It is easy to check that this definition is independent of the orthonormal basis $|1\rangle, \ldots, |k\rangle$ used for W. From the definition it can be shown that $|v\rangle\langle v|$ is Hermitian for any vector $|v\rangle$, so P is Hermitian, $P^\dagger = P$. We will often refer to the 'vector space' P, as shorthand for the vector space onto which P is a projector. The *orthogonal complement* of P is the operator $Q \equiv I - P$. It is easy to see that Q is a projector onto the vector space spanned by $|k+1\rangle, \ldots, |d\rangle$, which we also refer to as the *orthogonal complement* of P, and may denote by Q.

Exercise 2.16: Show that any projector P satisfies the equation $P^2 = P$.

An operator A is said to be *normal* if $AA^\dagger = A^\dagger A$. Clearly, an operator which is Hermitian is also normal. There is a remarkable representation theorem for normal operators known as the *spectral decomposition*, which states that an operator is a normal operator if and only if it is diagonalizable. This result is proved in Box 2.2 on page 72, which you should read closely.

Exercise 2.17: Show that a normal matrix is Hermitian if and only if it has real eigenvalues.

A matrix U is said to be *unitary* if $U^\dagger U = I$. Similarly an operator U is unitary if $U^\dagger U = I$. It is easily checked that an operator is unitary if and only if each of its matrix representations is unitary. A unitary operator also satisfies $UU^\dagger = I$, and therefore U is normal and has a spectral decomposition. Geometrically, unitary operators are important because they preserve inner products between vectors. To see this, let $|v\rangle$ and $|w\rangle$ be any two vectors. Then the inner product of $U|v\rangle$ and $U|w\rangle$ is the same as the inner product of $|v\rangle$ and $|w\rangle$,

$$(U|v\rangle, U|w\rangle) = \langle v|U^\dagger U|w\rangle = \langle v|I|w\rangle = \langle v|w\rangle. \tag{2.36}$$

This result suggests the following elegant outer product representation of any unitary U. Let $|v_i\rangle$ be any orthonormal basis set. Define $|w_i\rangle \equiv U|v_i\rangle$, so $|w_i\rangle$ is also an orthonormal basis set, since unitary operators preserve inner products. Note that $U = \sum_i |w_i\rangle\langle v_i|$. Conversely, if $|v_i\rangle$ and $|w_i\rangle$ are any two orthonormal bases, then it is easily checked that the operator U defined by $U \equiv \sum_i |w_i\rangle\langle v_i|$ is a unitary operator.

Exercise 2.18: Show that all eigenvalues of a unitary matrix have modulus 1, that is, can be written in the form $e^{i\theta}$ for some real θ.

Exercise 2.19: (**Pauli matrices: Hermitian and unitary**) Show that the Pauli matrices are Hermitian and unitary.

Exercise 2.20: (**Basis changes**) Suppose A' and A'' are matrix representations of an operator A on a vector space V with respect to two different orthonormal bases, $|v_i\rangle$ and $|w_i\rangle$. Then the elements of A' and A'' are $A'_{ij} = \langle v_i|A|v_j\rangle$ and $A''_{ij} = \langle w_i|A|w_j\rangle$. Characterize the relationship between A' and A''.

A special subclass of Hermitian operators is extremely important. This is the *positive operators*. A positive operator A is defined to be an operator such that for any vector $|v\rangle$, $(|v\rangle, A|v\rangle)$ is a real, non-negative number. If $(|v\rangle, A|v\rangle)$ is *strictly* greater than zero for all $|v\rangle \neq 0$ then we say that A is *positive definite*. In Exercise 2.24 on this page you will show that any positive operator is automatically Hermitian, and therefore by the spectral decomposition has diagonal representation $\sum_i \lambda_i|i\rangle\langle i|$, with non-negative eigenvalues λ_i.

Exercise 2.21: Repeat the proof of the spectral decomposition in Box 2.2 for the case when M is Hermitian, simplifying the proof wherever possible.

Exercise 2.22: Prove that two eigenvectors of a Hermitian operator with different eigenvalues are necessarily orthogonal.

Exercise 2.23: Show that the eigenvalues of a projector P are all either 0 or 1.

Exercise 2.24: (**Hermiticity of positive operators**) Show that a positive operator is necessarily Hermitian. (*Hint*: Show that an arbitrary operator A can be written $A = B + iC$ where B and C are Hermitian.)

Exercise 2.25: Show that for any operator A, $A^\dagger A$ is positive.

2.1.7 Tensor products

The *tensor product* is a way of putting vector spaces together to form larger vector spaces. This construction is crucial to understanding the quantum mechanics of multiparticle systems. The following discussion is a little abstract, and may be difficult to follow if you're not already familiar with the tensor product, so feel free to skip ahead now and revisit later when you come to the discussion of tensor products in quantum mechanics.

Suppose V and W are vector spaces of dimension m and n respectively. For convenience we also suppose that V and W are Hilbert spaces. Then $V \otimes W$ (read 'V tensor

---

Box 2.2: The spectral decomposition – important!

The *spectral decomposition* is an extremely useful representation theorem for normal operators.

Theorem 2.1: (**Spectral decomposition**) Any normal operator M on a vector
space V is diagonal with respect to some orthonormal basis for V.
Conversely, any diagonalizable operator is normal.

Proof

The converse is a simple exercise, so we prove merely the forward implication,
by induction on the dimension d of V. The case $d = 1$ is trivial. Let λ be an
eigenvalue of M, P the projector onto the λ eigenspace, and Q the projector onto
the orthogonal complement. Then $M = (P + Q)M(P + Q) = PMP + QMP +
PMQ + QMQ$. Obviously $PMP = \lambda P$. Furthermore, $QMP = 0$, as M takes
the subspace P into itself. We claim that $PMQ = 0$ also. To see this, let $|v\rangle$
be an element of the subspace P. Then $MM^\dagger|v\rangle = M^\dagger M|v\rangle = \lambda M^\dagger|v\rangle$. Thus,
$M^\dagger|v\rangle$ has eigenvalue λ and therefore is an element of the subspace P. It follows
that $QM^\dagger P = 0$. Taking the adjoint of this equation gives $PMQ = 0$. Thus
$M = PMP + QMQ$. Next, we prove that QMQ is normal. To see this, note that
$QM = QM(P + Q) = QMQ$, and $QM^\dagger = QM^\dagger(P + Q) = QM^\dagger Q$. Therefore,
by the normality of M, and the observation that $Q^2 = Q$,

$$QMQ\,QM^\dagger Q = QMQM^\dagger Q \tag{2.37}$$
$$= QMM^\dagger Q \tag{2.38}$$
$$= QM^\dagger MQ \tag{2.39}$$
$$= QM^\dagger QMQ \tag{2.40}$$
$$= QM^\dagger Q\,QMQ\,, \tag{2.41}$$

so QMQ is normal. By induction, QMQ is diagonal with respect to some orthonormal basis for the subspace Q, and PMP is already diagonal with respect
to some orthonormal basis for P. It follows that $M = PMP + QMQ$ is diagonal
with respect to some orthonormal basis for the total vector space.   □

In terms of the outer product representation, this means that M can be written as
$M = \sum_i \lambda_i |i\rangle\langle i|$, where λ_i are the eigenvalues of M, $|i\rangle$ is an orthonormal basis
for V, and each $|i\rangle$ an eigenvector of M with eigenvalue λ_i. In terms of projectors,
$M = \sum_i \lambda_i P_i$, where λ_i are again the eigenvalues of M, and P_i is the projector
onto the λ_i eigenspace of M. These projectors satisfy the completeness relation
$\sum_i P_i = I$, and the orthonormality relation $P_i P_j = \delta_{ij} P_i$.

W') is an mn dimensional vector space. The elements of $V \otimes W$ are linear combinations
of 'tensor products' $|v\rangle \otimes |w\rangle$ of elements $|v\rangle$ of V and $|w\rangle$ of W. In particular, if $|i\rangle$ and
$|j\rangle$ are orthonormal bases for the spaces V and W then $|i\rangle \otimes |j\rangle$ is a basis for $V \otimes W$. We
often use the abbreviated notations $|v\rangle|w\rangle$, $|v, w\rangle$ or even $|vw\rangle$ for the tensor product

$|v\rangle \otimes |w\rangle$. For example, if V is a two-dimensional vector space with basis vectors $|0\rangle$ and $|1\rangle$ then $|0\rangle \otimes |0\rangle + |1\rangle \otimes |1\rangle$ is an element of $V \otimes V$.

By definition the tensor product satisfies the following basic properties:

(1) For an arbitrary scalar z and elements $|v\rangle$ of V and $|w\rangle$ of W,

$$z \left(|v\rangle \otimes |w\rangle\right) = (z|v\rangle) \otimes |w\rangle = |v\rangle \otimes (z|w\rangle) . \tag{2.42}$$

(2) For arbitrary $|v_1\rangle$ and $|v_2\rangle$ in V and $|w\rangle$ in W,

$$\left(|v_1\rangle + |v_2\rangle\right) \otimes |w\rangle = |v_1\rangle \otimes |w\rangle + |v_2\rangle \otimes |w\rangle. \tag{2.43}$$

(3) For arbitrary $|v\rangle$ in V and $|w_1\rangle$ and $|w_2\rangle$ in W,

$$|v\rangle \otimes \left(|w_1\rangle + |w_2\rangle\right) = |v\rangle \otimes |w_1\rangle + |v\rangle \otimes |w_2\rangle. \tag{2.44}$$

What sorts of linear operators act on the space $V \otimes W$? Suppose $|v\rangle$ and $|w\rangle$ are vectors in V and W, and A and B are linear operators on V and W, respectively. Then we can define a linear operator $A \otimes B$ on $V \otimes W$ by the equation

$$(A \otimes B)(|v\rangle \otimes |w\rangle) \equiv A|v\rangle \otimes B|w\rangle. \tag{2.45}$$

The definition of $A \otimes B$ is then extended to all elements of $V \otimes W$ in the natural way to ensure linearity of $A \otimes B$, that is,

$$(A \otimes B) \left(\sum_i a_i|v_i\rangle \otimes |w_i\rangle\right) \equiv \sum_i a_i A|v_i\rangle \otimes B|w_i\rangle. \tag{2.46}$$

It can be shown that $A \otimes B$ defined in this way is a well-defined linear operator on $V \otimes W$. This notion of the tensor product of two operators extends in the obvious way to the case where $A : V \to V'$ and $B : W \to W'$ map between different vector spaces. Indeed, an arbitrary linear operator C mapping $V \otimes W$ to $V' \otimes W'$ can be represented as a linear combination of tensor products of operators mapping V to V' and W to W',

$$C = \sum_i c_i A_i \otimes B_i, \tag{2.47}$$

where by definition

$$\left(\sum_i c_i A_i \otimes B_i\right) |v\rangle \otimes |w\rangle \equiv \sum_i c_i A_i|v\rangle \otimes B_i|w\rangle. \tag{2.48}$$

The inner products on the spaces V and W can be used to define a natural inner product on $V \otimes W$. Define

$$\left(\sum_i a_i|v_i\rangle \otimes |w_i\rangle, \sum_j b_j|v_j'\rangle \otimes |w_j'\rangle\right) \equiv \sum_{ij} a_i^* b_j \langle v_i|v_j'\rangle \langle w_i|w_j'\rangle. \tag{2.49}$$

It can be shown that the function so defined is a well-defined inner product. From this inner product, the inner product space $V \otimes W$ inherits the other structure we are familiar with, such as notions of an adjoint, unitarity, normality, and Hermiticity.

All this discussion is rather abstract. It can be made much more concrete by moving to a convenient matrix representation known as the *Kronecker product*. Suppose A is an m by n matrix, and B is a p by q matrix. Then we have the matrix representation:

$$
A \otimes B \equiv
\overbrace{
\left.
\begin{bmatrix}
A_{11}B & A_{12}B & \cdots & A_{1n}B \\
A_{21}B & A_{22}B & \cdots & A_{2n}B \\
\vdots & \vdots & \vdots & \vdots \\
A_{m1}B & A_{m2}B & \cdots & A_{mn}B
\end{bmatrix}
\right\} mp\,.
}^{nq}
\tag{2.50}
$$

In this representation terms like $A_{11}B$ denote p by q submatrices whose entries are proportional to B, with overall proportionality constant A_{11}. For example, the tensor product of the vectors $(1, 2)$ and $(2, 3)$ is the vector

$$
\begin{bmatrix} 1 \\ 2 \end{bmatrix} \otimes \begin{bmatrix} 2 \\ 3 \end{bmatrix} =
\begin{bmatrix} 1 \times 2 \\ 1 \times 3 \\ 2 \times 2 \\ 2 \times 3 \end{bmatrix} =
\begin{bmatrix} 2 \\ 3 \\ 4 \\ 6 \end{bmatrix}.
\tag{2.51}
$$

The tensor product of the Pauli matrices X and Y is

$$
X \otimes Y =
\begin{bmatrix} 0 \cdot Y & 1 \cdot Y \\ 1 \cdot Y & 0 \cdot Y \end{bmatrix} =
\begin{bmatrix}
0 & 0 & 0 & -i \\
0 & 0 & i & 0 \\
0 & -i & 0 & 0 \\
i & 0 & 0 & 0
\end{bmatrix}.
\tag{2.52}
$$

Finally, we mention the useful notation $|\psi\rangle^{\otimes k}$, which means $|\psi\rangle$ tensored with itself k times. For example $|\psi\rangle^{\otimes 2} = |\psi\rangle \otimes |\psi\rangle$. An analogous notation is also used for operators on tensor product spaces.

Exercise 2.26: Let $|\psi\rangle = (|0\rangle + |1\rangle)/\sqrt{2}$. Write out $|\psi\rangle^{\otimes 2}$ and $|\psi\rangle^{\otimes 3}$ explicitly, both in terms of tensor products like $|0\rangle|1\rangle$, and using the Kronecker product.

Exercise 2.27: Calculate the matrix representation of the tensor products of the Pauli operators (a) X and Z; (b) I and X; (c) X and I. Is the tensor product commutative?

Exercise 2.28: Show that the transpose, complex conjugation, and adjoint operations distribute over the tensor product,

$$
(A \otimes B)^* = A^* \otimes B^*; \quad (A \otimes B)^T = A^T \otimes B^T; \quad (A \otimes B)^\dagger = A^\dagger \otimes B^\dagger. \tag{2.53}
$$

Exercise 2.29: Show that the tensor product of two unitary operators is unitary.

Exercise 2.30: Show that the tensor product of two Hermitian operators is Hermitian.

Exercise 2.31: Show that the tensor product of two positive operators is positive.

Exercise 2.32: Show that the tensor product of two projectors is a projector.

Exercise 2.33: The Hadamard operator on one qubit may be written as

$$
H = \frac{1}{\sqrt{2}} \left[(|0\rangle + |1\rangle)\langle 0| + (|0\rangle - |1\rangle)\langle 1| \right].
\tag{2.54}
$$

Show explicitly that the Hadamard transform on n qubits, $H^{\otimes n}$, may be written as

$$H^{\otimes n} = \frac{1}{\sqrt{2^n}} \sum_{x,y} (-1)^{x \cdot y} |x\rangle\langle y|. \tag{2.55}$$

Write out an explicit matrix representation for $H^{\otimes 2}$.

2.1.8 Operator functions

There are many important functions which can be defined for operators and matrices. Generally speaking, given a function f from the complex numbers to the complex numbers, it is possible to define a corresponding matrix function on normal matrices (or some subclass, such as the Hermitian matrices) by the following construction. Let $A = \sum_a a|a\rangle\langle a|$ be a spectral decomposition for a normal operator A. Define $f(A) \equiv \sum_a f(a)|a\rangle\langle a|$. A little thought shows that $f(A)$ is uniquely defined. This procedure can be used, for example, to define the square root of a positive operator, the logarithm of a positive-definite operator, or the exponential of a normal operator. As an example,

$$\exp(\theta Z) = \begin{bmatrix} e^\theta & 0 \\ 0 & e^{-\theta} \end{bmatrix}, \tag{2.56}$$

since Z has eigenvectors $|0\rangle$ and $|1\rangle$.

Exercise 2.34: Find the square root and logarithm of the matrix

$$\begin{bmatrix} 4 & 3 \\ 3 & 4 \end{bmatrix}. \tag{2.57}$$

Exercise 2.35: (Exponential of the Pauli matrices) Let $\vec{v}$ be any real, three-dimensional unit vector and θ a real number. Prove that

$$\exp(i\theta \vec{v} \cdot \vec{\sigma}) = \cos(\theta)I + i\sin(\theta)\vec{v} \cdot \vec{\sigma}, \tag{2.58}$$

where $\vec{v} \cdot \vec{\sigma} \equiv \sum_{i=1}^{3} v_i \sigma_i$. This exercise is generalized in Problem 2.1 on page 117.

Another important matrix function is the *trace* of a matrix. The trace of A is defined to be the sum of its diagonal elements,

$$\mathrm{tr}(A) \equiv \sum_i A_{ii}. \tag{2.59}$$

The trace is easily seen to be *cyclic*, $\mathrm{tr}(AB) = \mathrm{tr}(BA)$, and *linear*, $\mathrm{tr}(A + B) = \mathrm{tr}(A) + \mathrm{tr}(B)$, $\mathrm{tr}(zA) = z\,\mathrm{tr}(A)$, where A and B are arbitrary matrices, and z is a complex number. Furthermore, from the cyclic property it follows that the trace of a matrix is invariant under the unitary *similarity transformation* $A \rightarrow UAU^\dagger$, as $\mathrm{tr}(UAU^\dagger) = \mathrm{tr}(U^\dagger UA) = \mathrm{tr}(A)$. In light of this result, it makes sense to define the trace of an *operator* A to be the trace of any matrix representation of A. The invariance of the trace under unitary similarity transformations ensures that the trace of an operator is well defined. As an example of the trace, suppose $|\psi\rangle$ is a unit vector and A is an arbitrary operator. To evaluate $\mathrm{tr}(A|\psi\rangle\langle\psi|)$ use the Gram–Schmidt procedure to extend $|\psi\rangle$ to an orthonormal basis $|i\rangle$ which includes $|\psi\rangle$ as the first element. Then we have

$$\text{tr}(A|\psi\rangle\langle\psi|) = \sum_i \langle i|A|\psi\rangle\langle\psi|i\rangle \tag{2.60}$$

$$= \langle\psi|A|\psi\rangle. \tag{2.61}$$

This result, that $\text{tr}(A|\psi\rangle\langle\psi|) = \langle\psi|A|\psi\rangle$ is extremely useful in evaluating the trace of an operator.

Exercise 2.36: Show that the Pauli matrices except for I have trace zero.

Exercise 2.37: (Cyclic property of the trace) If A and B are two linear operators show that

$$\text{tr}(AB) = \text{tr}(BA). \tag{2.62}$$

Exercise 2.38: (Linearity of the trace) If A and B are two linear operators, show that

$$\text{tr}(A + B) = \text{tr}(A) + \text{tr}(B) \tag{2.63}$$

and if z is an arbitrary complex number show that

$$\text{tr}(zA) = z\text{tr}(A). \tag{2.64}$$

Exercise 2.39: (The Hilbert–Schmidt inner product on operators) The set L_V of linear operators on a Hilbert space V is obviously a vector space – the sum of two linear operators is a linear operator, zA is a linear operator if A is a linear operator and z is a complex number, and there is a zero element 0. An important additional result is that the vector space L_V can be given a natural inner product structure, turning it into a Hilbert space.

(1) Show that the function $(\cdot, \cdot)$ on $L_V \times L_V$ defined by

$$(A, B) \equiv \text{tr}(A^\dagger B) \tag{2.65}$$

is an inner product function. This inner product is known as the *Hilbert–Schmidt* or *trace* inner product.
(2) If V has d dimensions show that L_V has dimension d^2.
(3) Find an orthonormal basis of Hermitian matrices for the Hilbert space L_V.

2.1.9 The commutator and anti-commutator

The *commutator* between two operators A and B is defined to be

$$[A, B] \equiv AB - BA. \tag{2.66}$$

If $[A, B] = 0$, that is, $AB = BA$, then we say A *commutes* with B. Similarly, the *anti-commutator* of two operators A and B is defined by

$$\{A, B\} \equiv AB + BA; \tag{2.67}$$

we say A *anti-commutes* with B if $\{A, B\} = 0$. It turns out that many important properties of pairs of operators can be deduced from their commutator and anti-commutator. Perhaps the most useful relation is the following connection between the commutator and the property of being able to *simultaneously diagonalize* Hermitian operators A and B, that is, write $A = \sum_i a_i |i\rangle\langle i|$, $B = \sum_i b_i |i\rangle\langle i|$, where $|i\rangle$ is some common orthonormal set of eigenvectors for A and B.

Theorem 2.2: (**Simultaneous diagonalization theorem**) Suppose A and B are Hermitian operators. Then $[A, B] = 0$ if and only if there exists an orthonormal basis such that both A and B are diagonal with respect to that basis. We say that A and B are *simultaneously diagonalizable* in this case.

This result connects the commutator of two operators, which is often easy to compute, to the property of being simultaneously diagonalizable, which is *a priori* rather difficult to determine. As an example, consider that

$$[X, Y] = \begin{bmatrix} 0 & 1 \\ 1 & 0 \end{bmatrix} \begin{bmatrix} 0 & -i \\ i & 0 \end{bmatrix} - \begin{bmatrix} 0 & -i \\ i & 0 \end{bmatrix} \begin{bmatrix} 0 & 1 \\ 1 & 0 \end{bmatrix} \tag{2.68}$$

$$= 2i \begin{bmatrix} 1 & 0 \\ 0 & -1 \end{bmatrix} \tag{2.69}$$

$$= 2iZ, \tag{2.70}$$

so X and Y do not commute. You have already shown, in Exercise 2.11, that X and Y do not have common eigenvectors, as we expect from the simultaneous diagonalization theorem.

Proof
You can (and should!) easily verify that if A and B are diagonal in the same orthonormal basis then $[A, B] = 0$. To show the converse, let $|a, j\rangle$ be an orthonormal basis for the eigenspace V_a of A with eigenvalue a; the index j is used to label possible degeneracies. Note that

$$AB|a, j\rangle = BA|a, j\rangle = aB|a, j\rangle, \tag{2.71}$$

and therefore $B|a, j\rangle$ is an element of the eigenspace V_a. Let P_a denote the projector onto the space V_a and define $B_a \equiv P_a B P_a$. It is easy to see that the restriction of B_a to the space V_a is Hermitian on V_a, and therefore has a spectral decomposition in terms of an orthonormal set of eigenvectors which span the space V_a. Let's call these eigenvectors $|a, b, k\rangle$, where the indices a and b label the eigenvalues of A and B_a, and k is an extra index to allow for the possibility of a degenerate B_a. Note that $B|a, b, k\rangle$ is an element of V_a, so $B|a, b, k\rangle = P_a B|a, b, k\rangle$. Moreover we have $P_a|a, b, k\rangle = |a, b, k\rangle$, so

$$B|a, b, k\rangle = P_a B P_a|a, b, k\rangle = b|a, b, k\rangle. \tag{2.72}$$

It follows that $|a, b, k\rangle$ is an eigenvector of B with eigenvalue b, and therefore $|a, b, k\rangle$ is an orthonormal set of eigenvectors of both A and B, spanning the entire vector space on which A and B are defined. That is, A and B are simultaneously diagonalizable.   □

Exercise 2.40: (**Commutation relations for the Pauli matrices**)   Verify the commutation relations

$$[X, Y] = 2iZ; \quad [Y, Z] = 2iX; \quad [Z, X] = 2iY. \tag{2.73}$$

There is an elegant way of writing this using ϵ_{jkl}, the antisymmetric tensor on three indices, for which $\epsilon_{jkl} = 0$ except for $\epsilon_{123} = \epsilon_{231} = \epsilon_{312} = 1$, and $\epsilon_{321} = \epsilon_{213} = \epsilon_{132} = -1$:

$$[\sigma_j, \sigma_k] = 2i \sum_{l=1}^{3} \epsilon_{jkl} \sigma_l. \tag{2.74}$$

Exercise 2.41: (Anti-commutation relations for the Pauli matrices) Verify the anti-commutation relations

$$\{\sigma_i, \sigma_j\} = 0 \tag{2.75}$$

where $i \neq j$ are both chosen from the set $1, 2, 3$. Also verify that $(i = 0, 1, 2, 3)$

$$\sigma_i^2 = I. \tag{2.76}$$

Exercise 2.42: Verify that

$$AB = \frac{[A, B] + \{A, B\}}{2}. \tag{2.77}$$

Exercise 2.43: Show that for $j, k = 1, 2, 3$,

$$\sigma_j \sigma_k = \delta_{jk} I + i \sum_{l=1}^{3} \epsilon_{jkl} \sigma_l. \tag{2.78}$$

Exercise 2.44: Suppose $[A, B] = 0$, $\{A, B\} = 0$, and A is invertible. Show that B must be 0.

Exercise 2.45: Show that $[A, B]^\dagger = [B^\dagger, A^\dagger]$.

Exercise 2.46: Show that $[A, B] = -[B, A]$.

Exercise 2.47: Suppose A and B are Hermitian. Show that $i[A, B]$ is Hermitian.

2.1.10 The polar and singular value decompositions

The *polar* and *singular value* decompositions are useful ways of breaking linear operators up into simpler parts. In particular, these decompositions allow us to break general linear operators up into products of unitary operators and positive operators. While we don't understand the structure of general linear operators terribly well, we do understand unitary operators and positive operators in quite some detail. The polar and singular value decompositions allow us to apply this understanding to better understand general linear operators.

Theorem 2.3: **(Polar decomposition)** Let A be a linear operator on a vector space V. Then there exists unitary U and positive operators J and K such that

$$A = UJ = KU, \tag{2.79}$$

where the unique positive operators J and K satisfying these equations are defined by $J \equiv \sqrt{A^\dagger A}$ and $K \equiv \sqrt{AA^\dagger}$. Moreover, if A is invertible then U is unique.

We call the expression $A = UJ$ the *left polar decomposition* of A, and $A = KU$ the *right polar decomposition* of A. Most often, we'll omit the 'right' or 'left' nomenclature, and use the term 'polar decomposition' for both expressions, with context indicating which is meant.

Proof

$J \equiv \sqrt{A^\dagger A}$ is a positive operator, so it can be given a spectral decomposition, $J = \sum_i \lambda_i |i\rangle\langle i|$ ($\lambda_i \geq 0$). Define $|\psi_i\rangle \equiv A|i\rangle$. From the definition, we see that $\langle \psi_i | \psi_i \rangle = \lambda_i^2$. Consider for now only those i for which $\lambda_i \neq 0$. For those i define $|e_i\rangle \equiv |\psi_i\rangle / \lambda_i$, so the $|e_i\rangle$ are normalized. Moreover, they are orthogonal, since if $i \neq j$ then $\langle e_i | e_j \rangle = \langle i | A^\dagger A | j \rangle / \lambda_i \lambda_j = \langle i | J^2 | j \rangle / \lambda_i \lambda_j = 0$.

We have been considering i such that $\lambda_i \neq 0$. Now use the Gram–Schmidt procedure to extend the orthonormal set $|e_i\rangle$ so it forms an orthonormal basis, which we also label $|e_i\rangle$. Define a unitary operator $U \equiv \sum_i |e_i\rangle\langle i|$. When $\lambda_i \neq 0$ we have $UJ|i\rangle = \lambda_i |e_i\rangle = |\psi_i\rangle = A|i\rangle$. When $\lambda_i = 0$ we have $UJ|i\rangle = 0 = |\psi_i\rangle$. We have proved that the action of A and UJ agree on the basis $|i\rangle$, and thus that $A = UJ$.

J is unique, since multiplying $A = UJ$ on the left by the adjoint equation $A^\dagger = JU^\dagger$ gives $J^2 = A^\dagger A$, from which we see that $J = \sqrt{A^\dagger A}$, uniquely. A little thought shows that if A is invertible, then so is J, so U is uniquely determined by the equation $U = AJ^{-1}$. The proof of the right polar decomposition follows, since $A = UJ = UJU^\dagger U = KU$, where $K \equiv UJU^\dagger$ is a positive operator. Since $AA^\dagger = KUU^\dagger K = K^2$ we must have $K = \sqrt{AA^\dagger}$, as claimed. $\qquad\square$

The singular value decomposition combines the polar decomposition and the spectral theorem.

Corollary 2.4: (**Singular value decomposition**) Let A be a square matrix. Then there exist unitary matrices U and V, and a diagonal matrix D with non-negative entries such that

$$A = UDV. \tag{2.80}$$

The diagonal elements of D are called the *singular values* of A.

Proof

By the polar decomposition, $A = SJ$, for unitary S, and positive J. By the spectral theorem, $J = TDT^\dagger$, for unitary T and diagonal D with non-negative entries. Setting $U \equiv ST$ and $V \equiv T^\dagger$ completes the proof. $\qquad\square$

Exercise 2.48: What is the polar decomposition of a positive matrix P? Of a unitary matrix U? Of a Hermitian matrix, H?

Exercise 2.49: Express the polar decomposition of a normal matrix in the outer product representation.

Exercise 2.50: Find the left and right polar decompositions of the matrix

$$\begin{bmatrix} 1 & 0 \\ 1 & 1 \end{bmatrix}. \tag{2.81}$$

2.2   The postulates of quantum mechanics

All understanding begins with our not accepting the world as it appears.
– Alan Kay

The most incomprehensible thing about the world is that it is comprehensible.
– Albert Einstein

Quantum mechanics is a mathematical framework for the development of physical theories. On its own quantum mechanics doesn't tell you what laws a physical system must obey, but it does provide a mathematical and conceptual framework for the development of such laws. In the next few sections we give a complete description of the basic postulates of quantum mechanics. These postulates provide a connection between the physical world and the mathematical formalism of quantum mechanics.

The postulates of quantum mechanics were derived after a long process of trial and (mostly) error, which involved a considerable amount of guessing and fumbling by the originators of the theory. Don't be surprised if the motivation for the postulates is not always clear; even to experts the basic postulates of quantum mechanics appear surprising. What you should expect to gain in the next few sections is a good working grasp of the postulates – how to apply them, and when.

2.2.1   State space

The first postulate of quantum mechanics sets up the arena in which quantum mechanics takes place. The arena is our familiar friend from linear algebra, Hilbert space.

> **Postulate 1**: Associated to any isolated physical system is a complex vector space with inner product (that is, a Hilbert space) known as the *state space* of the system. The system is completely described by its *state vector*, which is a unit vector in the system's state space.

Quantum mechanics does *not* tell us, for a given physical system, what the state space of that system is, nor does it tell us what the state vector of the system is. Figuring that out for a *specific* system is a difficult problem for which physicists have developed many intricate and beautiful rules. For example, there is the wonderful theory of quantum electrodynamics (often known as QED), which describes how atoms and light interact. One aspect of QED is that it tells us what state spaces to use to give quantum descriptions of atoms and light. We won't be much concerned with the intricacies of theories like QED (except in so far as they apply to physical realizations, in Chapter 7), as we are mostly interested in the general framework provided by quantum mechanics. For our purposes it will be sufficient to make some very simple (and reasonable) assumptions about the state spaces of the systems we are interested in, and stick with those assumptions.

The simplest quantum mechanical system, and the system which we will be most concerned with, is the *qubit*. A qubit has a two-dimensional state space. Suppose $|0\rangle$ and $|1\rangle$ form an orthonormal basis for that state space. Then an arbitrary state vector in the state space can be written

$$|\psi\rangle = a|0\rangle + b|1\rangle, \tag{2.82}$$

where a and b are complex numbers. The condition that $|\psi\rangle$ be a unit vector, $\langle\psi|\psi\rangle = 1$, is therefore equivalent to $|a|^2 + |b|^2 = 1$. The condition $\langle\psi|\psi\rangle = 1$ is often known as the *normalization condition* for state vectors.

We will take the qubit as our fundamental quantum mechanical system. Later, in Chapter 7, we will see that there are real physical systems which may be described in terms of qubits. For now, though, it is sufficient to think of qubits in abstract terms, without reference to a specific realization. Our discussions of qubits will always be referred to some orthonormal set of basis vectors, $|0\rangle$ and $|1\rangle$, which should be thought of as being fixed in advance. Intuitively, the states $|0\rangle$ and $|1\rangle$ are analogous to the two values 0 and 1 which a bit may take. The way a qubit differs from a bit is that *superpositions* of these two states, of the form $a|0\rangle + b|1\rangle$, can also exist, in which it is not possible to say that the qubit is definitely in the state $|0\rangle$, or definitely in the state $|1\rangle$.

We conclude with some useful terminology which is often used in connection with the description of quantum states. We say that any linear combination $\sum_i \alpha_i|\psi_i\rangle$ is a superposition of the states $|\psi_i\rangle$ with *amplitude* α_i for the state $|\psi_i\rangle$. So, for example, the state

$$\frac{|0\rangle - |1\rangle}{\sqrt{2}} \tag{2.83}$$

is a superposition of the states $|0\rangle$ and $|1\rangle$ with amplitude $1/\sqrt{2}$ for the state $|0\rangle$, and amplitude $-1/\sqrt{2}$ for the state $|1\rangle$.

2.2.2 Evolution

How does the state, $|\psi\rangle$, of a quantum mechanical system change with time? The following postulate gives a prescription for the description of such state changes.

> **Postulate 2**: The evolution of a *closed* quantum system is described by a *unitary transformation*. That is, the state $|\psi\rangle$ of the system at time t_1 is related to the state $|\psi'\rangle$ of the system at time t_2 by a unitary operator U which depends only on the times t_1 and t_2,
>
> $$|\psi'\rangle = U|\psi\rangle. \tag{2.84}$$

Just as quantum mechanics does not tell us the state space or quantum state of a *particular* quantum system, it does not tell us which unitary operators U describe real-world quantum dynamics. Quantum mechanics merely assures us that the evolution of any closed quantum system may be described in such a way. An obvious question to ask is: what unitary operators are natural to consider? In the case of single qubits, it turns out that *any* unitary operator at all can be realized in realistic systems.

Let's look at a few examples of unitary operators on a single qubit which are important in quantum computation and quantum information. We have already seen several examples of such unitary operators – the Pauli matrices, defined in Section 2.1.3, and the quantum gates described in Chapter 1. As remarked in Section 1.3.1, the X matrix is often known as the quantum NOT gate, by analogy to the classical NOT gate. The X and Z Pauli matrices are also sometimes referred to as the *bit flip* and *phase flip* matrices: the X matrix takes $|0\rangle$ to $|1\rangle$, and $|1\rangle$ to $|0\rangle$, thus earning the name bit flip; and the Z matrix leaves $|0\rangle$ invariant, and takes $|1\rangle$ to $-|1\rangle$, with the extra factor of -1 added known as a *phase factor*, thus justifying the term phase flip. We will not use the term phase flip for

Z very often, since it is easily confused with the phase gate to be defined in Chapter 4. (Section 2.2.7 contains more discussion of the many uses of the term 'phase'.)

Another interesting unitary operator is the *Hadamard gate*, which we denote H. This has the action $H|0\rangle \equiv (|0\rangle + |1\rangle)/\sqrt{2}$, $H|1\rangle \equiv (|0\rangle - |1\rangle)/\sqrt{2}$, and corresponding matrix representation

$$H = \frac{1}{\sqrt{2}} \begin{bmatrix} 1 & 1 \\ 1 & -1 \end{bmatrix}. \tag{2.85}$$

Exercise 2.51: Verify that the Hadamard gate H is unitary.

Exercise 2.52: Verify that $H^2 = I$.

Exercise 2.53: What are the eigenvalues and eigenvectors of H?

Postulate 2 requires that the system being described be closed. That is, it is not interacting in any way with other systems. In reality, of course, all systems (except the Universe as a whole) interact at least somewhat with other systems. Nevertheless, there are interesting systems which can be described to a good approximation as being closed, and which are described by unitary evolution to some good approximation. Furthermore, at least in principle every open system can be described as part of a larger closed system (the Universe) which is undergoing unitary evolution. Later, we'll introduce more tools which allow us to describe systems which are not closed, but for now we'll continue with the description of the evolution of closed systems.

Postulate 2 describes how the quantum states of a closed quantum system at two different times are related. A more refined version of this postulate can be given which describes the evolution of a quantum system in *continuous time*. From this more refined postulate we will recover Postulate 2. Before we state the revised postulate, it is worth pointing out two things. First, a notational remark. The operator H appearing in the following discussion is not the same as the Hadamard operator, which we just introduced. Second, the following postulate makes use of the apparatus of differential equations. Readers with little background in the study of differential equations should be reassured that they will not be necessary for much of the book, with the exception of parts of Chapter 7, on real physical implementations of quantum information processing.

> **Postulate 2′:** The time evolution of the state of a closed quantum system is described by the *Schrödinger equation*,
>
> $$i\hbar \frac{d|\psi\rangle}{dt} = H|\psi\rangle. \tag{2.86}$$
>
> In this equation, $\hbar$ is a physical constant known as *Planck's constant* whose value must be experimentally determined. The exact value is not important to us. In practice, it is common to absorb the factor $\hbar$ into H, effectively setting $\hbar = 1$. H is a fixed Hermitian operator known as the *Hamiltonian* of the closed system.

If we know the Hamiltonian of a system, then (together with a knowledge of $\hbar$) we understand its dynamics completely, at least in principle. In general figuring out the Hamiltonian needed to describe a particular physical system is a very difficult problem – much of twentieth century physics has been concerned with this problem – which requires substantial input from experiment in order to be answered. From our point of view this is a problem of *detail* to be addressed by physical theories built within the framework of quantum mechanics – what Hamiltonian do we need to describe atoms in such-and-such a configuration – and is not a question that needs to be addressed by the theory of quantum mechanics itself. Most of the time in our discussion of quantum computation and quantum information we won't need to discuss Hamiltonians, and when we do, we will usually just posit that some matrix is the Hamiltonian as a starting point, and proceed from there, without attempting to justify the use of that Hamiltonian.

Because the Hamiltonian is a Hermitian operator it has a spectral decomposition

$$H = \sum_E E|E\rangle\langle E|, \tag{2.87}$$

with eigenvalues E and corresponding normalized eigenvectors $|E\rangle$. The states $|E\rangle$ are conventionally referred to as *energy eigenstates*, or sometimes as *stationary states*, and E is the *energy* of the state $|E\rangle$. The lowest energy is known as the *ground state energy* for the system, and the corresponding energy eigenstate (or eigenspace) is known as the *ground state*. The reason the states $|E\rangle$ are sometimes known as stationary states is because their only change in time is to acquire an overall numerical factor,

$$|E\rangle \rightarrow \exp(-iEt/\hbar)|E\rangle. \tag{2.88}$$

As an example, suppose a single qubit has Hamiltonian

$$H = \hbar\omega X. \tag{2.89}$$

In this equation ω is a parameter that, in practice, needs to be experimentally determined. We won't worry about the parameter overly much here – the point is to give you a feel for the sort of Hamiltonians that are sometimes written down in the study of quantum computation and quantum information. The energy eigenstates of this Hamiltonian are obviously the same as the eigenstates of X, namely $(|0\rangle + |1\rangle)/\sqrt{2}$ and $(|0\rangle - |1\rangle)/\sqrt{2}$, with corresponding energies $\hbar\omega$ and $-\hbar\omega$. The ground state is therefore $(|0\rangle - |1\rangle)/\sqrt{2}$, and the ground state energy is $-\hbar\omega$.

What is the connection between the Hamiltonian picture of dynamics, Postulate 2′, and the unitary operator picture, Postulate 2? The answer is provided by writing down the solution to Schrödinger's equation, which is easily verified to be:

$$|\psi(t_2)\rangle = \exp\left[\frac{-iH(t_2 - t_1)}{\hbar}\right]|\psi(t_1)\rangle = U(t_1, t_2)|\psi(t_1)\rangle, \tag{2.90}$$

where we define

$$U(t_1, t_2) \equiv \exp\left[\frac{-iH(t_2 - t_1)}{\hbar}\right]. \tag{2.91}$$

You will show in the exercises that this operator is unitary, and furthermore, that any unitary operator U can be realized in the form $U = \exp(iK)$ for some Hermitian operator K. There is therefore a one-to-one correspondence between the discrete-time description of dynamics using unitary operators, and the continuous time description using Hamiltonians. For most of the book we use the unitary formulation of quantum dynamics.

Exercise 2.54: Suppose A and B are commuting Hermitian operators. Prove that $\exp(A)\exp(B) = \exp(A + B)$. (*Hint:* Use the results of Section 2.1.9.)

Exercise 2.55: Prove that $U(t_1, t_2)$ defined in Equation (2.91) is unitary.

Exercise 2.56: Use the spectral decomposition to show that $K \equiv -i \log(U)$ is Hermitian for any unitary U, and thus $U = \exp(iK)$ for some Hermitian K.

In quantum computation and quantum information we often speak of *applying* a unitary operator to a particular quantum system. For example, in the context of quantum circuits we may speak of applying the unitary gate X to a single qubit. Doesn't this contradict what we said earlier, about unitary operators describing the evolution of a *closed* quantum system? After all, if we are 'applying' a unitary operator, then that implies that there is an external 'we' who is interacting with the quantum system, and the system is not closed.

An example of this occurs when a laser is focused on an atom. After a lot of thought and hard work it is possible to write down a Hamiltonian describing the total atom–laser system. The interesting thing is that when we write down the Hamiltonian for the atom–laser system and consider the effects on the atom alone, the behavior of the state vector of the atom turns out to be almost but not quite perfectly described by another Hamiltonian, the *atomic Hamiltonian*. The atomic Hamiltonian contains terms related to laser intensity, and other parameters of the laser, which we can vary at will. It is *as if* the evolution of the atom were being described by a Hamiltonian which we can vary at will, despite the atom not being a closed system.

More generally, for many systems like this it turns out to be possible to write down a *time-varying* Hamiltonian for a quantum system, in which the Hamiltonian for the system is not a constant, but varies according to some parameters which are under an experimentalist's control, and which may be changed during the course of an experiment. The system is not, therefore, closed, but it does evolve according to Schrödinger's equation with a time-varying Hamiltonian, to some good approximation.

The upshot is that to begin we will often describe the evolution of quantum systems – even systems which aren't closed – using unitary operators. The main exception to this, quantum measurement, will be described in the next section. Later on we will investigate in more detail possible deviations from unitary evolution due to the interaction with other systems, and understand more precisely the dynamics of realistic quantum systems.

2.2.3 Quantum measurement

We postulated that closed quantum systems evolve according to unitary evolution. The evolution of systems which don't interact with the rest of the world is all very well, but there must also be times when the experimentalist and their experimental equipment – an external physical system in other words – observes the system to find out what is going on inside the system, an interaction which makes the system no longer closed, and thus not necessarily subject to unitary evolution. To explain what happens when this is done, we introduce Postulate 3, which provides a means for describing the effects of measurements on quantum systems.

> **Postulate 3**: Quantum measurements are described by a collection $\{M_m\}$ of *measurement operators*. These are operators acting on the state space of the system being measured. The index m refers to the measurement outcomes that may occur in the experiment. If the state of the quantum system is $|\psi\rangle$ immediately before the measurement then the probability that result m occurs is

given by

$$p(m) = \langle\psi|M_m^\dagger M_m|\psi\rangle\,, \tag{2.92}$$

and the state of the system after the measurement is

$$\frac{M_m|\psi\rangle}{\sqrt{\langle\psi|M_m^\dagger M_m|\psi\rangle}}. \tag{2.93}$$

The measurement operators satisfy the *completeness equation,*

$$\sum_m M_m^\dagger M_m = I\,. \tag{2.94}$$

The completeness equation expresses the fact that probabilities sum to one:

$$1 = \sum_m p(m) = \sum_m \langle\psi|M_m^\dagger M_m|\psi\rangle\,. \tag{2.95}$$

This equation being satisfied for all $|\psi\rangle$ is equivalent to the completeness equation. However, the completeness equation is much easier to check directly, so that's why it appears in the statement of the postulate.

A simple but important example of a measurement is the *measurement of a qubit in the computational basis.* This is a measurement on a single qubit with two outcomes defined by the two measurement operators $M_0 = |0\rangle\langle0|$, $M_1 = |1\rangle\langle1|$. Observe that each measurement operator is Hermitian, and that $M_0^2 = M_0, M_1^2 = M_1$. Thus the completeness relation is obeyed, $I = M_0^\dagger M_0 + M_1^\dagger M_1 = M_0 + M_1$. Suppose the state being measured is $|\psi\rangle = a|0\rangle + b|1\rangle$. Then the probability of obtaining measurement outcome 0 is

$$p(0) = \langle\psi|M_0^\dagger M_0|\psi\rangle = \langle\psi|M_0|\psi\rangle = |a|^2. \tag{2.96}$$

Similarly, the probability of obtaining the measurement outcome 1 is $p(1) = |b|^2$. The state after measurement in the two cases is therefore

$$\frac{M_0|\psi\rangle}{|a|} = \frac{a}{|a|}|0\rangle \tag{2.97}$$

$$\frac{M_1|\psi\rangle}{|b|} = \frac{b}{|b|}|1\rangle. \tag{2.98}$$

We will see in Section 2.2.7 that multipliers like $a/|a|$, which have modulus one, can effectively be ignored, so the two post-measurement states are effectively $|0\rangle$ and $|1\rangle$, just as described in Chapter 1.

The status of Postulate 3 as a fundamental postulate intrigues many people. Measuring devices are quantum mechanical systems, so the quantum system being measured and the measuring device together are part of a larger, isolated, quantum mechanical system. (It may be necessary to include quantum systems other than the system being measured and the measuring device to obtain a completely isolated system, but the point is that this can be done.) According to Postulate 2, the evolution of this larger isolated system can be described by a unitary evolution. Might it be possible to *derive* Postulate 3 as a consequence of this picture? Despite considerable investigation along these lines there is still disagreement between physicists about whether or not this is possible. We, however, are going to take the very pragmatic approach that in practice it is clear when to apply

Postulate 2 and when to apply Postulate 3, and not worry about deriving one postulate from the other.

Over the next few sections we apply Postulate 3 to several elementary but important measurement scenarios. Section 2.2.4 examines the problem of *distinguishing* a set of quantum states. Section 2.2.5 explains a special case of Postulate 3, the *projective* or *von Neumann* measurements. Section 2.2.6 explains another special case of Postulate 3, known as *POVM* measurements. Many introductions to quantum mechanics only discuss projective measurements, omitting a full discussion of Postulate 3 or of POVM elements. For this reason we have included Box 2.5 on page 91 which comments on the relationship between the different classes of measurement we describe.

Exercise 2.57: (Cascaded measurements are single measurements) Suppose $\{L_l\}$ and $\{M_m\}$ are two sets of measurement operators. Show that a measurement defined by the measurement operators $\{L_l\}$ followed by a measurement defined by the measurement operators $\{M_m\}$ is physically equivalent to a single measurement defined by measurement operators $\{N_{lm}\}$ with the representation $N_{lm} \equiv M_m L_l$.

2.2.4 Distinguishing quantum states

An important application of Postulate 3 is to the problem of *distinguishing quantum states*. In the classical world, distinct states of an object are usually distinguishable, at least in principle. For example, we can always identify whether a coin has landed heads or tails, at least in the ideal limit. Quantum mechanically, the situation is more complicated. In Section 1.6 we gave a plausible argument that non-orthogonal quantum states cannot be distinguished. With Postulate 3 as a firm foundation we can now give a much more convincing demonstration of this fact.

Distinguishability, like many ideas in quantum computation and quantum information, is most easily understood using the metaphor of a game involving two parties, Alice and Bob. Alice chooses a state $|\psi_i\rangle$ ($1 \le i \le n$) from some fixed set of states known to both parties. She gives the state $|\psi_i\rangle$ to Bob, whose task it is to identify the index i of the state Alice has given him.

Suppose the states $|\psi_i\rangle$ are orthonormal. Then Bob can do a quantum measurement to *distinguish* these states, using the following procedure. Define measurement operators $M_i \equiv |\psi_i\rangle\langle\psi_i|$, one for each possible index i, and an additional measurement operator M_0 defined as the positive square root of the positive operator $I - \sum_{i \ne 0} |\psi_i\rangle\langle\psi_i|$. These operators satisfy the completeness relation, and if the state $|\psi_i\rangle$ is prepared then $p(i) = \langle\psi_i|M_i|\psi_i\rangle = 1$, so the result i occurs with certainty. Thus, it is possible to reliably distinguish the orthonormal states $|\psi_i\rangle$.

By contrast, if the states $|\psi_i\rangle$ are not orthonormal then we can prove that there is *no quantum measurement capable of distinguishing the states*. The idea is that Bob will do a measurement described by measurement operators M_j, with outcome j. Depending on the outcome of the measurement Bob tries to guess what the index i was using some rule, $i = f(j)$, where $f(\cdot)$ represents the rule he uses to make the guess. The key to why Bob can't distinguish non-orthogonal states $|\psi_1\rangle$ and $|\psi_2\rangle$ is the observation that $|\psi_2\rangle$ can be decomposed into a (non-zero) component parallel to $|\psi_1\rangle$, and a component orthogonal to $|\psi_1\rangle$. Suppose j is a measurement outcome such that $f(j) = 1$, that is, Bob guesses that the state was $|\psi_1\rangle$ when he observes j. But because of the component of $|\psi_2\rangle$ parallel to $|\psi_1\rangle$, there is a non-zero probability of getting outcome j when $|\psi_2\rangle$ is prepared, so sometimes Bob will make an error identifying which state was prepared. A more rigorous argument that non-orthogonal states can't be distinguished is given in Box 2.3, but this captures the essential idea.

Box 2.3: Proof that non-orthogonal states can't be reliably distinguished

A proof by contradiction shows that no measurement distinguishing the non-orthogonal states $|\psi_1\rangle$ and $|\psi_2\rangle$ is possible. Suppose such a measurement is possible. If the state $|\psi_1\rangle$ ($|\psi_2\rangle$) is prepared then the probability of measuring j such that $f(j) = 1$ ($f(j) = 2$) must be 1. Defining $E_i \equiv \sum_{j:f(j)=i} M_j^\dagger M_j$, these observations may be written as:

$$\langle\psi_1|E_1|\psi_1\rangle = 1; \quad \langle\psi_2|E_2|\psi_2\rangle = 1. \tag{2.99}$$

Since $\sum_i E_i = I$ it follows that $\sum_i \langle\psi_1|E_i|\psi_1\rangle = 1$, and since $\langle\psi_1|E_1|\psi_1\rangle = 1$ we must have $\langle\psi_1|E_2|\psi_1\rangle = 0$, and thus $\sqrt{E_2}|\psi_1\rangle = 0$. Suppose we decompose $|\psi_2\rangle = \alpha|\psi_1\rangle + \beta|\varphi\rangle$, where $|\varphi\rangle$ is orthonormal to $|\psi_1\rangle$, $|\alpha|^2 + |\beta|^2 = 1$, and $|\beta| < 1$ since $|\psi_1\rangle$ and $|\psi_2\rangle$ are not orthogonal. Then $\sqrt{E_2}|\psi_2\rangle = \beta\sqrt{E_2}|\varphi\rangle$, which implies a contradiction with (2.99), as

$$\langle\psi_2|E_2|\psi_2\rangle = |\beta|^2\langle\varphi|E_2|\varphi\rangle \leq |\beta|^2 < 1, \tag{2.100}$$

where the second last inequality follows from the observation that

$$\langle\varphi|E_2|\varphi\rangle \leq \sum_i \langle\varphi|E_i|\varphi\rangle = \langle\varphi|\varphi\rangle = 1. \tag{2.101}$$

2.2.5 Projective measurements

In this section we explain an important special case of the general measurement postulate, Postulate 3. This special class of measurements is known as *projective measurements*. For many applications of quantum computation and quantum information we will be concerned primarily with projective measurements. Indeed, projective measurements actually turn out to be *equivalent* to the general measurement postulate, when they are augmented with the ability to perform unitary transformations, as described in Postulate 2. We will explain this equivalence in detail in Section 2.2.8, as the statement of the measurement postulate for projective measurements is superficially rather different from the general postulate, Postulate 3.

Projective measurements: A projective measurement is described by an *observable*, M, a Hermitian operator on the state space of the system being observed. The observable has a spectral decomposition,

$$M = \sum_m m P_m, \tag{2.102}$$

where P_m is the projector onto the eigenspace of M with eigenvalue m. The possible outcomes of the measurement correspond to the eigenvalues, m, of the observable. Upon measuring the state $|\psi\rangle$, the probability of getting result m is given by

$$p(m) = \langle\psi|P_m|\psi\rangle . \tag{2.103}$$

Given that outcome m occurred, the state of the quantum system immediately after the measurement is

$$\frac{P_m|\psi\rangle}{\sqrt{p(m)}} . \tag{2.104}$$

Projective measurements can be understood as a special case of Postulate 3. Suppose the measurement operators in Postulate 3, in addition to satisfying the completeness relation $\sum_m M_m^\dagger M_m = I$, also satisfy the conditions that M_m are orthogonal projectors, that is, the M_m are Hermitian, and $M_m M_{m'} = \delta_{m,m'} M_m$. With these additional restrictions, Postulate 3 reduces to a projective measurement as just defined.

Projective measurements have many nice properties. In particular, it is very easy to calculate average values for projective measurements. By definition, the average (see Appendix 1 for elementary definitions and results in probability theory) value of the measurement is

$$\mathbf{E}(M) = \sum_m m\, p(m) \tag{2.110}$$

$$= \sum_m m\langle\psi|P_m|\psi\rangle \tag{2.111}$$

$$= \langle\psi| \left(\sum_m m P_m \right) |\psi\rangle \tag{2.112}$$

$$= \langle\psi|M|\psi\rangle. \tag{2.113}$$

This is a useful formula, which simplifies many calculations. The average value of the observable M is often written $\langle M\rangle \equiv \langle\psi|M|\psi\rangle$. From this formula for the average follows a formula for the standard deviation associated to observations of M,

$$[\Delta(M)]^2 = \langle(M - \langle M\rangle)^2\rangle \tag{2.114}$$

$$= \langle M^2\rangle - \langle M\rangle^2. \tag{2.115}$$

The standard deviation is a measure of the typical spread of the observed values upon measurement of M. In particular, if we perform a large number of experiments in which the state $|\psi\rangle$ is prepared and the observable M is measured, then the standard deviation $\Delta(M)$ of the observed values is determined by the formula $\Delta(M) = \sqrt{\langle M^2\rangle - \langle M\rangle^2}$. This formulation of measurement and standard deviations in terms of observables gives rise in an elegant way to results such as the *Heisenberg uncertainty principle* (see Box 2.4).

Exercise 2.58: Suppose we prepare a quantum system in an eigenstate $|\psi\rangle$ of some observable M, with corresponding eigenvalue m. What is the average observed value of M, and the standard deviation?

Two widely used nomenclatures for measurements deserve emphasis. Rather than giving an observable to describe a projective measurement, often people simply list a complete set of orthogonal projectors P_m satisfying the relations $\sum_m P_m = I$ and $P_m P_{m'} =$

Box 2.4: The Heisenberg uncertainty principle

Perhaps the best known result of quantum mechanics is the *Heisenberg uncertainty principle*. Suppose A and B are two Hermitian operators, and $|\psi\rangle$ is a quantum state. Suppose $\langle\psi|AB|\psi\rangle = x + iy$, where x and y are real. Note that $\langle\psi|[A,B]|\psi\rangle = 2iy$ and $\langle\psi|\{A,B\}|\psi\rangle = 2x$. This implies that

$$|\langle\psi|[A,B]|\psi\rangle|^2 + |\langle\psi|\{A,B\}|\psi\rangle|^2 = 4\,|\langle\psi|AB|\psi\rangle|^2 \,. \qquad (2.105)$$

By the Cauchy–Schwarz inequality

$$|\langle\psi|AB|\psi\rangle|^2 \leq \langle\psi|A^2|\psi\rangle\langle\psi|B^2|\psi\rangle, \qquad (2.106)$$

which combined with Equation (2.105) and dropping a non–negative term gives

$$|\langle\psi|[A,B]|\psi\rangle|^2 \leq 4\langle\psi|A^2|\psi\rangle\langle\psi|B^2|\psi\rangle. \qquad (2.107)$$

Suppose C and D are two observables. Substituting $A = C - \langle C\rangle$ and $B = D - \langle D\rangle$ into the last equation, we obtain Heisenberg's uncertainty principle as it is usually stated:

$$\Delta(C)\Delta(D) \geq \frac{|\langle\psi|[C,D]|\psi\rangle|}{2} \,. \qquad (2.108)$$

You should be wary of a common misconception about the uncertainty principle, that measuring an observable C to some 'accuracy' $\Delta(C)$ causes the value of D to be 'disturbed' by an amount $\Delta(D)$ in such a way that some sort of inequality similar to (2.108) is satisfied. While it is true that measurements in quantum mechanics cause disturbance to the system being measured, this is most emphatically *not* the content of the uncertainty principle.

The correct interpretation of the uncertainty principle is that if we prepare a large number of quantum systems in identical states, $|\psi\rangle$, and then perform measurements of C on some of those systems, and of D in others, then the standard deviation $\Delta(C)$ of the C results times the standard deviation $\Delta(D)$ of the results for D will satisfy the inequality (2.108).

As an example of the uncertainty principle, consider the observables X and Y when measured for the quantum state $|0\rangle$. In Equation (2.70) we showed that $[X,Y] = 2iZ$, so the uncertainty principle tells us that

$$\Delta(X)\Delta(Y) \geq \langle 0|Z|0\rangle = 1 \,. \qquad (2.109)$$

One elementary consequence of this is that $\Delta(X)$ and $\Delta(Y)$ must both be strictly greater than 0, as can be verified by direct calculation.

$\delta_{mm'}P_m$. The corresponding observable implicit in this usage is $M = \sum_m mP_m$. Another widely used phrase, to 'measure in a basis $|m\rangle$', where $|m\rangle$ form an orthonormal basis, simply means to perform the projective measurement with projectors $P_m = |m\rangle\langle m|$.

Let's look at an example of projective measurements on single qubits. First is the measurement of the observable Z. This has eigenvalues $+1$ and -1 with corresponding eigenvectors $|0\rangle$ and $|1\rangle$. Thus, for example, measurement of Z on the state $|\psi\rangle = (|0\rangle + |1\rangle)/\sqrt{2}$ gives the result $+1$ with probability $\langle\psi|0\rangle\langle 0|\psi\rangle = 1/2$, and similarly the result -1 with probability $1/2$. More generally, suppose $\vec{v}$ is any real three-dimensional unit vector. Then we can define an observable:

$$\vec{v} \cdot \vec{\sigma} \equiv v_1 \sigma_1 + v_2 \sigma_2 + v_3 \sigma_3. \tag{2.116}$$

Measurement of this observable is sometimes referred to as a 'measurement of spin along the $\vec{v}$ axis', for historical reasons. The following two exercises encourage you to work out some elementary but important properties of such a measurement.

Exercise 2.59: Suppose we have qubit in the state $|0\rangle$, and we measure the observable X. What is the average value of X? What is the standard deviation of X?

Exercise 2.60: Show that $\vec{v} \cdot \vec{\sigma}$ has eigenvalues ± 1, and that the projectors onto the corresponding eigenspaces are given by $P_\pm = (I \pm \vec{v} \cdot \vec{\sigma})/2$.

Exercise 2.61: Calculate the probability of obtaining the result $+1$ for a measurement of $\vec{v} \cdot \vec{\sigma}$, given that the state prior to measurement is $|0\rangle$. What is the state of the system after the measurement if $+1$ is obtained?

2.2.6 POVM measurements

The quantum measurement postulate, Postulate 3, involves two elements. First, it gives a rule describing the measurement statistics, that is, the respective probabilities of the different possible measurement outcomes. Second, it gives a rule describing the post-measurement state of the system. However, for some applications the post-measurement state of the system is of little interest, with the main item of interest being the probabilities of the respective measurement outcomes. This is the case, for example, in an experiment where the system is measured only once, upon conclusion of the experiment. In such instances there is a mathematical tool known as the *POVM formalism* which is especially well adapted to the analysis of the measurements. (The acronym POVM stands for 'Positive Operator-Valued Measure', a technical term whose historical origins we won't worry about.) This formalism is a simple consequence of the general description of measurements introduced in Postulate 3, but the theory of POVMs is so elegant and widely used that it merits a separate discussion here.

Suppose a measurement described by measurement operators M_m is performed upon a quantum system in the state $|\psi\rangle$. Then the probability of outcome m is given by $p(m) = \langle \psi | M_m^\dagger M_m | \psi \rangle$. Suppose we define

$$E_m \equiv M_m^\dagger M_m. \tag{2.117}$$

Then from Postulate 3 and elementary linear algebra, E_m is a positive operator such that $\sum_m E_m = I$ and $p(m) = \langle \psi | E_m | \psi \rangle$. Thus the set of operators E_m are sufficient to determine the probabilities of the different measurement outcomes. The operators E_m are known as the *POVM elements* associated with the measurement. The complete set $\{E_m\}$ is known as a *POVM*.

As an example of a POVM, consider a projective measurement described by measurement operators P_m, where the P_m are projectors such that $P_m P_{m'} = \delta_{mm'} P_m$ and $\sum_m P_m = I$. In this instance (and only this instance) all the POVM elements are the same as the measurement operators themselves, since $E_m \equiv P_m^\dagger P_m = P_m$.

Box 2.5: General measurements, projective measurements, and POVMs

Most introductions to quantum mechanics describe only projective measurements, and consequently the general description of measurements given in Postulate 3 may be unfamiliar to many physicists, as may the POVM formalism described in Section 2.2.6. The reason most physicists don't learn the general measurement formalism is because most physical systems can only be measured in a very coarse manner. In quantum computation and quantum information we aim for an exquisite level of control over the measurements that may be done, and consequently it helps to use a more comprehensive formalism for the description of measurements.

Of course, when the other axioms of quantum mechanics are taken into account, projective measurements augmented by unitary operations turn out to be completely *equivalent* to general measurements, as shown in Section 2.2.8. So a physicist trained in the use of projective measurements might ask to what end we start with the general formalism, Postulate 3? There are several reasons for doing so. First, mathematically general measurements are in some sense simpler than projective measurements, since they involve fewer restrictions on the measurement operators; there is, for example, no requirement for general measurements analogous to the condition $P_i P_j = \delta_{ij} P_i$ for projective measurements. This simpler structure also gives rise to many useful properties for general measurements that are not possessed by projective measurements. Second, it turns out that there are important problems in quantum computation and quantum information – such as the optimal way to distinguish a set of quantum states – the answer to which involves a general measurement, rather than a projective measurement.

A third reason for preferring Postulate 3 as a starting point is related to a property of projective measurements known as *repeatability*. Projective measurements are repeatable in the sense that if we perform a projective measurement once, and obtain the outcome m, repeating the measurement gives the outcome m again and does not change the state. To see this, suppose $|\psi\rangle$ was the initial state. After the first measurement the state is $|\psi_m\rangle = \left(P_m|\psi\rangle\right)/\sqrt{\langle\psi|P_m|\psi\rangle}$. Applying P_m to $|\psi_m\rangle$ does not change it, so we have $\langle\psi_m|P_m|\psi_m\rangle = 1$, and therefore repeated measurement gives the result m each time, without changing the state.

This repeatability of projective measurements tips us off to the fact that many important measurements in quantum mechanics are not projective measurements. For instance, if we use a silvered screen to measure the position of a photon we destroy the photon in the process. This certainly makes it impossible to repeat the measurement of the photon's position! Many other quantum measurements are also not repeatable in the same sense as a projective measurement. For such measurements, the general measurement postulate, Postulate 3, must be employed. Where do POVMs fit in this picture? POVMs are best viewed as a special case of the general measurement formalism, providing the simplest means by which one can study general measurement statistics, without the necessity for knowing the post-measurement state. They are a mathematical convenience that sometimes gives extra insight into quantum measurements.

Exercise 2.62: Show that any measurement where the measurement operators and the POVM elements coincide is a projective measurement.

Above we noticed that the POVM operators are positive and satisfy $\sum_m E_m = I$. Suppose now that $\{E_m\}$ is some arbitrary set of positive operators such that $\sum_m E_m = I$. We will show that there exists a set of measurement operators M_m defining a measurement described by the POVM $\{E_m\}$. Defining $M_m \equiv \sqrt{E_m}$ we see that $\sum_m M_m^\dagger M_m = \sum_m E_m = I$, and therefore the set $\{M_m\}$ describes a measurement with POVM $\{E_m\}$. For this reason it is convenient to *define* a POVM to be any set of operators $\{E_m\}$ such that: (a) each operator E_m is positive; and (b) the *completeness relation* $\sum_m E_m = I$ is obeyed, expressing the fact that probabilities sum to one. To complete the description of POVMs, we note again that given a POVM $\{E_m\}$, the probability of outcome m is given by $p(m) = \langle \psi | E_m | \psi \rangle$.

We've looked at projective measurements as an example of the use of POVMs, but it wasn't very exciting since we didn't learn much that was new. The following more sophisticated example illustrates the use of the POVM formalism as a guide for our intuition in quantum computation and quantum information. Suppose Alice gives Bob a qubit prepared in one of two states, $|\psi_1\rangle = |0\rangle$ or $|\psi_2\rangle = (|0\rangle + |1\rangle)/\sqrt{2}$. As explained in Section 2.2.4 it is impossible for Bob to determine whether he has been given $|\psi_1\rangle$ or $|\psi_2\rangle$ with perfect reliability. However, it is possible for him to perform a measurement which distinguishes the states some of the time, but *never* makes an error of mis-identification. Consider a POVM containing three elements,

$$E_1 \equiv \frac{\sqrt{2}}{1+\sqrt{2}}|1\rangle\langle 1|, \tag{2.118}$$

$$E_2 \equiv \frac{\sqrt{2}}{1+\sqrt{2}} \frac{(|0\rangle - |1\rangle)\,(\langle 0| - \langle 1|)}{2}, \tag{2.119}$$

$$E_3 \equiv I - E_1 - E_2. \tag{2.120}$$

It is straightforward to verify that these are positive operators which satisfy the completeness relation $\sum_m E_m = I$, and therefore form a legitimate POVM.

Suppose Bob is given the state $|\psi_1\rangle = |0\rangle$. He performs the measurement described by the POVM $\{E_1, E_2, E_3\}$. There is zero probability that he will observe the result E_1, since E_1 has been cleverly chosen to ensure that $\langle \psi_1 | E_1 | \psi_1 \rangle = 0$. Therefore, if the result of his measurement is E_1 then Bob can safely conclude that the state he received must have been $|\psi_2\rangle$. A similar line of reasoning shows that if the measurement outcome E_2 occurs then it must have been the state $|\psi_1\rangle$ that Bob received. Some of the time, however, Bob will obtain the measurement outcome E_3, and he can infer nothing about the identity of the state he was given. The key point, however, is that Bob *never* makes a mistake identifying the state he has been given. This infallibility comes at the price that sometimes Bob obtains no information about the identity of the state.

This simple example demonstrates the utility of the POVM formalism as a simple and intuitive way of gaining insight into quantum measurements in instances where only the measurement statistics matter. In many instances later in the book we will only be concerned with measurement statistics, and will therefore use the POVM formalism rather than the more general formalism for measurements described in Postulate 3.

Exercise 2.63: Suppose a measurement is described by measurement operators M_m.

Show that there exist unitary operators U_m such that $M_m = U_m\sqrt{E_m}$, where E_m is the POVM associated to the measurement.

Exercise 2.64: Suppose Bob is given a quantum state chosen from a set $|\psi_1\rangle, \dots, |\psi_m\rangle$ of linearly independent states. Construct a POVM $\{E_1, E_2, \dots, E_{m+1}\}$ such that if outcome E_i occurs, $1 \le i \le m$, then Bob knows with certainty that he was given the state $|\psi_i\rangle$. (The POVM must be such that $\langle\psi_i|E_i|\psi_i\rangle > 0$ for each i.)

2.2.7 Phase

'Phase' is a commonly used term in quantum mechanics, with several different meanings dependent upon context. At this point it is convenient to review a couple of these meanings. Consider, for example, the state $e^{i\theta}|\psi\rangle$, where $|\psi\rangle$ is a state vector, and θ is a real number. We say that the state $e^{i\theta}|\psi\rangle$ is equal to $|\psi\rangle$, up to the *global phase factor* $e^{i\theta}$. It is interesting to note that the *statistics of measurement* predicted for these two states are the same. To see this, suppose M_m is a measurement operator associated to some quantum measurement, and note that the respective probabilities for outcome m occurring are $\langle\psi|M_m^\dagger M_m|\psi\rangle$ and $\langle\psi|e^{-i\theta}M_m^\dagger M_m e^{i\theta}|\psi\rangle = \langle\psi|M_m^\dagger M_m|\psi\rangle$. Therefore, from an observational point of view these two states are identical. For this reason we may ignore global phase factors as being irrelevant to the observed properties of the physical system.

There is another kind of phase known as the *relative phase*, which has quite a different meaning. Consider the states

$$\frac{|0\rangle + |1\rangle}{\sqrt{2}} \qquad \text{and} \qquad \frac{|0\rangle - |1\rangle}{\sqrt{2}}. \qquad (2.121)$$

In the first state the amplitude of $|1\rangle$ is $1/\sqrt{2}$. For the second state the amplitude is $-1/\sqrt{2}$. In each case the *magnitude* of the amplitudes is the same, but they differ in sign. More generally, we say that two amplitudes, a and b, *differ by a relative phase* if there is a real θ such that $a = \exp(i\theta)b$. More generally still, two states are said to *differ by a relative phase* in some basis if each of the amplitudes in that basis is related by such a phase factor. For example, the two states displayed above are the same up to a relative phase shift because the $|0\rangle$ amplitudes are identical (a relative phase factor of 1), and the $|1\rangle$ amplitudes differ only by a relative phase factor of -1. The difference between relative phase factors and global phase factors is that for relative phase the phase factors may vary from amplitude to amplitude. This makes the relative phase a basis-dependent concept unlike global phase. As a result, states which differ only by relative phases in some basis give rise to physically observable differences in measurement statistics, and it is not possible to regard these states as physically equivalent, as we do with states differing by a global phase factor

Exercise 2.65: Express the states $(|0\rangle + |1\rangle)/\sqrt{2}$ and $(|0\rangle - |1\rangle)/\sqrt{2}$ in a basis in which they are *not* the same up to a relative phase shift.

2.2.8 Composite systems

Suppose we are interested in a composite quantum system made up of two (or more) distinct physical systems. How should we describe states of the composite system? The following postulate describes how the state space of a composite system is built up from the state spaces of the component systems.

Postulate 4: The state space of a composite physical system is the tensor product of the state spaces of the component physical systems. Moreover, if we have systems numbered 1 through n, and system number i is prepared in the state $|\psi_i\rangle$, then the joint state of the total system is $|\psi_1\rangle \otimes |\psi_2\rangle \otimes \cdots \otimes |\psi_n\rangle$.

Why is the tensor product the mathematical structure used to describe the state space of a composite physical system? At one level, we can simply accept it as a basic postulate, not reducible to something more elementary, and move on. After all, we certainly expect that there be *some canonical way* of describing composite systems in quantum mechanics. Is there some other way we can arrive at this postulate? Here is one heuristic that is sometimes used. Physicists sometimes like to speak of the *superposition principle of quantum mechanics*, which states that if $|x\rangle$ and $|y\rangle$ are two states of a quantum system, then any superposition $\alpha|x\rangle + \beta|y\rangle$ should also be an allowed state of a quantum system, where $|\alpha|^2 + |\beta|^2 = 1$. For composite systems, it seems natural that if $|A\rangle$ is a state of system A, and $|B\rangle$ is a state of system B, then there should be some corresponding state, which we might denote $|A\rangle|B\rangle$, of the joint system AB. Applying the superposition principle to product states of this form, we arrive at the tensor product postulate given above. This is not a derivation, since we are not taking the superposition principle as a fundamental part of our description of quantum mechanics, but it gives you the flavor of the various ways in which these ideas are sometimes reformulated.

A variety of different notations for composite systems appear in the literature. Part of the reason for this proliferation is that different notations are better adapted for different applications, and we will also find it convenient to introduce some specialized notations on occasion. At this point it suffices to mention a useful subscript notation to denote states and operators on different systems, when it is not clear from context. For example, in a system containing three qubits, X_2 is the Pauli σ_x operator acting on the second qubit.

Exercise 2.66:　Show that the average value of the observable $X_1 Z_2$ for a two qubit system measured in the state $(|00\rangle + |11\rangle)/\sqrt{2}$ is zero.

In Section 2.2.5 we claimed that projective measurements together with unitary dynamics are sufficient to implement a general measurement. The proof of this statement makes use of composite quantum systems, and is a nice illustration of Postulate 4 in action. Suppose we have a quantum system with state space Q, and we want to perform a measurement described by measurement operators M_m on the system Q. To do this, we introduce an *ancilla system*, with state space M, having an orthonormal basis $|m\rangle$ in one-to-one correspondence with the possible outcomes of the measurement we wish to implement. This ancilla system can be regarded as merely a mathematical device appearing in the construction, or it can be interpreted physically as an extra quantum system introduced into the problem, which we assume has a state space with the required properties.

Letting $|0\rangle$ be any fixed state of M, define an operator U on products $|\psi\rangle|0\rangle$ of states $|\psi\rangle$ from Q with the state $|0\rangle$ by

$$U|\psi\rangle|0\rangle \equiv \sum_m M_m|\psi\rangle|m\rangle. \tag{2.122}$$

Using the orthonormality of the states $|m\rangle$ and the completeness relation $\sum_m M_m^\dagger M_m =$

I, we can see that U preserves inner products between states of the form $|\psi\rangle|0\rangle$,

$$\langle\varphi|\langle 0|U^\dagger U|\psi\rangle|0\rangle = \sum_{m,m'}\langle\varphi|M_m^\dagger M_{m'}|\psi\rangle\langle m|m'\rangle \tag{2.123}$$

$$= \sum_m\langle\varphi|M_m^\dagger M_m|\psi\rangle \tag{2.124}$$

$$= \langle\varphi|\psi\rangle. \tag{2.125}$$

By the results of Exercise 2.67 it follows that U can be extended to a unitary operator on the space $Q \otimes M$, which we also denote by U.

Exercise 2.67: Suppose V is a Hilbert space with a subspace W. Suppose $U : W \to V$ is a linear operator which preserves inner products, that is, for any $|w_1\rangle$ and $|w_2\rangle$ in W,

$$\langle w_1|U^\dagger U|w_2\rangle = \langle w_1|w_2\rangle. \tag{2.126}$$

Prove that there exists a unitary operator $U' : V \to V$ which *extends* U. That is, $U'|w\rangle = U|w\rangle$ for all $|w\rangle$ in W, but U' is defined on the entire space V. Usually we omit the prime symbol $'$ and just write U to denote the extension.

Next, suppose we perform a projective measurement on the two systems described by projectors $P_m \equiv I_Q \otimes |m\rangle\langle m|$. Outcome m occurs with probability

$$p(m) = \langle\psi|\langle 0|U^\dagger P_m U|\psi\rangle|0\rangle \tag{2.127}$$

$$= \sum_{m',m''}\langle\psi|M_{m'}^\dagger\langle m'|(I_Q \otimes |m\rangle\langle m|)M_{m''}|\psi\rangle|m''\rangle \tag{2.128}$$

$$= \langle\psi|M_m^\dagger M_m|\psi\rangle, \tag{2.129}$$

just as given in Postulate 3. The joint state of the system QM after measurement, conditional on result m occurring, is given by

$$\frac{P_m U|\psi\rangle|0\rangle}{\sqrt{\langle\psi|U^\dagger P_m U|\psi\rangle}} = \frac{M_m|\psi\rangle|m\rangle}{\sqrt{\langle\psi|M_m^\dagger M_m|\psi\rangle}}. \tag{2.130}$$

It follows that the state of system M after the measurement is $|m\rangle$, and the state of system Q is

$$\frac{M_m|\psi\rangle}{\sqrt{\langle\psi|M_m^\dagger M_m|\psi\rangle}}, \tag{2.131}$$

just as prescribed by Postulate 3. Thus unitary dynamics, projective measurements, and the ability to introduce ancillary systems, together allow any measurement of the form described in Postulate 3 to be realized.

Postulate 4 also enables us to define one of the most interesting and puzzling ideas associated with composite quantum systems – *entanglement*. Consider the two qubit state

$$|\psi\rangle = \frac{|00\rangle + |11\rangle}{\sqrt{2}}. \tag{2.132}$$

This state has the remarkable property that there are no single qubit states $|a\rangle$ and $|b\rangle$ such that $|\psi\rangle = |a\rangle|b\rangle$, a fact which you should now convince yourself of:

Exercise 2.68: Prove that $|\psi\rangle \neq |a\rangle |b\rangle$ for all single qubit states $|a\rangle$ and $|b\rangle$.

We say that a state of a composite system having this property (that it can't be written as a product of states of its component systems) is an *entangled* state. For reasons which nobody fully understands, entangled states play a crucial role in quantum computation and quantum information, and arise repeatedly through the remainder of this book. We have already seen entanglement play a crucial role in quantum teleportation, as described in Section 1.3.7. In this chapter we give two examples of the strange effects enabled by entangled quantum states, superdense coding (Section 2.3), and the violation of Bell's inequality (Section 2.6).

2.2.9 Quantum mechanics: a global view

We have now explained *all* the fundamental postulates of quantum mechanics. Most of the rest of the book is taken up with deriving consequences of these postulates. Let's quickly review the postulates and try to place them in some kind of global perspective.

Postulate 1 sets the arena for quantum mechanics, by specifying how the state of an isolated quantum system is to be described. Postulate 2 tells us that the dynamics of *closed* quantum systems are described by the Schrödinger equation, and thus by unitary evolution. Postulate 3 tells us how to extract information from our quantum systems by giving a prescription for the description of measurement. Postulate 4 tells us how the state spaces of different quantum systems may be combined to give a description of the composite system.

What's odd about quantum mechanics, at least by our classical lights, is that we can't directly observe the state vector. It's a little bit like a game of chess where you can never find out exactly where each piece is, but only know the rank of the board they are on. Classical physics – and our intuition – tells us that the fundamental properties of an object, like energy, position, and velocity, are directly accessible to observation. In quantum mechanics these quantities no longer appear as fundamental, being replaced by the state vector, which can't be directly observed. It is as though there is a *hidden world* in quantum mechanics, which we can only indirectly and imperfectly access. Moreover, merely observing a classical system does not necessarily change the state of the system. Imagine how difficult it would be to play tennis if each time you looked at the ball its position changed! But according to Postulate 3, observation in quantum mechanics is an invasive procedure that typically changes the state of the system.

What conclusions should we draw from these strange features of quantum mechanics? Might it be possible to reformulate quantum mechanics in a mathematically equivalent way so that it had a structure more like classical physics? In Section 2.6 we'll prove *Bell's inequality*, a surprising result that shows any attempt at such a reformulation is doomed to failure. We're stuck with the counter-intuitive nature of quantum mechanics. Of course, the proper reaction to this is glee, not sorrow! It gives us an opportunity to develop tools of thought that make quantum mechanics intuitive. Moreover, we can exploit the hidden nature of the state vector to do information processing tasks beyond what is possible in the classical world. Without this counter-intuitive behavior, quantum computation and quantum information would be a lot less interesting.

We can also turn this discussion about, and ask ourselves: 'If quantum mechanics is so different from classical physics, then how come the everyday world looks so classical?' Why do we see no evidence of a hidden state vector in our everyday lives? It turns out that the classical world we see can be *derived* from quantum mechanics as an approximate description of the world that will be valid on the sort of time, length and mass scales we commonly encounter in our everyday lives. Explaining the details of how quantum mechanics gives rise to classical physics is beyond the scope of this book, but the interested reader should check out the discussion of this topic in 'History and further reading'at the end of Chapter 8.

2.3 Application: superdense coding

Superdense coding is a simple yet surprising application of elementary quantum mechanics. It combines in a concrete, non-trivial way all the basic ideas of elementary quantum mechanics, as covered in the previous sections, and is therefore an ideal example of the information processing tasks that can be accomplished using quantum mechanics.

Superdense coding involves two parties, conventionally known as 'Alice' and 'Bob', who are a long way away from one another. Their goal is to transmit some classical information from Alice to Bob. Suppose Alice is in possession of two classical bits of information which she wishes to send Bob, but is only allowed to send a single qubit to Bob. Can she achieve her goal?

Superdense coding tells us that the answer to this question is yes. Suppose Alice and Bob initially share a pair of qubits in the entangled state

$$|\psi\rangle = \frac{|00\rangle + |11\rangle}{\sqrt{2}}. \tag{2.133}$$

Alice is initially in possession of the first qubit, while Bob has possession of the second qubit, as illustrated in Figure 2.3. Note that $|\psi\rangle$ is a fixed state; there is no need for Alice to have sent Bob any qubits in order to prepare this state. Instead, some third party may prepare the entangled state ahead of time, sending one of the qubits to Alice, and the other to Bob.

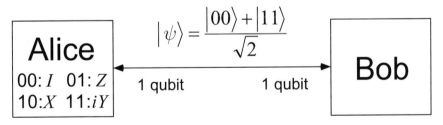

Figure 2.3. The initial setup for superdense coding, with Alice and Bob each in possession of one half of an entangled pair of qubits. Alice can use superdense coding to transmit two classical bits of information to Bob, using only a single qubit of communication and this preshared entanglement.

By sending the single qubit in her possession to Bob, it turns out that Alice can communicate two bits of classical information to Bob. Here is the procedure she uses. If she wishes to send the bit string '00' to Bob then she does nothing at all to her qubit. If she wishes to send '01' then she applies the phase flip Z to her qubit. If she wishes to send '10' then she applies the quantum NOT gate, X, to her qubit. If she wishes to send '11' then she applies the iY gate to her qubit. The four resulting states are easily seen to be:

$$00 : |\psi\rangle \rightarrow \frac{|00\rangle + |11\rangle}{\sqrt{2}} \tag{2.134}$$

$$01 : |\psi\rangle \rightarrow \frac{|00\rangle - |11\rangle}{\sqrt{2}} \tag{2.135}$$

$$10 : |\psi\rangle \rightarrow \frac{|10\rangle + |01\rangle}{\sqrt{2}} \tag{2.136}$$

$$11 : |\psi\rangle \rightarrow \frac{|01\rangle - |10\rangle}{\sqrt{2}}. \tag{2.137}$$

As we noted in Section 1.3.6, these four states are known as the *Bell basis*, *Bell states*, or *EPR pairs*, in honor of several of the pioneers who first appreciated the novelty of entanglement. Notice that the Bell states form an orthonormal basis, and can therefore be distinguished by an appropriate quantum measurement. If Alice sends her qubit to Bob, giving Bob possession of both qubits, then by doing a measurement in the Bell basis Bob can determine which of the four possible bit strings Alice sent.

Summarizing, Alice, interacting with only a single qubit, is able to transmit two bits of information to Bob. Of course, two qubits are involved in the protocol, but Alice never need interact with the second qubit. Classically, the task Alice accomplishes would have been impossible had she only transmitted a single classical bit, as we will show in Chapter 12. Furthermore, this remarkable superdense coding protocol has received partial verification in the laboratory. (See 'History and further reading' for references to the experimental verification.) In later chapters we will see many other examples, some of them much more spectacular than superdense coding, of quantum mechanics being harnessed to perform information processing tasks. However, a key point can already be seen in this beautiful example: information is physical, and surprising physical theories such as quantum mechanics may predict surprising information processing abilities.

Exercise 2.69: Verify that the Bell basis forms an orthonormal basis for the two qubit state space.

Exercise 2.70: Suppose E is any positive operator acting on Alice's qubit. Show that $\langle \psi | E \otimes I | \psi \rangle$ *takes the same value* when $|\psi\rangle$ is any of the four Bell states. Suppose some malevolent third party ('Eve') intercepts Alice's qubit on the way to Bob in the superdense coding protocol. Can Eve infer anything about which of the four possible bit strings $00, 01, 10, 11$ Alice is trying to send? If so, how, or if not, why not?

2.4 The density operator

We have formulated quantum mechanics using the language of state vectors. An alternate formulation is possible using a tool known as the *density operator* or *density matrix*. This alternate formulation is mathematically equivalent to the state vector approach, but it provides a much more convenient language for thinking about some commonly encountered scenarios in quantum mechanics. The next three sections describe the density operator formulation of quantum mechanics. Section 2.4.1 introduces the density operator using the concept of an ensemble of quantum states. Section 2.4.2 develops some general properties of the density operator. Finally, Section 2.4.3 describes an application where the density operator really shines – as a tool for the description of *individual subsystems* of a composite quantum system.

2.4.1 Ensembles of quantum states

The density operator language provides a convenient means for describing quantum systems whose state is not completely known. More precisely, suppose a quantum system is in one of a number of states $|\psi_i\rangle$, where i is an index, with respective probabilities p_i. We shall call $\{p_i, |\psi_i\rangle\}$ an *ensemble of pure states*. The density operator for the system is defined by the equation

$$\rho \equiv \sum_i p_i |\psi_i\rangle\langle\psi_i|. \tag{2.138}$$

The density operator is often known as the *density matrix*; we will use the two terms interchangeably. It turns out that all the postulates of quantum mechanics can be re-formulated in terms of the density operator language. The purpose of this section and the next is to explain how to perform this reformulation, and explain when it is useful. Whether one uses the density operator language or the state vector language is a matter of taste, since both give the same results; however it is sometimes much easier to approach problems from one point of view rather than the other.

Suppose, for example, that the evolution of a closed quantum system is described by the unitary operator U. If the system was initially in the state $|\psi_i\rangle$ with probability p_i then after the evolution has occurred the system will be in the state $U|\psi_i\rangle$ with probability p_i. Thus, the evolution of the density operator is described by the equation

$$\rho = \sum_i p_i |\psi_i\rangle\langle\psi_i| \xrightarrow{U} \sum_i p_i U|\psi_i\rangle\langle\psi_i|U^\dagger = U\rho U^\dagger. \tag{2.139}$$

Measurements are also easily described in the density operator language. Suppose we perform a measurement described by measurement operators M_m. If the initial state was $|\psi_i\rangle$, then the probability of getting result m is

$$p(m|i) = \langle\psi_i|M_m^\dagger M_m|\psi_i\rangle = \text{tr}(M_m^\dagger M_m|\psi_i\rangle\langle\psi_i|), \tag{2.140}$$

where we have used Equation (2.61) to obtain the last equality. By the law of total probability (see Appendix 1 for an explanation of this and other elementary notions of probability theory) the probability of obtaining result m is

$$p(m) = \sum_i p(m|i)p_i \tag{2.141}$$

$$= \sum_i p_i \text{tr}(M_m^\dagger M_m|\psi_i\rangle\langle\psi_i|) \tag{2.142}$$

$$= \text{tr}(M_m^\dagger M_m\rho). \tag{2.143}$$

What is the density operator of the system after obtaining the measurement result m? If the initial state was $|\psi_i\rangle$ then the state after obtaining the result m is

$$|\psi_i^m\rangle = \frac{M_m|\psi_i\rangle}{\sqrt{\langle\psi_i|M_m^\dagger M_m|\psi_i\rangle}}. \tag{2.144}$$

Thus, after a measurement which yields the result m we have an ensemble of states $|\psi_i^m\rangle$ with respective probabilities $p(i|m)$. The corresponding density operator ρ_m is therefore

$$\rho_m = \sum_i p(i|m)|\psi_i^m\rangle\langle\psi_i^m| = \sum_i p(i|m)\frac{M_m|\psi_i\rangle\langle\psi_i|M_m^\dagger}{\langle\psi_i|M_m^\dagger M_m|\psi_i\rangle}. \tag{2.145}$$

But by elementary probability theory, $p(i|m) = p(m,i)/p(m) = p(m|i)p_i/p(m)$. Substituting from (2.143) and (2.140) we obtain

$$\rho_m = \sum_i p_i \frac{M_m|\psi_i\rangle\langle\psi_i|M_m^\dagger}{\mathrm{tr}(M_m^\dagger M_m \rho)} \tag{2.146}$$

$$= \frac{M_m \rho M_m^\dagger}{\mathrm{tr}(M_m^\dagger M_m \rho)}. \tag{2.147}$$

What we have shown is that the basic postulates of quantum mechanics related to unitary evolution and measurement can be rephrased in the language of density operators. In the next section we complete this rephrasing by giving an intrinsic characterization of the density operator that does not rely on the idea of a state vector.

Before doing so, however, it is useful to introduce some more language, and one more fact about the density operator. First, the language. A quantum system whose state $|\psi\rangle$ is known exactly is said to be in a *pure state*. In this case the density operator is simply $\rho = |\psi\rangle\langle\psi|$. Otherwise, ρ is in a *mixed state*; it is said to be a *mixture* of the different pure states in the ensemble for ρ. In the exercises you will be asked to demonstrate a simple criterion for determining whether a state is pure or mixed: a pure state satisfies $\mathrm{tr}(\rho^2) = 1$, while a mixed state satisfies $\mathrm{tr}(\rho^2) < 1$. A few words of warning about the nomenclature: sometimes people use the term 'mixed state' as a catch-all to include both pure and mixed quantum states. The origin for this usage seems to be that it implies that the writer is not necessarily *assuming* that a state is pure. Second, the term 'pure state' is often used in reference to a state vector $|\psi\rangle$, to distinguish it from a density operator ρ.

Finally, imagine a quantum system is prepared in the state ρ_i with probability p_i. It is not difficult to convince yourself that the system may be described by the density matrix $\sum_i p_i \rho_i$. A proof of this is to suppose that ρ_i arises from some ensemble $\{p_{ij}, |\psi_{ij}\rangle\}$ (note that i is fixed) of pure states, so the probability for being in the state $|\psi_{ij}\rangle$ is $p_i p_{ij}$. The density matrix for the system is thus

$$\rho = \sum_{ij} p_i p_{ij} |\psi_{ij}\rangle\langle\psi_{ij}| \tag{2.148}$$

$$= \sum_i p_i \rho_i, \tag{2.149}$$

where we have used the definition $\rho_i = \sum_j p_{ij}|\psi_{ij}\rangle\langle\psi_{ij}|$. We say that ρ is a *mixture* of the states ρ_i with probabilities p_i. This concept of a mixture comes up repeatedly in the analysis of problems like quantum noise, where the effect of the noise is to introduce ignorance into our knowledge of the quantum state. A simple example is provided by the measurement scenario described above. Imagine that, for some reason, our record of the result m of the measurement was lost. We would have a quantum system in the state ρ_m with probability $p(m)$, but would no longer know the actual value of m. The state of such a quantum system would therefore be described by the density operator

$$\rho = \sum_m p(m)\rho_m \tag{2.150}$$

$$= \sum_m \text{tr}(M_m^\dagger M_m \rho) \frac{M_m \rho M_m^\dagger}{\text{tr}(M_m^\dagger M_m \rho)} \tag{2.151}$$

$$= \sum_m M_m \rho M_m^\dagger, \tag{2.152}$$

a nice compact formula which may be used as the starting point for analysis of further operations on the system.

2.4.2 General properties of the density operator

The density operator was introduced as a means of describing ensembles of quantum states. In this section we move away from this description to develop an intrinsic characterization of density operators that does not rely on an ensemble interpretation. This allows us to complete the program of giving a description of quantum mechanics that does not take as its foundation the state vector. We also take the opportunity to develop numerous other elementary properties of the density operator.

The class of operators that are density operators are characterized by the following useful theorem:

Theorem 2.5: (**Characterization of density operators**) An operator ρ is the density operator associated to some ensemble $\{p_i, |\psi_i\rangle\}$ if and only if it satisfies the conditions:

(1) (**Trace condition**) ρ has trace equal to one.
(2) (**Positivity condition**) ρ is a positive operator.

Proof
Suppose $\rho = \sum_i p_i |\psi_i\rangle\langle\psi_i|$ is a density operator. Then

$$\text{tr}(\rho) = \sum_i p_i \text{tr}(|\psi_i\rangle\langle\psi_i|) = \sum_i p_i = 1, \tag{2.153}$$

so the trace condition $\text{tr}(\rho) = 1$ is satisfied. Suppose $|\varphi\rangle$ is an arbitrary vector in state space. Then

$$\langle\varphi|\rho|\varphi\rangle = \sum_i p_i \langle\varphi|\psi_i\rangle\langle\psi_i|\varphi\rangle \tag{2.154}$$

$$= \sum_i p_i |\langle\varphi|\psi_i\rangle|^2 \tag{2.155}$$

$$\geq 0, \tag{2.156}$$

so the positivity condition is satisfied.

Conversely, suppose ρ is any operator satisfying the trace and positivity conditions. Since ρ is positive, it must have a spectral decomposition

$$\rho = \sum_j \lambda_j |j\rangle\langle j|, \tag{2.157}$$

where the vectors $|j\rangle$ are orthogonal, and λ_j are real, non-negative eigenvalues of ρ.

From the trace condition we see that $\sum_j \lambda_j = 1$. Therefore, a system in state $|j\rangle$ with probability λ_j will have density operator ρ. That is, the ensemble $\{\lambda_j, |j\rangle\}$ is an ensemble of states giving rise to the density operator ρ.   □

This theorem provides a characterization of density operators that is intrinsic to the operator itself: we can *define* a density operator to be a positive operator ρ which has trace equal to one. Making this definition allows us to reformulate the postulates of quantum mechanics in the density operator picture. For ease of reference we state all the reformulated postulates here:

> **Postulate 1**: Associated to any isolated physical system is a complex vector space with inner product (that is, a Hilbert space) known as the *state space* of the system. The system is completely described by its *density operator*, which is a positive operator ρ with trace one, acting on the state space of the system. If a quantum system is in the state ρ_i with probability p_i, then the density operator for the system is $\sum_i p_i \rho_i$.

> **Postulate 2**: The evolution of a *closed* quantum system is described by a *unitary transformation*. That is, the state ρ of the system at time t_1 is related to the state ρ' of the system at time t_2 by a unitary operator U which depends only on the times t_1 and t_2,
>
> $$\rho' = U\rho U^\dagger. \tag{2.158}$$

> **Postulate 3**: Quantum measurements are described by a collection $\{M_m\}$ of *measurement operators*. These are operators acting on the state space of the system being measured. The index m refers to the measurement outcomes that may occur in the experiment. If the state of the quantum system is ρ immediately before the measurement then the probability that result m occurs is given by
>
> $$p(m) = \text{tr}(M_m^\dagger M_m \rho), \tag{2.159}$$
>
> and the state of the system after the measurement is
>
> $$\frac{M_m \rho M_m^\dagger}{\text{tr}(M_m^\dagger M_m \rho)}. \tag{2.160}$$
>
> The measurement operators satisfy the *completeness equation*,
>
> $$\sum_m M_m^\dagger M_m = I. \tag{2.161}$$

> **Postulate 4**: The state space of a composite physical system is the tensor product of the state spaces of the component physical systems. Moreover, if we have systems numbered 1 through n, and system number i is prepared in the state ρ_i, then the joint state of the total system is $\rho_1 \otimes \rho_2 \otimes \ldots \rho_n$.

These reformulations of the fundamental postulates of quantum mechanics in terms of the density operator are, of course, mathematically equivalent to the description in terms of the state vector. Nevertheless, as a way of thinking about quantum mechanics, the density operator approach really shines for two applications: the description of quantum systems whose state is not known, and the description of subsystems of a composite quantum system, as will be described in the next section. For the remainder of this section we flesh out the properties of the density matrix in more detail.

Exercise 2.71: (Criterion to decide if a state is mixed or pure) Let ρ be a density operator. Show that $\text{tr}(\rho^2) \le 1$, with equality if and only if ρ is a pure state.

It is a tempting (and surprisingly common) fallacy to suppose that the eigenvalues and eigenvectors of a density matrix have some special significance with regard to the ensemble of quantum states represented by that density matrix. For example, one might suppose that a quantum system with density matrix

$$\rho = \frac{3}{4}|0\rangle\langle 0| + \frac{1}{4}|1\rangle\langle 1|. \tag{2.162}$$

must be in the state $|0\rangle$ with probability $3/4$ and in the state $|1\rangle$ with probability $1/4$. However, this is not necessarily the case. Suppose we define

$$|a\rangle \equiv \sqrt{\frac{3}{4}}|0\rangle + \sqrt{\frac{1}{4}}|1\rangle \tag{2.163}$$

$$|b\rangle \equiv \sqrt{\frac{3}{4}}|0\rangle - \sqrt{\frac{1}{4}}|1\rangle, \tag{2.164}$$

and the quantum system is prepared in the state $|a\rangle$ with probability $1/2$ and in the state $|b\rangle$ with probability $1/2$. Then it is easily checked that the corresponding density matrix is

$$\rho = \frac{1}{2}|a\rangle\langle a| + \frac{1}{2}|b\rangle\langle b| = \frac{3}{4}|0\rangle\langle 0| + \frac{1}{4}|1\rangle\langle 1|. \tag{2.165}$$

That is, these two *different* ensembles of quantum states give rise to the *same* density matrix. In general, the eigenvectors and eigenvalues of a density matrix just indicate *one* of many possible ensembles that may give rise to a specific density matrix, and there is no reason to suppose it is an especially privileged ensemble.

A natural question to ask in the light of this discussion is what class of ensembles does give rise to a particular density matrix? The solution to this problem, which we now give, has surprisingly many applications in quantum computation and quantum information, notably in the understanding of quantum noise and quantum error-correction (Chapters 8 and 10). For the solution it is convenient to make use of vectors $|\tilde{\psi}_i\rangle$ which may not be normalized to unit length. We say the set $|\tilde{\psi}_i\rangle$ *generates* the operator $\rho \equiv \sum_i |\tilde{\psi}_i\rangle\langle\tilde{\psi}_i|$, and thus the connection to the usual ensemble picture of density operators is expressed by the equation $|\tilde{\psi}_i\rangle = \sqrt{p_i}|\psi_i\rangle$. When do two sets of vectors, $|\tilde{\psi}_i\rangle$ and $|\tilde{\varphi}_j\rangle$ generate the same operator ρ? The solution to this problem will enable us to answer the question of what ensembles give rise to a given density matrix.

Theorem 2.6: **(Unitary freedom in the ensemble for density matrices)** The sets $|\tilde{\psi}_i\rangle$ and $|\tilde{\varphi}_j\rangle$ generate the same density matrix if and only if

$$|\tilde{\psi}_i\rangle = \sum_j u_{ij}|\tilde{\varphi}_j\rangle, \tag{2.166}$$

where u_{ij} is a unitary matrix of complex numbers, with indices i and j, and we 'pad' whichever set of vectors $|\tilde{\psi}_i\rangle$ or $|\tilde{\varphi}_j\rangle$ is smaller with additional vectors 0 so that the two sets have the same number of elements.

As a consequence of the theorem, note that $\rho = \sum_i p_i|\psi_i\rangle\langle\psi_i| = \sum_j q_j|\varphi_j\rangle\langle\varphi_j|$ for *normalized* states $|\psi_i\rangle, |\varphi_j\rangle$ and probability distributions p_i and q_j if and only if

$$\sqrt{p_i}|\psi_i\rangle = \sum_j u_{ij}\sqrt{q_j}|\varphi_j\rangle, \qquad (2.167)$$

for some unitary matrix u_{ij}, and we may pad the smaller ensemble with entries having probability zero in order to make the two ensembles the same size. Thus, Theorem 2.6 characterizes the freedom in ensembles $\{p_i, |\psi_i\rangle\}$ giving rise to a given density matrix ρ. Indeed, it is easily checked that our earlier example of a density matrix with two different decompositions, (2.162), arises as a special case of this general result. Let's turn now to the proof of the theorem.

Proof
Suppose $|\tilde{\psi}_i\rangle = \sum_j u_{ij}|\tilde{\varphi}_j\rangle$ for some unitary u_{ij}. Then

$$\sum_i |\tilde{\psi}_i\rangle\langle\tilde{\psi}_i| = \sum_{ijk} u_{ij}u_{ik}^*|\tilde{\varphi}_j\rangle\langle\tilde{\varphi}_k| \qquad (2.168)$$

$$= \sum_{jk}\left(\sum_i u_{ki}^\dagger u_{ij}\right)|\tilde{\varphi}_j\rangle\langle\tilde{\varphi}_k| \qquad (2.169)$$

$$= \sum_{jk} \delta_{kj}|\tilde{\varphi}_j\rangle\langle\tilde{\varphi}_k| \qquad (2.170)$$

$$= \sum_j |\tilde{\varphi}_j\rangle\langle\tilde{\varphi}_j|, \qquad (2.171)$$

which shows that $|\tilde{\psi}_i\rangle$ and $|\tilde{\varphi}_j\rangle$ generate the same operator.

Conversely, suppose

$$A = \sum_i |\tilde{\psi}_i\rangle\langle\tilde{\psi}_i| = \sum_j |\tilde{\varphi}_j\rangle\langle\tilde{\varphi}_j|. \qquad (2.172)$$

Let $A = \sum_k \lambda_k|k\rangle\langle k|$ be a decomposition for A such that the states $|k\rangle$ are orthonormal, and the λ_k are strictly positive. Our strategy is to relate the states $|\tilde{\psi}_i\rangle$ to the states $|\tilde{k}\rangle \equiv \sqrt{\lambda_k}|k\rangle$, and similarly relate the states $|\tilde{\varphi}_j\rangle$ to the states $|\tilde{k}\rangle$. Combining the two relations will give the result. Let $|\psi\rangle$ be any vector orthonormal to the space spanned by the $|\tilde{k}\rangle$, so $\langle\psi|\tilde{k}\rangle\langle\tilde{k}|\psi\rangle = 0$ for all k, and thus we see that

$$0 = \langle\psi|A|\psi\rangle = \sum_i \langle\psi|\tilde{\psi}_i\rangle\langle\tilde{\psi}_i|\psi\rangle = \sum_i |\langle\psi|\tilde{\psi}_i\rangle|^2. \qquad (2.173)$$

Thus $\langle\psi|\tilde{\psi}_i\rangle = 0$ for all i and all $|\psi\rangle$ orthonormal to the space spanned by the $|\tilde{k}\rangle$. It follows that each $|\tilde{\psi}_i\rangle$ can be expressed as a linear combination of the $|\tilde{k}\rangle$, $|\tilde{\psi}_i\rangle = \sum_k c_{ik}|\tilde{k}\rangle$. Since $A = \sum_k |\tilde{k}\rangle\langle\tilde{k}| = \sum_i |\tilde{\psi}_i\rangle\langle\tilde{\psi}_i|$ we see that

$$\sum_k |\tilde{k}\rangle\langle\tilde{k}| = \sum_{kl}\left(\sum_i c_{ik}c_{il}^*\right)|\tilde{k}\rangle\langle\tilde{l}|. \qquad (2.174)$$

The operators $|\tilde{k}\rangle\langle\tilde{l}|$ are easily seen to be linearly independent, and thus it must be that

$\sum_i c_{ik}c_{il}^* = \delta_{kl}$. This ensures that we may append extra columns to c to obtain a unitary matrix v such that $|\tilde{\psi}_i\rangle = \sum_k v_{ik}|\tilde{k}\rangle$, where we have appended zero vectors to the list of $|\tilde{k}\rangle$. Similarly, we can find a unitary matrix w such that $|\tilde{\varphi}_j\rangle = \sum_k w_{jk}|\tilde{k}\rangle$. Thus $|\tilde{\psi}_i\rangle = \sum_j u_{ij}|\tilde{\varphi}_j\rangle$, where $u = vw^\dagger$ is unitary. $\qquad\square$

Exercise 2.72: (Bloch sphere for mixed states) The Bloch sphere picture for pure states of a single qubit was introduced in Section 1.2. This description has an important generalization to mixed states as follows.

(1) Show that an arbitrary density matrix for a mixed state qubit may be written as

$$\rho = \frac{I + \vec{r} \cdot \vec{\sigma}}{2}, \tag{2.175}$$

where $\vec{r}$ is a real three-dimensional vector such that $\|\vec{r}\| \leq 1$. This vector is known as the *Bloch vector* for the state ρ.

(2) What is the Bloch vector representation for the state $\rho = I/2$?

(3) Show that a state ρ is pure if and only if $\|\vec{r}\| = 1$.

(4) Show that for pure states the description of the Bloch vector we have given coincides with that in Section 1.2.

Exercise 2.73: Let ρ be a density operator. A *minimal ensemble* for ρ is an ensemble $\{p_i, |\psi_i\rangle\}$ containing a number of elements equal to the rank of ρ. Let $|\psi\rangle$ be any state in the support of ρ. (The *support* of a Hermitian operator A is the vector space spanned by the eigenvectors of A with non-zero eigenvalues.) Show that there is a minimal ensemble for ρ that contains $|\psi\rangle$, and moreover that in any such ensemble $|\psi\rangle$ must appear with probability

$$p_i = \frac{1}{\langle\psi_i|\rho^{-1}|\psi_i\rangle}, \tag{2.176}$$

where ρ^{-1} is defined to be the inverse of ρ, when ρ is considered as an operator acting only on the support of ρ. (This definition removes the problem that ρ may not have an inverse.)

2.4.3 The reduced density operator

Perhaps the deepest application of the density operator is as a descriptive tool for *subsystems* of a composite quantum system. Such a description is provided by the *reduced density operator*, which is the subject of this section. The reduced density operator is so useful as to be virtually indispensable in the analysis of composite quantum systems.

Suppose we have physical systems A and B, whose state is described by a density operator ρ^{AB}. The reduced density operator for system A is defined by

$$\rho^A \equiv \text{tr}_B(\rho^{AB}), \tag{2.177}$$

where tr_B is a map of operators known as the *partial trace* over system B. The partial trace is defined by

$$\text{tr}_B\left(|a_1\rangle\langle a_2| \otimes |b_1\rangle\langle b_2|\right) \equiv |a_1\rangle\langle a_2| \, \text{tr}(|b_1\rangle\langle b_2|), \tag{2.178}$$

where $|a_1\rangle$ and $|a_2\rangle$ are any two vectors in the state space of A, and $|b_1\rangle$ and $|b_2\rangle$ are any two vectors in the state space of B. The trace operation appearing on the right hand side is the usual trace operation for system B, so $\text{tr}(|b_1\rangle\langle b_2|) = \langle b_2|b_1\rangle$. We have defined the partial trace operation only on a special subclass of operators on AB; the specification is completed by requiring in addition to Equation (2.178) that the partial trace be linear in its input.

It is not obvious that the reduced density operator for system A is in any sense a description for the state of system A. The physical justification for making this identification is that the reduced density operator provides the correct measurement statistics for measurements made on system A. This is explained in more detail in Box 2.6 on page 107. The following simple example calculations may also help understand the reduced density operator. First, suppose a quantum system is in the product state $\rho^{AB} = \rho \otimes \sigma$, where ρ is a density operator for system A, and σ is a density operator for system B. Then

$$\rho^A = \text{tr}_B(\rho \otimes \sigma) = \rho\,\text{tr}(\sigma) = \rho, \tag{2.184}$$

which is the result we intuitively expect. Similarly, $\rho^B = \sigma$ for this state. A less trivial example is the Bell state $(|00\rangle + |11\rangle)/\sqrt{2}$. This has density operator

$$\rho = \left(\frac{|00\rangle + |11\rangle}{\sqrt{2}}\right)\left(\frac{\langle 00| + \langle 11|}{\sqrt{2}}\right) \tag{2.185}$$

$$= \frac{|00\rangle\langle 00| + |11\rangle\langle 00| + |00\rangle\langle 11| + |11\rangle\langle 11|}{2}. \tag{2.186}$$

Tracing out the second qubit, we find the reduced density operator of the first qubit,

$$\rho^1 = \text{tr}_2(\rho) \tag{2.187}$$

$$= \frac{\text{tr}_2(|00\rangle\langle 00|) + \text{tr}_2(|11\rangle\langle 00|) + \text{tr}_2(|00\rangle\langle 11|) + \text{tr}_2(|11\rangle\langle 11|)}{2} \tag{2.188}$$

$$= \frac{|0\rangle\langle 0|\langle 0|0\rangle + |1\rangle\langle 0|\langle 0|1\rangle + |0\rangle\langle 1|\langle 1|0\rangle + |1\rangle\langle 1|\langle 1|1\rangle}{2} \tag{2.189}$$

$$= \frac{|0\rangle\langle 0| + |1\rangle\langle 1|}{2} \tag{2.190}$$

$$= \frac{I}{2}. \tag{2.191}$$

Notice that this state is a *mixed state*, since $\text{tr}((I/2)^2) = 1/2 < 1$. This is quite a remarkable result. The state of the joint system of two qubits is a pure state, that is, it is known *exactly*; however, the first qubit is in a mixed state, that is, a state about which we apparently do not have maximal knowledge. This strange property, that the joint state of a system can be completely known, yet a subsystem be in mixed states, is another hallmark of quantum entanglement.

Exercise 2.74: Suppose a composite of systems A and B is in the state $|a\rangle|b\rangle$, where $|a\rangle$ is a pure state of system A, and $|b\rangle$ is a pure state of system B. Show that the reduced density operator of system A alone is a pure state.

Exercise 2.75: For each of the four Bell states, find the reduced density operator for each qubit.

Quantum teleportation and the reduced density operator

A useful application of the reduced density operator is to the analysis of quantum teleportation. Recall from Section 1.3.7 that quantum teleportation is a procedure for sending

---

Box 2.6: Why the partial trace?

Why is the partial trace used to describe part of a larger quantum system? The reason for doing this is because the partial trace operation is the *unique* operation which gives rise to the correct description of *observable* quantities for subsystems of a composite system, in the following sense.

Suppose M is any observable on system A, and we have some measuring device which is capable of realizing measurements of M. Let $\tilde{M}$ denote the corresponding observable for the same measurement, performed on the composite system AB. Our immediate goal is to argue that $\tilde{M}$ is necessarily equal to $M \otimes I_B$. Note that if the system AB is prepared in the state $|m\rangle|\psi\rangle$, where $|m\rangle$ is an eigenstate of M with eigenvalue m, and $|\psi\rangle$ is any state of B, then the measuring device must yield the result m for the measurement, with probability one. Thus, if P_m is the projector onto the m eigenspace of the observable M, then the corresponding projector for $\tilde{M}$ is $P_m \otimes I_B$. We therefore have

$$\tilde{M} = \sum_m m P_m \otimes I_B = M \otimes I_B. \tag{2.179}$$

The next step is to show that the partial trace procedure gives the correct measurement statistics for observations on part of a system. Suppose we perform a measurement on system A described by the observable M. Physical consistency requires that any prescription for associating a 'state', ρ^A, to system A, must have the property that measurement averages be the same whether computed via ρ^A or ρ^{AB},

$$\text{tr}(M\rho^A) = \text{tr}(\tilde{M}\rho^{AB}) = \text{tr}((M \otimes I_B)\rho^{AB}). \tag{2.180}$$

This equation is certainly satisfied if we choose $\rho^A \equiv \text{tr}_B(\rho^{AB})$. In fact, the partial trace turns out to be the *unique* function having this property. To see this uniqueness property, let $f(\cdot)$ be any map of density operators on AB to density operators on A such that

$$\text{tr}(Mf(\rho^{AB})) = \text{tr}((M \otimes I_B)\rho^{AB}), \tag{2.181}$$

for all observables M. Let M_i be an orthonormal basis of operators for the space of Hermitian operators with respect to the Hilbert–Schmidt inner product $(X, Y) \equiv \text{tr}(XY)$ (compare Exercise 2.39 on page 76). Then expanding $f(\rho^{AB})$ in this basis gives

$$f(\rho^{AB}) = \sum_i M_i \text{tr}(M_i f(\rho^{AB})) \tag{2.182}$$

$$= \sum_i M_i \text{tr}((M_i \otimes I_B)\rho^{AB}). \tag{2.183}$$

It follows that f is uniquely determined by Equation (2.180). Moreover, the partial trace satisfies (2.180), so it is the unique function having this property.

---

quantum information from Alice to Bob, given that Alice and Bob share an EPR pair, and have a classical communications channel.

At first sight it appears as though teleportation can be used to do faster than light communication, a big no-no according to the theory of relativity. We surmised in Section 1.3.7 that what prevents faster than light communication is the need for Alice to communicate her measurement result to Bob. The reduced density operator allows us to make this rigorous.

Recall that immediately before Alice makes her measurement the quantum state of the three qubits is (Equation (1.32)):

$$|\psi_2\rangle = \frac{1}{2}\Big[|00\rangle\,(\alpha|0\rangle + \beta|1\rangle) + |01\rangle\,(\alpha|1\rangle + \beta|0\rangle)$$
$$+ |10\rangle\,(\alpha|0\rangle - \beta|1\rangle) + |11\rangle\,(\alpha|1\rangle - \beta|0\rangle)\Big]. \tag{2.192}$$

Measuring in Alice's computational basis, the state of the system after the measurement is:

$$|00\rangle\Big[\alpha|0\rangle + \beta|1\rangle\Big] \text{ with probability } \frac{1}{4} \tag{2.193}$$

$$|01\rangle\Big[\alpha|1\rangle + \beta|0\rangle\Big] \text{ with probability } \frac{1}{4} \tag{2.194}$$

$$|10\rangle\Big[\alpha|0\rangle - \beta|1\rangle\Big] \text{ with probability } \frac{1}{4} \tag{2.195}$$

$$|11\rangle\Big[\alpha|1\rangle - \beta|0\rangle\Big] \text{ with probability } \frac{1}{4}. \tag{2.196}$$

The density operator of the system is thus

$$\rho = \frac{1}{4}\Big[|00\rangle\langle00|(\alpha|0\rangle + \beta|1\rangle)(\alpha^*\langle0| + \beta^*\langle1|) + |01\rangle\langle01|(\alpha|1\rangle + \beta|0\rangle)(\alpha^*\langle1| + \beta^*\langle0|)$$
$$+ |10\rangle\langle10|(\alpha|0\rangle - \beta|1\rangle)(\alpha^*\langle0| - \beta^*\langle1|) + |11\rangle\langle11|(\alpha|1\rangle - \beta|0\rangle)(\alpha^*\langle1| - \beta^*\langle0|)\Big]. \tag{2.197}$$

Tracing out Alice's system, we see that the reduced density operator of Bob's system is

$$\rho^B = \frac{1}{4}\Big[(\alpha|0\rangle + \beta|1\rangle)(\alpha^*\langle0| + \beta^*\langle1|) + (\alpha|1\rangle + \beta|0\rangle)(\alpha^*\langle1| + \beta^*\langle0|)$$
$$+ (\alpha|0\rangle - \beta|1\rangle)(\alpha^*\langle0| - \beta^*\langle1|) + (\alpha|1\rangle - \beta|0\rangle)(\alpha^*\langle1| - \beta^*\langle0|)\Big] \tag{2.198}$$

$$= \frac{2(|\alpha|^2 + |\beta|^2)|0\rangle\langle0| + 2(|\alpha|^2 + |\beta|^2)|1\rangle\langle1|}{4} \tag{2.199}$$

$$= \frac{|0\rangle\langle0| + |1\rangle\langle1|}{2} \tag{2.200}$$

$$= \frac{I}{2}, \tag{2.201}$$

where we have used the completeness relation in the last line. Thus, the state of Bob's system *after* Alice has performed the measurement but *before* Bob has learned the measurement result is $I/2$. This state has no dependence upon the state $|\psi\rangle$ being teleported, and thus any measurements performed by Bob will contain no information about $|\psi\rangle$, thus preventing Alice from using teleportation to transmit information to Bob faster than light.

2.5 The Schmidt decomposition and purifications

Density operators and the partial trace are just the beginning of a wide array of tools useful for the study of composite quantum systems, which are at the heart of quantum computation and quantum information. Two additional tools of great value are the *Schmidt decomposition* and *purifications*. In this section we present both these tools, and try to give the flavor of their power.

Theorem 2.7: (**Schmidt decomposition**) Suppose $|\psi\rangle$ is a pure state of a composite system, AB. Then there exist orthonormal states $|i_A\rangle$ for system A, and orthonormal states $|i_B\rangle$ of system B such that

$$|\psi\rangle = \sum_i \lambda_i |i_A\rangle |i_B\rangle, \tag{2.202}$$

where λ_i are non-negative real numbers satisfying $\sum_i \lambda_i^2 = 1$ known as *Schmidt co-efficients*.

This result is very useful. As a taste of its power, consider the following consequence: let $|\psi\rangle$ be a pure state of a composite system, AB. Then by the Schmidt decomposition $\rho^A = \sum_i \lambda_i^2 |i_A\rangle\langle i_A|$ and $\rho^B = \sum_i \lambda_i^2 |i_B\rangle\langle i_B|$, so the eigenvalues of ρ^A and ρ^B are identical, namely λ_i^2 for both density operators. Many important properties of quantum systems are completely determined by the eigenvalues of the reduced density operator of the system, so for a pure state of a composite system such properties will be the same for both systems. As an example, consider the state of two qubits, $(|00\rangle + |01\rangle + |11\rangle)/\sqrt{3}$. This has no obvious symmetry property, yet if you calculate $\mathrm{tr}\left((\rho^A)^2\right)$ and $\mathrm{tr}\left((\rho^B)^2\right)$ you will discover that they have the same value, $7/9$ in each case. This is but one small consequence of the Schmidt decomposition.

Proof
We give the proof for the case where systems A and B have state spaces of the same dimension, and leave the general case to Exercise 2.76. Let $|j\rangle$ and $|k\rangle$ be any fixed orthonormal bases for systems A and B, respectively. Then $|\psi\rangle$ can be written

$$|\psi\rangle = \sum_{jk} a_{jk} |j\rangle |k\rangle, \tag{2.203}$$

for some matrix a of complex numbers a_{jk}. By the singular value decomposition, $a = udv$, where d is a diagonal matrix with non-negative elements, and u and v are unitary matrices. Thus

$$|\psi\rangle = \sum_{ijk} u_{ji} d_{ii} v_{ik} |j\rangle |k\rangle. \tag{2.204}$$

Defining $|i_A\rangle \equiv \sum_j u_{ji}|j\rangle$, $|i_B\rangle \equiv \sum_k v_{ik}|k\rangle$, and $\lambda_i \equiv d_{ii}$, we see that this gives

$$|\psi\rangle = \sum_i \lambda_i |i_A\rangle |i_B\rangle. \tag{2.205}$$

It is easy to check that $|i_A\rangle$ forms an orthonormal set, from the unitarity of u and the orthonormality of $|j\rangle$, and similarly that the $|i_B\rangle$ form an orthonormal set.   □

Exercise 2.76: Extend the proof of the Schmidt decomposition to the case where A and B may have state spaces of different dimensionality.

Exercise 2.77: Suppose ABC is a three component quantum system. Show by example that there are quantum states $|\psi\rangle$ of such systems which can not be written in the form

$$|\psi\rangle = \sum_i \lambda_i |i_A\rangle |i_B\rangle |i_C\rangle, \tag{2.206}$$

where λ_i are real numbers, and $|i_A\rangle, |i_B\rangle, |i_C\rangle$ are orthonormal bases of the respective systems.

The bases $|i_A\rangle$ and $|i_B\rangle$ are called the *Schmidt bases* for A and B, respectively, and the number of non-zero values λ_i is called the *Schmidt number* for the state $|\psi\rangle$. The Schmidt number is an important property of a composite quantum system, which in some sense quantifies the 'amount' of entanglement between systems A and B. To get some idea of why this is the case, consider the following obvious but important property: the Schmidt number is preserved under unitary transformations on system A or system B alone. To see this, notice that if $\sum_i \lambda_i |i_A\rangle |i_B\rangle$ is the Schmidt decomposition for $|\psi\rangle$ then $\sum_i \lambda_i (U|i_A\rangle)|i_B\rangle$ is the Schmidt decomposition for $U|\psi\rangle$, where U is a unitary operator acting on system A alone. Algebraic invariance properties of this type make the Schmidt number a very useful tool.

Exercise 2.78: Prove that a state $|\psi\rangle$ of a composite system AB is a product state if and only if it has Schmidt number 1. Prove that $|\psi\rangle$ is a product state if and only if ρ^A (and thus ρ^B) are pure states.

A second, related technique for quantum computation and quantum information is *purification*. Suppose we are given a state ρ^A of a quantum system A. It is possible to introduce another system, which we denote R, and *define a pure state* $|AR\rangle$ for the joint system AR such that $\rho^A = \mathrm{tr}_R(|AR\rangle\langle AR|)$. That is, the pure state $|AR\rangle$ reduces to ρ^A when we look at system A alone. This is a purely mathematical procedure, known as *purification*, which allows us to associate pure states with mixed states. For this reason we call system R a *reference* system: it is a fictitious system, without a direct physical significance.

To prove that purification can be done for *any* state, we explain how to construct a system R and purification $|AR\rangle$ for ρ^A. Suppose ρ^A has orthonormal decomposition $\rho^A = \sum_i p_i |i^A\rangle\langle i^A|$. To purify ρ^A we introduce a system R which has the same state space as system A, with orthonormal basis states $|i^R\rangle$, and define a pure state for the combined system

$$|AR\rangle \equiv \sum_i \sqrt{p_i} |i^A\rangle |i^R\rangle. \tag{2.207}$$

We now calculate the reduced density operator for system A corresponding to the state $|AR\rangle$:

$$\mathrm{tr}_R(|AR\rangle\langle AR|) = \sum_{ij} \sqrt{p_i p_j} |i^A\rangle\langle j^A| \, \mathrm{tr}(|i^R\rangle\langle j^R|) \tag{2.208}$$

$$= \sum_{ij} \sqrt{p_i p_j} |i^A\rangle\langle j^A| \, \delta_{ij} \tag{2.209}$$

$$= \sum_i p_i |i^A\rangle \langle i^A| \tag{2.210}$$

$$= \rho^A. \tag{2.211}$$

Thus $|AR\rangle$ is a purification of ρ^A.

Notice the close relationship of the Schmidt decomposition to purification: the procedure used to purify a mixed state of system A is to define a pure state whose Schmidt basis for system A is just the basis in which the mixed state is diagonal, with the Schmidt coefficients being the square root of the eigenvalues of the density operator being purified.

In this section we've explained two tools for studying composite quantum systems, the Schmidt decomposition and purifications. These tools will be indispensable to the study of quantum computation and quantum information, especially quantum information, which is the subject of Part III of this book.

Exercise 2.79: Consider a composite system consisting of two qubits. Find the Schmidt decompositions of the states

$$\frac{|00\rangle + |11\rangle}{\sqrt{2}}; \quad \frac{|00\rangle + |01\rangle + |10\rangle + |11\rangle}{2}; \quad \text{and} \quad \frac{|00\rangle + |01\rangle + |10\rangle}{\sqrt{3}}. \tag{2.212}$$

Exercise 2.80: Suppose $|\psi\rangle$ and $|\varphi\rangle$ are two pure states of a composite quantum system with components A and B, with identical Schmidt coefficients. Show that there are unitary transformations U on system A and V on system B such that $|\psi\rangle = (U \otimes V)|\varphi\rangle$.

Exercise 2.81: (Freedom in purifications) Let $|AR_1\rangle$ and $|AR_2\rangle$ be two purifications of a state ρ^A to a composite system AR. Prove that there exists a unitary transformation U_R acting on system R such that $|AR_1\rangle = (I_A \otimes U_R)|AR_2\rangle$.

Exercise 2.82: Suppose $\{p_i, |\psi_i\rangle\}$ is an ensemble of states generating a density matrix $\rho = \sum_i p_i |\psi_i\rangle \langle \psi_i|$ for a quantum system A. Introduce a system R with orthonormal basis $|i\rangle$.

(1) Show that $\sum_i \sqrt{p_i}|\psi_i\rangle|i\rangle$ is a purification of ρ.

(2) Suppose we measure R in the basis $|i\rangle$, obtaining outcome i. With what probability do we obtain the result i, and what is the corresponding state of system A?

(3) Let $|AR\rangle$ be *any* purification of ρ to the system AR. Show that there exists an orthonormal basis $|i\rangle$ in which R can be measured such that the corresponding post-measurement state for system A is $|\psi_i\rangle$ with probability p_i.

2.6 EPR and the Bell inequality

Anybody who is not shocked by quantum theory has not understood it.
– Niels Bohr

I recall that during one walk Einstein suddenly stopped, turned to me and asked whether I really believed that the moon exists only when I look at it. The rest of this walk was devoted to a discussion of what a physicist should mean by the term 'to exist'.
– Abraham Pais

...quantum phenomena do not occur in a Hilbert space, they occur in a laboratory.
– Asher Peres

...what is proved by impossibility proofs is lack of imagination.
– John Bell

This chapter has focused on introducing the tools and mathematics of quantum mechanics. As these techniques are applied in the following chapters of this book, an important recurring theme is the unusual, *non-classical* properties of quantum mechanics. But what exactly is the difference between quantum mechanics and the classical world? Understanding this difference is vital in learning how to perform information processing tasks that are difficult or impossible with classical physics. This section concludes the chapter with a discussion of the Bell inequality, a compelling example of an essential difference between quantum and classical physics.

When we speak of an object such as a person or a book, we assume that the physical properties of that object have an existence independent of observation. That is, measurements merely act to *reveal* such physical properties. For example, a tennis ball has as one of its physical properties its *position*, which we typically measure using light scattered from the surface of the ball. As quantum mechanics was being developed in the 1920s and 1930s a strange point of view arose that differs markedly from the classical view. As described earlier in the chapter, according to quantum mechanics, an unobserved particle does not possess physical properties that exist independent of observation. Rather, such physical properties arise as a consequence of measurements performed upon the system. For example, according to quantum mechanics a qubit does not possess definite properties of 'spin in the z direction, σ_z', and 'spin in the x direction, σ_x', each of which can be revealed by performing the appropriate measurement. Rather, quantum mechanics gives a set of rules which specify, given the state vector, the probabilities for the possible measurement outcomes when the observable σ_z is measured, or when the observable σ_x is measured.

Many physicists rejected this new view of Nature. The most prominent objector was Albert Einstein. In the famous 'EPR paper', co-authored with Nathan Rosen and Boris Podolsky, Einstein proposed a thought experiment which, he believed, demonstrated that quantum mechanics is not a complete theory of Nature.

The essence of the EPR argument is as follows. EPR were interested in what they termed 'elements of reality'. Their belief was that any such element of reality *must* be represented in any complete physical theory. The goal of the argument was to show that quantum mechanics is not a complete physical theory, by identifying elements of reality that were not included in quantum mechanics. The way they attempted to do this was by introducing what they claimed was a *sufficient condition* for a physical property to be an element of reality, namely, that it be possible to predict with certainty the value that property will have, immediately before measurement.

Box 2.7: Anti-correlations in the EPR experiment

Suppose we prepare the two qubit state

$$|\psi\rangle = \frac{|01\rangle - |10\rangle}{\sqrt{2}}, \tag{2.213}$$

a state sometimes known as the *spin singlet* for historical reasons. It is not difficult to show that this state is an entangled state of the two qubit system. Suppose we perform a measurement of spin along the $\vec{v}$ axis on both qubits, that is, we measure the observable $\vec{v} \cdot \vec{\sigma}$ (defined in Equation (2.116) on page 90) on each qubit, getting a result of $+1$ or -1 for each qubit. It turns out that no matter what choice of $\vec{v}$ we make, the results of the two measurements are always opposite to one another. That is, if the measurement on the first qubit yields $+1$, then the measurement on the second qubit will yield -1, and vice versa. It is as though the second qubit knows the result of the measurement on the first, no matter how the first qubit is measured. To see why this is true, suppose $|a\rangle$ and $|b\rangle$ are the eigenstates of $\vec{v} \cdot \vec{\sigma}$. Then there exist complex numbers $\alpha, \beta, \gamma, \delta$ such that

$$|0\rangle = \alpha|a\rangle + \beta|b\rangle \tag{2.214}$$
$$|1\rangle = \gamma|a\rangle + \delta|b\rangle. \tag{2.215}$$

Substituting we obtain

$$\frac{|01\rangle - |10\rangle}{\sqrt{2}} = (\alpha\delta - \beta\gamma)\frac{|ab\rangle - |ba\rangle}{\sqrt{2}}. \tag{2.216}$$

But $\alpha\delta - \beta\gamma$ is the determinant of the unitary matrix $\begin{bmatrix} \alpha & \beta \\ \gamma & \delta \end{bmatrix}$, and thus is equal to a phase factor $e^{i\theta}$ for some real θ. Thus

$$\frac{|01\rangle - |10\rangle}{\sqrt{2}} = \frac{|ab\rangle - |ba\rangle}{\sqrt{2}}, \tag{2.217}$$

up to an unobservable global phase factor. As a result, if a measurement of $\vec{v} \cdot \vec{\sigma}$ is performed on both qubits, then we can see that a result of $+1$ (-1) on the first qubit implies a result of -1 $(+1)$ on the second qubit.

Consider, for example, an entangled pair of qubits belonging to Alice and Bob, respectively:

$$\frac{|01\rangle - |10\rangle}{\sqrt{2}}. \tag{2.218}$$

Suppose Alice and Bob are a long way away from one another. Alice performs a measurement of spin along the $\vec{v}$ axis, that is, she measures the observable $\vec{v} \cdot \vec{\sigma}$ (defined in Equation (2.116) on page 90). Suppose Alice receives the result $+1$. Then a simple quantum mechanical calculation, given in Box 2.7, shows that she can predict with certainty that Bob will measure -1 on his qubit if he also measures spin along the $\vec{v}$ axis. Similarly, if Alice measured -1, then she can predict with certainty that Bob will measure $+1$ on his qubit. Because it is always possible for Alice to predict the value of the measurement result recorded when Bob's qubit is measured in the $\vec{v}$ direction, that physical property must correspond to an element of reality, by the EPR criterion, and should be represented in any complete physical theory. However, standard quantum mechanics, as we have presented it, merely tells one how to calculate the probabilities of the respective measurement outcomes if $\vec{v} \cdot \vec{\sigma}$ is measured. Standard quantum mechanics certainly does not include any fundamental element intended to represent the value of $\vec{v} \cdot \vec{\sigma}$, for all unit vectors $\vec{v}$.

The goal of EPR was to show that quantum mechanics is incomplete, by demonstrating that quantum mechanics lacked some essential 'element of reality', by their criterion. They hoped to force a return to a more classical view of the world, one in which systems could be ascribed properties which existed independently of measurements performed on those systems. Unfortunately for EPR, most physicists did not accept the above reasoning as convincing. The attempt to impose on Nature *by fiat* properties which she must obey seems a most peculiar way of studying her laws.

Indeed, Nature has had the last laugh on EPR. Nearly thirty years after the EPR paper was published, an *experimental test* was proposed that could be used to check whether or not the picture of the world which EPR were hoping to force a return to is valid or not. It turns out that Nature *experimentally invalidates* that point of view, while agreeing with quantum mechanics.

The key to this experimental invalidation is a result known as *Bell's inequality*. Bell's inequality is *not* a result about quantum mechanics, so the first thing we need to do is momentarily *forget* all our knowledge of quantum mechanics. To obtain Bell's inequality, we're going to do a thought experiment, which we will analyze using our common sense notions of how the world works – the sort of notions Einstein and his collaborators thought Nature ought to obey. After we have done the common sense analysis, we will perform a quantum mechanical analysis which we can show *is not consistent with the common sense analysis*. Nature can then be asked, by means of a real experiment, to decide between our common sense notions of how the world works, and quantum mechanics.

Imagine we perform the following experiment, illustrated in Figure 2.4. Charlie prepares two particles. It doesn't matter how he prepares the particles, just that he is capable of repeating the experimental procedure which he uses. Once he has performed the preparation, he sends one particle to Alice, and the second particle to Bob.

Once Alice receives her particle, she performs a measurement on it. Imagine that she has available two different measurement apparatuses, so she could choose to do one of two different measurements. These measurements are of physical properties which we shall label P_Q and P_R, respectively. Alice doesn't know in advance which measurement she will choose to perform. Rather, when she receives the particle she flips a coin or uses some other random method to decide which measurement to perform. We suppose for simplicity that the measurements can each have one of two outcomes, $+1$ or -1. Suppose Alice's particle has a value Q for the property P_Q. Q is assumed to be an *objective property* of Alice's particle, which is merely revealed by the measurement, much as we imagine the position of a tennis ball to be revealed by the particles of light being scattered off it. Similarly, let R denote the value revealed by a measurement of the property P_R.

Similarly, suppose that Bob is capable of measuring one of two properties, P_S or P_T, once again revealing an objectively existing value S or T for the property, each taking value $+1$ or -1. Bob does not decide beforehand which property he will measure, but waits until he has received the particle and then chooses randomly. The timing of the experiment is arranged so that Alice and Bob do their measurements *at the same time* (or, to use the more precise language of relativity, in a causally disconnected manner). Therefore, the measurement which Alice performs cannot disturb the result of Bob's measurement (or vice versa), since physical influences cannot propagate faster than light.

Figure 2.4. Schematic experimental setup for the Bell inequalities. Alice can choose to measure either Q or R, and Bob chooses to measure either S or T. They perform their measurements simultaneously. Alice and Bob are assumed to be far enough apart that performing a measurement on one system can not have any effect on the result of measurements on the other.

We are going to do some simple algebra with the quantity $QS + RS + RT - QT$. Notice that

$$QS + RS + RT - QT = (Q + R)S + (R - Q)T. \tag{2.219}$$

Because $R, Q = \pm 1$ it follows that either $(Q + R)S = 0$ or $(R - Q)T = 0$. In either case, it is easy to see from (2.219) that $QS + RS + RT - QT = \pm 2$. Suppose next that $p(q, r, s, t)$ is the probability that, before the measurements are performed, the system is in a state where $Q = q, R = r, S = s$, and $T = t$. These probabilities may depend on how Charlie performs his preparation, and on experimental noise. Letting $\mathbf{E}(\cdot)$ denote the mean value of a quantity, we have

$$\mathbf{E}(QS + RS + RT - QT) = \sum_{qrst} p(q, r, s, t)(qs + rs + rt - qt) \tag{2.220}$$

$$\leq \sum_{qrst} p(q, r, s, t) \times 2 \tag{2.221}$$

$$= 2. \tag{2.222}$$

Also,

$$\mathbf{E}(QS + RS + RT - QT) = \sum_{qrst} p(q, r, s, t)qs + \sum_{qrst} p(q, r, s, t)rs$$

$$+ \sum_{qrst} p(q, r, s, t)rt - \sum_{qrst} p(q, r, s, t)qt \tag{2.223}$$

$$= \mathbf{E}(QS) + \mathbf{E}(RS) + \mathbf{E}(RT) - \mathbf{E}(QT). \tag{2.224}$$

Comparing (2.222) and (2.224) we obtain the *Bell inequality*,

$$\mathbf{E}(QS) + \mathbf{E}(RS) + \mathbf{E}(RT) - \mathbf{E}(QT) \leq 2. \tag{2.225}$$

This result is also often known as the *CHSH inequality* after the initials of its four discoverers. It is part of a larger set of inequalities known generically as Bell inequalities, since the first was found by John Bell.

By repeating the experiment many times, Alice and Bob can determine each quantity on the left hand side of the Bell inequality. For example, after finishing a set of experiments, Alice and Bob get together to analyze their data. They look at all the experiments where Alice measured P_Q and Bob measured P_S. By multiplying the results of their experiments together, they get a sample of values for QS. By averaging over this sample, they can estimate $\mathbf{E}(QS)$ to an accuracy only limited by the number of experiments which they perform. Similarly, they can estimate all the other quantities on the left hand side of the Bell inequality, and thus check to see whether it is obeyed in a real experiment.

It's time to put some quantum mechanics back in the picture. Imagine we perform the following quantum mechanical experiment. Charlie prepares a quantum system of two qubits in the state

$$|\psi\rangle = \frac{|01\rangle - |10\rangle}{\sqrt{2}}. \tag{2.226}$$

He passes the first qubit to Alice, and the second qubit to Bob. They perform measurements of the following observables:

$$Q = Z_1 \qquad S = \frac{-Z_2 - X_2}{\sqrt{2}} \tag{2.227}$$

$$R = X_1 \qquad T = \frac{Z_2 - X_2}{\sqrt{2}}. \tag{2.228}$$

Simple calculations show that the average values for these observables, written in the quantum mechanical $\langle \cdot \rangle$ notation, are:

$$\langle QS \rangle = \frac{1}{\sqrt{2}}; \; \langle RS \rangle = \frac{1}{\sqrt{2}}; \; \langle RT \rangle = \frac{1}{\sqrt{2}}; \; \langle QT \rangle = -\frac{1}{\sqrt{2}}. \tag{2.229}$$

Thus,

$$\langle QS \rangle + \langle RS \rangle + \langle RT \rangle - \langle QT \rangle = 2\sqrt{2}. \tag{2.230}$$

Hold on! We learned back in (2.225) that the average value of QS plus the average value of RS plus the average value of RT minus the average value of QT can never exceed two. Yet here, quantum mechanics predicts that this sum of averages yields $2\sqrt{2}$!

Fortunately, we can ask Nature to resolve the apparent paradox for us. Clever experiments using photons – particles of light – have been done to check the prediction (2.230) of quantum mechanics versus the Bell inequality (2.225) which we were led to by our common sense reasoning. The details of the experiments are outside the scope of the book, but the results were resoundingly in favor of the quantum mechanical prediction. The Bell inequality (2.225) is *not* obeyed by Nature.

What does this mean? It means that one or more of the assumptions that went into the derivation of the Bell inequality must be incorrect. Vast tomes have been written analyzing the various forms in which this type of argument can be made, and analyzing the subtly different assumptions which must be made to reach Bell-like inequalities. Here we merely summarize the main points.

There are two assumptions made in the proof of (2.225) which are questionable:

(1) The assumption that the physical properties P_Q, P_R, P_S, P_T have definite values Q, R, S, T which exist independent of observation. This is sometimes known as the assumption of *realism*.

(2) The assumption that Alice performing her measurement does not influence the result of Bob's measurement. This is sometimes known as the assumption of *locality*.

These two assumptions together are known as the assumptions of *local realism*. They are certainly intuitively plausible assumptions about how the world works, and they fit our everyday experience. Yet the Bell inequalities show that at least one of these assumptions is not correct.

What can we learn from Bell's inequality? For physicists, the most important lesson is that their deeply held commonsense intuitions about how the world works are wrong. The world is *not* locally realistic. Most physicists take the point of view that it is the assumption of realism which needs to be dropped from our worldview in quantum mechanics, although others have argued that the assumption of locality should be dropped instead. Regardless, Bell's inequality together with substantial experimental evidence now points to the conclusion that either or both of locality and realism must be dropped from our view of the world if we are to develop a good intuitive understanding of quantum mechanics.

What lessons can the fields of quantum computation and quantum information learn from Bell's inequality? Historically the most useful lesson has perhaps also been the most vague: there is something profoundly 'up' with entangled states like the EPR state. A lot of mileage in quantum computation and, especially, quantum information, has come from asking the simple question: 'what would some entanglement buy me in this problem?' As we saw in teleportation and superdense coding, and as we will see repeatedly later in the book, by throwing some entanglement into a problem we open up a new world of possibilities unimaginable with classical information. The bigger picture is that Bell's inequality teaches us that entanglement is a fundamentally new resource in the world that goes essentially *beyond* classical resources; iron to the classical world's bronze age. A major task of quantum computation and quantum information is to exploit this new resource to do information processing tasks impossible or much more difficult with classical resources.

Problem 2.1: (Functions of the Pauli matrices)   Let $f(\cdot)$ be any function from complex numbers to complex numbers. Let $\vec{n}$ be a normalized vector in three dimensions, and let θ be real. Show that

$$f(\theta\vec{n}\cdot\vec{\sigma}) = \frac{f(\theta) + f(-\theta)}{2}I + \frac{f(\theta) - f(-\theta)}{2}\vec{n}\cdot\vec{\sigma}. \qquad (2.231)$$

Problem 2.2: (Properties of the Schmidt number)   Suppose $|\psi\rangle$ is a pure state of a composite system with components A and B.

(1) Prove that the Schmidt number of $|\psi\rangle$ is equal to the rank of the reduced density matrix $\rho_A \equiv \mathrm{tr}_B(|\psi\rangle\langle\psi|)$. (Note that the rank of a Hermitian operator is equal to the dimension of its support.)

(2) Suppose $|\psi\rangle = \sum_j |\alpha_j\rangle|\beta_j\rangle$ is a representation for $|\psi\rangle$, where $|\alpha_j\rangle$ and $|\beta_j\rangle$ are (un-normalized) states for systems A and B, respectively. Prove that the

number of terms in such a decomposition is greater than or equal to the Schmidt number of $|\psi\rangle$, $\text{Sch}(\psi)$.

(3) Suppose $|\psi\rangle = \alpha|\varphi\rangle + \beta|\gamma\rangle$. Prove that

$$\text{Sch}(\psi) \geq |\text{Sch}(\varphi) - \text{Sch}(\gamma)| \, . \tag{2.232}$$

Problem 2.3: (Tsirelson's inequality) Suppose $Q = \vec{q} \cdot \vec{\sigma}, R = \vec{r} \cdot \vec{\sigma}, S = \vec{s} \cdot \vec{\sigma}, T = \vec{t} \cdot \vec{\sigma}$, where $\vec{q}, \vec{r}, \vec{s}$ and $\vec{t}$ are real unit vectors in three dimensions. Show that

$$(Q \otimes S + R \otimes S + R \otimes T - Q \otimes T)^2 = 4I + [Q, R] \otimes [S, T]. \tag{2.233}$$

Use this result to prove that

$$\langle Q \otimes S \rangle + \langle R \otimes S \rangle + \langle R \otimes T \rangle - \langle Q \otimes T \rangle \leq 2\sqrt{2}, \tag{2.234}$$

so the violation of the Bell inequality found in Equation (2.230) is the maximum possible in quantum mechanics.

History and further reading

There are an enormous number of books on linear algebra at levels ranging from High School through to Graduate School. Perhaps our favorites are the two volume set by Horn and Johnson[HJ85, HJ91], which cover an extensive range of topics in an accessible manner. Other useful references include Marcus and Minc[MM92], and Bhatia[Bha97]. Good introductions to linear algebra include Halmos[Hal58], Perlis[Per52], and Strang[Str76].

There are many excellent books on quantum mechanics. Unfortunately, most of these books focus on topics of tangential interest to quantum information and computation. Perhaps the most relevant in the existing literature is Peres' superb book[Per93]. Beside an extremely clear exposition of elementary quantum mechanics, Peres gives an extensive discussion of the Bell inequalities and related results. Good introductory level texts include Sakurai's book[Sak95], Volume III of the superb series by Feynman, Leighton, and Sands[FLS65a], and the two volume work by Cohen-Tannoudji, Diu and Laloë[CTDL77a, CTDL77b]. All three of these works are somewhat closer in spirit to quantum computation and quantum information than are most other quantum mechanics texts, although the great bulk of each is still taken up by applications far removed from quantum computation and quantum information. As a result, none of these texts need be read in detail by someone interested in learning about quantum computation and quantum information. However, any one of these texts may prove handy as a reference, especially when reading articles by physicists. References for the history of quantum mechanics may be found at the end of Chapter 1.

Many texts on quantum mechanics deal only with projective measurements. For applications to quantum computing and quantum information it is more convenient – and, we believe, easier for novices – to start with the general description of measurements, of which projective measurements can be regarded as a special case. Of course, ultimately, as we have shown, the two approaches are equivalent. The theory of generalized measurements which we have employed was developed between the 1940s and 1970s. Much of the history can be distilled from the book of Kraus[Kra83]. Interesting discussion of quantum measurements may be found in Section 2.2 of Gardiner[Gar91], and in the book by Braginsky and Khahili[BK92]. The POVM measurement for distinguishing non–orthogonal states described in Section 2.2.6 is due to Peres[Per88]. The extension described in Exercise 2.64 appeared in Duan and Guo[DG98].

Superdense coding was invented by Bennett and Wiesner[BW92]. An experiment implementing a variant of superdense coding using entangled photon pairs was performed by Mattle, Weinfurter, Kwiat, and Zeilinger[MWKZ96].

The density operator formalism was introduced independently by Landau[Lan27] and by von Neumann[von27]. The unitary freedom in the ensemble for density matrices, Theorem 2.6, was first pointed out by Schrödinger[Sch36], and was later rediscovered and extended by Jaynes[Jay57] and by Hughston, Jozsa and Wootters[HJW93]. The result of Exercise 2.73 is from the paper by Jaynes, and the results of Exercises 2.81 and 2.82 appear in the paper by Hughston, Jozsa and Wootters. The class of probability distributions which may appear in a density matrix decomposition for a given density matrix has been studied by Uhlmann[Uhl70] and by Nielsen[Nie99b]. Schmidt's eponymous decomposition appeared in[Sch06]. The result of Exercise 2.77 was noted by Peres[Per95].

The EPR thought experiment is due to Einstein, Podolsky and Rosen[EPR35], and was recast in essentially the form we have given here by Bohm[Boh51]. It is sometimes misleadingly referred to as the EPR 'paradox'. The Bell inequality is named in honour of Bell[Bel64], who first derived inequalities of this type. The form we have presented is due to Clauser, Horne, Shimony, and Holt[CHSH69], and is often known as the CHSH inequality. This inequality was derived independently by Bell, who did not publish the result.

Part 3 of Problem 2.2 is due to Thapliyal (private communication). Tsirelson's inequality is due to Tsirelson[Tsi80].

3 Introduction to computer science

In natural science, Nature has given us a world and we're just to discover its laws. In computers, we can stuff laws into it and create a world.
– Alan Kay

Our field is still in its embryonic stage. It's great that we haven't been around for 2000 years. We are still at a stage where very, very important results occur in front of our eyes.
– Michael Rabin, on computer science

Algorithms are the key concept of computer science. An algorithm is a precise recipe for performing some task, such as the elementary algorithm for adding two numbers which we all learn as children. This chapter outlines the modern theory of algorithms developed by computer science. Our fundamental model for algorithms will be the *Turing machine*. This is an idealized computer, rather like a modern personal computer, but with a simpler set of basic instructions, and an idealized unbounded memory. The apparent simplicity of Turing machines is misleading; they are very powerful devices. We will see that they can be used to execute any algorithm whatsoever, even one running on an apparently much more powerful computer.

The fundamental question we are trying to address in the study of algorithms is: what resources are required to perform a given computational task? This question splits up naturally into two parts. First, we'd like to understand what computational tasks are possible, preferably by giving explicit algorithms for solving specific problems. For example, we have many excellent examples of algorithms that can quickly sort a list of numbers into ascending order. The second aspect of this question is to demonstrate *limitations* on what computational tasks may be accomplished. For example, lower bounds can be given for the number of operations that must be performed by any algorithm which sorts a list of numbers into ascending order. Ideally, these two tasks – the finding of algorithms for solving computational problems, and proving limitations on our ability to solve computational problems – would dovetail perfectly. In practice, a significant gap often exists between the best techniques known for solving a computational problem, and the most stringent limitations known on the solution. The purpose of this chapter is to give a broad overview of the tools which have been developed to aid in the analysis of computational problems, and in the construction and analysis of algorithms to solve such problems.

Why should a person interested in *quantum* computation and *quantum* information spend time investigating *classical* computer science? There are three good reasons for this effort. First, classical computer science provides a vast body of concepts and techniques which may be reused to great effect in quantum computation and quantum information. Many of the triumphs of quantum computation and quantum information have come by combining existing ideas from computer science with novel ideas from quantum mechanics. For example, some of the fast algorithms for quantum computers are based upon the Fourier transform, a powerful tool utilized by many classical algorithms. Once it was realized that quantum computers could perform a type of Fourier transform much more quickly than classical computers this enabled the development of many important quantum algorithms.

Second, computer scientists have expended great effort understanding what resources are required to perform a given computational task on a classical computer. These results can be used as the basis for a comparison with quantum computation and quantum information. For example, much attention has been focused on the problem of finding the prime factors of a given number. On a classical computer this problem is believed to have no 'efficient' solution, where 'efficient' has a meaning we'll explain later in the chapter. What is interesting is that an efficient solution to this problem *is* known for quantum computers. The lesson is that, for this task of finding prime factors, there appears to be a *gap* between what is possible on a classical computer and what is possible on a quantum computer. This is both intrinsically interesting, and also interesting in the broader sense that it suggests such a gap may exist for a wider class of computational problems than merely the finding of prime factors. By studying this specific problem further, it may be possible to discern features of the problem which make it more tractable on a quantum computer than on a classical computer, and then act on these insights to find interesting quantum algorithms for the solution of other problems.

Third, and most important, there is learning to *think* like a computer scientist. Computer scientists think in a rather different style than does a physicist or other natural scientist. Anybody wanting a deep understanding of quantum computation and quantum information must learn to think like a computer scientist at least some of the time; they must instinctively know what problems, what techniques, and most importantly what problems are of greatest interest to a computer scientist.

The structure of this chapter is as follows. In Section 3.1 we introduce two models for computation: the Turing machine model, and the circuit model. The Turing machine model will be used as our fundamental model for computation. In practice, however, we mostly make use of the circuit model of computation, and it is this model which is most useful in the study of quantum computation. With our models for computation in hand, the remainder of the chapter discusses resource requirements for computation. Section 3.2 begins by overviewing the computational tasks we are interested in as well as discusing some associated resource questions. It continues with a broad look at the key concepts of *computational complexity*, a field which examines the time and space requirements necessary to solve particular computational problems, and provides a broad classification of problems based upon their difficulty of solution. Finally, the section concludes with an examination of the energy resources required to perform computations. Surprisingly, it turns out that the energy required to perform a computation can be made vanishingly small, provided one can make the computation reversible. We explain how to construct reversible computers, and explain some of the reasons they are important both for computer science and for quantum computation and quantum information. Section 3.3 concludes the chapter with a broad look at the entire field of computer science, focusing on issues of particular relevance to quantum computation and quantum information.

3.1 Models for computation

...algorithms are concepts that have existence apart from any programming language.
– Donald Knuth

What does it mean to have an *algorithm* for performing some task? As children we all learn a procedure which enables us to add together any two numbers, no matter how large those numbers are. This is an example of an algorithm. Finding a mathematically precise formulation of the concept of an algorithm is the goal of this section.

Historically, the notion of an algorithm goes back centuries; undergraduates learn Euclid's two thousand year old algorithm for finding the greatest common divisor of two positive integers. However, it wasn't until the 1930s that the fundamental notions of the modern theory of algorithms, and thus of computation, were introduced, by Alonzo Church, Alan Turing, and other pioneers of the computer era. This work arose in response to a profound challenge laid down by the great mathematician David Hilbert in the early part of the twentieth century. Hilbert asked whether or not there existed some algorithm which could be used, in principle, to solve all the problems of mathematics. Hilbert expected that the answer to this question, sometimes known as the *entscheidungsproblem*, would be yes.

Amazingly, the answer to Hilbert's challenge turned out to be no: there is no algorithm to solve all mathematical problems. To prove this, Church and Turing had to solve the deep problem of capturing in a mathematical definition what we mean when we use the intuitive concept of an algorithm. In so doing, they laid the foundations for the modern theory of algorithms, and consequently for the modern theory of computer science.

In this chapter, we use two ostensibly different approaches to the theory of computation. The first approach is that proposed by Turing. Turing defined a class of machines, now known as *Turing machines*, in order to capture the notion of an algorithm to perform a computational task. In Section 3.1.1, we describe Turing machines, and then discuss some of the simpler variants of the Turing machine model. The second approach is via the *circuit model* of computation, an approach that is especially useful as preparation for our later study of quantum computers. The circuit model is described in Section 3.1.2. Although these models of computation appear different on the surface, it turns out that they are equivalent. Why introduce more than one model of computation, you may ask? We do so because different models of computation may yield different insights into the solution of specific problems. Two (or more) ways of thinking about a concept are better than one.

3.1.1 Turing machines

The basic elements of a Turing machine are illustrated in Figure 3.1. A Turing machine contains four main elements: (a) a *program*, rather like an ordinary computer; (b) a *finite state control*, which acts like a stripped-down microprocessor, co-ordinating the other operations of the machine; (c) a *tape*, which acts like a computer memory; and (d) a *read-write tape-head*, which points to the position on the tape which is currently readable or writable. We now describe each of these four elements in more detail.

The *finite state control* for a Turing machine consists of a finite set of *internal states*,

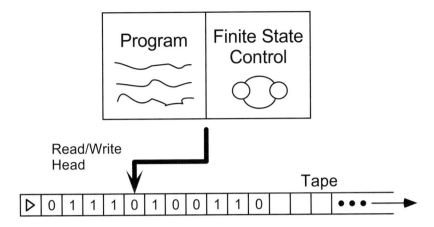

Figure 3.1. Main elements of a Turing machine. In the text, blanks on the tape are denoted by a 'b'. Note the ▷ marking the left hand end of the tape.

$q_1, \ldots, q_m$. The number m is allowed to be varied; it turns out that for m sufficiently large this does not affect the power of the machine in any essential way, so without loss of generality we may suppose that m is some fixed constant. The best way to think of the finite state control is as a sort of microprocessor, co-ordinating the Turing machine's operation. It provides temporary storage off-tape, and a central place where all processing for the machine may be done. In addition to the states $q_1, \ldots, q_m$, there are also two special internal states, labelled q_s and q_h. We call these the *starting state* and the *halting state*, respectively. The idea is that at the beginning of the computation, the Turing machine is in the starting state q_s. The execution of the computation causes the Turing machine's internal state to change. If the computation ever finishes, the Turing machine ends up in the state q_h to indicate that the machine has completed its operation.

The Turing machine *tape* is a one-dimensional object, which stretches off to infinity in one direction. The tape consists of an infinite sequence of *tape squares*. We number the tape squares $0, 1, 2, 3, \ldots$. The tape squares each contain one symbol drawn from some *alphabet*, Γ, which contains a finite number of distinct symbols. For now, it will be convenient to assume that the alphabet contains four symbols, which we denote by $0, 1, b$ (the 'blank' symbol), and $\triangleright$, to mark the left hand edge of the tape. Initially, the tape contains a $\triangleright$ at the left hand end, a finite number of 0s and 1s, and the rest of the tape contains blanks. The *read-write tape-head* identifies a single square on the Turing machine tape as the square that is currently being accessed by the machine.

Summarizing, the machine starts its operation with the finite state control in the state q_s, and with the read-write head at the leftmost tape square, the square numbered 0. The computation then proceeds in a step by step manner according to the *program*, to be defined below. If the current state is q_h, then the computation has halted, and the *output* of the computation is the current (non-blank) contents of the tape.

A *program* for a Turing machine is a finite ordered list of *program lines* of the form $\langle q, x, q', x', s \rangle$. The first item in the program line, q, is a state from the set of internal states of the machine. The second item, x, is taken from the alphabet of symbols which may appear on the tape, Γ. The way the program works is that on each machine cycle, the Turing machine looks through the list of program lines in order, searching for a line $\langle q, x, \cdot, \cdot, \cdot \rangle$, such that the current internal state of the machine is q, and the symbol being read on the tape is x. If it doesn't find such a program line, the internal state of the machine is changed to q_h, and the machine halts operation. If such a line is found, then that program line is *executed*. Execution of a program line involves the following steps: the internal state of the machine is changed to q'; the symbol x on the tape is overwritten by the symbol x', and the tape-head moves left, right, or stands still, depending on whether s is $-1, +1$, or 0, respectively. The only exception to this rule is if the tape-head is at the leftmost tape square, and $s = -1$, in which case the tape-head stays put.

Now that we know what a Turing machine is, let's see how it may be used to compute a simple function. Consider the following example of a Turing machine. The machine starts with a binary number, x, on the tape, followed by blanks. The machine has three internal states, q_1, q_2, and q_3, in addition to the starting state q_s and halting state q_h. The program contains the following program lines (the numbers on the left hand side are for convenience in referring to the program lines in later discussion, and do not form part of the program):

$$1: \langle q_s, \triangleright, q_1, \triangleright, +1 \rangle$$
$$2: \langle q_1, 0, q_1, b, +1 \rangle$$
$$3: \langle q_1, 1, q_1, b, +1 \rangle$$
$$4: \langle q_1, b, q_2, b, -1 \rangle$$
$$5: \langle q_2, b, q_2, b, -1 \rangle$$
$$6: \langle q_2, \triangleright, q_3, \triangleright, +1 \rangle$$
$$7: \langle q_3, b, q_h, 1, 0 \rangle.$$

What function does this program compute? Initially the machine is in the state q_s and at the left-most tape position so line 1, $\langle q_s, \triangleright, q_1, \triangleright, +1 \rangle$, is executed, which causes the tape-head to move right without changing what is written on the tape, but changing the internal state of the machine to q_1. The next three lines of the program ensure that while the machine is in the state q_1 the tape-head will continue moving right while it reads either 0s (line 2) or 1s (line 3) on the tape, over-writing the tape contents with blanks as it goes and remaining in the state q_1, until it reaches a tape square that is already blank, at which point the tape-head is moved one position to the left, and the internal state is changed to q_2 (line 4). Line 5 then ensures that the tape-head keeps moving left while blanks are being read by the tape-head, without changing the contents of the tape. This keeps up until the tape-head returns to its starting point, at which point it reads a $\triangleright$ on the tape, changes the internal state to q_3, and moves one step to the right (line 6). Line 7 completes the program, simply printing the number 1 onto the tape, and then halting.

The preceding analysis shows that this program computes the constant function $f(x) = 1$. That is, regardless of what number is input on the tape the number 1 is output. More generally, a Turing machine can be thought of as computing functions from the non-negative integers to the non-negative integers; the initial state of the tape is used to represent the input to the function, and the final state of the tape is used to represent the output of the function.

It seems as though we have gone to a very great deal of trouble to compute this simple function using our Turing machines. Is it possible to build up more complicated functions using Turing machines? For example, could we construct a machine such that when two numbers, x and y, are input on the tape with a blank to demarcate them, it will output the sum $x + y$ on the tape? More generally, what class of functions is it possible to compute using a Turing machine?

It turns out that the Turing machine model of computation can be used to compute an enormous variety of functions. For example, it can be used to do all the basic arithmetical operations, to search through text represented as strings of bits on the tape, and many other interesting operations. Surprisingly, it turns out that a Turing machine can be used to simulate all the operations performed on a modern computer! Indeed, according to a thesis put forward independently by Church and by Turing, the Turing machine model of computation *completely captures* the notion of computing a function using an algorithm. This is known as the *Church–Turing thesis*:

> The class of functions computable by a Turing machine corresponds exactly to the class of functions which we would naturally regard as being computable by an algorithm.

The Church–Turing thesis asserts an equivalence between a rigorous mathematical concept – function computable by a Turing machine – and the intuitive concept of what it means for a function to be computable by an algorithm. The thesis derives its importance from the fact that it makes the study of real-world algorithms, prior to 1936 a rather vague concept, amenable to rigorous mathematical study. To understand the significance of this point it may be helpful to consider the definition of a *continuous function* from real analysis. Every child can tell you what it means for a line to be continuous on a piece of paper, but it is far from obvious how to capture that intuition in a rigorous definition. Mathematicians in the nineteenth century spent a great deal of time arguing about the merits of various definitions of continuity before the modern definition of continuity came to be accepted. When making fundamental definitions like that of continuity or of computability it is important that good definitions be chosen, ensuring that one's intuitive notions closely match the precise mathematical definition. From this point of view the Church–Turing thesis is simply the assertion that the Turing machine model of computation provides a good foundation for computer science, capturing the intuitive notion of an algorithm in a rigorous definition.

A priori it is not obvious that every function which we would intuitively regard as computable by an algorithm can be computed using a Turing machine. Church, Turing and many other people have spent a great deal of time gathering evidence for the Church–Turing thesis, and in sixty years no evidence to the contrary has been found. Nevertheless, it is possible that in the future we will discover in Nature a process which computes a function not computable on a Turing machine. It would be wonderful if that ever happened, because we could then harness that process to help us perform new computations which could not be performed before. Of course, we would also need to overhaul the definition of computability, and with it, computer science.

Exercise 3.1: (Non-computable processes in Nature) How might we recognize that a process in Nature computes a function not computable by a Turing machine?

Exercise 3.2: (Turing numbers) Show that single-tape Turing machines can each be given a number from the list $1, 2, 3, \ldots$ in such a way that the number uniquely specifies the corresponding machine. We call this number the *Turing number* of the corresponding Turing machine. (*Hint:* Every positive integer has a unique prime factorization $p_1^{a_1} p_2^{a_2} \ldots p_k^{a_k}$, where p_i are distinct prime numbers, and $a_1, \ldots, a_k$ are non-negative integers.)

In later chapters, we will see that quantum computers also obey the Church–Turing thesis. That is, quantum computers can compute the same class of functions as is computable by a Turing machine. The difference between quantum computers and Turing machines turns out to lie in the *efficiency* with which the computation of the function may be performed – there are functions which can be computed much more efficiently on a quantum computer than is believed to be possible with a classical computing device such as a Turing machine.

Demonstrating in complete detail that the Turing machine model of computation can be used to build up all the usual concepts used in computer programming languages is beyond the scope of this book (see 'History and further reading' at the end of the chapter for more information). When specifying algorithms, instead of explicitly specifying the Turing machine used to compute the algorithm, we shall usually use a much higher level *pseudocode*, trusting in the Church–Turing thesis that this pseudocode can be translated into the Turing machine model of computation. We won't give any sort of rigorous definition for pseudocode. Think of it as a slightly more formal version of English or, if you like, a sloppy version of a high-level programming language such as C++ or BASIC. Pseudocode provides a convenient way of expressing algorithms, without going into the extreme level of detail required by a Turing machine. An example use of pseudocode may be found in Box 3.2 on page 130; it is also used later in the book to describe quantum algorithms.

There are many variants on the basic Turing machine model. We might imagine Turing machines with different kinds of tapes. For example, one could consider two-way infinite tapes, or perhaps computation with tapes of more than one dimension. So far as is presently known, it is not possible to change any aspect of the Turing model in a way that is physically reasonable, and which manages to extend the class of functions computable by the model.

As an example consider a Turing machine equipped with multiple tapes. For simplicity we consider the two-tape case, as the generalization to more than two tapes is clear from this example. Like the basic Turing machine, a two-tape Turing machine has a finite number of internal states $q_1, \ldots, q_m$, a start state q_s, and a halt state q_h. It has two tapes, each of which contain symbols from some finite alphabet of symbols, Γ. As before we find it convenient to assume that the alphabet contains four symbols, $0, 1, b$ and $\triangleright$, where $\triangleright$ marks the left hand edge of each tape. The machine has two tape-heads, one for each tape. The main difference between the two-tape Turing machine and the basic Turing machine is in the program. Program lines are of the form $\langle q, x_1, x_2, q', x_1', x_2', s_1, s_2 \rangle$, meaning that if the internal state of the machine is q, tape one is reading x_1 at its current position, and tape two is reading x_2 at its current position, then the internal state of the machine should be changed to q', x_1 overwritten with x_1', x_2 overwritten with x_2', and the tape-heads for tape one and tape two moved according to whether s_1 or s_2 are equal to $+1$, -1 or 0, respectively.

In what sense are the basic Turing machine and the two-tape Turing machine equivalent models of computation? They are equivalent in the sense that each computational model is able to *simulate* the other. Suppose we have a two-tape Turing machine which takes as input a bit string x on the first tape and blanks on the remainder of both tapes, except the endpoint marker ▷. This machine computes a function $f(x)$, where $f(x)$ is defined to be the contents of the first tape after the Turing machine has halted. Rather remarkably, it turns out that given a two-tape Turing machine to compute f, there exists an equivalent single-tape Turing machine that is also able to compute f. We won't explain how to do this explicitly, but the basic idea is that the single-tape Turing machine *simulates* the two-tape Turing machine, using its single tape to store the contents of both tapes of the two-tape Turing machine. There is some computational overhead required to do this simulation, but the important point is that in principle it can always be done. In fact, there exists a Universal Turing machine (see Box 3.1) which can simulate any other Turing machine!

Another interesting variant of the Turing machine model is to introduce randomness into the model. For example, imagine that the Turing machine can execute a program line whose effect is the following: if the internal state is q and the tape-head reads x, then flip an unbiased coin. If the coin lands heads, change the internal state to q_{i_H}, and if it lands tails, change the internal state to q_{i_T}, where q_{i_H} and q_{i_T} are two internal states of the Turing machine. Such a program line can be represented as $\langle q, x, q_{i_H}, q_{i_T} \rangle$. However, even this variant doesn't change the essential power of the Turing machine model of computation. It is not difficult to see that we can simulate the effect of the above algorithm on a deterministic Turing machine by explicitly 'searching out' all the possible computational paths corresponding to different values of the coin tosses. Of course, this deterministic simulation may be far less efficient than the random model, but the key point for the present discussion is that the class of functions computable is not changed by introducing randomness into the underlying model.

Exercise 3.3: (Turing machine to reverse a bit string)   Describe a Turing machine which takes a binary number x as input, and outputs the bits of x in reverse order. (*Hint*: In this exercise and the next it may help to use a multi-tape Turing machine and/or symbols other than ▷, 0, 1 and the blank.)

Exercise 3.4: (Turing machine to add modulo 2)   Describe a Turing machine to add two binary numbers x and y modulo 2. The numbers are input on the Turing machine tape in binary, in the form x, followed by a single blank, followed by a y. If one number is not as long as the other then you may assume that it has been padded with leading 0s to make the two numbers the same length.

Let us return to Hilbert's entscheidungsproblem, the original inspiration for the founders of computer science. Is there an algorithm to decide all the problems of mathematics? The answer to this question was shown by Church and Turing to be no. In Box 3.2, we explain Turing's proof of this remarkable fact. This phenomenon of *undecidability* is now known to extend far beyond the examples which Church and Turing constructed. For example, it is known that the problem of deciding whether two topological spaces are equivalent ('homeomorphic') is undecidable. There are simple problems related to the behavior of dynamical systems which are undecidable, as you will show in Problem 3.4. References for these and other examples are given in the end of chapter 'History and further reading'.

Besides its intrinsic interest, undecidability foreshadows a topic of great concern in computer science, and also to quantum computation and quantum information: the dis-

---

Box 3.1: The Universal Turing Machine

We've described Turing machines as containing three elements which may vary from machine to machine – the initial configuration of the tape, the internal states of the finite state control, and the program for the machine. A clever idea known as the *Universal Turing Machine* (UTM) allows us to fix the program and finite state control once and for all, leaving the initial contents of the tape as the only part of the machine which needs to be varied.

The Universal Turing Machine (see the figure below) has the following property. Let M be any Turing machine, and let T_M be the Turing number associated to machine M. Then on input of the binary representation for T_M followed by a blank, followed by any string of symbols x on the remainder of the tape, the Universal Turing Machine gives as output whatever machine M would have on input of x. Thus, the Universal Turing Machine is capable of simulating any other Turing machine!

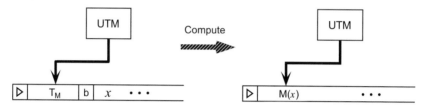

The Universal Turing Machine is similar in spirit to a modern programmable computer, in which the action to be taken by the computer – the 'program' – is stored in memory, analogous to the bit string T_M stored at the beginning of the tape by the Universal Turing Machine. The data to be processed by the program is stored in a separate part of memory, analogous to the role of x in the Universal Turing Machine. Then some fixed hardware is used to run the program, producing the output. This fixed hardware is analogous to the internal states and the (fixed) program being executed by the Universal Turing Machine.

Describing the detailed construction of a Universal Turing Machine is beyond the scope of this book. (Though industrious readers may like to attempt the construction.) The key point is the existence of such a machine, showing that a single fixed machine can be used to run any algorithm whatsoever. The existence of a Universal Turing Machine also explains our earlier statement that the number of internal states in a Turing machine does not matter much, for provided that number m exceeds the number needed for a Universal Turing Machine, such a machine can be used to simulate a Turing machine with any number of internal states.

---

tinction between problems which are easy to solve, and problems which are hard to solve. Undecidability provides the ultimate example of problems which are hard to solve – so hard that they are in fact impossible to solve.

Exercise 3.5: (Halting problem with no inputs) Show that given a Turing machine M there is no algorithm to determine whether M halts when the input to the machine is a blank tape.

Exercise 3.6: (Probabilistic halting problem)   Suppose we number the probabilistic Turing machines using a scheme similar to that found in Exercise 3.2 and define the probabilistic halting function $h_p(x)$ to be 1 if machine x halts on input of x with probability at least $1/2$ and 0 if machine x halts on input of x with probability less than $1/2$. Show that there is no probabilistic Turing machine which can output $h_p(x)$ with probability of correctness strictly greater than $1/2$ for all x.

Exercise 3.7: (Halting oracle)   Suppose a *black box* is made available to us which takes a non-negative integer x as input, and then outputs the value of $h(x)$, where $h(\cdot)$ is the halting function defined in Box 3.2 on page 130. This type of black box is sometimes known as an *oracle* for the halting problem. Suppose we have a regular Turing machine which is augmented by the power to call the oracle. One way of accomplishing this is to use a two-tape Turing machine, and add an extra program instruction to the Turing machine which results in the oracle being called, and the value of $h(x)$ being printed on the second tape, where x is the current contents of the second tape. It is clear that this model for computation is more powerful than the conventional Turing machine model, since it can be used to compute the halting function. Is the halting problem for this model of computation undecidable? That is, can a Turing machine aided by an oracle for the halting problem decide whether a program for the Turing machine with oracle will halt on a particular input?

3.1.2 Circuits

Turing machines are rather idealized models of computing devices. Real computers are *finite* in size, whereas for Turing machines we assumed a computer of unbounded size. In this section we investigate an alternative model of computation, the *circuit model*, that is equivalent to the Turing machine in terms of computational power, but is more convenient and realistic for many applications. In particular the circuit model of computation is especially important as preparation for our investigation of quantum computers.

A circuit is made up of *wires* and *gates*, which carry information around, and perform simple computational tasks, respectively. For example, Figure 3.2 shows a simple circuit which takes as input a single bit, a. This bit is passed through a NOT gate, which flips the bit, taking 1 to 0 and 0 to 1. The wires before and after the NOT gate serve merely to carry the bit to and from the NOT gate; they can represent movement of the bit through space, or perhaps just through time.

More generally, a circuit may involve many input and output bits, many wires, and many logical gates. A *logic gate* is a function $f : \{0,1\}^k \rightarrow \{0,1\}^l$ from some fixed number k of *input bits* to some fixed number l of *output bits*. For example, the NOT gate is a gate with one input bit and one output bit which computes the function $f(a) = 1 \oplus a$, where a is a single bit, and $\oplus$ is modulo 2 addition. It is also usual to make the convention that no loops are allowed in the circuit, to avoid possible instabilities, as illustrated in Figure 3.3. We say such a circuit is *acyclic*, and we adhere to the convention that circuits in the circuit model of computation be acyclic.

Box 3.2: The halting problem

In Exercise 3.2 you showed that each Turing machine can be uniquely associated with a number from the list $1, 2, 3, \ldots$. To solve Hilbert's problem, Turing used this numbering to pose the *halting problem*: does the machine with Turing number x halt upon input of the number y? This is a well posed and interesting mathematical problem. After all, it is a matter of some considerable interest to us whether our algorithms halt or not. Yet it turns out that there is no algorithm which is capable of solving the halting problem. To see this, Turing asked whether there is an algorithm to solve an even more specialized problem: does the machine with Turing number x halt upon input of the same number x? Turing defined the *halting function*,

$$h(x) \equiv \begin{cases} 0 & \text{if machine number } x \text{ does not halt upon input of } x \\ 1 & \text{if machine number } x \text{ halts upon input of } x. \end{cases}$$

If there is an algorithm to solve the halting problem, then there surely is an algorithm to evaluate $h(x)$. We will try to reach a contradiction by supposing such an algorithm exists, denoted by HALT(x). Consider an algorithm computing the function TURING(x), with pseudocode

```
TURING(x)

y = HALT(x)
if y = 0 then
          halt
else
          loop forever
end if
```

Since HALT is a valid program, TURING must also be a valid program, with some Turing number, t. By definition of the halting function, $h(t) = 1$ if and only if TURING halts on input of t. But by inspection of the program for TURING, we see that TURING halts on input of t if and only if $h(t) = 0$. Thus $h(t) = 1$ if and only if $h(t) = 0$, a contradiction. Therefore, our initial assumption that there is an algorithm to evaluate $h(x)$ must have been wrong. We conclude that there is no algorithm allowing us to solve the halting problem.

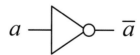

Figure 3.2. Elementary circuit performing a single NOT gate on a single input bit.

There are many other elementary logic gates which are useful for computation. A partial list includes the AND gate, the OR gate, the XOR gate, the NAND gate, and the NOR gate. Each of these gates takes two bits as input, and produces a single bit as output. The AND gate outputs 1 if and only if both of its inputs are 1. The OR gate outputs 1 if

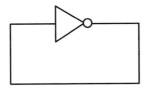

Figure 3.3. Circuits containing cycles can be unstable, and are not usually permitted in the circuit model of computation.

and only if at least one of its inputs is 1. The XOR gate outputs the sum, modulo 2, of its inputs. The NAND and NOR gates take the AND and OR, respectively, of their inputs, and then apply a NOT to whatever is output. The action of these gates is illustrated in Figure 3.4.

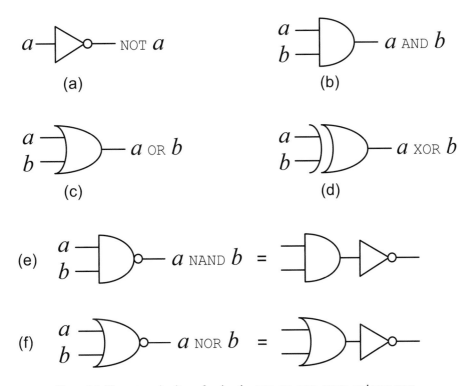

Figure 3.4. Elementary circuits performing the AND, OR, XOR, NAND, and NOR gates.

There are two important 'gates' missing from Figure 3.4, namely the FANOUT gate and the CROSSOVER gate. In circuits we often allow bits to 'divide', replacing a bit with two copies of itself, an operation referred to as FANOUT. We also allow bits to CROSSOVER, that is, the value of two bits are interchanged. A third operation missing from Figure 3.4, not really a logic gate at all, is to allow the preparation of extra *ancilla* or *work* bits, to allow extra working space during the computation.

These simple circuit elements can be put together to perform an enormous variety of computations. Below we'll show that these elements can be used to compute any function whatsoever. In the meantime, let's look at a simple example of a circuit which adds two n bit integers, using essentially the same algorithm taught to school-children around the world. The basic element in this circuit is a smaller circuit known as a *half-adder*, shown in Figure 3.5. A half-adder takes two bits, x and y, as input, and outputs the sum of the bits $x \oplus y$ modulo 2, together with a carry bit set to 1 if x and y are both 1, or 0 otherwise.

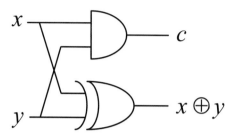

Figure 3.5. Half-adder circuit. The carry bit c is set to 1 when x and y are both 1, otherwise it is 0.

Two cascaded half-adders may be used to build a *full-adder*, as shown in Figure 3.6. A full-adder takes as input three bits, x, y, and c. The bits x and y should be thought of as data to be added, while c is a carry bit from an earlier computation. The circuit outputs two bits. One output bit is the modulo 2 sum, $x \oplus y \oplus c$ of all three input bits. The second output bit, c', is a carry bit, which is set to 1 if two or more of the inputs is 1, and is 0 otherwise.

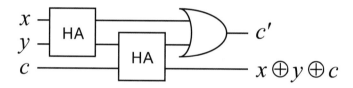

Figure 3.6. Full-adder circuit.

By cascading many of these full-adders together we obtain a circuit to add two n-bit integers, as illustrated in Figure 3.7 for the case $n = 3$.

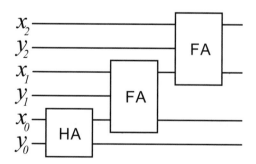

Figure 3.7. Addition circuit for two three-bit integers, $x = x_2 x_1 x_0$ and $y = y_2 y_1 y_0$, using the elementary algorithm taught to school-children.

We claimed earlier that just a few *fixed* gates can be used to compute *any* function $f : \{0, 1\}^n \rightarrow \{0, 1\}^m$ whatsoever. We will now prove this for the simplified case of a function $f : \{0, 1\}^n \rightarrow \{0, 1\}$ with n input bits and a single output bit. Such a function is known as a *Boolean function*, and the corresponding circuit is a *Boolean circuit*. The general universality proof follows immediately from the special case of Boolean functions. The proof is by induction on n. For $n = 1$ there are four possible functions: the identity, which has a circuit consisting of a single wire; the bit flip, which is implemented using a single NOT gate; the function which replaces the input bit with a 0, which can be obtained by ANDing the input with a work bit initially in the 0 state; and the function which replaces the input with a 1, which can be obtained by ORing the input with a work bit initially in the 1 state.

To complete the induction, suppose that any function on n bits may be computed by a circuit, and let f be a function on $n + 1$ bits. Define n-bit functions f_0 and f_1 by $f_0(x_1, \ldots, x_n) \equiv f(0, x_1, \ldots, x_n)$ and $f_1(x_1, \ldots, x_n) \equiv f(1, x_1, \ldots, x_n)$. These are both n-bit functions, so by the inductive hypothesis there are circuits to compute these functions.

It is now an easy matter to design a circuit which computes f. The circuit computes both f_0 and f_1 on the last n bits of the input. Then, depending on whether the first bit of the input was a 0 or a 1 it outputs the appropriate answer. A circuit to do this is shown in Figure 3.8. This completes the induction.

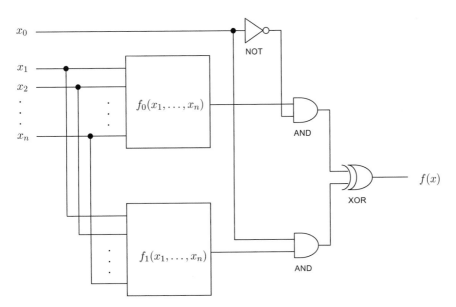

Figure 3.8. Circuit to compute an arbitrary function f on $n + 1$ bits, assuming by induction that there are circuits to compute the n-bit functions f_0 and f_1.

Five elements may be identified in the universal circuit construction: (1) wires, which preserve the states of the bits; (2) ancilla bits prepared in standard states, used in the $n = 1$ case of the proof; (3) the FANOUT operation, which takes a single bit as input and outputs two copies of that bit; (4) the CROSSOVER operation, which interchanges the value of two bits; and (5) the AND, XOR, and NOT gates. In Chapter 4 we'll define the quantum circuit model of computation in a manner analogous to classical circuits. It is interesting to note that many of these five elements pose some interesting challenges when extending to the quantum case: it is not necessarily obvious that good quantum wires for the preservation of qubits can be constructed, even in principle, the FANOUT operation cannot be performed in a straightforward manner in quantum mechanics, due to the no–cloning theorem (as explained in Section 1.3.5), and the AND and XOR gates are not invertible, and thus can't be implemented in a straightforward manner as unitary quantum gates. There is certainly plenty to think about in defining a quantum circuit model of computation!

Exercise 3.8: (Universality of NAND) Show that the NAND gate can be used to simulate the AND, XOR and NOT gates, provided wires, ancilla bits and FANOUT are available.

Let's return from our brief quantum digression, to the properties of classical circuits. We claimed earlier that the Turing machine model is equivalent to the circuit model of computation. In what sense do we mean the two models are equivalent? On the face of it, the two models appear quite different. The unbounded nature of a Turing machine makes them more useful for abstractly specifying what it is we mean by an algorithm, while circuits more closely capture what an actual physical computer does.

The two models are connected by introducing the notion of a *uniform circuit family*. A *circuit family* consists of a collection of circuits, $\{C_n\}$, indexed by a positive integer n. The circuit C_n has n input bits, and may have any finite number of extra work bits, and output bits. The output of the circuit C_n, upon input of a number x of at most n bits in length, is denoted by $C_n(x)$. We require that the circuits be *consistent*, that is, if $m < n$ and x is at most m bits in length, then $C_m(x) = C_n(x)$. The function computed by the circuit family $\{C_n\}$ is the function $C(\cdot)$ such that if x is n bits in length then $C(x) = C_n(x)$. For example, consider a circuit C_n that squares an n-bit number. This defines a family of circuits $\{C_n\}$ that computes the function, $C(x) = x^2$, where x is any positive integer.

It's not enough to consider unrestricted families of circuits, however. In practice, we need an algorithm to build the circuit. Indeed, if we don't place any restrictions on the circuit family then it becomes possible to compute all sorts of functions which we do not expect to be able to compute in a reasonable model of computation. For example, let $h_n(x)$ denote the halting function, restricted to values of x which are n bits in length. Thus h_n is a function from n bits to 1 bit, and we have proved there exists a circuit C_n to compute $h_n(\cdot)$. Therefore the circuit family $\{C_n\}$ computes the halting function! However, what prevents us from using this circuit family to solve the halting problem is that we haven't specified an algorithm which will allow us to build the circuit C_n for all values of n. Adding this requirement results in the notion of a uniform circuit family.

That is, a family of circuits $\{C_n\}$ is said to be a *uniform circuit family* if there is some algorithm running on a Turing machine which, upon input of n, generates a *description* of C_n. That is, the algorithm outputs a description of what gates are in the circuit C_n, how those gates are connected together to form a circuit, any ancilla bits needed by the circuit, FANOUT and CROSSOVER operations, and where the output from the circuit should be read out. For example, the family of circuits we described earlier for squaring n-bit numbers is certainly a uniform circuit family, since there is an algorithm which, given n, outputs a description of the circuit needed to square an n-bit number. You can think of this algorithm as the means by which an engineer is able to generate a description of (and thus build) the circuit for any n whatsoever. By contrast, a circuit family that is not uniform is said to be a *non–uniform* circuit family. There is no algorithm to construct the circuit for arbitrary n, which prevents our engineer from building circuits to compute functions like the halting function.

Intuitively, a uniform circuit family is a family of circuits that can be generated by some reasonable algorithm. It can be shown that the class of functions computable by uniform circuit families is exactly the same as the class of functions which can be computed on a Turing machine. With this uniformity restriction, results in the Turing machine model of computation can usually be given a straightforward translation into the circuit model of computation, and vice versa. Later we give similar attention to issues of uniformity in the quantum circuit model of computation.

3.2 The analysis of computational problems

The analysis of computational problems depends upon the answer to three fundamental questions:

(1) **What is a computational problem?** Multiplying two numbers together is a computational problem; so is programming a computer to exceed human abilities in the writing of poetry. In order to make progress developing a general theory for the analysis of computational problems we are going to isolate a special class of problems known as *decision problems*, and concentrate our analysis on those. Restricting ourselves in this way enables the development of a theory which is both elegant and rich in structure. Most important, it is a theory whose principles have application far beyond decision problems.

(2) **How may we design algorithms to solve a given computational problem?** Once a problem has been specified, what algorithms can be used to solve the problem? Are there general techniques which can be used to solve wide classes of problems? How can we be sure an algorithm behaves as claimed?

(3) **What are the minimal resources required to solve a given computational problem?** Running an algorithm requires the consumption of *resources*, such as time, space, and energy. In different situations it may be desirable to minimize consumption of one or more resource. Can we classify problems according to the resource requirements needed to solve them?

In the next few sections we investigate these three questions, especially questions 1 and 3. Although question 1, 'what is a computational problem?', is perhaps the most fundamental of the questions, we shall defer answering it until Section 3.2.3, pausing first to establish some background notions related to resource quantification in Section 3.2.1, and then reviewing the key ideas of *computational complexity* in Section 3.2.2.

Question 2, how to design good algorithms, is the subject of an enormous amount of ingenious work by many researchers. So much so that in this brief introduction we cannot even begin to describe the main ideas employed in the design of good algorithms. If you are interested in this beautiful subject, we refer you to the end of chapter 'History and further reading'. Our closest direct contact with this subject will occur later in the book, when we study quantum algorithms. The techniques involved in the creation of quantum algorithms have typically involved a blend of deep existing ideas in algorithm design for classical computers, and the creation of new, wholly quantum mechanical techniques for algorithm design. For this reason, and because the spirit of quantum algorithm design is so similar in many ways to classical algorithm design, we encourage you to become familiar with at least the basic ideas of algorithm design.

Question 3, what are the minimal resources required to solve a given computational problem, is the main focus of the next few sections. For example, suppose we are given two numbers, each n bits in length, which we wish to multiply. If the multiplication is performed on a single-tape Turing machine, how many computational steps must be executed by the Turing machine in order to complete the task? How much space is used on the Turing machine while completing the task?

These are examples of the type of resource questions we may ask. Generally speaking, computers make use of many different kinds of resources, however we will focus most of our attention on time, space, and energy. Traditionally in computer science, time and space have been the two major resource concerns in the study of algorithms, and we study these issues in Sections 3.2.2 through 3.2.4. Energy has been a less important consideration; however, the study of energy requirements motivates the subject of reversible classical computation, which in turn is a prerequisite for quantum computation, so we examine energy requirements for computation in some considerable detail in Section 3.2.5.

3.2.1 How to quantify computational resources

Different models of computation lead to different resource requirements for computation. Even something as simple as changing from a single-tape to a two-tape Turing machine may change the resources required to solve a given computational problem. For a computational task which is extremely well understood, like addition of integers, for example, such differences between computational models may be of interest. However, for a first pass at understanding a problem, we would like a way of quantifying resource requirements that is independent of relatively trivial changes in the computational model. One of the tools which has been developed to do this is the *asymptotic notation*, which can be used to summarize the *essential* behavior of a function. This asymptotic notation can be used, for example, to summarize the essence of how many time steps it takes a given algorithm to run, without worrying too much about the exact time count. In this section we describe this notation in detail, and apply it to a simple problem illustrating the quantification of computational resources – the analysis of algorithms for sorting a list of names into alphabetical order.

Suppose, for example, that we are interested in the number of gates necessary to add together two n-bit numbers. Exact counts of the number of gates required obscure the big picture: perhaps a specific algorithm requires $24n + 2\lceil \log n \rceil + 16$ gates to perform this task. However, in the limit of large problem size the only term which matters is the $24n$ term. Furthermore, we disregard constant factors as being of secondary importance to the analysis of the algorithm. The *essential* behavior of the algorithm is summed up by saying that the number of operations required scales like n, where n is the number of bits in the numbers being added. The asymptotic notation consists of three tools which make this notion precise.

The O ('big O') notation is used to set *upper bounds* on the behavior of a function. Suppose $f(n)$ and $g(n)$ are two functions on the non-negative integers. We say '$f(n)$ is in the class of functions $O(g(n))$', or just '$f(n)$ is $O(g(n))$', if there are constants c and n_0 such that for all values of n greater than n_0, $f(n) \leq cg(n)$. That is, for sufficiently large n, the function $g(n)$ is an upper bound on $f(n)$, up to an unimportant constant factor. The big O notation is particularly useful for studying the worst–case behavior of *specific* algorithms, where we are often satisfied with an upper bound on the resources consumed by an algorithm.

When studying the behaviors of a *class* of algorithms – say the entire class of algorithms which can be used to multiply two numbers – it is interesting to set lower bounds on the resources required. For this the Ω ('big Omega') notation is used. A function $f(n)$ is said to be $\Omega(g(n))$ if there exist constants c and n_0 such that for all n greater than n_0, $cg(n) \le f(n)$. That is, for sufficiently large n, $g(n)$ is a lower bound on $f(n)$, up to an unimportant constant factor.

Finally, the Θ ('big Theta') notation is used to indicate that $f(n)$ behaves the same as $g(n)$ asymptotically, up to unimportant constant factors. That is, we say $f(n)$ is $\Theta(g(n))$ if it is both $O(g(n))$ and $\Omega(g(n))$.

Asymptotic notation: examples

Let's consider a few simple examples of the asymptotic notation. The function $2n$ is in the class $O(n^2)$, since $2n \le 2n^2$ for all positive n. The function 2^n is $\Omega(n^3)$, since $n^3 \le 2^n$ for sufficiently large n. Finally, the function $7n^2 + \sqrt{n}\log(n)$ is $\Theta(n^2)$, since $7n^2 \le 7n^2 + \sqrt{n}\log(n) \le 8n^2$ for all sufficiently large values of n. In the following few exercises you will work through some of the elementary properties of the asymptotic notation that make it a useful tool in the analysis of algorithms.

Exercise 3.9:   Prove that $f(n)$ is $O(g(n))$ if and only if $g(n)$ is $\Omega(f(n))$. Deduce that $f(n)$ is $\Theta(g(n))$ if and only if $g(n)$ is $\Theta(f(n))$.

Exercise 3.10:   Suppose $g(n)$ is a polynomial of degree k. Show that $g(n)$ is $O(n^l)$ for any $l \ge k$.

Exercise 3.11:   Show that $\log n$ is $O(n^k)$ for any $k > 0$.

Exercise 3.12: ($n^{\log n}$ **is super-polynomial**)   Show that n^k is $O(n^{\log n})$ for any k, but that $n^{\log n}$ is never $O(n^k)$.

Exercise 3.13: ($n^{\log n}$ **is sub-exponential**)   Show that c^n is $\Omega(n^{\log n})$ for any $c > 1$, but that $n^{\log n}$ is never $\Omega(c^n)$.

Exercise 3.14:   Suppose $e(n)$ is $O(f(n))$ and $g(n)$ is $O(h(n))$. Show that $e(n)g(n)$ is $O(f(n)h(n))$.

An example of the use of the asymptotic notation in quantifying resources is the following simple application to the problem of sorting an n element list of names into alphabetical order. Many sorting algorithms are based upon the 'compare-and-swap' operation: two elements of an n element list are compared, and swapped if they are in the wrong order. If this compare-and-swap operation is the only means by which we can access the list, how many such operations are required in order to ensure that the list has been correctly sorted?

A simple compare-and-swap algorithm for solving the sorting problem is as follows: (note that compare-and-swap(j,k) compares the list entries numbered j and k, and swaps them if they are out of order)

```
for j = 1 to n-1
    for k = j+1 to n
        compare-and-swap(j,k)
    end k
end j
```

It is clear that this algorithm correctly sorts a list of n names into alphabetical order. Note that the number of compare-and-swap operations executed by the algorithm is $(n-1) + (n-2) + \cdots + 1 = n(n-1)/2$. Thus the number of compare-and-swap operations used by the algorithm is $\Theta(n^2)$. Can we do better than this? It turns out that we can. Algorithms such as 'heapsort' are known which run using $O(n \log n)$ compare-and-swap operations. Furthermore, in Exercise 3.15 you'll work through a simple counting argument that shows any algorithm based upon the compare-and-swap operation requires $\Omega(n \log n)$ such operations. Thus, the sorting problem requires $\Theta(n \log n)$ compare-and-swap operations, in general.

Exercise 3.15: (Lower bound for compare-and-swap based sorts) Suppose an n element list is sorted by applying some sequence of compare-and-swap operations to the list. There are $n!$ possible initial orderings of the list. Show that after k of the compare-and-swap operations have been applied, at most 2^k of the possible initial orderings will have been sorted into the correct order. Conclude that $\Omega(n \log n)$ compare-and-swap operations are required to sort all possible initial orderings into the correct order.

3.2.2 Computational complexity

> *The idea that there won't be an algorithm to solve it – this is something fundamental that won't ever change – that idea appeals to me.*
> – Stephen Cook

> *Sometimes it is good that some things are impossible. I am happy there are many things that nobody can do to me.*
> – Leonid Levin

> *It should not come as a surprise that our choice of polynomial algorithms as the mathematical concept that is supposed to capture the informal notion of 'practically efficient computation' is open to criticism from all sides. [...] Ultimately, our argument for our choice must be this: **Adopting polynomial worst-case performance as our criterion of efficiency results in an elegant and useful theory that says something meaningful about practical computation, and would be impossible without this simplification.**
> – Christos Papadimitriou

What time and space resources are required to perform a computation? In many cases these are the most important questions we can ask about a computational problem. Problems like addition and multiplication of numbers are regarded as efficiently solvable because we have *fast* algorithms to perform addition and multiplication, which consume little *space* when running. Many other problems have no known fast algorithm, and are effectively impossible to solve, not because we can't find an algorithm to solve the problem, but because all known algorithms consume such vast quantities of space or time as to render them practically useless.

Computational complexity is the study of the time and space resources required to solve computational problems. The task of computational complexity is to prove *lower bounds* on the resources required by the best possible algorithm for solving a problem, even if that algorithm is not explicitly known. In this and the next two sections, we give an overview of computational complexity, its major concepts, and some of the more important results of the field. Note that computational complexity is in a sense complementary to the field of algorithm design; ideally, the most efficient algorithms we could design would match perfectly with the lower bounds proved by computational complexity. Unfortunately, this is often not the case. As already noted, in this book we won't examine classical algorithm design in any depth.

One difficulty in formulating a theory of computational complexity is that different computational models may require different resources to solve the same problem. For instance, multiple-tape Turing machines can solve many problems substantially faster than single-tape Turing machines. This difficulty is resolved in a rather coarse way. Suppose a problem is specified by giving n bits as input. For instance, we might be interested in whether a particular n-bit number is prime or not. The chief distinction made in computational complexity is between problems which can be solved using resources which are bounded by a *polynomial* in n, or which require resources which grow faster than any polynomial in n. In the latter case we usually say that the resources required are *exponential* in the problem size, abusing the term exponential, since there are functions like $n^{\log n}$ which grow faster than any polynomial (and thus are 'exponential' according to this convention), yet which grow slower than any true exponential. A problem is regarded as *easy*, *tractable* or *feasible* if an algorithm for solving the problem using polynomial resources exists, and as *hard*, *intractable* or *infeasible* if the best possible algorithm requires exponential resources.

As a simple example, suppose we have two numbers with binary expansions $x_1 \ldots x_{m_1}$ and $y_1 \ldots y_{m_2}$, and we wish to determine the sum of the two numbers. The total size of the input is $n \equiv m_1 + m_2$. It's easy to see that the two numbers can be added using a number of elementary operations that scales as $\Theta(n)$; this algorithm uses a polynomial (indeed, linear) number of operations to perform its tasks. By contrast, it is believed (though it has never been proved!) that the problem of factoring an integer into its prime factors is intractable. That is, the belief is that there is no algorithm which can factor an arbitrary n-bit integer using $O(p(n))$ operations, where p is some fixed polynomial function of n. We will later give many other examples of problems which are believed to be intractable in this sense.

The polynomial versus exponential classification is rather coarse. In practice, an algorithm that solves a problem using $2^{n/1000}$ operations is probably more useful than one which runs in n^{1000} operations. Only for very large input sizes ($n \approx 10^8$) will the 'efficient' polynomial algorithm be preferable to the 'inefficient' exponential algorithm, and for many purposes it may be more practical to prefer the 'inefficient' algorithm.

Nevertheless, there are many reasons to base computational complexity primarily on the polynomial versus exponential classification. First, historically, with few exceptions, polynomial resource algorithms have been much faster than exponential algorithms. We might speculate that the reason for this is lack of imagination: coming up with algorithms requiring n, n^2 or some other low degree polynomial number of operations is often much easier than finding a natural algorithm which requires n^{1000} operations, although examples like the latter do exist. Thus, the predisposition for the human mind to come up with relatively simple algorithms has meant that in practice polynomial algorithms usually do perform much more efficiently than their exponential cousins.

A second and more fundamental reason for emphasizing the polynomial versus exponential classification is derived from the *strong Church–Turing thesis*. As discussed in Section 1.1, it was observed in the 1960s and 1970s that probabilistic Turing machines appear to be the strongest 'reasonable' model of computation. More precisely, researchers consistently found that if it was possible to compute a function using k elementary operations in some model that was *not* the probabilistic Turing machine model of computation, then it was always possible to compute the same function in the probabilistic Turing machine model, using at most $p(k)$ elementary operations, where $p(\cdot)$ is some *polynomial* function. This statement is known as the *strong Church–Turing thesis*:

> **Strong Church–Turing thesis**: *Any model of computation can be simulated on a probabilistic Turing machine with at most a polynomial increase in the number of elementary operations required.*

The strong Church–Turing thesis is great news for the theory of computational complexity, for it implies that attention may be restricted to the probabilistic Turing machine model of computation. After all, if a problem has no polynomial resource solution on a probabilistic Turing machine, then the strong Church–Turing thesis implies that it has no efficient solution on any computing device. Thus, the strong Church–Turing thesis implies that the entire theory of computational complexity will take on an elegant, model-independent form if the notion of efficiency is identified with polynomial resource algorithms, and this elegance has provided a strong impetus towards acceptance of the identification of 'solvable with polynomial resources' and 'efficiently solvable'. Of course, one of the prime reasons for interest in quantum computers is that they cast into doubt the strong Church–Turing thesis, by enabling the efficient solution of a problem which is believed to be intractable on all classical computers, including probabilistic Turing machines! Nevertheless, it is useful to understand and appreciate the role the strong Church–Turing thesis has played in the search for a model-independent theory of computational complexity.

Finally, we note that, in practice, computer scientists are not only interested in the polynomial versus exponential classification of problems. This is merely the first and coarsest way of understanding how difficult a computational problem is. However, it is an exceptionally important distinction, and illustrates many broader points about the nature of resource questions in computer science. For most of this book, it will be our central concern in evaluating the efficiency of a given algorithm.

Having examined the merits of the polynomial versus exponential classification, we now have to confess that the theory of computational complexity has one remarkable outstanding failure: it seems very hard to prove that there are interesting classes of problems which require exponential resources to solve. It is quite easy to give non-constructive proofs that *most* problems require exponential resources (see Exercise 3.16, below), and furthermore many interesting problems are *conjectured* to require exponential resources for their solution, but rigorous proofs seem very hard to come by, at least with the present state of knowledge. This failure of computational complexity has important implications for quantum computation, because it turns out that the computational power of quantum computers can be related to some major open problems in *classical* computational complexity theory. Until these problems are resolved, it cannot be stated with certainty how computationally powerful a quantum computer is, or even whether it is more powerful than a classical computer!

Exercise 3.16: (Hard-to-compute functions exist) Show that there exist Boolean functions on n inputs which require at least $2^n / \log n$ logic gates to compute.

3.2.3 Decision problems and the complexity classes P and NP

Many computational problems are most cleanly formulated as *decision problems* – problems with a yes or no answer. For example, is a given number m a prime number or not? This is the *primality* decision problem. The main ideas of computational complexity are most easily and most often formulated in terms of decision problems, for two reasons: the theory takes its simplest and most elegant form in this form, while still generalizing in a natural way to more complex scenarios; and historically computational complexity arose primarily from the study of decision problems.

Although most decision problems can easily be stated in simple, familiar language, discussion of the general properties of decision problems is greatly helped by the terminology of *formal languages*. In this terminology, a *language* L over the *alphabet* Σ is a subset of the set Σ^* of all (finite) strings of symbols from Σ. For example, if $\Sigma = \{0, 1\}$, then the set of binary representations of even numbers, $L = \{0, 10, 100, 110, \ldots\}$ is a language over Σ.

Decision problems may be encoded in an obvious way as problems about languages. For instance, the primality decision problem can be encoded using the binary alphabet $\Sigma = \{0, 1\}$. Strings from Σ^* can be interpreted in a natural way as non-negative integers. For example, 0010 can be interpreted as the number 2. The language L is defined to consist of all binary strings such that the corresponding number is prime.

To solve the primality decision problem, what we would like is a Turing machine which, when started with a given number n on its input tape, eventually outputs some equivalent of 'yes' if n is prime, and outputs 'no' if n is not prime. To make this idea precise, it is convenient to modify our old Turing machine definition (of Section 3.1.1) slightly, replacing the halting state q_h with two states q_Y and q_N to represent the answers 'yes' and 'no' respectively. In all other ways the machine behaves in the same way, and it still halts when it enters the state q_Y or q_N. More generally, a language L is *decided* by a Turing machine if the machine is able to decide whether an input x on its tape is a member of the language of L or not, eventually halting in the state q_Y if $x \in L$, and eventually halting in the state q_N if $x \notin L$. We say that the machine has *accepted* or *rejected* x depending on which of these two cases comes about.

How quickly can we determine whether or not a number is prime? That is, what is the fastest Turing machine which decides the language representing the primality decision problem? We say that a problem is in **TIME**$(f(n))$ if there exists a Turing machine which decides whether a candidate x is in the language in time $O(f(n))$, where n is the length of x. A problem is said to be solvable in *polynomial time* if it is in **TIME**(n^k) for some finite k. The collection of all languages which are in **TIME**(n^k), for some k, is denoted **P**. **P** is our first example of a *complexity class*. More generally, a complexity class is defined to be a collection of languages. Much of computational complexity theory is concerned with the definition of various complexity classes, and understanding the relationship between different complexity classes.

Not surprisingly, there are problems which cannot be solved in polynomial time. Unfortunately, proving that any given problem can't be solved in polynomial time seems to be very difficult, although conjectures abound! A simple example of an interesting decision problem which is believed not to be in **P** is the *factoring decision problem*:

> FACTORING: Given a composite integer m and $l < m$, does m have a non-trivial factor less than l?

An interesting property of factoring is that if somebody claims that the answer is 'yes, m does have a non-trivial factor less than l' then they can establish this by exhibiting such a factor, which can then be efficiently checked by other parties, simply by doing long-division. We call such a factor a *witness* to the fact that m has a factor less than l. This idea of an easily checkable witness is the key idea in the definition of the complexity class **NP**, below. We have phrased factoring as a decision problem, but you can easily verify that the decision problem is equivalent to finding the factors of a number:

Exercise 3.17:   Prove that a polynomial-time algorithm for finding the factors of a number m exists if and only if the factoring decision problem is in **P**.

Factoring is an example of a problem in an important complexity class known as **NP**. What distinguishes problems in **NP** is that 'yes' instances of a problem can easily be verified with the aid of an appropriate witness. More rigorously, a language L is in **NP** if there is a Turing machine M with the following properties:

(1) If $x \in L$ then there exists a witness string w such that M halts in the state q_Y after a time polynomial in $|x|$ when the machine is started in the state x-blank-w.
(2) If $x \notin L$ then for all strings w which attempt to play the role of a witness, the machine halts in state q_N after a time polynomial in $|x|$ when M is started in the state x-blank-w.

There is an interesting asymmetry in the definition of **NP**. While we have to be able to quickly decide whether a possible witness to $x \in L$ is truly a witness, there is no such need to produce a witness to $x \notin L$. For instance, in the factoring problem, we have an easy way of proving that a given number has a factor less than m, but exhibiting a witness to prove that a number has no factors less than m is more daunting. This suggests defining co**NP**, the class of languages which have witnesses to 'no' instances; obviously the languages in co**NP** are just the complements of languages in **NP**.

How are **P** and **NP** related? It is clear that **P** is a subset of **NP**. The most famous open problem in computer science is *whether or not there are problems in* **NP** *which are not in* **P**, often abbreviated as the **P** $\neq$ **NP** problem. Most computer scientists believe that **P** $\neq$ **NP**, but despite decades of work nobody has been able to prove this, and the possibility remains that **P** = **NP**.

Exercise 3.18:   Prove that if co**NP** $\neq$ **NP** then **P** $\neq$ **NP**.

Upon first acquaintance it's tempting to conclude that the conjecture **P** $\neq$ **NP** ought to be pretty easy to resolve. To see why it's actually rather subtle it helps to see couple of examples of problems that are in **P** and **NP**. We'll draw the examples from *graph theory*, a rich source of decision problems with surprisingly many practical applications. A *graph* is a finite collection of *vertices* $\{v_1, \ldots, v_n\}$ connected by *edges*, which are pairs (v_i, v_j) of vertices. For now, we are only concerned with *undirected graphs*, in which the order of the vertices (in each edge pair) does not matter; similar ideas can be investigated for *directed graphs* in which the order of vertices does matter. A typical graph is illustrated in Figure 3.9.

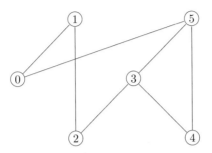

Figure 3.9. A graph.

A *cycle* in a graph is a sequence $v_1, \ldots, v_m$ of vertices such that each pair (v_j, v_{j+1}) is an edge, as is (v_1, v_m). A *simple cycle* is a cycle in which none of the vertices is repeated, except for the first and last vertices. A *Hamiltonian cycle* is a simple cycle which visits every vertex in the graph. Examples of graphs with and without Hamiltonian cycles are shown in Figure 3.10.

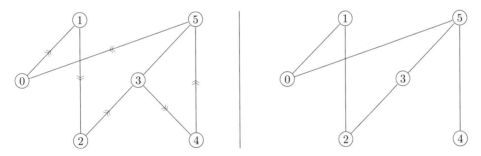

Figure 3.10. The graph on the left contains a Hamiltonian cycle, $0, 1, 2, 3, 4, 5, 0$. The graph on the right contains no Hamiltonian cycle, as can be verified by inspection.

The *Hamiltonian cycle problem* (HC) is to determine whether a given graph contains a Hamiltonian cycle or not. HC is a decision problem in **NP**, since if a given graph has a Hamiltonian cycle, then that cycle can be used as an easily checkable witness. Moreover, HC has no known polynomial time algorithm. Indeed, HC is a problem in the class of so-called **NP**-complete problems, which can be thought of as the 'hardest' problems in **NP**, in the sense that solving HC in time t allows any other problem in **NP** to be solved in time $O(\text{poly}(t))$. This also means that if any **NP**-complete problem has a polynomial time solution then it will follow that **P** = **NP**.

There is a problem, the Euler cycle decision problem, which is superficially similar to HC, but which has astonishingly different properties. An *Euler cycle* is an ordering of the edges of a graph G so that every edge in the graph is visited exactly once. The Euler cycle decision problem (EC) is to determine, given a graph G on n vertices, whether that graph contains an Euler cycle or not. EC is, in fact, exactly the same problem as HC, only the path visits edges, rather than vertices. Consider the following remarkable theorem, to be proven in Exercise 3.20:

Theorem 3.1: (**Euler's theorem**) A connected graph contains an Euler cycle if and only if every vertex has an even number of edges incident upon it.

Euler's theorem gives us a method for efficiently solving EC. First, check to see whether the graph is connected; this is easily done with $O(n^2)$ operations, as shown in Exercise 3.19. If the graph is not connected, then obviously no Euler cycle exists. If the graph is connected then for each vertex check whether there is an even number of edges incident upon the vertex. If a vertex is found for which this is not the case, then there is no Euler cycle, otherwise an Euler cycle exists. Since there are n vertices, and at most $n(n-1)/2$ edges, this algorithm requires $O(n^3)$ elementary operations. Thus EC is in **P**! Somehow, there is a structure present in the problem of visiting each edge that can be exploited to provide an efficient algorithm for EC, yet which does not seem to be reflected in the problem of visiting each vertex; it is not at all obvious why such a structure should be present in one case, but not in the other, if indeed it is absent for the HC problem.

Exercise 3.19: The REACHABILITY problem is to determine whether there is a path between two specified vertices in a graph. Show that REACHABILITY can be solved using $O(n)$ operations if the graph has n vertices. Use the solution to REACHABILITY to show that it is possible to decide whether a graph is connected in $O(n^2)$ operations.

Exercise 3.20: (**Euler's theorem**) Prove Euler's theorem. In particular, if each vertex has an even number of incident edges, give a constructive procedure for finding an Euler cycle.

The equivalence between the factoring decision problem and the factoring problem proper is a special instance of one of the most important ideas in computer science, an idea known as *reduction*. Intuitively, we know that some problems can be viewed as special instances of other problems. A less trivial example of reduction is the reduction of HC to the *traveling salesman* decision problem (TSP). The traveling salesman decision problem is as follows: we are given n cities $1, 2, \ldots, n$ and a non-negative integer distance d_{ij} between each pair of cities. Given a distance d the problem is to determine if there is a tour of all the cities of distance less than d.

The reduction of HC to TSP goes as follows. Suppose we have a graph containing n vertices. We turn this into an instance of TSP by thinking of each vertex of the graph as a 'city' and defining the distance d_{ij} between cities i and j to be one if vertices i and j are connected, and the distance to be two if the vertices are unconnected. Then a tour of the cities of distance less than $n+1$ must be of distance n, and be a Hamiltonian cycle for the graph. Conversely, if a Hamiltonian cycle exists then a tour of the cities of distance less than $n+1$ must exist. In this way, given an algorithm for solving TSP, we can convert it into an algorithm for solving HC without much overhead. Two consequences can be inferred from this. First, if TSP is a tractable problem, then HC is also tractable. Second, if HC is hard then TSP must also be hard. This is an example of a general technique known as *reduction*: we've reduced the problem HC to the problem TSP. This is a technique we will use repeatedly throughout this book.

A more general notion of reduction is illustrated in Figure 3.11. A language B is said to be *reducible* to another language A if there exists a Turing machine operating in polynomial time such that given as input x it outputs $R(x)$, and $x \in B$ if and only if $R(x) \in A$. Thus, if we have an algorithm for deciding A, then with a little extra overhead we can decide the language B. In this sense, the language B is essentially no more difficult to decide than the language A.

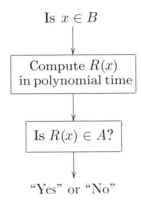

Figure 3.11. Reduction of B to A.

Exercise 3.21: (Transitive property of reduction)   Show that if a language L_1 is reducible to the language L_2 and the language L_2 is reducible to L_3 then the language L_1 is reducible to the language L_3.

Some complexity classes have problems which are *complete* with respect to that complexity class, meaning there is a language L in the complexity class which is the 'most difficult' to decide, in the sense that every other language in the complexity class can be reduced to L. Not all complexity classes have complete problems, but many of the complexity classes we are concerned with do have complete problems. A trivial example is provided by **P**. Let L be any language in **P** which is not empty or equal to the set of all words. That is, there exists a string x_1 such that $x_1 \notin L$ and a string x_2 such that $x_2 \in L$. Then any other language L' in **P** can be reduced to L using the following reduction: given an input x, use the polynomial time decision procedure to determine whether $x \in L'$ or not. If it is not, then set $R(x) = x_1$, otherwise set $R(x) = x_2$.

Exercise 3.22:   Suppose L is complete for a complexity class, and L' is another language in the complexity class such that L reduces to L'. Show that L' is complete for the complexity class.

Less trivially, **NP** also contains complete problems. An important example of such a problem and the prototype for all other **NP**-complete problems is the *circuit satisfiability* problem or CSAT: given a Boolean circuit composed of AND, OR and NOT gates, is there an assignment of values to the inputs to the circuit that results in the circuit outputting 1, that is, is the circuit *satisfiable* for some input? The **NP**-completeness of CSAT is known as the *Cook–Levin theorem*, for which we now outline a proof.

Theorem 3.2: (**Cook–Levin**) CSAT is **NP**-complete.

Proof

The proof has two parts. The first part of the proof is to show that CSAT is in **NP**, and the second part is to show that any language in **NP** can be reduced to CSAT. Both parts of the proof are based on *simulation* techniques: the first part of the proof is essentially showing that a Turing machine can efficiently simulate a circuit, while the second part of the proof is essentially showing that a circuit can efficiently simulate a Turing machine. Both parts of the proof are quite straightforward; for the purposes of illustration we give the second part in some detail.

The first part of the proof is to show that CSAT is in **NP**. Given a circuit containing n circuit elements, and a potential witness w, it is obviously easy to check in polynomial time on a Turing machine whether or not w satisfies the circuit, which establishes that CSAT is in **NP**.

The second part of the proof is to show that any language $L \in$ **NP** can be reduced to CSAT. That is, we aim to show that there is a polynomial time computable reduction R such that $x \in L$ if and only if $R(x)$ is a satisfiable circuit. The idea of the reduction is to find a circuit which simulates the action of the machine M which is used to check instance-witness pairs, (x, w), for the language L. The input variables for the circuit will represent the witness; the idea is that finding a witness which satisfies the circuit is equivalent to M accepting (x, w) for some specific witness w. Without loss of generality we may make the following assumptions about M to simplify the construction:

(1) M's tape alphabet is $\triangleright,0,1$ and the blank symbol.
(2) M runs using time at most $t(n)$ and total space at most $s(n)$ where $t(n)$ and $s(n)$ are polynomials in n.
(3) Machine M can actually be assumed to run using time *exactly* $t(n)$ for all inputs of size n. This is done by adding the lines $\langle q_Y, x, q_Y, x, 0\rangle$, and $\langle q_N, x, q_N, x, 0\rangle$ for each of $x = \triangleright, 0, 1$ and the blank, artificially halting the machine after exactly $t(n)$ steps.

The basic idea of the construction to simulate M is outlined in Figure 3.12. Each internal state of the Turing machine is represented by a single bit in the circuit. We name the corresponding bits $\tilde{q}_s, \tilde{q}_1, \ldots, \tilde{q}_m, \tilde{q}_Y, \tilde{q}_N$. Initially, $\tilde{q}_s$ is set to one, and all the other bits representing internal states are set to zero. Each square on the Turing machine tape is represented by three bits: two bits to represent the letter of the alphabet ($\triangleright, 0, 1$ or blank) currently residing on the tape, and a single 'flag' bit which is set to one if the read-write head is pointing to the square, and set to zero otherwise. We denote the bits representing the tape contents by $(u_1, v_1), \ldots, (u_{s(n)}, v_{s(n)})$ and the corresponding flag bits by $f_1, \ldots, f_{s(n)}$. Initially the u_j and v_j bits are set to represent the inputs x and w, as appropriate, while $f_1 = 1$ and all other $f_j = 0$. There is also a lone extra 'global flag' bit, F, whose function will be explained later. F is initially set to zero. We regard all the bits input to the circuit as fixed, except for those representing the witness w, which are the variable bits for the circuit.

The action of M is obtained by repeating $t(n)$ times a 'simulation step' which simulates the execution of a single program line for the Turing machine. Each simulation step may be broken up into a sequence of steps corresponding in turn to the respective program lines, with a final step which resets the global flag F to zero, as

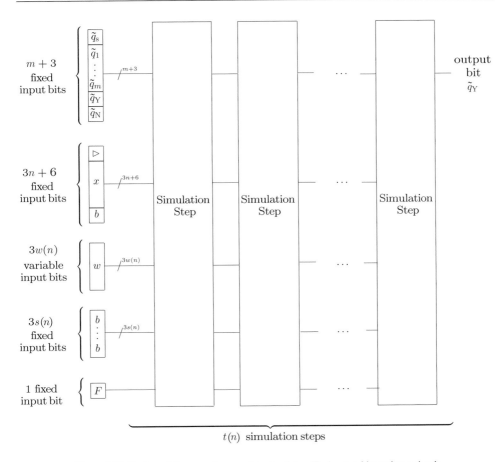

Figure 3.12. Outline of the procedure used to simulate a Turing machine using a circuit.

illustrated in Figure 3.13. To complete the simulation, we only need to simulate a program line of the form $\langle q_i, x, q_j, x', s \rangle$. For convenience, we assume $q_i \neq q_j$, but a similar construction works in the case when $q_i = q_j$. The procedure is as follows:

(1) Check to see that $\tilde{q}_i = 1$, indicating that the current state of the machine is q_i.
(2) For each tape square:

    (a) Check to see that the global flag bit is set to zero, indicating that no action has yet been taken by the Turing machine.

    (b) Check that the flag bit is set to one, indicating that the tape head is at this tape square.

    (c) Check that the simulated tape contents at this point are x.

    (d) If all conditions check out, then perform the following steps:

        1. Set $\tilde{q}_i = 0$ and $\tilde{q}_j = 1$.

        2. Update the simulated tape contents at this tape square to x'.

        3. Update the flag bit of this and adjacent 'squares' as appropriate, depending on whether $s = +1, 0, -1$, and whether we are at the left hand end of the tape.

        4. Set the global flag bit to one, indicating that this round of computation has been completed.

This is a fixed procedure which involves a constant number of bits, and by the universality result of Section 3.1.2 can be performed using a circuit containing a constant number of gates.

Figure 3.13. Outline of the simulation step used to simulate a Turing machine using a circuit.

The total number of gates in the entire circuit is easily seen to be $O(t(n)(s(n) + n))$, which is polynomial in size. At the end of the circuit, it is clear that $\tilde{q}_Y = 1$ if and only if the machine M accepts (x, w). Thus, the circuit is satisfiable if and only if there exists w such that machine M accepts (x, w), and we have found the desired reduction from L to CSAT. □

CSAT gives us a foot in the door which enables us to easily prove that many other problems are **NP**-complete. Instead of directly proving that a problem is **NP**-complete, we can instead prove that it is in **NP** and that CSAT reduces to it, so by Exercise 3.22 the problem must be **NP**-complete. A small sample of **NP**-complete problems is discussed in Box 3.3. An example of another **NP**-complete problem is the *satisfiability* problem (SAT), which is phrased in terms of a Boolean formula. Recall that a Boolean formula φ is composed of the following elements: a set of Boolean variables, $x_1, x_2, \ldots$; Boolean connectives, that is, a Boolean function with one or two inputs and one output, such as $\wedge$ (AND), $\vee$ (OR), and $\neg$ (NOT); and parentheses. The truth or falsity of a Boolean formula for a given set of Boolean variables is decided according to the usual laws of Boolean algebra. For example, the formula $\varphi = x_1 \vee \neg x_2$ has the satisfying assignment $x_1 = 0$ and $x_2 = 0$, while $x_1 = 0$ and $x_2 = 1$ is not a satisfying assignment. The satisfiability problem is to determine, given a Boolean formula φ, whether or not it is satisfiable by any set of possible inputs.

Exercise 3.23:   Show that SAT is **NP**-complete by first showing that SAT is in **NP**, and then showing that CSAT reduces to SAT. (*Hint*: for the reduction it may help to represent each distinct wire in an instance of CSAT by different variables in a Boolean formula.)

An important restricted case of SAT is also **NP**-complete, the 3-satisfiability problem (3SAT), which is concerned with formulae in *3-conjunctive normal form*. A formula is said to be in *conjunctive normal form* if it is the AND of a collection of *clauses*, each of which is the OR of one or more *literals*, where a literal is an expression is of the form x or $\neg x$. For example, the formula $(x_1 \vee \neg x_2) \wedge (x_2 \vee x_3 \vee \neg x_4)$ is in conjunctive normal form. A formula is in *3-conjunctive normal form* or *3-CNF* if each clause has exactly three literals. For example, the formula $(\neg x_1 \vee x_2 \vee \neg x_2) \wedge (\neg x_1 \vee x_3 \vee \neg x_4) \wedge (x_2 \vee x_3 \vee x_4)$ is in 3-conjunctive normal form. The 3-satisfiability problem is to determine whether a formula in 3-conjunctive normal form is satisfiable or not.

The proof that 3SAT is **NP**-complete is straightforward, but is a little too lengthy to justify inclusion in this overview. Even more than CSAT and SAT, 3SAT is in some sense *the* NP-complete problem, and it is the basis for countless proofs that other problems are NP-complete. We conclude our discussion of NP-completeness with the surprising fact that 2SAT, the analogue of 3SAT in which every clause has two literals, can be solved in polynomial time:

Exercise 3.24: (2SAT **has an efficient solution**)   Suppose φ is a Boolean formula in conjunctive normal form, in which each clause contains only two literals.

(1) Construct a (directed) graph $G(\varphi)$ with directed edges in the following way: the vertices of G correspond to variables x_j and their negations $\neg x_j$ in φ. There is a (directed) edge (α, β) in G if and only if the clause $(\neg \alpha \vee \beta)$ or the clause $(\beta \vee \neg \alpha)$ is present in φ. Show that φ is not satisfiable if and only if there exists a variable x such that there are paths from x to $\neg x$ and from $\neg x$ to x in $G(\varphi)$.

(2) Show that given a directed graph G containing n vertices it is possible to determine whether two vertices v_1 and v_2 are connected in polynomial time.

(3) Find an efficient algorithm to solve 2SAT.

Box 3.3: A zoo of NP-complete problems

The importance of the class **NP** derives, in part, from the enormous number of computational problems that are known to be **NP**-complete. We can't possibly hope to survey this topic here (see 'History and further reading'), but the following examples, taken from many distinct areas of mathematics, give an idea of the delicious melange of problems known to be **NP**-complete.

- CLIQUE (*graph theory*): A clique in an undirected graph G is a subset of vertices, each pair of which is connected by an edge. The *size* of a clique is the number of vertices it contains. Given an integer m and a graph G, does G have a clique of size m?
- SUBSET-SUM (*arithmetic*): Given a finite collection S of positive integers and a *target* t, is there any subset of S which sums to t?
- 0-1 INTEGER PROGRAMMING (*linear programming*): Given an integer $m \times n$ matrix A and an m-dimensional vector y with integer values, does there exist an n-dimensional vector x with entries in the set $\{0, 1\}$ such that $Ax \leq y$?
- VERTEX COVER (*graph theory*): A *vertex cover* for an undirected graph G is a set of vertices V' such that every edge in the graph has one or both vertices contained in V'. Given an integer m and a graph G, does G have a vertex cover V' containing m vertices?

Assuming that **P** $\neq$ **NP** it is possible to prove that there is a *non-empty* class of problems **NPI** (**NP** intermediate) which are neither solvable with polynomial resources, nor are **NP**-complete. Obviously, there are no problems known to be in **NPI** (otherwise we would know that **P** $\neq$ **NP**) but there are several problems which are regarded as being likely candidates. Two of the strongest candidates are the factoring and graph isomorphism problems:

GRAPH ISOMORPHISM: Suppose G and G' are two undirected graphs over the vertices $V \equiv \{v_1, \ldots, v_n\}$. Are G and G' isomorphic? That is, does there exist a one-to-one function $\varphi : V \to V$ such that the edge (v_i, v_j) is contained in G if and only if $(\varphi(v_i), \varphi(v_j))$ is contained in G?

Problems in **NPI** are interesting to researchers in quantum computation and quantum information for two reasons. First, it is desirable to find fast quantum algorithms to solve problems which are not in **P**. Second, many suspect that quantum computers will not be able to efficiently solve all problems in **NP**, ruling out **NP**-complete problems. Thus, it is natural to focus on the class **NPI**. Indeed, a fast quantum algorithm for factoring has been discovered (Chapter 5), and this has motivated the search for fast quantum algorithms for other problems suspected to be in **NPI**.

3.2.4 A plethora of complexity classes

We have investigated some of the elementary properties of some important complexity classes. A veritable pantheon of complexity classes exists, and there are many non-trivial relationships known or suspected between these classes. For quantum computation and quantum information, it is not necessary to understand all the different complexity classes that have been defined. However, it is useful to have some appreciation for the more important of the complexity classes, many of which have natural analogues in the study of quantum computation and quantum information. Furthermore, if we are to understand how powerful quantum computers are, then it behooves us to understand how the class of problems solvable on a quantum computer fits into the zoo of complexity classes which may be defined for classical computers.

There are essentially three properties that may be varied in the definition of a complexity class: the resource of interest (time, space, ...), the type of problem being considered (decision problem, optimization problem, ...), and the underlying computational model (deterministic Turing machine, probabilistic Turing machine, quantum computer, ...). Not surprisingly, this gives us an enormous range to define complexity classes. In this section, we briefly review a few of the more important complexity classes and some of their elementary properties. We begin with a complexity class defined by changing the resource of interest from time to *space*.

The most natural space-bounded complexity class is the class **PSPACE** of decision problems which may be solved on a Turing machine using a polynomial number of working bits, with no limitation on the amount of time that may be used by the machine (see Exercise 3.25). Obviously, **P** is included in **PSPACE**, since a Turing machine that halts after polynomial time can only traverse polynomially many squares, but it is also true that **NP** is a subset of **PSPACE**. To see this, suppose L is any language in **NP**. Suppose problems of size n have witnesses of size at most $p(n)$, where $p(n)$ is some polynomial in n. To determine whether or not the problem has a solution, we may sequentially test all $2^{p(n)}$ possible witnesses. Each test can be run in polynomial time, and therefore polynomial space. If we erase all the intermediate working between tests then we can test all the possibilities using polynomial space.

Unfortunately, at present it is not even known whether **PSPACE** contains problems which are not in **P**! This is a pretty remarkable situation – it seems fairly obvious that having unlimited time and polynomial spatial resources must be more powerful than having only a polynomial amount of time. However, despite considerable effort and ingenuity, this has never been shown. We will see later that the class of problems solvable on a quantum computer in polynomial time is a subset of **PSPACE**, so proving that a problem efficiently solvable on a quantum computer is not efficiently solvable on a classical computer would establish that **P** $\neq$ **PSPACE**, and thus solve a major outstanding problem of computer science. An optimistic way of looking at this result is that ideas from quantum computation might be useful in proving that **P** $\neq$ **PSPACE**. Pessimistically, one might conclude that it will be a long time before anyone rigorously proves that quantum computers can be used to efficiently solve problems that are intractable on a classical computer. Even more pessimistically, it is possible that **P** = **PSPACE**, in which case quantum computers offer no advantage over classical computers! However, very few (if any) computational complexity theorists believe that **P** = **PSPACE**.

Exercise 3.25: (PSPACE $\subseteq$ EXP) The complexity class **EXP** (for *exponential time*) contains all decision problems which may be decided by a Turing machine running in exponential time, that is time $O(2^{n^k})$, where k is any constant. Prove that **PSPACE** $\subseteq$ **EXP**. (*Hint*: If a Turing machine has l internal states, an m letter alphabet, and uses space $p(n)$, argue that the machine can exist in one of at most $l m^{p(n)}$ different states, and that if the Turing machine is to avoid infinite loops then it must halt before revisiting a state.)

Exercise 3.26: (L $\subseteq$ P) The complexity class **L** (for *logarithmic space*) contains all decision problems which may be decided by a Turing machine running in logarithmic space, that is, in space $O(\log(n))$. More precisely, the class **L** is defined using a two-tape Turing machine. The first tape contains the problem instance, of size n, and is a read-only tape, in the sense that only program lines which don't change the contents of the first tape are allowed. The second tape is a working tape which initially contains only blanks. The logarithmic space requirement is imposed on the second, working tape only. Show that **L** $\subseteq$ **P**.

Does allowing more time or space give greater computational power? The answer to this question is yes in both cases. Roughly speaking, the *time hierarchy theorem* states that **TIME**$(f(n))$ is a proper subset of **TIME**$(f(n) \log^2(f(n)))$. Similarly, the *space hierarchy theorem* states that **SPACE**$(f(n))$ is a proper subset of **SPACE**$(f(n) \log(f(n)))$, where **SPACE**$(f(n))$ is, of course, the complexity class consisting of all languages that can be decided with spatial resources $O(f(n))$. The hierarchy theorems have interesting implications with respect to the equality of complexity classes. We know that

$$\mathbf{L} \subseteq \mathbf{P} \subseteq \mathbf{NP} \subseteq \mathbf{PSPACE} \subseteq \mathbf{EXP}. \tag{3.1}$$

Unfortunately, although each of these inclusions is widely believed to be strict, none of them has ever been proved to be strict. However, the time hierarchy theorem implies that **P** is a strict subset of **EXP**, and the space hierarchy theorem implies that **L** is a strict subset of **PSPACE**! So we can conclude that at least one of the inclusions in (3.1) must be strict, although we do not know which one.

What should we do with a problem once we know that it is **NP**-complete, or that some other hardness criterion holds? It turns out that this is far from being the end of the story in problem analysis. One possible line of attack is to identify special cases of the problem which may be amenable to attack. For example, in Exercise 3.24 we saw that the 2SAT problem has an efficient solution, despite the **NP**-completeness of SAT.

Another approach is to change the type of problem which is being considered, a tactic which typically results in the definition of new complexity classes. For example, instead of finding exact solutions to an **NP**-complete problem, we can instead try to find good algorithms for finding *approximate* solutions to a problem. For example, the VERTEX COVER problem is an **NP**-complete problem, yet in Exercise 3.27 we show that it is possible to efficiently find an approximation to the minimal vertex cover which is correct to within a factor two! On the other hand, in Problem 3.6 we show that it is not possible to find approximations to solutions of TSP correct to within any factor, *unless* **P** = **NP**!

Exercise 3.27: (Approximation algorithm for VERTEX COVER**)** Let $G = (V, E)$ be an undirected graph. Prove that the following algorithm finds a vertex cover for G that is within a factor two of being a minimal vertex cover:

$$VC = \emptyset$$
$$E' = E$$
do until $E' = \emptyset$
> let (α, β) be any edge of E'
> $VC = VC \cup \{\alpha, \beta\}$
> remove from E' every edge incident on α or β

return VC.

Why is it possible to approximate the solution of one **NP**-complete problem, but not another? After all, isn't it possible to efficiently transform from one problem to another? This is certainly true, however it is not necessarily true that this transformation preserves the notion of a 'good approximation' to a solution. As a result, the computational complexity theory of approximation algorithms for problems in **NP** has a structure that goes beyond the structure of **NP** proper. An entire complexity theory of approximation algorithms exists, which unfortunately is beyond the scope of this book. The basic idea, however, is to define a notion of reduction that corresponds to being able to efficiently reduce one approximation problem to another, in such a way that the notion of good approximation is preserved. With such a notion, it is possible to define complexity classes such as **MAXSNP** by analogy to the class **NP**, as the set of problems for which it is possible to efficiently verify approximate solutions to the problem. Complete problems exist for **MAXSNP**, just as for **NP**, and it is an interesting open problem to determine how the class **MAXSNP** compares to the class of approximation problems which are efficiently solvable.

We conclude our discussion with a complexity class that results when the underlying model of computation itself is changed. Suppose a Turing machine is endowed with the ability to flip coins, using the results of the coin tosses to decide what actions to take during the computation. Such a Turing machine may only accept or reject inputs with a certain probability. The complexity class **BPP** (for *bounded-error probabilistic time*) contains all languages L with the property that there exists a probabilistic Turing machine M such that if $x \in L$ then M accepts x with probability at least $3/4$, and if $x \notin L$, then M rejects x with probability at least $3/4$. The following exercise shows that the choice of the constant $3/4$ is essentially arbitrary:

Exercise 3.28: (Arbitrariness of the constant in the definition of BPP) Suppose
k is a fixed constant, $1/2 < k \leq 1$. Suppose L is a language such that there
exists a Turing machine M with the property that whenever $x \in L$, M accepts
x with probability at least k, and whenever $x \notin L$, M rejects x with probability
at least k. Show that $L \in$ **BPP**.

Indeed, the *Chernoff bound*, discussed in Box 3.4, implies that with just a few repetitions
of an algorithm deciding a language in **BPP** the probability of success can be amplified
to the point where it is essentially equal to one, for all intents and purposes. For this
reason, **BPP** even more than **P** is the class of decision problems which is usually regarded
as being efficiently solvable on a classical computer, and it is the quantum analogue of
BPP, known as **BQP**, that is most interesting in our study of quantum algorithms.

3.2.5 Energy and computation

Computational complexity studies the amount of time and space required to solve a
computational problem. Another important computational resource is *energy*. In this
section, we study the energy requirements for computation. Surprisingly, it turns out that
computation, both classical and quantum, can in principle be done without expending
any energy! Energy consumption in computation turns out to be deeply linked to the
reversibility of the computation. Consider a gate like the NAND gate, which takes as
input two bits, and produces a single bit as output. This gate is intrinsically *irreversible*
because, given the output of the gate, the input is not uniquely determined. For example,
if the output of the NAND gate is 1, then the input could have been any one of $00, 01$,
or 10. On the other hand, the NOT gate is an example of a *reversible* logic gate because,
given the output of the NOT gate, it is possible to infer what the input must have been.

   Another way of understanding irreversibility is to think of it in terms of information
erasure. If a logic gate is irreversible, then some of the information input to the gate is lost
irretrievably when the gate operates – that is, some of the information has been erased by
the gate. Conversely, in a reversible computation, no information is ever erased, because
the input can always be recovered from the output. Thus, saying that a computation is
reversible is equivalent to saying that no information is erased during the computation.

   What is the connection between energy consumption and irreversibility in computation? *Landauer's principle* provides the connection, stating that, in order to erase
information, it is necessary to dissipate energy. More precisely, Landauer's principle
may be stated as follows:

> **Landauer's principle (first form):** Suppose a computer erases a single bit of
> information. The amount of energy dissipated into the environment is *at least*
> $k_B T \ln 2$, where k_B is a universal constant known as *Boltzmann's constant*, and T
> is the temperature of the environment of the computer.

According to the laws of thermodynamics, Landauer's principle can be given an alternative form stated not in terms of energy dissipation, but rather in terms of entropy:

> **Landauer's principle (second form):** Suppose a computer erases a single bit of
> information. The entropy of the environment increases by *at least* $k_B \ln 2$, where
> k_B is Boltzmann's constant.

Justifying Landauer's principle is a problem of physics that lies beyond the scope of this

Box 3.4: BPP and the Chernoff bound

Suppose we have an algorithm for a decision problem which gives the correct answer with probability $1/2 + \epsilon$, and the wrong answer with probability $1/2 - \epsilon$. If we run the algorithm n times, then it seems reasonable to guess that the correct answer is whichever appeared most frequently. How reliably does this procedure work? The *Chernoff bound* is a simple result from elementary probability which answers this question.

Theorem 3.3: (**The Chernoff bound**) Suppose $X_1, \ldots, X_n$ are independent and identically distributed random variables, each taking the value 1 with probability $1/2 + \epsilon$, and the value 0 with probability $1/2 - \epsilon$. Then

$$p\left(\sum_{i=1}^{n} X_i \leq n/2\right) \leq e^{-2\epsilon^2 n}. \tag{3.2}$$

Proof

Consider any sequence $(x_1, \ldots, x_n)$ containing at most $n/2$ ones. The probability of such a sequence occurring is maximized when it contains $\lfloor n/2 \rfloor$ ones, so

$$p(X_1 = x_1, \ldots, X_n = x_n) \leq \left(\frac{1}{2} - \epsilon\right)^{\frac{n}{2}} \left(\frac{1}{2} + \epsilon\right)^{\frac{n}{2}} \tag{3.3}$$

$$= \frac{(1 - 4\epsilon^2)^{\frac{n}{2}}}{2^n}. \tag{3.4}$$

There can be at most 2^n such sequences, so

$$p\left(\sum_{i} X_i \leq n/2\right) \leq 2^n \times \frac{(1 - 4\epsilon^2)^{\frac{n}{2}}}{2^n} = (1 - 4\epsilon^2)^{\frac{n}{2}}. \tag{3.5}$$

Finally, by calculus, $1 - x \leq \exp(-x)$, so

$$p\left(\sum_{i} X_i \leq n/2\right) \leq e^{-4\epsilon^2 n/2} = e^{-2\epsilon^2 n}. \tag{3.6}$$

$\square$

What this tells us is that for fixed ϵ, the probability of making an error decreases *exponentially quickly* in the number of repetitions of the algorithm. In the case of **BPP** we have $\epsilon = 1/4$, so it takes only a few hundred repetitions of the algorithm to reduce the probability of error below 10^{-20}, at which point an error in one of the computer's components becomes much more likely than an error due to the probabilistic nature of the algorithm.

book – see the end of chapter 'History and further reading' if you wish to understand why Landauer's principle holds. However, if we accept Landauer's principle as given, then it raises a number of interesting questions. First of all, Landauer's principle only provides a *lower bound* on the amount of energy that must be dissipated to erase information.

How close are existing computers to this lower bound? Not very, turns out to be the answer – computers circa the year 2000 dissipate roughly $500k_BT \ln 2$ in energy for each elementary logical operation.

Although existing computers are far from the limit set by Landauer's principle, it is still an interesting problem of principle to understand how much the energy consumption can be reduced. Aside from the intrinsic interest of the problem, a practical reason for the interest follows from Moore's law: if computer power keeps increasing then the amount of energy dissipated must also increase, unless the energy dissipated per operation drops at least as fast as the rate of increase in computing power.

If all computations could be done reversibly, then Landauer's principle would imply no lower bound on the amount of energy dissipated by the computer, since no bits at all are erased during a reversible computation. Of course, it is possible that some other physical principle might require that energy be dissipated during the computation; fortunately, this turns out not to be the case. But is it possible to perform universal computation without erasing any information? Physicists can cheat on this problem to see in advance that the answer to this question *must* be yes, because our present understanding of the laws of physics is that they are fundamentally reversible. That is, if we know the final state of a closed physical system, then the laws of physics allow us to work out the initial state of the system. If we believe that those laws are correct, then we must conclude that hidden in the irreversible logic gates like AND and OR, there must be some underlying reversible computation. But where is this hidden reversibility, and can we use it to construct manifestly reversible computers?

We will use two different techniques to give reversible circuit-based models capable of universal computation. The first model, a computer built entirely of billiard balls and mirrors, gives a beautiful concrete realization of the principles of reversible computation. The second model, based on a reversible logic gate known as the *Toffoli gate* (which we first encountered in Section 1.4.1), is a more abstract view of reversible computation that will later be of great use in our discussion of quantum computation. It is also possible to build reversible Turing machines that are universal for computation; however, we won't study these here, since the reversible circuit models turn out to be much more useful for quantum computation.

The basic idea of the billiard ball computer is illustrated in Figure 3.14. Billiard ball 'inputs' enter the computer from the left hand side, bouncing off mirrors and each other, before exiting as 'outputs' on the right hand side. The presence or absence of a billiard ball at a possible input site is used to indicate a logical 1 or a logical 0, respectively. The fascinating thing about this model is that it is manifestly reversible, insofar as its operation is based on the laws of classical mechanics. Furthermore, this model of computation turns out to be *universal* in the sense that it can be used to simulate an arbitrary computation in the standard circuit model of computation.

Of course, if a billiard ball computer were ever built it would be highly unstable. As any billiards player can attest, a billiard ball rolling frictionlessly over a smooth surface is easily knocked off course by small perturbations. The billiard ball model of computation depends on perfect operation, and the absence of external perturbations such as those caused by thermal noise. Periodic corrections can be performed, but information gained by doing this would have to be erased, requiring work to be performed. Expenditure of energy thus serves the purpose of reducing this susceptibility to noise, which is necessary for a practical, real-world computational machine. For the purposes of this introduction,

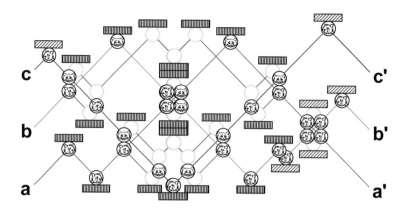

Figure 3.14. A simple billiard ball computer, with three input bits and three output bits, shown entering on the left and leaving on the right, respectively. The presence or absence of a billiard ball indicates a 1 or a 0, respectively. Empty circles illustrate potential paths due to collisions. This particular computer implements the Fredkin classical reversible logic gate, discussed in the text.

we will ignore the effects of noise on the billiard ball computer, and concentrate on understanding the essential elements of reversible computation.

The billiard ball computer provides an elegant means for implementing a reversible universal logic gate known as the *Fredkin gate*. Indeed, the properties of the Fredkin gate provide an informative overview of the general principles of reversible logic gates and circuits. The Fredkin gate has three input bits and three output bits, which we refer to as a, b, c and a', b', c', respectively. The bit c is a *control bit*, whose value is not changed by the action of the Fredkin gate, that is, $c' = c$. The reason c is called the control bit is because it controls what happens to the other two bits, a and b. If c is set to 0 then a and b are left alone, $a' = a$, $b' = b$. If c is set to 1, a and b are swapped, $a' = b$, $b' = a$. The explicit truth table for the Fredkin gate is shown in Figure 3.15. It is easy to see that the Fredkin gate is reversible, because given the output a', b', c', we can determine the inputs a, b, c. In fact, to recover the original inputs a, b and c we need only apply another Fredkin gate to a', b', c':

Exercise 3.29: (Fredkin gate is self-inverse)   Show that applying two consecutive Fredkin gates gives the same outputs as inputs.

Examining the paths of the billiard balls in Figure 3.14, it is not difficult to verify that this billiard ball computer implements the Fredkin gate:

Exercise 3.30:   Verify that the billiard ball computer in Figure 3.14 computes the Fredkin gate.

In addition to reversibility, the Fredkin gate also has the interesting property that the number of 1s is *conserved* between the input and output. In terms of the billiard ball computer, this corresponds to the number of billiard balls going into the Fredkin gate being equal to the number coming out. Thus, it is sometimes referred to as being a *conservative* reversible logic gate. Such reversibility and conservative properties are interesting to a physicist because they can be motivated by fundamental physical princi-

Inputs			Outputs		
a	b	c	a'	b'	c'
0	0	0	0	0	0
0	0	1	0	0	1
0	1	0	0	1	0
0	1	1	1	0	1
1	0	0	1	0	0
1	0	1	0	1	1
1	1	0	1	1	0
1	1	1	1	1	1

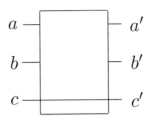

Figure 3.15. Fredkin gate truth table and circuit representation. The bits a and b are swapped if the control bit c is set, and otherwise are left alone.

ples. The laws of Nature are reversible, with the possible exception of the measurement postulate of quantum mechanics, discussed in Section 2.2.3 on page 84. The conservative property can be thought of as analogous to properties such as conservation of mass, or conservation of energy. Indeed, in the billiard ball model of computation the conservative property corresponds exactly to conservation of mass.

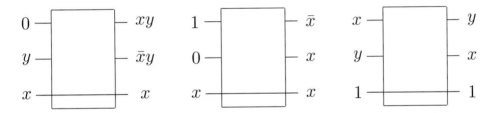

Figure 3.16. Fredkin gate configured to perform the elementary gates AND (left), NOT (middle), and a primitive routing function, the CROSSOVER (right). The middle gate also serves to perform the FANOUT operation, since it produces two copies of x at the output. Note that each of these configurations requires the use of extra 'ancilla' bits prepared in standard states – for example, the 0 input on the first line of the AND gate – and in general the output contains 'garbage' not needed for the remainder of the computation.

The Fredkin gate is not only reversible and conservative, it's a universal logic gate as well! As illustrated in Figure 3.16, the Fredkin gate can be configured to simulate AND, NOT, CROSSOVER and FANOUT functions, and thus can be cascaded to simulate any classical circuit whatsoever.

To simulate irreversible gates such as AND using the Fredkin gate, we made use of two ideas. First, we allowed the input of 'ancilla' bits to the Fredkin gate, in specially prepared states, either 0 or 1. Second, the output of the Fredkin gate contained extraneous 'garbage' not needed for the remainder of the computation. These ancilla and garbage bits are not directly important to the computation. Their importance lies in the fact that they make the computation reversible. Indeed the irreversibility of gates like the AND and OR may be viewed as a consequence of the ancilla and garbage bits being 'hidden'. Summarizing, given any classical circuit computing a function $f(x)$, we can build a reversible circuit made entirely of Fredkin gates, which on input of x, together with some ancilla bits in a standard state a, computes $f(x)$, together with some extra 'garbage' output, $g(x)$. Therefore, we represent the action of the computation as $(x, a) \rightarrow (f(x), g(x))$.

We now know how to compute functions reversibly. Unfortunately, this computation produces unwanted garbage bits. With some modifications it turns out to be possible to perform the computation so that any garbage bits produced are in a *standard state*. This construction is crucial for quantum computation, because garbage bits whose value depends upon x will in general destroy the interference properties crucial to quantum computation. To understand how this works it is convenient to assume that the NOT gate is available in our repertoire of reversible gates, so we may as well assume that the ancilla bits a all start out as 0s, with NOT gates being added where necessary to turn the ancilla 0s into 1s. It will also be convenient to assume that the classical controlled-NOT gate is available, defined in a manner analogous to the quantum definition of Section 1.3.2, that is, the inputs (c, t) are taken to $(c, t \oplus c)$, where $\oplus$ denotes addition modulo 2. Notice that $t = 0$ gives $(c, 0) \rightarrow (c, c)$, so the controlled-NOT can be thought of as a reversible copying gate or FANOUT, which leaves no garbage bits at the output.

With the additional NOT gates appended at the beginning of the circuit, the action of the computation may be written as $(x, 0) \rightarrow (f(x), g(x))$. We could also have added CNOT gates to the beginning of the circuit, in order to create a copy of x which is not changed during the subsequent computation. With this modification, the action of the circuit may be written

$$(x, 0, 0) \rightarrow (x, f(x), g(x)). \tag{3.7}$$

Equation (3.7) is a very useful way of writing the action of the reversible circuit, because it allows an idea known as *uncomputation* to be used to get rid of the garbage bits, for a small cost in the running time of the computation. The idea is the following. Suppose we start with a four register computer in the state $(x, 0, 0, y)$. The second register is used to store the result of the computation, and the third register is used to provide workspace for the computation, that is, the garbage bits $g(x)$. The use of the fourth register is described shortly, and we assume it starts in an arbitrary state y.

We begin as before, by applying a reversible circuit to compute f, resulting in the state $(x, f(x), g(x), y)$. Next, we use CNOTs to add the result $f(x)$ bitwise to the fourth register, leaving the machine in the state $(x, f(x), g(x), y \oplus f(x))$. However, all the steps used to compute $f(x)$ were reversible and did not affect the fourth register, so by applying the reverse of the circuit used to compute f we come to the state $(x, 0, 0, y \oplus f(x))$. Typically, we omit the ancilla 0s from the description of the function evaluation, and just write the action of the circuit as

$$(x, y) \rightarrow (x, y \oplus f(x)). \tag{3.8}$$

In general we refer to this modified circuit computing f as *the* reversible circuit computing f, even though in principle there are many other reversible circuits which could be used to compute f.

What resource overhead is involved in doing reversible computation? To analyze this question, we need to count the number of extra ancilla bits needed in a reversible circuit, and compare the gate counts with classical models. It ought to be clear that the number of gates in a reversible circuit is the same as in an irreversible circuit to within the constant factor which represents the number of Fredkin gates needed to simulate a single element of the irreversible circuit, and an additional factor of two for uncomputation, with an overhead for the extra CNOT operations used in reversible computation which is linear in the number of bits involved in the circuit. Similarly, the number of ancilla bits required scales at most linearly with the number of gates in the irreversible circuit, since each element in the irreversible circuit can be simulated using a constant number of ancilla bits. As a result, natural complexity classes such as **P** and **NP** are the same no matter whether a reversible or irreversible model of computation is used. For more elaborate complexity classes like **PSPACE** the situation is not so immediately clear; see Problem 3.9 and 'History and further reading' for a discussion of some such subtleties.

Exercise 3.31: (Reversible half-adder) Construct a reversible circuit which, when two bits x and y are input, outputs $(x, y, c, x \oplus y)$, where c is the carry bit when x and y are added.

The Fredkin gate and its implementation using the billiard ball computer offers a beautiful paradigm for reversible computation. There is another reversible logic gate, the *Toffoli gate*, which is also universal for classical computation. While the Toffoli gate does not have quite the same elegant physical simplicity as the billiard ball implementation of the Fredkin gate, it will be more useful in the study of quantum computation. We have already met the Toffoli gate in Section 1.4.1, but for convenience we review its properties here.

The Toffoli gate has three input bits, a, b and c. a and b are known as the first and second *control bits*, while c is the *target bit*. The gate leaves both control bits unchanged, flips the target bit if both control bits are set, and otherwise leaves the target bit alone. The truth table and circuit representation for the Toffoli gate are shown in Figure 3.17.

Inputs			Outputs		
a	b	c	a'	b'	c'
0	0	0	0	0	0
0	0	1	0	0	1
0	1	0	0	1	0
0	1	1	0	1	1
1	0	0	1	0	0
1	0	1	1	0	1
1	1	0	1	1	1
1	1	1	1	1	0

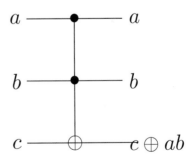

Figure 3.17. Truth table and circuit representation of the Toffoli gate.

How can the Toffoli gate be used to do universal computation? Suppose we wish to NAND the bits a and b. To do this using the Toffoli gate, we input a and b as control bits, and send in an ancilla bit set to 1 as the target bit, as shown in Figure 3.18. The NAND of a and b is output as the target bit. As expected from our study of the Fredkin gate, the Toffoli gate simulation of a NAND requires the use of a special ancilla input, and some of the outputs from the simulation are garbage bits.

The Toffoli gate can also be used to implement the FANOUT operation by inputting an ancilla 1 to the first control bit, and a to the second control bit, producing the output $1, a, a$. This is illustrated in Figure 3.19. Recalling that NAND and FANOUT are together

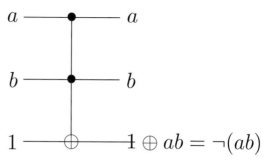

Figure 3.18. Implementing a NAND gate using a Toffoli gate. The top two bits represent the input to the NAND, while the third bit is prepared in the standard state 1, sometimes known as an *ancilla* state. The output from the NAND is on the third bit.

universal for computation, we see that an arbitrary circuit can be efficiently simulated using a reversible circuit consisting only of Toffoli gates and ancilla bits, and that useful additional techniques such as uncomputation may be achieved using the same methods as were employed with the Fredkin gate.

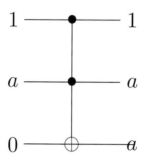

Figure 3.19. FANOUT with the Toffoli gate, with the second bit being the input to the FANOUT, and the other two bits standard ancilla states. The output from FANOUT appears on the second and third bits.

Our interest in reversible computation was motivated by our desire to understand the energy requirements for computation. It is clear that the noise-free billiard ball model of computation requires no energy for its operation; what about models based upon the Toffoli gate? This can only be determined by examining specific models for the computation of the Toffoli gate. In Chapter 7, we examine several such implementations, and it turns out that, indeed, the Toffoli gate can be implemented in a manner which does not require the expenditure of energy.

There is a significant caveat attached to the idea that computation can be done without the expenditure of energy. As we noted earlier, the billiard ball model of computation is highly sensitive to noise, and this is true of many other models of reversible computation. To nullify the effects of noise, some form of error-correction needs to be done. Such error-correction typically involves the performance of measurements on the system to determine whether the system is behaving as expected, or if an error has occurred. Because the computer's memory is finite, the bits used to store the measurement results utilized in error-correction must eventually be erased to make way for new measurement results. According to Landauer's principle, this erasure carries an associated energy cost that must be accounted for when tallying the total energy cost of the computation. We analyze the energy cost associated with error-correction in more detail in Section 12.4.4.

What can we conclude from our study of reversible computation? There are three key ideas. First, reversibility stems from keeping track of every bit of information; irreversibility occurs only when information is lost or erased. Second, by doing computation reversibly, we obviate the need for energy expenditure during computation. All computations can be done, in principle, for zero cost in energy. Third, reversible computation can be done efficiently, without the production of garbage bits whose value depends upon the input to the computation. That is, if there is an irreversible circuit computing a function f, then there is an efficient simulation of this circuit by a reversible circuit with action $(x, y) \rightarrow (x, y \oplus f(x))$.

What are the implications of these results for physics, computer science, and for quantum computation and quantum information? From the point of view of a physicist or hardware engineer worried about heat dissipation, the good news is that, in principle, it is possible to make computation dissipation-free by making it reversible, although in practice energy dissipation is required for system stability and immunity from noise. At an even more fundamental level, the ideas leading to reversible computation also lead to the resolution of a century-old problem in the foundations of physics, the famous problem of *Maxwell's demon*. The story of this problem and its resolution is outlined in Box 3.5 on page 162. From the point of view of a computer scientist, reversible computation validates the use of irreversible elements in models of computation such as the Turing machine (since using them or not gives polynomially equivalent models). Moreover, since the physical world is fundamentally reversible, one can argue that complexity classes based upon reversible models of computation are more natural than complexity classes based upon irreversible models, a point revisited in Problem 3.9 and 'History and further reading'. From the point of view of quantum computation and quantum information, reversible computation is enormously important. To harness the full power of quantum computation, any classical subroutines in a quantum computation must be performed reversibly and without the production of garbage bits depending on the classical input.

Exercise 3.32: (From Fredkin to Toffoli and back again) What is the smallest number of Fredkin gates needed to simulate a Toffoli gate? What is the smallest number of Toffoli gates needed to simulate a Fredkin gate?

3.3 Perspectives on computer science

In a short introduction such as this chapter, it is not remotely possible to cover in detail all the great ideas of a field as rich as computer science. We hope to have conveyed to you something of what it means to *think* like a computer scientist, and provided a basic vocabulary and overview of some of the fundamental concepts important in the understanding of computation. To conclude this chapter, we briefly touch on some more general issues, in order to provide some perspective on how quantum computation and quantum information fits into the overall picture of computer science.

Our discussion has revolved around the Turing machine model of computation. How does the computational power of unconventional models of computation such as massively parallel computers, DNA computers and analog computers compare with the standard

Box 3.5: Maxwell's demon

The laws of thermodynamics govern the amount of work that can be performed by a physical system at thermodynamic equilibrium. One of these laws, the second law of thermodynamics, states that the *entropy* in a closed system can never decrease. In 1871, James Clerk Maxwell proposed the existence of a machine that apparently violated this law. He envisioned a miniature little 'demon', like that shown in the figure below, which could reduce the entropy of a gas cylinder initially at equilibrium by individually separating the fast and slow molecules into the two halves of the cylinder. This demon would sit at a little door at the middle partition. When a fast molecule approaches from the left side the demon opens a door between the partitions, allowing the molecule through, and then closes the door. By doing this many times the total entropy of the cylinder can be *decreased*, in apparent violation of the second law of thermodynamics.

The resolution to the Maxwell's demon paradox lies in the fact that the demon must perform *measurements* on the molecules moving between the partitions, in order to determine their velocities. The result of this measurement must be stored in the demon's memory. Because any memory is finite, the demon must eventually begin erasing information from its memory, in order to have space for new measurement results. By Landauer's principle, this act of erasing information increases the total entropy of the combined system – demon, gas cylinder, and their environments. In fact, a complete analysis shows that Landauer's principle implies that the entropy of the combined system is increased *at least as much* by this act of erasing information as the entropy of the combined system is decreased by the actions of the demon, thus ensuring that the second law of thermodynamics is obeyed.

Turing machine model of computation and, implicitly, with quantum computation? Let's begin with parallel computing architectures. The vast majority of computers in existence are serial computers, processing instructions one at a time in some central processing unit. By contrast, parallel computers can process more than one instruction at a time, leading to a substantial savings in time and money for some applications. Nevertheless, parallel processing does not offer any fundamental advantage over the standard Turing machine model when issues of efficiency are concerned, because a Turing machine can simulate a parallel computer with polynomially equivalent total physical resources – the total space and time used by the computation. What a parallel computer gains in time, it loses in the total spatial resources required to perform the computation, resulting in a net of no essential change in the power of the computing model.

An interesting specific example of massively parallel computing is the technique of *DNA computing*. A strand of DNA, deoxyribonucleic acid, is a molecule composed of a sequence (a polymer) of four kinds of nucleotides distinguished by the bases they carry, denoted by the letter A (adenine), C (cytosine), G (guanine) and T (thymine). Two strands, under certain circumstances, can anneal to form a double strand, if the respective base pairs form complements of each other (A matches T and G matches C). The ends are also distinct and must match appropriately. Chemical techniques can be used to amplify the number of strands beginning or ending with specific sequences (polymerase chain reaction), separate the strands by length (gel electrophoresis), dissolve double strands into single strands (changing temperature and pH), read the sequence on a strand, cut strands at a specific position (restriction enzymes), and detect if a certain sequence of DNA is in a test tube. The procedure for using these mechanisms in a robust manner is rather involved, but the basic idea can be appreciated from an example.

The directed Hamiltonian path problem is a simple and equivalently hard variant of the Hamiltonian cycle problem of Section 3.2.2, in which the goal is to determine if a path exists or not between two specified vertices j_1 and j_N in a directed graph G of N vertices, entering each vertex exactly once, and following only allowed edge directions. This problem can be solved with a DNA computer using the following five steps, in which x_j are chosen to be unique sequences of bases (and $\bar{x}_j$ their complements), DNA strands $x_j x_k$ encode edges, and strands $\bar{x}_j \bar{x}_j$ encode vertices. (1) Generate random paths through G, by combining a mixture of all possible vertex and edge DNA strands, and waiting for the strands to anneal. (2) Keep only the paths beginning with j_1 and ending with j_N, by amplifying only the double strands beginning with $\bar{x}_{j_1}$ and ending with $\bar{x}_{j_N}$. (3) Select only paths of length N, by separating the strands according to their length. (4) Select only paths which enter each vertex at least once, by dissolving the DNA into single strands, and annealing with all possible vertex strands one at a time and filtering out only those strands which anneal. And (5) detect if any strands have survived the selection steps; if so, then a path exists, and otherwise, it does not. To ensure the answer is correct with sufficiently high probability, x_j may be chosen to contain many (≈ 30) bases, and a large number ($\approx 10^{14}$ or more are feasible) of strands are used in the reaction.

Heuristic methods are available to improve upon this basic idea. Of course, exhaustive search methods such as this only work as long as all possible paths can be generated efficiently, and thus the number of molecules used must grow exponentially as the size of the problem (the number of vertices in the example above). DNA molecules are relatively small and readily synthesized, and the huge number of DNA combinations one can fit into a test tube can stave off the exponential complexity cost increase for a while – up to a few dozen vertices – but eventually the exponential cost limits the applicability of this method. Thus, while DNA computing offers an attractive and physically realizable model of computation for the solution of certain problems, it is a classical computing technique and offers no essential improvement in principle over a Turing machine.

Analog computers offer a yet another paradigm for performing computation. A computer is analog when the physical representation of information it uses for computation is based on continuous degrees of freedom, instead of zeroes and ones. For example, a thermometer is an analog computer. Analog circuitry, using resistors, capacitors, and amplifiers, is also said to perform analog computation. Such machines have an infinite resource to draw upon in the ideal limit, since continuous variables like position and voltage can store an unlimited amount of information. But this is only true in the absence of noise. The presence of a finite amount of noise reduces the number of *distinguishable states* of a continuous variable to a finite number – and thus restricts analog computers to the representation of a finite amount of information. In practice, noise reduces analog computers to being no more powerful than conventional digital computers, and through them Turing machines. One might suspect that quantum computers are just analog computers, because of the use of continuous parameters in describing qubit states; however, it turns out that the effects of noise on a quantum computer can effectively be *digitized*. As a result, their computational advantages remain even in the presence of a finite amount of noise, as we shall see in Chapter 10.

What of the effects of noise on digital computers? In the early days of computation, noise was a very real problem for computers. In some of the original computers a vacuum tube would malfunction every few minutes. Even today, noise is a problem for computational devices such as modems and hard drives. Considerable effort was devoted to the problem of understanding how to construct reliable computers from unreliable components. It was proven by von Neumann that this is possible with only a polynomial increase in the resources required for computation. Ironically, however, modern computers use none of those results, because the components of modern computers are fantastically reliable. Failure rates of 10^{-17} and even less are common in modern electronic components. For this reason, failures happen so rarely that the extra effort required to protect against them is not regarded as being worth making. On the other hand, we shall find that quantum computers are very delicate machines, and will likely require substantial application of error-correction techniques.

Different architectures may change the effects of noise. For example, if the effect of noise is ignored, then changing to a computer architecture in which many operations are performed in parallel may not change the number of operations which need to be done. However, a parallel system may be substantially more resistant to noise, because the effects of noise have less time to accumulate. Therefore, in a realistic analysis, the parallel version of an algorithm may have some substantial advantages over a serial implementation. Architecture design is a well developed field of study for classical computers. Hardly anything similar has been developed along the same lines for quantum computers, but the study of noise already suggests some desirable traits for future quantum computer architectures, such as a high level of parallelism.

A fourth model of computation is *distributed computation*, in which two or more spatially separated computational units are available to solve a computational problem. Obviously, such a model of computation is no more powerful than the Turing machine model in the sense that it can be efficiently simulated on a Turing machine. However, distributed computation gives rise to an intriguing new resource challenge: how best to utilize multiple computational units when the cost of *communication* between the units is high. This problem of distributed computation becomes especially interesting as computers are connected through high speed networks; although the total computational capacity of all the computers on a network might be extremely large, utilization of that potential is difficult. Most interesting problems do not divide easily into independent chunks that can be solved separately, and may frequently require global communication between different computational subsystems to exchange intermediate results or synchronize status. The field of *communication complexity* has been developed to address such issues, by quantifying the cost of communication requirements in solving problems. When quantum resources are available and can be exchanged between distributed computers, the communication costs can sometimes be greatly reduced.

A recurring theme through these concluding thoughts and through the entire book is that despite the traditional independence of computer science from physical constraints, ultimately physical laws have tremendous impact not only upon how computers are realized, but also the class of problems they are capable of solving. The success of quantum computation and quantum information as a physically reasonable alternative model of computation questions closely held tenets of computer science, and thrusts notions of computer science into the forefront of physics. The task of the remainder of this book is to stir together ideas from these disparate fields, and to delight in what results!

Problem 3.1: (Minsky machines) A *Minsky machine* consists of a finite set of registers, $r_1, r_2, \ldots, r_k$, each capable of holding an arbitrary non-negative integer, and a *program*, made up of *orders* of one of two types. The first type has the form:

The interpretation is that at point m in the program register r_j is incremented by one, and execution proceeds to point n in the program. The second type of order has the form:

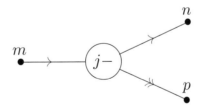

The interpretation is that at point m in the program, register r_j is decremented if it contains a positive integer, and execution proceeds to point n in the program. If register r_j is zero then execution simply proceeds to point p in the program. The *program* for the Minsky machine consists of a collection of such orders, of a form like:

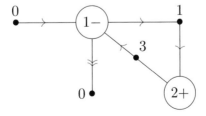

The starting and all possible halting points for the program are conventionally labeled zero. This program takes the contents of register r_1 and adds them to register r_2, while decrementing r_1 to zero.

(1) Prove that all (Turing) computable functions can be computed on a Minsky machine, in the sense that given a computable function $f(\cdot)$ there is a Minsky machine program that when the registers start in the state $(n, 0, \ldots, 0)$ gives as output $(f(n), 0, \ldots, 0)$.

(2) Sketch a proof that any function which can be computed on a Minsky machine, in the sense just defined, can also be computed on a Turing machine.

Problem 3.2: (Vector games) A *vector game* is specified by a finite list of vectors, all of the same dimension, and with integer co-ordinates. The game is to start with a vector x of non-negative integer co-ordinates and to add to x the first vector from the list which preserves the non-negativity of all the components, and to repeat this process until it is no longer possible. Prove that for any computable function $f(\cdot)$ there is a vector game which when started with the vector $(n, 0, \ldots, 0)$ reaches $(f(n), 0, \ldots, 0)$. (*Hint*: Show that a vector game in $k + 2$ dimensions can simulate a Minsky machine containing k registers.)

Problem 3.3: (Fractran) A *Fractran* program is defined by a list of positive rational numbers $q_1, \ldots, q_n$. It acts on a positive integer m by replacing it by $q_i m$, where i is the least number such that $q_i m$ is an integer. If there is ever a time when there is no i such that $q_i m$ is an integer, then execution stops. Prove that for any computable function $f(\cdot)$ there is a Fractran program which when started with 2^n reaches $2^{f(n)}$ without going through any intermediate powers of 2. (*Hint*: use the previous problem.)

Problem 3.4: (Undecidability of dynamical systems) A Fractran program is essentially just a very simple dynamical system taking positive integers to positive integers. Prove that there is no algorithm to decide whether such a dynamical system ever reaches 1.

Problem 3.5: (Non-universality of two bit reversible logic) Suppose we are trying to build circuits using only one and two bit reversible logic gates, and ancilla bits. Prove that there are Boolean functions which cannot be computed in this fashion. Deduce that the Toffoli gate cannot be simulated using one and two bit reversible gates, even with the aid of ancilla bits.

Problem 3.6: (Hardness of approximation of TSP) Let $r \geq 1$ and suppose that there is an approximation algorithm for TSP which is guaranteed to find the shortest tour among n cities to within a factor r. Let $G = (V, E)$ be any graph on n vertices. Define an instance of TSP by identifying cities with vertices in V, and defining the distance between cities i and j to be 1 if (i, j) is an edge of G, and to be $\lceil r \rceil |V| + 1$ otherwise. Show that if the approximation algorithm is applied to this instance of TSP then it returns a Hamiltonian cycle for G if one exists, and otherwise returns a tour of length more than $\lceil r \rceil |V|$. From the **NP**-completeness of HC it follows that no such approximation algorithm can exist unless **P = NP**.

Problem 3.7: (Reversible Turing machines)

(1) Explain how to construct a reversible Turing machine that can compute the same class of functions as is computable on an ordinary Turing machine. (*Hint*: It may be helpful to use a multi-tape construction.)

(2) Give general space and time bounds for the operation of your reversible Turing machine, in terms of the time $t(x)$ and space $s(x)$ required on an ordinary single-tape Turing machine to compute a function $f(x)$.

Problem 3.8: (Find a hard-to-compute class of functions (Research)) Find a natural class of functions on n inputs which requires a super-polynomial number of Boolean gates to compute.

Problem 3.9: (Reversible PSPACE = PSPACE) It can be shown that the problem 'quantified satisfiability', or QSAT, is **PSPACE**-complete. That is, every other language in **PSPACE** can be reduced to QSAT in polynomial time. The language QSAT is defined to consist of all Boolean formulae φ in n variables $x_1, \ldots, x_n$, and in conjunctive normal form, such that:

$$\exists_{x_1} \forall_{x_2} \exists_{x_3} \ldots \forall_{x_n} \varphi \text{ if } n \text{ is even;} \tag{3.9}$$

$$\exists_{x_1} \forall_{x_2} \exists_{x_3} \ldots \exists_{x_n} \varphi \text{ if } n \text{ is odd.} \tag{3.10}$$

Prove that a reversible Turing machine operating in polynomial space can be used to solve QSAT. Thus, the class of languages decidable by a computer operating reversibly in polynomial space is equal to **PSPACE**.

Problem 3.10: (Ancilla bits and efficiency of reversible computation) Let p_m be the mth prime number. Outline the construction of a reversible circuit which, upon input of m and n such that $n > m$, outputs the product $p_m p_n$, that is $(m, n) \to (p_m p_n, g(m, n))$, where $g(m, n)$ is the final state of the ancilla bits used by the circuit. Estimate the number of ancilla qubits your circuit requires. Prove that if a polynomial (in $\log n$) size reversible circuit can be found that uses $O(\log(\log n))$ ancilla bits then the problem of factoring a product of two prime numbers is in **P**.

History and further reading

Computer science is a huge subject with many interesting subfields. We cannot hope for any sort of completeness in this brief space, but instead take the opportunity to recommend a few titles of general interest, and some works on subjects of specific interest in relation to topics covered in this book, with the hope that they may prove stimulating.

Modern computer science dates to the wonderful 1936 paper of Turing[Tur36]. The Church–Turing thesis was first stated by Church[Chu36] in 1936, and was then given a more complete discussion from a different point of view by Turing. Several other researchers found their way to similar conclusions at about the same time. Many of these contributions and a discussion of the history may be found in a volume edited by Davis[Dav65]. Provocative discussions of the Church–Turing thesis and undecidability may be found in Hofstadter[Hof79] and Penrose[Pen89].

There are many excellent books on algorithm design. We mention only three. First, there is the classic series by Knuth[Knu97, Knu98a, Knu98b] which covers an enormous portion of computer science. Second, there is the marvelous book by Cormen, Leiserson, and Rivest[CLR90]. This huge book contains a plethora of well-written material on many areas of algorithm design. Finally, the book of Motwani and Raghavan[MR95] is an excellent survey of the field of randomized algorithms.

The modern theory of computational complexity was especially influenced by the papers of Cook[Coo71] and Karp[Kar72]. Many similar ideas were arrived at independently in Russia by Levin[Lev73], but unfortunately took time to propagate to the West. The classic book by Garey and Johnson[GJ79] has also had an enormous influence on the field. More recently, Papadimitriou[Pap94] has written a beautiful book that surveys many of the main ideas of computational complexity theory. Much of the material in this chapter is based upon Papadimitriou's book. In this chapter we considered only one type of reducibility between languages, polynomial time reducibility. There are many other notions of reductions between languages. An early survey of these notions was given by Ladner, Lynch and Selman[LLS75]. The study of different notions of reducibility later blossomed into a subfield of research known as *structural complexity*, which has been reviewed by Balcázar, Diaz, and Gabarró[BDG88a, BDG88b].

The connection between information, energy dissipation, and computation has a long history. The modern understanding is due to a 1961 paper by Landauer[Lan61], in which Landauer's principle was first formulated. A paper by Szilard[Szi29] and a 1949 lecture by von Neumann[von66] (page 66) arrive at conclusions close to Landauer's principle, but do not fully grasp the essential point that it is the *erasure* of information that requires dissipation.

Reversible Turing machines were invented by Lecerf[Lec63] and later, but independently, in an influential paper by Bennett[Ben73]. Fredkin and Toffoli[FT82] introduced reversible circuit models of computation. Two interesting historical documents are Barton's May, 1978 MIT 6.895 term paper[Bar78], and Ressler's 1981 Master's thesis[Res81], which contain designs for a reversible PDP-10! Today, reversible logic is potentially important in implementations of low-power CMOS circuitry[YK95].

Maxwell's demon is a fascinating subject, with a long and intricate history. Maxwell proposed his demon in 1871[Max71]. Szilard published a key paper in 1929[Szi29] which anticipated many of the details of the final resolution of the problem of Maxwell's demon. In 1965 Feynman[FLS65b] resolved a special case of Maxwell's demon. Bennett, building on Landauer's work[Lan61], wrote two beautiful papers on the subject[BBBW82, Ben87] which completed the resolution of the problem. An interesting book about the history of Maxwell's demon and its exorcism is the collection of papers by Leff and Rex[LR90].

DNA computing was invented by Adleman, and the solution of the directed Hamiltonian path problem we describe is his[Adl94]. Lipton has also shown how 3SAT and circuit satisfiability can be solved in this model[Lip95]. A good general article is Adleman's *Scientific American* article[Adl98]; for an insightful look into the universality of DNA operations, see Winfree[Win98]. An interesting place to read about performing reliable computation in the presence of noise is the book by Winograd and Cowan[WC67]. This topic will be addressed again in Chapter 10. A good textbook on computer architecture is by Hennessey, Goldberg, and Patterson.[HGP96].

Problems 3.1 through 3.4 explore a line of thought originated by Minsky (in his beautiful book on computational machines[Min67]) and developed by Conway[Con72, Con86]. The Fractran programming language is certainly one of the most beautiful and elegant universal computational models known, as demonstrated by the following example, known as PRIMEGAME[Con86]. PRIMEGAME is defined by the list of rational numbers:

$$\frac{17}{91}, \frac{78}{85}, \frac{19}{51}, \frac{23}{38}, \frac{29}{33}, \frac{77}{29}, \frac{95}{23}, \frac{77}{19}, \frac{1}{17}, \frac{11}{13}, \frac{13}{11}, \frac{15}{2}, \frac{1}{7}, \frac{55}{1}. \tag{3.11}$$

Amazingly, when PRIMEGAME is started at 2, the other powers of 2 that appear, namely, $2^2, 2^3, 2^5, 2^7, 2^{11}, 2^{13}, \ldots$, are precisely the prime powers of 2, with the powers stepping through the prime numbers, in order. Problem 3.9 is a special case of the more general subject of the spatial requirements for reversible computation. See the papers by Bennett[Ben89], and by Li, Tromp and Vitanyi[LV96, LTV98].

II Quantum computation

4 Quantum circuits

The theory of computation has traditionally been studied almost entirely in the abstract, as a topic in pure mathematics. This is to miss the point of it. Computers are physical objects, and computations are physical processes. What computers can or cannot compute is determined by the laws of physics alone, and not by pure mathematics.
– David Deutsch

Like mathematics, computer science will be somewhat different from the other sciences, in that it deals with artificial laws that can be proved, instead of natural laws that are never known with certainty.
– Donald Knuth

The opposite of a profound truth may well be another profound truth.
– Niels Bohr

This chapter begins Part II of the book, in which we explore quantum computation in detail. The chapter develops the fundamental principles of quantum computation, and establishes the basic building blocks for quantum circuits, a universal language for describing sophisticated quantum computations. The two fundamental quantum algorithms known to date are constructed from these circuits in the following two chapters. Chapter 5 presents the quantum Fourier transform and its applications to phase estimation, order-finding and factoring. Chapter 6 describes the quantum search algorithm, and its applications to database search, counting and speedup of solutions to **NP**-complete problems. Chapter 7 concludes Part II with a discussion of how quantum computation may one day be experimentally realized. Two other topics of great interest for quantum computation, quantum noise and quantum error-correction, are deferred until Part III of the book, in view of their wide interest also *outside* quantum computation.

There are two main ideas introduced in this chapter. First, we explain in detail the fundamental model of quantum computation, the quantum circuit model. Second, we demonstrate that there exists a small set of gates which are *universal*, that is, any quantum computation whatsoever can be expressed in terms of those gates. Along the way we also have occasion to describe many other basic results of quantum computation. Section 4.1 begins the chapter with an overview of quantum algorithms, focusing on what algorithms are known, and the unifying techniques underlying their construction. Section 4.2 is a detailed study of single qubit operations. Despite their simplicity, single qubit operations offer a rich playground for the construction of examples and techniques, and it is essential to understand them in detail. Section 4.3 shows how to perform multi-qubit *controlled unitary* operations, and Section 4.4 discusses the description of measurement in the quantum circuits model. These elements are then brought together in Section 4.5 for the statement and proof of the universality theorem. We summarize all the basic elements of quantum computation in Section 4.6, and discuss possible variants of the model, and the important question of the relationship in computational power between classical and quantum computers. Section 4.7 concludes the chapter with an important and instructive application of quantum computation to the *simulation* of real quantum systems.

This chapter is perhaps the most reader-intensive of all the chapters in the book, with a high density of exercises for you to complete, and it is worth explaining the reason for this intensity. Obtaining facility with the basic elements of the quantum circuit model of computation is quite easy, but requires assimilating a large number of simple results and techniques that must become second nature if one is to progress to the more difficult problem of designing quantum algorithms. For this reason we take an example-oriented approach in this chapter, and ask you to fill in many of the details, in order to acquire such a facility. A less intensive, but somewhat superficial overview of the basic elements of quantum computation may be obtained by skipping to Section 4.6.

4.1 Quantum algorithms

What is a quantum computer good for? We're all familiar with the frustration of needing more computer resources to solve a computational problem. Practically speaking, many interesting problems are impossible to solve on a classical computer, not because they are in principle insoluble, but because of the astronomical resources required to solve realistic cases of the problem.

The spectacular promise of quantum computers is to enable new algorithms which render feasible problems requiring exorbitant resources for their solution on a classical computer. At the time of writing, two broad classes of quantum algorithms are known which fulfill this promise. The first class of algorithms is based upon Shor's *quantum Fourier transform*, and includes remarkable algorithms for solving the factoring and discrete logarithm problems, providing a striking *exponential* speedup over the best known classical algorithms. The second class of algorithms is based upon Grover's algorithm for performing *quantum searching*. These provide a less striking but still remarkable *quadratic* speedup over the best possible classical algorithms. The quantum searching algorithm derives its importance from the widespread use of search-based techniques in classical algorithms, which in many instances allows a straightforward adaptation of the classical algorithm to give a faster quantum algorithm.

Figure 4.1 sketches the state of knowledge about quantum algorithms at the time of writing, including some sample applications of those algorithms. Naturally, at the core of the diagram are the quantum Fourier transform and the quantum searching algorithm. Of particular interest in the figure is the quantum counting algorithm. This algorithm is a clever combination of the quantum searching and Fourier transform algorithms, which can be used to estimate the number of solutions to a search problem more quickly than is possible on a classical computer.

The quantum searching algorithm has many potential applications, of which but a few are illustrated. It can be used to extract statistics, such as the minimal element, from an unordered data set, more quickly than is possible on a classical computer. It can be used to speed up algorithms for some problems in **NP** – specifically, those problems for which a straightforward search for a solution is the best algorithm known. Finally, it can be used to speed up the search for keys to cryptosystems such as the widely used Data Encryption Standard (DES). These and other applications are explained in Chapter 6.

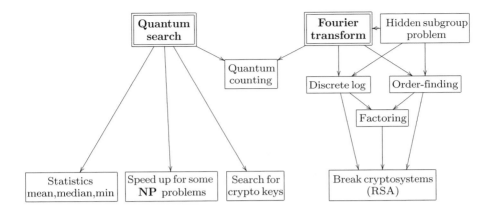

Figure 4.1. The main quantum algorithms and their relationships, including some notable applications.

The quantum Fourier transform also has many interesting applications. It can be used to solve the discrete logarithm and factoring problems. These results, in turn, enable a quantum computer to break many of the most popular cryptosystems now in use, including the RSA cryptosystem. The Fourier transform also turns out to be closely related to an important problem in mathematics, finding a hidden subgroup (a generalization of finding the period of a periodic function). The quantum Fourier transform and several of its applications, including fast quantum algorithms for factoring and discrete logarithm, are explained in Chapter 5.

Why are there so few quantum algorithms known which are better than their classical counterparts? The answer is that coming up with good quantum algorithms seems to be a difficult problem. There are at least two reasons for this. First, algorithm design, be it classical or quantum, is not an easy business! The history of algorithms shows us that considerable ingenuity is often required to come up with near optimal algorithms, even for apparently very simple problems, like the multiplication of two numbers. Finding good quantum algorithms is made doubly difficult because of the additional constraint that we want our quantum algorithms to be *better* than the best known classical algorithms. A second reason for the difficulty of finding good quantum algorithms is that our intuitions are much better adapted to the classical world than they are to the quantum world. If we think about problems using our native intuition, then the algorithms which we come up with are going to be classical algorithms. It takes special insights and special tricks to come up with good quantum algorithms.

Further study of quantum algorithms will be postponed until the next chapter. In this chapter we provide an efficient and powerful language for describing quantum algorithms, the language of quantum circuits – assemblies of discrete sets of components which describe computational procedures. This construction will enable us to quantify the cost of an algorithm in terms of things like the total number of gates required, or the circuit depth. The circuit language also comes with a toolbox of tricks that simplifies algorithm design and provides ready conceptual understanding.

4.2 Single qubit operations

The development of our quantum computational toolkit begins with operations on the simplest quantum system of all – a single qubit. Single qubit gates were introduced in Section 1.3.1. Let us quickly summarize what we learned there; you may find it useful to refer to the notes on notation on page xxiii as we go along.

A single qubit is a vector $|\psi\rangle = a|0\rangle + b|1\rangle$ parameterized by two complex numbers satisfying $|a|^2 + |b|^2 = 1$. Operations on a qubit must preserve this norm, and thus are described by 2×2 unitary matrices. Of these, some of the most important are the Pauli matrices; it is useful to list them again here:

$$X \equiv \begin{bmatrix} 0 & 1 \\ 1 & 0 \end{bmatrix} ; \quad Y \equiv \begin{bmatrix} 0 & -i \\ i & 0 \end{bmatrix} ; \quad Z \equiv \begin{bmatrix} 1 & 0 \\ 0 & -1 \end{bmatrix}. \tag{4.1}$$

Three other quantum gates will play a large part in what follows, the Hadamard gate (denoted H), phase gate (denoted S), and $\pi/8$ gate (denoted T):

$$H = \frac{1}{\sqrt{2}} \begin{bmatrix} 1 & 1 \\ 1 & -1 \end{bmatrix} ; \quad S = \begin{bmatrix} 1 & 0 \\ 0 & i \end{bmatrix} ; \quad T = \begin{bmatrix} 1 & 0 \\ 0 & \exp(i\pi/4) \end{bmatrix}. \tag{4.2}$$

A couple of useful algebraic facts to keep in mind are that $H = (X + Z)/\sqrt{2}$ and $S = T^2$. You might wonder why the T gate is called the $\pi/8$ gate when it is $\pi/4$ that appears in the definition. The reason is that the gate has historically often been referred to as the $\pi/8$ gate, simply because up to an unimportant global phase T is equal to a gate which has $\exp(\pm i\pi/8)$ appearing on its diagonals.

$$T = \exp(i\pi/8) \begin{bmatrix} \exp(-i\pi/8) & 0 \\ 0 & \exp(i\pi/8) \end{bmatrix}. \tag{4.3}$$

Nevertheless, the nomenclature is in some respects rather unfortunate, and we often refer to this gate as the T gate.

Recall also that a single qubit in the state $a|0\rangle + b|1\rangle$ can be visualized as a point (θ, φ) on the unit sphere, where $a = \cos(\theta/2)$, $b = e^{i\varphi} \sin(\theta/2)$, and a can be taken to be real because the overall phase of the state is unobservable. This is called the Bloch sphere representation, and the vector $(\cos \varphi \sin \theta, \sin \varphi \sin \theta, \cos \theta)$ is called the Bloch vector. We shall return to this picture often as an aid to intuition.

Exercise 4.1: In Exercise 2.11, which you should do now if you haven't already done it, you computed the eigenvectors of the Pauli matrices. Find the points on the Bloch sphere which correspond to the normalized eigenvectors of the different Pauli matrices.

The Pauli matrices give rise to three useful classes of unitary matrices when they are exponentiated, the *rotation operators* about the $\hat{x}$, $\hat{y}$, and $\hat{z}$ axes, defined by the equations:

$$R_x(\theta) \equiv e^{-i\theta X/2} = \cos\frac{\theta}{2}I - i\sin\frac{\theta}{2}X = \begin{bmatrix} \cos\frac{\theta}{2} & -i\sin\frac{\theta}{2} \\ -i\sin\frac{\theta}{2} & \cos\frac{\theta}{2} \end{bmatrix} \tag{4.4}$$

$$R_y(\theta) \equiv e^{-i\theta Y/2} = \cos\frac{\theta}{2}I - i\sin\frac{\theta}{2}Y = \begin{bmatrix} \cos\frac{\theta}{2} & -\sin\frac{\theta}{2} \\ \sin\frac{\theta}{2} & \cos\frac{\theta}{2} \end{bmatrix} \tag{4.5}$$

$$R_z(\theta) \equiv e^{-i\theta Z/2} = \cos\frac{\theta}{2}I - i\sin\frac{\theta}{2}Z = \begin{bmatrix} e^{-i\theta/2} & 0 \\ 0 & e^{i\theta/2} \end{bmatrix}. \tag{4.6}$$

Exercise 4.2: Let x be a real number and A a matrix such that $A^2 = I$. Show that

$$\exp(iAx) = \cos(x)I + i\sin(x)A. \tag{4.7}$$

Use this result to verify Equations (4.4) through (4.6).

Exercise 4.3: Show that, up to a global phase, the $\pi/8$ gate satisfies $T = R_z(\pi/4)$.

Exercise 4.4: Express the Hadamard gate H as a product of R_x and R_z rotations and $e^{i\varphi}$ for some φ.

If $\hat{n} = (n_x, n_y, n_z)$ is a real unit vector in three dimensions then we generalize the previous definitions by defining a rotation by θ about the $\hat{n}$ axis by the equation

$$R_{\hat{n}}(\theta) \equiv \exp(-i\theta\,\hat{n}\cdot\vec{\sigma}/2) = \cos\left(\frac{\theta}{2}\right)I - i\sin\left(\frac{\theta}{2}\right)(n_x X + n_y Y + n_z Z), \quad (4.8)$$

where $\vec{\sigma}$ denotes the three component vector (X, Y, Z) of Pauli matrices.

Exercise 4.5: Prove that $(\hat{n}\cdot\vec{\sigma})^2 = I$, and use this to verify Equation (4.8).

Exercise 4.6: (**Bloch sphere interpretation of rotations**) One reason why the $R_{\hat{n}}(\theta)$ operators are referred to as rotation operators is the following fact, which you are to prove. Suppose a single qubit has a state represented by the Bloch vector $\vec{\lambda}$. Then the effect of the rotation $R_{\hat{n}}(\theta)$ on the state is to rotate it by an angle θ about the $\hat{n}$ axis of the Bloch sphere. This fact explains the rather mysterious looking factor of two in the definition of the rotation matrices.

Exercise 4.7: Show that $XYX = -Y$ and use this to prove that $XR_y(\theta)X = R_y(-\theta)$.

Exercise 4.8: An arbitrary single qubit unitary operator can be written in the form

$$U = \exp(i\alpha)R_{\hat{n}}(\theta) \tag{4.9}$$

for some real numbers α and θ, and a real three-dimensional unit vector $\hat{n}$.

1. Prove this fact.
2. Find values for α, θ, and $\hat{n}$ giving the Hadamard gate H.
3. Find values for α, θ, and $\hat{n}$ giving the phase gate

$$S = \begin{bmatrix} 1 & 0 \\ 0 & i \end{bmatrix}. \tag{4.10}$$

An arbitrary unitary operator on a single qubit can be written in many ways as a combination of rotations, together with global phase shifts on the qubit. The following theorem provides a means of expressing an arbitrary single qubit rotation that will be particularly useful in later applications to controlled operations.

Theorem 4.1: (Z-Y decomposition for a single qubit) Suppose U is a unitary operation on a single qubit. Then there exist real numbers α, β, γ and δ such that

$$U = e^{i\alpha} R_z(\beta)R_y(\gamma)R_z(\delta). \tag{4.11}$$

Proof

Since U is unitary, the rows and columns of U are orthonormal, from which it follows that there exist real numbers α, β, γ,and δ such that

$$U = \begin{bmatrix} e^{i(\alpha-\beta/2-\delta/2)} \cos\frac{\gamma}{2} & -e^{i(\alpha-\beta/2+\delta/2)} \sin\frac{\gamma}{2} \\ e^{i(\alpha+\beta/2-\delta/2)} \sin\frac{\gamma}{2} & e^{i(\alpha+\beta/2+\delta/2)} \cos\frac{\gamma}{2} \end{bmatrix}. \tag{4.12}$$

Equation (4.11) now follows immediately from the definition of the rotation matrices and matrix multiplication. $\qquad\square$

Exercise 4.9: Explain why any single qubit unitary operator may be written in the form (4.12).

Exercise 4.10: (*X-Y decomposition of rotations*) Give a decomposition analogous to Theorem 4.1 but using R_x instead of R_z.

Exercise 4.11: Suppose $\hat{m}$ and $\hat{n}$ are non-parallel real unit vectors in three dimensions. Use Theorem 4.1 to show that an arbitrary single qubit unitary U may be written

$$U = e^{i\alpha} R_{\hat{n}}(\beta) R_{\hat{m}}(\gamma) R_{\hat{n}}(\delta), \tag{4.13}$$

for appropriate choices of α, β, γ and δ.

The utility of Theorem 4.1 lies in the following mysterious looking corollary, which is the key to the construction of controlled multi-qubit unitary operations, as explained in the next section.

Corollary 4.2: Suppose U is a unitary gate on a single qubit. Then there exist unitary operators A, B, C on a single qubit such that $ABC = I$ and $U = e^{i\alpha} AXBXC$, where α is some overall phase factor.

Proof

In the notation of Theorem 4.1, set $A \equiv R_z(\beta)R_y(\gamma/2)$, $B \equiv R_y(-\gamma/2)R_z(-(\delta+\beta)/2)$ and $C \equiv R_z((\delta - \beta)/2)$. Note that

$$ABC = R_z(\beta)R_y\left(\frac{\gamma}{2}\right) R_y\left(-\frac{\gamma}{2}\right) R_z\left(-\frac{\delta+\beta}{2}\right) R_z\left(\frac{\delta-\beta}{2}\right) = I. \tag{4.14}$$

Since $X^2 = I$, and using Exercise 4.7, we see that

$$XBX = XR_y\left(-\frac{\gamma}{2}\right) XXR_z\left(-\frac{\delta+\beta}{2}\right) X = R_y\left(\frac{\gamma}{2}\right) R_z\left(\frac{\delta+\beta}{2}\right). \tag{4.15}$$

Thus

$$AXBXC = R_z(\beta)R_y\left(\frac{\gamma}{2}\right) R_y\left(\frac{\gamma}{2}\right) R_z\left(\frac{\delta+\beta}{2}\right) R_z\left(\frac{\delta-\beta}{2}\right) \tag{4.16}$$

$$= R_z(\beta)R_y(\gamma)R_z(\delta). \tag{4.17}$$

Thus $U = e^{i\alpha} AXBXC$ and $ABC = I$, as required. $\qquad\square$

Exercise 4.12: Give A, B, C, and α for the Hadamard gate.

Exercise 4.13: (Circuit identities) It is useful to be able to simplify circuits by inspection, using well-known identities. Prove the following three identities:

$$HXH = Z; \quad HYH = -Y; \quad HZH = X. \tag{4.18}$$

Exercise 4.14: Use the previous exercise to show that $HTH = R_x(\pi/4)$, up to a global phase.

Exercise 4.15: (Composition of single qubit operations) The Bloch representation gives a nice way to visualize the effect of composing two rotations.

(1) Prove that if a rotation through an angle β_1 about the axis $\hat{n}_1$ is followed by a rotation through an angle β_2 about an axis $\hat{n}_2$, then the overall rotation is through an angle β_{12} about an axis $\hat{n}_{12}$ given by

$$c_{12} = c_1 c_2 - s_1 s_2 \, \hat{n}_1 \cdot \hat{n}_2 \tag{4.19}$$
$$s_{12}\hat{n}_{12} = s_1 c_2 \hat{n}_1 + c_1 s_2 \hat{n}_2 - s_1 s_2 \, \hat{n}_2 \times \hat{n}_1 , \tag{4.20}$$

where $c_i = \cos(\beta_i/2)$, $s_i = \sin(\beta_i/2)$, $c_{12} = \cos(\beta_{12}/2)$, and $s_{12} = \sin(\beta_{12}/2)$.

(2) Show that if $\beta_1 = \beta_2$ and $\hat{n}_1 = \hat{z}$ these equations simplify to

$$c_{12} = c^2 - s^2 \, \hat{z} \cdot \hat{n}_2 \tag{4.21}$$
$$s_{12}\hat{n}_{12} = sc(\hat{z} + \hat{n}_2) - s^2 \, \hat{n}_2 \times \hat{z} , \tag{4.22}$$

where $c = c_1$ and $s = s_1$.

Symbols for the common single qubit gates are shown in Figure 4.2. Recall the basic properties of quantum circuits: time proceeds from left to right; wires represent qubits, and a '/' may be used to indicate a bundle of qubits.

Hadamard	$-\boxed{H}-$	$\dfrac{1}{\sqrt{2}}\begin{bmatrix} 1 & 1 \\ 1 & -1 \end{bmatrix}$
Pauli-X	$-\boxed{X}-$	$\begin{bmatrix} 0 & 1 \\ 1 & 0 \end{bmatrix}$
Pauli-Y	$-\boxed{Y}-$	$\begin{bmatrix} 0 & -i \\ i & 0 \end{bmatrix}$
Pauli-Z	$-\boxed{Z}-$	$\begin{bmatrix} 1 & 0 \\ 0 & -1 \end{bmatrix}$
Phase	$-\boxed{S}-$	$\begin{bmatrix} 1 & 0 \\ 0 & i \end{bmatrix}$
$\pi/8$	$-\boxed{T}-$	$\begin{bmatrix} 1 & 0 \\ 0 & e^{i\pi/4} \end{bmatrix}$

Figure 4.2. Names, symbols, and unitary matrices for the common single qubit gates.

4.3 Controlled operations

'If A is true, then do B'. This type of *controlled operation* is one of the most useful in computing, both classical and quantum. In this section we explain how complex controlled operations may be implemented using quantum circuits built from elementary operations.

The prototypical controlled operation is the controlled-NOT, which we met in Section 1.2.1. Recall that this gate, which we'll often refer to as CNOT, is a quantum gate with two input qubits, known as the *control qubit* and *target qubit*, respectively. It is drawn as shown in Figure 4.3. In terms of the computational basis, the action of the CNOT is given by $|c\rangle|t\rangle \rightarrow |c\rangle|t \oplus c\rangle$; that is, if the control qubit is set to $|1\rangle$ then the target qubit is flipped, otherwise the target qubit is left alone. Thus, in the computational basis $|\text{control}, \text{target}\rangle$ the matrix representation of CNOT is

$$\begin{bmatrix} 1 & 0 & 0 & 0 \\ 0 & 1 & 0 & 0 \\ 0 & 0 & 0 & 1 \\ 0 & 0 & 1 & 0 \end{bmatrix}. \tag{4.23}$$

Figure 4.3. Circuit representation for the controlled-NOT gate. The top line represents the control qubit, the bottom line the target qubit.

More generally, suppose U is an arbitrary single qubit unitary operation. A *controlled-U* operation is a two qubit operation, again with a control and a target qubit. If the control qubit is set then U is applied to the target qubit, otherwise the target qubit is left alone; that is, $|c\rangle|t\rangle \rightarrow |c\rangle U^c|t\rangle$. The controlled-$U$ operation is represented by the circuit shown in Figure 4.4.

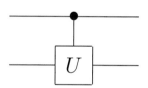

Figure 4.4. Controlled-U operation. The top line is the control qubit, and the bottom line is the target qubit. If the control qubit is set then U is applied to the target, otherwise it is left alone.

Exercise 4.16: (Matrix representation of multi-qubit gates) What is the 4×4 unitary matrix for the circuit

$$x_2 \; -\boxed{H}-$$

$$x_1 \; -\!\!-\!\!-$$

in the computational basis? What is the unitary matrix for the circuit

$$x_2 \; -\!\!-\!\!-$$

$$x_1 \; -\boxed{H}-$$

in the computational basis?

Exercise 4.17: (Building CNOT **from controlled-**Z **gates)** Construct a CNOT gate from one controlled-Z gate, that is, the gate whose action in the computational basis is specified by the unitary matrix

$$\begin{bmatrix} 1 & 0 & 0 & 0 \\ 0 & 1 & 0 & 0 \\ 0 & 0 & 1 & 0 \\ 0 & 0 & 0 & -1 \end{bmatrix},$$

and two Hadamard gates, specifying the control and target qubits.

Exercise 4.18: Show that

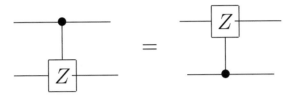

Exercise 4.19: (CNOT **action on density matrices)** The CNOT gate is a simple permutation whose action on a density matrix ρ is to rearrange the elements in the matrix. Write out this action explicitly in the computational basis.

Exercise 4.20: (CNOT **basis transformations)** Unlike ideal classical gates, ideal quantum gates do not have (as electrical engineers say) 'high-impedance' inputs. In fact, the role of 'control' and 'target' are arbitrary – they depend on what basis you think of a device as operating in. We have described how the CNOT behaves with respect to the computational basis, and in this description the state of the control qubit is not changed. However, if we work in a different basis then the control qubit *does* change: we will show that its phase is flipped depending on the state of the 'target' qubit! Show that

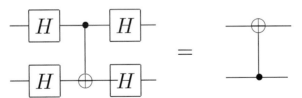

Introducing basis states $|\pm\rangle \equiv (|0\rangle \pm |1\rangle)/\sqrt{2}$, use this circuit identity to show that the effect of a CNOT with the first qubit as control and the second qubit as target is as follows:

$$|+\rangle|+\rangle \rightarrow |+\rangle|+\rangle \tag{4.24}$$
$$|-\rangle|+\rangle \rightarrow |-\rangle|+\rangle \tag{4.25}$$
$$|+\rangle|-\rangle \rightarrow |-\rangle|-\rangle \tag{4.26}$$
$$|-\rangle|-\rangle \rightarrow |+\rangle|-\rangle. \tag{4.27}$$

Thus, with respect to this new basis, the state of the target qubit is not changed, while the state of the control qubit is flipped if the target starts as $|-\rangle$, otherwise it is left alone. That is, in this basis, the target and control have essentially interchanged roles!

Our immediate goal is to understand how to implement the controlled-U operation for arbitrary single qubit U, using only single qubit operations and the CNOT gate. Our strategy is a two-part procedure based upon the decomposition $U = e^{i\alpha}AXBXC$ given in Corollary 4.2 on page 176.

Our first step will be to apply the phase shift $\exp(i\alpha)$ on the target qubit, controlled by the control qubit. That is, if the control qubit is $|0\rangle$, then the target qubit is left alone, while if the control qubit is $|1\rangle$, a phase shift $\exp(i\alpha)$ is applied to the target. A circuit implementing this operation using just a single qubit unitary gate is depicted on the right hand side of Figure 4.5. To verify that this circuit works correctly, note that the effect of the circuit on the right hand side is

$$|00\rangle \rightarrow |00\rangle, \quad |01\rangle \rightarrow |01\rangle, \quad |10\rangle \rightarrow e^{i\alpha}|10\rangle, \quad |11\rangle \rightarrow e^{i\alpha}|11\rangle, \qquad (4.28)$$

which is exactly what is required for the controlled operation on the left hand side.

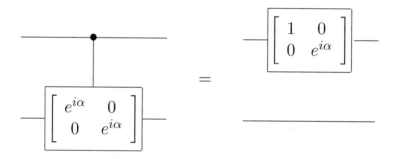

Figure 4.5. Controlled phase shift gate and an equivalent circuit for two qubits.

We may now complete the construction of the controlled-U operation, as shown in Figure 4.6. To understand why this circuit works, recall from Corollary 4.2 that U may be written in the form $U = e^{i\alpha}AXBXC$, where A, B and C are single qubit operations such that $ABC = I$. Suppose that the control qubit is set. Then the operation $e^{i\alpha}AXBXC = U$ is applied to the second qubit. If, on the other hand, the control qubit is not set, then the operation $ABC = I$ is applied to the second qubit; that is, no change is made. That is, this circuit implements the controlled-U operation.

Now that we know how to condition on a single qubit being set, what about conditioning on multiple qubits? We've already met one example of multiple qubit conditioning, the Toffoli gate, which flips the third qubit, the target qubit, conditioned on the first two qubits, the control qubits, being set to one. More generally, suppose we have $n + k$ qubits, and U is a k qubit unitary operator. Then we define the controlled operation $C^n(U)$ by the equation

$$C^n(U)|x_1 x_2 \ldots x_n\rangle|\psi\rangle = |x_1 x_2 \ldots x_n\rangle U^{x_1 x_2 \ldots x_n}|\psi\rangle, \qquad (4.29)$$

where $x_1 x_2 \ldots x_n$ in the exponent of U means the *product* of the bits $x_1, x_2, \ldots, x_n$. That is, the operator U is applied to the last k qubits if the first n qubits are all equal to one, and otherwise, nothing is done. Such conditional operations are so useful that we

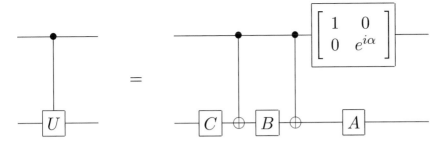

Figure 4.6. Circuit implementing the controlled-U operation for single qubit U. α, A, B and C satisfy $U = \exp(i\alpha)AXBXC$, $ABC = I$.

introduce a special circuit notation for them, illustrated in Figure 4.7. For the following we assume that $k = 1$, for simplicity. Larger k can be dealt with using essentially the same methods, however for $k \geq 2$ there is the added complication that we don't (yet) know how to perform arbitrary operations on k qubits.

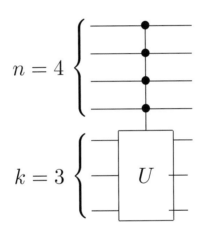

Figure 4.7. Sample circuit representation for the $C^n(U)$ operation, where U is a unitary operator on k qubits, for $n = 4$ and $k = 3$.

Suppose U is a single qubit unitary operator, and V is a unitary operator chosen so that $V^2 = U$. Then the operation $C^2(U)$ may be implemented using the circuit shown in Figure 4.8.

Exercise 4.21: Verify that Figure 4.8 implements the $C^2(U)$ operation.

Exercise 4.22: Prove that a $C^2(U)$ gate (for any single qubit unitary U) can be constructed using at most eight one-qubit gates, and six controlled-NOTs.

Exercise 4.23: Construct a $C^1(U)$ gate for $U = R_x(\theta)$ and $U = R_y(\theta)$, using only CNOT and single qubit gates. Can you reduce the number of single qubit gates needed in the construction from three to two?

The familiar Toffoli gate is an especially important special case of the $C^2(U)$ operation,

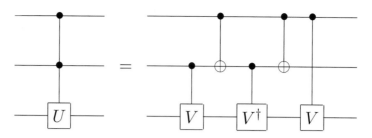

Figure 4.8. Circuit for the $C^2(U)$ gate. V is any unitary operator satisfying $V^2 = U$. The special case $V \equiv (1 - i)(I + iX)/2$ corresponds to the Toffoli gate.

the case $C^2(X)$. Defining $V \equiv (1 - i)(I + iX)/2$ and noting that $V^2 = X$, we see that Figure 4.8 gives an implementation of the Toffoli gate in terms of one and two qubit operations. From a classical viewpoint this is a remarkable result; recall from Problem 3.5 that one and two bit classical reversible gates are not sufficient to implement the Toffoli gate, or, more generally, universal computation. By contrast, in the quantum case we see that one and two qubit reversible gates are sufficient to implement the Toffoli gate, and will eventually prove that they suffice for universal computation.

Ultimately we will show that any unitary operation can be composed to an arbitrarily good approximation from just the Hadamard, phase, controlled-NOT and $\pi/8$ gates. Because of the great usefulness of the Toffoli gate it is interesting to see how it can be built from just this gate set. Figure 4.9 illustrates a simple circuit for the Toffoli gate made up of just Hadamard, phase, controlled-NOT and $\pi/8$ gates.

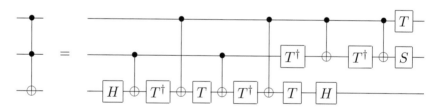

Figure 4.9. Implementation of the Toffoli gate using Hadamard, phase, controlled-NOT and $\pi/8$ gates.

Exercise 4.24: Verify that Figure 4.9 implements the Toffoli gate.

Exercise 4.25: (Fredkin gate construction) Recall that the Fredkin (controlled-swap) gate performs the transform

$$
\begin{bmatrix}
1 & 0 & 0 & 0 & 0 & 0 & 0 & 0 \\
0 & 1 & 0 & 0 & 0 & 0 & 0 & 0 \\
0 & 0 & 1 & 0 & 0 & 0 & 0 & 0 \\
0 & 0 & 0 & 1 & 0 & 0 & 0 & 0 \\
0 & 0 & 0 & 0 & 1 & 0 & 0 & 0 \\
0 & 0 & 0 & 0 & 0 & 0 & 1 & 0 \\
0 & 0 & 0 & 0 & 0 & 1 & 0 & 0 \\
0 & 0 & 0 & 0 & 0 & 0 & 0 & 1
\end{bmatrix}. \tag{4.30}
$$

(1) Give a quantum circuit which uses three Toffoli gates to construct the Fredkin gate (*Hint*: think of the swap gate construction – you can control each gate, one at a time).

(2) Show that the first and last Toffoli gates can be replaced by CNOT gates.

(3) Now replace the middle Toffoli gate with the circuit in Figure 4.8 to obtain a Fredkin gate construction using only six two-qubit gates.

(4) Can you come up with an even simpler construction, with only five two-qubit gates?

Exercise 4.26: Show that the circuit:

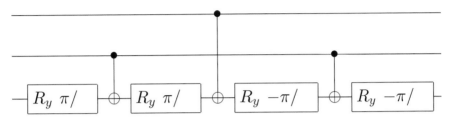

differs from a Toffoli gate only by relative phases. That is, the circuit takes $|c_1, c_2, t\rangle$ to $e^{i\theta(c_1, c_2, t)}|c_1, c_2, t \oplus c_1 \cdot c_2\rangle$, where $e^{i\theta(c_1, c_2, t)}$ is some relative phase factor. Such gates can sometimes be useful in experimental implementations, where it may be much easier to implement a gate that is the same as the Toffoli up to relative phases than it is to do the Toffoli directly.

Exercise 4.27: Using just CNOTs and Toffoli gates, construct a quantum circuit to perform the transformation

$$\begin{bmatrix} 1 & 0 & 0 & 0 & 0 & 0 & 0 & 0 \\ 0 & 0 & 0 & 0 & 0 & 0 & 0 & 1 \\ 0 & 1 & 0 & 0 & 0 & 0 & 0 & 0 \\ 0 & 0 & 1 & 0 & 0 & 0 & 0 & 0 \\ 0 & 0 & 0 & 1 & 0 & 0 & 0 & 0 \\ 0 & 0 & 0 & 0 & 1 & 0 & 0 & 0 \\ 0 & 0 & 0 & 0 & 0 & 1 & 0 & 0 \\ 0 & 0 & 0 & 0 & 0 & 0 & 1 & 0 \end{bmatrix}. \qquad (4.31)$$

This kind of partial cyclic permutation operation will be useful later, in Chapter 7.

How may we implement $C^n(U)$ gates using our existing repertoire of gates, where U is an arbitrary single qubit unitary operation? A particularly simple circuit for achieving this task is illustrated in Figure 4.10. The circuit divides up into three stages, and makes use of a small number $(n - 1)$ of working qubits, which all start and end in the state $|0\rangle$. Suppose the control qubits are in the computational basis state $|c_1, c_2, \ldots, c_n\rangle$. The first stage of the circuit is to reversibly AND all the control bits $c_1, \ldots, c_n$ together to produce the product $c_1 \cdot c_2 \ldots c_n$. To do this, the first gate in the circuit ANDs c_1 and c_2 together, using a Toffoli gate, changing the state of the first work qubit to $|c_1 \cdot c_2\rangle$. The next Toffoli gate ANDs c_3 with the product $c_1 \cdot c_2$, changing the state of the second work qubit to $|c_1 \cdot c_2 \cdot c_3\rangle$. We continue applying Toffoli gates in this fashion, until the final work qubit is in the state $|c_1 \cdot c_2 \ldots c_n\rangle$. Next, a U operation on the target qubit is performed, conditional on the final work qubit being set to one. That is, U is applied if and only if all of c_1 through c_n are set. Finally, the last part of the circuit just reverses the steps of the first stage, returning all the work qubits to their initial state, $|0\rangle$. The combined result, therefore, is to apply the unitary operator U to the target qubit, if and only if all the control bits c_1 through c_n are set, as desired.

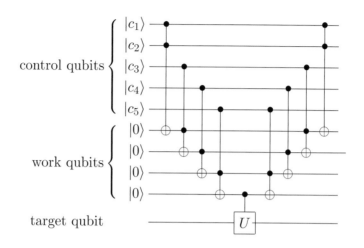

Figure 4.10. Network implementing the $C^n(U)$ operation, for the case $n = 5$.

Exercise 4.28: For $U = V^2$ with V unitary, construct a $C^5(U)$ gate analogous to that in Figure 4.10, but using no work qubits. You may use controlled-V and controlled-$V^\dagger$ gates.

Exercise 4.29: Find a circuit containing $O(n^2)$ Toffoli, CNOT and single qubit gates which implements a $C^n(X)$ gate (for $n > 3$), using no work qubits.

Exercise 4.30: Suppose U is a single qubit unitary operation. Find a circuit containing $O(n^2)$ Toffoli, CNOT and single qubit gates which implements a $C^n(U)$ gate (for $n > 3$), using no work qubits.

In the controlled gates we have been considering, conditional dynamics on the target qubit occurs if the control bits are set to *one*. Of course, there is nothing special about one, and it is often useful to consider dynamics which occur conditional on the control bit being set to zero. For instance, suppose we wish to implement a two qubit gate in which the second ('target') qubit is flipped, conditional on the first ('control') qubit being set to zero. In Figure 4.11 we introduce a circuit notation for this gate, together with an equivalent circuit in terms of the gates we have already introduced. Generically we shall use the open circle notation to indicate conditioning on the qubit being set to zero, while a closed circle indicates conditioning on the qubit being set to one.

A more elaborate example of this convention, involving three control qubits, is illustrated in Figure 4.12. The operation U is applied to the target qubit if the first and third qubits are set to zero, and the second qubit is set to one. It is easy to verify by inspection that the circuit on the right hand side of the figure implements the desired operation. More generally, it is easy to move between circuits which condition on qubits being set

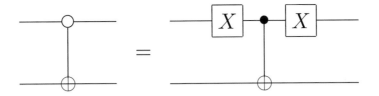

Figure 4.11. Controlled operation with a NOT gate being performed on the second qubit, conditional on the first qubit being set to zero.

to one and circuits which condition on qubits being set to zero, by insertion of X gates in appropriate locations, as illustrated in Figure 4.12.

Another convention which is sometimes useful is to allow controlled-NOT gates to have multiple targets, as shown in Figure 4.13. This natural notation means that when the control qubit is 1, then all the qubits marked with a $\oplus$ are flipped, and otherwise nothing happens. It is convenient to use, for example, in constructing classical functions such as permutations, or in encoders and decoders for quantum error-correction circuits, as we shall see in Chapter 10.

Exercise 4.31: (More circuit identities) Let subscripts denote which qubit an operator acts on, and let C be a CNOT with qubit 1 the control qubit and qubit 2 the target qubit. Prove the following identities:

$$CX_1C = X_1X_2 \tag{4.32}$$
$$CY_1C = Y_1X_2 \tag{4.33}$$
$$CZ_1C = Z_1 \tag{4.34}$$
$$CX_2C = X_2 \tag{4.35}$$
$$CY_2C = Z_1Y_2 \tag{4.36}$$
$$CZ_2C = Z_1Z_2 \tag{4.37}$$
$$R_{z,1}(\theta)C = CR_{z,1}(\theta) \tag{4.38}$$
$$R_{x,2}(\theta)C = CR_{x,2}(\theta). \tag{4.39}$$

4.4 Measurement

A final element used in quantum circuits, almost implicitly sometimes, is measurement. In our circuits, we shall denote a projective measurement in the computational basis (Section 2.2.5) using a 'meter' symbol, illustrated in Figure 4.14. In the theory of quantum circuits it is conventional to not use any special symbols to denote more general measurements, because, as explained in Chapter 2, they can always be represented by unitary transforms with ancilla qubits followed by projective measurements.

There are two important principles that it is worth bearing in mind about quantum circuits. Both principles are rather obvious; however, they are of such great utility that they are worth emphasizing early. The first principle is that classically conditioned operations can be replaced by quantum conditioned operations:

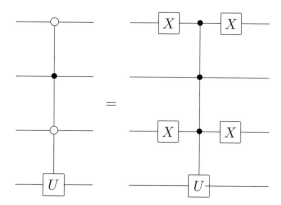

Figure 4.12. Controlled-*U* operation and its equivalent in terms of circuit elements we already know how to implement. The fourth qubit has *U* applied if the first and third qubits are set to zero, and the second qubit is set to one.

Figure 4.13. Controlled-NOT gate with multiple targets.

> **Principle of deferred measurement:** Measurements can always be moved from an intermediate stage of a quantum circuit to the end of the circuit; if the measurement results are used at any stage of the circuit then the classically controlled operations can be replaced by conditional quantum operations.

Often, quantum measurements are performed as an intermediate step in a quantum circuit, and the measurement results are used to conditionally control subsequent quantum gates. This is the case, for example, in the teleportation circuit of Figure 1.13 on page 27. However, such measurements can *always* be moved to the end of the circuit. Figure 4.15 illustrates how this may be done by replacing all the classical conditional operations by corresponding quantum conditional operations. (Of course, some of the interpretation of this circuit as performing 'teleportation' is lost, because no classical information is transmitted from Alice to Bob, but it is clear that the overall action of the two quantum circuits is the same, which is the key point.)

The second principle is even more obvious – and surprisingly useful!

Figure 4.14. Symbol for projective measurement on a single qubit. In this circuit nothing further is done with the measurement result, but in more general quantum circuits it is possible to change later parts of the quantum circuit, *conditional* on measurement outcomes in earlier parts of the circuit. Such a usage of classical information is depicted using wires drawn with double lines (not shown here).

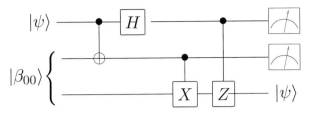

Figure 4.15. Quantum teleportation circuit in which measurements are done at the end, instead of in the middle of the circuit. As in Figure 1.13, the top two qubits belong to Alice, and the bottom one to Bob.

> **Principle of implicit measurement:** Without loss of generality, any unterminated quantum wires (qubits which are not measured) at the end of a quantum circuit may be assumed to be measured.

To understand why this is true, imagine you have a quantum circuit containing just two qubits, and only the first qubit is measured at the end of the circuit. Then the measurement statistics observed at this time are completely determined by the reduced density matrix of the first qubit. However, if a measurement had also been performed on the second qubit, then it would be highly surprising if that measurement could change the statistics of measurement on the first qubit. You'll prove this in Exercise 4.32 by showing that the reduced density matrix of the first qubit is not affected by performing a measurement on the second.

As you consider the role of measurements in quantum circuits, it is important to keep in mind that in its role as an interface between the quantum and classical worlds, measurement is generally considered to be an irreversible operation, destroying quantum information and replacing it with classical information. In certain carefully designed cases, however, this need not be true, as is vividly illustrated by teleportation and quantum error-correction (Chapter 10). What teleportation and quantum error-correction have in common is that in neither instance does the measurement result reveal any information about the identity of the quantum state being measured. Indeed, we will see in Chapter 10 that this is a more general feature of measurement – in order for a measurement to be reversible, it must reveal no information about the quantum state being measured!

Exercise 4.32: Suppose ρ is the density matrix describing a two qubit system. Suppose we perform a projective measurement in the computational basis of the second qubit. Let $P_0 = |0\rangle\langle0|$ and $P_1 = |1\rangle\langle1|$ be the projectors onto the $|0\rangle$ and $|1\rangle$ states of the second qubit, respectively. Let ρ' be the density matrix which would be assigned to the system after the measurement by an observer who did not learn the measurement result. Show that

$$\rho' = P_0\rho P_0 + P_1\rho P_1. \tag{4.40}$$

Also show that the reduced density matrix for the first qubit is not affected by the measurement, that is, $\text{tr}_2(\rho) = \text{tr}_2(\rho')$.

Exercise 4.33: (Measurement in the Bell basis) The measurement model we have specified for the quantum circuit model is that measurements are performed only in the computational basis. However, often we want to perform a measurement in some other basis, defined by a complete set of orthonormal states. To perform this measurement, simply unitarily transform from the basis we wish to perform the measurement in to the computational basis, then measure. For example, show that the circuit

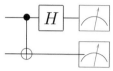

performs a measurement in the basis of the Bell states. More precisely, show that this circuit results in a measurement being performed with corresponding POVM elements the four projectors onto the Bell states. What are the corresponding measurement operators?

Exercise 4.34: (Measuring an operator) Suppose we have a single qubit operator U with eigenvalues ± 1, so that U is both Hermitian and unitary, so it can be regarded both as an observable and a quantum gate. Suppose we wish to measure the observable U. That is, we desire to obtain a measurement result indicating one of the two eigenvalues, and leaving a post-measurement state which is the corresponding eigenvector. How can this be implemented by a quantum circuit? Show that the following circuit implements a measurement of U:

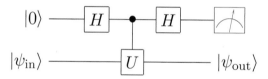

Exercise 4.35: (Measurement commutes with controls) A consequence of the principle of deferred measurement is that measurements commute with quantum gates when the qubit being measured is a control qubit, that is:

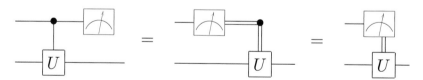

(Recall that the double lines represent classical bits in this diagram.) Prove the first equality. The rightmost circuit is simply a convenient notation to depict the use of a measurement result to classically control a quantum gate.

4.5 Universal quantum gates

A small set of gates (e.g. AND, OR, NOT) can be used to compute an arbitrary classical function, as we saw in Section 3.1.2. We say that such a set of gates is *universal* for classical computation. In fact, since the Toffoli gate is universal for classical computation, quantum circuits subsume classical circuits. A similar universality result is true for quantum computation, where a set of gates is said to be *universal for quantum computation* if any unitary operation may be approximated to arbitrary accuracy by a quantum circuit involving only those gates. We now describe three universality constructions for quantum computation. These constructions build upon each other, and culminate in a proof that any unitary operation can be approximated to arbitrary accuracy using Hadamard, phase, CNOT, and $\pi/8$ gates. You may wonder why the phase gate appears in this list, since it can be constructed from two $\pi/8$ gates; it is included because of its natural role in the fault-tolerant constructions described in Chapter 10.

The first construction shows that an arbitrary unitary operator may be expressed *exactly* as a product of unitary operators that each acts non-trivially only on a subspace spanned by two computational basis states. The second construction combines the first construction with the results of the previous section to show that an arbitrary unitary operator may be expressed *exactly* using single qubit and CNOT gates. The third construction combines the second construction with a proof that single qubit operation may be approximated to arbitrary accuracy using the Hadamard, phase, and $\pi/8$ gates. This in turn implies that any unitary operation can be approximated to arbitrary accuracy using Hadamard, phase, CNOT, and $\pi/8$ gates.

Our constructions say little about efficiency – how many (polynomially or exponentially many) gates must be composed in order to create a given unitary transform. In Section 4.5.4 we show that there *exist* unitary transforms which require exponentially many gates to approximate. Of course, the goal of quantum computation is to find interesting families of unitary transformations that *can* be performed efficiently.

Exercise 4.36: Construct a quantum circuit to add two two-bit numbers x and y modulo 4. That is, the circuit should perform the transformation $|x, y\rangle \rightarrow |x, x + y \bmod 4\rangle$.

4.5.1 Two-level unitary gates are universal

Consider a unitary matrix U which acts on a d-dimensional Hilbert space. In this section we explain how U may be decomposed into a product of *two-level unitary matrices*; that is, unitary matrices which act non-trivially only on two-or-fewer vector components. The essential idea behind this decomposition may be understood by considering the case when U is 3×3, so suppose that U has the form

$$U = \begin{bmatrix} a & d & g \\ b & e & h \\ c & f & j \end{bmatrix}. \tag{4.41}$$

We will find two-level unitary matrices $U_1, \ldots, U_3$ such that

$$U_3 U_2 U_1 U = I. \tag{4.42}$$

It follows that

$$U = U_1^\dagger U_2^\dagger U_3^\dagger. \tag{4.43}$$

U_1, U_2 and U_3 are all two-level unitary matrices, and it is easy to see that their inverses, $U_1^\dagger, U_2^\dagger$ and $U_3^\dagger$ are also two-level unitary matrices. Thus, if we can demonstrate (4.42), then we will have shown how to break U up into a product of two-level unitary matrices.

Use the following procedure to construct U_1: if $b = 0$ then set

$$U_1 \equiv \begin{bmatrix} 1 & 0 & 0 \\ 0 & 1 & 0 \\ 0 & 0 & 1 \end{bmatrix}. \tag{4.44}$$

If $b \neq 0$ then set

$$U_1 \equiv \begin{bmatrix} \dfrac{a^*}{\sqrt{|a|^2+|b|^2}} & \dfrac{b^*}{\sqrt{|a|^2+|b|^2}} & 0 \\ \dfrac{b}{\sqrt{|a|^2+|b|^2}} & \dfrac{-a}{\sqrt{|a|^2+|b|^2}} & 0 \\ 0 & 0 & 1 \end{bmatrix}. \tag{4.45}$$

Note that in either case U_1 is a two-level unitary matrix, and when we multiply the matrices out we get

$$U_1 U = \begin{bmatrix} a' & d' & g' \\ 0 & e' & h' \\ c' & f' & j' \end{bmatrix}. \tag{4.46}$$

The key point to note is that the middle entry in the left hand column is zero. We denote the other entries in the matrix with a generic prime $'$; their actual values do not matter.

Now apply a similar procedure to find a two-level matrix U_2 such that $U_2 U_1 U$ has no entry in the *bottom left* corner. That is, if $c' = 0$ we set

$$U_2 \equiv \begin{bmatrix} a'^* & 0 & 0 \\ 0 & 1 & 0 \\ 0 & 0 & 1 \end{bmatrix}, \tag{4.47}$$

while if $c' \neq 0$ then we set

$$U_2 \equiv \begin{bmatrix} \dfrac{a'^*}{\sqrt{|a'|^2+|c'|^2}} & 0 & \dfrac{c'^*}{\sqrt{|a'|^2+|c'|^2}} \\ 0 & 1 & 0 \\ \dfrac{c'}{\sqrt{|a'|^2+|c'|^2}} & 0 & \dfrac{-a'}{\sqrt{|a'|^2+|c'|^2}} \end{bmatrix}. \tag{4.48}$$

In either case, when we carry out the matrix multiplication we find that

$$U_2 U_1 U = \begin{bmatrix} 1 & d'' & g'' \\ 0 & e'' & h'' \\ 0 & f'' & j'' \end{bmatrix}. \tag{4.49}$$

Since U, U_1 and U_2 are unitary, it follows that $U_2 U_1 U$ is unitary, and thus $d'' = g'' = 0$, since the first row of $U_2 U_1 U$ must have norm 1. Finally, set

$$U_3 \equiv \begin{bmatrix} 1 & 0 & 0 \\ 0 & e''^* & f''^* \\ 0 & h''^* & j''^* \end{bmatrix}. \tag{4.50}$$

It is now easy to verify that $U_3 U_2 U_1 U = I$, and thus $U = U_1^\dagger U_2^\dagger U_3^\dagger$, which is a decomposition of U into two-level unitaries.

More generally, suppose U acts on a d-dimensional space. Then, in a similar fashion to the 3×3 case, we can find two-level unitary matrices $U_1, \ldots, U_{d-1}$ such that the matrix

$U_{d-1}U_{d-2}\ldots U_1U$ has a one in the top left hand corner, and all zeroes elsewhere in the first row and column. We then repeat this procedure for the $d-1$ by $d-1$ unitary submatrix in the lower right hand corner of $U_{d-1}U_{d-2}\ldots U_1U$, and so on, with the end result that an arbitrary $d\times d$ unitary matrix may be written

$$U = V_1\ldots V_k, \tag{4.51}$$

where the matrices V_i are two-level unitary matrices, and $k \le (d-1)+(d-2)+\cdots+1 = d(d-1)/2$.

Exercise 4.37: Provide a decomposition of the transform

$$\frac{1}{2}\begin{bmatrix} 1 & 1 & 1 & 1 \\ 1 & i & -1 & -i \\ 1 & -1 & 1 & -1 \\ 1 & -i & -1 & i \end{bmatrix} \tag{4.52}$$

into a product of two-level unitaries. This is a special case of the quantum Fourier transform, which we study in more detail in the next chapter.

A corollary of the above result is that an arbitrary unitary matrix on an n qubit system may be written as a product of at most $2^{n-1}(2^n - 1)$ two-level unitary matrices. For specific unitary matrices, it may be possible to find much more efficient decompositions, but as you will now show there exist matrices which *cannot* be decomposed as a product of fewer than $d-1$ two-level unitary matrices!

Exercise 4.38: Prove that there exists a $d\times d$ unitary matrix U which cannot be decomposed as a product of fewer than $d-1$ two-level unitary matrices.

4.5.2 Single qubit and CNOT gates are universal

We have just shown that an arbitrary unitary matrix on a d-dimensional Hilbert space may be written as a product of two-level unitary matrices. Now we show that single qubit and CNOT gates together can be used to implement an arbitrary two-level unitary operation on the state space of n qubits. Combining these results we see that single qubit and CNOT gates can be used to implement an arbitrary unitary operation on n qubits, and therefore are universal for quantum computation.

Suppose U is a two-level unitary matrix on an n qubit quantum computer. Suppose in particular that U acts non-trivially on the space spanned by the computational basis states $|s\rangle$ and $|t\rangle$, where $s = s_1\ldots s_n$ and $t = t_1\ldots t_n$ are the binary expansions for s and t. Let $\tilde{U}$ be the non-trivial 2×2 unitary submatrix of U; $\tilde{U}$ can be thought of as a unitary operator on a single qubit.

Our immediate goal is to construct a circuit implementing U, built from single qubit and CNOT gates. To do this, we need to make use of *Gray codes*. Suppose we have distinct binary numbers, s and t. A *Gray code* connecting s and t is a sequence of binary numbers, starting with s and concluding with t, such that adjacent members of the list differ in exactly one bit. For instance, with $s = 101001$ and $t = 110011$ we have the Gray code

$$
\begin{matrix}
1 & 0 & 1 & 0 & 0 & 1 \\
1 & 0 & 1 & 0 & 1 & 1 \\
1 & 0 & 0 & 0 & 1 & 1 \\
1 & 1 & 0 & 0 & 1 & 1
\end{matrix}
\tag{4.53}
$$

Let g_1 through g_m be the elements of a Gray code connecting s and t, with $g_1 = s$ and $g_m = t$. Note that we can always find a Gray code such that $m \le n + 1$ since s and t can differ in at most n locations.

The basic idea of the quantum circuit implementing U is to perform a sequence of gates effecting the state changes $|g_1\rangle \to |g_2\rangle \to \cdots \to |g_{m-1}\rangle$, then to perform a controlled-$\tilde{U}$ operation, with the target qubit located at the single bit where g_{m-1} and g_m differ, and then to undo the first stage, transforming $|g_{m-1}\rangle \to |g_{m-2}\rangle \to \cdots \to |g_1\rangle$. Each of these steps can be easily implemented using operations developed earlier in this chapter, and the final result is an implementation of U.

A more precise description of the implementation is as follows. The first step is to swap the states $|g_1\rangle$ and $|g_2\rangle$. Suppose g_1 and g_2 differ at the ith digit. Then we accomplish the swap by performing a controlled bit flip on the ith qubit, conditional on the values of the other qubits being identical to those in both g_1 and g_2. Next we use a controlled operation to swap $|g_2\rangle$ and $|g_3\rangle$. We continue in this fashion until we swap $|g_{m-2}\rangle$ with $|g_{m-1}\rangle$. The effect of this sequence of $m - 2$ operations is to achieve the operation

$$|g_1\rangle \to |g_{m-1}\rangle \tag{4.54}$$

$$|g_2\rangle \to |g_1\rangle \tag{4.55}$$

$$|g_3\rangle \to |g_2\rangle \tag{4.56}$$

$$\cdots\cdots\cdots$$

$$|g_{m-1}\rangle \to |g_{m-2}\rangle. \tag{4.57}$$

All other computational basis states are left unchanged by this sequence of operations. Next, suppose g_{m-1} and g_m differ in the jth bit. We apply a controlled-$\tilde{U}$ operation with the jth qubit as target, conditional on the other qubits having the same values as appear in both g_m and g_{m-1}. Finally, we complete the U operation by undoing the swap operations: we swap $|g_{m-1}\rangle$ with $|g_{m-2}\rangle$, then $|g_{m-2}\rangle$ with $|g_{m-3}\rangle$ and so on, until we swap $|g_2\rangle$ with $|g_1\rangle$.

A simple example illuminates the procedure further. Suppose we wish to implement the two-level unitary transformation

$$
U = \begin{bmatrix}
a & 0 & 0 & 0 & 0 & 0 & 0 & c \\
0 & 1 & 0 & 0 & 0 & 0 & 0 & 0 \\
0 & 0 & 1 & 0 & 0 & 0 & 0 & 0 \\
0 & 0 & 0 & 1 & 0 & 0 & 0 & 0 \\
0 & 0 & 0 & 0 & 1 & 0 & 0 & 0 \\
0 & 0 & 0 & 0 & 0 & 1 & 0 & 0 \\
0 & 0 & 0 & 0 & 0 & 0 & 1 & 0 \\
b & 0 & 0 & 0 & 0 & 0 & 0 & d
\end{bmatrix}.
\tag{4.58}
$$

Here, a, b, c and d are any complex numbers such that $\tilde{U} \equiv \begin{bmatrix} a & c \\ b & d \end{bmatrix}$ is a unitary matrix.

Notice that U acts non-trivially only on the states $|000\rangle$ and $|111\rangle$. We write a Gray code connecting 000 and 111:

$$
\begin{array}{ccc}
A & B & C \\
0 & 0 & 0 \\
0 & 0 & 1 \\
0 & 1 & 1 \\
1 & 1 & 1
\end{array}
\qquad (4.59)
$$

From this we read off the required circuit, shown in Figure 4.16. The first two gates shuffle the states so that $|000\rangle$ gets swapped with $|011\rangle$. Next, the operation $\tilde{U}$ is applied to the first qubit of the states $|011\rangle$ and $|111\rangle$, conditional on the second and third qubits being in the state $|11\rangle$. Finally, we unshuffle the states, ensuring that $|011\rangle$ gets swapped back with the state $|000\rangle$.

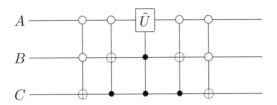

Figure 4.16. Circuit implementing the two-level unitary operation defined by (4.58).

Returning to the general case, we see that implementing the two-level unitary operation U requires at most $2(n-1)$ controlled operations to swap $|g_1\rangle$ with $|g_{m-1}\rangle$ and then back again. Each of these controlled operations can be realized using $O(n)$ single qubit and CNOT gates; the controlled-$\tilde{U}$ operation also requires $O(n)$ gates. Thus, implementing U requires $O(n^2)$ single qubit and CNOT gates. We saw in the previous section that an arbitrary unitary matrix on the 2^n-dimensional state space of n qubits may be written as a product of $O(2^{2n}) = O(4^n)$ two-level unitary operations. Combining these results, we see that an arbitrary unitary operation on n qubits can be implemented using a circuit containing $O(n^2 4^n)$ single qubit and CNOT gates. Obviously, this construction does not provide terribly efficient quantum circuits! However, we show in Section 4.5.4 that the construction is close to optimal in the sense that there are unitary operations that require an exponential number of gates to implement. Thus, to find fast quantum algorithms we will clearly need a different approach than is taken in the universality construction.

Exercise 4.39: Find a quantum circuit using single qubit operations and CNOTs to implement the transformation

$$
\begin{bmatrix}
1 & 0 & 0 & 0 & 0 & 0 & 0 & 0 \\
0 & 1 & 0 & 0 & 0 & 0 & 0 & 0 \\
0 & 0 & a & 0 & 0 & 0 & 0 & c \\
0 & 0 & 0 & 1 & 0 & 0 & 0 & 0 \\
0 & 0 & 0 & 0 & 1 & 0 & 0 & 0 \\
0 & 0 & 0 & 0 & 0 & 1 & 0 & 0 \\
0 & 0 & 0 & 0 & 0 & 0 & 1 & 0 \\
0 & 0 & b & 0 & 0 & 0 & 0 & d
\end{bmatrix},
\qquad (4.60)
$$

where $\tilde{U} = \begin{bmatrix} a & c \\ b & d \end{bmatrix}$ is an arbitrary 2×2 unitary matrix.

4.5.3 A discrete set of universal operations

In the previous section we proved that the CNOT and single qubit unitaries together form a universal set for quantum computation. Unfortunately, no straightforward method is known to implement all these gates in a fashion which is resistant to errors. Fortunately, in this section we'll find a discrete set of gates which can be used to perform universal quantum computation, and in Chaper 10 we'll show how to perform these gates in an error-resistant fashion, using quantum error-correcting codes.

Approximating unitary operators

Obviously, a discrete set of gates can't be used to implement an arbitrary unitary operation *exactly*, since the set of unitary operations is continuous. Rather, it turns out that a discrete set can be used to *approximate* any unitary operation. To understand how this works, we first need to study what it means to approximate a unitary operation. Suppose U and V are two unitary operators on the same state space. U is the target unitary operator that we wish to implement, and V is the unitary operator that is actually implemented in practice. We define the *error* when V is implemented instead of U by

$$E(U, V) \equiv \max_{|\psi\rangle} \|(U - V)|\psi\rangle\|, \tag{4.61}$$

where the maximum is over all normalized quantum states $|\psi\rangle$ in the state space. In Box 4.1 on page 195 we show that this measure of error has the interpretation that if $E(U, V)$ is small, then any measurement performed on the state $V|\psi\rangle$ will give approximately the same measurement statistics as a measurement of $U|\psi\rangle$, for any initial state $|\psi\rangle$. More precisely, we show that if M is a POVM element in an arbitrary POVM, and P_U (or P_V) is the probability of obtaining this outcome if U (or V) were performed with a starting state $|\psi\rangle$, then

$$|P_U - P_V| \le 2E(U, V). \tag{4.62}$$

Thus, if $E(U, V)$ is small, then measurement outcomes occur with similar probabilities, regardless of whether U or V were performed. Also shown in Box 4.1 is that if we perform a sequence of gates $V_1, \ldots, V_m$ intended to approximate some other sequence of gates $U_1, \ldots, U_m$, then the errors add at most linearly,

$$E(U_m U_{m-1} \ldots U_1, V_m V_{m-1} \ldots V_1) \le \sum_{j=1}^{m} E(U_j, V_j). \tag{4.63}$$

The approximation results (4.62) and (4.63) are extremely useful. Suppose we wish to perform a quantum circuit containing m gates, U_1 through U_m. Unfortunately, we are only able to approximate the gate U_j by the gate V_j. In order that the probabilities of different measurement outcomes obtained from the approximate circuit be within a tolerance $\Delta > 0$ of the correct probabilities, it suffices that $E(U_j, V_j) \le \Delta/(2m)$, by the results (4.62) and (4.63).

Universality of Hadamard + phase + CNOT + π/8 gates

We're now in a good position to study the approximation of arbitrary unitary operations by discrete sets of gates. We're going to consider two different discrete sets of gates, both

---

Box 4.1: Approximating quantum circuits

Suppose a quantum system starts in the state $|\psi\rangle$, and we perform either the unitary operation U, or the unitary operation V. Following this, we perform a measurement. Let M be a POVM element associated with the measurement, and let P_U (or P_V) be the probability of obtaining the corresponding measurement outcome if the operation U (or V) was performed. Then

$$|P_U - P_V| = |\langle\psi|U^\dagger MU|\psi\rangle - \langle\psi|V^\dagger MV|\psi\rangle|. \qquad (4.64)$$

Let $|\Delta\rangle \equiv (U - V)|\psi\rangle$. Simple algebra and the Cauchy–Schwarz inequality show that

$$|P_U - P_V| = |\langle\psi|U^\dagger M|\Delta\rangle + \langle\Delta|MV|\psi\rangle|. \qquad (4.65)$$
$$\leq |\langle\psi|U^\dagger M|\Delta\rangle| + |\langle\Delta|MV|\psi\rangle| \qquad (4.66)$$
$$\leq \||\Delta\rangle\| + \||\Delta\rangle\| \qquad (4.67)$$
$$\leq 2E(U, V). \qquad (4.68)$$

The inequality $|P_U - P_V| \leq 2E(U, V)$ gives quantitative expression to the idea that when the error $E(U, V)$ is small, the difference in probabilities between measurement outcomes is also small.

Suppose we perform a sequence $V_1, V_2, \ldots, V_m$ of gates intended to approximate some other sequence of gates, $U_1, U_2, \ldots, U_m$. Then it turns out that the error caused by the entire sequence of imperfect gates is at most the sum of the errors in the individual gates,

$$E(U_m U_{m-1} \ldots U_1, V_m V_{m-1} \ldots V_1) \leq \sum_{j=1}^{m} E(U_j, V_j). \qquad (4.69)$$

To prove this we start with the case $m = 2$. Note that for some state $|\psi\rangle$ we have

$$E(U_2 U_1, V_2 V_1) = \|(U_2 U_1 - V_2 V_1)|\psi\rangle\| \qquad (4.70)$$
$$= \|(U_2 U_1 - V_2 U_1)|\psi\rangle + (V_2 U_1 - V_2 V_1)|\psi\rangle\|. \qquad (4.71)$$

Using the triangle inequality $\||a\rangle + |b\rangle\| \leq \||a\rangle\| + \||b\rangle\|$, we obtain

$$E(U_2 U_1, V_2 V_1) \leq \|(U_2 - V_2)U_1|\psi\rangle\| + \|V_2(U_1 - V_1)|\psi\rangle\| \qquad (4.72)$$
$$\leq E(U_2, V_2) + E(U_1, V_1), \qquad (4.73)$$

which was the desired result. The result for general m follows by induction.

---

of which are universal. The first set, the *standard set* of universal gates, consists of the Hadamard, phase, controlled-NOT and $\pi/8$ gates. We provide fault-tolerant constructions for these gates in Chapter 10; they also provide an exceptionally simple universality construction. The second set of gates we consider consists of the Hadamard gate, phase gate, the controlled-NOT gate, and the Toffoli gate. These gates can also all be done fault-tolerantly; however, the universality proof and fault-tolerance construction for these gates is a little less appealing.

We begin the universality proof by showing that the Hadamard and $\pi/8$ gates can be used to approximate any single qubit unitary operation to arbitrary accuracy. Consider the gates T and HTH. T is, up to an unimportant global phase, a rotation by $\pi/4$ radians around the $\hat{z}$ axis on the Bloch sphere, while HTH is a rotation by $\pi/4$ radians around the $\hat{x}$ axis on the Bloch sphere (Exercise 4.14). Composing these two operations gives, up to a global phase,

$$
\exp\left(-i\frac{\pi}{8}Z\right)\exp\left(-i\frac{\pi}{8}X\right) = \left[\cos\frac{\pi}{8}I - i\sin\frac{\pi}{8}Z\right]\left[\cos\frac{\pi}{8}I - i\sin\frac{\pi}{8}X\right] \quad (4.74)
$$

$$
= \cos^2\frac{\pi}{8}I - i\left[\cos\frac{\pi}{8}(X+Z) + \sin\frac{\pi}{8}Y\right]\sin\frac{\pi}{8}. \quad (4.75)
$$

This is a rotation of the Bloch sphere about an axis along $\vec{n} = (\cos\frac{\pi}{8}, \sin\frac{\pi}{8}, \cos\frac{\pi}{8})$ with corresponding unit vector $\hat{n}$, and through an angle θ defined by $\cos(\theta/2) \equiv \cos^2\frac{\pi}{8}$. That is, using only the Hadamard and $\pi/8$ gates we can construct $R_{\hat{n}}(\theta)$. Moreover, this θ can be shown to be an irrational multiple of 2π. Proving this latter fact is a little beyond our scope; see the end of chapter 'History and further reading'.

Next, we show that repeated iteration of $R_{\hat{n}}(\theta)$ can be used to approximate to arbitrary accuracy any rotation $R_{\hat{n}}(\alpha)$. To see this, let $\delta > 0$ be the desired accuracy, and let N be an integer larger than $2\pi/\delta$. Define θ_k so that $\theta_k \in [0, 2\pi)$ and $\theta_k = (k\theta) \bmod 2\pi$. Then the pigeonhole principle implies that there are distinct j and k in the range $1, \ldots, N$ such that $|\theta_k - \theta_j| \leq 2\pi/N < \delta$. Without loss of generality assume that $k > j$, so we have $|\theta_{k-j}| < \delta$. Since $j \neq k$ and θ is an irrational multiple of 2π we must have $\theta_{k-j} \neq 0$. It follows that the sequence $\theta_{l(k-j)}$ fills up the interval $[0, 2\pi)$ as l is varied, so that adjacent members of the sequence are no more than δ apart. It follows that for any $\epsilon > 0$ there exists an n such that

$$
E(R_{\hat{n}}(\alpha), R_{\hat{n}}(\theta)^n) < \frac{\epsilon}{3}. \quad (4.76)
$$

Exercise 4.40: For arbitrary α and β show that

$$
E(R_{\hat{n}}(\alpha), R_{\hat{n}}(\alpha + \beta)) = |1 - \exp(i\beta/2)|, \quad (4.77)
$$

and use this to justify (4.76).

We are now in position to verify that any single qubit operation can be approximated to arbitrary accuracy using the Hadamard and $\pi/8$ gates. Simple algebra implies that for any α

$$
HR_{\hat{n}}(\alpha)H = R_{\hat{m}}(\alpha), \quad (4.78)
$$

where $\hat{m}$ is a unit vector in the direction $(\cos\frac{\pi}{8}, -\sin\frac{\pi}{8}, \cos\frac{\pi}{8})$, from which it follows that

$$
E(R_{\hat{m}}(\alpha), R_{\hat{m}}(\theta)^n) < \frac{\epsilon}{3}. \quad (4.79)
$$

But by Exercise 4.11 an arbitrary unitary U on a single qubit may be written as

$$
U = R_{\hat{n}}(\beta)R_{\hat{m}}(\gamma)R_{\hat{n}}(\delta), \quad (4.80)
$$

up to an unimportant global phase shift. The results (4.76) and (4.79), together with the chaining inequality (4.63) therefore imply that for suitable positive integers n_1, n_2, n_3,

$$E(U, R_{\hat{n}}(\theta)^{n_1} H R_{\hat{n}}(\theta)^{n_2} H R_{\hat{n}}(\theta)^{n_3}) < \epsilon. \tag{4.81}$$

That is, given any single qubit unitary operator U and any $\epsilon > 0$ it is possible to approximate U to within ϵ using a circuit composed of Hadamard gates and $\pi/8$ gates alone.

Since the $\pi/8$ and Hadamard gates allow us to approximate any single qubit unitary operator, it follows from the arguments of Section 4.5.2 that we can approximate any m gate quantum circuit, as follows. Given a quantum circuit containing m gates, either CNOTs or single qubit unitary gates, we may approximate it using Hadamard, controlled-NOT and $\pi/8$ gates (later, we will find that phase gates make it possible to do the appoximation fault-tolerantly, but for the present universality argument they are not strictly necessary). If we desire an accuracy of ϵ for the entire circuit, then this may be achieved by approximating each single qubit unitary using the above procedure to within ϵ/m and applying the chaining inequality (4.63) to obtain an accuracy of ϵ for the entire circuit.

How efficient is this procedure for approximating quantum circuits using a discrete set of gates? This is an important question. Suppose, for example, that approximating an arbitrary single qubit unitary to within a distance ϵ were to require $\Omega(2^{1/\epsilon})$ gates from the discrete set. Then to approximate the m gate quantum circuit considered in the previous paragraph would require $\Omega(m2^{m/\epsilon})$ gates, an exponential increase over the original circuit size! Fortunately, the rate of convergence is much better than this. Intuitively, it is plausible that the sequence of angles θ_k 'fills in' the interval $[0, 2\pi)$ in a more or less uniform fashion, so that to approximate an arbitrary single qubit gate ought to take roughly $\Theta(1/\epsilon)$ gates from the discrete set. If we use this estimate for the number of gates required to approximate an arbitrary single qubit gate, then the number required to approximate an m gate circuit to accuracy ϵ becomes $\Theta(m^2/\epsilon)$. This is a quadratic increase over the original size of the circuit, m, which for many applications may be sufficient.

Rather remarkably, however, a much faster rate of convergence can be proved. The *Solovay–Kitaev theorem*, proved in Appendix 3, implies that an arbitrary single qubit gate may be approximated to an accuracy ϵ using $O(\log^c(1/\epsilon))$ gates from our discrete set, where c is a constant approximately equal to 2. The Solovay–Kitaev theorem therefore implies that to approximate a circuit containing m CNOTs and single qubit unitaries to an accuracy ϵ requires $O(m \log^c(m/\epsilon))$ gates from the discrete set, a polylogarithmic increase over the size of the original circuit, which is likely to be acceptable for virtually all applications.

To sum up, we have shown that the Hadamard, phase, controlled-NOT and $\pi/8$ gates are universal for quantum computation in the sense that given a circuit containing CNOTs and arbitrary single qubit unitaries it is possible to simulate this circuit to good accuracy using only this discrete set of gates. Moreover, the simulation can be performed efficiently, in the sense that the overhead required to perform the simulation is polynomial in $\log(m/\epsilon)$, where m is the number of gates in the original circuit, and ϵ is the desired accuracy of the simulation.

Exercise 4.41: This and the next two exercises develop a construction showing that the Hadamard, phase, controlled-NOT and Toffoli gates are universal. Show that

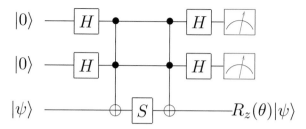

Figure 4.17. Provided both measurement outcomes are 0 this circuit applies $R_z(\theta)$ to the target, where $\cos\theta = 3/5$. If some other measurement outcome occurs then the circuit applies Z to the target.

the circuit in Figure 4.17 applies the operation $R_z(\theta)$ to the third (target) qubit if the measurement outcomes are both 0, where $\cos\theta = 3/5$, and otherwise applies Z to the target qubit. Show that the probability of both measurement outcomes being 0 is 5/8, and explain how repeated use of this circuit and $Z = S^2$ gates may be used to apply a $R_z(\theta)$ gate with probability approaching 1.

Exercise 4.42: (Irrationality of θ) Suppose $\cos\theta = 3/5$. We give a proof by contradiction that θ is an irrational multiple of 2π.

(1) Using the fact that $e^{i\theta} = (3 + 4i)/5$, show that if θ is rational, then there must exist a positive integer m such that $(3 + 4i)^m = 5^m$.

(2) Show that $(3 + 4i)^m = 3 + 4i \pmod 5$ for all $m > 0$, and conclude that no m such that $(3 + 4i)^m = 5^m$ can exist.

Exercise 4.43: Use the results of the previous two exercises to show that the Hadamard, phase, controlled-NOT and Toffoli gates are universal for quantum computation.

Exercise 4.44: Show that the three qubit gate G defined by the circuit:

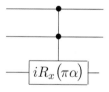

is universal for quantum computation whenever α is irrational.

Exercise 4.45: Suppose U is a unitary transform implemented by an n qubit quantum circuit constructed from H, S, CNOT and Toffoli gates. Show that U is of the form $2^{-k/2}M$, for some integer k, where M is a $2^n \times 2^n$ matrix with only complex integer entries. Repeat this exercise with the Toffoli gate replaced by the $\pi/8$ gate.

4.5.4 Approximating arbitrary unitary gates is generically hard

We've seen that any unitary transformation on n qubits can be built up out of a small set of elementary gates. Is it always possible to do this efficiently? That is, given a unitary transformation U on n qubits does there always exist a circuit of size polynomial in n approximating U? The answer to this question turns out to be a resounding no: in fact, most unitary transformations can only be implemented very inefficiently. One way to see this is to consider the question: how many gates does it take to generate an arbitrary state of n qubits? A simple counting argument shows that this requires exponentially many operations, in general; it immediately follows that there are unitary operations requiring exponentially many operations. To see this, suppose we have g different types of gates available, and each gate works on at most f input qubits. These numbers, f and g, are fixed by the computing hardware we have available, and may be considered to be constants. Suppose we have a quantum circuit containing m gates, starting from the computational basis state $|0\rangle^{\otimes n}$. For any particular gate in the circuit there are therefore

at most $\begin{bmatrix} n \\ f \end{bmatrix}^g = O(n^{fg})$ possible choices. It follows that at most $O(n^{fgm})$ different

states may be computed using m gates.

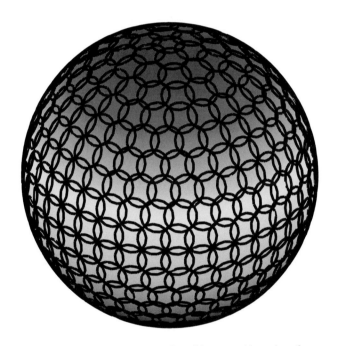

Figure 4.18. Visualization of covering the set of possible states with patches of constant radius.

Suppose we wish to approximate a particular state, $|\psi\rangle$, to within a distance ϵ. The idea of the proof is to cover the set of all possible states with a collection of 'patches,' each of radius ϵ (Figure 4.18), and then to show that the number of patches required rises doubly exponentially in n; comparing with the exponential number of different states that may be computed using m gates will imply the result. The first observation we need is that the space of state vectors of n qubits can be regarded as just the unit $(2^{n+1} - 1)$-sphere. To see this, suppose the n qubit state has amplitudes $\psi_j = X_j + iY_j$, where X_j and Y_j are the real and imaginary parts, respectively, of the jth amplitude. The normalization condition for quantum states can be written $\sum_j (X_j^2 + Y_j^2) = 1$, which is just the condition for a point to be on the unit sphere in 2^{n+1} real dimensions, that is, the unit $(2^{n+1} - 1)$-sphere. Similarly, the surface area of radius ϵ near $|\psi\rangle$ is approximately the same as the volume of a $(2^{n+1} - 2)$-sphere of radius ϵ. Using the formula $S_k(r) = 2\pi^{(k+1)/2} r^k / \Gamma((k+1)/2)$ for the surface area of a k-sphere of radius r, and $V_k(r) = 2\pi^{(k+1)/2} r^{k+1} / [(k+1)\Gamma((k+1)/2)]$ for the volume of a k-sphere of radius r, we see that the number of patches needed to cover the state space goes like

$$\frac{S_{2^{n+1}-1}(1)}{V_{2^{n+1}-2}(\epsilon)} = \frac{\sqrt{\pi}\Gamma(2^n - \frac{1}{2})(2^{n+1} - 1)}{\Gamma(2^n)\epsilon^{2^{n+1}-1}}, \tag{4.82}$$

where Γ is the usual generalization of the factorial function. But $\Gamma(2^n - 1/2) \geq \Gamma(2^n)/2^n$, so the number of patches required to cover the space is at least

$$\Omega\left(\frac{1}{\epsilon^{2^{n+1}-1}}\right). \tag{4.83}$$

Recall that the number of patches which can be reached in m gates is $O(n^{fgm})$, so in order to reach all the ϵ-patches we must have

$$O\left(n^{fgm}\right) \geq \Omega\left(\frac{1}{\epsilon^{2^{n+1}-1}}\right) \tag{4.84}$$

which gives us

$$m = \Omega\left(\frac{2^n \log(1/\epsilon)}{\log(n)}\right). \tag{4.85}$$

That is, there are states of n qubits which take $\Omega(2^n \log(1/\epsilon)/\log(n))$ operations to approximate to within a distance ϵ. This is exponential in n, and thus is 'difficult', in the sense of computational complexity introduced in Chapter 3. Furthermore, this immediately implies that there are unitary transformations U on n qubits which take $\Omega(2^n \log(1/\epsilon)/\log(n))$ operations to approximate by a quantum circuit implementing an operation V such that $E(U, V) \leq \epsilon$. By contrast, using our universality constructions and the Solovay–Kitaev theorem it follows that an arbitrary unitary operation U on n qubits may be approximated to within a distance ϵ using $O(n^2 4^n \log^c(n^2 4^n/\epsilon))$ gates. Thus, to within a polynomial factor the construction for universality we have given is optimal; unfortunately, it does not address the problem of determining which families of unitary operations can be computed efficiently in the quantum circuits model.

4.5.5 Quantum computational complexity

In Chapter 3 we described a theory of 'computational complexity' for classical computers that classified the resource requirements to solve computational problems on classical computers. Not surprisingly there is considerable interest in developing a theory of quantum computational complexity, and relating it to classical computational complexity theory. Although only first steps have been taken in this direction, it will doubtless be an enormously fruitful direction for future researchers. We content ourselves with presenting one result about quantum complexity classes, relating the quantum complexity class **BQP** to the classical complexity class **PSPACE**. Our discussion of this result is rather informal; for more details you are referred to the paper of Bernstein and Vazirani referenced in the end of chapter 'History and further reading'.

Recall that **PSPACE** was defined in Chapter 3 as the class of decision problems which can be solved on a Turing machine using space polynomial in the problem size and an arbitrary amount of time. **BQP** is an essentially quantum complexity class consisting of those decision problems that can be solved with bounded probability of error using a polynomial size quantum circuit. Slightly more formally, we say a language L is in **BQP** if there is a family of polynomial size quantum circuits which decides the language, accepting strings in the language with probability at least $3/4$, and rejecting strings which aren't in the language with probability at least $3/4$. In practice, what this means is that the quantum circuit takes as input binary strings, and tries to determine whether they are elements of the language or not. At the conclusion of the circuit one qubit is measured, with 0 indicating that the string has been accepted, and 1 indicating rejection. By testing the string a few times to determine whether it is in L, we can determine with very high probability whether a given string is in L.

Of course, a quantum circuit is a fixed entity, and any given quantum circuit can only decide whether strings up to some finite length are in L. For this reason, we use an entire family of circuits in the definition of **BQP**; for every possible input length there is a different circuit in the family. We place two restrictions on the circuit in addition to the acceptance / rejection criterion already described. First, the size of the circuits should only grow polynomially with the size of the input string x for which we are trying to determine whether $x \in L$. Second, we require that the circuits be *uniformly generated*, in a sense similar to that described in Section 3.1.2. This uniformity requirement arises because, in practice, given a string x of some length n, somebody will have to build a quantum circuit capable of deciding whether x is in L. To do so, they will need to have a clear set of instructions – an algorithm – for building the circuit. For this reason, we require that our quantum circuits be uniformly generated, that is, there is a Turing machine capable of efficiently outputting a description of the quantum circuit. This restriction may seem rather technical, and in practice is nearly always satisfied trivially, but it does save us from pathological examples such as that described in Section 3.1.2. (You might also wonder if it matters whether the Turing machine used in the uniformity requirement is a quantum or classical Turing machine; it turns out that it doesn't matter – see 'History and further reading'.)

One of the most significant results in quantum computational complexity is that **BQP** $\subseteq$ **PSPACE**. It is clear that **BPP** $\subseteq$ **BQP**, where **BPP** is the classical complexity class of decision problems which can be solved with bounded probability of error using poly-nomial time on a classical Turing machine. Thus we have the chain of inclusions **BPP** $\subseteq$ **BQP** $\subseteq$ **PSPACE**. Proving that **BQP** $\neq$ **BPP** – intuitively the statement that quantum computers are more powerful than classical computers – will therefore imply that **BPP** $\neq$ **PSPACE**. However, it is not presently known whether **BPP** $\neq$ **PSPACE**, and proving this would represent a major breakthrough in classical computer science! So proving that quantum computers are more powerful than classical computers would have some very interesting implications for classical computational complexity! Unfortunately, it also means that providing such a proof may be quite difficult.

Why is it that **BQP** $\subseteq$ **PSPACE**? Here is an intuitive outline of the proof (a rigorous proof is left to the references in 'History and further reading'). Suppose we have an n qubit quantum computer, and do a computation involving a sequence of $p(n)$ gates, where $p(n)$ is some polynomial in n. Supposing the quantum circuit starts in the state $|0\rangle$ we will explain how to evaluate in polynomial space on a classical computer the probability that it ends up in the state $|y\rangle$. Suppose the gates that are executed on the quantum computer are, in order, $U_1, U_2, \ldots, U_{p(n)}$. Then the probability of ending up in the state $|y\rangle$ is the modulus squared of

$$\langle y|U_{p(n)} \cdots U_2 U_1|0\rangle . \tag{4.86}$$

This quantity may be estimated in polynomial space on a classical computer. The basic idea is to insert the completeness relation $\sum_x |x\rangle\langle x| = I$ between each term in (4.86), obtaining

$$\langle y|U_{p(n)}\cdots U_2 U_1|0\rangle = \sum_{x_1,\ldots,x_{p(n)-1}} \langle y|U_{p(n)}|x_{p(n)-1}\rangle\langle x_{p(n)-1}|U_{p(n)-2}\cdots U_2|x_1\rangle\langle x_1|U_1|0\rangle .$$

(4.87)

Given that the individual unitary gates appearing in this sum are operations such as the Hadamard gate, CNOT, and so on, it is clear that each term in the sum can be calculated to high accuracy using only polynomial space on a classical computer, and thus the sum as a whole can be calculated using polynomial space, since individual terms in the sum can be erased after being added to the running total. Of course, this algorithm is rather slow, since there are exponentially many terms in the sum which need to be calculated and added to the total; however, only polynomially much space is consumed, and thus **BQP $\subseteq$ PSPACE**, as we set out to show.

   A similar procedure can be used to simulate an *arbitrary* quantum computation on a classical computer, no matter the length of the quantum computation. Therefore, the class of problems solvable on a quantum computer with unlimited time and space resources is no larger than the class of problems solvable on a classical computer. Stated another way, this means that quantum computers do not violate the Church–Turing thesis that any algorithmic process can be simulated efficiently using a Turing machine. Of course, quantum computers may be much more *efficient* than their classical counterparts, thereby challenging the *strong* Church–Turing thesis that any algorithmic process can be simulated *efficiently* using a probabilistic Turing machine.

4.6 Summary of the quantum circuit model of computation

In this book the term 'quantum computer' is synonymous with the quantum circuit model of computation. This chapter has provided a detailed look at quantum circuits, their basic elements, universal families of gates, and some applications. Before we move on to more sophisticated applications, let us summarize the key elements of the quantum circuit model of computation:

(1) **Classical resources:** A quantum computer consists of two parts, a classical part and a quantum part. In principle, there is no need for the classical part of the computer, but in practice certain tasks may be made much easier if parts of the computation can be done classically. For example, many schemes for quantum error-correction (Chapter 10) are likely to involve classical computations in order to maximize efficiency. While classical computations can always be done, in principle, on a quantum computer, it may be more convenient to perform the calculations on a classical computer.

(2) **A suitable state space:** A quantum circuit operates on some number, n, of qubits. The state space is thus a 2^n-dimensional complex Hilbert space. Product states of the form $|x_1,\ldots,x_n\rangle$, where $x_i = 0,1$, are known as *computational basis states* of the computer. $|x\rangle$ denotes a computational basis state, where x is the number whose binary representation is $x_1\ldots x_n$.

(3) **Ability to prepare states in the computational basis:** It is assumed that any computational basis state $|x_1,\ldots,x_n\rangle$ can be prepared in at most n steps.

(4) **Ability to perform quantum gates:** Gates can be applied to any subset of qubits as desired, and a universal family of gates can be implemented. For example, it should be possible to apply the CNOT gate to any pair of qubits in the quantum computer. The Hadamard, phase, CNOT and $\pi/8$ gates form a family of gates from which any unitary operation can be approximated, and thus is a universal set of gates. Other universal families exist.

(5) **Ability to perform measurements in the computational basis:**
Measurements may be performed in the computational basis of one or more of the qubits in the computer.

The quantum circuit model of quantum computation is equivalent to many other models of computation which have been proposed, in the sense that other models result in essentially the same resource requirements for the same problems. As a simple example which illustrates the basic idea, one might wonder whether moving to a design based on three-level quantum systems, rather than the two-level qubits, would confer any computational advantage. Of course, although there may be some slight advantage in using three-level quantum systems (*qutrits*) over two-level systems, any difference will be essentially negligible from the theoretical point of view. At a less trivial level, the 'quantum Turing machine' model of computation, a quantum generalization of the classical Turing machine model, has been shown to be equivalent to the model based upon quantum circuits. We do not consider that model of computation in this book, but the reader interested in learning more about quantum Turing machines may consult the references given in the end of chapter 'History and further reading'.

Despite the simplicity and attraction of the quantum circuit model, it is useful to keep in mind possible criticisms, modifications, and extensions. For example, it is by no means clear that the basic assumptions underlying the state space and starting conditions in the quantum circuit model are justified. Everything is phrased in terms of finite dimensional state spaces. Might there be anything to be gained by using systems whose state space is infinite dimensional? Assuming that the starting state of the computer is a computational basis state is also not necessary; we know that many systems in Nature 'prefer' to sit in highly entangled states of many systems; might it be possible to exploit this preference to obtain extra computational power? It might be that having access to certain states allows particular computations to be done much more easily than if we are constrained to start in the computational basis. Likewise, the ability to efficiently perform entangling measurements in multi-qubit bases might be as useful as being able to perform just entangling unitary operations. Indeed, it may be possible to harness such measurements to perform tasks intractable within the quantum circuit model.

A detailed examination and attempted justification of the physics underlying the quantum circuit model is outside the scope of the present discussion, and, indeed, outside the scope of present knowledge! By raising these issues we wish to introduce the question of the completeness of the quantum circuit model, and re-emphasize the fundamental point that information is physical. In our attempts to formulate models for information processing we should always attempt to go back to fundamental physical laws. For the purposes of this book, we shall stay within the quantum circuit model of computation. It offers a rich and powerful model of computation that exploits the properties of quantum mechanics to perform amazing feats of information processing, without classical precedent. Whether physically reasonable models of computation exist which go beyond the quantum circuit model is a fascinating question which we leave open for you.

4.7 Simulation of quantum systems

Perhaps [...] we need a mathematical theory of quantum automata. [...] the quantum state space has far greater capacity than the classical one: for a classical system with N states, its quantum version allowing superposition accommodates c^N states. When we join two classical systems, their number of states N_1 and N_2 are multiplied, and in the quantum case we get the exponential growth $c^{N_1 N_2}$. [...] These crude estimates show that the quantum behavior of the system might be much more complex than its classical simulation.
– Yu Manin (1980)[Man80], as translated in [Man99]

The quantum-mechanical computation of one molecule of methane requires 10^{42} grid points. Assuming that at each point we have to perform only 10 elementary operations, and that the computation is performed at the extremely low temperature $T = 3 \times 10^{-3} K$, we would still have to use all the energy produced on Earth during the last century.
– R. P. Poplavskii (1975)[Pop75], as quoted by Manin

Can physics be simulated by a universal computer? [...] the physical world is quantum mechanical, and therefore the proper problem is the simulation of quantum physics [...] the full description of quantum mechanics for a large system with R particles [...] has too many variables, it **cannot be simulated** *with a normal computer with a number of elements proportional to R [... but it can be simulated with] quantum computer elements. [...] Can a quantum system be probabilistically simulated by a classical (probabilistic, I'd assume) universal computer? [...] If you take the computer to be the classical kind I've described so far [..] the answer is certainly, No!*
– Richard P. Feynman (1982)[Fey82]

Let us close out this chapter by providing an interesting and useful application of the quantum circuit model. One of the most important practical applications of computation is the simulation of physical systems. For example, in the engineering design of a new building, finite element analysis and modeling is used to ensure safety while minimizing cost. Cars are made lightweight, structurally sound, attractive, and inexpensive, by using computer aided design. Modern aeronautical engineering depends heavily on computational fluid dynamics simulations for aircraft designs. Nuclear weapons are no longer exploded (for the most part), but rather, tested by exhaustive computational modeling. Examples abound, because of the tremendous practical applications of predictive simulations. We begin by describing some instances of the simulation problem, then we present a quantum algorithm for simulation and an illustrative example, concluding with some perspective on this application.

4.7.1 Simulation in action

The heart of simulation is the solution of differential equations which capture the physical laws governing the dynamical behavior of a system. Some include Newton's law,

$$\frac{d}{dt}\left(m\frac{dx}{dt}\right) = F\,,\tag{4.88}$$

Poisson's equation,

$$-\vec{\nabla}\cdot(k\,\vec{\nabla}\vec{u}) = \vec{Q}\,,\tag{4.89}$$

the electromagnetic vector wave equation,

$$\vec{\nabla}\cdot\vec{\nabla}\vec{E} = \epsilon_0\mu_0\frac{\partial^2\vec{E}}{\partial t^2}\,,\tag{4.90}$$

and the diffusion equation,

$$\vec{\nabla}^2\psi = \frac{1}{a^2}\frac{\partial\psi}{\partial t}\,,\tag{4.91}$$

just to name a very few. The goal is generally: given an initial state of the system, what is the state at some other time and/or position? Solutions are usually obtained by *approximating* the state with a digital representation, then *discretizing* the differential equation in space and time such that an iterative application of a procedure carries the state from the initial to the final conditions. Importantly, the error in this procedure is bounded, and known not to grow faster than some small power of the number of iterations. Furthermore, *not* all dynamical systems can be simulated *efficiently*: generally, only those systems which can be described efficiently can be simulated efficiently.

Simulation of quantum systems by classical computers is possible, but generally only very inefficiently. The dynamical behavior of many simple quantum systems is governed by Schrödinger's equation,

$$i\hbar\frac{d}{dt}|\psi\rangle = H|\psi\rangle\,.\tag{4.92}$$

We will find it convenient to absorb $\hbar$ into H, and use this convention for the rest of this section. For a typical Hamiltonian of interest to physicists dealing with real particles in space (rather than abstract systems such as qubits, which we have been dealing with!), this reduces to

$$i\frac{\partial}{\partial t}\psi(x) = \left[-\frac{1}{2m}\frac{\partial^2}{\partial x^2} + V(x)\right]\psi(x)\,,\tag{4.93}$$

using a convention known as the position representation $\langle x|\psi\rangle = \psi(x)$. This is an elliptical equation very much like Equation (4.91). So just simulating Schrödinger's equation is not the especial difficulty faced in simulating quantum systems. What is the difficulty?

The key challenge in simulating quantum systems is the *exponential* number of differential equations which must be solved. For one qubit evolving according to the Schrödinger equation, a system of two differential equations must be solved; for two qubits, four equations; and for n qubits, 2^n equations. Sometimes, insightful approximations can be made which reduce the effective number of equations involved, thus making classical simulation of the quantum system feasible. However, there are many physically interesting quantum systems for which no such approximations are known.

Exercise 4.46: (Exponential complexity growth of quantum systems) Let ρ be a density matrix describing the state of n qubits. Show that describing ρ requires $4^n - 1$ independent real numbers.

The reader with a physics background may appreciate that there are many important quantum systems for which classical simulation is intractable. These include the Hubbard model, a model of interacting fermionic particles with the Hamiltonian

$$H = \sum_{k=1}^{n} V_0 n_{k\uparrow} n_{k\downarrow} + \sum_{k,j \text{ neighbors},\sigma} t_0 c_{k\sigma}^* c_{j\sigma} , \qquad (4.94)$$

which is useful in the study of superconductivity and magnetism, the Ising model,

$$H = \sum_{k=1}^{n} \vec{\sigma}_k \cdot \vec{\sigma}_{k+1} , \qquad (4.95)$$

and many others. Solutions to such models give many physical properties such as the dielectric constant, conductivity, and magnetic susceptibility of materials. More sophisticated models such as quantum electrodynamics (QED) and quantum chromodynamics (QCD) can be used to compute constants such as the mass of the proton.

Quantum computers can efficiently simulate quantum systems for which there is no known efficient classical simulation. Intuitively, this is possible for much the same reason any quantum circuit can be constructed from a universal set of quantum gates. Moreover, just as there exist unitary operations which cannot be *efficiently* approximated, it is possible in principle to imagine quantum systems with Hamiltonians which cannot be efficiently simulated on a quantum computer. Of course, we believe that such systems aren't actually realized in Nature, otherwise we'd be able to exploit them to do information processing beyond the quantum circuit model.

4.7.2 The quantum simulation algorithm

Classical simulation begins with the realization that in solving a simple differential equation such as $dy/dt = f(y)$, to first order, it is known that $y(t + \Delta t) \approx y(t) + f(y)\Delta t$. Similarly, the quantum case is concerned with the solution of $id|\psi\rangle/dt = H|\psi\rangle$, which, for a time-independent H, is just

$$|\psi(t)\rangle = e^{-iHt}|\psi(0)\rangle . \qquad (4.96)$$

Since H is usually extremely difficult to exponentiate (it may be sparse, but it is also exponentially large), a good beginning is the first order solution $|\psi(t + \Delta t)\rangle \approx (I - iH\Delta t)|\psi(t)\rangle$. This is tractable, because for many Hamiltonians H it is straightforward to compose quantum gates to efficiently approximate $I - iH\Delta t$. However, such first order solutions are generally not very satisfactory.

Efficient approximation of the solution to Equation (4.96), to high order, is possible for many classes of Hamiltonian. For example, in most physical systems, the Hamiltonian can be written as a sum over many local interactions. Specifically, for a system of n particles,

$$H = \sum_{k=1}^{L} H_k , \qquad (4.97)$$

where each H_k acts on at most a constant c number of systems, and L is a polynomial in n. For example, the terms H_k are often just two-body interactions such as $X_i X_j$ and one-body Hamiltonians such as X_i. Both the Hubbard and Ising models have Hamiltonians of this form. Such locality is quite physically reasonable, and originates in many systems from the fact that most interactions fall off with increasing distance or difference in energy. There are sometimes additional global symmetry constraints such as particle statistics; we shall come to those shortly. The important point is that although e^{-iHt} is difficult to compute, $e^{-iH_k t}$ acts on a much smaller subsystem, and is straightforward to approximate using quantum circuits. But because $[H_j, H_k] \neq 0$ in general, $e^{-iHt} \neq \prod_k e^{-iH_k t}$! How, then, can $e^{-iH_k t}$ be useful in constructing e^{-iHt}?

Exercise 4.47: For $H = \sum_k^L H_k$, prove that $e^{-iHt} = e^{-iH_1 t} e^{-iH_2 t} \ldots e^{-iH_L t}$ for all t if $[H_j, H_k] = 0$, for all j, k.

Exercise 4.48: Show that the restriction of H_k to involve at most c particles *implies* that in the sum (4.97), L is upper bounded by a polynomial in n.

The heart of quantum simulation algorithms is the following asymptotic approximation theorem:

Theorem 4.3: **(Trotter formula)** Let A and B be Hermitian operators. Then for any real t,

$$\lim_{n \to \infty} (e^{iAt/n} e^{iBt/n})^n = e^{i(A+B)t} . \tag{4.98}$$

Note that (4.98) is true even if A and B do not commute. Even more interestingly, perhaps, it can be generalized to hold for A and B which are generators of certain kinds of semigroups, which correspond to general quantum operations; we shall describe such generators (the 'Lindblad form') in Section 8.4.1 of Chapter 8. For now, we only consider the case of A and B being Hermitian matrices.

Proof
By definition,

$$e^{iAt/n} = I + \frac{1}{n} iAt + O\left(\frac{1}{n^2}\right) , \tag{4.99}$$

and thus

$$e^{iAt/n} e^{iBt/n} = I + \frac{1}{n} i(A + B)t + O\left(\frac{1}{n^2}\right) . \tag{4.100}$$

Taking products of these gives us

$$(e^{iAt/n} e^{iBt/n})^n = I + \sum_{k=1}^{n} \binom{n}{k} \frac{1}{n^k} \left[i(A + B)t \right]^k + O\left(\frac{1}{n}\right) , \tag{4.101}$$

and since $\binom{n}{k} \frac{1}{n^k} = \left(1 + O\left(\frac{1}{n}\right)\right) / k!$, this gives

$$\lim_{n \to \infty} (e^{iAt/n} e^{iBt/n})^n = \lim_{n \to \infty} \sum_{k=0}^{n} \frac{(i(A + B)t)^k}{k!} \left(1 + O\left(\frac{1}{n}\right)\right) + O\left(\frac{1}{n}\right) = e^{i(A+B)t} . \tag{4.102}$$

$\square$

Modifications of the Trotter formula provide the methods by which higher order approximations can be derived for performing quantum simulations. For example, using similar reasoning to the proof above, it can be shown that

$$e^{i(A+B)\Delta t} = e^{iA\Delta t}e^{iB\Delta t} + O(\Delta t^2).$$

(4.103)

Similarly,

$$e^{i(A+B)\Delta t} = e^{iA\Delta t/2}e^{iB\Delta t}e^{iA\Delta t/2} + O(\Delta t^3).$$

(4.104)

An overview of the quantum simulation algorithm is given below, and an explicit example of simulating the one-dimensional non-relativistic Schrödinger equation is shown in Box 4.2.

Algorithm: Quantum simulation

Inputs: (1) A Hamiltonian $H = \sum_k H_k$ acting on an N-dimensional system, where each H_k acts on a small subsystem of size independent of N, (2) an initial state $|\psi_0\rangle$, of the system at $t = 0$, (3) a positive, non-zero accuracy δ, and (3) a time t_f at which the evolved state is desired.

Outputs: A state $|\tilde{\psi}(t_f)\rangle$ such that $|\langle\tilde{\psi}(t_f)|e^{-iHt_f}|\psi_0\rangle|^2 \geq 1 - \delta$.

Runtime: $O(\text{poly}(1/\delta))$ operations.

Procedure: Choose a representation such that the state $|\tilde{\psi}\rangle$ of $n = \text{poly}(\log N)$ qubits approximates the system and the operators $e^{-iH_k\Delta t}$ have efficient quantum circuit approximations. Select an approximation method (see for example Equations (4.103)–(4.105)) and Δt such that the expected error is acceptable (and $j\Delta t = t_f$ for an integer j), construct the corresponding quantum circuit $U_{\Delta t}$ for the iterative step, and do:

1.	$\|\tilde{\psi}_0\rangle \leftarrow \|\psi_0\rangle$; $j = 0$	initialize state
2.	$\rightarrow \|\tilde{\psi}_{j+1}\rangle = U_{\Delta t}\|\tilde{\psi}_j\rangle$	iterative update
3.	$\rightarrow j = j + 1$; goto 2 until $j\Delta t \geq t_f$	loop
4.	$\rightarrow \|\tilde{\psi}(t_f)\rangle = \|\tilde{\psi}_j\rangle$	final result

Exercise 4.49: (**Baker–Campbell–Hausdorf formula**) Prove that

$$e^{(A+B)\Delta t} = e^{A\Delta t}e^{B\Delta t}e^{-\frac{1}{2}[A,B]\Delta t^2} + O(\Delta t^3),$$

(4.105)

and also prove Equations (4.103) and (4.104).

Exercise 4.50: Let $H = \sum_k^L H_k$, and define

$$U_{\Delta t} = \left[e^{-iH_1\Delta t}e^{-iH_2\Delta t}\ldots e^{-iH_L\Delta t}\right]\left[e^{-iH_L\Delta t}e^{-iH_{L-1}\Delta t}\ldots e^{-iH_1\Delta t}\right].$$

(4.106)

(a) Prove that $U_{\Delta t} = e^{-2iH\Delta t} + O(\Delta t^3)$.

(b) Use the results in Box 4.1 to prove that for a positive integer m,

$$E(U_{\Delta t}^m, e^{-2miH\Delta t}) \leq m\alpha\Delta t^3,$$

(4.107)

for some constant α.

Box 4.2: Quantum simulation of Schrödinger's equation

The methods and limitations of quantum simulation may be illustrated by the following example, drawn from the conventional models studied by physicists, rather than the abstract qubit model. Consider a single particle living on a line, in a one-dimensional potential $V(x)$, governed by the Hamiltonian

$$H = \frac{p^2}{2m} + V(x), \tag{4.108}$$

where p is the momentum operator and x is the position operator. The eigenvalues of x are continuous, and the system state $|\psi\rangle$ resides in an infinite dimensional Hilbert space; in the x basis, it can be written as

$$|\psi\rangle = \int_{-\infty}^{\infty} |x\rangle\langle x|\psi\rangle \, dx. \tag{4.109}$$

In practice, only some finite region is of interest, which we may take to be the range $-d \le x \le d$. Furthermore, it is possible to choose a differential step size Δx sufficiently small compared to the shortest wavelength in the system such that

$$|\tilde{\psi}\rangle = \sum_{k=-d/\Delta x}^{d/\Delta x} a_k |k\Delta x\rangle \tag{4.110}$$

provides a good physical approximation of $|\psi\rangle$. This state can be represented using $n = \lceil \log(2d/\Delta x + 1) \rceil$ qubits; we simply replace the basis $|k\Delta x\rangle$ (an eigenstate of the x operator) with $|k\rangle$, a computational basis state of n qubits. Note that only n qubits are required for this simulation, whereas classically 2^n complex numbers would have to be kept track of, thus leading to an exponential resource saving when performing the simulation on a quantum computer.

Computation of $|\tilde{\psi}(t)\rangle = e^{-iHt}|\tilde{\psi}(0)\rangle$ must utilize one of the approximations of Equations (4.103)–(4.105) because in general $H_1 = V(x)$ does not commute with $H_0 = p^2/2m$. Thus, we must be able to compute $e^{-iH_1\Delta t}$ and $e^{-iH_0\Delta t}$. Because $|\tilde{\psi}\rangle$ is expressed in the eigenbasis of H_1, $e^{-iH_1\Delta t}$ is a diagonal transformation of the form

$$|k\rangle \rightarrow e^{-iV(k\Delta x)\Delta t}|k\rangle. \tag{4.111}$$

It is straightforward to compute this, since we can compute $V(k\Delta x)\Delta t$. (See also Problem 4.1.) The second term is also simple, because x and p are conjugate variables related by a quantum Fourier transform $U_{FFT}xU_{FFT}^\dagger = p$, and thus $e^{-iH_0\Delta t} = U_{FFT}e^{-ix^2\Delta t/2m}U_{FFT}^\dagger$; to compute $e^{-iH_0\Delta t}$, do

$$|k\rangle \rightarrow U_{FFT}e^{-ix^2/2m}U_{FFT}^\dagger|k\rangle. \tag{4.112}$$

The construction of U_{FFT} is discussed in Chapter 5.

4.7.3 An illustrative example

The procedure we have described for quantum simulations has concentrated on simulating Hamiltonians which are sums of local interations. However, this is not a fundamental requirement! As the following example illustrates, efficient quantum simulations are possible even for Hamiltonians which act non-trivially on all or nearly all parts of a large system.

Suppose we have the Hamiltonian

$$H = Z_1 \otimes Z_2 \otimes \cdots \otimes Z_n , \tag{4.113}$$

which acts on an n qubit system. Despite this being an interaction involving all of the system, indeed, it can be simulated efficiently. What we desire is a simple quantum circuit which implements $e^{-iH\Delta t}$, for arbitrary values of Δt. A circuit doing precisely this, for $n = 3$, is shown in Figure 4.19. The main insight is that although the Hamiltonian involves all the qubits in the system, it does so in a *classical* manner: the phase shift applied to the system is $e^{-i\Delta t}$ if the *parity* of the n qubits in the computational basis is even; otherwise, the phase shift should be $e^{i\Delta t}$. Thus, simple simulation of H is possible by first classically computing the parity (storing the result in an ancilla qubit), then applying the appropriate phase shift conditioned on the parity, then uncomputing the parity (to erase the ancilla). This strategy clearly works not only for $n = 3$, but also for arbitrary values of n.

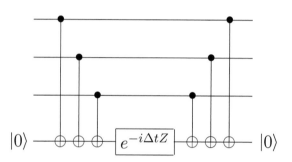

Figure 4.19. Quantum circuit for simulating the Hamiltonian $H = Z_1 \otimes Z_2 \otimes Z_3$ for time Δt.

Furthermore, extending the same procedure allows us to simulate more complicated extended Hamiltonians. Specifically, we can efficiently simulate any Hamiltonian of the form

$$H = \bigotimes_{k=1}^{n} \sigma_{c(k)}^{k} , \tag{4.114}$$

where $\sigma_{c(k)}^{k}$ is a Pauli matrix (or the identity) acting on the kth qubit, with $c(k) \in \{0, 1, 2, 3\}$ specifying one of $\{I, X, Y, Z\}$. The qubits upon which the identity operation is performed can be disregarded, and X or Y terms can be transformed by single qubit gates to Z operations. This leaves us with a Hamiltonian of the form of (4.113), which is simulated as described above.

Exercise 4.51: Construct a quantum circuit to simulate the Hamiltonian

$$H = X_1 \otimes Y_2 \otimes Z_3 , \tag{4.115}$$

performing the unitary transform $e^{-i\Delta t H}$ for any Δt.

Using this procedure allows us to simulate a wide class of Hamiltonians containing terms which are not local. In particular, it is possible to simulate a Hamiltonian of the form

$H = \sum_{k=1}^{L} H_k$ where the only restriction is that the individual H_k have a tensor product structure, and that L is polynomial in the total number of particles n. More generally, all that is required is that there be an efficient circuit to simulate each H_k separately. As an example, the Hamiltonian $H = \sum_{k=1}^{n} X_k + Z^{\otimes n}$ can easily be simulated using the above techniques. Such Hamiltonians typically do not arise in Nature. However, they provide a new and possibly valuable vista on the world of quantum algorithms.

4.7.4 Perspectives on quantum simulation

The quantum simulation algorithm is very similar to classical methods, but also differs in a fundamental way. Each iteration of the quantum algorithm must completely replace the old state with the new one; there is no way to obtain (non-trivial) information from an intermediate step without significantly changing the algorithm, because the state is a quantum one. Furthermore, the final measurement must be chosen cleverly to provide the desired result, because it disturbs the quantum state. Of course, the quantum simulation can be repeated to obtain statistics, but it is desirable to repeat the algorithm only at most a polynomial number of times. It may be that even though the simulation can be performed efficiently, there is no way to efficiently perform a desired measurement.

Also, there are Hamiltonians which simply can't be simulated efficiently. In Section 4.5.4, we saw that there exist unitary transformations which quantum computers cannot efficiently approximate. As a corollary, not all Hamiltonian evolutions can be efficiently simulated on a quantum computer, for if this were possible, then all unitary transformations could be efficiently approximated!

Another difficult problem – one which is very interesting – is the simulation of equilibration processes. A system with Hamiltonian H in contact with an environment at temperature T will generally come to thermal equilibrium in a state known as the *Gibbs state*, $\rho_{\text{therm}} = e^{-H/k_B T}/\mathcal{Z}$, where k_B is Boltzmann's constant, and $\mathcal{Z} = \text{tr}\, e^{-H/k_B T}$ is the usual partition function normalization, which ensures that $\text{tr}(\rho) = 1$. The process by which this equilibration occurs is not very well understood, although certain requirements are known: the environment must be large, it must have non-zero population in states with energies matching the eigenstates of H, and its coupling with the system should be weak. Obtaining ρ_{therm} for arbitrary H and T is generally an exponentially difficult problem for a classical computer. Might a quantum computer be able to solve this efficiently? We do not yet know.

On the other hand, as we discussed above many interesting quantum problems can indeed be simulated efficiently with a quantum computer, even when they have extra constraints beyond the simple algorithms presented here. A particular class of these involve global symmetries originating from particle statistics. In the everyday world, we are used to being able to identify different particles; tennis balls can be followed around a tennis court, keeping track of which is which. This ability to keep track of which object is which is a general feature of classical objects – by continuously measuring the position of a classical particle it can be tracked at all times, and thus uniquely distinguished from other particles. However, this breaks down in quantum mechanics, which prevents us from following the motion of individual particles exactly. If the two particles are inherently different, say a proton and an electron, then we can distinguish them by measuring the sign of the charge to tell which particle is which. But in the case of identical particles, like two electrons, it is found that they are truly indistinguishable.

Indistinguishability of particles places a constraint on the state vector of a system which manifests itself in two ways. Experimentally, particles in Nature are found to come in two distinct flavors, known as bosons and fermions. The state vector of a system of bosons remains unchanged under permutation of any two constituents, reflecting their fundamental indistinguishability. Systems of fermions, in contrast, experience a sign change in their state vector under interchange of any two constituents. Both kinds of systems can be simulated efficiently on a quantum computer. The detailed description of how this is done is outside the scope of this book; suffice it to say the procedure is fairly straightforward. Given an initial state of the wrong symmetry, it can be properly symmetrized before the simulation begins. And the operators used in the simulation can be constructed to respect the desired symmetry, even allowing for the effects of higher order error terms. The reader who is interested in pursuing this and other topics further will find pointers to the literature in 'History and further reading,' at the end of the chapter.

Problem 4.1: (Computable phase shifts) Let m and n be positive integers. Suppose $f : \{0, \ldots, 2^m - 1\} \to \{0, \ldots, 2^n - 1\}$ is a classical function from m to n bits which may be computed reversibly using T Toffoli gates, as described in Section 3.2.5. That is, the function $(x, y) \to (x, y \oplus f(x))$ may be implemented using T Toffoli gates. Give a quantum circuit using $2T + n$ (or fewer) one, two, and three qubit gates to implement the unitary operation defined by

$$|x\rangle \to \exp\left(\frac{-2i\pi f(x)}{2^n}\right)|x\rangle. \tag{4.116}$$

Problem 4.2: Find a depth $O(\log n)$ construction for the $C^n(X)$ gate. (*Comment:* The depth of a circuit is the number of distinct timesteps at which gates are applied; the point of this problem is that it is possible to parallelize the $C^n(X)$ construction by applying many gates in parallel during the same timestep.)

Problem 4.3: (Alternate universality construction) Suppose U is a unitary matrix on n qubits. Define $H \equiv i \ln(U)$. Show that

(1) H is Hermitian, with eigenvalues in the range 0 to 2π.

(2) H can be written

$$H = \sum_g h_g g, \tag{4.117}$$

where h_g are real numbers and the sum is over all n-fold tensor products g of the Pauli matrices $\{I, X, Y, Z\}$.

(3) Let $\Delta = 1/k$, for some positive integer k. Explain how the unitary operation $\exp(-ih_g g \Delta)$ may be implemented using $O(n)$ one and two qubit operations.

(4) Show that

$$\exp(-iH\Delta) = \prod_g \exp(-ih_g g \Delta) + O(4^n \Delta^2), \tag{4.118}$$

where the product is taken with respect to any fixed ordering of the n-fold tensor products of Pauli matrices, g.

(5) Show that

$$U = \left[\prod_g \exp(-ih_g g\Delta) \right]^k + O(4^n \Delta).$$

(4.119)

(6) Explain how to approximate U to within a distance $\epsilon > 0$ using $O(n16^n/\epsilon)$ one and two qubit unitary operations.

Problem 4.4: (Minimal Toffoli construction) (Research)

(1) What is the smallest number of two qubit gates that can be used to implement the Toffoli gate?

(2) What is the smallest number of one qubit gates and CNOT gates that can be used to implement the Toffoli gate?

(3) What is the smallest number of one qubit gates and controlled-Z gates that can be used to implement the Toffoli gate?

Problem 4.5: (Research) Construct a family of Hamiltonians, $\{H_n\}$, on n qubits, such that simulating H_n requires a number of operations super-polynomial in n. (*Comment:* This problem seems to be quite difficult.)

Problem 4.6: (Universality with prior entanglement) Controlled-NOT gates and single qubit gates form a universal set of quantum logic gates. Show that an alternative universal set of resources is comprised of single qubit unitaries, the ability to perform measurements of pairs of qubits in the Bell basis, and the ability to prepare arbitrary four qubit entangled states.

Summary of Chapter 4: Quantum circuits

- **Universality:** Any unitary operation on n qubits may be implemented exactly by composing single qubit and controlled-NOT gates.

- **Universality with a discrete set:** The Hadamard gate, phase gate, controlled-NOT gate, and $\pi/8$ gate are *universal* for quantum computation, in the sense that an arbitrary unitary operation on n qubits can be approximated to an arbitrary accuracy $\epsilon > 0$ using a circuit composed of only these gates. Replacing the $\pi/8$ gate in this list with the Toffoli gate also gives a universal family.

- **Not all unitary operations can be efficiently implemented:** There are unitary operations on n qubits which require $\Omega(2^n \log(1/\epsilon)/\log(n))$ gates to approximate to within a distance ϵ using any finite set of gates.

- **Simulation:** For a Hamiltonian $H = \sum_k H_k$ which is a sum of polynomially many terms H_k such that efficient quantum circuits for H_k can be constructed, a quantum computer can efficiently simulate the evolution e^{-iHt} and approximate $|\psi(t)\rangle = e^{-iHt}|\psi(0)\rangle$, given $|\psi(0)\rangle$.

History and further reading

The gate constructions in this chapter are drawn from a wide variety of sources. The paper by Barenco, Bennett, Cleve, DiVincenzo, Margolus, Shor, Sleator, Smolin, and Weinfurter[BBC+95] was the source of many of the circuit constructions in this chapter, and for the universality proof for single qubit and controlled-NOT gates. Another useful source of insights about quantum circuits is the paper by Beckman, Chari, Devabhaktuni, and Preskill[BCDP96]. A gentle and accessible introduction has been provided by DiVincenzo[DiV98]. The fact that measurements commute with control qubit terminals was pointed out by Griffiths and Niu[GN96].

The universality proof for two-level unitaries is due to Reck, Zeilinger, Bernstein, and Bertani[RZBB94]. The universality of the controlled-NOT and single qubit gates was proved by DiVincenzo[DiV95b]. The universal gate G in Exercise 4.44 is sometimes known as the Deutsch gate[Deu89]. Deutsch, Barenco, and Ekert[DBE95] and Lloyd[Llo95] independently proved that almost any two qubit quantum logic gate is universal. That errors caused by sequences of gates is at most the sum of the errors of the individual gates was proven by Bernstein and Vazirani [BV97]. The specific universal set of gates we have focused on – the Hadamard, phase, controlled-NOT and $\pi/8$ gates, was proved universal in Boykin, Mor, Pulver, Roychowdhury, and Vatan[BMP+99], which also contains a proof that θ defined by $\cos(\theta/2) \equiv \cos^2(\pi/8)$ is an irrational multiple of π. The bound in Section 4.5.4 is based on a paper by Knill[Kni95], which does a much more detailed investigation of the hardness of approximating arbitrary unitary operations using quantum circuits. In particular, Knill obtains tighter and more general bounds than we do, and his analysis applies also to cases where the universal set is a continuum of gates, not just a finite set, as we have considered.

The quantum circuit model of computation is due to Deutsch[Deu89], and was further developed by Yao[Yao93]. The latter paper showed that the quantum circuit model of computation is equivalent to the quantum Turing machine model. Quantum Turing machines were introduced in 1980 by Benioff[Ben80], further developed by Deutsch[Deu85] and Yao[Yao93], and their modern definition given by Bernstein and Vazirani[BV97]. The latter two papers also take first steps towards setting up a theory of quantum computational complexity, analogous to classical computational complexity theory. In particular, the inclusion **BQP** $\subseteq$ **PSPACE** and some slightly stronger results was proved by Bernstein and Vazirani. Knill and Laflamme[KL99] develop some fascinating connections between quantum and classical computational complexity. Other interesting work on quantum computational complexity includes the paper by Adleman, Demarrais and Huang[ADH97], and the paper by Watrous[Wat99]. The latter paper gives intriguing evidence to suggest that quantum computers are more powerful than classical computers in the setting of 'interactive proof systems'.

The suggestion that non-computational basis starting states may be used to obtain computational power beyond the quantum circuits model was made by Daniel Gottesman and Michael Nielson.

That quantum computers might simulate quantum systems more efficiently than classical computers was intimated by Manin[Man80] in 1980, and independently developed in more detail by Feynman[Fey82] in 1982. Much more detailed investigations were subsequently carried out by Abrams and Lloyd[AL97], Boghosian and Taylor[BT97], Sornborger and Stewart[SS99], Wiesner[Wie96], and Zalka[Zal98]. The Trotter formula is attributed to Trotter[Tro59], and was also proven by Chernoff[Che68], although the simpler form for unitary operators is much older, and goes back to the time of Sophus Lie. The third order version of the Baker–Campbell–Hausdorff formula, Equation (4.104), was given by Sornborger and Stewart[SS99]. Abrams and Lloyd[AL97] give a procedure for simulating many-body Fermi systems on a quantum computer. Terhal and DiVincenzo address the problem of simulating the equilibration of quantum systems to the Gibbs state[TD98]. The method used to simulate the Schrödinger equation in Box 4.2 is due to Zalka[Zal98] and Wiesner[Wie96].

Exercise 4.25 is due to Vandersypen, and is related to work by Chau and Wilczek[CW95]. Exercise 4.45 is due to Boykin, Mor, Pulver, Roychowdhury, and Vatan[BMP+99]. Problem 4.2 is due to Gottesman. Problem 4.6 is due to Gottesman and Chuang[GC99].

5 The quantum Fourier transform and its applications

If computers that you build are quantum,
Then spies everywhere will all want 'em.
Our codes will all fail,
And they'll read our email,
Till we get crypto that's quantum, and daunt 'em.
  – Jennifer and Peter Shor

To read our E-mail, how mean
of the spies and their quantum machine;
be comforted though,
they do not yet know
how to factorize twelve or fifteen.
  – Volker Strassen

Computer programming is an art form, like the creation of poetry or music.
  – Donald Knuth

The most spectacular discovery in quantum computing to date is that quantum computers can efficiently perform some tasks which are not feasible on a classical computer. For example, finding the prime factorization of an n-bit integer is thought to require $\exp(\Theta(n^{1/3} \log^{2/3} n))$ operations using the best classical algorithm known at the time of writing, the so-called *number field sieve*. This is exponential in the size of the number being factored, so factoring is generally considered to be an intractable problem on a classical computer: it quickly becomes impossible to factor even modest numbers. In contrast, a quantum algorithm can accomplish the same task using $O(n^2 \log n \log \log n)$ operations. That is, a quantum computer can factor a number *exponentially* faster than the best known classical algorithms. This result is important in its own right, but perhaps the most exciting aspect is the question it raises: what other problems can be done efficiently on a quantum computer which are infeasible on a classical computer?

In this chapter we develop the *quantum Fourier transform*, which is the key ingredient for quantum factoring and many other interesting quantum algorithms. The quantum Fourier transform, with which we begin in Section 5.1, is an efficient quantum algorithm for performing a Fourier transform of quantum mechanical amplitudes. It does *not* speed up the classical task of computing Fourier transforms of classical data. But one important task which it does enable is *phase estimation*, the approximation of the eigenvalues of a unitary operator under certain circumstances, as described in Section 5.2. This allows us to solve several other interesting problems, including the *order-finding problem* and the *factoring problem*, which are covered in Section 5.3. Phase estimation can also be combined with the quantum search algorithm to solve the problem of *counting solutions* to a search problem, as described in the next chapter. Section 5.4 concludes the chapter with a discussion of how the quantum Fourier transform may be used to solve the *hidden subgroup problem*, a generalization of the phase estimation and order-finding problems that has among its special cases an efficient quantum algorithm for the *discrete logarithm* problem, another problem thought to be intractable on a classical computer.

5.1 The quantum Fourier transform

> *A good idea has a way of becoming simpler and solving problems other than that for which it was intended.*
> – Robert Tarjan

One of the most useful ways of solving a problem in mathematics or computer science is to *transform* it into some other problem for which a solution is known. There are a few transformations of this type which appear so often and in so many different contexts that the transformations are studied for their own sake. A great discovery of quantum computation has been that some such transformations can be computed much faster on a quantum computer than on a classical computer, a discovery which has enabled the construction of fast algorithms for quantum computers.

One such transformation is the *discrete Fourier transform*. In the usual mathematical notation, the discrete Fourier transform takes as input a vector of complex numbers, $x_0, \ldots, x_{N-1}$ where the length N of the vector is a fixed parameter. It outputs the transformed data, a vector of complex numbers $y_0, \ldots, y_{N-1}$, defined by

$$y_k \equiv \frac{1}{\sqrt{N}} \sum_{j=0}^{N-1} x_j e^{2\pi i j k / N} . \tag{5.1}$$

The *quantum Fourier transform* is exactly the same transformation, although the conventional notation for the quantum Fourier transform is somewhat different. The quantum Fourier transform on an orthonormal basis $|0\rangle, \ldots, |N-1\rangle$ is defined to be a linear operator with the following action on the basis states,

$$|j\rangle \longrightarrow \frac{1}{\sqrt{N}} \sum_{k=0}^{N-1} e^{2\pi i j k / N} |k\rangle . \tag{5.2}$$

Equivalently, the action on an arbitrary state may be written

$$\sum_{j=0}^{N-1} x_j |j\rangle \longrightarrow \sum_{k=0}^{N-1} y_k |k\rangle , \tag{5.3}$$

where the amplitudes y_k are the discrete Fourier transform of the amplitudes x_j. It is not obvious from the definition, but this transformation is a unitary transformation, and thus can be implemented as the dynamics for a quantum computer. We shall demonstrate the unitarity of the Fourier transform by constructing a manifestly unitary quantum circuit computing the Fourier transform. It is also easy to prove directly that the Fourier transform is unitary:

Exercise 5.1: Give a direct proof that the linear transformation defined by Equation (5.2) is unitary.

Exercise 5.2: Explicitly compute the Fourier transform of the n qubit state $|00\ldots0\rangle$.

In the following, we take $N = 2^n$, where n is some integer, and the basis $|0\rangle, \ldots, |2^n - 1\rangle$ is the computational basis for an n qubit quantum computer. It is helpful to write the state $|j\rangle$ using the binary representation $j = j_1 j_2 \ldots j_n$. More formally, $j = j_1 2^{n-1} + j_2 2^{n-2} + \cdots + j_n 2^0$. It is also convenient to adopt the notation $0.j_l j_{l+1} \ldots j_m$ to represent the *binary fraction* $j_l/2 + j_{l+1}/4 + \cdots + j_m/2^{m-l+1}$.

With a little algebra the quantum Fourier transform can be given the following useful *product representation*:

$$|j_1, \ldots, j_n\rangle \rightarrow \frac{\left(|0\rangle + e^{2\pi i 0.j_n}|1\rangle\right)\left(|0\rangle + e^{2\pi i 0.j_{n-1}j_n}|1\rangle\right)\cdots\left(|0\rangle + e^{2\pi i 0.j_1 j_2\cdots j_n}|1\rangle\right)}{2^{n/2}}.$$

$$(5.4)$$

This product representation is so useful that you may even wish to consider this to be the *definition* of the quantum Fourier transform. As we explain shortly this representation allows us to construct an efficient quantum circuit computing the Fourier transform, a proof that the quantum Fourier transform is unitary, and provides insight into algorithms based upon the quantum Fourier transform. As an incidental bonus we obtain the classical fast Fourier transform, in the exercises!

The equivalence of the product representation (5.4) and the definition (5.2) follows from some elementary algebra:

$$|j\rangle \rightarrow \frac{1}{2^{n/2}} \sum_{k=0}^{2^n-1} e^{2\pi i jk/2^n} |k\rangle \tag{5.5}$$

$$= \frac{1}{2^{n/2}} \sum_{k_1=0}^{1} \cdots \sum_{k_n=0}^{1} e^{2\pi i j\left(\sum_{l=1}^{n} k_l 2^{-l}\right)} |k_1 \ldots k_n\rangle \tag{5.6}$$

$$= \frac{1}{2^{n/2}} \sum_{k_1=0}^{1} \cdots \sum_{k_n=0}^{1} \bigotimes_{l=1}^{n} e^{2\pi i j k_l 2^{-l}} |k_l\rangle \tag{5.7}$$

$$= \frac{1}{2^{n/2}} \bigotimes_{l=1}^{n} \left[\sum_{k_l=0}^{1} e^{2\pi i j k_l 2^{-l}} |k_l\rangle\right] \tag{5.8}$$

$$= \frac{1}{2^{n/2}} \bigotimes_{l=1}^{n} \left[|0\rangle + e^{2\pi i j 2^{-l}}|1\rangle\right] \tag{5.9}$$

$$= \frac{\left(|0\rangle + e^{2\pi i 0.j_n}|1\rangle\right)\left(|0\rangle + e^{2\pi i 0.j_{n-1}j_n}|1\rangle\right)\cdots\left(|0\rangle + e^{2\pi i 0.j_1 j_2\cdots j_n}|1\rangle\right)}{2^{n/2}}. \tag{5.10}$$

The product representation (5.4) makes it easy to derive an efficient circuit for the quantum Fourier transform. Such a circuit is shown in Figure 5.1. The gate R_k denotes the unitary transformation

$$R_k \equiv \begin{bmatrix} 1 & 0 \\ 0 & e^{2\pi i/2^k} \end{bmatrix}. \tag{5.11}$$

To see that the pictured circuit computes the quantum Fourier transform, consider what happens when the state $|j_1 \ldots j_n\rangle$ is input. Applying the Hadamard gate to the first bit produces the state

$$\frac{1}{2^{1/2}} \left(|0\rangle + e^{2\pi i 0.j_1}|1\rangle\right) |j_2 \ldots j_n\rangle, \tag{5.12}$$

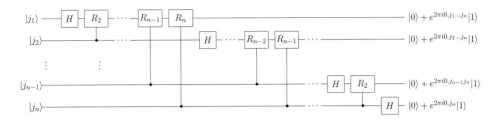

Figure 5.1. Efficient circuit for the quantum Fourier transform. This circuit is easily derived from the product representation (5.4) for the quantum Fourier transform. Not shown are swap gates at the end of the circuit which reverse the order of the qubits, or normalization factors of $1/\sqrt{2}$ in the output.

since $e^{2\pi i 0.j_1} = -1$ when $j_1 = 1$, and is $+1$ otherwise. Applying the controlled-R_2 gate produces the state

$$\frac{1}{2^{1/2}} \left(|0\rangle + e^{2\pi i 0.j_1 j_2} |1\rangle \right) |j_2 \ldots j_n\rangle . \tag{5.13}$$

We continue applying the controlled-R_3, R_4 through R_n gates, each of which adds an extra bit to the phase of the co-efficient of the first $|1\rangle$. At the end of this procedure we have the state

$$\frac{1}{2^{1/2}} \left(|0\rangle + e^{2\pi i 0.j_1 j_2 \ldots j_n} |1\rangle \right) |j_2 \ldots j_n\rangle . \tag{5.14}$$

Next, we perform a similar procedure on the second qubit. The Hadamard gate puts us in the state

$$\frac{1}{2^{2/2}} \left(|0\rangle + e^{2\pi i 0.j_1 j_2 \ldots j_n} |1\rangle \right) \left(|0\rangle + e^{2\pi i 0.j_2} |1\rangle \right) |j_3 \ldots j_n\rangle , \tag{5.15}$$

and the controlled-R_2 through R_{n-1} gates yield the state

$$\frac{1}{2^{2/2}} \left(|0\rangle + e^{2\pi i 0.j_1 j_2 \ldots j_n} |1\rangle \right) \left(|0\rangle + e^{2\pi i 0.j_2 \ldots j_n} |1\rangle \right) |j_3 \ldots j_n\rangle . \tag{5.16}$$

We continue in this fashion for each qubit, giving a final state

$$\frac{1}{2^{n/2}} \left(|0\rangle + e^{2\pi i 0.j_1 j_2 \ldots j_n} |1\rangle \right) \left(|0\rangle + e^{2\pi i 0.j_2 \ldots j_n} |1\rangle \right) \ldots \left(|0\rangle + e^{2\pi i 0.j_n} |1\rangle \right) . \tag{5.17}$$

Swap operations (see Section 1.3.4 for a description of the circuit), omitted from Figure 5.1 for clarity, are then used to reverse the order of the qubits. After the swap operations, the state of the qubits is

$$\frac{1}{2^{n/2}} \left(|0\rangle + e^{2\pi i 0.j_n} |1\rangle \right) \left(|0\rangle + e^{2\pi i 0.j_{n-1} j_n} |1\rangle \right) \ldots \left(|0\rangle + e^{2\pi i 0.j_1 j_2 \ldots j_n} |1\rangle \right) . \tag{5.18}$$

Comparing with Equation (5.4) we see that this is the desired output from the quantum Fourier transform. This construction also proves that the quantum Fourier transform is unitary, since each gate in the circuit is unitary. An explicit example showing a circuit for the quantum Fourier transform on three qubits is given in Box 5.1.

How many gates does this circuit use? We start by doing a Hadamard gate and $n - 1$ conditional rotations on the first qubit – a total of n gates. This is followed by a Hadamard gate and $n - 2$ conditional rotations on the second qubit, for a total of $n + (n - 1)$ gates. Continuing in this way, we see that $n + (n - 1) + \cdots + 1 = n(n + 1)/2$ gates are required,

Box 5.1: Three qubit quantum Fourier transform

For concreteness it may help to look at the explicit circuit for the three qubit quantum Fourier transform:

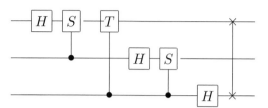

Recall that S and T are the phase and $\pi/8$ gates (see page xxiii). As a matrix the quantum Fourier transform in this instance may be written out explicitly, using $\omega = e^{2\pi i/8} = \sqrt{i}$, as

$$\frac{1}{\sqrt{8}}\begin{bmatrix} 1 & 1 & 1 & 1 & 1 & 1 & 1 & 1 \\ 1 & \omega & \omega^2 & \omega^3 & \omega^4 & \omega^5 & \omega^6 & \omega^7 \\ 1 & \omega^2 & \omega^4 & \omega^6 & 1 & \omega^2 & \omega^4 & \omega^6 \\ 1 & \omega^3 & \omega^6 & \omega^1 & \omega^4 & \omega^7 & \omega^2 & \omega^5 \\ 1 & \omega^4 & 1 & \omega^4 & 1 & \omega^4 & 1 & \omega^4 \\ 1 & \omega^5 & \omega^2 & \omega^7 & \omega^4 & \omega^1 & \omega^6 & \omega^3 \\ 1 & \omega^6 & \omega^4 & \omega^2 & 1 & \omega^6 & \omega^4 & \omega^2 \\ 1 & \omega^7 & \omega^6 & \omega^5 & \omega^4 & \omega^3 & \omega^2 & \omega^1 \end{bmatrix}. \tag{5.19}$$

plus the gates involved in the swaps. At most $n/2$ swaps are required, and each swap can be accomplished using three controlled-NOT gates. Therefore, this circuit provides a $\Theta(n^2)$ algorithm for performing the quantum Fourier transform.

In contrast, the best classical algorithms for computing the discrete Fourier transform on 2^n elements are algorithms such as the *Fast Fourier Transform (FFT)*, which compute the discrete Fourier transform using $\Theta(n2^n)$ gates. That is, it requires exponentially more operations to compute the Fourier transform on a classical computer than it does to implement the quantum Fourier transform on a quantum computer.

At face value this sounds terrific, since the Fourier transform is a crucial step in so many real-world data processing applications. For example, in computer speech recognition, the first step in phoneme recognition is to Fourier transform the digitized sound. Can we use the quantum Fourier transform to speed up the computation of these Fourier transforms? Unfortunately, the answer is that there is no known way to do this. The problem is that the amplitudes in a quantum computer cannot be directly accessed by measurement. Thus, there is no way of determining the Fourier transformed amplitudes of the original state. Worse still, there is in general no way to efficiently prepare the original state to be Fourier transformed. Thus, finding uses for the quantum Fourier transform is more subtle than we might have hoped. In this and the next chapter we develop several algorithms based upon a more subtle application of the quantum Fourier transform.

Exercise 5.3: (Classical fast Fourier transform) Suppose we wish to perform a Fourier transform of a vector containing 2^n complex numbers on a classical computer. Verify that the straightforward method for performing the Fourier transform, based upon direct evaluation of Equation (5.1) requires $\Theta(2^{2n})$ elementary arithmetic operations. Find a method for reducing this to $\Theta(n2^n)$ operations, based upon Equation (5.4).

Exercise 5.4: Give a decomposition of the controlled-R_k gate into single qubit and CNOT gates.

Exercise 5.5: Give a quantum circuit to perform the inverse quantum Fourier transform.

Exercise 5.6: (Approximate quantum Fourier transform) The quantum circuit construction of the quantum Fourier transform apparently requires gates of exponential precision in the number of qubits used. However, such precision is never required in any quantum circuit of polynomial size. For example, let U be the ideal quantum Fourier transform on n qubits, and V be the transform which results if the controlled-R_k gates are performed to a precision $\Delta = 1/p(n)$ for some polynomial $p(n)$. Show that the error $E(U, V) \equiv \max_{|\psi\rangle} \|(U - V)|\psi\rangle\|$ scales as $\Theta(n^2/p(n))$, and thus polynomial precision in each gate is sufficient to guarantee polynomial accuracy in the output state.

5.2 Phase estimation

The Fourier transform is the key to a general procedure known as *phase estimation*, which in turn is the key for many quantum algorithms. Suppose a unitary operator U has an eigenvector $|u\rangle$ with eigenvalue $e^{2\pi i\varphi}$, where the value of φ is unknown. The goal of the phase estimation algorithm is to estimate φ. To perform the estimation we assume that we have available *black boxes* (sometimes known as *oracles*) capable of preparing the state $|u\rangle$ and performing the controlled-U^{2^j} operation, for suitable non-negative integers j. The use of black boxes indicates that the phase estimation procedure is not a complete quantum algorithm in its own right. Rather, you should think of phase estimation as a kind of 'subroutine' or 'module' that, when combined with other subroutines, can be used to perform interesting computational tasks. In specific applications of the phase estimation procedure we shall do exactly this, describing how these black box operations are to be performed, and combining them with the phase estimation procedure to do genuinely useful tasks. For the moment, though, we will continue to imagine them as black boxes.

The quantum phase estimation procedure uses two registers. The first register contains t qubits initially in the state $|0\rangle$. How we choose t depends on two things: the number of digits of accuracy we wish to have in our estimate for φ, and with what probability we wish the phase estimation procedure to be successful. The dependence of t on these quantities emerges naturally from the following analysis.

The second register begins in the state $|u\rangle$, and contains as many qubits as is necessary to store $|u\rangle$. Phase estimation is performed in two stages. First, we apply the circuit shown in Figure 5.2. The circuit begins by applying a Hadamard transform to the first register, followed by application of controlled-U operations on the second register, with U raised to successive powers of two. The final state of the first register is easily seen to be:

$$\frac{1}{2^{t/2}} \left(|0\rangle + e^{2\pi i 2^{t-1}\varphi}|1\rangle \right) \left(|0\rangle + e^{2\pi i 2^{t-2}\varphi}|1\rangle \right) \cdots \left(|0\rangle + e^{2\pi i 2^0 \varphi}|1\rangle \right)$$

$$= \frac{1}{2^{t/2}} \sum_{k=0}^{2^t-1} e^{2\pi i \varphi k}|k\rangle . \qquad (5.20)$$

We omit the second register from this description, since it stays in the state $|u\rangle$ throughout the computation.

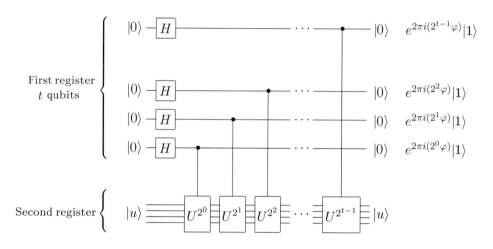

Figure 5.2. The first stage of the phase estimation procedure. Normalization factors of $1/\sqrt{2}$ have been omitted, on the right.

Exercise 5.7:   Additional insight into the circuit in Figure 5.2 may be obtained by showing, as you should now do, that the effect of the sequence of controlled-U operations like that in Figure 5.2 is to take the state $|j\rangle|u\rangle$ to $|j\rangle U^j|u\rangle$. (Note that this does not depend on $|u\rangle$ being an eigenstate of U.)

The second stage of phase estimation is to apply the *inverse* quantum Fourier transform on the first register. This is obtained by reversing the circuit for the quantum Fourier transform in the previous section (Exercise 5.5), and can be done in $\Theta(t^2)$ steps. The third and final stage of phase estimation is to read out the state of the first register by doing a measurement in the computational basis. We will show that this provides a pretty good estimate of φ. An overall schematic of the algorithm is shown in Figure 5.3.

To sharpen our intuition as to why phase estimation works, suppose φ may be expressed exactly in t bits, as $\varphi = 0.\varphi_1 \ldots \varphi_t$. Then the state (5.20) resulting from the first stage of phase estimation may be rewritten

$$\frac{1}{2^{t/2}} \left(|0\rangle + e^{2\pi i 0.\varphi_t}|1\rangle \right) \left(|0\rangle + e^{2\pi i 0.\varphi_{t-1}\varphi_t}|1\rangle \right) \cdots \left(|0\rangle + e^{2\pi i 0.\varphi_1\varphi_2\cdots\varphi_t}|1\rangle \right) . \quad (5.21)$$

The second stage of phase estimation is to apply the inverse quantum Fourier transform. But comparing the previous equation with the product form for the Fourier transform, Equation (5.4), we see that the output state from the second stage is the product state $|\varphi_1 \ldots \varphi_t\rangle$. A measurement in the computational basis therefore gives us φ exactly!

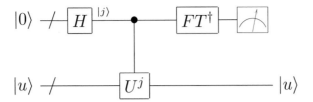

Figure 5.3. Schematic of the overall phase estimation procedure. The top t qubits (the '/' denotes a bundle of wires, as usual) are the first register, and the bottom qubits are the second register, numbering as many as required to perform U. $|u\rangle$ is an eigenstate of U with eigenvalue $e^{2\pi i\varphi}$. The output of the measurement is an approximation to φ accurate to $t - \left\lceil \log\left(2 + \frac{1}{2\epsilon}\right) \right\rceil$ bits, with probability of success at least $1 - \epsilon$.

Summarizing, the phase estimation algorithm allows one to estimate the phase φ of an eigenvalue of a unitary operator U, given the corresponding eigenvector $|u\rangle$. An essential feature at the heart of this procedure is the ability of the inverse Fourier transform to perform the transformation

$$\frac{1}{2^{t/2}} \sum_{j=0}^{2^t-1} e^{2\pi i\varphi j}|j\rangle|u\rangle \rightarrow |\tilde{\varphi}\rangle|u\rangle, \tag{5.22}$$

where $|\tilde{\varphi}\rangle$ denotes a state which is a good estimator for φ when measured.

5.2.1 Performance and requirements

The above analysis applies to the ideal case, where φ can be written exactly with a t bit binary expansion. What happens when this is not the case? It turns out that the procedure we have described will produce a pretty good approximation to φ with high probability, as foreshadowed by the notation used in (5.22). Showing this requires some careful manipulations.

Let b be the integer in the range 0 to $2^t - 1$ such that $b/2^t = 0.b_1 \ldots b_t$ is the best t bit approximation to φ which is less than φ. That is, the difference $\delta \equiv \varphi - b/2^t$ between φ and $b/2^t$ satisfies $0 \le \delta \le 2^{-t}$. We aim to show that the observation at the end of the phase estimation procedure produces a result which is close to b, and thus enables us to estimate φ accurately, with high probability. Applying the inverse quantum Fourier transform to the state (5.20) produces the state

$$\frac{1}{2^t} \sum_{k,l=0}^{2^t-1} e^{\frac{-2\pi ikl}{2^t}} e^{2\pi i\varphi k}|l\rangle. \tag{5.23}$$

Let α_l be the amplitude of $|(b+l)(\bmod 2^t)\rangle$,

$$\alpha_l \equiv \frac{1}{2^t} \sum_{k=0}^{2^t-1} \left(e^{2\pi i(\varphi-(b+l)/2^t)}\right)^k. \tag{5.24}$$

This is the sum of a geometric series, so

$$\alpha_l = \frac{1}{2^t}\left(\frac{1 - e^{2\pi i(2^t\varphi-(b+l))}}{1 - e^{2\pi i(\varphi-(b+l)/2^t)}}\right) \tag{5.25}$$

$$= \frac{1}{2^t} \left(\frac{1 - e^{2\pi i(2^t\delta - l)}}{1 - e^{2\pi i(\delta - l/2^t)}} \right).$$ (5.26)

Suppose the outcome of the final measurement is m. We aim to bound the probability of obtaining a value of m such that $|m - b| > e$, where e is a positive integer characterizing our desired tolerance to error. The probability of observing such an m is given by

$$p(|m - b| > e) = \sum_{-2^{t-1} < l \le -(e+1)} |\alpha_l|^2 + \sum_{e+1 \le l \le 2^{t-1}} |\alpha_l|^2.$$ (5.27)

But for any real θ, $|1 - \exp(i\theta)| \le 2$, so

$$|\alpha_l| \le \frac{2}{2^t |1 - e^{2\pi i(\delta - l/2^t)}|}.$$ (5.28)

By elementary geometry or calculus $|1 - \exp(i\theta)| \ge 2|\theta|/\pi$ whenever $-\pi \le \theta \le \pi$. But when $-2^{t-1} < l \le 2^{t-1}$ we have $-\pi \le 2\pi(\delta - l/2^t) \le \pi$. Thus

$$|\alpha_l| \le \frac{1}{2^{t+1}(\delta - l/2^t)}.$$ (5.29)

Combining (5.27) and (5.29) gives

$$p(|m - b| > e) \le \frac{1}{4} \left[\sum_{l=-2^{t-1}+1}^{-(e+1)} \frac{1}{(l - 2^t\delta)^2} + \sum_{l=e+1}^{2^{t-1}} \frac{1}{(l - 2^t\delta)^2} \right].$$ (5.30)

Recalling that $0 \le 2^t\delta \le 1$, we obtain

$$p(|m - b| > e) \le \frac{1}{4} \left[\sum_{l=-2^{t-1}+1}^{-(e+1)} \frac{1}{l^2} + \sum_{l=e+1}^{2^{t-1}} \frac{1}{(l - 1)^2} \right]$$ (5.31)

$$\le \frac{1}{2} \sum_{l=e}^{2^{t-1}-1} \frac{1}{l^2}$$ (5.32)

$$\le \frac{1}{2} \int_{e-1}^{2^{t-1}-1} dl \frac{1}{l^2}$$ (5.33)

$$= \frac{1}{2(e - 1)}.$$ (5.34)

Suppose we wish to approximate φ to an accuracy 2^{-n}, that is, we choose $e = 2^{t-n} - 1$. By making use of $t = n + p$ qubits in the phase estimation algorithm we see from (5.34) that the probability of obtaining an approximation correct to this accuracy is at least $1 - 1/2(2^p - 2)$. Thus to successfully obtain φ accurate to n bits with probability of success at least $1 - \epsilon$ we choose

$$t = n + \left\lceil \log \left(2 + \frac{1}{2\epsilon} \right) \right\rceil.$$ (5.35)

In order to make use of the phase estimation algorithm, we need to be able to prepare an eigenstate $|u\rangle$ of U. What if we do not know how to prepare such an eigenstate? Suppose that we prepare some other state $|\psi\rangle$ in place of $|u\rangle$. Expanding this state in terms of eigenstates $|u\rangle$ of U gives $|\psi\rangle = \sum_u c_u |u\rangle$. Suppose the eigenstate $|u\rangle$ has eigenvalue $e^{2\pi i\varphi_u}$. Intuitively, the result of running the phase estimation algorithm will be to give as output a state close to $\sum_u c_u |\widetilde{\varphi_u}\rangle |u\rangle$, where $\widetilde{\varphi_u}$ is a pretty good approximation to the phase φ_u. Therefore, we expect that reading out the first register will give us a good approximation to φ_u, where u is chosen at random with probability $|c_u|^2$. Making this argument rigorous is left for Exercise 5.8. This procedure allows us to avoid preparing a (possibly unknown) eigenstate, at the cost of introducing some additional randomness into the algorithm.

Exercise 5.8: Suppose the phase estimation algorithm takes the state $|0\rangle |u\rangle$ to the state $|\widetilde{\varphi_u}\rangle |u\rangle$, so that given the input $|0\rangle \left(\sum_u c_u |u\rangle \right)$, the algorithm outputs $\sum_u c_u |\widetilde{\varphi_u}\rangle |u\rangle$. Show that if t is chosen according to (5.35), then the probability for measuring φ_u accurate to n bits at the conclusion of the phase estimation algorithm is at least $|c_u|^2 (1 - \epsilon)$.

Why is phase estimation interesting? For its own sake, phase estimation solves a problem which is both non-trivial and interesting from a physical point of view: how to estimate the eigenvalue associated to a given eigenvector of a unitary operator. Its real use, though, comes from the observation that other interesting problems can be reduced to phase estimation, as will be shown in subsequent sections. The phase estimation algorithm is summarized below.

Algorithm: Quantum phase estimation

   Inputs: (1) A black box wich performs a controlled-U^j operation, for integer j, (2) an eigenstate $|u\rangle$ of U with eigenvalue $e^{2\pi i \varphi_u}$, and (3) $t = n + \left\lceil \log \left(2 + \frac{1}{2\epsilon} \right) \right\rceil$ qubits initialized to $|0\rangle$.

   Outputs: An n-bit approximation $\widetilde{\varphi_u}$ to φ_u.

   Runtime: $O(t^2)$ operations and one call to controlled-U^j black box. Succeeds with probability at least $1 - \epsilon$.

   Procedure:

   1. $|0\rangle |u\rangle$ initial state

   2. $\longrightarrow \dfrac{1}{\sqrt{2^t}} \displaystyle\sum_{j=0}^{2^t-1} |j\rangle |u\rangle$ create superposition

   3. $\longrightarrow \dfrac{1}{\sqrt{2^t}} \displaystyle\sum_{j=0}^{2^t-1} |j\rangle U^j |u\rangle$ apply black box

         $= \dfrac{1}{\sqrt{2^t}} \displaystyle\sum_{j=0}^{2^t-1} e^{2\pi i j \varphi_u} |j\rangle |u\rangle$ result of black box

   4. $\longrightarrow |\widetilde{\varphi_u}\rangle |u\rangle$ apply inverse Fourier transform

   5. $\longrightarrow \widetilde{\varphi_u}$ measure first register

Exercise 5.9: Let U be a unitary transform with eigenvalues ± 1, which acts on a state $|\psi\rangle$. Using the phase estimation procedure, construct a quantum circuit to collapse $|\psi\rangle$ into one or the other of the two eigenspaces of U, giving also a classical indicator as to which space the final state is in. Compare your result with Exercise 4.34.

5.3 Applications: order-finding and factoring

The phase estimation procedure can be used to solve a variety of interesting problems. We now describe two of the most interesting of these problems: the *order-finding problem*, and the *factoring problem*. These two problems are, in fact, equivalent to one another, so in Section 5.3.1 we explain a quantum algorithm for solving the order-finding problem, and in Section 5.3.2 we explain how the order-finding problem implies the ability to factor as well.

To understand the quantum algorithms for factoring and order-finding requires a little background in number theory. All the required materials are collected together in Appendix 4. The description we give over the next two sections focuses on the quantum aspects of the problem, and requires only a little familiarity with modular arithmetic to be readable. Detailed proofs of the number-theoretic results we quote here may be found in Appendix 4.

The fast quantum algorithms for order-finding and factoring are interesting for at least three reasons. First, and most important in our opinion, they provide evidence for the idea that quantum computers may be inherently more powerful than classical computers, and provide a credible challenge to the strong Church–Turing thesis. Second, both problems are of sufficient intrinsic worth to justify interest in any novel algorithm, be it classical or quantum. Third, and most important from a practical standpoint, efficient algorithms for order-finding and factoring can be used to break the RSA public-key cryptosystem (Appendix 5).

5.3.1 Application: order-finding

For positive integers x and N, $x < N$, with no common factors, the *order* of x modulo N is defined to be the least positive integer, r, such that $x^r = 1 \pmod{N}$. The order-finding problem is to determine the order for some specified x and N. Order-finding is believed to be a hard problem on a classical computer, in the sense that no algorithm is known to solve the problem using resources polynomial in the $O(L)$ bits needed to specify the problem, where $L \equiv \lceil \log(N) \rceil$ is the number of bits needed to specify N. In this section we explain how phase estimation may be used to obtain an efficient quantum algorithm for order-finding.

Exercise 5.10: Show that the order of $x = 5$ modulo $N = 21$ is 6.

Exercise 5.11: Show that the order of x satisfies $r \leq N$.

The quantum algorithm for order-finding is just the phase estimation algorithm applied to the unitary operator

$$U|y\rangle \equiv |xy(\bmod N)\rangle, \tag{5.36}$$

with $y \in \{0, 1\}^L$. (Note that here and below, when $N \leq y \leq 2^L - 1$, we use the convention that $xy(\bmod N)$ is just y again. That is, U only acts non-trivially when

$0 \leq y \leq N - 1$.) A simple calculation shows that the states defined by

$$|u_s\rangle \equiv \frac{1}{\sqrt{r}} \sum_{k=0}^{r-1} \exp\left[\frac{-2\pi isk}{r}\right] |x^k \bmod N\rangle, \tag{5.37}$$

for integer $0 \leq s \leq r - 1$ are eigenstates of U, since

$$U|u_s\rangle = \frac{1}{\sqrt{r}} \sum_{k=0}^{r-1} \exp\left[\frac{-2\pi isk}{r}\right] |x^{k+1} \bmod N\rangle \tag{5.38}$$

$$= \exp\left[\frac{2\pi is}{r}\right] |u_s\rangle. \tag{5.39}$$

Using the phase estimation procedure allows us to obtain, with high accuracy, the corresponding eigenvalues $\exp(2\pi is/r)$, from which we can obtain the order r with a little bit more work.

Exercise 5.12: Show that U is unitary (*Hint:* x is co-prime to N, and therefore has an inverse modulo N).

There are two important requirements for us to be able to use the phase estimation procedure: we must have efficient procedures to implement a controlled-U^{2^j} operation for any integer j, and we must be able to efficiently prepare an eigenstate $|u_s\rangle$ with a non-trivial eigenvalue, or at least a superposition of such eigenstates. The first requirement is satisfied by using a procedure known as *modular exponentiation*, with which we can implement the entire sequence of controlled-U^{2^j} operations applied by the phase estimation procedure using $O(L^3)$ gates, as described in Box 5.2.

The second requirement is a little tricker: preparing $|u_s\rangle$ requires that we know r, so this is out of the question. Fortunately, there is a clever observation which allows us to circumvent the problem of preparing $|u_s\rangle$, which is that

$$\frac{1}{\sqrt{r}} \sum_{s=0}^{r-1} |u_s\rangle = |1\rangle. \tag{5.44}$$

In performing the phase estimation procedure, if we use $t = 2L + 1 + \lceil \log\left(2 + \frac{1}{2\epsilon}\right) \rceil$ qubits in the first register (referring to Figure 5.3), and prepare the second register in the state $|1\rangle$ – which is trivial to construct – it follows that for each s in the range 0 through $r - 1$, we will obtain an estimate of the phase $\varphi \approx s/r$ accurate to $2L + 1$ bits, with probability at least $(1 - \epsilon)/r$. The order-finding algorithm is schematically depicted in Figure 5.4.

Exercise 5.13: Prove (5.44). (*Hint:* $\sum_{s=0}^{r-1} \exp(-2\pi isk/r) = r\delta_{k0}$.) In fact, prove that

$$\frac{1}{\sqrt{r}} \sum_{s=0}^{r-1} e^{2\pi isk/r} |u_s\rangle = |x^k \bmod N\rangle. \tag{5.45}$$

Exercise 5.14: The quantum state produced in the order-finding algorithm, before the inverse Fourier transform, is

$$|\psi\rangle = \sum_{j=0}^{2^t-1} |j\rangle U^j |1\rangle = \sum_{j=0}^{2^t-1} |j\rangle |x^j \bmod N\rangle, \tag{5.46}$$

Box 5.2: Modular exponentiation

How can we compute the sequence of controlled-U^{2^j} operations used by the phase estimation procedure as part of the order-finding algorithm? That is, we wish to compute the transformation

$$|z\rangle|y\rangle \rightarrow |z\rangle U^{z_t 2^{t-1}} \dots U^{z_1 2^0}|y\rangle \tag{5.40}$$

$$= |z\rangle|x^{z_t 2^{t-1}} \times \dots \times x^{z_1 2^0} y (\bmod N)\rangle \tag{5.41}$$

$$= |z\rangle|x^z y (\bmod N)\rangle. \tag{5.42}$$

Thus the sequence of controlled-U^{2^j} operations used in phase estimation is equivalent to multiplying the contents of the second register by the *modular exponential* $x^z (\bmod N)$, where z is the contents of the first register. This operation may be accomplished easily using the techniques of reversible computation. The basic idea is to reversibly compute the function $x^z (\bmod N)$ of z in a third register, and then to reversibly multiply the contents of the second register by $x^z (\bmod N)$, using the trick of uncomputation to erase the contents of the third register upon completion. The algorithm for computing the modular exponential has two stages. The first stage uses modular multiplication to compute $x^2 (\bmod N)$, by squaring x modulo N, then computes $x^4 (\bmod N)$ by squaring $x^2 (\bmod N)$, and continues in this way, computing $x^{2^j} (\bmod N)$ for all j up to $t-1$. We use $t = 2L + 1 + \lceil \log(2 + 1/(2\epsilon)) \rceil = O(L)$, so a total of $t - 1 = O(L)$ squaring operations is performed at a cost of $O(L^2)$ each (this cost assumes the circuit used to do the squaring implements the familiar algorithm we all learn as children for multiplication), for a total cost of $O(L^3)$ for the first stage. The second stage of the algorithm is based upon the observation we've already noted,

$$x^z (\bmod N) = \left(x^{z_t 2^{t-1}} (\bmod N) \right) \left(x^{z_{t-1} 2^{t-2}} (\bmod N) \right) \dots \left(x^{z_1 2^0} (\bmod N) \right). \tag{5.43}$$

Performing $t - 1$ modular multiplications with a cost $O(L^2)$ each, we see that this product can be computed using $O(L^3)$ gates. This is sufficiently efficient for our purposes, but more efficient algorithms are possible based on more efficient algorithms for multiplication (see 'History and further reading'). Using the techniques of Section 3.2.5, it is now straightforward to construct a reversible circuit with a t bit register and an L bit register which, when started in the state (z, y) outputs $(z, x^z y (\bmod N))$, using $O(L^3)$ gates, which can be translated into a quantum circuit using $O(L^3)$ gates computing the transformation $|z\rangle|y\rangle \rightarrow |z\rangle|x^z y (\bmod N)\rangle$.

---

if we initialize the second register as $|1\rangle$. Show that the same state is obtained if we replace U^j with a *different* unitary transform V, which computes

$$V|j\rangle|k\rangle = |j\rangle|k + x^j \bmod N\rangle, \tag{5.47}$$

and start the second register in the state $|0\rangle$. Also show how to construct V using $O(L^3)$ gates.

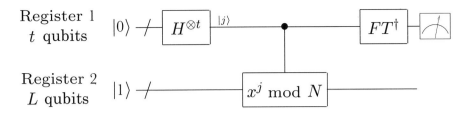

Figure 5.4. Quantum circuit for the order-finding algorithm. The second register is shown as being initialized to the $|1\rangle$ state, but if the method of Exercise 5.14 is used, it can be initialized to $|0\rangle$ instead. This circuit can also be used for factoring, using the reduction given in Section 5.3.2.

The continued fraction expansion

The reduction of order-finding to phase estimation is completed by describing how to obtain the desired answer, r, from the result of the phase estimation algorithm, $\varphi \approx s/r$. We only know φ to $2L + 1$ bits, but we also know *a priori* that it is a *rational* number – the ratio of two bounded integers – and if we could compute the nearest such fraction to φ we might obtain r.

Remarkably, there is an algorithm which accomplishes this task efficiently, known as the *continued fractions algorithm*. An example of how this works is described in Box 5.3. The reason this algorithm satisfies our needs is the following theorem, which is proved in Appendix 4:

Theorem 5.1: Suppose s/r is a rational number such that

$$\left| \frac{s}{r} - \varphi \right| \le \frac{1}{2r^2}. \tag{5.48}$$

Then s/r is a convergent of the continued fraction for φ, and thus can be computed in $O(L^3)$ operations using the continued fractions algorithm.

Since φ is an approximation of s/r accurate to $2L + 1$ bits, it follows that $|s/r - \varphi| \le 2^{-2L-1} \le 1/2r^2$, since $r \le N \le 2^L$. Thus, the theorem applies.

Summarizing, given φ the continued fractions algorithm *efficiently* produces numbers s' and r' with no common factor, such that $s'/r' = s/r$. The number r' is our candidate for the order. We can check to see whether it is the order by calculating $x^{r'} \bmod N$, and seeing if the result is 1. If so, then r' is the order of x modulo N, and we are done!

Performance

How can the order-finding algorithm fail? There are two possibilities. First, the phase estimation procedure might produce a bad estimate to s/r. This occurs with probability at most ϵ, and can be made small with a negligible increase in the size of the circuit. More seriously, it might be that s and r have a common factor, in which case the number r' returned by the continued fractions algorithm be a factor of r, and not r itself. Fortunately, there are at least three ways around this problem.

Perhaps the most straightforward way is to note that for randomly chosen s in the range 0 through $r - 1$, it's actually pretty likely that s and r are co-prime, in which case the continued fractions algorithm must return r. To see that this is the case, note that by Problem 4.1 on page 638 the number of prime numbers less than r is at least

Box 5.3: The continued fractions algorithm

The idea of the continued fractions algorithm is to describe real numbers in terms of integers alone, using expressions of the form

$$[a_0, \ldots, a_M] \equiv a_0 + \cfrac{1}{a_1 + \cfrac{1}{a_2 + \cfrac{1}{\ddots + \cfrac{1}{a_M}}}}, \tag{5.49}$$

where $a_0, \ldots, a_M$ are positive integers. (For applications to quantum computing it is convenient to allow $a_0 = 0$ as well.) We define the *mth convergent* $(0 \leq m \leq M)$ to this continued fraction to be $[a_0, \ldots, a_m]$. The *continued fractions algorithm* is a method for determining the continued fraction expansion of an arbitrary real number. It is easily understood by example. Suppose we are trying to decompose $31/13$ as a continued fraction. The first step of the continued fractions algorithm is to split $31/13$ into its integer and fractional part,

$$\frac{31}{13} = 2 + \frac{5}{13}. \tag{5.50}$$

Next we invert the fractional part, obtaining

$$\frac{31}{13} = 2 + \frac{1}{\frac{13}{5}}. \tag{5.51}$$

These steps – split then invert – are now applied to $13/5$, giving

$$\frac{31}{13} = 2 + \cfrac{1}{2 + \frac{3}{5}} = 2 + \cfrac{1}{2 + \frac{1}{\frac{5}{3}}}. \tag{5.52}$$

Next we split and invert $5/3$:

$$\frac{31}{13} = 2 + \cfrac{1}{2 + \cfrac{1}{1 + \frac{2}{3}}} = 2 + \cfrac{1}{2 + \cfrac{1}{1 + \frac{1}{\frac{3}{2}}}}. \tag{5.53}$$

The decomposition into a continued fraction now terminates, since

$$\frac{3}{2} = 1 + \frac{1}{2} \tag{5.54}$$

may be written with a 1 in the numerator without any need to invert, giving a final continued fraction representation of $31/13$ as

$$\frac{31}{13} = 2 + \cfrac{1}{2 + \cfrac{1}{1 + \frac{1}{1 + \frac{1}{2}}}}. \tag{5.55}$$

It's clear that the continued fractions algorithm terminates after a finite number of 'split and invert' steps for any rational number, since the numerators which appear ($31, 5, 3, 2, 1$ in the example) are strictly decreasing. How quickly does this termination occur? It turns out that if $\varphi = s/r$ is a rational number, and s and r are L bit integers, then the continued fraction expansion for φ can be computed using $O(L^3)$ operations – $O(L)$ 'split and invert' steps, each using $O(L^2)$ gates for elementary arithmetic.

$r/2 \log r$, and thus the chance that s is prime (and therefore, co-prime to r) is at least $1/2 \log(r) > 1/2 \log(N)$. Thus, repeating the algorithm $2 \log(N)$ times we will, with high probability, observe a phase s/r such that s and r are co-prime, and therefore the continued fractions algorithm produces r, as desired.

A second method is to note that if $r' \neq r$, then r' is guaranteed to be a factor of r, unless $s = 0$, which possibility occurs with probability $1/r \leq 1/2$, and which can be discounted further by a few repetitions. Suppose we replace a by $a' \equiv a^{r'} \pmod{N}$. Then the order of a' is r/r'. We can now repeat the algorithm, and try to compute the order of a', which, if we succeed, allows us to compute the order of a, since $r = r' \times r/r'$. If we fail, then we obtain r'' which is a factor of r/r', and we now try to compute the order of $a'' \equiv (a')^{r''} \pmod{N}$. We iterate this procedure until we determine the order of a. At most $\log(r) = O(L)$ iterations are required, since each repetition reduces the order of the current candidate $a'^{''\cdots}$ by a factor of at least two.

The third method is better than the first two methods, in that it requires only a constant number of trials, rather than $O(L)$ repetitions. The idea is to repeat the phase estimation–continued fractions procedure twice, obtaining r_1', s_1' the first time, and r_2', s_2' the second time. Provided s_1' and s_2' have no common factors, r may be extracted by taking the least common multiple of r_1 and r_2. The probability that s_1' and s_2' have no common factors is given by

$$1 - \sum_q p(q|s_1')p(q|s_2'), \tag{5.56}$$

where the sum is over all prime numbers q, and $p(x|y)$ here means the probability of x dividing y. If q divides s_1' then it must also divide the true value of s, s_1, on the first iteration, so to upper bound $p(q|s_1')$ it suffices to upper bound $p(q|s_1)$, where s_1 is chosen uniformly at random from 0 through $r - 1$. It is easy to see that $p(q|s_1) \leq 1/q$, and thus $p(q|s_1') \leq 1/q$. Similarly, $p(q|s_2') \leq 1/q$, and thus the probability that s_1' and s_2' have no common factors satisfies

$$1 - \sum_q p(q|s_1')p(q|s_2') \geq 1 - \sum_q \frac{1}{q^2}. \tag{5.57}$$

The right hand side can be upper bounded in a number of ways; a simple technique is provided in Exercise 5.16, which gives

$$1 - \sum_q p(q|s_1')p(q|s_2') \geq \frac{1}{4}, \tag{5.58}$$

and thus the probability of obtaining the correct r is at least $1/4$.

Exercise 5.15: Show that the least common multiple of positive integers x and y is $xy/\gcd(x,y)$, and thus may be computed in $O(L^2)$ operations if x and y are L bit numbers.

Exercise 5.16: For all $x \geq 2$ prove that $\int_x^{x+1} 1/y^2 \, dy \geq 2/3x^2$. Show that

$$\sum_q \frac{1}{q^2} \leq \frac{3}{2} \int_2^\infty \frac{1}{y^2} dy = \frac{3}{4}, \tag{5.59}$$

and thus that (5.58) holds.

What resource requirements does this algorithm consume? The Hadamard transform requires $O(L)$ gates, and the inverse Fourier transform requires $O(L^2)$ gates. The major cost in the quantum circuit proper actually comes from the modular exponentiation, which uses $O(L^3)$ gates, for a total of $O(L^3)$ gates in the quantum circuit proper. The continued fractions algorithm adds $O(L^3)$ more gates, for a total of $O(L^3)$ gates to obtain r'. Using the third method for obtaining r from r' we need only repeat this procedure a constant number of times to obtain the order, r, for a total cost of $O(L^3)$. The algorithm is summarized below.

Algorithm: Quantum order-finding

> **Inputs:** (1) A black box $U_{x,N}$ which performs the transformation $|j\rangle|k\rangle \to |j\rangle|x^j k \bmod N\rangle$, for x co-prime to the L-bit number N, (2) $t = 2L + 1 + \lceil \log \left(2 + \frac{1}{2\epsilon}\right) \rceil$ qubits initialized to $|0\rangle$, and (3) L qubits initialized to the state $|1\rangle$.
>
> **Outputs:** The least integer $r > 0$ such that $x^r = 1 \pmod N$.
>
> **Runtime:** $O(L^3)$ operations. Succeeds with probability $O(1)$.
>
> **Procedure:**
>
> 1. $\quad |0\rangle|1\rangle$ $\hfill$ initial state
>
> 2. $\quad \to \dfrac{1}{\sqrt{2^t}} \displaystyle\sum_{j=0}^{2^t-1} |j\rangle|1\rangle$ $\hfill$ create superposition
>
> 3. $\quad \to \dfrac{1}{\sqrt{2^t}} \displaystyle\sum_{j=0}^{2^t-1} |j\rangle|x^j \bmod N\rangle$ $\hfill$ apply $U_{x,N}$
>
> $\quad \approx \dfrac{1}{\sqrt{r2^t}} \displaystyle\sum_{s=0}^{r-1}\sum_{j=0}^{2^t-1} e^{2\pi i s j/r}|j\rangle|u_s\rangle$
>
> 4. $\quad \to \dfrac{1}{\sqrt{r}} \displaystyle\sum_{s=0}^{r-1} |\widetilde{s/r}\rangle|u_s\rangle$ $\hfill$ apply inverse Fourier transform to first register
>
> 5. $\quad \to \widetilde{s/r}$ $\hfill$ measure first register
>
> 6. $\quad \to r$ $\hfill$ apply continued fractions algorithm

5.3.2 Application: factoring

> *The problem of distinguishing prime numbers from composites, and of resolving composite numbers into their prime factors, is one of the most important and useful in all of arithmetic. [...] The dignity of science seems to demand that every aid to the solution of such an elegant and celebrated problem be zealously cultivated.*
> – Carl Friedrich Gauss, as quoted by Donald Knuth

Given a positive composite integer N, what prime numbers when multiplied together equal it? This *factoring problem* turns out to be equivalent to the order-finding problem we just studied, in the sense that a fast algorithm for order-finding can easily be turned into a fast algorithm for factoring. In this section we explain the method used to reduce factoring to order-finding, and give a simple example of this reduction.

The reduction of factoring to order-finding proceeds in two basic steps. The first step is to show that we can compute a factor of N if we can find a non-trivial solution $x \neq \pm 1 \pmod{N}$ to the equation $x^2 = 1 \pmod{N}$. The second step is to show that a randomly chosen y co-prime to N is quite likely to have an order r which is even, and such that $y^{r/2} \neq \pm 1 \pmod{N}$, and thus $x \equiv y^{r/2} \pmod{N}$ is a non-trivial solution to $x^2 = 1 \pmod{N}$. These two steps are embodied in the following theorems, whose proofs may be found in Section A4.3 of Appendix 4.

Theorem 5.2: Suppose N is an L bit composite number, and x is a non-trivial solution to the equation $x^2 = 1 \pmod{N}$ in the range $1 \leq x \leq N$, that is, neither $x = 1 \pmod{N}$ nor $x = N - 1 = -1 \pmod{N}$. Then at least one of $\gcd(x - 1, N)$ and $\gcd(x + 1, N)$ is a non-trivial factor of N that can be computed using $O(L^3)$ operations.

Theorem 5.3: Suppose $N = p_1^{\alpha_1} \ldots p_m^{\alpha_m}$ is the prime factorization of an odd composite positive integer. Let x be an integer chosen uniformly at random, subject to the requirements that $1 \leq x \leq N - 1$ and x is co-prime to N. Let r be the order of x modulo N. Then

$$p(r \text{ is even and } x^{r/2} \neq - 1 \pmod{N}) \geq 1 - \frac{1}{2^m} . \tag{5.60}$$

Theorems 5.2 and 5.3 can be combined to give an algorithm which, with high probability, returns a non-trivial factor of any composite N. All the steps in the algorithm can be performed efficiently on a classical computer except (so far as is known today) an order-finding 'subroutine' which is used by the algorithm. By repeating the procedure we may find a complete prime factorization of N. The algorithm is summarized below.

Algorithm: Reduction of factoring to order-finding

    Inputs: A composite number N

    Outputs: A non-trivial factor of N.

    Runtime: $O((\log N)^3)$ operations. Succeeds with probability $O(1)$.

    Procedure:

    1.   If N is even, return the factor 2.

    2.   Determine whether $N = a^b$ for integers $a \geq 1$ and $b \geq 2$, and if so return the factor a (uses the classical algorithm of Exercise 5.17).

    3.   Randomly choose x in the range 1 to $N - 1$. If $\gcd(x, N) > 1$ then return the factor $\gcd(x, N)$.

    4.   Use the order-finding subroutine to find the order r of x modulo N.

5.   If r is even and $x^{r/2} \neq -1 \pmod N$ then compute $\gcd(x^{r/2} - 1, N)$ and $\gcd(x^{r/2} + 1, N)$, and test to see if one of these is a non-trivial factor, returning that factor if so. Otherwise, the algorithm fails.

Steps **1** and **2** of the algorithm either return a factor, or else ensure that N is an odd integer with more than one prime factor. These steps may be performed using $O(1)$ and $O(L^3)$ operations, respectively. Step **3** either returns a factor, or produces a randomly chosen element x of $\{0, 1, 2, \ldots, N - 1\}$. Step **4** calls the order-finding subroutine, computing the order r of x modulo N. Step **5** completes the algorithm, since Theorem 5.3 guarantees that with probability at least one-half r will be even and $x^{r/2} \neq -1 \pmod N$, and then Theorem 5.2 guarantees that either $\gcd(x^{r/2} - 1, N)$ or $\gcd(x^{r/2} + 1, N)$ is a non-trivial factor of N. An example illustrating the use of this algorithm with the quantum order-finding subroutine is shown in Box 5.4.

Exercise 5.17: Suppose N is L bits long. The aim of this exercise is to find an efficient classical algorithm to determine whether $N = a^b$ for some integers $a \geq 1$ and $b \geq 2$. This may be done as follows:

(1) Show that b, if it exists, satisfies $b \leq L$.
(2) Show that it takes at most $O(L^2)$ operations to compute $\log_2 N$, $x = y/b$ for $b \leq L$, and the two integers u_1 and u_2 nearest to 2^x.
(3) Show that it takes at most $O(L^2)$ operations to compute u_1^b and u_2^b (use repeated squaring) and check to see if either is equal to N.
(4) Combine the previous results to give an $O(L^3)$ operation algorithm to determine whether $N = a^b$ for integers a and b.

Exercise 5.18: (Factoring 91) Suppose we wish to factor $N = 91$. Confirm that steps **1** and **2** are passed. For step **3**, suppose we choose $x = 4$, which is co-prime to 91. Compute the order r of x with respect to N, and show that $x^{r/2} \bmod 91 = 64 \neq -1 \pmod{91}$, so the algorithm succeeds, giving $\gcd(64 - 1, 19) = 7$.
It is unlikely that this is the most efficient method you've seen for factoring 91. Indeed, if all computations had to be carried out on a classical computer, this reduction would not result in an efficient factoring algorithm, as no efficient method is known for solving the order-finding problem on a classical computer.

Exercise 5.19: Show that $N = 15$ is the smallest number for which the order-finding subroutine is required, that is, it is the smallest composite number that is not even or a power of some smaller integer.

5.4 General applications of the quantum Fourier transform

The main applications of the quantum Fourier transform we have described so far in this chapter are phase estimation and order-finding. What other problems can be solved with these techniques? In this section, we define a very general problem known as the *hidden subgroup problem*, and describe an efficient quantum algorithm for solving it. This problem, which encompasses all known 'exponentially fast' applications of the quantum Fourier transform, can be thought of as a generalization of the task of finding the unknown period of a periodic function, in a context where the structure of the domain and range

Box 5.4: Factoring 15 quantum-mechanically

The use of order-finding, phase estimation, and continued fraction expansions in the quantum factoring algorithm is illustrated by applying it to factor $N = 15$. First, we choose a random number which has no common factors with N; suppose we choose $x = 7$. Next, we compute the order r of x with respect to N, using the quantum order-finding algorithm: begin with the state $|0\rangle|0\rangle$ and create the state

$$\frac{1}{\sqrt{2^t}} \sum_{k=0}^{2^t-1} |k\rangle|0\rangle = \frac{1}{\sqrt{2^t}} \left[|0\rangle + |1\rangle + |2\rangle + \cdots + |2^t - 1\rangle \right]|0\rangle \qquad (5.61)$$

by applying $t = 11$ Hadamard transforms to the first register. Choosing this value of t ensures an error probability ϵ of at most $1/4$. Next, compute $f(k) = x^k \bmod N$, leaving the result in the second register,

$$\frac{1}{\sqrt{2^t}} \sum_{k=0}^{2^t-1} |k\rangle|x^k \bmod N\rangle \qquad (5.62)$$

$$= \frac{1}{\sqrt{2^t}} \left[|0\rangle|1\rangle + |1\rangle|7\rangle + |2\rangle|4\rangle + |3\rangle|13\rangle + |4\rangle|1\rangle + |5\rangle|7\rangle + |6\rangle|4\rangle + \cdots \right].$$

We now apply the inverse Fourier transform $FT^\dagger$ to the first register and measure it. One way of analyzing the distribution of outcomes obtained is to calculate the reduced density matrix for the first register, and apply $FT^\dagger$ to it, and calculate the measurement statistics. However, since no further operation is applied to the second register, we can instead apply the principle of implicit measurement (Section 4.4) and assume that the second register is measured, obtaining a *random* result from 1, 7, 4, or 13. Suppose we get 4 (any of the results works); this means the state input to $FT^\dagger$ would have been $\sqrt{\frac{4}{2^t}} \left[|2\rangle + |6\rangle + |10\rangle + |14\rangle + \cdots \right]$. After applying $FT^\dagger$ we obtain some state $\sum_\ell \alpha_\ell |\ell\rangle$, with the probability distribution

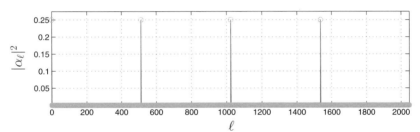

shown for $2^t = 2048$. The final measurement therefore gives either 0, 512, 1024, or 1536, each with probability almost exactly $1/4$. Suppose we obtain $\ell = 1536$ from the measurement; computing the continued fraction expansion thus gives $1536/2048 = 1/(1 + (1/3))$, so that $3/4$ occurs as a convergent in the expansion, giving $r = 4$ as the order of $x = 7$. By chance, r is even, and moreover, $x^{r/2} \bmod N = 7^2 \bmod 15 = 4 \neq -1 \bmod 15$, so the algorithm works: computing the greatest common divisor $\gcd(x^2 - 1, 15) = 3$ and $\gcd(x^2 + 1, 15) = 5$ tells us that $15 = 3{\times}5$.

of the function may be very intricate. In order to present this problem in the most approachable manner, we begin with two more specific applications: period-finding (of a one-dimensional function), and discrete logarithms. We then return to the general hidden subgroup problem. Note that the presentation in this section is rather more schematic and conceptual than earlier sections in this chapter; of necessity, this means that the reader interested in understanding all the details will have to work much harder!

5.4.1 Period-finding

Consider the following problem. Suppose f is a periodic function producing a single bit as output and such that $f(x + r) = f(x)$, for some unknown $0 < r < 2^L$, where $x, r \in \{0, 1, 2, \ldots\}$. Given a quantum black box U which performs the unitary transform $U|x\rangle|y\rangle \rightarrow |x\rangle|y \oplus f(x)\rangle$ (where $\oplus$ denotes addition modulo 2) how many black box queries and other operations are required to determine r? Note that in practice U operates on a finite domain, whose size is determined by the desired accuracy for r. Here is a quantum algorithm which solves this problem using *one* query, and $O(L^2)$ other operations:

Algorithm: Period-finding

    Inputs: (1) A black box which performs the operation $U|x\rangle|y\rangle = |x\rangle|y \oplus f(x)\rangle$, (2) a state to store the function evaluation, initialized to $|0\rangle$, and (3) $t = O(L + \log(1/\epsilon))$ qubits initialized to $|0\rangle$.

    Outputs: The least integer $r > 0$ such that $f(x + r) = f(x)$.

    Runtime: One use of U, and $O(L^2)$ operations. Succeeds with probability $O(1)$.

    Procedure:

1.	$\|0\rangle\|0\rangle$	initial state
2.	$\rightarrow \dfrac{1}{\sqrt{2^t}} \sum\limits_{x=0}^{2^t-1} \|x\rangle\|0\rangle$	create superposition
3.	$\rightarrow \dfrac{1}{\sqrt{2^t}} \sum\limits_{x=0}^{2^t-1} \|x\rangle\|f(x)\rangle$	apply U
	$\approx \dfrac{1}{\sqrt{r2^t}} \sum\limits_{\ell=0}^{r-1} \sum\limits_{x=0}^{2^t-1} e^{2\pi i\ell x/r} \|x\rangle\|\hat{f}(\ell)\rangle$	
4.	$\rightarrow \dfrac{1}{\sqrt{r}} \sum\limits_{\ell=0}^{r-1} \|\widetilde{\ell/r}\rangle\|\hat{f}(\ell)\rangle$	apply inverse Fourier transform to first register
5.	$\rightarrow \widetilde{\ell/r}$	measure first register
6.	$\rightarrow r$	apply continued fractions algorithm

    The key to understanding this algorithm, which is based on phase estimation, and is nearly identical to the algorithm for quantum order-finding, is step **3**, in which we introduce the state

$$|\hat{f}(\ell)\rangle \equiv \frac{1}{\sqrt{r}} \sum_{x=0}^{r-1} e^{-2\pi i\ell x/r} |f(x)\rangle \,, \tag{5.63}$$

the Fourier transform of $|f(x)\rangle$. The identity used in step **3** is based on

$$|f(x)\rangle = \frac{1}{\sqrt{r}} \sum_{\ell=0}^{r-1} e^{2\pi i\ell x/r} |\hat{f}(\ell)\rangle , \tag{5.64}$$

which is easy to verify by noting that $\sum_{\ell=0}^{r-1} e^{2\pi i\ell x/r} = r$ for x an integer multiple of r, and zero otherwise. The approximate equality in step **3** is required because 2^t may not be an integer multiple of r in general (it need not be: this is taken account of by the phase estimation bounds). By Equation (5.22), applying the inverse Fourier transform to the first register, in step **4**, gives an estimate of the phase ℓ/r, where ℓ is chosen randomly. r can be efficiently obtained in the final step using a continued fraction expansion.

Box 5.5: The shift-invariance property of the Fourier transform

The Fourier transform, Equation (5.1), has an interesting and very useful property, known as *shift invariance*. Using notation which is useful in describing the general application of this property, let us describe the quantum Fourier transform as

$$\sum_{h \in H} \alpha_h |h\rangle \rightarrow \sum_{g \in G} \tilde{\alpha}_g |g\rangle , \tag{5.65}$$

where $\tilde{\alpha}_g = \sum_{h \in H} \alpha_h \exp(2\pi igh/|G|)$, H is some subset of G, and G indexes the states in an orthonormal basis of the Hilbert space. For example, G may be the set of numbers from 0 to $2^n - 1$ for an n qubit system. $|G|$ denotes the number of elements in G. Suppose we apply to the initial state an operator U_k which performs the unitary transform

$$U_k |g\rangle = |g + k\rangle , \tag{5.66}$$

then apply the Fourier transform. The result,

$$U_k \sum_{h \in H} \alpha_h |h\rangle = \sum_{h \in H} \alpha_h |h + k\rangle \rightarrow \sum_{g \in G} e^{2\pi igk/|G|} \tilde{\alpha}_g |g\rangle \tag{5.67}$$

has the property that the magnitude of the amplitude for $|g\rangle$ does not change, no matter what k is, that is: $|\exp(2\pi igk/|G|)\tilde{\alpha}_g| = |\tilde{\alpha}_g|$.

In the language of group theory, G is a group, H a subgroup of G, and we say that if a function f on G is constant on cosets of H, then the Fourier transform of f is invariant over cosets of H.

Why does this work? One way to understand this is to realize that (5.63) is approximately the Fourier transform over $\{0, 1, \ldots, 2^L - 1\}$ of $|f(x)\rangle$ (see Exercise 5.20), and the Fourier transform has an interesting and very useful property, known as *shift invariance*, described in Box 5.5. Another is to realize that what the order-finding algorithm does is just to find the period of the function $f(k) = x^k \mod N$, so the ability to find the period of a general periodic function is not unexpected. Yet another way is to realize that the implementation of the black box U is naturally done using a certain unitary operator whose eigenvectors are precisely $|\hat{f}(\ell)\rangle$, as described in Exercise 5.21 below, so that the phase estimation procedure of Section 5.2 can be applied.

Exercise 5.20: Suppose $f(x + r) = f(x)$, and $0 \leq x < N$, for N an integer multiple of r. Compute

$$\hat{f}(\ell) \equiv \frac{1}{\sqrt{N}} \sum_{x=0}^{N-1} e^{-2\pi i \ell x / N} f(x),\qquad(5.68)$$

and relate the result to (5.63). You will need to use the fact that

$$\sum_{k \in \{0, r, 2r, \ldots, N-r\}} e^{2\pi i k \ell / N} = \begin{cases} \sqrt{N/r} & \text{if } \ell \text{ is an integer multiple of } N/r \\ 0 & \text{otherwise.} \end{cases}$$

$$(5.69)$$

Exercise 5.21: (Period-finding and phase estimation) Suppose you are given a unitary operator U_y which performs the transformation $U_y |f(x)\rangle = |f(x + y)\rangle$, for the periodic function described above.

(1) Show that the eigenvectors of U_y are $|\hat{f}(\ell)\rangle$, and calculate their eigenvalues.
(2) Show that given $|f(x_0)\rangle$ for some x_0, U_y can be used to realize a black box which is as useful as U in solving the period-finding problem.

5.4.2 Discrete logarithms

The period finding problem we just considered is a simple one, in that the domain and range of the periodic function were integers. What happens when the function is more complex? Consider the function $f(x_1, x_2) = a^{s x_1 + x_2} \bmod N$, where all the variables are integers, and r is the smallest positive integer for which $a^r \bmod N = 1$. This function is periodic, since $f(x_1 + \ell, x_2 - \ell s) = f(x_1, x_2)$, but now the period is a 2-tuple, $(\ell, -\ell s)$, for integer ℓ. This may seem to be a strange function, but it is very useful in cryptography, since determining s allows one to solve what is known as the *discrete logarithm* problem: given a and $b = a^s$, what is s? Here is a quantum algorithm which solves this problem using *one* query of a quantum black box U which performs the unitary transform $U|x_1\rangle|x_2\rangle|y\rangle \rightarrow |x_1\rangle|x_2\rangle|y \oplus f(x)\rangle$ (where $\oplus$ denotes bitwise addition modulo 2), and $O(\lceil \log r \rceil^2)$ other operations. We assume knowledge of the minimum $r > 0$ such that $a^r \bmod N = 1$, which can be obtained using the order-finding algorithm described previously.

Algorithm: Discrete logarithm

> **Inputs:** (1) A black box which performs the operation
> $U|x_1\rangle|x_2\rangle|y\rangle = |x_1\rangle|x_2\rangle|y \oplus f(x_1, x_2)\rangle$, for $f(x_1, x_2) = b^{x_1} a^{x_2}$, (2) a state to store the function evaluation, initialized to $|0\rangle$, and (3) two $t = O(\lceil \log r \rceil + \log(1/\epsilon))$ qubit registers initialized to $|0\rangle$.
>
> **Outputs:** The least positive integer s such that $a^s = b$.
>
> **Runtime:** One use of U, and $O(\lceil \log r \rceil^2)$ operations. Succeeds with probability $O(1)$.
>
> **Procedure:**
>
> 1. $|0\rangle|0\rangle|0\rangle$ initial state

2. $\quad \to \dfrac{1}{2^t} \sum\limits_{x_1=0}^{2^t-1} \sum\limits_{x_2=0}^{2^t-1} |x_1\rangle|x_2\rangle|0\rangle$ \hfill create superposition

3. $\quad \to \dfrac{1}{2^t} \sum\limits_{x_1=0}^{2^t-1} \sum\limits_{x_2=0}^{2^t-1} |x_1\rangle|x_2\rangle|f(x_1,x_2)\rangle$ \hfill apply U

$$\approx \dfrac{1}{2^t\sqrt{r}} \sum_{\ell_2=0}^{r-1}\sum_{x_1=0}^{2^t-1}\sum_{x_2=0}^{2^t-1} e^{2\pi i(s\ell_2 x_1+\ell_2 x_2)/r}|x_1\rangle|x_2\rangle|\hat{f}(s\ell_2,\ell_2)\rangle$$

$$= \dfrac{1}{2^t\sqrt{r}} \sum_{\ell_2=0}^{r-1}\left[\sum_{x_1=0}^{2^t-1} e^{2\pi i(s\ell_2 x_1)/r}|x_1\rangle\right]\left[\sum_{x_2=0}^{2^t-1} e^{2\pi i(\ell_2 x_2)/r}|x_2\rangle\right]|\hat{f}(s\ell_2,\ell_2)\rangle$$

4. $\quad \to \dfrac{1}{\sqrt{r}} \sum\limits_{\ell_2=0}^{r-1} |\widetilde{s\ell_2/r}\rangle|\widetilde{\ell_2/r}\rangle|\hat{f}(s\ell_2,\ell_2)\rangle$ \hfill apply inverse Fourier transform to first two registers

5. $\quad \to \left(\widetilde{s\ell_2/r},\ \widetilde{\ell_2/r}\right)$ \hfill measure first two registers

6. $\quad \to s$ \hfill apply generalized continued fractions algorithm

Again, the key to understanding this algorithm is step 3, in which we introduce the state

$$|\hat{f}(\ell_1,\ell_2)\rangle = \dfrac{1}{\sqrt{r}} \sum_{j=0}^{r-1} e^{-2\pi i\ell_2 j/r}|f(0,j)\rangle, \qquad (5.70)$$

the Fourier transform of $|f(x_1,x_2)\rangle$ (see Exercise 5.22). In this equation, the values of ℓ_1 and ℓ_2 must satisfy

$$\sum_{k=0}^{r-1} e^{2\pi i k(\ell_1/s-\ell_2)/r} = r. \qquad (5.71)$$

Otherwise, the amplitude of $|\hat{f}(\ell_1,\ell_2)\rangle$ is nearly zero. The generalized continued fraction expansion used in the final step to determine s is analogous to the procedures used in Section 5.3.1, and is left as a simple exercise for you to construct.

Exercise 5.22: Show that

$$|\hat{f}(\ell_1,\ell_2)\rangle = \sum_{x_1=0}^{r-1}\sum_{x_2=0}^{r-1} e^{-2\pi i(\ell_1 x_1+\ell_2 x_2)/r}|f(x_1,x_2)\rangle = \dfrac{1}{\sqrt{r}} \sum_{j=0}^{r-1} e^{-2\pi i\ell_2 j/r}|f(0,j)\rangle,$$

$$(5.72)$$

and we are constrained to have $\ell_1/s - \ell_2$ be an integer multiple of r for this expression to be non-zero.

Exercise 5.23: Compute

$$\dfrac{1}{r}\sum_{\ell_1=0}^{r-1}\sum_{\ell_2=0}^{r-1} e^{-2\pi i(\ell_1 x_1+\ell_2 x_2)/r}|\hat{f}(\ell_1,\ell_2)\rangle \qquad (5.73)$$

using (5.70), and show that the result is $f(x_1,x_2)$.

Exercise 5.24: Construct the generalized continued fractions algorithm needed in step **6** of the discrete logarithm algorithm to determine s from estimates of $s\ell_2/r$ and ℓ_2/r.

Exercise 5.25: Construct a quantum circuit for the black box U used in the quantum discrete logarithm algorithm, which takes a and b as parameters, and performs the unitary transform $|x_1\rangle|x_2\rangle|y\rangle \rightarrow |x_1\rangle|x_2\rangle|y \oplus b^{x_1}a^{x_2}\rangle$. How many elementary operations are required?

5.4.3 The hidden subgroup problem

By now, a pattern should be coming clear: if we are given a periodic function, even when the structure of the periodicity is quite complicated, we can often use a quantum algorithm to determine the period efficiently. Importantly, however, not *all* periods of periodic functions can be determined. The general problem which defines a broad framework for these questions can be succinctly expressed in the language of group theory (see Appendix 2 for a quick review) as follows:

> Let f be a function from a finitely generated group G to a finite set X such that f is constant on the cosets of a subgroup K, and distinct on each coset. Given a quantum black box for performing the unitary transform $U|g\rangle|h\rangle = |g\rangle|h \oplus f(g)\rangle$, for $g \in G$, $h \in X$, and $\oplus$ an appropriately chosen binary operation on X, find a generating set for K.

Order-finding, period-finding, discrete logarithms, and many other problems are instances of this *hidden subgroup problem*; some interesting ones are listed in Figure 5.5.

For G a finite *Abelian* group, a quantum computer can solve the hidden subgroup problem using a number of operations polynomial in $\log |G|$, and one use of the black box function evaluation, using an algorithm very similar to the others in this section. (In fact, solution for a finitely generated Abelian group is also possible, along similar lines, but we'll stick to the finite case here.) We shall leave detailed specification of the algorithm to you as an exercise, which should be simple after we explain the basic idea. Many things remain essentially the same, because finite Abelian groups are isomorphic to products of additive groups over the integers in modular arithmetic. This means that the quantum Fourier transform of f over G is well defined (see Section A2.3), and can still be done efficiently. The first non-trivial step of the algorithm is to use a Fourier transform (generalizing the Hadamard operation) to create a superposition over group elements, which is then transformed by applying the quantum black box for f in the next step, to give

$$\frac{1}{\sqrt{|G|}} \sum_{g \in G} |g\rangle|f(g)\rangle . \tag{5.74}$$

As before, we would now like to rewrite $|f(g)\rangle$ in the Fourier basis. We start with

$$|f(g)\rangle = \frac{1}{\sqrt{|G|}} \sum_{\ell=0}^{|G|-1} e^{2\pi i\ell g/|G|}|\hat{f}(\ell)\rangle , \tag{5.75}$$

where we have chosen $\exp[-2\pi i\ell g/|G|]$ as a representation (see Exercise A2.13) of $g \in G$ indexed by ℓ (the Fourier transform maps between group elements and representations: see Exercise A2.23). The key is to recognize that this expression can be simplified because

Name	G	X	K	Function
Deutsch	$\{0,1\},\oplus$	$\{0,1\}$	$\{0\}$ or $\{0,1\}$	$K = \{0,1\} : \begin{cases} f(x) = 0 \\ f(x) = 1 \end{cases}$ $K = \{0\} : \begin{cases} f(x) = x \\ f(x) = 1 - x \end{cases}$
Simon	$\{0,1\}^n,\oplus$	any finite set	$\{0,s\}$ $s \in \{0,1\}^n$	$f(x \oplus s) = f(x)$
Period-finding	$\mathbf{Z},+$	any finite set	$\{0,r,2r,\ldots\}$ $r \in G$	$f(x+r) = f(x)$
Order-finding	$\mathbf{Z},+$	$\{a^j\}$ $j \in Z_r$ $a^r = 1$	$\{0,r,2r,\ldots\}$ $r \in G$	$f(x) = a^x$ $f(x+r) = f(x)$
Discrete logarithm	$\mathbf{Z}_r \times \mathbf{Z}_r$ $+ \pmod r$	$\{a^j\}$ $j \in Z_r$ $a^r = 1$	$(\ell, -\ell s)$ $\ell, s \in Z_r$	$f(x_1, x_2) = a^{kx_1 + x_2}$ $f(x_1 + \ell, x_2 - \ell s) = f(x_1, x_2)$
Order of a permutation	$\mathbf{Z}_{2^m} \times \mathbf{Z}_{2^n}$ $+ \pmod{2^m}$	$\mathbf{Z}_{2^n}$	$\{0,r,2r,\ldots\}$ $r \in X$	$f(x,y) = \pi^x(y)$ $f(x+r,y) = f(x,y)$ $\pi = $ permutation on X
Hidden linear function	$\mathbf{Z} \times \mathbf{Z}, +$	$\mathbf{Z}_N$	$(\ell, -\ell s)$ $\ell, s \in X$	$f(x_1, x_2) =$ $\pi(sx_1 + x_2 \bmod N)$ $\pi = $ permutation on X
Abelian stabilizer	(H, X) $H = $ any Abelian group	any finite set	$\{s \in H \mid$ $f(s,x) = x,$ $\forall x \in X\}$	$f(gh, x) = f(g, f(h, x))$ $f(gs, x) = f(g, x)$

Figure 5.5. Hidden subgroup problems. The function f maps from the group G to the finite set X, and is promised to be constant on cosets of the hidden subgroup $K \subseteq G$. $\mathbf{Z}_N$ represents the set $\{0, 1, \ldots, N-1\}$ in this table, and $\mathbf{Z}$ is the integers. The problem is to find K (or a generating set for it), given a black box for f.

f is constant and distinct on cosets of the subgroup K, so that

$$|\hat{f}(\ell)\rangle = \frac{1}{\sqrt{|G|}} \sum_{g \in G} e^{-2\pi i \ell g / |G|} |f(g)\rangle \tag{5.76}$$

has nearly zero amplitude for all values of ℓ except those which satisfy

$$\sum_{h \in K} e^{-2\pi i \ell h / |G|} = |K| . \tag{5.77}$$

If we can determine ℓ, then using the linear constraints given by this expression allows us to determine elements of K, and since K is Abelian, this allows us to eventually determine a generating set for the whole hidden subgroup, solving the problem.

However, life is not so simple. An important reason why the period-finding and discrete logarithm algorithms work is because of the success of the continued fraction expansion in obtaining ℓ from $\ell/|G|$. In those problems, ℓ and $|G|$ are arranged to not have any common factors, with high probability. In the general case, however, this may not be true, since $|G|$ is free to be a composite number with many factors, and we have no useful prior information about ℓ.

Fortunately, this problem can be solved: as mentioned above, any finite Abelian group G is isomorphic to a product of cyclic groups of prime power order, that is, $G = \mathbf{Z}_{p_1} \times \mathbf{Z}_{p_2} \times \cdots \times \mathbf{Z}_{p_M}$, where p_i are primes, and $\mathbf{Z}_{p_i}$ is the group over integers $\{0, 1, \ldots, p_i - 1\}$ with addition modulo p_i being the group operation. We can thus re-express the phase which appears in (5.75) as

$$e^{2\pi i \ell g/|G|} = \prod_{i=1}^{M} e^{2\pi i \ell_i' g_i/p_i} \tag{5.78}$$

for $g_i \in \mathbf{Z}_{p_i}$. The phase estimation procedure now gives us ℓ_i', from which we determine ℓ, and thus, sample K as described above, to solve the hidden subgroup problem.

Exercise 5.26: Since K is a subgroup of G, when we decompose G into a product of cyclic groups of prime power order, this also decomposes K. Re-express (5.77) to show that determining ℓ_i' allows one to sample from the corresponding cyclic subgroup K_{p_i} of K.

Exercise 5.27: Of course, the decomposition of a general finite Abelian group G into a product of cyclic groups of prime power order is usually a difficult problem (at least as hard as factoring integers, for example). Here, quantum algorithms come to the rescue again: explain how the algorithms in this chapter can be used to efficiently decompose G as desired.

Exercise 5.28: Write out a detailed specification of the quantum algorithm to solve the hidden subgroup problem, complete with runtime and success probability estimates, for finite Abelian groups.

Exercise 5.29: Give quantum algorithms to solve the Deutsch and Simon problems listed in Figure 5.5, using the framework of the hidden subgroup problem.

5.4.4 Other quantum algorithms?

One of the most intriguing aspects of this framework for describing quantum algorithms in terms of the hidden subgroup problem is the suggestion that more difficult problems might be solvable by considering various groups G and functions f. We have only described the solution of this problem for Abelian groups. What about *non-Abelian* groups? They are quite interesting (see Appendix 2 for a discussion of general Fourier transforms over non-Abelian groups): for example, the problem of *graph isomorphism* is to determine if two given graphs are the same under some permutation of the labels of the n vertices (Section 3.2.3). These permutations can be described as transformations under the symmetric group S_n, and algorithms for performing fast Fourier transforms over these groups exists. However, a quantum algorithm for efficiently solving the graph isomporphism problem remains unknown.

Even if more general cases of the hidden subgroup problem remain unsolvable by quantum computers, having this unifying framework is useful, because it allows us to ask questions about how one might be able to step outside its limitations. It is difficult to believe that all fast quantum algorithms that will ever be discovered will be just ways to solve the hidden subgroup problem. If one thinks of these problems as being based on the coset invariance property of the Fourier transform, in searching for new algorithms, perhaps the thing to do then is to investigate other transforms with different invariances. Going in another direction, one might ask: what difficult hidden subgroup problems might be efficiently solved given an arbitrary (but specified independently of the problem) quantum state as a helper? After all, as discussed in Chapter 4, most quantum states are actually exponentially hard to construct. Such a state might be a useful resource (a real 'quantum oracle'), if quantum algorithms existed to utilize them to solve hard problems!

The hidden subgroup problem also captures an important constraint underlying the class of quantum algorithms which are exponentially faster than their (known) classical counterparts: this is a *promise* problem, meaning that it is of the form '$F(X)$ is promised to have such and such property: characterize that property.' Rather disappointingly, perhaps, we shall show at the end of the next chapter that, in solving problems without some sort of promise, quantum computers *cannot* achieve an exponential speedup over classical computers; the best speedup is polynomial. On the other hand, this gives us an important clue as to what kinds of problems quantum computers might be good at: in retrospect, the hidden subgroup problem might be thought of as a natural candidate for quantum computation. What other natural problems are there? Think about it!

Problem 5.1:   Construct a quantum circuit to perform the quantum Fourier transform

$$|j\rangle \longrightarrow \frac{1}{\sqrt{p}} \sum_{k=0}^{p-1} e^{2\pi ijk/p}|k\rangle \tag{5.79}$$

where p is prime.

Problem 5.2: (Measured quantum Fourier transform)   Suppose the quantum Fourier transform is performed as the last step of a quantum computation, followed by a measurement in the computational basis. Show that the combination of quantum Fourier transform and measurement is equivalent to a circuit consisting entirely of *one* qubit gates and measurement, with classical control, and no two qubit gates. You may find the discussion of Section 4.4 useful.

Problem 5.3: (Kitaev's algorithm)   Consider the quantum circuit

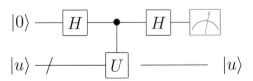

where $|u\rangle$ is an eigenstate of U with eigenvalue $e^{2\pi i\varphi}$. Show that the top qubit is measured to be 0 with probability $p \equiv \cos^2(\pi\varphi)$. Since the state $|u\rangle$ is unaffected by the circuit it may be reused; if U can be replaced by U^k, where k is an arbitrary integer under your control, show that by repeating this circuit and increasing k appropriately, you can efficiently obtain as many bits of p as desired, and thus, of φ. This is an alternative to the phase estimation algorithm.

Problem 5.4: The runtime bound $O(L^3)$ we have given for the factoring algorithm is not tight. Show that a better upper bound of $O(L^2 \log L \log \log L)$ operations can be achieved.

Problem 5.5: (Non-Abelian hidden subgroups – Research) Let f be a function on a finite group G to an arbitrary finite range X, which is promised to be constant and distinct on distinct left cosets of a subgroup K. Start with the state

$$\frac{1}{\sqrt{|G|^m}} \sum_{g_1,\ldots,g_m} |g_1,\ldots,g_m\rangle |f(g_1),\ldots,f(g_m)\rangle , \tag{5.80}$$

and prove that picking $m = 4\log|G| + 2$ allows K to be identified with probability at least $1 - 1/|G|$. Note that G does not necessarily have to be Abelian, and being able to perform a Fourier transform over G is not required. This result shows that one can produce (using only $O(\log|G|)$ oracle calls) a final result in which the pure state outcomes corresponding to different possible hidden subgroups are nearly orthogonal. However, it is unknown whether a POVM exists or not which allows the hidden subgroup to be identified *efficiently* (i.e. using poly($\log|G|$) operations) from this final state.

Problem 5.6: (Addition by Fourier transforms) Consider the task of constructing a quantum circuit to compute $|x\rangle \rightarrow |x + y \bmod 2^n\rangle$, where y is a fixed constant, and $0 \leq x < 2^n$. Show that one efficient way to do this, for values of y such as 1, is to first perform a quantum Fourier transform, then to apply single qubit phase shifts, then an inverse Fourier transform. What values of y can be added easily this way, and how many operations are required?

Summary of Chapter 5: The quantum Fourier transform and its applications

- When $N = 2^n$ the quantum Fourier transform

$$|j\rangle = |j_1, \ldots, j_n\rangle \longrightarrow \frac{1}{\sqrt{N}} \sum_{k=0}^{N-1} e^{2\pi i \frac{jk}{N}} |k\rangle \qquad (5.81)$$

may be written in the form

$$|j\rangle \rightarrow \frac{1}{2^{n/2}} \left(|0\rangle + e^{2\pi i 0.j_n} |1\rangle\right) \left(|0\rangle + e^{2\pi i 0.j_{n-1}j_n} |1\rangle\right) \cdots \left(|0\rangle + e^{2\pi i 0.j_1 j_2 \cdots j_n} |1\rangle\right),$$
$$(5.82)$$

and may be implemented using $\Theta(n^2)$ gates.

- **Phase estimation**: Let $|u\rangle$ be an eigenstate of the operator U with eigenvalue $e^{2\pi i \varphi}$. Starting from the initial state $|0\rangle^{\otimes t} |u\rangle$, and given the ability to efficiently perform U^{2^k} for integer k, this algorithm (shown in Figure 5.3) can be used to efficiently obtain the state $|\tilde{\varphi}\rangle |u\rangle$, where $\tilde{\varphi}$ accurately approximates φ to $t - \lceil \log\left(2 + \frac{1}{2\epsilon}\right) \rceil$ bits with probability at least $1 - \epsilon$.

- **Order-finding**: The order of x modulo N is the least positive integer r such that $x^r \bmod N = 1$. This number can be computed in $O(L^3)$ operations using the quantum phase estimation algorithm, for L-bit integers x and N.

- **Factoring**: The prime factors of an L-bit integer N can be determined in $O(L^3)$ operations by reducing this problem to finding the order of a random number x co-prime with N.

- **Hidden subgroup problem**: All the known fast quantum algorithms can be described as solving the following problem: Let f be a function from a finitely generated group G to a finite set X such that f is constant on the cosets of a subgroup K, and distinct on each coset. Given a quantum black box for performing the unitary transform $U|g\rangle|h\rangle = |g\rangle|h \oplus f(g)\rangle$, for $g \in G$ and $h \in X$, find a generating set for K.

History and further reading

The definition of the Fourier transform may be generalized beyond what we have considered in this chapter. In the general scenario a Fourier transform is defined on a set of complex numbers α_g, where the index g is chosen from some group, G. In this chapter we have chosen G to be the additive group of integers modulo 2^n, often denoted $\mathbf{Z}_{2^n}$. Deutsch[Deu85] showed that the Fourier transform over the group $\mathbf{Z}_2^n$ could be implemented efficiently on a quantum computer – this is the Hadamard transform of earlier chapters. Shor [Sho94] realized to spectacular effect that quantum computers could efficiently implement the quantum Fourier transform over groups $\mathbf{Z}_m$ for certain special values of m. Inspired by this result Coppersmith[Cop94], Deutsch (unpublished), and Cleve (unpublished) gave the simple quantum circuits for computing the quantum Fourier transform over $\mathbf{Z}_{2^n}$ which we have used in this chapter. Cleve, Ekert, Macchiavello and Mosca[CEMM98] and Griffiths and Niu[GN96] independently discovered the product formula (5.4); in fact, this result had been realized much earlier by Danielson and Lanczos. The simplified proof starting in Equation (5.5) was suggested by Zhou. Griffiths and Niu[GN96] are responsible for the measured quantum Fourier transform found in Problem 5.2.

The Fourier transform over Z_{2^n} was generalized to obtain a Fourier transform over an arbitrary finite Abelian group by Kitaev[Kit95], who also introduced the phase estimation procedure in the form given in Problem 5.3. Cleve, Ekert, Macchiavello and Mosca[CEMM98] also integrated several of the techniques of Shor and Kitaev into one nice picture, upon which Section 5.2 is based. A good description of the phase estimation algorithm can be found in Mosca's Ph.D. thesis[Mos99].

Shor announced the quantum order-finding algorithm in a seminal paper in 1994[Sho94], and noted that the problems of performing discrete logarithms and factoring could be reduced to order-finding. The final paper, including extended discussion and references, was published in 1997[Sho97]. This paper also contains a discussion of clever multiplication methods that may be used to speed up the algorithm even further than in our description, which uses relatively naive multiplication techniques. With these faster multiplication methods the resources required to factor a composite integer n scale as $O(n^2 \log n \log \log n)$, as claimed in the introduction to the chapter. In 1995 Kitaev[Kit95] announced an algorithm for finding the stabilizer of a general Abelian group, which he showed could be used to solve discrete logarithm and factoring as special cases. In addition, this algorithm contained several elements not present in Shor's algorithm. A good review of the factoring algorithm was written by Ekert and Jozsa [EJ96]; also see DiVincenzo [DiV95a]. The discussion of continued fractions is based upon Chapter 10 of Hardy and Wright[HW60]. At the time of writing, the most efficient classical algorithm for factoring on a classical computer is the number field sieve. This is described in a collection edited by A. K. Lenstra and H. W. Lenstra, Jr.[LL93].

The generalization of quantum algorithms to solving the hidden subgroup problem has been considered by many authors. Historically, Simon was first to note that a quantum computer could find a hidden period of a function satisfying $f(x \oplus s) = f(x)$[Sim94, Sim97]. In fact, Shor found his result by generalizing Simon's result, and by applying a Fourier transform over Z_N instead of Simon's Hadamard transforms (a Fourier transform over Z_2^k). Boneh and Lipton then noted the connection to the hidden subgroup problem, and described a quantum algorithm for solving the hidden linear function problem[BL95]. Jozsa was the first to explicitly provide a uniform description of the Deutsch–Jozsa, Simon, and Shor algorithms in terms of the hidden subgroup problem[Joz97]. Ekert and Jozsa's work in studying the role of the Abelian and non-Abelian Fast Fourier Transform algorithms in speedup of quantum algorithms[EJ98] has also been insightful. Our description of the hidden subgroup problem in Section 5.4 follows the framework of Mosca and Ekert[ME99, Mos99]. Cleve has proven that the problem of finding an order of a permutation requires an exponential number of queries for a bounded-error probabilistic classical computer[Cle99]. Generalizations of this method to beyond Abelian groups have been attempted by Ettinger and Høyer[EH99], by Roetteler and Beth[RB98] and Pueschel, Roetteler, and Beth[PRB98], by Beals, who also described constructions of quantum Fourier transforms over the symmetric group[BBC+98], and by Ettinger, Høyer, and Knill[EHK99]. These results have shown, so far, that *there exists* a quantum algorithm to solve the hidden subgroup problem for non–Abelian groups using only $O(\log |G|)$ oracle calls, but whether this can be realized in polynomial time is unknown (Problem 5.5).

6 Quantum search algorithms

Suppose you are given a map containing many cities, and wish to determine the shortest route passing through all cities on the map. A simple algorithm to find this route is to search all possible routes through the cities, keeping a running record of which route has the shortest length. On a classical computer, if there are N possible routes, it obviously takes $O(N)$ operations to determine the shortest route using this method. Remarkably, there is a *quantum search algorithm*, sometimes known as *Grover's algorithm*, which enables this search method to be sped up substantially, requiring only $O(\sqrt{N})$ operations. Moreover, the quantum search algorithm is *general* in the sense that it can be applied far beyond the route-finding example just described to speed up many (though not all) classical algorithms that use search heuristics.

In this chapter we explain the fast quantum search algorithm. The basic algorithm is described in Section 6.1. In Section 6.2 we derive the algorithm from another point of view, based on the quantum simulation algorithm of Section 4.7. Three important applications of this algorithm are also described: quantum counting in Section 6.3, speedup of solution of **NP**-complete problems in Section 6.4, and search of unstructured databases in Section 6.5. One might hope to improve upon the search algorithm to do even better than a square root speedup but, as we show in Section 6.6, it turns out this is not possible. We conclude in Section 6.7 by showing that this speed limit applies to most unstructured problems.

6.1 The quantum search algorithm

Let us begin by setting the stage for the search algorithm in terms of an *oracle*, similar to that encountered in Section 3.1.1. This allows us to present a very general description of the search procedure, and a geometric way to visualize its action and see how it performs.

6.1.1 The oracle

Suppose we wish to search through a search space of N elements. Rather than search the elements directly, we concentrate on the *index* to those elements, which is just a number in the range 0 to $N - 1$. For convenience we assume $N = 2^n$, so the index can be stored in n bits, and that the search problem has exactly M solutions, with $1 \leq M \leq N$. A particular instance of the search problem can conveniently be represented by a function f, which takes as input an integer x, in the range 0 to $N - 1$. By definition, $f(x) = 1$ if x is a solution to the search problem, and $f(x) = 0$ if x is not a solution to the search problem.

Suppose we are supplied with a quantum *oracle* – a black box whose internal workings we discuss later, but which are not important at this stage – with the ability to *recognize* solutions to the search problem. This recognition is signalled by making use of an *oracle qubit*. More precisely, the oracle is a unitary operator, O, defined by its action on the computational basis:

$$|x\rangle|q\rangle \xrightarrow{O} |x\rangle|q \oplus f(x)\rangle, \tag{6.1}$$

where $|x\rangle$ is the index register, $\oplus$ denotes addition modulo 2, and the oracle qubit $|q\rangle$ is a single qubit which is flipped if $f(x) = 1$, and is unchanged otherwise. We can check whether x is a solution to our search problem by preparing $|x\rangle|0\rangle$, applying the oracle, and checking to see if the oracle qubit has been flipped to $|1\rangle$.

In the quantum search algorithm it is useful to apply the oracle with the oracle qubit initially in the state $(|0\rangle - |1\rangle)/\sqrt{2}$, just as was done in the Deutsch–Jozsa algorithm of Section 1.4.4. If x is not a solution to the search problem, applying the oracle to the state $|x\rangle(|0\rangle - |1\rangle)/\sqrt{2}$ does not change the state. On the other hand, if x is a solution to the search problem, then $|0\rangle$ and $|1\rangle$ are interchanged by the action of the oracle, giving a final state $-|x\rangle(|0\rangle - |1\rangle)/\sqrt{2}$. The action of the oracle is thus:

$$|x\rangle \left(\frac{|0\rangle - |1\rangle}{\sqrt{2}} \right) \xrightarrow{O} (-1)^{f(x)}|x\rangle \left(\frac{|0\rangle - |1\rangle}{\sqrt{2}} \right). \tag{6.2}$$

Notice that the state of the oracle qubit is not changed. It turns out that this remains $(|0\rangle - |1\rangle)/\sqrt{2}$ throughout the quantum search algorithm, and can therefore be omitted from further discussion of the algorithm, simplifying our description.

With this convention, the action of the oracle may be written:

$$|x\rangle \xrightarrow{O} (-1)^{f(x)}|x\rangle. \tag{6.3}$$

We say that the oracle *marks* the solutions to the search problem, by shifting the phase of the solution. For an N item search problem with M solutions, it turns out that we need only apply the search oracle $O(\sqrt{N/M})$ times in order to obtain a solution, on a quantum computer.

This discussion of the oracle without describing how it works in practice is rather abstract, and perhaps even puzzling. It seems as though the oracle already *knows* the answer to the search problem; what possible use could it be to have a quantum search algorithm based upon such oracle consultations?! The answer is that there is a distinction between *knowing* the solution to a search problem, and being able to *recognize* the solution; the crucial point is that it is possible to do the latter without necessarily being able to do the former.

A simple example to illustrate this is the problem of factoring. Suppose we have been given a large number, m, and told that it is a product of two primes, p and q – the same sort of situation as arises in trying to break the RSA public key cryptosystem (Appendix 5). To determine p and q, the obvious method on a classical computer is to *search* all numbers from 2 through $m^{1/2}$ for the smaller of the two prime factors. That is, we successively do a trial division of m by each number in the range 2 to $m^{1/2}$, until we find the smaller prime factor. The other prime factor can then be found by dividing m by the smaller prime. Obviously, this search-based method requires roughly $m^{1/2}$ trial divisions to find a factor on a classical computer.

The quantum search algorithm can be used to speed up this process. By definition, the action of the oracle upon input of the state $|x\rangle$ is to divide m by x, and check to see if the division is exact, flipping the oracle qubit if this is so. Applying the quantum search algorithm with this oracle yields the smaller of the two prime factors with high probability.

But to make the algorithm work, we need to construct an efficient circuit implementing the oracle. How to do this is an exercise in the techniques of reversible computation. We begin by defining the function $f(x) \equiv 1$ if x divides m, and $f(x) = 0$ otherwise; $f(x)$ tells us whether the trial division is successful or not. Using the techniques of reversible computation discussed in Section 3.2.5, construct a classical reversible circuit which takes (x, q) – representing an input register initially set to x and a one bit output register initially set to q – to $(x, q \oplus f(x))$, by modifying the usual (irreversible) classical circuit for doing trial division. The resource cost of this reversible circuit is the same to within a factor two as the irreversible classical circuit used for trial division, and therefore we regard the two circuits as consuming essentially the same resources. Furthermore, the classical reversible circuit can be immediately translated into a quantum circuit that takes $|x\rangle|q\rangle$ to $|x\rangle|q \oplus f(x)\rangle$, as required of the oracle. The key point is that *even without knowing the prime factors of m, we can explicitly construct an oracle which recognizes a solution to the search problem when it sees one.* Using this oracle and the quantum search algorithm we can search the range 2 to $m^{1/2}$ using $O(m^{1/4})$ oracle consultations. That is, we need only perform the trial division roughly $m^{1/4}$ times, instead of $m^{1/2}$ times, as with the classical algorithm!

The factoring example is conceptually interesting but not practical: there are classical algorithms for factoring which work much faster than searching through all possible divisors. However, it illustrates the general way in which the quantum search algorithm may be applied: classical algorithms which rely on search-based techniques may be sped up using the quantum search algorithm. Later in this chapter we examine scenarios where the quantum search algorithm offers a genuinely useful aid in speeding up the solution of **NP**-complete problems.

6.1.2 The procedure

Schematically, the search algorithm operates as shown in Figure 6.1. The algorithm proper makes use of a single n qubit register. The internal workings of the oracle, including the possibility of it needing extra work qubits, are not important to the description of the quantum search algorithm proper. The goal of the algorithm is to find a solution to the search problem, using the smallest possible number of applications of the oracle.

The algorithm begins with the computer in the state $|0\rangle^{\otimes n}$. The Hadamard transform is used to put the computer in the equal superposition state,

$$|\psi\rangle = \frac{1}{N^{1/2}} \sum_{x=0}^{N-1} |x\rangle . \tag{6.4}$$

The quantum search algorithm then consists of repeated application of a quantum subroutine, know as the *Grover iteration* or *Grover operator*, which we denote G. The Grover iteration, whose quantum circuit is illustrated in Figure 6.2, may be broken up into four steps:

(1) Apply the oracle O.
(2) Apply the Hadamard transform $H^{\otimes n}$.
(3) Perform a conditional phase shift on the computer, with every computational basis state except $|0\rangle$ receiving a phase shift of -1,

$$|x\rangle \rightarrow -(-1)^{\delta_{x0}}|x\rangle. \tag{6.5}$$

(4) Apply the Hadamard transform $H^{\otimes n}$.

Exercise 6.1:   Show that the unitary operator corresponding to the phase shift in the Grover iteration is $2|0\rangle\langle 0| - I$.

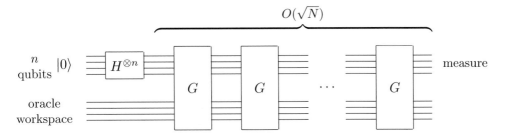

Figure 6.1. Schematic circuit for the quantum search algorithm. The oracle may employ work qubits for its implementation, but the analysis of the quantum search algorithm involves only the n qubit register.

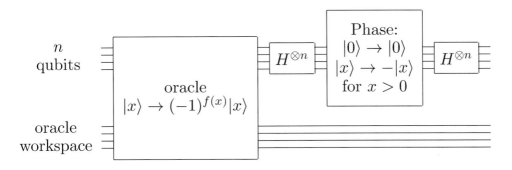

Figure 6.2. Circuit for the Grover iteration, G.

Each of the operations in the Grover iteration may be efficiently implemented on a quantum computer. Steps 2 and 4, the Hadamard transforms, require $n = \log(N)$ operations each. Step 3, the conditional phase shift, may be implemented using the techniques of Section 4.3, using $O(n)$ gates. The cost of the oracle call depends upon the specific application; for now, we merely need note that the Grover iteration requires only a single oracle call. It is useful to note that the combined effect of steps 2, 3, and 4 is

$$H^{\otimes n}(2|0\rangle\langle 0| - I)H^{\otimes n} = 2|\psi\rangle\langle\psi| - I, \qquad (6.6)$$

where $|\psi\rangle$ is the equally weighted superposition of states, (6.4). Thus the Grover iteration, G, may be written $G = (2|\psi\rangle\langle\psi| - I)O$.

Exercise 6.2:   Show that the operation $(2|\psi\rangle\langle\psi| - I)$ applied to a general state $\sum_k \alpha_k |k\rangle$ produces

$$\sum_k \left[-\alpha_k + 2\langle\alpha\rangle \right] |k\rangle, \qquad (6.7)$$

where $\langle a \rangle \equiv \sum_k a_k / N$ is the mean value of the a_k. For this reason, $(2|\psi\rangle\langle\psi| - I)$ is sometimes referred to as the *inversion about mean* operation.

6.1.3 Geometric visualization

What does the Grover iteration do? We have noted that $G = (2|\psi\rangle\langle\psi| - I)O$. In fact, we will show that the Grover iteration can be regarded as a *rotation* in the two-dimensional space spanned by the starting vector $|\psi\rangle$ and the state consisting of a uniform superposition of solutions to the search problem. To see this it is useful to adopt the convention that $\sum_x'$ indicates a sum over all x which are solutions to the search problem, and $\sum_x''$ indicates a sum over all x which are not solutions to the search problem. Define normalized states

$$|\alpha\rangle \equiv \frac{1}{\sqrt{N-M}} \sum_x'' |x\rangle \qquad (6.8)$$

$$|\beta\rangle \equiv \frac{1}{\sqrt{M}} \sum_x' |x\rangle . \qquad (6.9)$$

Simple algebra shows that the initial state $|\psi\rangle$ may be re-expressed as

$$|\psi\rangle = \sqrt{\frac{N-M}{N}}|\alpha\rangle + \sqrt{\frac{M}{N}}|\beta\rangle , \qquad (6.10)$$

so the initial state of the quantum computer is in the space spanned by $|\alpha\rangle$ and $|\beta\rangle$.

The effect of G can be understood in a beautiful way by realizing that the oracle operation O performs a *reflection* about the vector $|\alpha\rangle$ in the plane defined by $|\alpha\rangle$ and $|\beta\rangle$. That is, $O(a|\alpha\rangle + b|\beta\rangle) = a|\alpha\rangle - b|\beta\rangle$. Similarly, $2|\psi\rangle\langle\psi| - I$ also performs a reflection in the plane defined by $|\alpha\rangle$ and $|\beta\rangle$, about the vector $|\psi\rangle$. And the product of two reflections is a rotation! This tells us that the state $G^k|\psi\rangle$ remains in the space spanned by $|\alpha\rangle$ and $|\beta\rangle$ for all k. It also gives us the rotation angle. Let $\cos\theta/2 = \sqrt{(N-M)/N}$, so that $|\psi\rangle = \cos\theta/2|\alpha\rangle + \sin\theta/2|\beta\rangle$. As Figure 6.3 shows, the two reflections which comprise G take $|\psi\rangle$ to

$$G|\psi\rangle = \cos\frac{3\theta}{2}|\alpha\rangle + \sin\frac{3\theta}{2}|\beta\rangle , \qquad (6.11)$$

so the rotation angle is in fact θ. It follows that continued application of G takes the state to

$$G^k|\psi\rangle = \cos\left(\frac{2k+1}{2}\theta\right)|\alpha\rangle + \sin\left(\frac{2k+1}{2}\theta\right)|\beta\rangle . \qquad (6.12)$$

Summarizing, G is a rotation in the two-dimensional space spanned by $|\alpha\rangle$ and $|\beta\rangle$, rotating the space by θ radians per application of G. Repeated application of the Grover iteration rotates the state vector close to $|\beta\rangle$. When this occurs, an observation in the computational basis produces with high probability one of the outcomes superposed in $|\beta\rangle$, that is, a solution to the search problem! An example illustrating the search algorithm with $N = 4$ is given in Box 6.1.

Exercise 6.3: Show that in the $|\alpha\rangle$, $|\beta\rangle$ basis, we may write the Grover iteration as

$$G = \begin{bmatrix} \cos\theta & -\sin\theta \\ \sin\theta & \cos\theta \end{bmatrix} , \qquad (6.13)$$

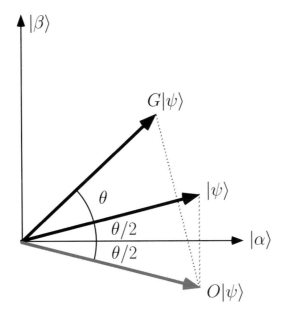

Figure 6.3. The action of a single Grover iteration, G: the state vector is rotated by θ towards the superposition $|\beta\rangle$ of all solutions to the search problem. Initially, it is inclined at angle $\theta/2$ from $|\alpha\rangle$, a state orthogonal to $|\beta\rangle$. An oracle operation O reflects the state about the state $|\alpha\rangle$, then the operation $2|\psi\rangle\langle\psi| - I$ reflects it about $|\psi\rangle$. In the figure $|\alpha\rangle$ and $|\beta\rangle$ are lengthened slightly to reduce clutter (all states should be unit vectors). After repeated Grover iterations, the state vector gets close to $|\beta\rangle$, at which point an observation in the computational basis outputs a solution to the search problem with high probability. The remarkable efficiency of the algorithm occurs because θ behaves like $\Omega(\sqrt{M/N})$, so only $O(\sqrt{N/M})$ applications of G are required to rotate the state vector close to $|\beta\rangle$.

where θ is a real number in the range 0 to $\pi/2$ (assuming for simplicity that $M \leq N/2$; this limitation will be lifted shortly), chosen so that

$$\sin\theta = \frac{2\sqrt{M(N-M)}}{N}. \tag{6.14}$$

6.1.4 Performance

How many times must the Grover iteration be repeated in order to rotate $|\psi\rangle$ near $|\beta\rangle$? The initial state of the system is $|\psi\rangle = \sqrt{(N-M)/N}|\alpha\rangle + \sqrt{M/N}|\beta\rangle$, so rotating through $\arccos\sqrt{M/N}$ radians takes the system to $|\beta\rangle$. Let $\text{CI}(x)$ denote the integer closest to the real number x, where by convention we round halves down, $\text{CI}(3.5) = 3$, for example. Then repeating the Grover iteration

$$R = \text{CI}\left(\frac{\arccos\sqrt{M/N}}{\theta}\right) \tag{6.15}$$

times rotates $|\psi\rangle$ to within an angle $\theta/2 \leq \pi/4$ of $|\beta\rangle$. Observation of the state in the computational basis then yields a solution to the search problem with probability at least one-half. Indeed, for specific values of M and N it is possible to achieve a much higher probability of success. For example, when $M \ll N$ we have $\theta \approx \sin\theta \approx 2\sqrt{M/N}$, and thus the angular error in the final state is at most $\theta/2 \approx \sqrt{M/N}$, giving a probability of error of at most M/N. Note that R depends on the number of solutions M, but not on the identity of those solutions, so provided we know M we can apply the quantum search algorithm as described. In Section 6.3 we will explain how to remove even the need for a knowledge of M in applying the search algorithm.

The form (6.15) is useful as an exact expression for the number of oracle calls used to perform the quantum search algorithm, but it would be useful to have a simpler expression summarizing the essential behavior of R. To achieve this, note from (6.15) that $R \leq \lceil \pi/2\theta \rceil$, so a lower bound on θ will give an upper bound on R. Assuming for the moment that $M \leq N/2$, we have

$$\frac{\theta}{2} \geq \sin\frac{\theta}{2} = \sqrt{\frac{M}{N}}, \tag{6.16}$$

from which we obtain an elegant upper bound on the number of iterations required,

$$R \leq \left\lceil \frac{\pi}{4}\sqrt{\frac{N}{M}} \right\rceil. \tag{6.17}$$

That is, $R = O(\sqrt{N/M})$ Grover iterations (and thus oracle calls) must be performed in order to obtain a solution to the search problem with high probability, a quadratic improvement over the $O(N/M)$ oracle calls required classically. The quantum search algorithm is summarized below, for the case $M = 1$.

Algorithm: Quantum search

>   **Inputs:** (1) a black box oracle O which performs the transformation $O|x\rangle|q\rangle = |x\rangle|q \oplus f(x)\rangle$, where $f(x) = 0$ for all $0 \leq x < 2^n$ except x_0, for which $f(x_0) = 1$; (2) $n + 1$ qubits in the state $|0\rangle$.
>
>   **Outputs:** x_0.
>
>   **Runtime:** $O(\sqrt{2^n})$ operations. Succeeds with probability $O(1)$.
>
>   **Procedure:**
>
>   1. $|0\rangle^{\otimes n}|0\rangle$ initial state
>
>   2. $\rightarrow \dfrac{1}{\sqrt{2^n}}\displaystyle\sum_{x=0}^{2^n-1}|x\rangle\left[\dfrac{|0\rangle - |1\rangle}{\sqrt{2}}\right]$ apply $H^{\otimes n}$ to the first n qubits, and HX to the last qubit
>
>   3. $\rightarrow \left[(2|\psi\rangle\langle\psi| - I)O\right]^R \dfrac{1}{\sqrt{2^n}}\displaystyle\sum_{x=0}^{2^n-1}|x\rangle\left[\dfrac{|0\rangle - |1\rangle}{\sqrt{2}}\right]$ apply the Grover iteration $R \approx \lceil\pi\sqrt{2^n}/4\rceil$ times.
>
> $\approx |x_0\rangle\left[\dfrac{|0\rangle - |1\rangle}{\sqrt{2}}\right]$
>
>   4. $\rightarrow x_0$ measure the first n qubits

Exercise 6.4: Give explicit steps for the quantum search algorithm, as above, but for the case of multiple solutions ($1 < M < N/2$).

What happens when more than half the items are solutions to the search problem, that is, $M \geq N/2$? From the expression $\theta = \arcsin(2\sqrt{M(N-M)}/N)$ (compare (6.14)) we see that the angle θ gets smaller as M varies from $N/2$ to N. As a result, the number of iterations needed by the search algorithm *increases* with M, for $M \geq N/2$. Intuitively, this is a silly property for a search algorithm to have: we expect that it should become easier to find a solution to the problem as the number of solutions increases. There are at least two ways around this problem. If M is known in advance to be larger than $N/2$ then we can just randomly pick an item from the search space, and then check that it is a solution using the oracle. This approach has a success probability at least one-half, and only requires one consultation with the oracle. It has the disadvantage that we may not know the number of solutions M in advance.

In the case where it isn't known whether $M \geq N/2$, another approach can be used. This approach is interesting in its own right, and has a useful application to simplify the analysis of the quantum algorithm for counting the number of solutions to the search problem, as presented in Section 6.3. The idea is to double the number of elements in the search space by adding N extra items to the search space, none of which are solutions. As a consequence, less than half the items in the new search space are solutions. This is effected by adding a single qubit $|q\rangle$ to the search index, doubling the number of items to be searched to $2N$. A new *augmented* oracle O' is constructed which marks an item only if it is a solution to the search problem and the extra bit is set to zero. In Exercise 6.5 you will explain how the oracle O' may be constructed using one call to O. The new search problem has only M solutions out of $2N$ entries, so running the search algorithm with the new oracle O' we see that at most $R = \pi/4\sqrt{2N/M}$ calls to O' are required, and it follows that $O(\sqrt{N/M})$ calls to O are required to perform the search.

Exercise 6.5: Show that the augmented oracle O' may be constructed using one application of O, and elementary quantum gates, using the extra qubit $|q\rangle$.

The quantum search algorithm may be used in a wide variety of ways, some of which will be explored in subsequent sections. The great utility of the algorithm arises because we do not assume any particular structure to the search problems being performed. This is the great advantage of posing the problem in terms of a 'black box' oracle, and we adopt this point of view whenever convenient through the remainder of this chapter. In practical applications, of course, it is necessary to understand how the oracle is being implemented, and in each of the practical problems we concern ourselves with an explicit description of the oracle implementation is given.

Exercise 6.6: Verify that the gates in the dotted box in the second figure of Box 6.1 perform the conditional phase shift operation $2|00\rangle\langle00| - I$, up to an unimportant global phase factor.

6.2 Quantum search as a quantum simulation

The correctness of the quantum search algorithm is easily verified, but it is by no means obvious how one would dream up such an algorithm from a state of ignorance. In this section we sketch a heuristic means by which one can 'derive' the quantum search algorithm, in the hope of lending some intuition as to the tricky task of quantum algorithm design. As a useful side effect we also obtain a *deterministic* quantum search algorithm. Because our goal is to obtain insight rather than generality, we assume for the sake of simplicity that the search problem has exactly one solution, which we label x.

Our method involves two steps. First, we make a guess as to a *Hamiltonian* which

---

Box 6.1: Quantum search: a two-bit example

Here is an explicit example illustrating how the quantum search algorithm works on a search space of size $N = 4$. The oracle, for which $f(x) = 0$ for all x except $x = x_0$, in which case $f(x_0) = 1$, can be taken to be one of the four circuits

corresponding to $x_0 = 0, 1, 2$, or 3 from left to right, where the top two qubits carry the query x, and the bottom qubit carries the oracle's response. The quantum circuit which performs the initial Hadamard transforms and a single Grover iteration G is

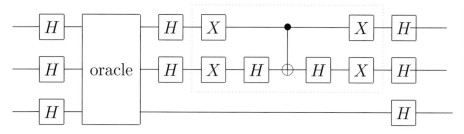

Initially, the top two qubits are prepared in the state $|0\rangle$, and the bottom one as $|1\rangle$. The gates in the dotted box perform the conditional phase shift operation $2|00\rangle\langle00| - I$. How many times must we repeat G to obtain x_0? From Equation (6.15), using $M = 1$, we find that less than one iteration is required. It turns out that because $\theta = \pi/3$ in (6.14), only *exactly one* iteration is required, to perfectly obtain x_0, in this special case. In the geometric picture of Figure 6.3, our initial state $|\psi\rangle = (|00\rangle + |01\rangle + |10\rangle + |11\rangle)/2$ is 30° from $|\alpha\rangle$, and a single rotation by $\theta = 60°$ moves $|\psi\rangle$ to $|\beta\rangle$. You can confirm for yourself directly, using the quantum circuits, that measurement of the top two qubits gives x_0, after using the oracle only once. In contrast, a classical computer – or classical circuit – trying to differentiate between the four oracles would require on average 2.25 oracle queries!

---

solves the search problem. More precisely, we write down a Hamiltonian H which depends on the solution x and an initial state $|\psi\rangle$ such that a quantum system evolving according to H will change from $|\psi\rangle$ to $|x\rangle$ after some prescribed time. Once we've found such a Hamiltonian and initial state, we can move on to the second step, which is to attempt to simulate the action of the Hamiltonian using a quantum circuit. Amazingly, following this procedure leads very quickly to the quantum search algorithm! We have already met this two-part procedure while studying universality in quantum circuits, in Problem 4.3, and it also serves well in the study of quantum searching.

We suppose that the algorithm starts with the quantum computer in a state $|\psi\rangle$. We'll tie down what $|\psi\rangle$ should be later on, but it is convenient to leave $|\psi\rangle$ undetermined until we understand the dynamics of the algorithm. The goal of quantum searching is to change $|\psi\rangle$ into $|x\rangle$ or some approximation thereof. What Hamiltonians might we guess do a good job of causing such an evolution? Simplicity suggests that we should guess a Hamiltonian constructed entirely from the terms $|\psi\rangle$ and $|x\rangle$. Thus, the Hamiltonian must be a sum of terms like $|\psi\rangle\langle\psi|$, $|x\rangle\langle x|$, $|\psi\rangle\langle x|$ and $|x\rangle\langle\psi|$. Perhaps the simplest choices along these lines are the Hamiltonians:

$$H = |x\rangle\langle x| + |\psi\rangle\langle\psi| \tag{6.18}$$
$$H = |x\rangle\langle\psi| + |\psi\rangle\langle x|. \tag{6.19}$$

It turns out that both these Hamiltonians lead to the quantum search algorithm! For now, however, we restrict ourselves to analyzing the Hamiltonian in Equation (6.18). Recall from Section 2.2.2 that after a time t, the state of a quantum system evolving according to the Hamiltonian H and initially in the state $|\psi\rangle$ is given by

$$\exp(-iHt)|\psi\rangle\,. \tag{6.20}$$

Intuitively it looks pretty good: for small t the effect of the evolution is to take $|\psi\rangle$ to $(I - itH)|\psi\rangle = (1 - it)|\psi\rangle - it\langle x|\psi\rangle|x\rangle$. That is, the $|\psi\rangle$ vector is rotated slightly, into the $|x\rangle$ direction. Let's actually do a full analysis, with the goal being to determine whether there is a t such that $\exp(-iHt)|\psi\rangle = |x\rangle$. Clearly we can restrict the analysis to the two-dimensional space spanned by $|x\rangle$ and $|\psi\rangle$. Performing the Gram–Schmidt procedure, we can find $|y\rangle$ such that $|x\rangle, |y\rangle$ forms an orthonormal basis for this space, and $|\psi\rangle = \alpha|x\rangle + \beta|y\rangle$, for some α, β such that $\alpha^2 + \beta^2 = 1$, and for convenience we have chosen the phases of $|x\rangle$ and $|y\rangle$ so that α and β are real and non-negative. In the $|x\rangle, |y\rangle$ basis we have

$$H = \begin{bmatrix} 1 & 0 \\ 0 & 0 \end{bmatrix} + \begin{bmatrix} \alpha^2 & \alpha\beta \\ \alpha\beta & \beta^2 \end{bmatrix} = \begin{bmatrix} 1+\alpha^2 & \alpha\beta \\ \alpha\beta & 1-\alpha^2 \end{bmatrix} = I + \alpha(\beta X + \alpha Z). \tag{6.21}$$

Thus

$$\exp(-iHt)|\psi\rangle = \exp(-it)\left[\cos(\alpha t)|\psi\rangle - i\sin(\alpha t)(\beta X + \alpha Z)|\psi\rangle\right]. \tag{6.22}$$

The global phase factor $\exp(-it)$ can be ignored, and simple algebra shows that $(\beta X + \alpha Z)|\psi\rangle = |x\rangle$, so the state of the system after a time t is

$$\cos(\alpha t)|\psi\rangle - i\sin(\alpha t)|x\rangle\,. \tag{6.23}$$

Thus, observation of the system at time $t = \pi/2\alpha$ yields the result $|x\rangle$ with probability *one*: we have found a solution to the search problem! Unfortunately, the time of the observation depends on α, the component of $|\psi\rangle$ in the $|x\rangle$ direction, and thus on x, which is what we are trying to determine. The obvious solution is to attempt to arrange α to be the same for all $|x\rangle$, that is, to choose $|\psi\rangle$ to be the uniform superposition state

$$|\psi\rangle = \frac{\sum_x |x\rangle}{\sqrt{N}}\,. \tag{6.24}$$

Making this choice gives $\alpha = 1/\sqrt{N}$ for all x, and thus the time of observation $t = \pi\sqrt{N}/2$ does not depend on knowing the value of x. Furthermore, the state (6.24) has the obvious advantage that we already know how to prepare such a state by doing a Hadamard transform.

We now know that the Hamiltonian (6.18) rotates the vector $|\psi\rangle$ to $|x\rangle$. Can we find a quantum circuit to simulate the Hamiltonian (6.18), and thus obtain a quantum search algorithm? Applying the method of Section 4.7, we see that a natural way of simulating H is to alternately simulate the Hamiltonians $H_1 \equiv |x\rangle\langle x|$ and $H_2 \equiv |\psi\rangle\langle\psi|$ for short time increments Δt. These Hamiltonians are easily simulated using the methods of Chapter 4, as illustrated in Figures 6.4 and 6.5.

Exercise 6.7: Verify that the circuits shown in Figures 6.4 and 6.5 implement the operations $\exp(-i|x\rangle\langle x|\Delta t)$ and $\exp(-i|\psi\rangle\langle\psi|\Delta t)$, respectively, with $|\psi\rangle$ as in (6.24).

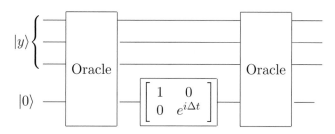

Figure 6.4. Circuit implementing the operation $\exp(-i|x\rangle\langle x|\Delta t)$ using two oracle calls.

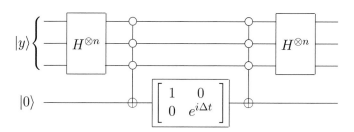

Figure 6.5. Circuit implementing the operation $\exp(-i|\psi\rangle\langle\psi|\Delta t)$, for $|\psi\rangle$ as in (6.24).

The number of oracle calls required by the quantum simulation is determined by how small a time-step is required to obtain reasonably accurate results. Suppose we use a simulation step of length Δt that is accurate to $O(\Delta t^2)$. The total number of steps required is $t/\Delta t = \Theta(\sqrt{N}/\Delta t)$, and thus the cumulative error is $O(\Delta t^2 \times \sqrt{N}/\Delta t) = O(\Delta t\sqrt{N})$. To obtain a reasonably high success probability we need the error to be $O(1)$, which means that we must choose $\Delta t = \Theta(1/\sqrt{N})$ which results in a number of oracle calls that scales like $O(N)$ – no better than the classical solution! What if we use a more accurate method of quantum simulation, say one that is accurate to $O(\Delta t^3)$? The cumulative error in this case is $O(\Delta t^2\sqrt{N})$, and thus to achieve a reasonable success probability we need to choose $\Delta t = \Theta(N^{-1/4})$, resulting in a total number of oracle calls $O(N^{3/4})$, which is a distinct improvement over the classical situation, although still not as good as achieved by the quantum search algorithm of Section 6.1! In general going to a more accurate quantum simulation technique results in a reduction in the number of oracle calls required to perform the simulation:

Exercise 6.8: Suppose the simulation step is performed to an accuracy $O(\Delta t^r)$. Show that the number of oracle calls required to simulate H to reasonable accuracy is $O(N^{r/2(r-1)})$. Note that as r becomes large the exponent of N approaches $1/2$.

We have been analyzing the accuracy of the quantum simulation of the Hamiltonian (6.18) using general results on quantum simulation from Section 4.7. Of course, in this instance we are dealing with a specific Hamiltonian, not the general case, which suggests that it might be interesting to calculate explicitly the effect of a simulation step of time Δt, rather than relying on the general analysis. We can do this for any specific choice of simulation method – it can be a little tedious to work out the effect of the simulation step, but it is essentially a straightforward calculation. The obvious starting point is to explicitly calculate the action of the lowest-order simulation techniques, that is, to calculate one or both of $\exp(-i|x\rangle\langle x|\Delta t)\exp(-i|\psi\rangle\langle\psi|\Delta t)$ or $\exp(-i|\psi\rangle\langle\psi|\Delta t)\exp(-i|x\rangle\langle x|\Delta t)$. The results are essentially the same in both instances; we will focus on the study of $U(\Delta t) \equiv \exp(-i|\psi\rangle\langle\psi|\Delta t)\exp(-i|x\rangle\langle x|\Delta t)$. $U(\Delta t)$ clearly acts non-trivially only in the space spanned by $|x\rangle\langle x|$ and $|\psi\rangle\langle\psi|$, so we restrict ourselves to that space, working in the basis $|x\rangle, |y\rangle$, where $|y\rangle$ is defined as before. Note that in this representation $|x\rangle\langle x| = (I + Z)/2 = (I + \hat{z} \cdot \vec{\sigma})/2$, where $\hat{z} \equiv (0,0,1)$ is the unit vector in the z direction, and $|\psi\rangle\langle\psi| = (I + \vec{\psi} \cdot \vec{\sigma})/2$, where $\vec{\psi} = (2\alpha\beta, 0, (\alpha^2 - \beta^2))$ (recall that this is the Bloch vector representation; see Section 4.2). A simple calculation shows that up to an unimportant global phase factor,

$$U(\Delta t) = \left(\cos^2\left(\frac{\Delta t}{2}\right) - \sin^2\left(\frac{\Delta t}{2}\right)\vec{\psi} \cdot \hat{z}\right) I$$

$$-2i\sin\left(\frac{\Delta t}{2}\right)\left(\cos\left(\frac{\Delta t}{2}\right)\frac{\vec{\psi} + \hat{z}}{2} + \sin\left(\frac{\Delta t}{2}\right)\frac{\vec{\psi} \times \hat{z}}{2}\right) \cdot \vec{\sigma}. \quad (6.25)$$

Exercise 6.9: Verify Equation (6.25). (*Hint*: see Exercise 4.15.)

Equation (6.25) implies that $U(\Delta t)$ is a rotation on the Bloch sphere about an axis of rotation $\vec{r}$ defined by

$$\vec{r} = \cos\left(\frac{\Delta t}{2}\right)\frac{\vec{\psi} + \hat{z}}{2} + \sin\left(\frac{\Delta t}{2}\right)\frac{\vec{\psi} \times \hat{z}}{2}, \quad (6.26)$$

and through an angle θ defined by

$$\cos\left(\frac{\theta}{2}\right) = \cos^2\left(\frac{\Delta t}{2}\right) - \sin^2\left(\frac{\Delta t}{2}\right)\vec{\psi} \cdot \hat{z}, \quad (6.27)$$

which simplifies upon substitution of $\vec{\psi} \cdot \hat{z} = \alpha^2 - \beta^2 = (2/N - 1)$ to

$$\cos\left(\frac{\theta}{2}\right) = 1 - \frac{2}{N}\sin^2\left(\frac{\Delta t}{2}\right). \quad (6.28)$$

Note that $\vec{\psi} \cdot \vec{r} = \hat{z} \cdot \vec{r}$, so both $|\psi\rangle\langle\psi|$ and $|x\rangle\langle x|$ lie on the same circle of revolution about the $\vec{r}$ axis on the Bloch sphere. Summarizing, the action of $U(\Delta t)$ is to rotate $|\psi\rangle\langle\psi|$ about the $\vec{r}$ axis, through an angle θ for each application of $U(\Delta t)$, as illustrated in Figure 6.6. We terminate the procedure when enough rotations have been performed to rotate $|\psi\rangle\langle\psi|$ near to the solution $|x\rangle\langle x|$. Now initially we imagined that Δt was small, since we were considering the case of quantum simulation, but Equation (6.28) shows that the smart thing to do is to choose $\Delta t = \pi$, in order to maximize the rotation angle θ. If we do this, then we obtain $\cos(\theta/2) = 1 - 2/N$, which for large N corresponds to $\theta \approx 4/\sqrt{N}$, and the number of oracle calls required to find the solution $|x\rangle$ is $O(\sqrt{N})$, just as for the original quantum search algorithm.

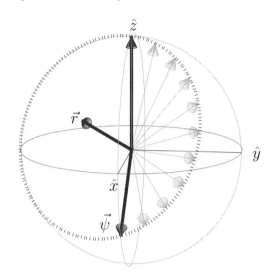

Figure 6.6. Bloch sphere diagram showing the initial state $\vec{\psi}$ rotating around the axis of rotation $\vec{r}$ going toward the final state $\hat{z}$.

Indeed, if we make the choice $\Delta t = \pi$, then this 'quantum simulation' is in fact identical with the original quantum search algorithm, since the operators applied in the quantum simulation are $\exp(-i\pi|\psi\rangle\langle\psi|) = I - 2|\psi\rangle\langle\psi|$ and $\exp(-i\pi|x\rangle\langle x|) = I - 2|x\rangle\langle x|$, and up to a global phase shift these are *identical* to the steps making up the Grover iteration. Viewed this way, the circuits shown in Figures 6.2 and 6.3 for the quantum search algorithm are simplifications of the circuits shown in Figures 6.4 and 6.5 for the simulation, in the special case $\Delta t = \pi$!

Exercise 6.10: Show that by choosing Δt appropriately we can obtain a quantum search algorithm which uses $O(\sqrt{N})$ queries, and for which the final state is $|x\rangle$ *exactly*, that is, the algorithm works with probability 1, rather than with some smaller probability.

We have re-derived the quantum search algorithm from a different point of view, the point of view of quantum simulation. Why did this approach work? Might it be used to find other fast quantum algorithms? We can't answer these questions in any definitive way, but the following few thoughts may be of some interest. The basic procedure used is four-fold: (1) specify the problem to be solved, including a description of the desired input and output from the quantum algorithm; (2) guess a Hamiltonian to solve the problem, and verify that it does in fact work; (3) find a procedure to simulate the Hamiltonian; and (4) analyze the resource costs of the simulation. This is different from the more conventional approach in two respects: we guess a Hamiltonian, rather than a quantum circuit, and there is no analogue to the simulation step in the conventional approach. The more important of these two differences is the first. There is a great deal of freedom in specifying a quantum circuit to solve a problem. While that freedom is, in part, responsible for the great power of quantum computation, it makes searching for good circuits rather difficult. By contrast, specifying a Hamiltonian is a much more constrained problem, and therefore affords less freedom in the solution of problems, but those same constraints may in fact make it much easier to find an efficient quantum algorithm to solve a problem. We've seen this happen for the quantum search algorithm, and perhaps other quantum algorithms will be discovered by this method; we don't know. What seems certain is that this 'quantum algorithms as quantum simulations' point of view offers a useful alternative viewpoint to stimulate in the development of quantum algorithms.

Exercise 6.11: (Multiple solution continuous quantum search) Guess a
Hamiltonian with which one may solve the continuous time search problem in the case where the search problem has M solutions.

Exercise 6.12: (Alternative Hamiltonian for quantum search) Suppose

$$H = |x\rangle\langle\psi| + |\psi\rangle\langle x| . \tag{6.29}$$

(1) Show that it takes time $O(1)$ to rotate from the state $|\psi\rangle$ to the state $|x\rangle$, given an evolution according to the Hamiltonian H.
(2) Explain how a quantum simulation of the Hamiltonian H may be performed, and determine the number of oracle calls your simulation technique requires to obtain the solution with high probability.

6.3 Quantum counting

How quickly can we determine the number of solutions, M, to an N item search problem, if M is not known in advance? Clearly, on a classical computer it takes $\Theta(N)$ consultations with an oracle to determine M. On a quantum computer it is possible to estimate the number of solutions much more quickly than is possible on a classical computer by combining the Grover iteration with the phase estimation technique based upon the quantum Fourier transform (Chapter 5). This has some important applications. First, if we can estimate the number of solutions quickly then it is also possible to find a solution quickly, even if the number of solutions is unknown, by first counting the number of solutions, and then applying the quantum search algorithm to find a solution. Second, quantum counting allows us to decide whether or not a solution even exists, depending on whether the number of solutions is zero, or non-zero. This has applications, for example, to the solution of **NP**-complete problems, which may be phrased in terms of the existence of a solution to a search problem.

Exercise 6.13: Consider a classical algorithm for the counting problem which samples uniformly and independently k times from the search space, and let $X_1, \ldots, X_k$ be the results of the oracle calls, that is, $X_j = 1$ if the jth oracle call revealed a solution to the problem, and $X_j = 0$ if the jth oracle call did not reveal a solution to the problem. This algorithm returns the estimate $S \equiv N \times \sum_j X_j/k$ for the number of solutions to the search problem. Show that the standard deviation in S is $\Delta S = \sqrt{M(N-M)/k}$. Prove that to obtain a probability at least $3/4$ of estimating M correctly to within an accuracy $\sqrt{M}$ for all values of M we must have $k = \Omega(N)$.

Exercise 6.14: Prove that *any* classical counting algorithm with a probability at least $3/4$ for estimating M correctly to within an accuracy $c\sqrt{M}$ for some constant c and for all values of M must make $\Omega(N)$ oracle calls.

Quantum counting is an application of the phase estimation procedure of Section 5.2 to estimate the eigenvalues of the Grover iteration G, which in turn enables us to determine the number of solutions M to the search problem. Suppose $|a\rangle$ and $|b\rangle$ are the two eigenvectors of the Grover iteration in the space spanned by $|\alpha\rangle$ and $|\beta\rangle$. Let θ be the angle of rotation determined by the Grover iteration. From Equation (6.13) we see that the corresponding eigenvalues are $e^{i\theta}$ and $e^{i(2\pi-\theta)}$. For ease of analysis it is convenient to assume that the oracle has been augmented, as described in Section 6.1, expanding the size of the search space to $2N$, and ensuring that $\sin^2(\theta/2) = M/2N$.

The phase estimation circuit used for quantum counting is shown in Figure 6.7. The function of the circuit is to estimate θ to m bits of accuracy, with a probability of success at least $1 - \epsilon$. The first register contains $t \equiv m + \lceil \log(2 + 1/2\epsilon) \rceil$ qubits, as per the phase estimation algorithm, and the second register contains $n + 1$ qubits, enough to implement the Grover iteration on the augmented search space. The state of the second register is initialized to an equal superposition of all possible inputs $\sum_x |x\rangle$ by a Hadamard transform. As we saw in Section 6.1 this state is a superposition of the eigenstates $|a\rangle$ and $|b\rangle$, so by the results of Section 5.2 the circuit in Figure 6.7 gives us an estimate of θ or $2\pi - \theta$ accurate to within $|\Delta\theta| \leq 2^{-m}$, with probability at least $1 - \epsilon$. Furthermore, an estimate for $2\pi - \theta$ is clearly equivalent to an estimate of θ with the same level of accuracy, so effectively the phase estimation algorithm determines θ to an accuracy 2^{-m} with probability $1 - \epsilon$.

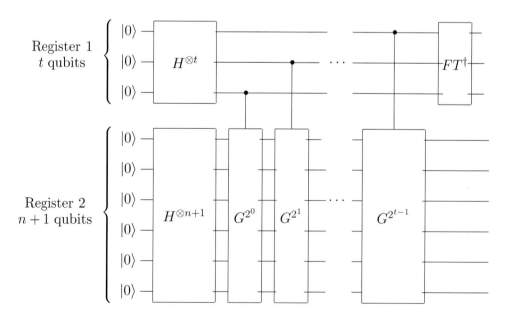

Figure 6.7. Circuit for performing approximate quantum counting on a quantum computer.

Using the equation $\sin^2(\theta/2) = M/2N$ and our estimate for θ we obtain an estimate of the number of solutions, M. How large an error, ΔM, is there in this estimate?

$$\frac{|\Delta M|}{2N} = \left| \sin^2\left(\frac{\theta + \Delta\theta}{2}\right) - \sin^2\left(\frac{\theta}{2}\right) \right| \tag{6.30}$$

$$= \left(\sin\left(\frac{\theta + \Delta\theta}{2}\right) + \sin\left(\frac{\theta}{2}\right) \right) \left| \sin\left(\frac{\theta + \Delta\theta}{2}\right) - \sin\left(\frac{\theta}{2}\right) \right|. \tag{6.31}$$

Calculus implies that $|\sin((\theta + \Delta\theta)/2) - \sin(\theta/2)| \le |\Delta\theta|/2$, and elementary trigonometry that $|\sin((\theta + \Delta\theta)/2)| < \sin(\theta/2) + |\Delta\theta|/2$, so

$$\frac{|\Delta M|}{2N} < \left(2\sin\left(\frac{\theta}{2}\right) + \frac{|\Delta\theta|}{2} \right) \frac{|\Delta\theta|}{2}. \tag{6.32}$$

Substituting $\sin^2(\theta/2) = M/2N$ and $|\Delta\theta| \le 2^{-m}$ gives our final estimate for the error in our estimate of M,

$$|\Delta M| < \left(\sqrt{2MN} + \frac{N}{2^{m+1}} \right) 2^{-m}. \tag{6.33}$$

As an example, suppose we choose $m = \lceil n/2 \rceil + 1$, and $\epsilon = 1/6$. Then we have $t = \lceil n/2 \rceil + 3$, so the algorithm requires $\Theta(\sqrt{N})$ Grover iterations, and thus $\Theta(\sqrt{N})$ oracle calls. By (6.33) our accuracy is $|\Delta M| < \sqrt{M/2} + 1/4 = O(\sqrt{M})$. Compare this with Exercise 6.14, according to which it would have required $O(N)$ oracle calls to obtain a similar accuracy on a classical computer.

Indeed, the example just described serves double duty as an algorithm for determining whether a solution to the search problem exists at all, that is, whether $M = 0$ or $M \ne 0$. If $M = 0$ then we have $|\Delta M| < 1/4$, so the algorithm must produce the estimate zero with probability at least $5/6$. Conversely, if $M \ne 0$ then it is easy to verify that the estimate for M is not equal to 0 with probability at least $5/6$.

Another application of quantum counting is to *find* a solution to a search problem when the number M of solutions is unknown. The difficulty in applying the quantum search algorithm as described in Section 6.1 is that the number of times to repeat the Grover iteration, Equation (6.15), depends on knowing the number of solutions M. This problem can be alleviated by using the quantum counting algorithm to first estimate θ and M to high accuracy using phase estimation, and then to apply the quantum search algorithm as in Section 6.1, repeating the Grover iteration a number of times determined by (6.15), with the estimates for θ and M obtained by phase estimation substituted to determine R. The angular error in this case is at most $\pi/4(1 + |\Delta\theta|/\theta)$, so choosing $m = \lceil n/2 \rceil + 1$ as before gives an angular error at most $\pi/4 \times 3/2 = 3\pi/8$, which corresponds to a success probability of at least $\cos^2(3\pi/8) = 1/2 - 1/2\sqrt{2} \approx 0.15$ for the search algorithm. If the probability of obtaining an estimate of θ this accurate is $5/6$, as in our earlier example, then the total probability of obtaining a solution to the search problem is $5/6 \times \cos^2(3\pi/8) \approx 0.12$, a probability which may quickly be boosted close to 1 by a few repetitions of the combined counting–search procedure.

6.4 Speeding up the solution of NP-complete problems

Quantum searching may be used to speed up the solution to problems in the complexity class **NP** (Section 3.2.2). We already saw, in Section 6.1.1, how factoring can be sped up; here, we illustrate how quantum search can be applied to assist the solution of the Hamiltonian cycle problem (HC). Recall that a Hamiltonian cycle of a graph is a simple cycle which visits every vertex of the graph. The HC problem is to determine whether a given graph has a Hamiltonian cycle or not. This problem belongs to the class of **NP**-complete problems, widely believed (but not yet proved) to be intractable on a classical computer.

A simple algorithm to solve HC is to perform a search through all possible orderings of the vertices:

(1) Generate each possible ordering $(v_1, \ldots, v_n)$ of vertices for the graph. Repetitions will be allowed, as they ease the analysis without affecting the essential result.
(2) For each ordering check to see whether it is a Hamiltonian cycle for the graph. If not, continue checking the orderings.

Since there are $n^n = 2^{n \log n}$ possible orderings of the vertices which must be searched, this algorithm requires $2^{n \log n}$ checks for the Hamiltonian cycle property in the worst case. Indeed, any problem in **NP** may be solved in a similar way: if a problem of size n has witnesses which can be specified using $w(n)$ bits, where $w(n)$ is some polynomial in n, then searching through all $2^{w(n)}$ possible witnesses will reveal a solution to the problem, if one exists.

The quantum search algorithm may be used to speed up this algorithm by increasing the speed of the search. Specifically, we use the algorithm described in Section 6.3 to determine whether a solution to the search problem exists. Let $m \equiv \lceil \log n \rceil$. The search space for the algorithm will be represented by a string of mn qubits, each block of m qubits being used to store the index to a single vertex. Thus we can write the computational basis states as $|v_1, \ldots, v_n\rangle$, where each $|v_i\rangle$ is represented by the appropriate string of m qubits, for a total of nm qubits. The oracle for the search algorithm must apply the transformation:

$$O|v_1, \ldots, v_n\rangle = \left\{ \begin{array}{ll} |v_1, \ldots, v_n\rangle & \text{if } v_1, \ldots, v_n \text{ is not a Hamiltonian cycle} \\ -|v_1, \ldots, v_n\rangle & \text{if } v_1, \ldots, v_n \text{ is a Hamiltonian cycle} \end{array} \right. \tag{6.34}$$

Such an oracle is easy to design and implement when one has a description of the graph. One takes a polynomial size classical circuit recognizing Hamiltonian cycles in the graph, and converts it to a reversible circuit, also of polynomial size, computing the transformation $(v_1, \ldots, v_n, q) \rightarrow (v_1, \ldots, v_n, q \oplus f(v_1, \ldots, v_n))$, where $f(v_1, \ldots, v_n) = 1$ if $v_1, \ldots, v_n$ is a Hamiltonian cycle, and is 0 otherwise. Implementing the corresponding circuit on a quantum computer with the final qubit starting in the state $(|0\rangle - |1\rangle)/\sqrt{2}$ gives the desired transformation. We won't explicitly describe the details here, except to note the key point: the oracle requires a number of gates polynomial in n, as a direct consequence of the fact that Hamiltonian cycles can be recognized using polynomially many gates classically. Applying the variant of the search algorithm which determines whether a solution to the search problem exists (Section 6.3) we see that it takes $O(2^{mn/2}) = O(2^{n\lceil \log n \rceil/2})$ applications of the oracle to determine whether a Hamiltonian cycle exists. When one does exist it is easy to apply the combined counting–search algorithm to find an example of such a cycle, which can then be exhibited as a witness for the problem.

To summarize:

- The classical algorithm requires $O\left(p(n)2^{n\lceil \log n \rceil}\right)$ operations to determine whether a

Hamiltonian cycle exists, where the polynomial factor $p(n)$ is overhead predominantly due to the implementation of the oracle, that is, the gates checking whether a candidate path is Hamiltonian or not. The dominant effect in determining the resources required is the exponent in $2^{n\lceil \log n \rceil}$. The classical algorithm is deterministic, that is, it succeeds with probability 1.

- The quantum algorithm requires $O\left(p(n)2^{n\lceil \log n \rceil/2}\right)$ operations to determine whether a Hamiltonian cycle exists. Once again, the polynomial $p(n)$ is overhead predominantly due to implementation of the oracle. The dominant effect in determining the resources required is the exponent in $2^{n\lceil \log n \rceil/2}$. There is a constant probability (say, $1/6$) of error for the algorithm, which may be reduced to $1/6^r$ by r repetitions of the algorithm.

- Asymptotically the quantum algorithm requires the *square root* of the number of operations the classical algorithm requires.

6.5 Quantum search of an unstructured database

Suppose somebody gives you a list containing one thousand flower names, and asks you where 'Perth Rose' appears on the list. If the flower appears exactly once on the list, and the list is not ordered in any obvious way, then you will need to examine five hundred names, on average, before you find the 'Perth Rose'. Might it be possible to speed up this kind of *database searching* using the quantum search algorithm? Indeed, the quantum search algorithm is sometimes referred to as a database search algorithm, but its usefulness for that application is limited, and based on certain assumptions. In this section we take a look at how the quantum search algorithm can *conceptually* be used to search an unstructured database, in a setting rather like that found on a conventional computer. The picture we construct will clarify what resources are required to enable a quantum computer to search classical databases.

Suppose we have a database containing $N \equiv 2^n$ items, each of length l bits. We will label these items $d_1, \ldots, d_N$. We want to determine where a particular l bit string, s, is in the database. A classical computer, used to solve this problem, is typically split into two parts, illustrated in Figure 6.8. One is the *central processing unit*, or CPU, where data manipulation takes place, using a small amount of temporary memory. The second part is a large *memory* which stores the database in a string of 2^n blocks of l bit cells. The memory is assumed to be passive, in the sense that it is not capable of processing data on its own. What is possible is to LOAD data from memory into the CPU, and STORE data from the CPU in memory, and to do manipulations of the data stored temporarily in the CPU. Of course, classical computers may be designed along different lines, but this CPU–memory split is a popular and common architecture.

To find out where a given string s is in the unstructured database, the most efficient classical algorithm is as follows. First, an n-bit index to the database elements is set up in the CPU. We assume that the CPU is large enough to store the $n \equiv \lceil \log N \rceil$ bit index. The index starts out at zero, and is incremented by one on each iteration of the algorithm. At each iteration, the database entry corresponding to the index is loaded into the CPU, and compared to the string which is being searched for. If they are the same, the algorithm outputs the value of the index and halts. If not, the algorithm continues incrementing the index. Obviously, this algorithm requires that items be loaded from

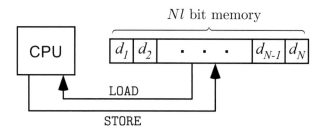

Figure 6.8. Classical database searching on a computer with distinct central processing unit (CPU) and memory. Only two operations may be directly performed on the memory – a memory element may be LOADed into the CPU, or an item from the CPU may be STOREd in memory.

memory 2^n times in the worst case. It is also clear that this is the most efficient possible algorithm for solving the problem in this model of computation.

How efficiently can an analogous algorithm be implemented on a quantum computer? And, even if a quantum speedup is possible, how useful is such an algorithm? We show first that a speedup is possible, and then return to the question of the utility of such an algorithm. Suppose our quantum computer consists of two units, just like the classical computer, a CPU and a memory. We assume that the CPU contains four registers: (1) an n qubit 'index' register initialized to $|0\rangle$; (2) an l qubit register initialized to $|s\rangle$ and remaining in that state for the entire computation; (3) an l qubit 'data' register initialized to $|0\rangle$; and (4) a 1 qubit register initialized to $(|0\rangle - |1\rangle)/\sqrt{2}$.

The memory unit can be implemented in one of two ways. The simplest is a quantum memory containing $N = 2^n$ cells of l qubits each, containing the database entries $|d_x\rangle$. The second implementation is to implement the memory as a *classical* memory with $N = 2^n$ cells of l bits each, containing the database entries d_x. Unlike a traditional classical memory, however, it can be addressed by an index x which can be in a superposition of multiple values. This quantum index allows a superposition of cell values to be LOADed from memory. Memory access works in the following way: if the CPU's index register is in the state $|x\rangle$ and the data register is in the state $|d\rangle$, then the contents d_x of the xth memory cell are added to the data register: $|d\rangle \rightarrow |d \oplus d_x\rangle$, where the addition is done bitwise, modulo 2. First, let us see how this capability is used to perform quantum search, then we shall discuss how such a memory might be physically constructed.

The key part of implementing the quantum search algorithm is realization of the oracle, which must flip the phase of the index which locates s in the memory. Suppose the CPU is in the state

$$|x\rangle|s\rangle|0\rangle \frac{|0\rangle - |1\rangle}{\sqrt{2}}. \tag{6.35}$$

Applying the LOAD operation puts the computer in the state

$$|x\rangle|s\rangle|d_x\rangle \frac{|0\rangle - |1\rangle}{\sqrt{2}}. \tag{6.36}$$

Now the second and third registers are compared, and if they are the same, then a bit flip is applied to register 4; otherwise nothing is changed. The effect of this operation is

$$|x\rangle|s\rangle|d_x\rangle\frac{|0\rangle - |1\rangle}{\sqrt{2}} \rightarrow \begin{cases} -|x\rangle|s\rangle|d_x\rangle\dfrac{|0\rangle - |1\rangle}{\sqrt{2}} & \text{if } d_x = s \\[2mm] |x\rangle|s\rangle|d_x\rangle\dfrac{|0\rangle - |1\rangle}{\sqrt{2}} & \text{if } d_x \neq s. \end{cases} \tag{6.37}$$

The data register is then restored to the state $|0\rangle$ by performing the LOAD operation again. The total action of the oracle thus leaves registers 2, 3 and 4 unaffected, and unentangled with register 1. Thus, the overall effect is to take the state of register 1 from $|x\rangle$ to $-|x\rangle$ if $d_x = s$, and to leave the register alone otherwise. Using the oracle implemented in this way, we may apply the quantum search algorithm to determine the location of s in the database, using $O(\sqrt{N})$ LOAD operations, compared to the N LOAD operations that were required classically.

In order for the oracle to function correctly on superpositions it seems at first glance as though the memory needs to be quantum mechanical. In fact, as we noted above, with some caveats the memory can actually be implemented classically, which likely makes it much more resistant to the effects of noise. But a quantum *addressing* scheme is still needed; a conceptual picture illustrating how this might be done is shown in Figure 6.9. The principle of operation is a means by which the binary encoded state of the quantum index (where 0 to $2^n - 1$ is represented by n qubits) is translated into a unary encoding (where 0 to $2^n - 1$ is represented by the position of a single probe within 2^n possible locations) which addresses the classical database. The database effects a change on a degree of freedom within the probe which is unrelated to its position. The binary to unary encoding is then reversed, leaving the data register with the desired contents.

Are there practical instances in which the quantum search algorithm could be useful for searching classical databases? Two distinct points may be made. First, databases are not ordinarily unstructured. Simple databases, like one containing flower names discussed in the introduction to this section, may be maintained in alphabetical order, such that a binary search can be used to locate an item in time which is $O(\log(N))$ for an N-element database. However, some databases may require a much more complex structure, and although sophisticated techniques exist to optimize classical searches, given queries of a sufficiently complex or unanticipated nature, a predetermined structure may not be of assistance, and the problem can be regarded as being essentially the unstructured database search problem we discussed.

Second, for a quantum computer to be able to search a classical database, a quantum addressing scheme is required. The scheme we depicted requires $O(N \log N)$ quantum switches – about the same amount of hardware as would be required to store the database itself. Presumably, these switches may someday be as simple and inexpensive as classical memory elements, but if that is not the case, then building a quantum computer to perform a quantum search may not be economically advantageous, compared with using classical computing hardware distributed over the memory elements.

Given these considerations, it appears that the principle use of quantum search algorithms will not be in searching classical databases. Rather, their use will probably be in searching for solutions to hard problems, as discussed in the last section, such as the Hamiltonian cycle, traveling salesman, and satisfiability problems.

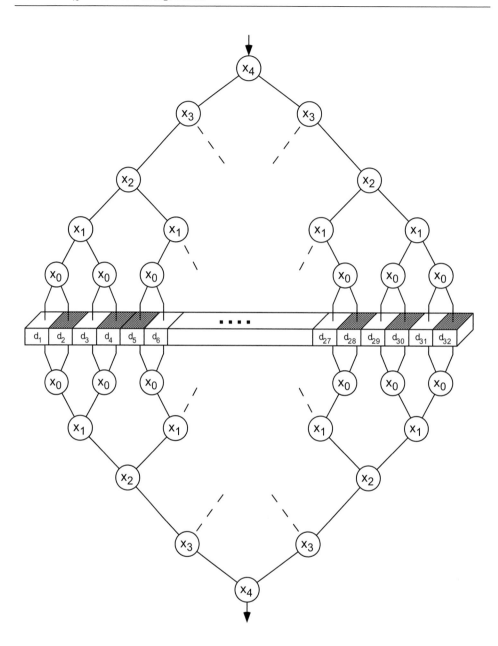

Figure 6.9. Conceptual diagram of a 32 cell classical memory with a five qubit quantum addressing scheme. Each circle represents a switch, addressed by the qubit inscribed within. For example, when $|x_4\rangle = |0\rangle$, the corresponding switch routes the input qubit towards the left; when $|x_4\rangle = |1\rangle$ the switch routes the input qubit to the right. If $|x_4\rangle = (|0\rangle + |1\rangle)/\sqrt{2}$, then an equal superposition of both routes is taken. The data register qubits enter at the top of the tree, and are routed down to the database, which changes their state according to the contents of the memory. The qubits are then routed back into a definite position, leaving them with the retrieved information. Physically, this could be realized using, for example, single photons for the data register qubits, which are steered using nonlinear interferometers (Chapter 7). The classical database could be just a simple sheet of plastic in which a 'zero' (illustrated as white squares) transmits light unchanged, and a 'one' (shaded squares) shifts the polarization of the incident light by $90°$.

6.6 Optimality of the search algorithm

We have shown that a quantum computer can search N items, consulting the search oracle only $O(\sqrt{N})$ times. We now prove that no quantum algorithm can perform this task using fewer than $\Omega(\sqrt{N})$ accesses to the search oracle, and thus the algorithm we have demonstrated is optimal.

Suppose the algorithm starts in the state $|\psi\rangle$. For simplicity, we prove the lower bound for the case where the search problem has a single solution, x. To determine x we are allowed to apply an oracle O_x which gives a phase shift of -1 to the solution $|x\rangle$ and leaves all other states invariant, $O_x = I - 2|x\rangle\langle x|$. We suppose the algorithm starts in a state $|\psi\rangle$ and applies the oracle O_x exactly k times, with unitary operations $U_1, U_2, \ldots, U_k$ interleaved between the oracle operations. Define

$$|\psi_k^x\rangle \equiv U_k O_x U_{k-1} O_x \ldots U_1 O_x |\psi\rangle \tag{6.38}$$
$$|\psi_k\rangle \equiv U_k U_{k-1} \ldots U_1 |\psi\rangle . \tag{6.39}$$

That is, $|\psi_k\rangle$ is the state that results when the sequence of unitary operations $U_1, \ldots, U_k$ is carried out, *without* the oracle operations. Let $|\psi_0\rangle = |\psi\rangle$. Our goal will be to bound the quantity

$$D_k \equiv \sum_x \| \psi_k^x - \psi_k \|^2 , \tag{6.40}$$

where we use the notation ψ for $|\psi\rangle$ as a convenience to simplify formulas. Intuitively, D_k is a measure of the *deviation after k steps* caused by the oracle, from the evolution that would otherwise have ensued. If this quantity is small, then all the states $|\psi_k^x\rangle$ are roughly the same, and it is not possible to correctly identify x with high probability. The strategy of the proof is to demonstrate two things: (a) a bound on D_k that shows it can grow no faster than $O(k^2)$; and (b) a proof that D_k must be $\Omega(N)$ if it is to be possible to distinguish N alternatives. Combining these two results gives the desired lower bound.

First, we give an inductive proof that $D_k \leq 4k^2$. This is clearly true for $k = 0$, where $D_k = 0$. Note that

$$D_{k+1} = \sum_x \| O_x \psi_k^x - \psi_k \|^2 \tag{6.41}$$

$$= \sum_x \| O_x(\psi_k^x - \psi_k) + (O_x - I)\psi_k \|^2. \tag{6.42}$$

Applying $\|b+c\|^2 \leq \|b\|^2 + 2\|b\| \, \|c\| + \|c\|^2$ with $b \equiv O_x(\psi_k^x - \psi_k)$ and $c \equiv (O_x - I)\psi_k = -2\langle x|\psi_k\rangle|x\rangle$, gives

$$D_{k+1} \leq \sum_x \left(\|\psi_k^x - \psi_k\|^2 + 4\|\psi_k^x - \psi_k\| \, |\langle x|\psi_k\rangle| + 4|\langle\psi_k|x\rangle|^2 \right) . \tag{6.43}$$

Applying the Cauchy–Schwarz inequality to the second term on the right hand side, and noting that $\sum_x |\langle x|\psi_k\rangle|^2 = 1$ gives

$$D_{k+1} \leq D_k + 4 \left(\sum_x \|\psi_k^x - \psi_k\|^2 \right)^{\frac{1}{2}} \left(\sum_{x'} |\langle\psi_k|x'\rangle|^2 \right)^{\frac{1}{2}} + 4 \tag{6.44}$$

$$\leq D_k + 4\sqrt{D_k} + 4. \tag{6.45}$$

By the inductive hypothesis $D_k \leq 4k^2$ we obtain

$$D_{k+1} \leq 4k^2 + 8k + 4 = 4(k+1)^2, \tag{6.46}$$

which completes the induction.

To complete the proof we need to show that the probability of success can only be high if D_k is $\Omega(N)$. We suppose $|\langle x|\psi_k^x\rangle|^2 \geq 1/2$ for all x, so that an observation yields a solution to the search problem with probability at least one-half. Replacing $|x\rangle$ by $e^{i\theta}|x\rangle$ does not change the probability of success, so without loss of generality we may assume that $\langle x|\psi_k^x\rangle = |\langle x|\psi_k^x\rangle|$, and therefore

$$\|\psi_k^x - x\|^2 = 2 - 2|\langle x|\psi_k^x\rangle| \leq 2 - \sqrt{2}. \tag{6.47}$$

Defining $E_k \equiv \sum_x \|\psi_k^x - x\|^2$ we see that $E_k \leq (2 - \sqrt{2})N$. We are now in position to prove that D_k is $\Omega(N)$. Defining $F_k \equiv \sum_x \|x - \psi_k\|^2$ we have

$$D_k = \sum_x \|(\psi_k^x - x) + (x - \psi_k)\|^2 \tag{6.48}$$

$$\geq \sum_x \|\psi_k^x - x\|^2 - 2\sum_x \|\psi_k^x - x\| \|x - \psi_k\| + \sum_x \|x - \psi_k\|^2 \tag{6.49}$$

$$= E_k + F_k - 2\sum_x \|\psi_k^x - x\| \|x - \psi_k\|. \tag{6.50}$$

Applying the Cauchy–Schwarz inequality gives $\sum_x \|\psi_k^x - x\| \|x - \psi_k\| \leq \sqrt{E_k F_k}$, so we have

$$D_k \geq E_k + F_k - 2\sqrt{E_k F_k} = (\sqrt{F_k} - \sqrt{E_k})^2. \tag{6.51}$$

In Exercise 6.15 you will show that $F_k \geq 2N - 2\sqrt{N}$. Combining this with the result $E_k \leq (2 - \sqrt{2})N$ gives $D_k \geq cN$ for sufficiently large N, where c is any constant less than $(\sqrt{2} - \sqrt{2 - \sqrt{2}})^2 \approx 0.42$. Since $D_k \leq 4k^2$ this implies that

$$k \geq \sqrt{cN/4}. \tag{6.52}$$

Summarizing, to achieve a probability of success at least one-half for finding a solution to the search problem we must call the oracle $\Omega(\sqrt{N})$ times.

Exercise 6.15: Use the Cauchy–Schwarz inequality to show that for any normalized state vector $|\psi\rangle$ and set of N orthonormal basis vectors $|x\rangle$,

$$\sum_x \|\psi - x\|^2 \geq 2N - 2\sqrt{N}. \tag{6.53}$$

Exercise 6.16: Suppose we merely require that the probability of an error being made is less than $1/2$ when averaged uniformly over the possible values for x, instead of for all values of x. Show that $O(\sqrt{N})$ oracle calls are still required to solve the search problem.

This result, that the quantum search algorithm is essentially optimal, is both exciting and disappointing. It is exciting because it tells us that for this problem, at least, we have fully plumbed the depths of quantum mechanics; no further improvement is possible. The disappointment arises because we might have hoped to do much better than the square root speedup offered by the quantum search algorithm. The sort of dream result we might have hoped for *a priori* is that it would be possible to search an N item search space using $O(\log N)$ oracle calls. If such an algorithm existed, it would allow us to solve **NP**-complete problems efficiently on a quantum computer, since it could search all $2^{w(n)}$ possible witnesses using roughly $w(n)$ oracle calls, where the polynomial $w(n)$ is the length of a witness in bits. Unfortunately, such an algorithm is not possible. This is useful information for would-be algorithm designers, since it indicates that a naive search-based method for attacking **NP**-complete problems is guaranteed to fail.

Venturing into the realm of opinion, we note that many researchers believe that the essential reason for the difficulty of **NP**-complete problems is that their search space has essentially no structure, and that (up to polynomial factors) the best possible method for solving such a problem is to adopt a search method. If one takes this point of view, then it is bad news for quantum computing, indicating that the class of problems efficiently soluble on a quantum computer, **BQP**, does not contain the **NP**-complete problems. Of course, this is merely opinion, and it is still possible that the **NP**-complete problems contain some unknown structure that allows them to be efficiently solved on a quantum computer, or perhaps even on a classical computer. A nice example to illustrate this point is the problem of factoring, widely believed to be in the class **NPI** of problems intermediate in difficulty between **P** and the **NP**-complete problems. The key to the efficient quantum mechanical solution of the factoring problem was the exploitation of a structure 'hidden' within the problem – a structure revealed by the reduction to order-finding. Even with this amazing structure revealed, it has not been found possible to exploit the structure to develop an efficient classical algorithm for factoring, although, of course, quantum mechanically the structure can be harnessed to give an efficient factoring algorithm! Perhaps a similar structure lurks in other problems suspected to be in **NPI**, such as the graph isomorphism problem, or perhaps even in the **NP**-complete problems themselves.

Exercise 6.17: (Optimality for multiple solutions) Suppose the search problem has M solutions. Show that $O(\sqrt{N/M})$ oracle applications are required to find a solution.

6.7 Black box algorithm limits

We conclude this chapter with a generalization of the quantum search algorithm which provides insightful bounds on the power of quantum computation. At the beginning of the chapter, we described the search problem as finding an n-bit integer x such that the function $f : \{0,1\}^n \to \{0,1\}$ evaluates to $f(x) = 1$. Related to this is the *decision* problem of whether or not there exists x such that $f(x) = 1$. Solving this decision problem is equivalently difficult, and can be expressed as computing the Boolean function $F(X) = X_0 \vee X_1 \vee \cdots \vee X_{N-1}$, where $\vee$ denotes the binary OR operation, $X_k \equiv f(k)$, and X denotes the set $\{X_0, X_1, \ldots, X_{N-1}\}$. More generally, we may wish to compute some function other than OR. For example, $F(X)$ could be the AND, PARITY (sum modulo two), or MAJORITY ($F(X) = 1$ if and only if more $X_k = 1$ than not) functions. In general, we can consider F to be any Boolean function. How fast (measured in number of queries) can a computer, classical or quantum, compute these functions, given an oracle for f?

It might seem difficult to answer such questions without knowing something about the function f, but in fact a great deal can be determined even in this 'black box' model, where the means by which the oracle accomplishes its task is taken for granted, and complexity is measured only in terms of the number of required oracle queries. The analysis of the search algorithm in the previous sections demonstrated one way to approach such problems, but a more powerful approach for obtaining query complexities is the *method of polynomials*, which we now briefly describe.

Let us begin with some useful definitions. The deterministic query complexity $D(F)$ is the minimum number of oracle queries a classical computer must perform to compute F with certainty. The quantum equivalent, $Q_E(F)$, is the minimum number of oracle queries a *quantum* computer requires to compute F with certainty. Since a quantum computer produces probabilistic outputs by nature, a more interesting quantity is the bounded error complexity $Q_2(F)$, the minimum number of oracle queries a quantum computer requires to produce an output which equals F with probability at least $2/3$. (The $2/3$ is an arbitrary number – the probability need only be bounded finitely away from $1/2$ in order to be boosted close to 1 by repetitions.) A related measure is the zero-error complexity $Q_0(F)$, the minimum number of oracle queries a quantum computer requires to produce an output which either equals F with certainty, or, with probability less than $1/2$, an admission of an inconclusive result. All these bounds must hold for any oracle function f (or in other words, any input X into F). Note that $Q_2(F) \leq Q_0(F) \leq Q_E(F) \leq D(F) \leq N$.

The method of polynomials is based upon the properties of minimum-degree multilinear polynomials (over the *real numbers*) which *represent* Boolean functions. All the polynomials we shall consider below are functions of $X_k \in \{0, 1\}$ and are thus multilinear, since $X_k^2 = X_k$. We say that a polynomial $p : \mathbf{R}^N \to \mathbf{R}$ represents F if $p(X) = F(X)$ for all $X \in \{0, 1\}^N$ (where $\mathbf{R}$ denotes the real numbers). Such a polynomial p always exists, since we can explicitly construct a suitable candidate:

$$p(X) = \sum_{Y \in \{0,1\}^N} F(Y) \prod_{k=0}^{N-1} \left[1 - (Y_k - X_k)^2\right] . \tag{6.54}$$

That the minimum degree p is unique is left as Exercise 6.18 for the reader. The minimum degree of such a representation for F, denoted as $\deg(F)$, is a useful measure of the complexity of F. For example, it is known that $\deg(\text{OR})$, $\deg(\text{AND})$, and $\deg(\text{PARITY})$ are all equal to N. In fact, it is known that the degree of most functions is of order N. Moreover, it has also been proven that

$$D(F) \leq 2 \deg(F)^4 . \tag{6.55}$$

This result places an upper bound on the performance of deterministic classical computation in calculating most Boolean functions. Extending this concept, if a polynomial satisfies $|p(X) - F(X)| \leq 1/3$ for all $X \in \{0, 1\}^N$, we say p *approximates* F, and $\widetilde{\deg}(F)$ denotes the minimum degree of such an approximating polynomial. Such measures are important in randomized classical computation and, as we shall see, in describing the quantum case. It is known that $\widetilde{\deg}(\text{PARITY}) = N$,

$$\widetilde{\deg}(\text{OR}) \in \Theta(\sqrt{N}) \quad \text{and} \quad \widetilde{\deg}(\text{AND}) \in \Theta(\sqrt{N}), \tag{6.56}$$

and

$$D(F) \leq 216 \, \widetilde{\deg}(F)^6 . \tag{6.57}$$

The bounds of Equations (6.55) and (6.57) are only the best known at the time of writing; their proof is outside the scope of this book, but you may find further information about them in 'History and further reading'. It is believed that tighter bounds are possible, but these will be good enough for our purposes.

Exercise 6.18: Prove that the minimum degree polynomial representing a Boolean function $F(X)$ is unique.

Exercise 6.19: Show that $P(X) = 1 - (1 - X_0)(1 - X_1) \ldots (1 - X_{N-1})$ represents OR.

Polynomials naturally arise in describing the results of quantum algorithms. Let us write the output of a quantum algorithm $\mathcal{Q}$ which performs T queries to an oracle O as

$$\sum_{k=0}^{2^n - 1} c_k |k\rangle \,. \tag{6.58}$$

We will show that the amplitudes c_k are polynomials of degree at most T in the variables $X_0, X_1, \ldots, X_{N-1}$. Any $\mathcal{Q}$ can be realized using the quantum circuit shown in Figure 6.10. The state $|\psi_0\rangle$ right before the first oracle query can be written as

$$|\psi_0\rangle = \sum_{ij} \left(a_{i0j} |i\rangle |0\rangle + a_{i1j} |i\rangle |1\rangle \right) |j\rangle \,, \tag{6.59}$$

where the first label corresponds to the n qubit oracle query, the next to a single qubit in which the oracle leaves its result, and the last to the $m - n - 1$ working qubits used by $\mathcal{Q}$. After the oracle query, we obtain the state

$$|\psi_1\rangle = \sum_{ij} \left(a_{i0j} |i\rangle |X_i\rangle + a_{i1j} |i\rangle |X_i \oplus 1\rangle \right) |j\rangle \,, \tag{6.60}$$

but since X_i is either 0 or 1, we can re-express this as

$$|\psi_1\rangle = \sum_{ij} \left[\left((1 - X_i) a_{i0j} + X_i a_{i1j} \right) |i0\rangle + \left((1 - X_i) a_{i1j} + X_i a_{i0j} \right) |i1\rangle \right] |j\rangle \,. \tag{6.61}$$

Note that in $|\psi_0\rangle$, the amplitudes of the computational basis states were of degree 0 in X, while those of $|\psi_1\rangle$ are of degree 1 (linear in X). The important observation is that any unitary operation which $\mathcal{Q}$ performs before or after the oracle query cannot change the degree of these polynomials, but each oracle call can increase the degree by at most one. Thus, after T queries, the amplitudes are polynomials of at most degree T. Moreover, measuring the final output (6.58) in the computational basis produces a result k with probability $P_k(X) = |c_k|^2$, which are real-valued polynomials in X of degree at most $2T$.

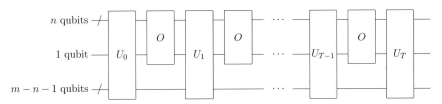

Figure 6.10. General quantum circuit for a quantum algorithm which performs T queries to an oracle O. $U_0, U_1, \ldots, U_T$ are arbitrary unitary transforms on m qubits, and the oracle acts on $n + 1$ qubits.

The total probability $P(X)$ of obtaining a one as the output from the algorithm is a sum over some subset of the polynomials $P_k(X)$, and thus also has degree at most $2T$. In the case that Q produces the correct answer with certainty we must have $P(X) = F(X)$, and thus $\deg(F) \leq 2T$, from which we deduce

$$Q_E(F) \geq \frac{\deg(F)}{2}. \tag{6.62}$$

In the case where Q produces an answer with bounded probability of error it follows that $P(X)$ approximates $F(X)$, and thus $\widetilde{\deg}(F) \leq 2T$, from which we deduce

$$Q_2(F) \geq \frac{\widetilde{\deg}(F)}{2}. \tag{6.63}$$

Combining (6.55) and (6.62), we find that

$$Q_E(F) \geq \left[\frac{D(F)}{32}\right]^{1/4}. \tag{6.64}$$

Similarly, combining (6.57) and (6.63), we find that

$$Q_2(F) \geq \left[\frac{D(F)}{13\,824}\right]^{1/6}. \tag{6.65}$$

This means that in computing Boolean functions using a black box, quantum algorithms may only provide a polynomial speedup over classical algorithms, *at best* – and even that is not generally possible (since $\deg(F)$ is $\Omega(N)$ for most functions). On the other hand, it is known that for $F = $ OR, $D(F) = N$, and the randomized classical query complexity $R(F) \in \Theta(N)$, whereas combining (6.63) and (6.56), and the known performance of the quantum search algorithm, shows that $Q_2(F) \in \Theta(\sqrt{N})$. This square root speedup is just what the quantum search algorithm achieves, and the method of polynomials indicates that the result can perhaps be generalized to a somewhat wider class of problems, but without extra information about the *structure* of the black box oracle function f, no exponential speedup over classical algorithms is possible.

Exercise 6.20: Show that $Q_0(\text{OR}) \geq N$ by constructing a polynomial which represents the OR function from the output of a quantum circuit which computes OR with zero error.

Problem 6.1: (Finding the minimum) Suppose $x_1, \ldots, x_N$ is a database of numbers held in memory, as in Section 6.5. Show that only $O(\log(N)\sqrt{N})$ accesses to the memory are required on a quantum computer, in order to find the smallest element on the list, with probability at least one-half.

Problem 6.2: (Generalized quantum searching) Let $|\psi\rangle$ be a quantum state, and define $U_{|\psi\rangle} \equiv I - 2|\psi\rangle\langle\psi|$. That is, $U_{|\psi\rangle}$ gives the state $|\psi\rangle$ a -1 phase, and leaves states orthogonal to $|\psi\rangle$ invariant.

(1) Suppose we have a quantum circuit implementing a unitary operator U such that $U|0\rangle^{\otimes n} = |\psi\rangle$. Explain how to implement $U_{|\psi\rangle}$.

(2) Let $|\psi_1\rangle = |1\rangle, |\psi_2\rangle = (|0\rangle - |1\rangle)/\sqrt{2}, |\psi_3\rangle = (|0\rangle - i|1\rangle)/\sqrt{2}$. Suppose an unknown oracle O is selected from the set $U_{|\psi_1\rangle}, U_{|\psi_2\rangle}, U_{|\psi_3\rangle}$. Give a quantum algorithm which identifies the oracle with just *one* application of the oracle. (*Hint*: consider superdense coding.)

(3) **Research**: More generally, given k states $|\psi_1\rangle, \ldots, |\psi_k\rangle$, and an unknown oracle O selected from the set $U_{|\psi_1\rangle}, \ldots, U_{|\psi_k\rangle}$, how many oracle applications are required to identify the oracle, with high probability?

Problem 6.3: (Database retrieval) Given a quantum oracle which returns $|k, y \oplus X_k\rangle$ given an n qubit query (and one scratchpad qubit) $|k, y\rangle$, show that with high probability, all $N = 2^n$ bits of X can be obtained using only $N/2 + \sqrt{N}$ queries. This implies the general upper bound $Q_2(F) \leq N/2 + \sqrt{N}$ for any F.

Problem 6.4: (Quantum searching and cryptography) Quantum searching can, potentially, be used to speed up the search for cryptographic keys. The idea is to search through the space of all possible keys for decryption, in each case trying the key, and checking to see whether the decrypted message makes 'sense'. Explain why this idea doesn't work for the Vernam cipher (Section 12.6). When might it work for cryptosystems such as DES? (For a description of DES see, for example, [MvOV96] or [Sch96a].)

Summary of Chapter 6: Quantum search algorithms

- **Quantum search algorithm**: For a search problem with M solutions out of $N = 2^n$ possibilities, prepare $\sum_x |x\rangle$ and then repeat $G \equiv H^{\otimes n} U H^{\otimes n} O$ a total of $O(\sqrt{N/M})$ times, where O is the search oracle, $|x\rangle \rightarrow -|x\rangle$ if x is a solution, no change otherwise, and U takes $|0\rangle \rightarrow -|0\rangle$ and leaves all other computational basis states alone. Measuring yields a solution to the search problem with high probability.

- **Quantum counting algorithm**: Suppose a search problem has an unknown number M of solutions. G has eigenvalues $\exp(\pm i\theta)$ where $\sin^2(\theta/2) = M/N$. The Fourier transform based phase estimation procedure enables us to estimate M to high accuracy using $O(\sqrt{N})$ oracle applications. Quantum counting, in turn, allows us to determine whether a given search problem has any solutions, and to find one if there are, even if the number of solutions is not known in advance.

- **Polynomial bounds**: For problems which are described as evaluations of total functions F (as opposed to partial functions, or 'promise' problems), quantum algorithms can give no more than a polynomial speedup over classical algorithms. Specifically, $Q_2(F) \geq \left[D(F)/13\,824\right]^{1/6}$. Moreover, the performance of the quantum search is optimal: it is $\Theta(\sqrt{N})$.

History and further reading

The quantum search algorithm and much of its further development and elaboration is due to Grover[Gro96, Gro97]. Boyer, Brassard, Høyer and Tapp[BBHT98] wrote an influential paper in which they developed the quantum search algorithm for cases where the number of solutions M is greater than one, and outlined the quantum counting algorithm, later developed in more detail by Brassard, Høyer, and Tapp[BHT98], and from the point of view of phase estimation by Mosca[Mos98]. That the Grover iteration can be understood as a product of two reflections was first pointed out in a review by Aharonov[Aha99b]. The continuous–time Hamiltonian (6.18) was first investigated by Farhi and Gutmann[FG98], from a rather different point of view than we take in Section 6.2. That Grover's algorithm is the best possible oracle-based search algorithm was proved by Bennett, Bernstein, Brassard and Vazirani[BBBV97]. The version of this proof we have presented is based upon that given by Boyer, Brassard, Høyer and Tapp[BBHT98]. Zalka[Zal99] has refined these proofs to show that the quantum search algorithm is, asymptotically, exactly optimal.

The method of polynomials for bounding the power of quantum algorithms was introduced into quantum computing by Beals, Buhrman, Cleve, Mosca, and de Wolf[BBC+98]. An excellent discussion is also available in Mosca's Ph.D. thesis[Mos99], on which much of the discussion in Section 6.7 is based. A number of results are quoted in that section without proof; here are the citations: Equation (6.55) is attributed to Nisan and Smolensky in [BBC+98], but otherwise is presently unpublished, (6.56) is derived from a theorem by Paturi[Pat92] and (6.57) is derived in [BBC+98]. A better bound than (6.65) is given in [BBC+98], but requires concepts such as *block sensitivity* which are outside the scope of this book. A completely different approach for bounding quantum black box algorithms, using arguments based on entanglement, was presented by Ambainis[Amb00].

Problem 6.1 is due to Dürr and Høyer[DH96]. Problem 6.3 is due to van Dam[van98a].

7 Quantum computers: physical realization

Computers in the future may weigh no more than 1.5 tons.
– Popular Mechanics, forecasting the relentless march of science, 1949

I think there is a world market for maybe five computers.
– Thomas Watson, chairman of IBM, 1943

Quantum computation and quantum information is a field of fundamental interest because we believe quantum information processing machines can actually be realized in Nature. Otherwise, the field would be just a mathematical curiosity! Nevertheless, experimental realization of quantum circuits, algorithms, and communication systems has proven extremely challenging. In this chapter we explore some of the guiding principles and model systems for physical implementation of quantum information processing devices and systems.

We begin in Section 7.1 with an overview of the tradeoffs in selecting a physical realization of a quantum computer. This discussion provides perspective for an elaboration of a set of conditions sufficient for the experimental realization of quantum computation in Section 7.2. These conditions are illustrated in Sections 7.3 through 7.7, through a series of case studies, which consider five different model physical systems: the simple harmonic oscillator, photons and nonlinear optical media, cavity quantum electrodynamics devices, ion traps, and nuclear magnetic resonance with molecules. For each system, we briefly describe the physical apparatus, the Hamiltonian which governs its dynamics, means for controlling the system to perform quantum computation, and its principal drawbacks. We do not go into much depth in describing the physics of these systems; as each of these are entire fields unto themselves, that would be outside the scope of this book! Instead, we summarize just the concepts relevant to quantum computation and quantum information such that both the experimental challenge and theoretical potential can be appreciated. On the other hand, analyzing these systems from the standpoint of quantum information also provides a fresh perspective which we hope you will find insightful and useful, as it also allows strikingly simple derivations of some important physics. We conclude the chapter in Section 7.8 by discussing aspects of some other physical systems – quantum dots, superconducting gates, and spins in semiconductors – which are also of interest to this field. For the benefit of the reader wishing to catch just the highlights of each implementation, a summary is provided at the end of each section.

7.1 Guiding principles

What are the experimental requirements for building a quantum computer? The elementary units of the theory are quantum bits – two-level quantum systems; in Section 1.5 we took a brief look at why it is believed that qubits exist in Nature, and what physical forms they may take on. To realize a quantum computer, we must not only give qubits some robust physical representation (in which they retain their quantum properties), but also select a system in which they can be made to evolve as desired. Furthermore, we must be able to prepare qubits in some specified set of initial states, and to measure the final output state of the system.

The challenge of experimental realization is that these basic requirements can often only be partially met. A coin has two states, and makes a good bit, but a poor qubit because it cannot remain in a superposition state (of 'heads' and 'tails') for very long. A single nuclear spin can be a very good qubit, because superpositions of being aligned with or against an external magnetic field can last a long time – even for days. But it can be difficult to build a quantum computer from nuclear spins because their coupling to the world is so small that it is hard to measure the orientation of single nuclei. The observation that the constraints are opposing is general: a quantum computer has to be well isolated in order to retain its quantum properties, but at the same time its qubits have to be accessible so that they can be manipulated to perform a computation and to read out the results. A realistic implementation must strike a delicate balance between these constraints, so that the relevant question is not *how* to build a quantum computer, but rather, *how good* a quantum computer can be built.

System	τ_Q	τ_{op}	$n_{op} = \lambda^{-1}$
Nuclear spin	$10^{-2} - 10^{8}$	$10^{-3} - 10^{-6}$	$10^{5} - 10^{14}$
Electron spin	10^{-3}	10^{-7}	10^{4}
Ion trap (In^+)	10^{-1}	10^{-14}	10^{13}
Electron – Au	10^{-8}	10^{-14}	10^{6}
Electron – GaAs	10^{-10}	10^{-13}	10^{3}
Quantum dot	10^{-6}	10^{-9}	10^{3}
Optical cavity	10^{-5}	10^{-14}	10^{9}
Microwave cavity	10^{0}	10^{-4}	10^{4}

Figure 7.1. Crude estimates for decoherence times τ_Q (seconds), operation times τ_{op} (seconds), and maximum number of operations $n_{op} = \lambda^{-1} = \tau_Q/\tau_{op}$ for various candidate physical realizations of interacting systems of quantum bits. Despite the number of entries in this table, only three fundamentally different qubit representations are given: spin, charge, and photon. The ion trap utilizes either fine or hyperfine transitions of a trapped atom (Section 7.6), which correspond to electron and nuclear spin flips. The estimates for electrons in gold and GaAs, and in quantum dots are given for a charge representation, with an electrode or some confined area either containing an electron or not. In optical and microwave cavities, photons (of frequencies from gigahertz to hundreds of terahertz) populating different modes of the cavities represent the qubit. Take these estimates with a grain of salt: they are only meant to give some perspective on the wide range of possibilities.

What physical systems are potentially good candidates for handling quantum information? A key concept in understanding the merit of a particular quantum computer realization is the notion of *quantum noise* (sometimes called *decoherence*) , the subject of Chapter 8: processes corrupting the desired evolution of the system. This is because the length of the longest possible quantum computation is roughly given by the ratio of τ_Q, the time for which a system remains quantum-mechanically coherent, to τ_{op}, the time it takes to perform elementary unitary transformations (which involve at least two qubits). These two times are actually related to each other in many systems, since they are both determined by the strength of coupling of the system to the external world. Nevertheless, $\lambda = \tau_{op}/\tau_Q$ can vary over a surprisingly wide range, as shown in Figure 7.1.

These estimates give some insight into the merits of different possible physical realizations of a quantum information processing machine, but many other important sources of noise and imperfections arise in actual implementations. For example, manipulations of a qubit represented by two electronic levels of an atom by using light to cause transitions between levels would also cause transitions to other electronic levels with some probability. These would also be considered noise processes, since they take the system out of the two states which define the qubit. Generally speaking, anything which causes loss of (quantum) information is a noise process – later, in Chapter 8, we discuss the theory of quantum noise in more depth.

7.2 Conditions for quantum computation

Let us return to discuss in detail the four basic requirements for quantum computation which were mentioned at the beginning of the previous section. These requirements are the abilities to:

1. Robustly represent quantum information
2. Perform a universal family of unitary transformations
3. Prepare a fiducial initial state
4. Measure the output result

7.2.1 Representation of quantum information

Quantum computation is based on transformation of quantum states. Quantum bits are two-level quantum systems, and as the simplest elementary building blocks for a quantum computer, they provide a convenient labeling for pairs of states and their physical realizations. Thus, for example, the four states of a spin-3/2 particle, $|m = +3/2\rangle, |m = +1/2\rangle, |m = -1/2\rangle, |m = -3/2\rangle$, could be used to represent two qubits.

For the purpose of computation, the crucial realization is that the set of accessible states should be *finite*. The position x of a particle along a one-dimensional line is not generally a good set of states for computation, even though the particle may be in a quantum state $|x\rangle$, or even some superposition $\sum_x c_x |x\rangle$. This is because x has a continuous range of possibilities, and the Hilbert space has infinite size, so that in the absence of noise the information capacity is infinite. For example, in a perfect world, the entire texts of Shakespeare could be stored in (and retrieved from) the infinite number of digits in the binary fraction $x = 0.010111011001 \ldots$. This is clearly unrealistic; what happens in reality is that the presence of noise reduces the number of distinguishable states to a finite number.

In fact, it is generally desirable to have some aspect of symmetry dictate the finiteness of the state space, in order to minimize decoherence. For example, a spin-1/2 particle lives in a Hilbert space spanned by the $|\uparrow\rangle$ and $|\downarrow\rangle$ states; the spin state cannot be anything outside this two-dimensional space, and thus is a nearly ideal quantum bit when well isolated.

If the choice of representation is poor, then decoherence will result. For example, as described in Box 7.1, a particle in a finite square well which is just deep enough to contain two bound states would make a mediocre quantum bit, because transitions from the bound states to the continuum of unbound states would be possible. These would lead to decoherence since they could destroy qubit superposition states. For single qubits, the figure of merit is the minimum lifetime of arbitrary superposition states; a good measure, used for spin states and atomic systems, is T_2, the ('transverse') relaxation time of states such as $(|0\rangle + |1\rangle)/\sqrt{2}$. Note that T_1, the ('longitudinal') relaxation time of the higher energy $|1\rangle$ state, is just a *classical* state lifetime, which is usually longer than T_2.

Box 7.1: Square wells and qubits

A prototypical quantum system is known as the 'square well,' which is a particle in a one-dimensional box, behaving according to Schrödinger's equation, (2.86). The Hamiltonian for this system is $H = p^2/2m + V(x)$, where $V(x) = 0$ for $0 < x < L$, and $V(x) = \infty$ otherwise. The energy eigenstates, expressed as wavefunctions in the position basis, are

$$|\psi_n\rangle = \sqrt{\frac{2}{L}} \sin\left(\frac{n\pi}{L}x\right) , \tag{7.1}$$

where n is an integer, and $|\psi_n(t)\rangle = e^{-iE_n t}|\psi_n\rangle$, with $E_n = n^2\pi^2 m/2L^2$. These states have a discrete spectrum. In particular, suppose that we arrange matters such that only the two lowest energy levels need be considered in an experiment. We define an arbitrary wavefunction of interest as $|\psi\rangle = a|\psi_1\rangle + b|\psi_2\rangle$. Since

$$|\psi(t)\rangle = e^{-i(E_1+E_2)/2t}\left[ae^{-i\omega t}|\psi_1\rangle + be^{i\omega t}|\psi_2\rangle\right] , \tag{7.2}$$

where $\omega = (E_1 - E_2)/2$, we can just forget about everything except a and b, and write our state *abstractly* as the two-component vector $|\psi\rangle = \begin{bmatrix} a \\ b \end{bmatrix}$. This two-level system represents a qubit! Does our two-level system transform like a qubit? Under time evolution, this qubit evolves under the effective Hamiltonian $H = \hbar\omega Z$, which can be disregarded by moving into the rotating frame. To perform operations to this qubit, we perturb H. Consider the effect of adding the additional term

$$\delta V(x) = -V_0(t)\frac{9\pi^2}{16L}\left(\frac{x}{L} - \frac{1}{2}\right) \tag{7.3}$$

to $V(x)$. In the basis of our two-level system, this can be rewritten by taking the matrix elements $V_{nm} = \langle\psi_n|\delta V(x)|\psi_m\rangle$, giving $V_{11} = V_{22} = 0$, and $V_{12} = V_{21} = V_0$, such that, to lowest order in V_0, the perturbation to H is $H_1 = V_0(t)X$. This generates rotations about the $\hat{x}$ axis. Similar techniques can be used to perform other single qubit operations, by manipulating the potential function.

This shows how a single qubit can be represented by the two lowest levels in a square well potential, and how simple perturbations of the potential can effect computational operations on the qubit. However, perturbations also introduce higher order effects, and in real physical systems boxes are not infinitely deep, other levels begin to enter the picture, and our two-level approximation begins to fail. Also, in reality, the controlling system is just another quantum system, and it couples to the one we are trying to do quantum computation with. These problems lead to decoherence.

7.2.2 Performance of unitary transformations

Closed quantum systems evolve unitarily as determined by their Hamiltonians, but to perform quantum computation one must be able to control the Hamiltonian to effect an *arbitrary* selection from a universal family of unitary transformations (as described in Section 4.5). For example, a single spin might evolve under the Hamiltonian $H = P_x(t)X + P_y(t)Y$, where $P_{\{x,y\}}$ are classically controllable parameters. From Exercise 4.10, we know that by manipulating P_x and P_y appropriately, one can perform arbitrary single spin rotations.

According to the theorems of Section 4.5, any unitary transform can be composed from single spin operations and controlled-NOT gates, and thus realization of those two kinds of quantum logic gates are natural goals for experimental quantum computation. However, implicitly required also is the ability to address individual qubits, and to apply these gates to select qubits or pairs of qubits. This is not simple to accomplish in many physical systems. For example, in an ion trap, one can direct a laser at one of many individual ions to selectively excite it, but only as long as the ions are spatially separated by a wavelength or more.

Unrecorded imperfections in unitary transforms can lead to decoherence. In Chapter 8 we shall see how the average effect of random kicks (small rotations to a single spin about its $\hat{z}$ axis) leads to loss of quantum information which is represented by the relative phases in a quantum state. Similarly, the cumulative effect of systematic errors is decoherence, when the information needed to be able to reverse them is lost. Furthermore, the control parameters in the Hamiltonian are only approximately classical controls: in reality, the controlling system is just another quantum system, and the true Hamiltonian should include the back-action of the control system upon the quantum computer. For example, instead of $P_x(t)$ in the above example, one actually has a Jaynes–Cummings type atom–photon interaction Hamiltonian (Section 7.5.2), with $P_x(t) = \sum_k \omega_k(t)(a_k + a_k^\dagger)$ or something similar being the cavity photon field. After interacting with a qubit, a photon can carry away information about the state of the qubit, and this is thus a decoherence process.

Two important figures of merit for unitary transforms are the minimum achievable fidelity $\mathcal{F}$ (Chapter 9), and the maximum time t_{op} required to perform elementary operations such as single spin rotations or a controlled-NOT gate.

7.2.3 Preparation of fiducial initial states

One of the most important requirements for being able to perform a useful computation, even classically, is to be able to prepare the desired input. If one has a box which can perform perfect computations, what use is it if numbers cannot be input? With classical machines, establishing a definite input state is rarely a difficulty – one merely sets some switches in the desired configuration and that defines the input state. However, with quantum systems this can be very difficult, depending on the realization of qubits.

Note that it is only necessary to be able to (repeatedly) produce one specific quantum state with high fidelity, since a unitary transform can turn it into any other desired input state. For example, being able to put n spins into the $|00\ldots0\rangle$ state is good enough. The fact that they may not stay there for very long due to thermal heating is a problem with the choice of representation.

Input state preparation is a significant problem for most physical systems. For example, ions can be prepared in good input states by physically cooling them into their ground state

(Section 7.6), but this is challenging. Moreover, for physical systems in which ensembles of quantum computers are involved, extra concerns arise. In nuclear magnetic resonance (Section 7.7), each molecule can be thought of as a single quantum computer, and a large number of molecules is needed to obtain a measurable signal strength. Although qubits can remain in arbitrary superposition states for relatively long times, it is difficult to put all of the qubits in all of the molecules into the same state, because the energy difference $\hbar\omega$ between the $|0\rangle$ and $|1\rangle$ states is much smaller than $k_B T$. On the other hand, simply letting the system equilibrate establishes it in a very well-known state, the thermal one, with the density matrix $\rho \approx e^{-\mathcal{H}/k_B T}/\mathcal{Z}$, where $\mathcal{Z}$ is a normalization factor required to maintain $\mathrm{tr}(\rho) = 1$.

Two figures of merit are relevant to input state preparation: the minimum fidelity with which the initial state can be prepared in a given state ρ_{in}, and the entropy of ρ_{in}. The entropy is important because, for example, it is very easy to prepare the state $\rho_{\mathrm{in}} = I/2^n$ with high fidelity, but that is a useless state for quantum computation, since it is invariant under unitary transforms! Ideally, the input state is a pure state, with zero entropy. Generally, input states with non-zero entropy reduce the accessibility of the answer from the output result.

7.2.4 Measurement of output result

What measurement capability is required for quantum computation? For the purpose of the present discussion, let us think of measurement as a process of coupling one or more qubits to a classical system such that after some interval of time, the state of the qubits is indicated by the state of the classical system. For example, a qubit state $a|0\rangle + b|1\rangle$, represented by the ground and excited states of a two-level atom, might be measured by pumping the excited state and looking for fluorescence. If an electrometer indicates that fluorescence had been detected by a photomultiplier tube, then the qubit would collapse into the $|1\rangle$ state; this would happen with probability $|b|^2$. Otherwise, the electrometer would detect no charge, and the qubit would collapse into the $|0\rangle$ state.

An important characteristic of the measurement process for quantum computation is the wavefunction collapse which describes what happens when a projective measurement is performed (Section 2.2.5). The output from a good quantum algorithm is a superposition state which gives a useful answer with high probability when measured. For example, one step in Shor's quantum factoring algorithm is to find an integer r from the measurement result, which is an integer close to qc/r, where q is the dimension of a Hilbert space. The output state is actually in a nearly uniform superposition of all possible values of c, but a measurement collapses this into a single, random integer, thus allowing r to be determined with high probability (using a continued fraction expansion, as was described in Chapter 5).

Many difficulties with measurement can be imagined; for example, inefficient photon counters and amplifier thermal noise can reduce the information obtained about measured qubit states in the scheme just described. Furthermore, projective measurements (sometimes called 'strong' measurements) are often difficult to implement. They require that the coupling between the quantum and classical systems be large, and switchable. Measurements should not occur when not desired; otherwise they can be a decoherence process.

Surprisingly, however, strong measurements are not necessary; weak measurements which are performed continuously and never switched off, are usable for quantum computation. This is made possible by completing the computation in time short compared with the measurement coupling, and by using large ensembles of quantum computers. These ensembles together give an aggregate signal which is macroscopically observable and indicative of the quantum state. Use of an ensemble introduces additional problems. For example, in the factoring algorithm, if the measurement output is $q\langle c\rangle/r$, the algorithm would fail because $\langle c\rangle$, the average value of c, is not necessarily an integer (and thus the continued fraction expansion would not be possible). Fortunately, it is possible to modify quantum algorithms to work with ensemble average readouts. This will be discussed further in Section 7.7.

A good figure of merit for measurement capability is the signal to noise ratio (SNR). This accounts for measurement inefficiency as well as inherent signal strength available from coupling a measurement apparatus to the quantum system.

7.3 Harmonic oscillator quantum computer

Before continuing on to describe a complete physical model for a realizable quantum computer, let us pause for a moment to consider a very elementary system – the simple harmonic oscillator – and discuss why it does not serve as a good quantum computer. The formalism used in this example will also serve as a basis for studying other physical systems.

7.3.1 Physical apparatus

An example of a simple harmonic oscillator is a particle in a parabolic potential well, $V(x) = m\omega^2 x^2/2$. In the classical world, this could be a mass on a spring, which oscillates back and forth as energy is transfered between the potential energy of the spring and the kinetic energy of the mass. It could also be a resonant electrical circuit, where the energy sloshes back and forth between the inductor and the capacitor. In these systems, the total energy of the system is a continuous parameter.

In the quantum domain, which is reached when the coupling to the external world becomes very small, the total energy of the system can only take on a discrete set of values. An example is given by a single mode of electromagnetic radiation trapped in a high Q cavity; the total amount of energy (up to a fixed offset) can only be integer multiples of $\hbar\omega$, an energy scale which is determined by the fundamental constant $\hbar$ and the frequency of the trapped radiation, ω.

The set of discrete energy eigenstates of a simple harmonic oscillator can be labeled as $|n\rangle$, where $n = 0, 1, \ldots, \infty$. The relationship to quantum computation comes by taking a finite subset of these states to represent qubits. These qubits will have lifetimes determined by physical parameters such as the cavity quality factor Q, which can be made very large by increasing the reflectivity of the cavity walls. Moreover, unitary transforms can be applied by simply allowing the system to evolve in time. However, there are problems with this scheme, as will become clear below. We begin by studying the system Hamiltonian, then discuss how one might implement simple quantum logic gates such as the controlled-NOT.

7.3.2 The Hamiltonian

The Hamiltonian for a particle in a one-dimensional parabolic potential is

$$H = \frac{p^2}{2m} + \frac{1}{2}m\omega^2 x^2 , \tag{7.4}$$

where p is the particle momentum operator, m is the mass, x is the position operator, and ω is related to the potential depth. Recall that x and p are operators in this expression (see Box 7.2), which can be rewritten as

$$H = \hbar\omega \left(a^\dagger a + \frac{1}{2} \right) , \tag{7.5}$$

where $a^\dagger$ and a are creation and annihilation operators, defined as

$$a = \frac{1}{\sqrt{2m\hbar\omega}} \left(m\omega x + ip \right) \tag{7.6}$$

$$a^\dagger = \frac{1}{\sqrt{2m\hbar\omega}} \left(m\omega x - ip \right) . \tag{7.7}$$

The zero point energy $\hbar\omega/2$ contributes an unobservable overall phase factor, which can be disregarded for our present purpose.

The eigenstates $|n\rangle$ of H, where $n = 0, 1, \ldots$, have the properties

$$a^\dagger a |n\rangle = n|n\rangle \tag{7.10}$$
$$a^\dagger |n\rangle = \sqrt{n+1}\,|n+1\rangle \tag{7.11}$$
$$a|n\rangle = \sqrt{n}\,|n-1\rangle . \tag{7.12}$$

Later, we will find it convenient to express interactions with a simple harmonic oscillator by introducing additional terms involving a and $a^\dagger$, and interactions between oscillators with terms such as $a_1^\dagger a_2 + a_1 a_2^\dagger$. For now, however, we confine our attention to a single oscillator.

Exercise 7.1: Using the fact that x and p do not commute, and that in fact $[x, p] = i\hbar$, explicitly show that $a^\dagger a = H/\hbar\omega - 1/2$.

Exercise 7.2: Given that $[x, p] = i\hbar$, compute $[a, a^\dagger]$.

Exercise 7.3: Compute $[H, a]$ and use the result to show that if $|\psi\rangle$ is an eigenstate of H with energy $E \geq n\hbar\omega$, then $a^n|\psi\rangle$ is an eigenstate with energy $E - n\hbar\omega$.

Exercise 7.4: Show that $|n\rangle = \frac{(a^\dagger)^n}{\sqrt{n!}}|0\rangle$.

Exercise 7.5: Verify that Equations (7.11) and (7.12) are consistent with (7.10) and the normalization condition $\langle n|n\rangle = 1$.

Time evolution of the eigenstates is given by solving the Schrödinger equation, (2.86), from which we find that the state $|\psi(0)\rangle = \sum_n c_n(0)|n\rangle$ evolves in time to become

$$|\psi(t)\rangle = e^{-iHt/\hbar}|\psi(0)\rangle = \sum_n c_n e^{-in\omega t}|n\rangle . \tag{7.13}$$

We will assume for the purpose of discussion that an arbitrary state can be perfectly prepared, and that the state of the system can be projectively measured (Section 2.2.3),

Box 7.2: The quantum harmonic oscillator

The harmonic oscillator is an extremely important and useful concept in the quantum description of the physical world, and a good way to begin to understand its properties is to determine the energy eigenstates of its Hamiltonian, (7.4). One way to do this is simply to solve the Schrödinger equation

$$\frac{\hbar^2}{2m}\frac{d^2\psi_n(x)}{dx^2} + \frac{1}{2}m\omega^2 x^2\psi_n(x) = E\psi_n(x) \tag{7.8}$$

for $\psi_n(x)$ and the eigenenergies E, subject to $\psi(x) \rightarrow 0$ at $x = \pm\infty$, and $\int |\psi(x)|^2 = 1$; the first five solutions are sketched here:

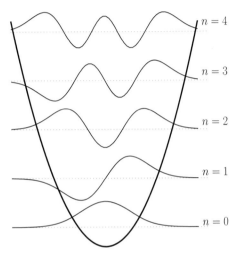

These wavefunctions describe the probability amplitudes that a particle in the harmonic oscillator will be found at different positions within the potential.

Although these pictures may give some intuition about what a physical system is doing in co-ordinate space, we will generally be more interested in the abstract *algebraic* properties of the states. Specifically, suppose $|\psi\rangle$ satisfies (7.8) with energy E. Then defining operators a and $a^\dagger$ as in (7.6)–(7.7), we find that since $[H, a^\dagger] = \hbar\omega a^\dagger$,

$$Ha^\dagger|\psi\rangle = \left([H, a^\dagger] + a^\dagger H\right)|\psi\rangle = (\hbar\omega + E)a^\dagger|\psi\rangle, \tag{7.9}$$

that is, $a^\dagger|\psi\rangle$ is an eigenstate of H, with energy $E + \hbar\omega$! Similarly, $a|\psi\rangle$ is an eigenstate with energy $E - \hbar\omega$. Because of this, $a^\dagger$ and a are called raising and lowering operators. It follows that $a^{\dagger n}|\psi\rangle$ are eigenstates for any integer n, with energies $E + n\hbar\omega$. There are thus an infinite number of energy eigenstates, whose energies are equally spaced apart, by $\hbar\omega$. Moreover, since H is positive definite, there must be some $|\psi_0\rangle$ for which $a|\psi_0\rangle = 0$; this is the ground state – the eigenstate of H with lowest energy. These results efficiently capture the essence of the quantum harmonic oscillator, and allow us to use a compact notation $|n\rangle$ for the eigenstates, where n is an integer, and $H|n\rangle = \hbar(n + 1/2)|n\rangle$. We shall often work with $|n\rangle$, a, and $a^\dagger$ in this chapter, as harmonic oscillators arise in the guise of many different physical systems.

but otherwise, there are no interactions with the external world, so that the system is perfectly closed.

7.3.3 Quantum computation

Suppose we want to perform quantum computation with the single simple harmonic oscillator described above. What can be done? The most natural choice for representation of qubits are the energy eigenstates $|n\rangle$. This choice allows us to perform a controlled-NOT gate in the following way. Recall that this transformation performs the mapping

$$
\begin{aligned}
|00\rangle_L &\rightarrow |00\rangle_L \\
|01\rangle_L &\rightarrow |01\rangle_L \\
|10\rangle_L &\rightarrow |11\rangle_L \\
|11\rangle_L &\rightarrow |10\rangle_L,
\end{aligned}
\tag{7.14}
$$

on two qubit states (here, the subscript L is used to clearly distinguish 'logical' states in contrast to the harmonic oscillator basis states). Let us *encode* these two qubits using the mapping

$$
\begin{aligned}
|00\rangle_L &= |0\rangle \\
|01\rangle_L &= |2\rangle \\
|10\rangle_L &= (|4\rangle + |1\rangle)/\sqrt{2} \\
|11\rangle_L &= (|4\rangle - |1\rangle)/\sqrt{2}.
\end{aligned}
\tag{7.15}
$$

Now suppose that at $t = 0$ the system is started in a state spanned by these basis states, and we simply evolve the system forward to time $t = \pi/\hbar\omega$. This causes the energy eigenstates to undergo the transformation $|n\rangle \rightarrow \exp(-i\pi a^\dagger a)|n\rangle = (-1)^n|n\rangle$, such that $|0\rangle$, $|2\rangle$, and $|4\rangle$ stay unchanged, but $|1\rangle \rightarrow -|1\rangle$. As a result, we obtain the desired controlled-NOT gate transformation.

In general, a necessary and sufficient condition for a physical system to be able to perform a unitary transform U is simply that the time evolution operator for the system, $T = \exp(-iHt)$, defined by its Hamiltonian H, has nearly the same eigenvalue spectrum as U. In the case above, the controlled-NOT gate was simple to implement because it only has eigenvalues $+1$ and -1; it was straightforward to arrange an encoding to obtain the same eigenvalues from the time evolution operator for the harmonic oscillator. The Hamiltonian for an oscillator could be perturbed to realize nearly any eigenvalue spectrum, and any number of qubits could be represented by simply mapping them into the infinite number of eigenstates of the system. This suggests that perhaps one might be able to realize an entire quantum computer in a single simple harmonic oscillator!

7.3.4 Drawbacks

Of course, there are many problems with the above scenario. Clearly, one will not always know the eigenvalue spectrum of the unitary operator for a certain quantum computation, even though one may know how to construct the operator from elementary gates. In fact, for most problems addressed by quantum algorithms, knowledge of the eigenvalue spectrum is tantamount to knowledge of the solution!

Another obvious problem is that the technique used above does not allow one computation to be cascaded with another, because in general, cascading two unitary transforms results in a new transform with unrelated eigenvalues.

Finally, the idea of using a single harmonic oscillator to perform quantum computation is flawed because it neglects the principle of digital representation of information. A Hilbert space of 2^n dimensions mapped into the state space of a single harmonic oscillator would have to allow for the possibility of states with energy $2^n \hbar \omega$. In contrast, the same Hilbert space could be obtained by using n two-level quantum systems, which has an energy of at most $n\hbar\omega$. Similar comparisons can be made between a classical dial with 2^n settings, and a register of n classical bits. Quantum computation builds upon *digital* computation, not analog computation.

The main features of the harmonic oscillator quantum computer are summarized below (each system we consider will be summarized similarly, at the end of the corresponding section). With this, we leave behind us the study of single oscillators, and turn next to systems of harmonic oscillators, made of photons and atoms.

Harmonic oscillator quantum computer

- **Qubit representation**: Energy levels $|0\rangle$, $|1\rangle$, ..., $|2^n\rangle$ of a single quantum oscillator give n qubits.
- **Unitary evolution**: Arbitrary transforms U are realized by matching their eigenvalue spectrums to that given by the Hamiltonian $H = a^\dagger a$.
- **Initial state preparation**: Not considered.
- **Readout**: Not considered.
- **Drawbacks**: Not a digital representation! Also, matching eigenvalues to realize transformations is not feasible for arbitrary U, which generally have unknown eigenvalues.

7.4 Optical photon quantum computer

An attractive physical system for representing a quantum bit is the optical photon. Photons are chargeless particles, and do not interact very strongly with each other, or even with most matter. They can be guided along long distances with low loss in optical fibers, delayed efficiently using phase shifters, and combined easily using beamsplitters. Photons exhibit signature quantum phenomena, such as the interference produced in two-slit experiments. Furthermore, in principle, photons *can* be made to interact with each other, using nonlinear optical media which mediate interactions. There are problems with this ideal scenario; nevertheless, many things can be learned from studying the components, architecture, and drawbacks of an optical photon quantum information processor, as we shall see in this section.

7.4.1 Physical apparatus

Let us begin by considering what single photons are, how they can represent quantum states, and the experimental components useful for manipulating photons. The classical behavior of phase shifters, beamsplitters, and nonlinear optical Kerr media is described.

Photons can represent qubits in the following manner. As we saw in the discussion of the simple harmonic oscillator, the energy in an electromagnetic cavity is quantized in units of $\hbar\omega$. Each such quantum is called a photon. It is possible for a cavity to contain a superposition of zero or one photon, a state which could be expressed as a qubit $c_0|0\rangle + c_1|1\rangle$, but we shall do something different. Let us consider two cavities, whose total energy is $\hbar\omega$, and take the two states of a qubit as being whether the photon is in one cavity ($|01\rangle$) or the other ($|10\rangle$). The physical state of a superposition would thus be written as $c_0|01\rangle + c_1|10\rangle$; we shall call this the *dual-rail* representation. Note that we shall focus on single photons traveling as a wavepacket through free space, rather than inside a cavity; one can imagine this as having a cavity moving along with the wavepacket. Each cavity in our qubit state will thus correspond to a different spatial mode.

One scheme for generating single photons in the laboratory is by attenuating the output of a laser. A laser outputs a state known as a coherent state, $|\alpha\rangle$, defined as

$$|\alpha\rangle = e^{-|\alpha|^2/2} \sum_{n=0}^{\infty} \frac{\alpha^n}{\sqrt{n!}}|n\rangle, \tag{7.16}$$

where $|n\rangle$ is an n-photon energy eigenstate. This state, which has been the subject of thorough study in the field of quantum optics, has many beautiful properties which we shall not describe here. It suffices to understand just that coherent states are naturally radiated from driven oscillators such as a laser when pumped high above its lasing threshold. Note that the mean energy is $\langle\alpha|n|\alpha\rangle = |\alpha|^2$. When attenuated, a coherent state just becomes a weaker coherent state, and a weak coherent state can be made to have just one photon, with high probability.

Exercise 7.6: (Eigenstates of photon annihilation)   Prove that a coherent state is an eigenstate of the photon annihilation operator, that is, show $a|\alpha\rangle = \lambda|\alpha\rangle$ for some constant λ.

For example, for $\alpha = \sqrt{0.1}$, we obtain the state $\sqrt{0.90}\,|0\rangle + \sqrt{0.09}\,|1\rangle + \sqrt{0.002}\,|2\rangle + \cdots$. Thus if light ever makes it through the attenuator, one knows it is a single photon with probability better than 95%; the failure probability is thus 5%. Note also that 90% of the time, no photons come through at all; this source thus has a rate of 0.1 photons per unit time. Finally, this source does not indicate (by means of some classical readout) when a photon has been output or not; two of these sources cannot be synchronized.

Better synchronicity can be achieved using parametric down-conversion. This involves sending photons of frequency ω_0 into a nonlinear optical medium such as KH_2PO_4 to generate photon pairs at frequencies $\omega_1 + \omega_2 = \omega_0$. Momentum is also conserved, such that $\vec{k}_1 + \vec{k}_2 = \vec{k}_3$, so that when a single ω_2 photon is (destructively) detected, then a single ω_1 photon is known to exist (Figure 7.2). By coupling this to a gate, which is opened only when a single photon (as opposed to two or more) is detected, and by appropriately delaying the outputs of multiple down-conversion sources, one can, in principle, obtain multiple single photons propagating in time synchronously, within the time resolution of the detector and gate.

Single photons can be detected with high quantum efficiency for a wide range of wavelengths, using a variety of technologies. For our purposes, the most important characteristic of a detector is its capability of determining, with high probability, whether zero or one photon exists in a particular spatial mode. For the dual-rail representation, this translates into a projective measurement in the computational basis. In practice, imperfections reduce the probability of being able to detect a single photon; the *quantum efficiency* η ($0 \leq \eta \leq 1$) of a photodetector is the probability that a single photon incident on the detector generates a photocarrier pair that contributes to detector current. Other important characteristics of a detector are its bandwidth (time responsivity), noise, and 'dark counts' which are photocarriers generated even when no photons are incident.

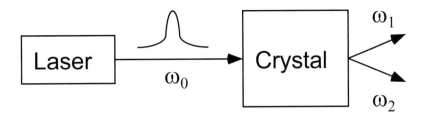

Figure 7.2. Parametric down-conversion scheme for generation of single photons.

Three of the most experimentally accessible devices for manipulating photon states are mirrors, phase shifters and beamsplitters. High reflectivity mirrors reflect photons and change their propagation direction in space. Mirrors with 0.01% loss are not unusual. We shall take these for granted in our scenario. A phase shifter is nothing more than a slab of transparent medium with index of refraction n different from that of free space, n_0; for example, ordinary borosilicate glass has $n \approx 1.5n_0$ at optical wavelengths. Propagation in such a medium through a distance L changes a photon's phase by e^{ikL}, where $k = n\omega/c_0$, and c_0 is the speed of light in vacuum. Thus, a photon propagating through a phase shifter will experience a phase shift of $e^{i(n-n_0)L\omega/c_0}$ compared to a photon going the same distance through free space.

Another useful component, the beamsplitter, is nothing more than a partially silvered piece of glass, which reflects a fraction R of the incident light, and transmits $1 - R$. In the laboratory, a beamsplitter is usually fabricated from two prisms, with a thin metallic layer sandwiched in-between, schematically drawn as shown in Figure 7.3. It is convenient to define the *angle* θ of a beamsplitter as $\cos \theta = R$; note that the angle parameterizes the amount of partial reflection, and does not necessarily have anything to do with the physical orientation of the beamsplitter. The two inputs and two outputs of this device are related by

$$a_{out} = a_{in} \cos \theta + b_{in} \sin \theta \tag{7.17}$$

$$b_{out} = -a_{in} \sin \theta + b_{in} \cos \theta , \tag{7.18}$$

where classically we think of a and b as being the electromagnetic fields of the radiation at the two ports. Note that in this definition we have chosen a non-standard phase convention convenient for our purposes. In the special case of a 50/50 beamsplitter, $\theta = 45°$.

Nonlinear optics provides one final useful component for this exercise: a material

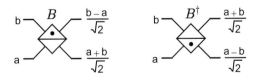

Figure 7.3. Schematic of an optical beamsplitter, showing the two input ports, the two output ports, and the phase conventions for a 50/50 beamsplitter ($\theta = \pi/4$). The beamsplitter on the right is the inverse of the one on the left (the two are distinguished by the dot drawn inside). The input–output relations for the mode operators a and b are given for $\theta = \pi/4$.

whose index of refraction n is proportional to the total intensity I of light going through it:

$$n(I) = n + n_2 I . \qquad (7.19)$$

This is known as the optical Kerr effect, and it occurs (very weakly) in materials as mundane as glass and sugar water. In doped glasses, n_2 ranges from 10^{-14} to 10^{-7} cm^2/W, and in semiconductors, from 10^{-10} to 10^2. Experimentally, the relevant behavior is that when two beams of light of equal intensity are nearly co-propagated through a Kerr medium, each beam will experience an extra phase shift of $e^{in_2 I L\omega/c_0}$ compared to what happens in the single beam case. This would be ideal if the length L could be arbitrarily long, but unfortunately that fails because most Kerr media are also highly absorptive, or scatter light out of the desired spatial mode. This is the primary reason why a single photon quantum computer is impractical, as we shall discuss in Section 7.4.3.

We turn next to a quantum description of these optical components.

7.4.2 Quantum computation

Arbitrary unitary transforms can be applied to quantum information, encoded with single photons in the $c_0|01\rangle + c_1|10\rangle$ dual-rail representation, using phase shifters, beamsplitters, and nonlinear optical Kerr media. How this works can be understood in the following manner, by giving a quantum-mechanical Hamiltonian description of each of these devices.

The time evolution of a cavity mode of electromagnetic radiation is modeled quantum-mechanically by a harmonic oscillator, as we saw in Section 7.3.2. $|0\rangle$ is the vacuum state, $|1\rangle = a^\dagger|0\rangle$ is a single photon state, and in general, $|n\rangle = \frac{a^{\dagger n}}{\sqrt{n!}}|0\rangle$ is an n-photon state, where $a^\dagger$ is the creation operator for the mode. Free space evolution is described by the Hamiltonian

$$H = \hbar\omega a^\dagger a , \qquad (7.20)$$

and applying (7.13), we find that the state $|\psi\rangle = c_0|0\rangle + c_1|1\rangle$ evolves in time to become $|\psi(t)\rangle = c_0|0\rangle + c_1 e^{-i\omega t}|1\rangle$. Note that the dual-rail representation is convenient because free evolution only changes $|\varphi\rangle = c_0|01\rangle + c_1|10\rangle$ by an overall phase, which is undetectable. Thus, for that manifold of states, the evolution Hamiltonian is zero.

Phase shifter. A phase shifter P acts just like normal time evolution, but at a different rate, and localized to only the modes going through it. That is because light slows down in a medium with larger index of refraction; specifically, it takes $\Delta \equiv (n - n_0)L/c_0$ more time to propagate a distance L in a medium with index of refraction n than in vacuum. For example, the action of P on the vacuum state is to do nothing: $P|0\rangle = |0\rangle$, but on a single photon state, one obtains $P|1\rangle = e^{i\Delta}|1\rangle$.

P performs a useful logical operation on a dual-rail state. Placing a phase shifter in one mode retards its phase evolution with respect to another mode, which travels the same distance but without going through the shifter. For dual-rail states this transforms $c_0|01\rangle + c_1|10\rangle$ to $c_0 e^{-i\Delta/2}|01\rangle + c_1 e^{i\Delta/2}|10\rangle$, up to an irrelevant overall phase. Recall from Section 4.2 that this operation is nothing more than a rotation,

$$R_z(\Delta) = e^{-iZ\Delta/2} , \qquad (7.21)$$

where we take as the logical zero $|0_L\rangle = |01\rangle$ and one $|1_L\rangle = |10\rangle$, and Z is the usual

Pauli operator. One can thus think of P as resulting from time evolution under the Hamiltonian

$$H = (n_0 - n)Z ,$$ (7.22)

where $P = \exp(-iHL/c_0)$.

Exercise 7.7: Show that the circuit below transforms a dual-rail state by

$$|\psi_{out}\rangle = \begin{bmatrix} e^{i\pi} & 0 \\ 0 & 1 \end{bmatrix} |\psi_{in}\rangle ,$$ (7.23)

if we take the top wire to represent the $|01\rangle$ mode, and $|10\rangle$ the bottom mode, and the boxed π to represent a phase shift by π:

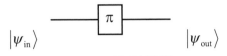

Note that in such 'optical circuits', propagation in space is explicitly represented by putting in lumped circuit elements such as in the above, to represent phase evolution. In the dual-rail representation, evolution according to (7.20) changes the logical state only by an unobservable global phase, and thus we are free to disregard it and keep only relative phase shifts.

Exercise 7.8: Show that $P|\alpha\rangle = |\alpha e^{i\Delta}\rangle$ where $|\alpha\rangle$ is a coherent state (note that, in general, α is a complex number!).

Beamsplitter. A similar Hamiltonian description of the beamsplitter also exists, but instead of motivating it phenomenologically, let us begin with the Hamiltonian and show how the expected classical behavior, Equations (7.17)–(7.18) arises from it. Recall that the beamsplitter acts on two modes, which we shall describe by the creation (annihilation) operators a ($a^\dagger$) and b ($b^\dagger$). The Hamiltonian is

$$H_{bs} = i\theta \left(ab^\dagger - a^\dagger b \right) ,$$ (7.24)

and the beamsplitter performs the unitary operation

$$B = \exp\left[\theta \left(a^\dagger b - ab^\dagger \right) \right] .$$ (7.25)

The transformations effected by B on a and b, which will later be useful, are found to be

$$BaB^\dagger = a \cos\theta + b \sin\theta \quad \text{and} \quad BbB^\dagger = -a \sin\theta + b \cos\theta .$$ (7.26)

We verify these relations using the Baker–Campbell–Hausdorf formula (also see Exercise 4.49)

$$e^{\lambda G} A e^{-\lambda G} = \sum_{n=0}^{\infty} \frac{\lambda^n}{n!} C_n ,$$ (7.27)

where λ is a complex number, A, G, and C_n are operators, and C_n is defined recursively as the sequence of commutators $C_0 = A$, $C_1 = [G, C_0]$, $C_2 = [G, C_1]$, $C_3 = [G, C_2]$, ..., $C_n = [G, C_{n-1}]$. Since it follows from $[a, a^\dagger] = 1$ and $[b, b^\dagger] = 1$ that $[G, a] = -b$ and $[G, b] = a$, for $G \equiv a^\dagger b - ab^\dagger$, we obtain for the expansion of $BaB^\dagger$ the series coefficients

$C_0 = a$, $C_1 = [G, a] = -b$, $C_2 = [G, C_1] = -a$, $C_3 = [G, C_2] = -[G, C_0] = b$, which in general are

$$C_{n \text{ even}} = i^n a \tag{7.28}$$
$$C_{n \text{ odd}} = i^{n+1} b. \tag{7.29}$$

From this, our desired result follows straightforwardly:

$$BaB^\dagger = e^{\theta G} a e^{-\theta G} \tag{7.30}$$
$$= \sum_{n=0}^\infty \frac{\theta^n}{n!} C_n \tag{7.31}$$
$$= \sum_{n \text{ even}} \frac{(i\theta)^n}{n!} a + i \sum_{n \text{ odd}} \frac{(i\theta)^n}{n!} b \tag{7.32}$$
$$= a \cos\theta - b \sin\theta. \tag{7.33}$$

The transform $BbB^\dagger$ is trivially found by swapping a and b in the above solution. Note that the beamsplitter operator arises from a deep relationship between the beamsplitter and the algebra of $SU(2)$, as explained in Box 7.3.

In terms of quantum logic gates, B performs a useful operation. First note that $B|00\rangle = |00\rangle$, that is, when no photons in either input mode exist, no photons will exist in either output mode. When one photon exists in mode a, recalling that $|1\rangle = a^\dagger|0\rangle$, we find that

$$B|01\rangle = Ba^\dagger|00\rangle = Ba^\dagger B^\dagger B|00\rangle = (a^\dagger \cos\theta + b^\dagger \sin\theta)|00\rangle = \cos\theta|01\rangle + \sin\theta|10\rangle. \tag{7.34}$$

Similarly, $B|10\rangle = \cos\theta|10\rangle - \sin\theta|01\rangle$. Thus, on the $|0_L\rangle$ and $|1_L\rangle$ manifold of states, we may write B as

$$B = \begin{bmatrix} \cos\theta & -\sin\theta \\ \sin\theta & \cos\theta \end{bmatrix} = e^{i\theta Y}. \tag{7.35}$$

Phase shifters and beamsplitters together allow arbitrary single qubit operations to be performed to our optical qubit. This a consequence of Theorem 4.1 on page 175, which states that all single qubit operations can be generated from $\hat{z}$-axis rotations $R_z(\alpha) = \exp(-i\alpha Z/2)$, and $\hat{y}$-axis rotations, $R_y(\alpha) = \exp(-i\alpha Y/2)$. A phase shifter performs R_z rotations, and a beamsplitter performs R_y rotations.

Exercise 7.9: (Optical Hadamard gate) Show that the following circuit acts as a Hadamard gate on dual-rail single photon states, that is, $|01\rangle \to (|01\rangle + |10\rangle)/\sqrt{2}$ and $|10\rangle \to (|01\rangle - |10\rangle)/\sqrt{2}$ up to an overall phase:

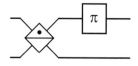

Exercise 7.10: (Mach–Zehnder interferometer) Interferometers are optical tools used to measure small phase shifts, which are constructed from two beamsplitters. Their basic principle of operation can be understood by this simple exercise.

1. Note that this circuit performs the identity operation:

Box 7.3: $SU(2)$ Symmetry and quantum beamsplitters

There is an interesting connection between the Lie group $SU(2)$ and the algebra of two coupled harmonic oscillators, which is useful for understanding the quantum beamsplitter transformation. Identify

$$a^\dagger a - b^\dagger b \to Z \tag{7.36}$$

$$a^\dagger b \to \sigma_+ \tag{7.37}$$

$$ab^\dagger \to \sigma_- , \tag{7.38}$$

where Z is the Pauli operator, and $\sigma_\pm = (X \pm iY)/2$ are raising and lowering operators defined in terms of Pauli X and Y. From the commutation relations for a, $a^\dagger$, b, and $b^\dagger$, it is easy to verify that these definitions satisfy the usual commutation relations for the Pauli operators, (2.40). Also note that the total number operator, $a^\dagger a + b^\dagger b$, commutes with σ_z, σ_+, and σ_-, as it should, being an invariant quantity under rotations in the $SU(2)$ space. Using $X = a^\dagger b + ab^\dagger$ and $Y = -i(a^\dagger b - ab^\dagger)$ in the traditional $SU(2)$ rotation operator

$$R(\hat{n}, \theta) = e^{-i\theta \vec{\sigma} \cdot \hat{n}/2} \tag{7.39}$$

gives us the desired beamsplitter operator when $\hat{n}$ is taken to be the $-\hat{y}$-axis.

2. Compute the rotation operation (on dual-rail states) which this circuit performs, as a function of the phase shift φ:

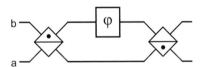

Exercise 7.11: What is $B|2, 0\rangle$ for $\theta = \pi/4$?

Exercise 7.12: (Quantum beamsplitter with classical inputs) What is $B|\alpha\rangle|\beta\rangle$ where $|\alpha\rangle$ and $|\beta\rangle$ are two coherent states as in Equation (7.16)? (*Hint:* recall that $|n\rangle = \frac{(a^\dagger)^n}{\sqrt{n!}}|0\rangle$.)

Nonlinear Kerr media. The most important effect of a Kerr medium is the *cross phase modulation* it provides between two modes of light. That is classically described by the n_2 term in (7.19), which is effectively an interaction between photons, mediated by atoms in the Kerr medium. Quantum-mechanically, this effect is described by the Hamiltonian

$$H_{xpm} = -\chi a^\dagger a b^\dagger b , \tag{7.40}$$

where a and b describe two modes propagating through the medium, and for a crystal of length L we obtain the unitary transform

$$K = e^{i\chi L a^\dagger a b^\dagger b} . \tag{7.41}$$

χ is a coefficient related to n_2, and the third order nonlinear susceptibility coefficient usually denoted as $\chi^{(3)}$. That the expected classical behavior arises from this Hamiltonian is left as Exercise 7.14 for the reader.

By combining Kerr media with beamsplitters, a controlled-NOT gate can be constructed in the following manner. For single photon states, we find that

$$K|00\rangle = |00\rangle \tag{7.42}$$
$$K|01\rangle = |01\rangle \tag{7.43}$$
$$K|10\rangle = |10\rangle \tag{7.44}$$
$$K|11\rangle = e^{i\chi L}|11\rangle , \tag{7.45}$$

and let us take $\chi L = \pi$, such that $K|11\rangle = -|11\rangle$. Now consider two dual-rail states, that is, four modes of light. These live in a space spanned by the four basis states $|e_{00}\rangle = |1001\rangle$, $|e_{01}\rangle = |1010\rangle$, $|e_{10}\rangle = |0101\rangle$, $|e_{11}\rangle = |0110\rangle$. Note that we have flipped the usual order of the two modes for the first pair, for convenience (physically, the two modes are easily swapped using mirrors). Now, if a Kerr medium is applied to act upon the two middle modes, then we find that $K|e_i\rangle = |e_i\rangle$ for all i except $K|e_{11}\rangle = -|e_{11}\rangle$. This is useful because the controlled-NOT operation can be factored into

$$\underbrace{\begin{bmatrix} 1 & 0 & 0 & 0 \\ 0 & 1 & 0 & 0 \\ 0 & 0 & 0 & 1 \\ 0 & 0 & 1 & 0 \end{bmatrix}}_{U_{CN}} = \frac{1}{\sqrt{2}} \underbrace{\begin{bmatrix} 1 & 1 & 0 & 0 \\ 1 & -1 & 0 & 0 \\ 0 & 0 & 1 & 1 \\ 0 & 0 & 1 & -1 \end{bmatrix}}_{I \otimes H} \underbrace{\begin{bmatrix} 1 & 0 & 0 & 0 \\ 0 & 1 & 0 & 0 \\ 0 & 0 & 1 & 0 \\ 0 & 0 & 0 & -1 \end{bmatrix}}_{K} \frac{1}{\sqrt{2}} \underbrace{\begin{bmatrix} 1 & 1 & 0 & 0 \\ 1 & -1 & 0 & 0 \\ 0 & 0 & 1 & 1 \\ 0 & 0 & 1 & -1 \end{bmatrix}}_{I \otimes H} ,$$

$$\tag{7.46}$$

where H is the single qubit Hadamard transform (simply implemented with beamsplitters and phase shifters), and K is the Kerr interaction we just considered, with $\chi L = \pi$. Such an apparatus has been considered before, for constructing a reversible classical optical logic gate, as described in Box 7.4; in the single photon regime, it also functions as a quantum logic gate.

Summarizing, the CNOT can be constructed from Kerr media, and arbitrary single qubit operations realized using beamsplitters and phase shifters. Single photons can be created using attenuated lasers, and detected with photodetectors. Thus, in theory, a quantum computer can be implemented using these optical components!

Exercise 7.13: (Optical Deutsch–Jozsa quantum circuit) In Section 1.4.4 (page 34), we described a quantum circuit for solving the one-bit Deutsch–Jozsa problem. Here is a version of that circuit for single photon states (in the dual-rail representation), using beamsplitters, phase shifters, and nonlinear Kerr media:

Box 7.4: The quantum optical Fredkin gate

An optical Fredkin gate can be built using two beamsplitters and a nonlinear Kerr medium as shown in this schematic diagram:

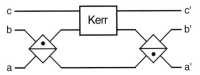

This performs the unitary transform $U = B^\dagger K B$, where B is a 50/50 beamsplitter, K is the Kerr cross phase modulation operator $K = e^{i\xi\, b^\dagger b\, c^\dagger c}$, and $\xi = \chi L$ is the product of the coupling constant and the interaction distance. This simplifies to give

$$U = \exp\left[i\xi c^\dagger c \left(\frac{b^\dagger - a^\dagger}{2} \right) \left(\frac{b - a}{2} \right) \right] \tag{7.47}$$

$$= e^{i\frac{\pi}{2} b^\dagger b}\, e^{\frac{\xi}{2} c^\dagger c (a^\dagger b - b^\dagger a)}\, e^{-i\frac{\pi}{2} b^\dagger b}\, e^{i\frac{\xi}{2} a^\dagger a\, c^\dagger c}\, e^{i\frac{\xi}{2} b^\dagger b\, c^\dagger c}. \tag{7.48}$$

The first and third exponentials are constant phase shifts, and the last two phase shifts come from cross phase modulation. All those effects are not fundamental, and can be compensated for. The interesting term is the second exponential, which is defines the quantum Fredkin operator

$$F(\xi) = \exp\left[\frac{\xi}{2} c^\dagger c (a^\dagger b - b^\dagger a) \right]. \tag{7.49}$$

The usual (classical) Fredkin gate operation is obtained for $\xi = \pi$, in which case when no photons are input at c, then $a' = a$ and $b' = b$, but when a single photon is input at c, then $a' = b$ and $b' = a$. This can be understood by realizing that $F(\chi)$ is like a controlled-beamsplitter operator, where the rotation angle is $\xi c^\dagger c$. Note that this description does not use the dual-rail representation; in that representation, this Fredkin gate corresponds to a controlled-NOT gate.

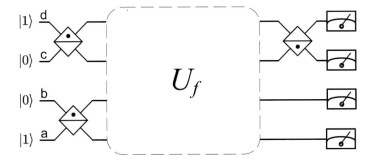

1. Construct circuits for the four possible classical functions U_f using Fredkin gates and beamsplitters.
2. Why are no phase shifters necessary in this construction?
3. For each U_f show explicitly how interference can be used to explain how the quantum algorithm works.

4. Does this implementation work if the single photon states are replaced by coherent states?

Exercise 7.14: (Classical cross phase modulation)   To see that the expected classical behavior of a Kerr medium is obtained from the definition of K, Equation (7.41), apply it to two modes, one with a coherent state and the other in state $|n\rangle$; that is, show that

$$K|\alpha\rangle|n\rangle = |\alpha e^{i\chi L n}\rangle|n\rangle . \tag{7.50}$$

Use this to compute

$$\rho_a = \text{Tr}_b\left[K|\alpha\rangle|\beta\rangle\langle\beta|\langle\alpha|K^\dagger\right] \tag{7.51}$$

$$= e^{-|\beta|^2} \sum_m \frac{|\beta|^{2m}}{m!}|\alpha e^{i\chi L m}\rangle\langle\alpha e^{i\chi L m}| , \tag{7.52}$$

and show that the main contribution to the sum is for $m = |\beta|^2$.

7.4.3 Drawbacks

The single photon representation of a qubit is attractive. Single photons are relatively simple to generate and measure, and in the dual-rail representation, arbitrary single qubit operations are possible. Unfortunately, interacting photons is difficult – the best nonlinear Kerr media available are very weak, and cannot provide a cross phase modulation of π between single photon states. In fact, because a nonlinear index of refraction is usually obtained by using a medium near an optical resonance, there is always some absorption associated with the nonlinearity, and it can theoretically be estimated that in the best such arrangement, approximately 50 photons must be absorbed for each photon which experiences a π cross phase modulation. This means that the outlook for building quantum computers from traditional nonlinear optics components is slim at best.

Nevertheless, from studying this optical quantum computer, we have gained some valuable insight into the nature of the *architecture* and system design of a quantum computer. We now can see what an actual quantum computer might look like in the laboratory (if only sufficiently good components were available to construct it), and a striking feature is that it is constructed nearly completely from optical interferometers. In the apparatus, information is encoded both in the photon number and the phase of the photon, and interferometers are used to convert between the two representations. Although it is feasible to construct stable optical interferometers, if an alternate, massive representation of a qubit were chosen, then it could rapidly become difficult to build stable interferometers because of the shortness of typical de Broglie wavelengths. Even with the optical representation, the multiple interlocked interferometers which would be needed to realize a large quantum algorithm would be a challenge to stabilize in the laboratory.

Historically, optical *classical* computers were once thought to be promising replacements for electronic machines, but they ultimately failed to live up to expectations when sufficiently nonlinear optical materials were not discovered, and when their speed and parallelism advantages did not sufficiently outweigh their alignment and power disadvantages. On the other hand, optical communications is a vital and important area; one reason for this is that for distances longer than one centimeter, the energy needed to transmit a bit using a photon over a fiber is smaller than the energy required to charge a typical 50 ohm electronic transmission line covering the same distance. Similarly, it may be that optical qubits may find a natural home in communication of quantum information, such as in quantum cryptography, rather than in computation.

Despite the drawbacks facing optical quantum computer realizations, the theoretical formalism which describes them is absolutely fundamental in all the other realizations we shall study in the remainder of this chapter. In fact, you may think of what we shall turn to next as being just another kind of optical quantum computer, but with a different (and better!) kind of nonlinear medium.

Optical photon quantum computer

- **Qubit representation**: Location of single photon between two modes, $|01\rangle$ and $|10\rangle$, or polarization.

- **Unitary evolution**: Arbitrary transforms are constructed from phase shifters (R_z rotations), beamsplitters (R_y rotations), and nonlinear Kerr media, which allow two single photons to cross phase modulate, performing $\exp\left[i\chi L|11\rangle\langle 11|\right]$.

- **Initial state preparation**: Create single photon states (e.g. by attenuating laser light).

- **Readout**: Detect single photons (e.g. using a photomultipler tube).

- **Drawbacks**: Nonlinear Kerr media with large ratio of cross phase modulation strength to absorption loss are difficult to realize.

7.5 Optical cavity quantum electrodynamics

Cavity quantum electrodynamics (QED) is a field of study which accesses an important regime involving coupling of single atoms to only a few optical modes. Experimentally, this is made possible by placing single atoms within optical cavities of very high Q; because only one or two electromagnetic modes exist within the cavity, and each of these has a very high electric field strength, the dipole coupling between the atom and the field is very high. Because of the high Q, photons within the cavity have an opportunity to interact many times with the atoms before escaping. Theoretically, this technique presents a unique opportunity to control and study single quantum systems, opening many opportunities in quantum chaos, quantum feedback control, and quantum computation.

In particular, single-atom cavity QED methods offer a potential solution to the dilemma with the optical quantum computer described in the previous section. Single photons can be good carriers of quantum information, but they require some other medium in order to interact with each other. Because they are bulk materials, traditional nonlinear optical Kerr media are unsatisfactory in satisfying this need. However, well isolated single atoms might not necessarily suffer from the same decoherence effects, and moreover, they could also provide cross phase modulation between photons. In fact, what if the state of single photons could be efficiently transfered to and from single atoms, whose interactions could be controlled? This potential scenario is the topic of this section.

7.5.1 Physical apparatus

The two main experimental components of a cavity QED system are the electromagnetic cavity and the atom. We begin by describing the basic physics of cavity modes, and then summarize basic ideas about atomic structure and the interaction of atoms with light.

Fabry–Perot cavity

The main interaction involved in cavity QED is the dipolar interaction $\vec{d} \cdot \vec{E}$ between an electric dipole moment $\vec{d}$ and an electric field $\vec{E}$. How large can this interaction be? It is difficult in practice to change the size of $\vec{d}$; however, $|\vec{E}|$ is experimentally accessible, and one of the most important tools for realizing a very large electric field in a narrow band of frequencies and in a small volume of space, is the Fabry–Perot cavity.

In the approximation that the electric field is monochromatic and occupies a single spatial mode, it can be given a very simple quantum-mechanical description:

$$\vec{E}(r) = i\vec{\epsilon}\, E_0 \left[a e^{ikr} - a^\dagger e^{-ikr} \right] . \tag{7.56}$$

As described in Box 7.5, these approximations are appropriate for the field in a Fabry–Perot cavity. Here, $k = \omega/c$ is the spatial frequency of the light, E_0 is the field strength, $\vec{\epsilon}$ is the polarization, and r is the position at which the field is desired. a and $a^\dagger$ are creation and annihilation operators for photons in the mode, and behave as described in Section 7.4.2. Note that the Hamiltonian governing the evolution of the field in the cavity is simply

$$H_{\text{field}} = \hbar\omega a^\dagger a , \tag{7.57}$$

and this is consistent with the semiclassical notion that the energy is the volume integral of $|\vec{E}|^2$ in the cavity.

Exercise 7.15: Plot (7.55) as a function of field detuning φ, for $R_1 = R_2 = 0.9$.

Two-level atoms

Until this section of the chapter, we have discussed only photons, or interactions such as the cross phase modulation between photons mediated by a semiclassical medium. Now, let us turn our attention to atoms, their electronic structure, and their interactions with photons. This is, of course, a very deep and well-developed field of study; we shall only describe a small part of it that touches upon quantum computation.

The electronic energy eigenstates of an atom can be very complicated (see Box 7.6), but for our purposes modeling an atom as having only two states is an excellent approximation. This *two-level atom* approximation can be valid because we shall be concerned with the interaction with monochromatic light and, in this case, the only relevant energy levels are those satisfying two conditions: their energy difference matches the energy of the incident photons, and symmetries ('selection rules') do not inhibit the transition. These conditions arise from basic conservation laws for energy, angular momentum, and parity. Energy conservation is no more than the condition that

$$\hbar\omega = E_2 - E_1 , \tag{7.58}$$

where E_2 and E_1 are two eigenenergies of the atom. Angular momentum and parity conservation requirements can be illustrated by considering the matrix element of $\hat{r}$ between two orbital wavefunctions, $\langle l_1, m_1 | \hat{r} | l_2, m_2 \rangle$. Without loss of generality, we can

Box 7.5: The Fabry–Perot cavity

A basic component of a Fabry–Perot cavity is a partially silvered mirror, off which incident light E_a and E_b partially reflect and partially transmit, producing the output fields $E_{a'}$ and $E_{b'}$. These are related by the unitary transform

$$\begin{bmatrix} E_{a'} \\ E_{b'} \end{bmatrix} = \begin{bmatrix} \sqrt{R} & \sqrt{1-R} \\ \sqrt{1-R} & -\sqrt{R} \end{bmatrix} \begin{bmatrix} E_a \\ E_b \end{bmatrix}, \tag{7.53}$$

where R is the reflectivity of the mirror, and the location of the '$-$' sign is a convention chosen as given here for convenience.

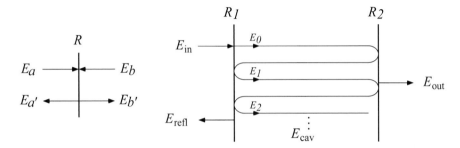

A Fabry–Perot cavity is made from two plane parallel mirrors of reflectivities R_1 and R_2, upon which light E_{in} is incident from the outside, as shown in the figure. Inside the cavity, light bounces back and forth between the two mirrors, such that the field acquires a phase shift $e^{i\varphi}$ on each round-trip; φ is a function of the path length and the frequency of the light. Thus, using (7.53), we find the cavity internal field to be

$$E_{cav} = \sum_k E_k = \frac{\sqrt{1-R_1}\,E_{in}}{1 + e^{i\varphi}\sqrt{R_1 R_2}}, \tag{7.54}$$

where $E_0 = \sqrt{1-R_1}\,E_{in}$, and $E_k = -e^{i\varphi}\sqrt{R_1 R_2}\,E_{k-1}$. Similarly, we find $E_{out} = e^{i\varphi/2}\sqrt{1-R_2}$, and $E_{refl} = \sqrt{R_1}\,E_{in} + \sqrt{1-R_1}\sqrt{R_2}\,e^{i\varphi}\,E_{cav}$.

One of the most important characteristics of a Fabry–Perot cavity for our purpose is the power in the cavity internal field as a function of the input power and field frequency,

$$\frac{P_{cav}}{P_{in}} = \left| \frac{E_{cav}}{E_{in}} \right|^2 = \frac{1-R_1}{|1 + e^{i\varphi}\sqrt{R_1 R_2}|^2}. \tag{7.55}$$

Two aspects are noteworthy. First, frequency selectivity is given by the fact that $\varphi = \omega d/c$, where d is the mirror separation, c is the speed of light, and ω is the frequency of the field. Physically, it comes about because of constructive and destructive interference between the cavity field and the front surface reflected light. And second, on resonance, the cavity field achieves a maximum value which is approximately $1/(1-R)$ times the incident field. This property is invaluable for cavity QED.

take $\hat{r}$ to be in the $\hat{x} - \hat{y}$ plane, such that it can be expressed in terms of spherical harmonics (Box 7.6) as

$$\hat{r} = \sqrt{\frac{3}{8\pi}} \left[(-r_x + ir_y)Y_{1,+1} + (r_x + ir_y)Y_{1,-1} \right] \tag{7.59}$$

In this basis, the relevant terms in $\langle l, m_1 | \hat{r} | l, m_2 \rangle$ are

$$\int Y^*_{l_1 m_1} Y_{1m} Y_{l_2 m_2} \, d\Omega . \tag{7.60}$$

Recall that $m = \pm 1$; this integral is non-zero only when $m_2 - m_1 = \pm 1$ and $\Delta l = \pm 1$. The first condition is the conservation of angular momentum, and the second, parity, under the dipole approximation where $\langle l_1, m_1 | \hat{r} | l_2, m_2 \rangle$ becomes relevant. These conditions are selection rules which are important in the two-level atom approximation.

Exercise 7.16: (Electric dipole selection rules) Show that (7.60) is non-zero only when $m_2 - m_1 = \pm 1$ and $\Delta l = \pm 1$.

In reality, light is never *perfectly* monochromatic; it is generated from some source such as a laser, in which longitudinal modes, pump noise, and other sources give rise to a finite linewidth. Also, an atom coupled to the external world never has *perfectly* defined energy eigenstates; small perturbations such as nearby fluctuating electric potentials, or even interaction with the vacuum, cause each energy level to be smeared out and become a distribution with finite width.

Nevertheless, by choosing an atom and excitation energy carefully, and by taking advantage of the selection rules, it is possible to arrange circumstances such that the two-level atom approximation is superb. The whole point of this procedure is that in this approximation, if $|\psi_1\rangle$ and $|\psi_2\rangle$ are the two selected levels, then the matrix elements of the $\hat{r}$ are

$$r_{ij} = \langle \psi_i | \hat{r} | \psi_j \rangle \approx r_0 Y , \tag{7.65}$$

where r_0 is some constant, and Y is a Pauli operator (Section 2.1.3; that we obtain Y as opposed to X doesn't really matter – it is a matter of convention, and convenience, for later calculations). This will be relevant in describing interactions between the atom and incident electric fields. The Hamiltonian of the atom itself, in this two-level subspace, is simply

$$H_{\text{atom}} = \frac{\hbar\omega_0}{2} Z , \tag{7.66}$$

where $\hbar\omega_0$ is the difference of the energies of the two levels, since the two states are energy eigenstates.

7.5.2 The Hamiltonian

The $\vec{d} \cdot \vec{E}$ interaction between an atom and a cavity confined electric field can be approximated quite well by a much simpler model, in the two-level approximation of the atom, using the quantization of the field in the cavity, and the minute size of the electron compared to the wavelength of the field. Using the fact that $\vec{d} \propto \hat{r}$ (the electric dipole size is charge times distance), we can combine (7.56) with (7.65) to obtain the interaction

$$H_I = -igY(a - a^\dagger) , \tag{7.67}$$

Box 7.6: Energy levels of an atom

The electrons of an atom behave like particles in a three dimensional box, with a Hamiltonian of the form

$$H_A = \sum_k \frac{|\vec{p}_k|^2}{2m} - \frac{Ze^2}{r_k} + H_{rel} + H_{ee} + H_{so} + H_{hf}, \qquad (7.61)$$

where the first two terms describe the balance of the electrons' kinetic energy with the Coulomb attraction of the negatively charged electrons to the positively charged nucleus, H_{rel} is a relativistic correction term, H_{ee} describes electron–electron couplings and contributions from the fermionic nature of the electrons, H_{so} is the spin orbit interaction, which can be interpreted as the spin of the electron interacting with a magnetic field generated by its orbit around the atom, and H_{hf} is the hyperfine interaction: the electron spin interacting with the magnetic field generated by the nucleus. The energy eigenstates of H_A are generally pretty well categorized according to three integers or half-integers (*quantum numbers*): n, the principle quantum number; l, the orbital angular momentum; and m, its $\hat{z}$ component. In addition, S, the total electron spin, and I, the nuclear spin, are often important. The eigenvalues of H are roughly determined to order α^2 by n, to slightly smaller order by H_{ee}, to order α^4 by H_{rel} and H_{so}, and to order $\approx 10^{-3}\alpha^4$ by H_{hf}, where $\alpha = 1/137$ is the dimensionless *fine structure constant*.

The derivation of n is simple and follows the usual one-dimensional Schrödinger equation solutions for a particle in a box, since the Coulomb confining potential is dependent on radial distance only. However, orbital angular momentum is a feature of being in three dimensions which deserves some explanation. The essential properties arise from the angular dependence of the coordinate representation of H_A, in which $\vec{p}$ becomes the Laplacian operator $\vec{\nabla}^2$, giving the Schrödinger equation

$$\frac{\Phi(\varphi)}{\sin\theta} \frac{d}{d\theta}\left(\sin\theta \frac{d\Theta}{d\theta}\right) + \frac{\Theta(\theta)}{\sin^2\theta} \frac{d\Phi(\varphi)}{d\varphi^2} + l(l+1)\Theta(\theta)\Phi(\varphi) = 0, \qquad (7.62)$$

where θ and φ are the usual spherical coordinates, and Φ and Θ are the eigenfunctions we desire. The solutions $Y_{lm}(\theta, \varphi) = \Theta_{lm}(\theta)\Phi_m(\varphi)$ are the *spherical harmonics*

$$Y_{lm}(\theta, \varphi) \equiv (-1)^m \sqrt{\frac{2l+1}{4\pi} \frac{(l-m)!}{(l+m)!}} P_{lm}(\cos\theta)e^{im\varphi}, \qquad (7.63)$$

where P_{lm} are the usual Legendre functions

$$P_{lm}(x) = \frac{(1-x^2)^{m/2}}{2^l l!} \frac{d^{m+l}}{dx^{m+l}}(x^2-1)^l. \qquad (7.64)$$

In these equations, $-l \leq m \leq l$, and it can be shown that m and l must be either integer or half-integer. l is known as the orbital angular momentum, and m is its component along the $\hat{z}$ axis. Similarly, the electron spins S and the nuclear spin I have components m_s and m_i. As you can see, the description of the energy states of an atom can be quite complicated! Summarizing: for our purposes we may think of the eigenenergies of an atom as being determined by seven numbers: n, l, m, S, m_s, I, and m_i.

where we have chosen $r = 0$ as the point to place the atom (and thus evaluate $\vec{E}$), and also oriented the atom such that $\hat{r}$ is aligned properly with the electric field vector. g is some constant (we need not be concerned about the specific values here, just the forms) which describes the strength of the interaction. The i is present simply to allow g to be real, since H_I must be Hermitian. H_I can be simplified further, by recognizing that it contains terms which are generally small; to see these, it is useful to define the Pauli raising and lowering operators,

$$\sigma_\pm = \frac{X \pm iY}{2},$$ (7.68)

such that we can re-express H_I as

$$H_I = g(\sigma_+ - \sigma_-)(a - a^\dagger).$$ (7.69)

The terms containing $\sigma_+ a^\dagger$ and $\sigma_- a$ oscillate at twice the frequencies of interest, which are ω and ω_0, and dropping them is a fairly good approximation (the *rotating wave approximation*) which leads us to the total Hamiltonian $H = H_{\text{atom}} + H_{\text{field}} + H_I$,

$$H = \frac{\hbar\omega_0}{2}Z + \hbar\omega a^\dagger a + g(a^\dagger \sigma_- + a\sigma_+).$$ (7.70)

where, again, just to recap: the Pauli operators act on the two-level atom, $a^\dagger$, a are raising and lowering operators on the single mode field, ω is the frequency of the field, ω_0 is the frequency of the atom, and g is the coupling constant for the interaction between atom and field. This is the fundamental theoretical tool in the study of cavity QED, the *Jaynes–Cummings* Hamiltonian, which describes interactions between two-level atoms and an electromagnetic field.

This Hamiltonian can be written in another convenient form by noting that $N = a^\dagger a + Z/2$ is a constant of the motion, that is $[H, N] = 0$, so that we find

$$H = \hbar\omega N + \delta Z + g(a^\dagger \sigma_- + a\sigma_+),$$ (7.71)

where $\delta = (\omega_0 - \omega)/2$ is known as the *detuning* – the frequency difference between the field and atomic resonance. This Hamiltonian, the Jaynes–Cummings Hamiltonian, is very important, and we shall be spending nearly all of the rest of the chapter studying its properties and guises in different physical systems.

Exercise 7.17: (Eigenstates of the Jaynes–Cummings Hamiltonian) Show that

$$|\chi_n\rangle = \frac{1}{\sqrt{2}}\left[|n, 1\rangle + |n + 1, 0\rangle\right]$$ (7.72)

$$|\overline{\chi}_n\rangle = \frac{1}{\sqrt{2}}\left[|n, 1\rangle - |n + 1, 0\rangle\right]$$ (7.73)

are eigenstates of the Jaynes–Cummings Hamiltonian (7.71) for $\omega = \delta = 0$, with the eigenvalues

$$H|\chi_n\rangle = g\sqrt{n + 1}|\chi_n\rangle$$ (7.74)
$$H|\overline{\chi}_n\rangle = -g\sqrt{n + 1}|\overline{\chi}_n\rangle,$$ (7.75)

where the labels in the ket are $|\text{field}, \text{atom}\rangle$.

7.5.3 Single-photon single-atom absorption and refraction

The most interesting regime in cavity QED, for our purposes, is that in which *single* photons interact with *single* atoms. This is an unusual regime, in which traditional concepts (such as index of refraction and permittivity) in classical theories of electromagnetism break down. In particular, we would like to utilize a single atom to obtain a nonlinear interaction between photons.

Let us begin by showing one striking and general characteristic of the atom–field system known as *Rabi oscillations*. Without loss of generality we may neglect N, since it only contributes a fixed phase. Recalling that time evolution is given by $U = e^{-iHt}$ (here and in the following, it will often be convenient to drop $\hbar$, and we shall do so freely), and focusing on the case of at most a single excitation in the field mode, where

$$
H = - \begin{bmatrix} \delta & 0 & 0 \\ 0 & \delta & g \\ 0 & g & -\delta \end{bmatrix} , \tag{7.76}
$$

(the basis states are $|00\rangle$, $|01\rangle$, $|10\rangle$, from left to right and top to bottom, where the left label corresponds to the field, and the right one to the atom), we find that

$$
\begin{aligned}
U = \ & e^{-i\delta t}|00\rangle\langle 00| \\
& + (\cos \Omega t + i\frac{\delta}{\Omega} \sin \Omega t)|01\rangle\langle 01| \\
& + (\cos \Omega t - i\frac{\delta}{\Omega} \sin \Omega t)|10\rangle\langle 10| \\
& - i\frac{g}{\Omega} \sin \Omega t \left(|01\rangle\langle 10| + |10\rangle\langle 01| \right) .
\end{aligned} \tag{7.77}
$$

The interesting behavior is in the last line of this equation, which shows that the atom and field oscillate back and forth exchanging a quantum of energy, at the *Rabi frequency* $\Omega = \sqrt{g^2 + \delta^2}$.

Exercise 7.18: (Rabi oscillations) Show that (7.77) is correct by using

$$
e^{i\hat{n}\cdot\vec{\sigma}} = \sin |n| + i\hat{n} \cdot \vec{\sigma} \cos |n| \tag{7.78}
$$

to exponentiate H. This is an unusually simple derivation of the Rabi oscillations and the Rabi frequency; ordinarily, one solves coupled differential equations to obtain Ω, but here we obtain the essential dynamics just by focusing on the single-atom, single-photon subspace!

The transformation of the photon, in interacting with a single atom, can be obtained by tracing over the atom's state (Section 2.4.3). The probability that an initial photon $|1\rangle$ is absorbed by the atom (which we assume starts in its ground state, $|0\rangle$) is simply

$$
\chi_r = \sum_k |\langle 0k|U|10\rangle|^2 = \frac{g^2}{g^2 + \delta^2} \sin^2 \Omega t . \tag{7.79}
$$

This has the usual Lorentzian profile expected for absorption as a function of detuning δ from resonance.

The refractive index (of the single atom!) is given by the matrix elements of U in which the atom stays in the ground state. The phase shift experienced by the photon is the difference in the angle of rotation experienced by the $|1\rangle$ and the $|0\rangle$ states of the field, tracing over the atom. This is found to be

$$\chi_i = \arg\left[e^{i\delta t}\left(\cos\Omega t - i\frac{\delta}{\Omega}\sin\Omega t\right)\right]. \qquad (7.80)$$

For fixed non-zero δ, as the coupling g is decreased, the absorption probability χ_r decreases as g^2, but the phase shift χ_i remains nearly constant. This is the origin of materials which can perform phase shifts without scattering much light.

Exercise 7.19: (Lorentzian absorption profile) Plot (7.79) for $t = 1$ and $g = 1.2$, as a function of the detuning δ, and (if you know it) the corresponding classical result. What are the oscillations due to?

Exercise 7.20: (Single photon phase shift) Derive (7.80) from U, and plot it for $t = 1$ and $g = 1.2$, as a function of the detuning δ. Compare with δ/Ω^2.

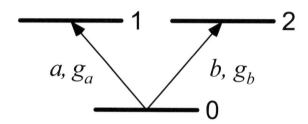

Figure 7.4. Three level atom (with levels 0, 1, and 2) interacting with two orthogonal polarizations of light, described by the operators a and b. The atom–photon couplings are respectively g_a and g_b. The energy differences between 0 and 1, and between 0 and 2 are assumed to be nearly equal.

A natural application of the atom–photon interaction is to study what happens when two different photon modes (each containing at most one photon) interact with the same atom. This can give rise to a nonlinear interaction between the two modes. Recall from Section 7.4.2 that nonlinear Kerr media can be described phenomenologically as media which induce a cross phase modulation with Hamiltonian of the form $H = \chi a^\dagger a b^\dagger b$. There, we did not see how that effect arises from fundamental interactions. Using the present formalism, the origin of the Kerr effect can be illustrated using a simple model, in which two polarizations of light interact with a three-level atom, as shown in Figure 7.4. This is described by a modified version of the Jaynes–Cummings Hamiltonian,

$$H = \delta\begin{bmatrix} -1 & 0 & 0 \\ 0 & 1 & 0 \\ 0 & 0 & 1 \end{bmatrix} + g_a\left(a\begin{bmatrix} 0 & 0 & 0 \\ 1 & 0 & 0 \\ 0 & 0 & 0 \end{bmatrix} + a^\dagger\begin{bmatrix} 0 & 1 & 0 \\ 0 & 0 & 0 \\ 0 & 0 & 0 \end{bmatrix}\right)$$
$$+ g_b\left(b\begin{bmatrix} 0 & 0 & 0 \\ 0 & 0 & 0 \\ 1 & 0 & 0 \end{bmatrix} + b^\dagger\begin{bmatrix} 0 & 0 & 1 \\ 0 & 0 & 0 \\ 0 & 0 & 0 \end{bmatrix}\right), \qquad (7.81)$$

where the basis elements for the 3×3 atom operators are $|0\rangle$, $|1\rangle$, and $|2\rangle$. In matrix form, the relevant terms in H are found to be the block-diagonal matrix

$$
H = \begin{bmatrix} H_0 & 0 & 0 \\ 0 & H_1 & 0 \\ 0 & 0 & H_2 \end{bmatrix} ,
\tag{7.82}
$$

where

$$
H_0 = -\delta
\tag{7.83}
$$

$$
H_1 = \begin{bmatrix} -\delta & g_a & 0 & 0 \\ g_a & \delta & 0 & 0 \\ 0 & 0 & -\delta & g_b \\ 0 & 0 & g_b & \delta \end{bmatrix}
\tag{7.84}
$$

$$
H_2 = \begin{bmatrix} -\delta & g_a & g_b \\ g_a & \delta & 0 \\ g_b & 0 & \delta \end{bmatrix} .
\tag{7.85}
$$

in the basis $|a, b, \text{atom}\rangle = |000\rangle$ for H_0, $|100\rangle, |001\rangle, |010\rangle, |002\rangle$ for H_1, and $|110\rangle$, $|011\rangle, |102\rangle$ for H_2, across the columns from left to right. Exponentiating to give $U = \exp(iHt)$ allows one to find the single photon phase shifts $\varphi_a = \arg(\langle 100|U|100\rangle) - \arg(\langle 000|U|000\rangle)$ and $\varphi_b = \arg(\langle 010|U|010\rangle) - \arg(\langle 000|U|000\rangle)$ and the two photon phase shift $\varphi_{ab} = \arg(\langle 110|U|110\rangle) - \arg(\langle 000|U|000\rangle)$. For linear media, one would expect that $\varphi_{ab} = \varphi_a + \varphi_b$, that is, the two photon state has twice the phase shift of the single photon state, since $\exp[-i\omega(a^\dagger a + b^\dagger b)]|11\rangle = \exp(-2i\omega)|11\rangle$. However, this system behaves nonlinearly, and gives $\chi_3 \equiv \varphi_{ab} - \varphi_a - \varphi_b$ as shown in Figure 7.5. In this physical system, this Kerr effect arises from the slight amplitude for the atom to exchange quanta between the two optical modes.

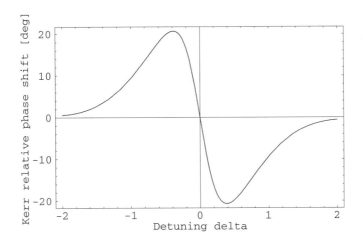

Figure 7.5. Kerr phase shift χ_3 in degrees, for $t = 0.98$ and $g_a = g_b = 1$, plotted as a function of the detuning δ, computed from (7.82) for single photons interacting with a single three-level atom.

Exercise 7.21: Explicitly exponentiate (7.82) and show that

$$
\varphi_{ab} = \arg \left[e^{i\delta t} \left(\cos \Omega' t - i \frac{\delta}{\Omega'} \sin \Omega' t \right) \right] ,
\tag{7.86}
$$

where $\Omega' = \sqrt{\delta^2 + g_a^2 + g_b^2}$. Use this to compute χ_3, the nonlinear Kerr phase shift. This is a very simple way to model and understand the Kerr interaction, which sidesteps much of the complication typically involved in classical nonlinear optics.

Exercise 7.22: Associated with the cross phase modulation is also a certain amount of loss, which is given by the probability that a photon is absorbed by the atom. Compute this probability, $1 - \langle 110|U|110 \rangle$, where $U = \exp(-iHt)$ for H as in (7.82); compare with $1 - \langle 100|U|100 \rangle$ as a function of δ, g_a, g_b, and t.

7.5.4 Quantum computation

Broadly speaking, cavity QED techniques can be used to perform quantum computation in a number of different ways, two of which are the following: quantum information can be represented by photon states, using cavities with atoms to provide nonlinear interactions between photons; or quantum information can be represented using atoms, using photons to communicate between the atoms. Let us now close out this subject by describing an experiment which demonstrates the first of these methods to realize a quantum logic gate.

As we saw in Section 7.4.2, a quantum computer can be constructed using single photon states, phase shifters, beamsplitters, and nonlinear Kerr media, but the π cross phase modulation required to produce a controlled-NOT gate is nearly infeasible with standard bulk nonlinear optics techniques. Cavity QED can be used to implement a Kerr interaction, as shown in Section 7.5.3; unlike for bulk media, this can have a very strong effect even at the single photon level, because of the strong field provided by a Fabry–Perot type cavity.

Figure 7.6 illustrates a cavity QED experiment which was performed (see 'History and further reading' at the end of the chapter) to demonstrate the potential for realizing a logic gate with the unitary transform

$$\begin{bmatrix} 1 & 0 & 0 & 0 \\ 0 & e^{i\varphi_a} & 0 & 0 \\ 0 & 0 & e^{i\varphi_b} & 0 \\ 0 & 0 & 0 & e^{i(\varphi_a+\varphi_b+\Delta)} \end{bmatrix}, \tag{7.87}$$

where $\Delta = 16°$, using single photons. In the experiment, two modes of light (distinguished by a very small frequency difference) with weak coherent states are prepared, one linearly polarized (the probe), and one circularly polarized (the pump), as input to the cavity. This state can be expressed as

$$|\psi_{\text{in}}\rangle = |\beta^+\rangle \left[\frac{|\alpha^+\rangle + |\alpha^-\rangle}{\sqrt{2}} \right], \tag{7.88}$$

recalling that linearly polarized light is an equal superposition of the two possible circularly polarized states, $+$ and $-$. Approximating the weak coherent states as $|\alpha\rangle \approx |0\rangle + \alpha|1\rangle$ and similarly for $|\beta\rangle$ (and leaving out normalizations for the moment) gives

$$|\psi_{\text{in}}\rangle \approx \left[|0^+\rangle + \beta|1^+\rangle \right] \left[|0^+\rangle + \alpha|1^+\rangle + |0^-\rangle + \alpha|1^-\rangle \right]. \tag{7.89}$$

These photons pass through the optical cavity and interact with the atom, which is modeled as causing a different phase shift to occur to states depending on the total number of photons in each polarization (independent of which mode the photons are in). Specifically, we assume that a photon in the $|1^+\rangle$ state experiences a $e^{i\varphi_a}$ phase shift if it is in the probe beam, and $e^{i\varphi_b}$ for the pump. In addition to this single photon phase shift, the state $|1^+1^+\rangle$ experiences an additional Kerr phase shift Δ, so it becomes $e^{i(\varphi_a+\varphi_b+\Delta)}|1^+1^+\rangle$. Other states (and in particular, other polarizations) remain unchanged. The physics which leads to this behavior is similar to that described in Section 7.5.3, and the end effect is the same: a cross phase modulation between the pump and the probe light. The output from the cavity is thus

$$|\psi_{\text{out}}\rangle \approx |0^+\rangle\left[|0^+\rangle + \alpha e^{i\varphi_a}|1^+\rangle + |0^-\rangle + \alpha|1^-\rangle\right]$$

$$+ e^{i\varphi_b}\beta|1^+\rangle\left[|0^+\rangle + \alpha e^{i(\varphi_a+\Delta)}|1^+\rangle + |0^-\rangle + \alpha|1^-\rangle\right] \quad (7.90)$$

$$\approx |0^+\rangle|\alpha, \varphi_a/2\rangle + e^{i\varphi_b}\beta|1^+\rangle|\alpha, (\varphi_a+\Delta)/2\rangle, \quad (7.91)$$

where $|\alpha, \varphi_a/2\rangle$ denotes a linearly polarized probe field rotated from the vertical by $\varphi_a/2$. The field polarizations are measured by the detector, giving $\varphi_a \approx 17.5°$, $\varphi_b \approx 12.5°$ and $\Delta \approx 16°$. Since Δ is a non-trivial value, this result suggests that a universal two qubit logic gate (Exercise 7.23) is possible using single photons, and a single atom in a cavity as a nonlinear optical Kerr medium to interact photons.

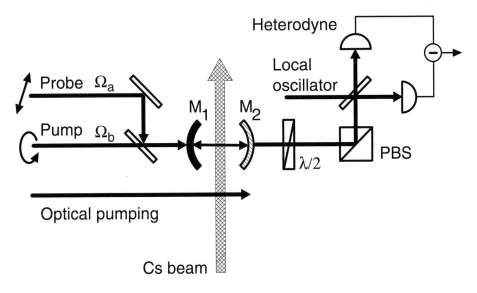

Figure 7.6. Schematic of an experimental apparatus used to demonstrate the possibility of using a single atom to provide cross phase modulation between single photons, as an elementary quantum logic gate. A linearly polarized weak probe beam of light Ω_a, and a stronger circularly polarized pump beam Ω_b are prepared and shone on an optical cavity with high reflectivity mirrors M_1 and M_2. Cesium atoms prepared in the electronic state $6S_{1/2}, F = 4, m = 4$ by optical pumping fall (the figure shows the atoms upside down) such that the average number of atoms in the cavity is around one. The light traverses the cavity, interacting with the atom; $\sigma+$ polarized light causes strong transitions to the $6P_{3/2}, F' = 5, m' = 5$ state, and the orthogonal $\sigma-$ polarized light causes weak transitions to the $6P_{3/2}, F' = 5, m' = 3$ state. The polarization of the output light is then measured, using a half wave plate, a polarizing beamsplitter (PBS), and a sensitive balanced heterodyne detector (which selectively detects light at a specific frequency, as determined by the local oscillator). Figure courtesy of Q. Turchette.

Several important caveats must be kept in mind in interpreting these experimental results. The incident photons are absorbed with non-trivial probability when traversing the cavity and atom, and thus the true quantum operation performed is not unitary; this problem would be aggravated if multiple gates were cascaded, which would be required, for example, to realize a controlled-NOT gate (which requires $\Delta = \pi$). In fact, reflection losses of the cavity arrangement used in this experiment would significantly impede cascading; to understand how to get around this, a proper time-dependent model would have to be developed and studied. Also, although the cross phase modulation model is consistent with data measured, the photon–atom interaction model used is an *ansatz*, and other models are not ruled out by the experiment. In fact, it would be possible in principle to use single photon states (as opposed to attenuated coherent states) in the experiment, and measurements of the resulting entanglement of the two modes in $|\psi_{out}\rangle$ would be a good test. At the time this experiment was carried out, no general procedure was known for fully characterizing a quantum operation and its suitability as a quantum logic gate. However, a method for doing this, known as *process tomography*, is now well understood (Chapter 8), and remarkably it even allows full characterization of dissipation and other non-unitary behavior. Performing such a test would unambiguously determine exactly the extent to which the experiment described here actually reflects a quantum computation.

Despite these drawbacks, the experiment does demonstrate fundamental concepts required for quantum information processing. It certifies that nonlinear optical behavior such as the Kerr interaction really does occur at the single photon level, thus validating the essence of the Jaynes–Cummings model. Also, this experiment is performed in what is called the *bad cavity regime*, where the atom's coherent coupling rate g^2/κ to the cavity mode dominates incoherent emission rate γ into free space, but this coupling is weaker than the rate κ at which input photons enter and leave the cavity. The *strong coupling* operating regime, in which $g > \kappa > \gamma$, offers an alternative in which larger conditional phase shifts Δ may be obtained.

Most importantly, perhaps, cavity QED opens the door to a wealth of additional interactions which are valuable for quantum information processing. We have also seen how the quantum information perspective – focusing on single photons and single atoms – has allowed us take the Jaynes–Cummings Hamiltonian, the basic cavity QED interaction, and construct from it some of the most fundamental physics of the interaction of electromagnetic waves with matter. We now leave the subject of cavity QED, but as we continue on next to ion traps, and then to magnetic resonance, we shall keep with us these notions of photon–atom interactions, single atoms and photons, and the Jaynes–Cummings Hamiltonian.

Exercise 7.23: Show that the two qubit gate of (7.87) can be used to realize a controlled-NOT gate, when augmented with arbitrary single qubit operations, for any φ_a and φ_b, and $\Delta = \pi$. It turns out that for nearly any value of Δ this gate is universal when augmented with single qubit unitaries.

Optical cavity quantum electrodynamics

- **Qubit representation**: Location of single photon between two modes, $|01\rangle$ and $|10\rangle$, or polarization.

- **Unitary evolution**: Arbitrary transforms are constructed from phase shifters (R_z

rotations), beamsplitters (R_y rotations), and a cavity QED system, comprised of a Fabry–Perot cavity containing a few atoms, to which the optical field is coupled.

- **Initial state preparation**: Create single photon states (e.g. by attenuating laser light).
- **Readout**: Detect single photons (e.g. using a photomultipler tube).
- **Drawbacks**: The coupling of two photons is mediated by an atom, and thus it is desirable to increase the atom–field coupling. However, coupling the photon into and out of the cavity then becomes difficult, and limits cascadibility.

7.6 Ion traps

Thus far in this chapter, we have focused mainly on representing qubits using photons. Let us now turn to representations which use atomic and nuclear states. Specifically, as we saw in Section 7.1, electron and nuclear spins provide potentially good representations for qubits. Spin is a strange (but very real!) concept (Box 7.7), but since the energy difference between different spin states is typically very small compared with other energy scales (such as the kinetic energy of typical atoms at room temperature), the spin states of an atom are usually difficult to observe, and even more difficult to control. In carefully crafted environments, however, exquisite control is possible. Such circumstances are provided by isolating and trapping small numbers of charged atoms in electromagnetic traps, then cooling the atoms until their kinetic energy is much lower than the spin energy contribution. After doing this, incident monochromatic light can be tuned to selectively cause transitions which change certain spin states depending on other spin states. This is the essence of how trapped ions can be made to perform quantum computation, as we describe in this section. We begin with an overview of the experimental apparatus and its main components, then we present a Hamiltonian modeling the system. We describe an experiment which has been performed to demonstrate a controlled-NOT gate with trapped ^{9}Be ions, and then close with a few comments on the potential and limitations of the method.

Exercise 7.24: The energy of a nuclear spin in a magnetic field is approximately $\mu_N B$, where $\mu_N = eh/4\pi m_p \approx 5\times10^{-27}$ joules per tesla is the nuclear Bohr magneton. Compute the energy of a nuclear spin in a $B = 10$ tesla field, and compare with the thermal energy $k_B T$ at $T = 300$ K.

7.6.1 Physical apparatus

An ion trap quantum computer has as its main components an electromagnetic trap with lasers and photodetectors, and ions.

Trap geometry and lasers

The main experimental apparatus, an electromagnetic trap constructed from four cylindrical electrodes, is shown in Figure 7.7. The end segments of the electrodes are biased at a different voltage U_0 than the middle, so that the ions are axially confined by a static potential $\Phi_{dc} = \kappa U_0 \left[z^2 - (x^2 + y^2) \right] /2$ along the $\hat{z}$ axis (κ is a geometrical factor). However, a result known as *Earnshaw's theorem* states that a charge cannot be confined in three dimensions by static potentials. Thus, to provide confinement, two of the electrodes

Box 7.7: Spin

Spin is a strange concept. When a particle has spin, it possesses a magnetic moment as if it were a composite particle with some current running around in a loop. But electrons are elementary particles, and the quarks which compose a nucleon are not known to produce spin by orbital motion. Furthermore, the spin of a particle is only ever either integer or half-integer.

Spin is nevertheless quite real, and an important part of everyday physics. Integer spin particles, known as *bosons*, include the photon. Being massless, it is somewhat special and only has spin ± 1 (and no spin zero) components; these correspond to the two familiar orthogonal polarization states. Sunglasses made from cheap plastic polarizers are effective when driving because sunlight becomes partially polarized in the opposite direction after reflecting off of surfaces such as roadways (light polarized with the electric field transverse to the interface always partially reflects no matter the angle of incidence, in contrast with the transverse magnetic polarization which does not reflect when the angle of incidence is at *Brewster's angle*). Half-integer spin particles, known as *fermions*, include the electron, proton, and neutron. These are 'spin-1/2' particles, in that their spin component can either be +1/2 (spin 'up') or −1/2 (spin 'down'). When we say 'spin' often what is meant is a spin-1/2 particle.

The energy eigenstates of an atom intimately involve spin, and the combination of multiple spins. For example, the nucleus of ^{9}Be has spin 3/2. Spins interact with a magnetic field just as magnetic moments do; in a magnetic field $\vec{B}$, an electron with spin $\vec{S}$ has energy $g_e \vec{S} \cdot \vec{B}$, and similarly, a nucleus $\vec{I}$ has energy $g_n \vec{I} \cdot \vec{B}$. Pictorially, for example, the spin contribution to an atom's energy levels can be viewed as:

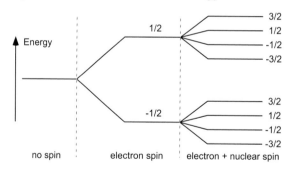

where we have assumed a spin-1/2 electron, and a spin-3/2 nucleus. By tuning the frequency of an incident laser just right, any of these transitions could be selected, as long as conservation laws (Section 7.5.1) are satisfied. In particular, angular momentum conversation implies that when a photon is absorbed by an atom, one unit of angular momentum or spin must change between the initial and final states. These states thus must have definite values of angular momenta; this can be taken into account.

Unlike continuous variables such as position and momentum, and other infinite Hilbert space systems which must be artificially truncated to represent quantum bits, spin states provide good representations for quantum information because they live in an inherently finite state space.

are grounded, while the other two are driven by a fast oscillating voltage which creates a radiofrequency (RF) potential $\Phi_{\mathrm{rf}} = (V_0 \cos \Omega_T t + U_r)(1 + (x^2 - y^2)/R^2)/2$, where R is a geometrical factor. The segments of the electrodes are capacitively coupled such that the RF potential is constant across them. The combination of Φ_{dc} and Φ_{rf} creates, on average (over Ω_T), a harmonic potential in x, y, and z. Together with the Coulomb repulsion of the ions, this gives a Hamiltonian governing the motion of the N ions in the trap,

$$H = \sum_{i=1}^{N} \frac{M}{2} \left(\omega_x^2 x_i^2 + \omega_y^2 y_i^2 + \omega_z^2 z_i^2 + \frac{|\vec{p}_i|^2}{M^2} \right) + \sum_{i=1}^{N} \sum_{j>i} \frac{e^2}{4\pi\epsilon_0 |\vec{r}_i - \vec{r}_j|}, \qquad (7.92)$$

where M is the mass of each ion. Typically, $\omega_x, \omega_y \gg \omega_z$ by design, so that the ions all lie generally along the $\hat{z}$ axis. As the number of ions becomes large, the geometrical configuration of the ions can become quite complicated, forming zig-zag and other patterns, but we shall focus on the simple case where just a few ions are trapped, in a string-like configuration.

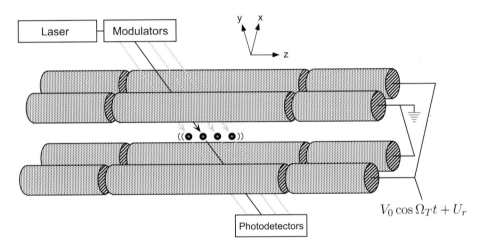

Figure 7.7. Schematic drawing (not to scale) of an ion trap quantum computer, depicting four ions trapped in the center of a potential created by four cylindrical electrodes. The apparatus is typically contained in a high vacuum ($\approx 10^{-8}$ Pa), and the ions are loaded from a nearby oven. Modulated laser light incident on the ions through windows in vacuum chamber perform operations on and are used to readout the atomic states.

Just as a mass on a spring can behave as a quantum system when the coupling to the external world becomes sufficiently small, the motion of the electromagnetically confined ion becomes quantized when it is sufficiently well isolated. Let us first understand what the quantization means, then consider the isolation criteria. As we saw in Section 7.3, the energy levels of a harmonic oscillator are equally spaced, in units of $\hbar\omega_z$. In the ion trap, in the regime which concerns us, these energy eigenstates represent different vibrational modes of the *entire* linear chain of ions moving together as one body, with mass NM. These are called the *center of mass* modes. Each $\hbar\omega_z$ quantum of vibrational energy is called a *phonon*, and can be thought of as a particle, just as a quantum of electromagnetic radiation in a cavity is a photon.

For the above phonon description to hold, certain criteria must hold. First, the coupling to the environment must be sufficiently small such that thermalization does not randomize the state of the system (and thus cause it to behave classically). Physically, what can happen is that nearby fluctuating electric and magnetic fields push on the ions, causing their motional state to randomly transition between energy eigenstates. Such noise sources are nearly inevitable, in a technical sense, since, for example, one cannot drive the confining electrodes from a perfect voltage source; the source will always have a finite resistance, and this resistance gives rise to Johnson noise, which has fluctuations on time scales the ions are sensitive to. The electric field on local patches of the electrodes can also fluctuate, randomly driving the ions' motion. As the randomness increases, the quantum properties of the ions' state is lost, and their behavior becomes well described by classical statistical averages. For example, both their momentum and position become well defined, which cannot be simultaneously true for a quantum system. Nevertheless, in practice, most technical noise sources can indeed be controlled quite well, to the extent that they do not heat or dephase the trapped ions too much on the time scale of most experiments. In part, one important reason this is possible is that as long as the harmonic approximation holds, the trapped ions are very selective about the frequency of the noise they are sensitive to; just as transitions between atomic levels can be selected by radiation tuned only to the correct frequency, only fluctuations which have high spectral power density around ω_z will affect the ions.

It is also quite important for the ions to be sufficiently cool so as to make the one-dimensional harmonic approximation valid. The true potential is non-quadratic for large displacements along any direction away from the trap center. And higher order vibrational modes in which the ions move relative to each other (instead of moving together) must have energies much higher than the center of mass mode. When this holds, and the ions are cooled to their motional ground state, their transition to the next higher energy state is through absorption of a center of mass phonon; this process is related to the *Mossbaüer effect*, in which a photon is absorbed by atoms in a crystal without generating local phonons because the entire crystal recoils together.

How are the ions cooled to their motional ground state? The goal is to satisfy $k_B T \ll \hbar\omega_z$, where T is the temperature reflecting the kinetic energy of the ions. Essentially, this can be done by using the fact that photons carry not only energy, but also momentum $p = h/\lambda$, where λ is the wavelength of the light. Just as the whistle of an approaching train has a higher pitch than a departing train, an atom moving toward a laser beam has transition frequencies which are slightly higher in energy than an atom moving away. If the laser is tuned such that it is absorbed only by approaching atoms, then the atoms slow down because the photons kick them in the opposite direction. This method is known as Doppler cooling. Shining a properly tuned laser (which has momentum vector components along each axis) at trapped atoms thus can cool the atoms down to the limit $k_B T \approx \hbar\Gamma/2$, where Γ is the radiative width of the transition used for the cooling. To cool beyond this limit, another method, known as *sideband cooling*, is then applied, as illustrated in Figure 7.8. This allows one to reach the $k_B T \ll \hbar\omega_z$ limit.

Another criterion which must be satisfied is that the width of the ion oscillation in the trap potential should be small compared to the wavelength of the incident light. This *Lamb–Dicke criterion* is conveniently expressed in terms of the *Lamb-Dicke parameter* $\eta \equiv 2\pi z_0/\lambda$, where λ is the wavelength, and $z_0 = \sqrt{\hbar/2NM\omega}$ is the characteristic length scale of the spacing between ions in the trap. The Lamb–Dicke criterion requires that $\eta \ll 1$; this does not strictly have to be met in order for ion traps to be useful for quantum computation, but it is desired to have that $\eta \approx 1$ at least, in order that the individual

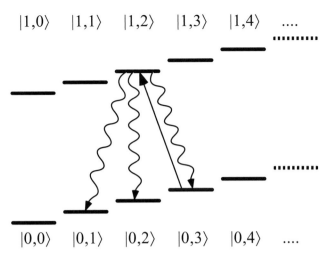

$|1,0\rangle$ $|1,1\rangle$ $|1,2\rangle$ $|1,3\rangle$ $|1,4\rangle$ ....

$|0,0\rangle$ $|0,1\rangle$ $|0,2\rangle$ $|0,3\rangle$ $|0,4\rangle$ ....

Figure 7.8. Sideband cooling method, showing transitions between $|0,n\rangle$ and $|1,m\rangle$, where 0 and 1 are two electronic levels, and n and m are phonon levels representing motional states of the ion. Laser light is tuned to have energy one phonon less than the electronic transition, such that, for example, the $|0,3\rangle$ state transitions to the $|1,2\rangle$ state, as shown. The atom then spontaneously decays into the lower energy 0 state (wiggly lines), randomly going to either $|0,1\rangle$, $|0,2\rangle$, or $|0,3\rangle$ (with nearly equal probabilities). Note that the laser light actually causes all possible transitions between $|0,n\rangle$ and $|1,n-1\rangle$, since these all have the same energy. However, this process does not touch the $|0,0\rangle$ state, and eventually that is the state in which the atom will be left.

ions can be resolved by different laser beams, but without making their motional state too difficult to optically excite in order to perform logic operations.

Atomic structure

The purpose of the trap apparatus described above is to allow ions to be cooled to the extent that their vibrational state is sufficiently close to having zero phonons ($|0\rangle$), an appropriate initial state for computation. Similarly, the internal states of the ions must be initialized appropriately, so they may be used to store quantum information. Let us now consider what these internal states are, and understand why they are good qubit representations by estimating their coherent lifetime.

The internal atomic states relevant to the trapped ion we shall consider result from the combination F of electron spin S and nuclear spin I, giving $F = S + I$. The formal piece of theory which describes this – known as the *addition of angular momenta* – not only describes important physics for understanding atomic structure, but also is an interesting mechanism for quantum information. A single photon interacting with an atom can provide or carry away one unit of angular momentum, as we saw in Section 7.5.1. But there are numerous possible sources of angular momenta in an atom: orbital, electron spin, and nuclear spin. Where it comes from is partly determined by the energy levels selected by the energy of the photon, but beyond that, the photon cannot distinguish between different sources, and to describe what happens we must select a basis in which *total* angular momentum becomes a uniquely defined property of the state.

Consider, for example, two spin-1/2 spins. The 'computational' basis for this two qubit space is $|00\rangle$, $|01\rangle$, $|10\rangle$, $|11\rangle$, but to span the state space we could equally well choose the basis

$$|0,0\rangle_J = \frac{|01\rangle - |10\rangle}{\sqrt{2}} \qquad (7.93)$$

$$|1,-1\rangle_J = |00\rangle \qquad (7.94)$$

$$|1,0\rangle_J = \frac{|01\rangle + |10\rangle}{\sqrt{2}} \qquad (7.95)$$

$$|1,1\rangle_J = |11\rangle . \qquad (7.96)$$

These basis states are special, because they are eigenstates of the total momentum operator, defined by $j_x = (X_1 + X_2)/2$, $j_y = (Y_1 + Y_2)/2$, $j_z = (Z_1 + Z_2)/2$, and

$$J^2 = j_x^2 + j_y^2 + j_z^2 . \qquad (7.97)$$

The states $|j, m_j\rangle_J$ are eigenstates of J^2 with eigenvalue $j(j + 1)$, and simultaneously eigenstates of j_z, with eigenvalue m_j. These states are the natural ones selected by many physical interactions; for example, in a $\hat{z}$ oriented magnetic field the magnetic moment μ in the Hamiltonian μB_z is proportional to m_j, the component of the total angular momentum in the $\hat{z}$ direction.

The theory of addition of angular momenta is sophisticated and well developed, and we have but scratched its surface (for the interested reader, some relevant exercises are provided below, and pointers to the literature are given in the 'History and further reading' section at the end of the chapter). Nevertheless, some interesting observations which concern quantum information can already be drawn from the above examples. Normally, we think of entangled states such as the Bell states (Section 1.3.6) as being unnatural states of matter, because they have strange, non-local properties. However, the state $|0,0\rangle_J$ is a Bell state! Why does Nature prefer this state here? It is because of a *symmetry* under which the interaction involving the magnetic moment is invariant under interchange of the two spins. Such symmetries actually occur widely in Nature, and are potentially quite useful for performing entangling measurements and operations.

Exercise 7.25: Show that the total angular momenta operators obey the commutation relations for $SU(2)$, that is, $[j_i, j_k] = i\epsilon_{ikl} j_l$.

Exercise 7.26: Verify the properties of $|j, m_j\rangle_J$ by explicitly writing the 4×4 matrices J^2 and j_z in the basis defined by $|j, m_j\rangle_J$.

Exercise 7.27: (**Three spin angular momenta states**) Three spin-1/2 spins can combine together to give states of total angular momenta with $j = 1/2$ and $j = 3/2$. Show that the states

$$|3/2, 3/2\rangle = |111\rangle \qquad (7.98)$$

$$|3/2, 1/2\rangle = \frac{1}{\sqrt{3}} \left[|011\rangle + |101\rangle + |110\rangle \right] \qquad (7.99)$$

$$|3/2, -1/2\rangle = \frac{1}{\sqrt{3}} \left[|100\rangle + |010\rangle + |001\rangle \right] \qquad (7.100)$$

$$|3/2, -3/2\rangle = |000\rangle \qquad (7.101)$$

$$|1/2, 1/2\rangle_1 = \frac{1}{\sqrt{2}} \left[-|001\rangle + |100\rangle \right] \qquad (7.102)$$

$$|1/2, -1/2\rangle_1 = \frac{1}{\sqrt{2}}\left[|110\rangle - |011\rangle\right] \tag{7.103}$$

$$|1/2, 1/2\rangle_2 = \frac{1}{\sqrt{6}}\left[|001\rangle - 2|010\rangle + |100\rangle\right] \tag{7.104}$$

$$|1/2, -1/2\rangle_2 = \frac{1}{\sqrt{6}}\left[-|110\rangle + 2|101\rangle - |011\rangle\right] \tag{7.105}$$

form a basis for the space, satisfying $J^2|j, m_j\rangle = j(j + 1)|j, m_j\rangle$ and $j_z|j, m_j\rangle = m_j|j, m_j\rangle$, for $j_z = (Z_1 + Z_2 + Z_3)/2$ (similarly for j_x and j_y) and $J^2 = j_x^2 + j_y^2 + j_z^2$. There are sophisticated ways to obtain these states, but a straightforward brute-force method is simply to simultaneously diagonalize the 8×8 matrices J^2 and j_z.

Exercise 7.28: (Hyperfine states) We shall be taking a look at beryllium in Section 7.6.4 – the total angular momenta states relevant there involve a nuclear spin $I = 3/2$ combining with an electron spin $S = 1/2$ to give $F = 2$ or $F = 1$. For a spin-$3/2$ particle, the angular momenta operators are

$$i_x = \frac{1}{2}\begin{bmatrix} 0 & \sqrt{3} & 0 & 0 \\ \sqrt{3} & 0 & 2 & 0 \\ 0 & 2 & 0 & \sqrt{3} \\ 0 & 0 & \sqrt{3} & 0 \end{bmatrix} \tag{7.106}$$

$$i_y = \frac{1}{2}\begin{bmatrix} 0 & i\sqrt{3} & 0 & 0 \\ -i\sqrt{3} & 0 & 2i & 0 \\ 0 & -2i & 0 & i\sqrt{3} \\ 0 & 0 & -i\sqrt{3} & 0 \end{bmatrix} \tag{7.107}$$

$$i_z = \frac{1}{2}\begin{bmatrix} -3 & 0 & 0 & 0 \\ 0 & -1 & 0 & 0 \\ 0 & 0 & 1 & 0 \\ 0 & 0 & 0 & 3 \end{bmatrix} \tag{7.108}$$

1. Show that i_x, i_y, and i_z satisfy $SU(2)$ commutation rules.
2. Give 8×8 matrix representations of $f_z = i_z \otimes I + I \otimes Z/2$ (where I here represents the identity operator on the appropriate subspace) and similarly f_x and f_y, and, $F^2 = f_x^2 + f_y^2 + f_z^2$. Simultaneously diagonalize f_z and F^2 to obtain basis states $|F, m_F\rangle$ for which $F^2|F, m_F\rangle = F(F + 1)|F, m_F\rangle$ and $f_z|F, m_F\rangle = m_F|F, m_F\rangle$.

How long can a superposition of different spin states exist? The limiting process, known as *spontaneous emission*, occurs when an atom transitions from its excited state to its ground state by emitting a photon. This happens at some random time, at a rate which we shall estimate. It might seem that spontaneously emitting a photon is a strange thing for an atom to do, if it is simply sitting in free space with nothing apparently disturbing it. But this process is actually a very natural consequence of the coupling of the atom to electromagnetic fields, described simply by the Jaynes–Cummings interaction,

$$H_I = g(a^\dagger \sigma_- + a\sigma_+), \tag{7.109}$$

as we recall from Section 7.5.2. Previously, we used this model to describe how a laser interacts with an atom, but the model also describes what happens to an atom even when no optical field is present! Consider an atom in its excited state coupled to a single mode which contains no photon, the state $|01\rangle$ (using $|\text{field}, \text{atom}\rangle$). *This is not an eigenstate of H_I*, and thus it cannot remain stationary as time evolves. What happens is described by the unitary operator U in (7.77), by which we find that there is a probability $p_{\text{decay}} = |\langle 10|U|01\rangle|^2$ for the atom to decay into its ground state and emit a photon, where

$$p_{\text{decay}} = g^2 \frac{4\sin^2 \frac{1}{2}(\omega - \omega_0)^2 t}{(\omega - \omega_0)^2}, \tag{7.110}$$

to lowest order in g, the atom–field coupling. ω is the frequency of the photon, and $\hbar\omega_0$ the energy difference between the two levels of the atom. An atom sitting in free space interacts with many different optical modes; inserting the coupling

$$g^2 = \frac{\omega_0^2}{2\hbar\omega\epsilon_0 c^2} |\langle 0|\vec{\mu}|1\rangle|^2, \tag{7.111}$$

where $\vec{\mu}$ is the atomic dipole operator, integrating over all the optical modes (Exercise 7.29) and taking a time derivative gives the probability per second of decay,

$$\gamma_{\text{rad}} = \frac{\omega_0^3 |\langle 0|\vec{\mu}|1\rangle|^2}{3\pi\hbar\epsilon_0 c^5}. \tag{7.112}$$

If we make the approximation that $|\langle 0|\vec{\mu}|1\rangle| \approx \mu_B \approx 9\times10^{-24}$ J/T, the Bohr magneton, and assume that $\omega_0/2\pi \approx 10$ GHz, then $\gamma_{\text{rad}} \approx 10^{-15}$ sec^{-1}, a spontaneous emission rate of less than one decay every 3 000 000 years. This calculation is representative of those done to estimate lifetimes of atomic states; as you can see, the hyperfine states can have remarkably long coherence times in theory, and this is generally consistent with experiments, in which lifetimes of tens of seconds to tens of hours have been observed.

Exercise 7.29: (Spontaneous emission)   The spontaneous emission rate (7.112) can be derived from (7.110)–(7.111) by the following steps.

1. Integrate

$$\frac{1}{(2\pi c)^3} \frac{8\pi}{3} \int_0^\infty \omega^2 \, p_{\text{decay}} \, d\omega, \tag{7.113}$$

   where the $8\pi/3$ comes from summing over polarizations and integrating over the solid angle $d\Omega$, and $\omega^2/(2\pi c)^3$ comes from the mode density in three-dimensional space. (*Hint:* you may want to extend the lower limit of the integral to $-\infty$.)

2. Differentiate the result with respect to t, to obtain γ_{rad}.

   The form of g^2 is a result of quantum electrodynamics; taking this for granted, the remainder of the calculation as presented here really stems from just the Jaynes–Cummings interaction. Again, we see how considering its properties in the single atom, single photon regime gives us a fundamental property of atoms, without resorting to perturbation theory!

Exercise 7.30: (Electronic state lifetimes)   A calculation similar to that for γ_{red} can be done to estimate the lifetimes expected for electronic transitions, that is, those which involve energy level changes $\Delta n \neq 0$. For such transitions, the relevant interaction couples the atom's electric dipole moment to the electromagnetic field, giving

$$g_{ed}^2 = \frac{\omega_0^2}{2\hbar\omega\epsilon_0}|\langle 0|\vec{\mu}_{ed}|1\rangle|^2 . \tag{7.114}$$

This gives a spontaneous emission rate

$$\gamma_{red}^{ed} = \frac{\omega_0^3|\langle 0|\vec{\mu}_{ed}|1\rangle|^2}{3\pi\hbar\epsilon_0 c^3} . \tag{7.115}$$

Give a value for γ_{red}^{ed}, taking $|\langle 0|\vec{\mu}_{ed}|1\rangle| \approx qa_0$, where q is the electric charge, and a_0 the Bohr radius, and assuming $\omega_0/2\pi \approx 10^{15}$ Hz. The result show how much faster electronic states can decay compared with hyperfine states.

7.6.2 The Hamiltonian

Combining the simplified models given in the previous section for the harmonic electromagnetic trap and the atomic structure provides us with the following simplified toy model for an ion trap quantum information processor. Imagine a single two-level spin interacting via the usual magnetic dipole interaction $H_I = -\vec{\mu} \cdot \vec{B}$ with an electromagnetic field, where the dipole moment $\vec{\mu} = \mu_m \vec{S}$ is proportional to the spin operator S, and the magnetic field is $\vec{B} = B_1\hat{x}\cos(kz - \omega t + \varphi)$, and B_1 is the field strength, k its momentum in the $\hat{z}$ direction, ω its frequency, and φ its phase. Note that in this section, we shall use $S_x = X/2$, $S_y = Y/2$, and $S_z = Z/2$ as the spin operators; they are related to the Pauli operators by a factor of two.

In addition to the usual electromagnetic interaction, there are interactions with the vibrational modes. The spin is physically confined within a harmonic potential of energy scale $\hbar\omega_z$ (Figure 7.9), such that its position becomes *quantized* and we must describe it by an operator $z = z_0(a^\dagger + a)$, where $a^\dagger$, a are raising and lowering operators for the vibrational modes of the particle, representing creation and annihilation of phonons.

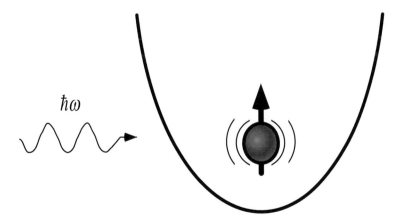

$\hbar\omega$

Figure 7.9. Toy model of a trapped ion: a single particle in a harmonic potential with two internal states, interacting with electromagnetic radiation.

Let us assume that the particle is cooled to near its lowest vibrational mode, such that the width of its oscillation in the well is small compared to the wavelength of the incident light, that is, the Lamb–Dicke parameter $\eta \equiv kz_0$ is small. Defining the Rabi frequency of the spin as $\Omega = \mu_m B_1/2\hbar$, and recalling that $S_x = (S_+ + S_-)/2$, we find that the interaction Hamiltonian simplifies in the small η limit to become

$$H_I = -\vec{\mu} \cdot \vec{B} \tag{7.116}$$

$$\approx \left[\frac{\hbar\Omega}{2} \left(S_+ e^{i(\varphi - \omega t)} + S_- e^{-i(\varphi - \omega t)} \right) \right]$$

$$+ \left[i\frac{\eta\hbar\Omega}{2} \{ S_+ a + S_- a^\dagger + S_+ a^\dagger + S_- a \} \left(e^{i(\varphi - \omega t)} - e^{-i(\varphi - \omega t)} \right) \right]. \tag{7.117}$$

The first term in brackets results from the usual Jaynes–Cummings Hamiltonian as we saw in Section 7.5.2, which occurs when the location z of the spin is a constant. However, it is simplified and does not contain photon operators because it turns out that as long as B_1 is a strong coherent state, we can neglect its quantum properties and leave ourselves with a Hamiltonian which describes just the evolution of the internal atomic state. It is in fact quite remarkable that a coherent state of the field does not become entangled with an atom after interacting with it (to an excellent degree of approximation); this is a deep result which you may explore further by looking at Problem 7.3 at the end of the chapter. We shall also touch on this fact in describing resonance in Section 7.7.2.

The second term in brackets describes the coupling of the motional state of the ion to its spin state, through the fact that the magnetic field it sees is dependent on its position. The four terms in braces correspond to four transitions (two up and two down) which are known as the red and blue motional sidebands, illustrated in Figure 7.10.

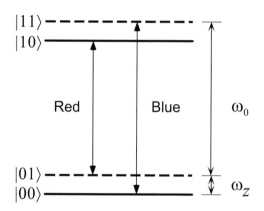

Figure 7.10. Energy levels of the toy model trapped ion showing the red and blue motional sideband transitions, which correspond to creation or annihilation of a single phonon. There is an infinite ladder of additional motional states, which are usually not involved. The states are labeled as $|n, m\rangle$ where n represents the spin state, and m the number of phonons.

Why these sideband transitions have frequencies $\omega_0 \pm \omega_z$ is easy to see, by including the free particle Hamiltonian

$$H_0 = \hbar\omega_0 S_z + \hbar\omega_z a^\dagger a, \tag{7.118}$$

which causes the spin and phonon operators to evolve as

$$S_+(t) = S_+ e^{i\omega_0 t} \qquad S_-(t) = S_- e^{-i\omega_0 t} \tag{7.119}$$

$$a^\dagger(t) = a^\dagger e^{i\omega_z t} \qquad a(t) = a e^{-i\omega_z t}. \tag{7.120}$$

Thus, in the frame of reference of H_0, the dominant terms of $H_I' = e^{iH_0t/\hbar}H_Ie^{-iH_0t/\hbar}$ are found to be

$$H_I' = \begin{cases} i\frac{\eta\hbar\Omega}{2}\left(S_+a^\dagger e^{i\varphi} - S_-ae^{-i\varphi}\right) & \omega = \omega_0 + \omega_z \\ i\frac{\eta\hbar\Omega}{2}\left(S_+ae^{i\varphi} - S_-a^\dagger e^{-i\varphi}\right) & \omega = \omega_0 - \omega_z \end{cases} \tag{7.121}$$

where the frequency of the electromagnetic field, ω, is as shown on the right.

Extending the above model from one spin to N spins confined within the same harmonic potential is simple if we assume that they share a single center of mass vibrational mode, whose energy is much lower than any other vibrational mode of the system. A straightforward extension of the theory shows that the only required modification is replacement of Ω by $\Omega/\sqrt{N}$, since all N particles move together collectively.

7.6.3 Quantum computation

Quantum computation with trapped ions requires one to be able to construct arbitrary unitary transforms on the internal states of the atoms. We now show how this is done, in three steps: we describe (1) how arbitrary single qubit operations are performed on the internal atomic (spin) state, (2) a method for performing a controlled two qubit gate between the spin and the phonon state, and (3) a way to swap quantum information between the spin and the phonon. Given these building blocks, we then describe an experiment which was performed to demonstrate a controlled-NOT gate, complete with state preparation and readout.

Single qubit operations

Applying an electromagnetic field tuned to frequency ω_0 turns on the internal Hamiltonian term

$$H_I^{\text{internal}} = \frac{\hbar\Omega}{2}\left(S_+e^{i\varphi} + S_-e^{-i\varphi}\right). \tag{7.122}$$

By choosing φ and the duration of the interaction appropriately, this allows us to perform rotation operations $R_x(\theta) = \exp(-i\theta S_x)$ and $R_y(\theta) = \exp(-i\theta S_y)$, which, by Theorem 4.1 on page 175, thereby allow us to perform any single qubit operation on the spin state. We shall denote rotations on the jth ion by a subscript, for example, $R_{xj}(\theta)$.

Exercise 7.31: Construct a Hadamard gate from R_y and R_x rotations.

Controlled phase-flip gate

Suppose, now, that one qubit is stored in the atom's internal spin state, and another qubit is stored using the $|0\rangle$ and $|1\rangle$ phonon states. If this is the case, we can perform a controlled phase-flip gate, with the unitary transform

$$\begin{bmatrix} 1 & 0 & 0 & 0 \\ 0 & 1 & 0 & 0 \\ 0 & 0 & 1 & 0 \\ 0 & 0 & 0 & -1 \end{bmatrix}. \tag{7.123}$$

It is easiest to explain how to do this with an atom that has a third energy level, as shown in Figure 7.11 (the extra level is not fundamentally necessary; see Problem 7.4). A laser is tuned to the frequency $\omega_{aux} + \omega_z$, to cause transitions between the $|20\rangle$ and $|11\rangle$ states; this turns on a Hamiltonian of the form

$$H_{aux} = i\frac{\eta\hbar\Omega'}{2}\left(S'_+ e^{i\varphi} + S'_- e^{-i\varphi}\right), \tag{7.124}$$

where S'_+ and S'_- denote transitions between $|20\rangle$ and $|11\rangle$, and we assume that higher order motional states are unoccupied. Note that because of the uniqueness of this frequency, no other transitions are excited. We apply the laser with phase and duration to perform a 2π pulse, that is, the rotation $R_x(2\pi)$ on the space spanned by $|11\rangle$ and $|20\rangle$, which is just the unitary transform $|11\rangle \rightarrow -|11\rangle$. All the other states remain unchanged, assuming that undesired states such as $|1,2\rangle$ have no probability amplitude. This realizes the transform of (7.123), as desired. We shall write this gate as $C_j(Z)$ (denoting a controlled-Z operation), where j indicates which ion the gate is applied to. Note that the same phonon is shared by all the ions, since it is a center-of-mass phonon; because of this, adopting engineering terminology, this has been called the phonon 'bus' qubit in the literature.

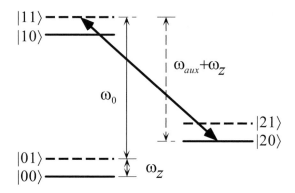

Figure 7.11. Energy levels of a three-level atom in an ion trap, with two phonon states each. The labels $|n, m\rangle$ indicate the atom's state n and the phonon state m. The $|20\rangle \leftrightarrow |11\rangle$ transition is used to perform a controlled phase-flip gate.

Swap gate

Finally, we need some way to swap qubits between the atom's internal spin state and the phonon state. This can be done by tuning a laser to the frequency $\omega_0 - \omega_z$, and arranging for the phase to be such that we perform the rotation $R_y(\pi)$ on the subspace spanned by $|01\rangle$ and $|10\rangle$, which is just the unitary transform

$$\begin{bmatrix} 1 & 0 & 0 & 0 \\ 0 & 0 & 1 & 0 \\ 0 & -1 & 0 & 0 \\ 0 & 0 & 0 & 1 \end{bmatrix} \tag{7.125}$$

on the $|00\rangle, |01\rangle, |10\rangle, |11\rangle$ space. If the initial state is $a|00\rangle + b|10\rangle$ (that is, the phonon is initially $|0\rangle$), then the state after the swap is $a|00\rangle + b|01\rangle$, so this accomplishes the desired swap operation. We shall write this as SWAP$_j$ when acting on ion j; the inverse operation $\overline{\text{SWAP}}_j$ corresponds to $R_y(-\pi)$. Technically, because of the minus sign in the $|10\rangle\langle01|$ entry of $R_y(\pi)$, this is not a perfect swap operation, but it is equivalent to one up relative phases (see Exercise 4.26). Thus, this is sometimes referred to as being a 'mapping operation' instead of as a swap.

*Controlled-*NOT *gate*

Putting these gates together allows us to construct a CNOT gate acting on ions j (control) and k (target) using the sequence of operations

$$\text{CNOT}_{jk} = H_k \,\overline{\text{SWAP}}_k\, C_j(Z)\,\text{SWAP}_k\, H_k , \qquad (7.126)$$

(time going from right to left, as usual for matrices) where H_k is a Hadamard gate (constructed from R_y and R_x rotations on ion k). This is very similar to how a controlled-NOT gate was constructed using beamsplitters and optical Kerr media, as in Equation (7.46).

7.6.4 Experiment

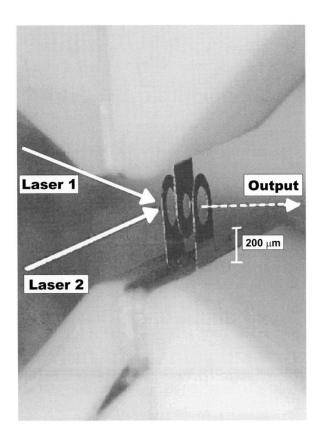

Figure 7.12. Photograph of a microfabricated elliptical electrode ion trap, in which ions have been confined. The ions in this trap are barium ions, rather than beryllium, but the basic principles are the same as described in the text. Reproduced courtesy of R. Devoe and C. Kurtsiefer, IBM Almaden Research Center.

A controlled-NOT gate using a single trapped ion has been demonstrated (see 'History and further reading' at the end of the chapter); precisely how this experiment is done is insightful. In the experiment, a single ion of $^9\text{Be}^+$ is trapped in a coaxial resonator RF ion trap, different in geometry from the linear ion trap of Figure 7.7, but functionally equivalent, and similar to the photograph of an actual ion trap shown in Figure 7.12. Beryllium was chosen for its convenient hyperfine and electronic level structure, shown in Figure 7.13. The $^2S_{1/2}(1,1)$ and $^2S_{1/2}(2,2)$ energy levels (Exercise 7.28) are used as the atom's internal qubit state, and the $|0\rangle$ and $|1\rangle$ phonon states as another qubit (labeled

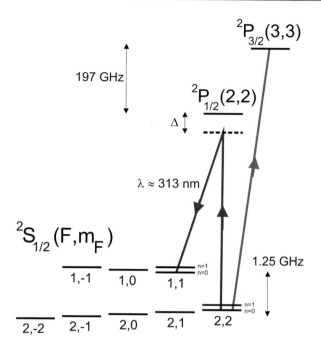

Figure 7.13. Energy levels of ^{9}Be$^+$ used in the ion trap experiment. Figure courtesy of C. Monroe at NIST.

in the figure as $n = 0$ and $n = 1$). The ≈ 313 nm transition between the $^2S_{1/2}(1, 1)$ and $^2S_{1/2}(2, 2)$ levels is accomplished not by tuning a single laser to the transition frequency, but rather two lasers whose *difference* frequency is that of the transition. This *Raman transition* method simplifies requirements for laser phase stability. The $^2S_{1/2}(2, 0)$ state is used as the auxiliary level; the $^2S_{1/2}$ states have different energies by virtue of a 0.18 millitesla magnetic field applied to the system. The trapped ion has vibrational frequencies $(\omega_x, \omega_y, \omega_z)/2\pi = (11.2, 18.2, 29.8)$ MHz in the trap, and a ground state $n_x = 0$ wavefunction spread of about 7 nm, giving a Lamb–Dicke parameter of about $\eta_x = 0.2$. The Rabi frequency of the on-resonance transition is $\Omega/2\pi = 140$ kHz, the two motional sidebands, $\eta_x\Omega/2\pi = 30$ kHz, and the auxiliary transition $\eta_x\Omega'/2\pi = 12$ kHz.

The state of the ion is initialized using Doppler and sideband cooling to obtain, with approximately 95% probability, the state $|00\rangle = |^2S_{1/2}(2, 2)\rangle|n_x = 0\rangle$. The internal and motional states of the ion are then prepared in one of the four basis states $|00\rangle$, $|01\rangle$, $|10\rangle$, or $|11\rangle$ using single qubit operations, then a controlled-NOT gate is performed using three pulses, which implement a $R_y(\pi/2)$ rotation on the internal state qubit, a controlled-Z operation between the two qubits, then a $R_y(-\pi/2)$ rotation on the internal state qubit. It is simple to show (Exercise 7.32) that this circuit, drawn in Figure 7.14, realizes a controlled-NOT gate.

Readout of the computational output is performed with two measurements. The first is to collect the fluorescence from the ion which occurs when + circularly polarized light tuned to the $^2S_{1/2}(2, 2) - \,^2P_{3/2}(3, 3)$ 'cycling' transition is applied. The light does not couple appreciably to the $^2S_{1/2}(1, 1)$ state, and thus the intensity of the observed fluorescence is proportional to the probability of the internal state qubit being in the $|0\rangle$

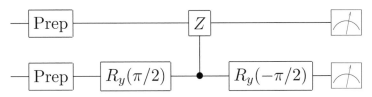

Figure 7.14. Quantum circuit modeling the ion trap controlled-NOT experiment. The top wire represents the phonon state, and the bottom, the ion's internal hyperfine state.

state; it is a projective measurement. This measurement technique is powerful because the transition cycles many times – the ion absorbs a photon, jumping to the $^2P_{3/2}(3,3)$ state, then emits a photon, decaying back into the $^2S_{1/2}(2,2)$ state where it started. Thousands or more cycles are possible, allowing good statistics to be accumulated. The second measurement is similar to the first, but a swap pulse is applied first to exchange the motional and internal state qubits; this projectively measures the motional state qubit.

The experiment as performed verifies the classical truth table of the controlled-NOT operation and, in principle, by preparing superposition input states and measuring output density matrices, the unitary transform could be completely characterized using process tomography (Chapter 8). The controlled-NOT gate requires about 50 microseconds to perform with the optical power used in the experiment. On the other hand, the coherence time was measured to be somewhere around hundreds to thousands of microseconds. The dominant decoherence mechanisms included instabilities in the laser beam power and the RF ion trap drive frequency and voltage amplitude, and fluctuating external magnetic fields. Moreover, the experiment involved only a single ion, and only two qubits, and thus was not useful for computation; to be useful, a controlled-NOT gate should generally be applied between different ions, and not just between a single ion and the motional state.

However, the technical limitations can probably be overcome, and lifetimes can be extended by using the short-lived motional state only intermittently, capitalizing on the much longer coherence times of the internal atomic states. And scaling to larger numbers of ions is conceptually viable. Shown in Figure 7.15 is a string of 40 mercury ions which have been trapped. There are many hurdles to making such systems behave as useful quantum information processing machines, but technological surprises are a never-ending saga. Someday, perhaps, trapped ions such as these could be registers of qubits in a quantum computer.

Figure 7.15. Image of fluorescence from about 40 trapped mercury ($^{199}Hg^+$) atomic ions. The ions are spaced by approximately 15 micrometers, and the two apparent gaps are different isotopes of mercury which do not respond to the probe laser. Reprinted courtesy of D. Wineland, at NIST.

Exercise 7.32: Show that the circuit in Figure 7.14 is equivalent (up to relative phases) to a controlled-NOT gate, with the phonon state as the control qubit.

Ion trap quantum computer

- **Qubit representation**: Hyperfine (nuclear spin) state of an atom, and lowest level vibrational modes (phonons) of trapped atoms.
- **Unitary evolution**: Arbitrary transforms are constructed from application of laser pulses which externally manipulate the atomic state, via the Jaynes–Cummings interaction. Qubits interact via a shared phonon state.
- **Initial state preparation**: Cool the atoms (by trapping and using optical pumping) into their motional ground state, and hyperfine ground state.
- **Readout**: Measure population of hyperfine states.
- **Drawbacks**: Phonon lifetimes are short, and ions are difficult to prepare in their motional ground states.

7.7 Nuclear magnetic resonance

Nuclear spin systems would be nearly ideal for quantum computation if only spin–spin couplings could be large and controllable; this is an important observation from our study of ion traps in the last section. The principal drawback of ion trap quantum computers is the weakness of the phonon mediated spin–spin coupling technique and its susceptibility to decoherence. One way this limitation could be circumvented would be to trap molecules instead of single atoms – the magnetic dipole and electron mediated Fermi contact interactions between neighboring nuclei would provide strong natural couplings. However, with their many vibrational modes, single molecules have been difficult to trap and cool, and thus optical manipulation and detection of nuclear spins in trapped molecules has not been feasible except in special circumstances.

On the other hand, direct manipulation and detection of nuclear spin states using radiofrequency electromagnetic waves is a well-developed field known as nuclear magnetic resonance (NMR). These techniques are widely used in chemistry, for example, to measure properties of liquids, solids, and gases, to determine the structure of molecules, and to image materials and even biological systems. These many applications have lead the technology of NMR to become quite sophisticated, allowing control and observation of tens to hundreds and thousands of nuclei in an experiment.

However, two problems arise in using NMR for quantum computation. First, because of the smallness of the nuclear magnetic moment, a large number (more than $\approx 10^8$) molecules must be present in order to produce a measurable induction signal. Conceptually, a single molecule might be a fine quantum computer, but how can this be true of an ensemble of molecules? In particular, the output of an NMR measurement is an *average* over all the molecule's signals; can the average output of an ensemble of quantum computers be meaningful? Second, NMR is typically applied to physical systems in equilibrium at room temperature, where the spin energy $\hbar\omega$ is much less than $k_B T$. This means that the initial state of the spins is nearly completely random. Traditional quantum computation requires the system be prepared in a pure state; how can quantum computation be performed with a system which is in a high entropy mixed state?

Solutions to these two problems have made NMR a particularly attractive and insightful method for implementing quantum computation, despite stringent limitations which arise from the thermal nature of typical systems. Many lessons can be learned from NMR: for example, techniques for controlling realistic Hamiltonians to perform arbitrary unitary transforms, methods for characterizing and circumventing decoherence (and systematic errors), and considerations which arise in assembling components in implementing full quantum algorithms on entire systems. We begin with a description of the physical apparatus and the main Hamiltonian involved, then we discuss how quantum information processing with NMR is possible despite the thermal input state and ensemble problems, concluding with some experiments which have been performed demonstrating quantum algorithms, and the drawbacks of this method.

7.7.1 Physical apparatus

Let us begin with a general description of the apparatus, whose workings will be mathematically modeled in detail later. The two main parts of a pulsed NMR system for liquid samples, which we shall focus on here, are the sample and the NMR spectrometer. A typical molecule which might be used would contain a number n of protons which have spin 1/2 (other possible nuclei include ^{13}C, ^{19}F, ^{15}N, and ^{31}P), and produce an NMR signal at about 500 MHz when placed in a magnetic field of about 11.7 tesla. The frequencies of different nuclei within a molecule can differ by a few kHz to hundreds of kHz because of differences in the local magnetic fields due to chemical environment shielding effects. The molecules are typically dissolved in a solvent, reducing the concentration to the extent that inter-molecular interactions become negligible, leaving a system that might well be described as an ensemble of n qubit quantum 'computers.'

The spectrometer itself is constructed from radiofrequency (RF) electronics and a large superconducting magnet, within the bore of which is held the sample in a glass tube, as shown in Figure 7.16. There, the static $\hat{z}$ oriented magnetic field B_0 is carefully trimmed to be uniform over approximately 1 cm^3 to better than one part in 10^9. Orthogonal saddle or Helmholtz coils lying in the transverse plane allow small, oscillating magnetic fields to be applied along the $\hat{x}$ and $\hat{y}$ directions. These fields can be rapidly pulsed on and off to manipulate nuclear spin states. The same coils are also part of tuned circuits which are used to pick up the RF signal generated by the precessing nuclei (much like how a spinning magnet inductively generates an alternating current in a nearby coil).

A typical experiment begins with a long waiting period in which the nuclei are allowed to thermalize to equilibrium; this can require several minutes for well-prepared liquid samples. Under control of a (classical) computer, RF pulses are then applied to effect the desired transformation on the state of the nuclei. The high power pulse amplifiers are then quickly switched off and a sensitive pre-amplifier is enabled, to measure the final state of the spins. This output, called the *free induction decay*, is Fourier transformed to obtain a frequency spectrum with peaks whose areas are functions of the spin states (Figure 7.17).

There are many important practical issues which lead to observable imperfections. For example, spatial inhomogeneities in the static magnetic field cause nuclei in different parts of the fields to precess at different frequencies. This broadens lines in the spectrum. An even more challenging problem is the homogeneity of the RF field, which is generated by a coil which must be orthogonal to the B_0 magnet; this geometric constraint and the requirement to simultaneously maintain high B_0 homogeneity usually forces the RF

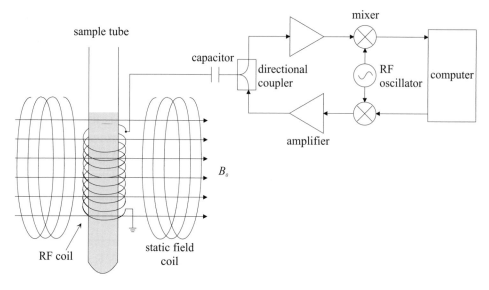

Figure 7.16. Schematic diagram of an NMR apparatus.

field to be inhomogeneous and generated by a small coil, leading to imperfect control of the nuclear system. Also, pulse timing, and stability of power, phase, and frequency are important issues; however, unlike the ion traps, because of the lower frequencies, good control of these parameters is more tractable. We shall return to imperfections in Section 7.7.4, after understanding the basic mathematical description of the system and the methodology for performing quantum information processing with NMR.

7.7.2 The Hamiltonian

The basic theory of NMR can be understood from an ideal model of one and two spins, which we describe here. The first step is to describe how electromagnetic radiation interacts with a single magnetic spin. We then consider the physical nature of couplings between spins which arise in molecules. These tools enable us to model readout of the magnetization which results from transformation of an initial state which is in thermal equilibrium. Finally, we describe a phenomenological model of decoherence, and how its T_1 and T_2 parameters can be experimentally determined.

Single spin dynamics

The magnetic interaction of a classical electromagnetic field with a two–state spin is described by the Hamiltonian $H = -\vec{\mu} \cdot \vec{B}$, where $\vec{\mu}$ is the spin, and $B = B_0\hat{z} + B_1(\hat{x}\cos\omega t + \hat{y}\sin\omega t)$ is a typical applied magnetic field. B_0 is static and very large, and B_1 is usually time varying and several orders of magnitude smaller than B_0 in strength, so that perturbation theory is traditionally employed to study this system. However, the Schrödinger equation for this system can be solved straightforwardly without perturbation theory, using the Pauli matrix techniques of Chapter 2, in terms of which the Hamiltonian can be written as

$$H = \frac{\omega_0}{2}Z + g(X\cos\omega t + Y\sin\omega t), \qquad (7.127)$$

where g is related to the strength of the B_1 field, and ω_0 to B_0, and X, Y, Z are the Pauli matrices as usual. Define $|\varphi(t)\rangle = e^{i\omega t Z/2}|\chi(t)\rangle$, such that the Schrödinger equation

$$i\partial_t|\chi(t)\rangle = H|\chi(t)\rangle \tag{7.128}$$

can be re-expressed as

$$i\partial_t|\varphi(t)\rangle = \left[e^{i\omega Z t/2}He^{-i\omega Z t/2} - \frac{\omega}{2}Z\right]|\varphi(t)\rangle. \tag{7.129}$$

Since

$$e^{i\omega Z t/2}Xe^{-i\omega Z t/2} = (X\cos\omega t - Y\sin\omega t), \tag{7.130}$$

(7.129) simplifies to become

$$i\partial_t|\varphi(t)\rangle = \left[\frac{\omega_0 - \omega}{2}Z + gX\right]|\varphi(t)\rangle, \tag{7.131}$$

where the terms on the right multiplying the state can be identified as the effective 'rotating frame' Hamiltonian. The solution to this equation is

$$|\varphi(t)\rangle = e^{i\left[\frac{\omega_0 - \omega}{2}Z + gX\right]t}|\varphi(0)\rangle. \tag{7.132}$$

The concept of *resonance* arises from the behavior of this solution, which can be understood using (4.8) to be a single qubit rotation about the axis

$$\hat{n} = \frac{\hat{z} + \frac{2g}{\omega_0 - \omega}\hat{x}}{\sqrt{1 + \left(\frac{2g}{\omega_0 - \omega}\right)^2}} \tag{7.133}$$

by an angle

$$|\vec{n}| = t\sqrt{\left(\frac{\omega_0 - \omega}{2}\right)^2 + g^2}. \tag{7.134}$$

When ω is far from ω_0, the spin is negligibly affected by the B_1 field; the axis of its rotation is nearly parallel with $\hat{z}$, and its time evolution is nearly exactly that of the free B_0 Hamiltonian. On the other hand, when $\omega_0 \approx \omega$, the B_0 contribution becomes negligible, and a small B_1 field can cause large changes in the state, corresponding to rotations about the $\hat{x}$ axis. The enormous effect a small perturbation can have on the spin system, when tuned to the appropriate frequency, is responsible for the 'resonance' in nuclear magnetic resonance. The same effect, of course, is also at the heart of the selectivity of two-level atoms for specifically tuned laser fields that was used (but not explained) in Section 7.5.1.

In general, when $\omega = \omega_0$, the single spin rotating frame Hamiltonian can be written as

$$H = g_1(t)X + g_2(t)Y, \tag{7.135}$$

where g_1 and g_2 are functions of the applied transverse RF fields.

Exercise 7.33: (Magnetic resonance) Show that (7.128) simplifies to become (7.129). What laboratory frame Hamiltonian gives rise to the rotating frame Hamiltonian (7.135)?

Exercise 7.34: (NMR frequencies) Starting with the nuclear Bohr magneton,
compute the precession frequency of a proton in a magnetic field of 11.8 tesla.
How many gauss should B_1 be to accomplish a 90° rotation in 10 microseconds?

Spin–spin couplings

More than one spin is usually present in systems of interest; ^1H, ^{13}C, ^{19}F, and ^{15}N all
have nuclear spin $1/2$. These spins interact through two dominant mechanisms: direct
dipolar coupling, and indirect through-bond electron mediated interactions. Through-
space dipolar coupling is described by an interaction Hamiltonian of the form

$$H_{1,2}^D = \frac{\gamma_1 \gamma_2 \hbar}{4r^3} \left[\vec{\sigma}_1 \cdot \vec{\sigma}_2 - 3(\vec{\sigma}_1 \cdot \hat{n})(\vec{\sigma}_2 \cdot \hat{n}) \right] , \qquad (7.136)$$

where $\hat{n}$ is the unit vector in the direction joining the two nuclei, and $\vec{\sigma}$ is the magnetic
moment vector (times two). In a low viscosity liquid, dipolar interactions are rapidly
averaged away; mathematically this is calculated by showing that the spherical average of
$H_{1,2}^D$ over $\hat{n}$ goes to zero as the averaging becomes rapid compared to the dipolar coupling
energy scale.

Through-bond interactions, also known simply as 'J-coupling,' are indirect interactions mediated by electrons shared through a chemical bond; the magnetic field seen by
one nucleus is perturbed by the state of its electronic cloud, which interacts with another
nucleus through the overlap of the electronic wavefunction with the nucleus (a Fermi
contact interaction). This coupling has the form

$$H_{1,2}^J = \frac{\hbar J}{4} \vec{\sigma}_1 \cdot \vec{\sigma}_2 = \frac{\hbar J}{4} Z_1 Z_2 + \frac{\hbar J}{8} \left[\sigma_+ \sigma_- + \sigma_- \sigma_+ \right] . \qquad (7.137)$$

We shall be interested in the case where J is a scalar (in general it may be a tensor),
which is an excellent approximation in liquids and when couplings are weak, or when
the interacting nuclear species have vastly different precession frequencies. This case is
described by

$$H_{12}^J \approx \frac{\hbar}{4} J Z_1 Z_2 . \qquad (7.138)$$

Exercise 7.35: (Motional narrowing) Show that the spherical average of $H_{1,2}^D$ over
$\hat{n}$ is zero.

Thermal equilibrium

NMR differs significantly from the other physical systems we have studied previously
in this chapter in that it uses an ensemble of systems, and the primary measurement is
an ensemble average. Furthermore, no extensive procedures are employed to prepare the
initial state in a special state such as the ground state; in fact, to do so is challenging with
present technology.

Rather, the initial state is the thermal equilibrium state,

$$\rho = \frac{e^{-\beta H}}{\mathcal{Z}} , \qquad (7.139)$$

where $\beta = 1/k_B T$, and $\mathcal{Z} = \text{tr}\, e^{-\beta H}$ is the usual partition function normalization, which
ensures that $\text{tr}(\rho) = 1$. Since $\beta \approx 10^{-4}$ at modest fields for typical nuclei at room

temperature, the high temperature approximation

$$\rho \approx 2^{-n} \left[1 - \beta H \right] \tag{7.140}$$

is appropriate, for a system of n spins.

Since spin–spin couplings are very small compared with the precession frequencies, the thermal state density matrix is very nearly diagonal in the Z basis, and thus it can be interpreted as being a mixture of the pure states $|00\ldots0\rangle$, $|00\ldots01\rangle$, ..., $|11\ldots1\rangle$. What is actually the true physical state of each ensemble member is a matter of debate, because an infinite number of unravelings exist for a given density matrix. In principle, with NMR the true physical state can be measured if the ensemble members (individual molecules) are accessible, but this is experimentally difficult.

Exercise 7.36: (Thermal equilibrium NMR state) For $n = 1$ show that the thermal equilibrium state is

$$\rho \approx 1 - \frac{\hbar\omega}{2k_{\rm B}T} \begin{bmatrix} 1 & 0 \\ 0 & -1 \end{bmatrix}, \tag{7.141}$$

and for $n = 2$ (and $\omega_A \approx 4\omega_B$),

$$\rho \approx 1 - \frac{\hbar\omega_B}{4k_{\rm B}T} \begin{bmatrix} 5 & 0 & 0 & 0 \\ 0 & 3 & 0 & 0 \\ 0 & 0 & -3 & 0 \\ 0 & 0 & 0 & -5 \end{bmatrix}. \tag{7.142}$$

Magnetization readout

The principal output of an experiment is the free induction decay signal, mathematically given as

$$V(t) = V_0 \text{tr} \left[e^{-iHt} \rho e^{iHt} (iX_k + Y_k) \right], \tag{7.143}$$

where X_k and Y_k operate only on the kth spin, and V_0 is a constant factor dependent on coil geometry, quality factor, and maximum magnetic flux from the sample volume. This signal originates from the pickup coils detecting the magnetization of the sample in the $\hat{x} - \hat{y}$ plane. In the laboratory frame, this signal will oscillate at a frequency equal to the precession frequency ω_0 of the nuclei; however, $V(t)$ is usually mixed down with an oscillator locked at ω_0, then Fourier transformed, such that the final signal appears as shown in Figure 7.17.

Exercise 7.37: (NMR spectrum of coupled spins) Calculate $V(t)$ for $H = JZ_1Z_2$ and $\rho = e^{i\pi Y_1/4}\frac{1}{4}[1 - \beta\hbar\omega_0(Z_1 + Z_2)]e^{-i\pi Y_1/4}$. How many lines would there be in the spectrum of the first spin if the Hamiltonian were $H = JZ_1(Z_2 + Z_3 + Z_4)$ (with a similar initial density matrix) and what would their relative magnitudes be?

Decoherence

A prominent characteristic of the free induction decay whose description lies outside the simple models presented so far for NMR is the exponential decay of the magnetization signal. One cause of this is inhomogeneity in the static magnetic field, which leads to

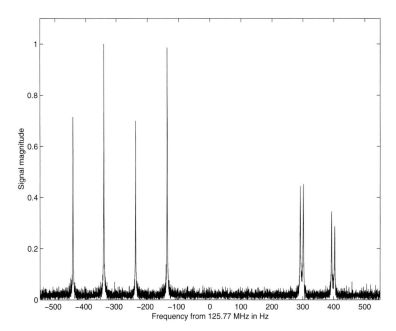

Figure 7.17. Carbon spectrum of ^{13}C labeled trichloroethylene. The four lines on the left come from the carbon nucleus directly bound to the proton; four lines appear because of couplings to the proton and to the second carbon nucleus, whose own signal gives the closely spaced four lines on the right. The second carbon nucleus is further away from the proton than the first, and thus has a much smaller coupling to it.

precessing spins in one part of the sample getting out of phase with those in another part. Effects due to inhomogeneities are reversible in principle, but there are other sources of phase randomization which are fundamentally irreversible, such as those originating from spin–spin couplings. Another irreversible mechanism is the thermalization of the spins to equilibrium at the temperature of their environment, a process which involves exchange of energy. For a single qubit state, these effects may be phenomenologically characterized with a density matrix transformation model,

$$
\begin{bmatrix} a & b \\ b^* & 1-a \end{bmatrix} \rightarrow \begin{bmatrix} (a-a_0)e^{-t/T_1}+a_0 & be^{-t/T_2} \\ b^*e^{-t/T_2} & (a_0-a)e^{-t/T_1}+1-a_0 \end{bmatrix}, \quad (7.144)
$$

where T_1 and T_2 are known as the spin–lattice (or 'longitudinal') and spin–spin (or 'transverse') relaxation rates, respectively, and a_0 characterizes the thermal equilibrium state. They define important time scales for the lifetimes of non-equilibrium classical states and quantum superpositions. Theoretical tools for calculating T_1 and T_2 in NMR systems are well-developed, and, in fact, measurements of these rates play an important role in using NMR to distinguish between different chemical species.

Experimental methods for measuring T_1 and T_2 are well known in NMR. Let $R_x = e^{-i\pi X/4}$ be a 90° pulse about the $\hat{x}$ axis. To measure T_1, apply R_x^2, wait time τ, then R_x. The first pulse flips the spin by 180°, after which it relaxes for time τ back towards equilibrium (visualize this as the Bloch vector shrinking back towards the top of the Bloch sphere, the ground state), then the final 90° pulse puts the magnetization in the $\hat{x}-\hat{y}$ plane, where it is detected. The measured magnetization M from this *inversion–recovery* experiment is found to decay exponentially with τ as $M = M_0\left[1-2\exp(-\tau/T_1)\right]$. To measure T_2, one can, to first order, simply measure the linewidth of a peak. A better way, the Carr–Purcell–Meiboom–Gill technique, is to apply an R_x operation, followed by k iterations of 'wait time $\tau/2$, apply R_x^2, wait time $\tau/2$, apply R_x^2'. This train of 180° pulses 'refocuses' couplings (Section 7.7.3) and partially cancels B_0 field inhomogeneities, so that one can better estimate the true T_2 of the system. The observed magnetization decays as $M = M_0 e^{-k\tau/T_2}$.

Multiple spin Hamiltonian

Summarizing our discussion of the NMR Hamiltonian, we can write H for an n spin coupled system as

$$H = \sum_k \omega_k Z_k + \sum_{j,k} H_{j,k}^J + H^{\mathrm{RF}} + \sum_{j,k} H_{j,k}^{\mathrm{D}} + H^{\mathrm{env}}, \tag{7.145}$$

where the first term is the free precession of the spins in the ambient magnetic field, H^{D} is the magnetic dipole coupling of (7.136), H^J is the 'J' coupling of (7.137), H^{RF} describes the effect of the externally applied radiofrequency magnetic fields of (7.135), and H^{env} describes interactions with the environment which lead to decoherence, as in (7.144).

For the sake of understanding the basic principles about how this Hamiltonian can be manipulated, we shall find it sufficient to consider the simplified Hamiltonian

$$H = \sum_k \omega_k Z_k + \sum_{j,k} Z_j \otimes Z_k + \sum_k g_k^x(t) X_k + g_k^y(t) Y_k, \tag{7.146}$$

in much of the following discussion. The treatment of the more general (7.145) follows the same ideas.

7.7.3 Quantum computation

Quantum information processing requires performing unitary transforms to a system prepared in a proper initial state. Three questions arise for the present system: First, how can arbitrary unitary transforms be implemented in a system of n coupled spins described by the Hamiltonian of (7.146)? And second, how can the thermal state (7.140) of an NMR system be used as a proper initial state for computation? Third, the quantum algorithms we have studied in the last three chapters ask for projective measurements to obtain output results, whereas with NMR, we can only easily perform ensemble average measurements. How can we deal with this ensemble readout problem? We answer these questions in this section.

Refocusing

Perhaps one of the most interesting techniques available to us in performing arbitrary unitary transforms using Hamiltonians of the sort of (7.146) is *refocusing*, as it is known in the art of NMR. Consider the simple two spin Hamiltonian $H = H^{\mathrm{sys}} + H^{\mathrm{RF}}$ where

$$H^{\mathrm{sys}} = a Z_1 + b Z_2 + c Z_1 Z_2. \tag{7.147}$$

As was shown in Section 7.7.2, when a large RF field is applied at the proper frequency, to a good approximation, we can approximate

$$e^{-iHt/\hbar} \approx e^{-iH^{\mathrm{RF}}t/\hbar}. \tag{7.148}$$

This allows arbitrary single qubit operations to be performed with excellent fidelity. Let us define

$$R_{x1} = e^{-i\pi X_1/4} \tag{7.149}$$

as a 90° rotation about $\hat{x}$ on spin 1, and similarly for spin 2. The 180° rotation R_{x1}^2 has the special property that

$$R_{x1}^2 e^{-iaZ_1 t} R_{x1}^2 = e^{iaZ_1 t}, \tag{7.150}$$

as can be easily verified. This is known as refocusing, because of the way it reverses time evolution such that different frequency spins starting together at some point on the Bloch sphere come back to the same point on the Bloch sphere. 180° pulses applied in this manner are known as refocusing pulses. Note that in the above expression, a can be an operator as well as a constant (as long as it contains no operators which act on spin 1), and thus

$$e^{-iH^{sys}t/\hbar} R_{x1}^2 e^{-iH^{sys}t/\hbar} R_{x1}^2 = e^{-2ibZ_2 t/\hbar}. \tag{7.151}$$

Using another set of refocusing pulses applied to spin 2 would remove even this remaining term. Refocusing is thus a useful technique for removing coupled evolution between spins, and for removing all evolution entirely.

Exercise 7.38: (Refocusing) Explicitly show that (7.150) is true (use the anti-commutativity of the Pauli matrices).

Exercise 7.39: (Three-dimensional refocusing) What set of pulses can be used to refocus evolution under *any* single spin Hamiltonian $H^{sys} = \sum_k c_k \sigma_k$?

Exercise 7.40: (Refocusing dipolar interactions) Give a sequence of pulses which can be used to turn two spin dipolar coupling $H_{1,2}^D$ into the much simpler form of (7.138).

Controlled-NOT gate

Realization of a controlled-NOT gate is simple using refocusing pulses and single qubit pulses. Let us show how this is done for a two spin system with the Hamiltonian of (7.147). From the construction of (7.46), we know that being able to realize the unitary transform

$$U_{CZ} = \begin{bmatrix} 1 & 0 & 0 & 0 \\ 0 & 1 & 0 & 0 \\ 0 & 0 & 1 & 0 \\ 0 & 0 & 0 & -1 \end{bmatrix} \tag{7.152}$$

is sufficient. Since $\sqrt{i}\, e^{iZ_1 Z_2 \pi/4} e^{-iZ_1 \pi/4} e^{-iZ_2 \pi/4} = U_{CZ}$, getting a controlled-NOT from one evolution period of time $\hbar\pi/4c$ together with several single qubit pulses is straightforward.

Exercise 7.41: (NMR controlled-NOT) Give an explicit sequence of single qubit rotations which realize a controlled-NOT between two spins evolving under the Hamiltonian of (7.147). You may start with (7.46), but the result can be simplified to reduce the number of single qubit rotations.

Temporal, spatial, and logical labeling

Being able to realize arbitrary unitary transforms on a spin system to good fidelity using RF pulses is one of the most attractive aspects of NMR for quantum computation. However, the major drawback is the fact that the initial state is usually the thermal state of (7.140). Despite the high entropy of this state, quantum computation can nevertheless be done, with some cost. Two techniques for achieving this are called temporal and logical labeling.

Temporal labeling, also sometimes called temporal averaging, is based on two important facts: quantum operations are linear, and the observables measured in NMR are traceless (see Section 2.2.5 for background on quantum measurements). Suppose a two spin system starts out with the density matrix

$$\rho_1 = \begin{bmatrix} a & 0 & 0 & 0 \\ 0 & b & 0 & 0 \\ 0 & 0 & c & 0 \\ 0 & 0 & 0 & d \end{bmatrix}, \tag{7.153}$$

where a, b, c, and d are arbitrary positive numbers satisfying $a+b+c+d = 1$. We can use a circuit P constructed from controlled-NOT gates to obtain a state with the permuted populations

$$\rho_2 = P\rho_1 P^\dagger = \begin{bmatrix} a & 0 & 0 & 0 \\ 0 & c & 0 & 0 \\ 0 & 0 & d & 0 \\ 0 & 0 & 0 & b \end{bmatrix}, \tag{7.154}$$

and similarly,

$$\rho_3 = P^\dagger \rho_1 P = \begin{bmatrix} a & 0 & 0 & 0 \\ 0 & d & 0 & 0 \\ 0 & 0 & b & 0 \\ 0 & 0 & 0 & c \end{bmatrix}. \tag{7.155}$$

A unitary quantum computation U is applied to each of these states, to obtain (in three separate experiments, which may be performed at different times) $C_k = U\rho_k U^\dagger$. By linearity,

$$\sum_{k=1,2,3} C_k = \sum_k U\rho_k U^\dagger \tag{7.156}$$

$$= U \left[\sum_k \rho_k \right] U^\dagger \tag{7.157}$$

$$= (4a - 1)U \begin{bmatrix} 1 & 0 & 0 & 0 \\ 0 & 0 & 0 & 0 \\ 0 & 0 & 0 & 0 \\ 0 & 0 & 0 & 0 \end{bmatrix} U^\dagger + (1 - a) \begin{bmatrix} 1 & 0 & 0 & 0 \\ 0 & 1 & 0 & 0 \\ 0 & 0 & 1 & 0 \\ 0 & 0 & 0 & 1 \end{bmatrix}. \tag{7.158}$$

In NMR, observables M (such as Pauli X and Y) for which $\text{tr}(M) = 0$ are the only ones ever measured; thus,

$$\text{tr}\left(\sum_k C_k M \right) = \sum_k \text{tr}\left(C_k M \right) \tag{7.159}$$

$$= (4a - 1) \operatorname{tr}\left(U \begin{bmatrix} 1 & 0 & 0 & 0 \\ 0 & 0 & 0 & 0 \\ 0 & 0 & 0 & 0 \\ 0 & 0 & 0 & 0 \end{bmatrix} U^\dagger M \right) \tag{7.160}$$

$$= (4a - 1) \operatorname{tr}\left(U|00\rangle\langle 00|U^\dagger \right) . \tag{7.161}$$

We find that the sum of the measured signals from the three experiments gives us a result which is proportional to what we would have obtained had the original system been prepared in a pure state $|00\rangle\langle 00|$ instead of in the arbitrary state of (7.153). This is always possible to accomplish for arbitrarily prepared systems of any size, given enough experiments which are summed over, and sufficiently long coherence time for unitary operations to be performed before decoherence sets in. Note that the different C_k experiments can actually also be done simultaneously with three different systems, or in different parts of a single system; this is experimentally feasible by applying gradient magnetic fields which vary systematically over a single sample, and that variant of this technique is called spatial labeling.

Exercise 7.42: (Permutations for temporal labeling) Give a quantum circuit to accomplish the permutations P and $P^\dagger$ necessary to transform ρ_1 of (7.153) to ρ_2 of (7.154).

Logical labeling is based on similar observations, but does not require multiple experiments to be performed. Suppose we have a system of three nearly identical spins in the state

$$\rho = \delta I + \alpha \begin{bmatrix} 6 & 0 & 0 & 0 & 0 & 0 & 0 & 0 \\ 0 & 2 & 0 & 0 & 0 & 0 & 0 & 0 \\ 0 & 0 & 2 & 0 & 0 & 0 & 0 & 0 \\ 0 & 0 & 0 & -2 & 0 & 0 & 0 & 0 \\ 0 & 0 & 0 & 0 & 2 & 0 & 0 & 0 \\ 0 & 0 & 0 & 0 & 0 & -2 & 0 & 0 \\ 0 & 0 & 0 & 0 & 0 & 0 & -2 & 0 \\ 0 & 0 & 0 & 0 & 0 & 0 & 0 & -6 \end{bmatrix} \tag{7.162}$$

$$\approx \left(\delta' I + \alpha' \begin{bmatrix} 2 & 0 \\ 0 & -2 \end{bmatrix} \right)^{\otimes 3}, \tag{7.163}$$

where δI refers to a background population which is unobservable (because of traceless measurement observables), and $\alpha \ll \delta$ is a small constant. We may then apply a unitary operation which performs a permutation $\tilde{P}$, giving

$$\rho' = \tilde{P}\rho\tilde{P}^\dagger = \delta I + \alpha \begin{bmatrix} 6 & 0 & 0 & 0 & 0 & 0 & 0 & 0 \\ 0 & -2 & 0 & 0 & 0 & 0 & 0 & 0 \\ 0 & 0 & -2 & 0 & 0 & 0 & 0 & 0 \\ 0 & 0 & 0 & -2 & 0 & 0 & 0 & 0 \\ 0 & 0 & 0 & 0 & -6 & 0 & 0 & 0 \\ 0 & 0 & 0 & 0 & 0 & 2 & 0 & 0 \\ 0 & 0 & 0 & 0 & 0 & 0 & 2 & 0 \\ 0 & 0 & 0 & 0 & 0 & 0 & 0 & 2 \end{bmatrix} . \tag{7.164}$$

Note that the upper 4×4 block of this matrix has the form

$$\begin{bmatrix} 6 & 0 & 0 & 0 \\ 0 & -2 & 0 & 0 \\ 0 & 0 & -2 & 0 \\ 0 & 0 & 0 & -2 \end{bmatrix} = 8|00\rangle\langle00| - 2I, \tag{7.165}$$

where I here denotes the 4×4 identity matrix. Just as for temporal labeling, we find that if a computation is performed on such a state, in this case the $|000\rangle, |001\rangle, |010\rangle, |011\rangle$ manifold, then it produces a result which is proportional to what we would have obtained had the original system been prepared in a pure state $|00\rangle\langle00|$! Experimentally, it is possible to perform $\tilde{P}$, and also to isolate the signal from just this manifold of states.

States which are of the form $\rho = 2^{-n}(1-\epsilon)I + \epsilon U|00\dots0\rangle\langle00\dots0|U^\dagger$ (where U is any unitary operator), are called 'effective pure states', or 'pseudopure' states. n is the number of qubits, but the dimension of the Hilbert space need not be a power of two in general. There are many strategies for preparing such states, and in general they all incur some cost. We shall return to discuss this later, in Section 7.7.4. Effective pure states make it possible to observe zero temperature dynamics from a system which equilibrates to a high temperature state, as long as the coupling of the system to its high temperature environment is sufficiently small. This is the way it is used in NMR quantum computation.

Exercise 7.43: (Permutations for logical labeling) Give a quantum circuit to accomplish the permutations $\tilde{P}$ necessary to transform ρ of (7.163) to ρ' of (7.165).

Exercise 7.44: (Logical labeling for n spins) Suppose we have a system of n nearly identical spins of Zeeman frequency $\hbar\omega$ in thermal equilibrium at temperature T with state ρ. What is the largest effective pure state that you can construct from ρ using logical labeling? (*Hint*: take advantage of states whose labels have Hamming weight of $n/2$.)

Ensemble readout of quantum algorithm results

We have seen how arbitrary unitary transforms can be performed on an n spin system with a Hamiltonian of (7.146), and we have learned how to prepare from a thermal state a well-defined input which behaves as a low-entropy ground state. However, to complete the requirements for quantum computation, we must have a way of performing measurements on the system to read out computational results. The difficulty is that the output of a typical quantum algorithm is a random number, whose distribution gives information which allows the problem to be solved. Unfortunately, the average value of the random variable does not necessarily give any relevant information. This would be the output if the quantum algorithm were executed without modification on an NMR quantum computer, since it is performed with a large *ensemble* of molecules, rather than with a single n spin molecule.

This difficulty is illustrated by the following example. The quantum factoring algorithm produces as its output a random rational number c/r, where c is an unknown integer, and r is the desired result (also an integer). Normally, a projective measurement is used to obtain c/r, then a classical continued fractions algorithm is performed to obtain c with high probability (Section 5.3.1). The answer is then checked by plugging in the result to the original problem, and if it fails the entire algorithm is repeated. Unfortunately, however, if only the ensemble average is available, since c is nearly uniformly distributed the average value $\langle c/r \rangle$ gives no meaningful information.

A simple resolution to this problem, which works for all quantum algorithms based on the hidden subgroup problem (Chapter 5), is simply to append any required classical post-processing step to the quantum computation. This is always possible because quantum computation subsumes classical computation. In the example given above, we simply ask each individual quantum computer (each molecule) to perform a continued fractions algorithm. The result is then checked on each quantum computer, then only the computers which succeed in the verification give an output. The final ensemble average measurement thus gives $\langle r \rangle$.

7.7.4 Experiment

One of the most exciting aspects of the NMR approach is the ready experimental realization of small instances of quantum computation and quantum information tasks. In this concluding section on NMR, we briefly describe three experiments which have been performed, demonstrating state tomography, elementary logic gates, and the quantum search algorithm. We also summarize the drawbacks of this method.

State tomography

How does one debug a quantum computer? A classical computer is analyzed by measuring its internal state at different points in time. Analogously, for a quantum computer, an essential technique is the ability to measure its density matrix – this is called *state tomography*.

Recall that the density matrix of a single qubit can be expressed as

$$\rho = \frac{1}{2}\left[1 + \sum_k r_k \sigma_k\right], \tag{7.166}$$

where σ_k are the Pauli matrices and r_b is a real, three-component vector. Because of the trace orthogonality property of Pauli matrices,

$$\mathrm{tr}(\sigma_k \sigma_j) = 2\delta_{kj}, \tag{7.167}$$

it follows that ρ can be reconstructed from the three measurement results

$$r_k = \langle \sigma_k \rangle = \mathrm{tr}(\rho \sigma_k). \tag{7.168}$$

Measurement of the usual observable in NMR, (7.143), preceded by the appropriate single qubit pulses, allows us to determine $\langle \sigma_k \rangle$, and thus obtain ρ. Similar results hold for larger numbers of spins. In practice, it is convenient to measure just the traceless deviation of ρ from the identity; this is called the deviation density matrix. Example results for two and three spin systems are shown in Figure 7.18.

Exercise 7.45: (State tomography with NMR) Let the voltage measurement
$$V_k(t) = V_0 \mathrm{tr}\left[e^{-iHt} M_k \rho M_k^\dagger e^{iHt}(iX_k + Y_k)\right]$$
be the result of experiment k.
Show that for two spins, nine experiments, with $M_0 = I$, $M_1 = R_{x1}$, $M_2 = R_{y1}$, $M_3 = R_{x2}$, $M_4 = R_{x2}R_{x1}$, $M_5 = R_{x2}R_{y1}$ etc. provide sufficient data from which ρ can be reconstructed.

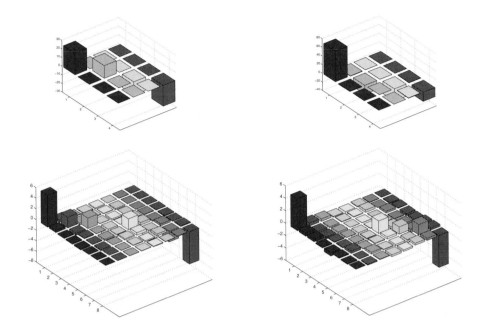

Figure 7.18. Experimentally measured deviation density matrices. Vertical scales are arbitrary, and only the real part is shown; the imaginary components are all relatively small. (top left) The two qubit thermal equilibrium state of the proton and the carbon nucleus in molecules of chloroform (^{13}CHCl$_3$) in a 11.78 tesla magnetic field. The 0.5 milliliter, 200 millimolar sample was diluted in acetone-d_6, degassed, and flame sealed in a thin walled, 5 mm glass tube. (top right) Two qubit effective pure state created using temporal labeling with the chloroform, as in Equation (7.161). (bottom left) The three qubit thermal equilibrium state of three fluorine nuclei in trifluoroethylene. (bottom right) An effective pure state created from the three spin system using logical labeling, as in Equation (7.164).

Exercise 7.46: How many experiments are sufficient for three spins? Necessary?

Quantum logic gates

The two qubit proton–carbon system in chloroform presents an excellent system for demonstration of single qubit and two qubit gates, for many reasons. At ≈ 500 and ≈ 125 MHz in an ≈ 11.8 tesla field, the frequencies of the two spins are well separated and easily addressed. The 215 Hz J-coupling frequency of the two nuclei is also convenient; it is much slower than the time scale required for single qubit RF pulses, but much faster than the relaxation time scales. In typical experiments, the T_1 of the proton and carbon are approximately 18 and 25 seconds, respectively, while T_2 are 7 and 0.3 seconds. The carbon T_2 is short because of interactions with the three quadrupolar chlorine nuclei, but taking the product of the shortest T_2 lifetime and the J-coupling indicates that roughly 60 gates should still be realizable before quantum coherence is lost.

The Hamiltonian of the two spin system is very well approximated by the expression in (7.147), but it can be simplified significantly using an experimental trick. By tuning two oscillators to exactly the rotating frequencies of the proton and carbon, we obtain, in the rotating frame defined by the oscillators, the simplified Hamiltonian

$$H = 2\pi\hbar J Z_1 Z_2 \,, \tag{7.169}$$

where $J = 215$ Hz. This Hamiltonian makes the realization of the controlled-NOT gate quite simple. A circuit which performs a CNOT transform equivalent up to single qubit phases is shown in Figure 7.19, as well as a circuit for creating a Bell state, and experimentally measured outputs.

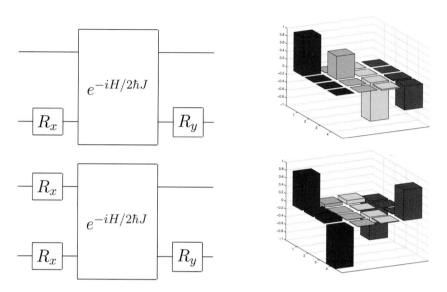

Figure 7.19. Quantum circuits implemented with NMR and the real part of the experimentally measured output deviation density matrices. In these circuits, R_x and R_y denote single qubit gates which perform $90°$ rotations about $\hat{x}$ and $\hat{y}$, implemented with RF pulses about 10 microseconds long, and the two qubit box with $e^{-iH/2\hbar J}$ is a free evolution period of time $1/2J \approx 2.3$ milliseconds. (top) Controlled-NOT circuit, and the output measured for a thermal state input, showing the exchange of the $|10\rangle$ and $|11\rangle$ diagonal elements, as expected from the classical truth table for the CNOT gate. (bottom) Circuit for creating the Bell state $(|00\rangle - |11\rangle)/\sqrt{2}$, and its output, when a $|00\rangle$ effective pure state is prepared as an input.

Exercise 7.47: (NMR controlled-NOT gate) Verify that the circuit shown in the top left of Figure 7.19 performs a controlled-NOT gate, up to single qubit phases; that is, it acts properly on classical input states, and furthermore can be turned into a proper controlled-NOT gate by applying additional single qubit R_z rotations. Give another circuit using the same building blocks to realize a proper CNOT gate.

Exercise 7.48: Verify that the circuit shown in the bottom left of Figure 7.19 creates the Bell state $(|00\rangle - |11\rangle)/\sqrt{2}$ as advertised.

Exercise 7.49: (NMR swap gate) An important chemical application of NMR is measurement of connectivity of spins, i.e. what protons, carbons, and phosphorus atoms are nearest neighbors in a molecule. One pulse sequence to do this is known as INADEQUATE (incredible natural abundance double quantum transfer experiment – the art of NMR is full of wonderfully creative acronyms). In the language of quantum computation, it can be understood as simply trying to apply a CNOT between any two resonances; if the CNOT works, the two nuclei must be neighbors. Another building block which is used in sequences such as

TOCSY (total correlation spectroscopy) is a swap operation, but not quite in the perfect form we can describe simply with quantum gates! Construct a quantum circuit using only $e^{-iH/2\hbar J}$, R_x, and R_y operations to implement a swap gate (you may start from the circuit in Figure 1.7).

Quantum algorithms

Grover's quantum search algorithm provides another simple example for NMR quantum computation. For a problem size of four elements ($n = 2$ qubits), we are given the set $x = \{0, 1, 2, 3\}$ for which $f(x) = 0$ except at one value x_0, where $f(x_0) = 1$. The goal is to find x_0, which can be classically accomplished by evaluating $f(x)$ an average of 2.25 times. In comparison, the quantum algorithm finds x_0 by evaluating $f(x)$ only once (Chapter 6; see, in particular, Box 6.1).

Three operators are required: the oracle operator O (which performs a phase flip based on the function $f(x)$), the Hadamard operator on two qubits $H^{\otimes 2}$, and the conditional phase shift operator P. The oracle O flips the sign of the basis element corresponding to x_0; for $x_0 = 3$, this is

$$O = \begin{bmatrix} 1 & 0 & 0 & 0 \\ 0 & 1 & 0 & 0 \\ 0 & 0 & 1 & 0 \\ 0 & 0 & 0 & -1 \end{bmatrix}, \tag{7.170}$$

Denoting a $t = 1/2J$ (2.3 millisecond for the chloroform) period evolution $e^{-iH/2\hbar J}$ as τ, we find that $O = R_{y1}\bar{R}_{x1}\bar{R}_{y1}R_{y2}\bar{R}_{x2}\bar{R}_{y2}\,\tau$ (up to an irrelevant overall phase factor) for the $x_0 = 3$ case. $H^{\otimes 2}$ is just two single qubit Hadamard operations, $H_1 \otimes H_2$, where $H_k = R_{xk}^2\bar{R}_{yk}$. And the operator P,

$$P = \begin{bmatrix} 1 & 0 & 0 & 0 \\ 0 & -1 & 0 & 0 \\ 0 & 0 & -1 & 0 \\ 0 & 0 & 0 & -1 \end{bmatrix} \tag{7.171}$$

is simply realized as $P = R_{y1}R_{x1}\bar{R}_{y1}R_{y2}R_{x2}\bar{R}_{y2}\,\tau$. From these, we construct the Grover iteration $G = H^{\otimes 2}PH^{\otimes 2}O$. This operator can be simplified straightforwardly by eliminating unnecessary operations which obviously cancel (see Exercise 7.51). Let $|\psi_k\rangle = G^k|00\rangle$ be the state after k applications of the Grover iteration to the initial state. We find that the amplitude $\langle x_0|\psi_k\rangle \approx \sin((2k + 1)\theta)$, where $\theta = \arcsin(1/\sqrt{2})$; this periodicity is a fundamental property of the quantum search algorithm, and is a natural feature to test in an experiment. For the two qubit case, and $x_0 = 3$, we expect $|11\rangle = |\psi_1\rangle = -|\psi_4\rangle = |\psi_7\rangle = -|\psi_{10}\rangle$, a period of 3 if the overall sign is disregarded.

Exercise 7.50: Find quantum circuits using just single qubit rotations and $e^{-iH/2\hbar J}$ to implement the oracle O for $x_0 = 0, 1, 2$.

Exercise 7.51: Show that the Grover iteration can be simplified, by canceling adjacent single qubit rotations appropriately, to obtain

$$
G = \begin{cases}
\bar{R}_{x1}\bar{R}_{y1}\bar{R}_{x2}\bar{R}_{y2}\, \tau\, R_{x1}\bar{R}_{y1}R_{x2}\bar{R}_{y2}\, \tau & (x_0 = 3) \\
\bar{R}_{x1}\bar{R}_{y1}\bar{R}_{x2}\bar{R}_{y2}\, \tau\, R_{x1}\bar{R}_{y1}\bar{R}_{x2}\bar{R}_{y2}\, \tau & (x_0 = 2) \\
\bar{R}_{x1}\bar{R}_{y1}\bar{R}_{x2}\bar{R}_{y2}\, \tau\, \bar{R}_{x1}\bar{R}_{y1}R_{x2}\bar{R}_{y2}\, \tau & (x_0 = 1) \\
\bar{R}_{x1}\bar{R}_{y1}\bar{R}_{x2}\bar{R}_{y2}\, \tau\, \bar{R}_{x1}\bar{R}_{y1}\bar{R}_{x2}\bar{R}_{y2}\, \tau & (x_0 = 0)
\end{cases} , \tag{7.172}
$$

for the four possible cases of x_0.

Figure 7.20 shows the theoretical and measured deviation density matrices $\rho_{\Delta n} = |\psi_n\rangle\langle\psi_n| - \mathrm{tr}(|\psi_n\rangle\langle\psi_n|)/4$ for the first seven iterations of U. As expected, $\rho_{\Delta 1}$ clearly reveals the $|11\rangle$ state corresponding to $x_0 = 3$. Analogous results were obtained for experiments repeated for the other possible values of x_0. Measuring each density matrix required $9 \times 3 = 27$ experimental repetitions, 9 for the tomographic reconstruction and 3 for the pure state preparation.

The longest computation, for $n = 7$, took less than 35 milliseconds, which was well within the coherence time. The periodicity of Grover's algorithm is clearly seen in Figure 7.20, with good agreement between theory and experiment. The large signal-to-noise (typically better than 10^4 to 1) was obtained with just single-shot measurements. Numerical simulations indicate that the 7–44% errors are primarily due to inhomogeneity of the magnetic field, magnetization decay during the measurement, and imperfect calibration of the rotations (in order of importance).

Drawbacks

The bulk-ensemble NMR implementation of quantum computation has been successful in demonstrating quantum algorithms and quantum information tasks with systems up to seven qubits, which is quite impressive. However, there are important limitations which arise from the temporal, spatial, and logical labeling techniques which are at the heart of the method.

The essential objective of these labeling techniques is to isolate the signal from the subset of spins which happen to be in the $|00\ldots0\rangle$ component (or any other standard, high probability state) of the thermal equilibrium state. Temporal and spatial labeling do this by adding up signals to cancel all but the desired state; logical labeling trades off Hilbert space for purity. Irrespective of the method used, however, nothing can increase the probability of the $|00\ldots0\rangle$ component of the thermal state,

$$
p_{00\ldots0} = \frac{1}{\mathcal{Z}}\langle 00\ldots0|e^{-\beta H}|00\ldots0\rangle . \tag{7.173}
$$

Taking $H = \sum_k \omega Z_k$, we find that $p_{00\ldots0}$ is proportional to $n2^{-n}$, for an n spin molecule. This means that the total signal *decreases exponentially* as the number of qubits 'distilled' into an effective pure state using labeling techniques, for constant initial state temperature.

Another limitation comes from using molecules as quantum computers. The structure of the molecule plays the role of the *architecture* of the computer, determining what pairs (or groups) of qubits interact with each other (analogously, the RF pulses serve as the software). Not all qubits are necessarily well connected! This is doubly important since interactions cannot be switched off, except by performing refocusing. Furthermore qubits are addressed by distinguishing them in frequency, but this rapidly becomes difficult as the number of nuclei is increased. A solution to this exists, which is to use a cellular automata style architecture, such as the one-dimensional chain $X - A - B - C - A -$

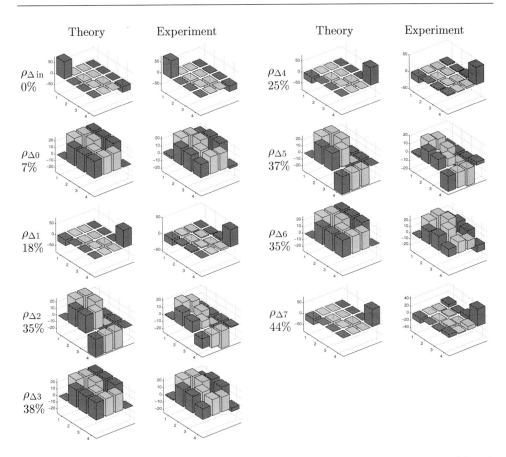

Figure 7.20. Theoretical and experimental deviation density matrices (in arbitrary units) for seven steps of Grover's algorithm performed on the hydrogen and carbon spins in chloroform. Three full cycles, with a periodicity of four iterations are clearly seen. Only the real component is plotted (the imaginary portion is theoretically zero and was found to contribute less than 12% to the experimental results). Relative errors $||\rho_{\text{theory}} - \rho_{\text{expt}}||/||\rho_{\text{theory}}||$ are shown.

$B - C - \cdots - A - B - C - Y$, in which the ends are distinguished but the middle is composed of a repeating regular sequence. In this architecture, only distinct letters are addressable, and it might seem as if this is a highly restrictive model of computation. However, it has been shown that in fact it is universal, with only polynomial slowdown. The precise amount of slowdown required will of course be important when performing tasks such as the quantum search algorithm, which only has a square root speedup.

Methods for circumventing the limitations of labeling techniques also exist. One possibility is to polarize the nuclei through some physical mechanism; this has been done by using optical pumping (similar to how ions are cooled as shown in Figure 7.8) to polarize the electronic state of rubidium atoms, which then transfer their polarization to the nuclei of xenon atoms through formation of short-lived van der Waals molecules. This has also been done for helium. Doing similarly for molecules is conceivable, albeit technologically challenging. Another possibility is the use of a different labeling scheme; logical labeling is essentially a compression algorithm, which increases the relative probability of one state in an ensemble by discarding ensemble members. An improved version of this method has been developed which achieves the entropic limit, giving $nH(p)$ pure qubits from

n-spin molecules originally at temperature T such that $p = (1 - e^{-\Delta E/k_B T})/2$, where ΔE is the spin flip energy. This scheme does *not* have any exponential cost; the compression can be achieved using only a polynomial number of basic operations. However, it is inefficient unless p is relatively small, and today $p \approx 0.4999$ in good solenoid magnet systems.

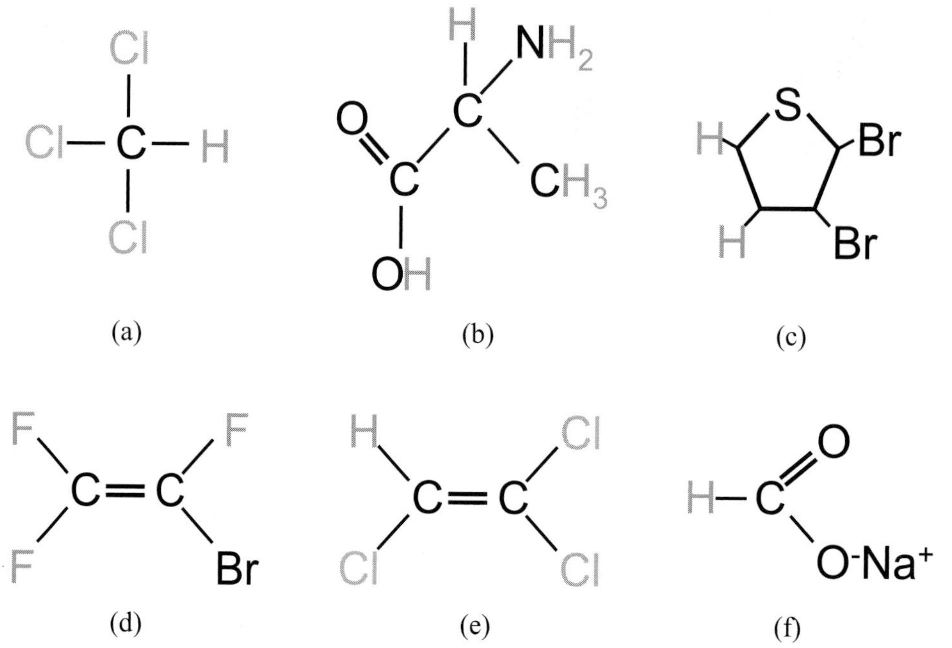

Figure 7.21. A selection of simple molecules which have been used to demonstrate various quantum computation and quantum information tasks with NMR. (a) Chloroform: two qubits, proton and carbon, have been used to implement the Deutsch–Jozsa algorithm, and a two qubit quantum search. (b) Alanine: three qubits, composed of the carbon backbone, have been used to demonstrate error-correction. Note how the three carbon nuclei have distinguishable frequencies because their surrounding chemical environments are different (for example, the electronegativity of the oxygen causes it to draw much of the nearby electrons away from the neighboring carbon). (c) 2,3-dibromothiophene: two qubits, composed of the two protons, have been used to simulate four levels of a truncated simple harmonic oscillator. Here, the two protons are at different distances from the sulphur atom, and thus have distinguishable frequencies. (d) Trifluorobromoethylene: three qubits, the three fluorines, have been used to demonstrate logical labeling and the creation of a $(|000\rangle + |111\rangle)/\sqrt{2}$ superposition state. (e) Trichloroethylene: three qubits, the proton and two carbons, were used to demonstrate teleportation, with the proton's state being teleported to the rightmost carbon. (f) Sodium formate: two qubits, proton and carbon, used to demonstrate the two qubit quantum error detection code. In this molecule, the sodium radical is used to tune the T_2 times of the two qubits to be nearly equal, by changing the ambient temperature to modify its exchange rate with the solvent.

Despite these drawbacks, NMR provides a testbed for quantum algorithms and illustrates basic techniques which other realizations will have to implement to perform quantum computation. Some of the molecules which have been used to demonstrate quantum computation and quantum information tasks are shown in Figure 7.21. The NMR idea is also a very rich area for innovation, combining chemistry, physics, engineering, and mathematics, and undoubtedly continued innovation between fields will further this technique.

NMR quantum computer

- **Qubit representation**: Spin of an atomic nucleus.

- **Unitary evolution**: Arbitrary transforms are constructed from magnetic field pulses applied to spins in a strong magnetic field. Couplings between spins are provided by chemical bonds between neighboring atoms.

- **Initial state preparation**: Polarize the spins by placing them in a strong magnetic field, then use 'effective pure state' preparation techniques.

- **Readout**: Measure voltage signal induced by precessing magnetic moment.

- **Drawbacks**: Effective pure state preparation schemes reduce the signal exponentially in the number of qubits, unless the initial polarization is sufficiently high.

7.8 Other implementation schemes

In this chapter we have described but a fraction of the number of ideas people have considered for implementation of quantum computers. Our selection illustrates basic requirements and challenges common to all implementations: robust representation of quantum information, application of unitary transforms, preparation of a fiducial input state, and measurement of the output.

The simple harmonic oscillator example emphasizes how a *digital* representation is crucial: each unit (qubit, qutrit, qudit, or whatever) of quantum information should reside in *physically separate* degrees of freedom; otherwise, some resource (such as energy) is inefficiently utilized. That example also provides the mathematical basis for studying representations of qubits through the remainder of the chapter. Single photons are nearly ideal qubits, but the nonlinear optical materials required to get them to interact are difficult to realize without causing coherence loss. Cavity-QED techniques can address this problem by using single atoms to interact photons, but even more importantly, they introduce the notion of two-level atoms, and the idea of guarding qubit representations using selection rules imposed by physical symmetries such as dipole selection.

A natural extension of this idea is to represent qubits using spin-$1/2$ particles, which inherently only have two states. This is the tack taken with trapped ions, which store qubits in electron and nuclear spins; the difficulty with this method, though, is that the center of mass oscillations – phonons – used to mediate interactions between spins have short coherence times. Molecules, in which nuclear spins can be strongly coupled by chemical bonds, can solve this difficulty, but the spin resonance signal from single molecules is too small to detect with present technology. NMR quantum computing solves this by creating 'effective pure states' with bulk ensembles of $O(10^{18})$ molecules, thereby demonstrating simple quantum algorithms in the laboratory. But without providing initial polarization, this capability comes at the cost of signal strength, which decreases exponentially with the number of qubits.

As these examples demonstrate, coming up with a good physical realization for a quantum computer is tricky business, fraught with tradeoffs. All of the above schemes are unsatisfactory, in that none allow a large-scale quantum computer to be realized anytime in the near future. However, that does not preclude the possibility, and in fact many other implementation schemes have been proposed, some of which we briefly touch upon in this final section here.

A good way to categorize realizations is in terms of the physical degree of freedom used to represent qubits. Recall Figure 7.1: just about anything which comes in quantum units could be a qubit, but, as we have seen, fundamental physical quanta such as photons and spin are particularly good choices.

Another fundamental quantum unit which could serve as a qubit representation is electric *charge*. Modern electronics provides excellent techniques to create, control, and measure charges, even at the level of single electrons. For example, *quantum dots*, fabricated from semiconductor materials, metals, or even small molecules, can serve as three-dimensional boxes with electrostatic potentials which confine charge quanta. This is verified by observation of the Coulomb blockade effect, in which electrical conductance through dots of capacitance C is found to increase in discrete steps as a function of bias voltage across the dot, reflecting the $e^2/2C$ energy required to add each additional electron. Unlike photons, (net) charge cannot be destroyed; charges can only be moved around, and thus a charge state qubit would have to use something like the dual-rail representation of Section 7.4.2, whereby the $|0\rangle$ and $|1\rangle$ states correspond to having charge located in either of two dots, or two states within a single dot.

Just as for single photons, single qubit operations on charge state qubits can be performed using electrostatic gates (the analogue of optical phase shifters) and either single mode waveguide couplers (the analogue of beamsplitters) for moving electrons, or tunnel junctions for quantum dots. Electrical charges experience long-range Coulomb interactions with other charges (the potential created by a single charge at distance r is $V(r) = e/4\pi\epsilon r$), and thus charges far away can modulate the phase of a local charge, much like the Kerr interaction between photons. Controlled Coulomb interactions can thus be used to perform two qubit operations. Finally, single electron charges are straightforward to measure; modern field effect transistors easily detect movements of single charges in their channels, and single electron transistors operating at ≈ 100 millikelvin temperatures can detect charge to better than $10^{-4}e/\sqrt{\text{Hz}}$ at frequencies over 200 MHz. Unfortunately, uncontrolled distant charge motion leads to dephasing; this, and other scattering mechanism such as those due to phonon interactions cause coherence times to be relatively short for charge states, on the order of hundreds of femtoseconds to hundreds of picoseconds.

Charge carriers in superconductors have also been suggested as qubit representations. At low temperatures in certain metals, two electrons can bind together through a phonon interaction to form a Cooper pair, which has charge $2e$. And just as electrons can be confined within quantum dots, Cooper pairs can be confined within an electrostatic box, such that the number of Cooper pairs in the box becomes a good quantum number, and can be used to represent quantum information. Single qubit operations can be realized by using electrostatic gates to modulate the box potential, and Josephson junctions between coupled boxes. These junctions can also be used to couple qubits, and their strength can be modulated using an external magnetic field by coupling appropriately to superconducting interferometric loops. Finally, qubits can be measured simply by measuring electric charge. This superconductor qubit representation is interesting because of the relative robustness of Cooper pairs; it is estimated that the main decoherence mechanism is spontaneous emission of electromagnetic photons, which may allow coherence times exceeding one microsecond, long compared with typical dynamical time scales of hundreds of picoseconds. Unfortunately, just as with the electronic charge representation, a fluctuating background of extraneous charges ('quasiparticles') is highly deleterious to qubit coherence. One means around this problem, using superconductor technology, is to choose instead a *magnetic flux* qubit representation, in which qubit states correspond to left and right hand orientations of flux localized through a superconducting loop device. Here, decoherence is caused by background magnetic fluctuations, which are expected to be much quieter than the electrostatic case.

The locality of magnetic interactions is a good feature for qubit representations, and thus we return to spin, for which schemes have also been proposed to take advantage of solid state technology. A fairly large quantum dot, even one containing many electrons, can behave as a spin-$1/2$ object, where the spin is carried by a single excess electron. This spin state can be prepared by equilibrating in a strong magnetic field, at low temperatures, such that the spin flip energy ΔE is much larger than $k_B T$. Manipulating a single spin, as we have seen in Section 7.7, can be done by applying pulsed local magnetic fields, and coupled qubit operations can be realized with a controlled Heisenberg coupling,

$$H(t) = J(t)\vec{S}_1 \cdot \vec{S}_2 = \frac{1}{4}\left[X_1 X_2 + Y_1 Y_2 + Z_1 Z_2\right], \qquad (7.174)$$

where $\vec{S}$ are the spin operators (Pauli operators divided by two), and $J(t) = 4\tau_0^2(t)/u$, u being the charging energy of the dot, and $\tau_0(t)$ being the tunneling matrix element controlled by local electrostatic gates placed between dots. This interaction is universal, in the sense that it is equivalent to the controlled-NOT gate (see exercise below). Spin states may, in theory, be measured by allowing the spin-carrying electron to tunnel to a readout paramagnetic dot, or to spin-dependently tunnel through a 'spin valve' to a readout electrometer. The challenge is to realize such measurements in practice; high fidelity single spin measurement in semiconductors has not been accomplished with present technology.

Exercise 7.52: (Universality of Heisenberg Hamiltonian) Show that a swap operation U can be implemented by turning on $J(t)$ for an appropriate amount of time in the Heisenberg coupling Hamiltonian of (7.174), to obtain $U = \exp(-i\pi\vec{S}_1 \cdot \vec{S}_2)$. The '$\sqrt{\text{SWAP}}$' gate obtained by turning on the interaction for half this time is universal; compute this transform and show how to obtain a controlled-NOT gate by composing it with single qubit operations.

Eventually, with sufficiently advanced technology, it may be possible to place, control, and measure single *nuclear* spins in semiconductors, making possible the following vision. Imagine being able to precisely place single phosphorus atoms (nuclear spin-$1/2$) within a crystalline wafer of ^{28}Si (nuclear spin 0), positioned beneath lithographically patterned electrostatic gates. These gates allow manipulation of the electron cloud surrounding the ^{31}P dopants, to perform single qubit operations via modulation of the magnetic field seen by the ^{31}P nuclei. Additional gates located above the region separating the ^{31}P dopants can be used to artificially create electron distributions connecting adjacent ^{31}P, much like a chemical bond, thus allowing two qubit operations to be performed. Although fabrication constraints of such a scheme are extremely challenging – for example, the gates should be separated by $100\,\text{Å}$ or less, and the ^{31}P dopants must be registered precisely and in an ordered array – at least this vision articulates a possible venue for marrying quantum computing with more conventional computing technologies.

Of the schemes we have described for realizing quantum computers, the ones which have most captivated the attention of technologists are the ones based on solid state technologies. Of course, atomic, molecular, and optical quantum computing schemes continue to be proposed, using systems such as optical lattices (artificial crystals made from atoms confined by crossed beams of light) and Bose condensates which are at the forefront of those fields; someday, we may even see quantum computing proposals using mesons, quarks, and gluons, or even black holes. But the motivation to envision *some* kind of solid state quantum computer is enormous. It has been estimated that the world has invested over US$1 trillion in silicon technology since the invention of the transistor in the late 1940s. Condensed matter systems have also been rich in new physics, such as superconductivity, the quantum Hall effect, and the Coulomb blockade (a classical effect discovered long after it was widely thought that everything about classical physics was well known!).

This chapter has concentrated mainly on the implementation of quantum computing machines, but the basic components which were presented are also useful in many other quantum applications. Quantum cryptography and its experimental realization are described in Section 12.6. And pointers to experimental work on quantum teleportation and superdense coding are given in 'History and further reading' at the end of this chapter. The general interface between quantum communication and computation includes challenges such as distributed quantum computation; development of new algorithms and experimental implementations of such systems will certainly continue to be a rich area of research for the future.

Much of the allure of quantum computing and communication machines is certainly their potential economic ramifications, as novel information technologies. But as we have seen in this chapter, quantum computation and quantum information also motivate new questions about physical systems, and provide different ways to understand their properties. These new ideas embody a need to move away from traditional many-body, statistical, and thermodynamic studies of physical systems, all the way from atoms to condensed matter systems. They represent a new opportunity to focus instead on *dynamical* properties of physical systems at the *single* quantum level. Hopefully, by giving a flavor of the richness of this approach, this chapter will motivate you to continue 'thinking algorithmically' about Physics.

Problem 7.1: (Efficient temporal labeling) Can you construct efficient circuits (which require only $O(\text{poly}(n))$ gates) to cyclically permute all diagonal elements in a $2^n \times 2^n$ diagonal density matrix except the $|0^n\rangle\langle 0^n|$ term?

Problem 7.2: (Computing with linear optics) In performing quantum computation with single photons, suppose that instead of the dual rail representation of Section 7.4.1 we use a *unary* representation of states, where $|00\ldots01\rangle$ is 0, $|00\ldots010\rangle$ is 1, $|00\ldots0100\rangle$ is 2, and so on, up to $|10\ldots0\rangle$ being $2^n - 1$.

1. Show that an arbitrary unitary transformation on these states can be constructed completely from just beamsplitters and phase shifters (and no nonlinear media).

2. Construct a circuit of beamsplitters and phase shifters to perform the one qubit Deutsch–Jozsa algorithm.

3. Construct a circuit of beamsplitters and phase shifters to perform the two qubit quantum search algorithm.

4. Prove that an arbitrary unitary transform will, in general, require an exponential number (in n) of components to realize.

Problem 7.3: (Control via Jaynes–Cummings interactions)   Robust and accurate control of small quantum systems – via an external *classical* degree of freedom – is important to the ability to perform quantum computation. It is quite remarkable that atomic states can be controlled by applying optical pulses, without causing superpositions of atomic states to decohere very much! In this problem, we see what approximations are necessary for this to be the case. Let us begin with the Jaynes–Cummings Hamiltonian for a single atom coupled to a single mode of an electromagnetic field,

$$H = a^\dagger \sigma_- + a\sigma_+ \,, \tag{7.175}$$

where $\sigma_\pm$ act on the atom, and a, $a^\dagger$ act on the field.

1. For $U = e^{i\theta H}$, compute

$$A_n = \langle n|U|\alpha\rangle \,, \tag{7.176}$$

where $|\alpha\rangle$ and $|n\rangle$ are coherent states and number eigenstates of the field, respectively; A_n is an *operator* on atomic states, and you should obtain

$$A_n = e^{-|\alpha|^2}\frac{|\alpha|^2}{n!}\left[\begin{array}{cc} \cos(\theta\sqrt{n}) & \frac{i\sqrt{n}}{\alpha}\sin(\theta\sqrt{n}) \\ \frac{i\alpha}{\sqrt{n+1}}\sin(\theta\sqrt{n+1}) & \cos(\theta\sqrt{n+1}) \end{array}\right]. \tag{7.177}$$

The results of Exercise 7.17 may be helpful.

2. It is useful to make an approximation that α is large (without loss of generality, we may choose α real). Consider the probability distribution

$$p_n = e^{-x}\frac{x^n}{n!} \,, \tag{7.178}$$

which has mean $\langle n\rangle = x$, and standard deviation $\sqrt{\langle n^2\rangle - \langle n\rangle^2} = \sqrt{x}$. Now change variables to $n = x - L\sqrt{x}$, and use Stirling's approximation

$$n! \approx \sqrt{2\pi n}\,n^n e^{-n} \tag{7.179}$$

to obtain

$$p_L \approx \frac{e^{-L^2/2}}{\sqrt{2\pi}}. \tag{7.180}$$

3. The most important term is A_n for $n = |\alpha|^2$. Define $n = \alpha^2 + L\alpha$, and for

$$a = y\sqrt{\frac{1}{y^2} + \frac{L}{y}} \quad\text{and}\quad b = y\sqrt{\frac{1}{y^2} + \frac{L}{y} + 1}\,, \tag{7.181}$$

where $y = 1/\alpha$, show that

$$A_L \approx \frac{e^{-L^2/4}}{(2\pi)^{1/4}}\left[\begin{array}{cc} \cos a\varphi & ia\sin a\varphi \\ (i/b)\sin b\varphi & \cos b\varphi \end{array}\right], \tag{7.182}$$

using $\theta = \varphi/\alpha$. Also verify that

$$\int_{-\infty}^{\infty} A_L^\dagger A_L \, dL = I \tag{7.183}$$

as expected.

4. The ideal unitary transform which occurs to the atom is

$$U = \begin{bmatrix} \cos \alpha\theta & i \sin \alpha\theta \\ i \sin \alpha\theta & \cos \alpha\theta \end{bmatrix}. \tag{7.184}$$

How close is A_L to U? See if you can estimate the *fidelity*

$$\mathcal{F} = \min_{|\psi\rangle} \int_{-\infty}^{\infty} |\langle\psi|U^\dagger A_L|\psi\rangle|^2 \, dL \tag{7.185}$$

as a Taylor series in y.

Problem 7.4: (Ion trap logic with two-level atoms) The controlled-NOT gate described in Section 7.6.3 used a three-level atom for simplicity. It is possible to do without this third level, with some extra complication, as this problem demonstrates.

Let $\Upsilon_{\hat{n}}^{\text{blue},j}(\theta)$ denote the operation accomplished by pulsing light at the blue sideband frequency, $\omega = \Omega + \omega_z$, of the jth particle for time $\theta\sqrt{N}/\eta\Omega$, and similarly for the red sideband. $\hat{n}$ denotes the axis of the rotation in the $\hat{x}$-$\hat{y}$ plane, controlled by setting the phase of the incident light. The superscript j may be omitted when it is clear which ion is being addressed. Specifically,

$$\Upsilon_{\hat{n}}^{\text{blue}}(\theta) = \exp\left[\left(e^{i\varphi}|00\rangle\langle 11| + e^{-i\varphi}|11\rangle\langle 00| \right.\right.$$
$$\left.\left. + e^{i\varphi}\sqrt{2}|01\rangle\langle 12| + e^{-i\varphi}\sqrt{2}|12\rangle\langle 01| + \cdots \right) \frac{i\theta}{2} \right], \tag{7.186}$$

where $\hat{n} = \hat{x}\cos\varphi + \hat{y}\sin\varphi$, and the two labels in the ket represent the internal and the motional states, respectively, from left to right. The $\sqrt{2}$ factor comes from the fact that $a^\dagger|n\rangle = \sqrt{n+1}|n+1\rangle$ for bosonic states.

(1) Show that $S^j = \Upsilon_{\hat{y}}^{\text{red},j}(\pi)$ performs a swap between the internal and motional states of ion j when the motional state is initially $|0\rangle$.

(2) Find a value of θ such that $\Upsilon_{\hat{n}}^{\text{blue}}(\theta)$ acting on any state in the computational subspace, spanned by $|00\rangle$, $|01\rangle$, $|10\rangle$, and $|11\rangle$, leaves it in that subspace. This should work for any axis $\hat{n}$.

(3) Show that if $\Upsilon_{\hat{n}}^{\text{blue}}(\varphi)$ stays within the computational subspace, then $U = \Upsilon_{\alpha}^{\text{blue}}(-\beta)\Upsilon_{\hat{n}}^{\text{blue}}(\theta)\Upsilon_{\alpha}^{\text{blue}}(\beta)$ also stays within the computational subspace, for any choice of rotation angle β and axis α.

(4) Find values of α and β such that U is diagonal. Specifically, it is useful to obtain an operator such as

$$\begin{bmatrix} e^{-i\pi/\sqrt{2}} & 0 & 0 & 0 \\ 0 & -1 & 0 & 0 \\ 0 & 0 & 1 & 0 \\ 0 & 0 & 0 & e^{i\pi/\sqrt{2}} \end{bmatrix}. \tag{7.187}$$

(5) Show that (7.187) describes a non-trivial gate, in that a controlled-NOT gate between the internal states of two ions can be constructed from it and single qubit operations. Can you come up with a composite pulse sequence for performing a CNOT without requiring the motional state to initially be $|0\rangle$?

Summary of Chapter 7: Quantum computers: physical realization

- **There are four basic requirements** for implementation of quantum computation: (1) Representation of qubits, (2) Controllable unitary evolution, (3) Preparation of initial qubit states, and (4) Measurement of final qubit states.

- **Single photons** can serve as good qubits, using $|01\rangle$ and $|10\rangle$ as logical 0 and 1, but conventional nonlinear optical materials which are sufficiently strong to allow single photons to interact inevitably absorb or scatter the photons.

- **Cavity-QED** is a technique by which single atoms can be made to interact strongly with single photons. It provides a mechanism for using an atom to mediate interactions between single photons.

- **Trapped ions** can be cooled to the extent that their electronic and nuclear spin states can be controlled by applying laser pulses. By coupling spin states through center-of-mass phonons, logic gates between different ions can be performed.

- **Nuclear spins** are nearly ideal qubits, and single molecules would be nearly ideal quantum computers if their spin states could only be controlled and measured. Nuclear magnetic resonance makes this possible using large ensembles of molecules at room temperature, but at the expense of signal loss due to an inefficient preparation procedure.

History and further reading

For an excellent discussion of why building a quantum computer is difficult, see the article by DiVincenzo[DiV95a], on which Figure 7.1 is based. DiVincenzo also presents five criteria for realizing a quantum computer, which are similar to those discussed in Section 7.2.

The quantum simple harmonic oscillator of Section 7.3 is a staple of quantum mechanics, and is treated in any standard textbook, such as [Sak95]. The general necessary and sufficient conditions for quantum computation given in Section 7.3.3 were discussed by Lloyd[Llo94].

The optical quantum computer of Section 7.4 uses as its main theoretical tools the formalism of quantum optics, which is described in many textbooks; two excellent ones are [Lou73, Gar91]. For more on basic optics and optical technology, such as polarizers, beamsplitters, photon detectors, and the like, see for example, the textbook [ST91]. The beamsplitter in the single photon regime was studied by Campos, Saleh, and Tiech[CST89], and related to that context, the elegant connection between $SU(2)$ and two coupled harmonic oscillators was first described by Schwinger[Sak95]. The concept of a 'dual-rail' representation of a qubit was suggested by Yurke and used by Chuang and Yamamoto[CY95] to describe a complete quantum computer (using nonlinear Kerr media) to perform the

Deutsch–Jozsa algorithm (as in Exercise 7.13). The quantum-optical Fredkin gate was first described by Yamamoto, Kitagawa, and Igeta[YKI88] and Milburn[Mil89a]. The single photon generation and detection technology required for an optical quantum computer has been discussed by Imamoglu and Yamamoto[IY94] and by Kwiat, Steinberg, Chiao, Eberhard, and Petroff[KSC+94]. An analogous mechanism using electron optics and the Coulomb interaction in place of the Kerr interaction has been discussed by Kitagawa and Ueda[KU91]. The fundamental limits of traditional, off-resonance nonlinear optical materials at single quanta levels have been studied by Watanabe and Yamamoto[WY90]. Cerf, Adami, and Kwiat have studied the simulation of quantum logic using (exponentially many) linear optical components[CAK98]. An influential earlier paper by Reck, Zeilinger, Bernstein and Bertani[RZBB94] described similar constructions, but did not make the explicit connection to quantum computation. Kwiat, Mitchell, Schwindt, and White have constructed an optical simulation of Grover's quantum search algorithm which uses linear optics, but appears to require exponential resources when scaling[KMSW99]. See Miller[Mil89b] for a discussion of the energetics of optical communication over different distances.

Allen and Eberly[AE75] have written a beautiful treatise on two-level atoms and optical resonance. The experiment described in Section 7.5.4 was performed by Turchette, Hood, Lange, Mabuchi, and Kimble at Caltech in 1995[THL+95]. A detailed explanation is also given in Turchette's Ph.D. thesis[Tur97]. The single photons used in this experiment were called 'flying qubits'. A different approach, in which atomic states are used as qubits and the atoms travel through optical cavities, was proposed by Domokos, Raimond, Brune, and Haroche[DRBH95]. It is based on the idea of using a single atom to switch a coherent state into a cavity, as described by Davidovich, Maali, Brune, Raimond, and Haroche[DRBH87, DMB+93].

The idea of ion trap quantum computation was proposed by Cirac and Zoller[CZ95]. Our discussion of this idea in Section 7.6.1 benefited greatly from the articles by Steane[Ste97], and by Wineland, Monroe, Itano, Leibfried, King, and Meekof[WMI+98]. Earnshaw's theorem can be derived from Laplace's equation, as described in his original paper[Ear42], or a modern electromagnetics textbook such as the one by Ramo, Whinnery, and van Duzer[RWvD84]. Figure 7.8 is drawn after Figure 6 in [Ste97]. Figure 7.7 is drawn after [WMI+98]. The experiment described in Section 7.6.4 was performed by Monroe, Meekhof, King, Itano, and Wineland at the National Institute of Standards and Technology in Boulder, Colorado[MMK+95]. Figure 7.15 is reprinted courtesy of Wineland[WMI+98]. Brewer, DeVoe, and Kallenbach have envisioned using large arrays of planar ion microtraps[BDK92] as a scalable quantum computer; this is the kind of trap shown in Figure 7.12. James has extensively studied the theory of heating and other decoherence mechanisms in ion traps[Jam98]. The impact of decoherence on ion trap quantum computation has also been studied in some depth by Plenio and Knight, who also consider effects such as the failure of the two-level approximation[PK96].

DiVincenzo first suggested the use of nuclear spins in quantum computation[DiV95b], and pointed out that the well known, and very old ENDOR (electron nucleon double resonance) pulse sequence is essentially an instance of the controlled-NOT gate. However, the problem of how to use an ensemble of nuclei at room temperature for quantum computation was not solved until Cory, Fahmy, and Havel[CFH97], and Gershenfeld and Chuang[GC97] realized that effective pure states could be prepared. It was also necessary to realize that quantum algorithms can be modified to allow ensemble readout; a solution to this problem was provided in [GC97], and is presented in Section 7.7.3. Excellent textbooks on NMR have been written by Ernst, Bodenhausen, and Wokaun[EBW87], and by Slichter[Sli96]. Warren has written a criticism of NMR quantum computation[War97]; interestingly, in the same paper he advocates using electron spin resonance (ESR) to do quantum computation. Temporal labeling was proposed by Knill, Chuang, and Laflamme[KCL98]. The NMR logic gates, and Bell state preparation circuit of Section 7.7.4 were discussed by Chuang, Gershenfeld, Kubinec, and Leung[CGKL98]. The realization of Grover's algorithm in Section 7.7.4 was by Chuang, Gershenfeld, and Kubinec[CGK98]; Figure 7.20 is taken from their paper. Linden, Kupce, and Freeman note that the swap gate may be a useful contribution of quantum computing to NMR, and provides a pulse sequence for it[LKF99]. The three spin data in Figure 7.18 showing logical labeling are from the article by Vandersypen, Yannoni, Sherwood and Chuang[VYSC99]. Yamaguchi and Yamamoto have creatively extended the NMR idea to use a crystal lattice[YY99]. The molecules shown in Figure 7.21 are attributed to (a) Chuang, Vandersypen, Zhou, Leung, and Lloyd[CVZ⁺98] (b) Cory, Mass, Price, Knill, Laflamme, Zurek, and Havel[CMP⁺98] (c) Somaroo, Tseng, Havel, Laflamme, and Cory[STH⁺99] (d) Vandersypen, Yannoni, Sherwood and Chuang[VYSC99] (e) Nielsen, Knill, and Laflamme[NKL98] (f) Leung, Vandersypen, Zhou, Sherwood, Yannoni, Kubinec and Chuang[LVZ⁺99]. Jones, Mosca, and Hansen have also realized various quantum algorithms on small molecules[JM98, JMH98]. The optimal labeling scheme which achieves the entropic limit was devised by Schulman and Vazirani[SV99].

Various criticisms of the NMR approach to quantum information processing have been leveled. Perhaps the most comprehensive discussion is due to Schack and Caves[SC99], building on earlier work by Braunstein, Caves, Jozsa, Linden, Popescu and Schack[BCJ⁺99], whose main technical conclusions (though not the possible connection with NMR) were obtained by Vidal and Tarrach[Vid99] and also by Zyczkowski, Horodecki, Sanpera and Lewenstein[ZHSL99]. See also the discussion by Linden and Popescu[LP99].

There are far too many proposals for implementations of quantum computers to all be mentioned here, so only a few are given; references in the given citations should provide good leads to additional articles. Lloyd has envisioned many implementations of quantum computers, including polymer systems[Llo93]. Nakamura, Pashkin, and Tsai have demonstrated control of single Cooper-pair qubits and observation of their Rabi oscillations[NPT99]. Mooij, Orlando, Levitov, Tian, van der Waal, and Lloyd have studied the representation of a qubit using a superconducting flux representation[MOL⁺99]. Platzman and Dykman have creatively proposed the use of electrons bound to the surface of liquid helium as qubits[PD99]. The description of the spin based quantum dot qubit realization in Section 7.8 was proposed by Loss and DiVincenzo[LD98]; Exercise 7.52 is due to them. An interesting entree into the literature on the coherence times of quantum dots is the article by Huibers, Switkes, Marcus, Campman, and Gossard[HSM⁺98]. Imamoglu, Awschalom, Burkard, DiVincenzo, Loss, Sherwin, and Small have proposed a quantum computer implementation using electron spins in quantum dots manipulated with cavity QED techniques[IAB⁺99]. The silicon-based nuclear spin quantum computer with ^{31}P dopants was proposed by Kane[Kan98], and Vrijen, Yablonovitch, Wang, Jiang, Balandin, Roychowdhury, Mor, and DiVincenzo[VYW⁺99] have extended this to use electron spins buried in a silicon-germanium heterostructure. Finally, Brennen, Caves, Jessen, and Deutsch[BCJD99] have proposed an implementation of quantum computation using neutral atoms trapped in a far off-resonance optical lattice.

Quantum teleportation has been experimentally realized using single photons and nuclear spins as qubits, as was described in 'History and further reading' at the end of Chapter 1. One of these implementations, by Furusawa, Sørensen, Braunstein, Fuchs, Kimble, and Polzik[FSB⁺98], is particularly of note in the context of this chapter, because it eschews the use of a finite Hilbert space representation of quantum information (such as qubits)! Instead, it utilizes infinite dimensional Hilbert spaces, where *continuous variables* (such as position and momentum, as in Section 7.3.2) parameterize quantum states. This approach to teleportation was originally suggested by Vaidman[Vai94], then further developed by Braunstein and Kimble[BK98a]. The continuous variable representation has also been extended to superdense coding by Braunstein and Kimble[BK99]; to quantum error correction, independently by Braunstein[Bra98] and by Lloyd and Slotine[LS98]; and to computation, by Lloyd and Braunstein[LB99].

III Quantum information

8 Quantum noise and quantum operations

Until now we have dealt almost solely with the dynamics of *closed* quantum systems, that is, with quantum systems that do not suffer any unwanted interactions with the outside world. Although fascinating conclusions can be drawn about the information processing tasks which may be accomplished *in principle* in such ideal systems, these observations are tempered by the fact that in the real world there are no perfectly closed systems, except perhaps the universe as a whole. Real systems suffer from unwanted interactions with the outside world. These unwanted interactions show up as *noise* in quantum information processing systems. We need to understand and control such noise processes in order to build useful quantum information processing systems. This is a central topic of the third part of this book, which begins in this chapter with the description of the *quantum operations formalism*, a powerful set of tools enabling us to describe quantum noise and the behavior of *open* quantum systems.

What is the distinction between an open and a closed system? A swinging pendulum like that found in some mechanical clocks can be a nearly ideal closed system. A pendulum interacts only very slightly with the rest of the world – its *environment* – mainly through friction. However, to properly describe the full dynamics of the pendulum and why it eventually ceases to move one must take into account the damping effects of air friction and imperfections in the suspension mechanism of the pendulum. Similarly, no quantum systems are ever perfectly closed, and especially not quantum computers, which must be delicately programmed by an external system to perform some desired set of operations. For example, if the state of a qubit is represented by two positions of a electron, then that electron will interact with other charged particles, which act as a source of uncontrolled *noise* affecting the state of the qubit. An open system is nothing more than one which has interactions with some other environment system, whose dynamics we wish to neglect, or average over.

The mathematical formalism of *quantum operations* is the key tool for our description of the dynamics of open quantum systems. This tool is very powerful, in that it simultaneously addresses a wide range of physical scenarios. It can be used to describe not only nearly closed systems which are weakly coupled to their environments, but also systems which are strongly coupled to their environments, and closed systems that are opened suddenly and subject to measurement. Another advantage of quantum operations in applications to quantum computation and quantum information is that they are especially well adapted to describe *discrete state changes*, that is, transformations between an initial state ρ and final state ρ', without explicit reference to the passage of time. This discrete-time analysis is rather different to the tools traditionally used by physicists for the description of open quantum systems (such as 'master equations', 'Langevin equations', and 'stochastic differential equations'), which tend to be continuous-time descriptions.

The chapter is structured as follows. We begin in Section 8.1 with a discussion of how noise is described in classical systems. The intuition gained by understanding classical noise is invaluable in learning how to think about quantum operations and quantum noise. Section 8.2 introduces the quantum operations formalism from three different points of view, enabling us to become thoroughly familiar with the basic theory of quantum operations. Section 8.3 illustrates several important examples of noise using quantum operations. These include such examples as depolarization, amplitude damping, and phase damping. A geometric approach to understanding quantum noise on single qubits is explained, using the Bloch sphere. Section 8.4 explains some miscellaneous applications of quantum operations: the connection between quantum operations and other tools conventionally used by physicists to describe quantum noise, such as *master equations*; how to *experimentally determine* the dynamics a quantum system undergoes using a procedure known as *quantum process tomography*; and an explanation of how quantum operations can be used to understand the fact that the world around us appears to obey the rules of classical physics, while it really follows quantum mechanical laws. The chapter concludes in Section 8.5 with a discussion of the limitations of the quantum operations formalism as a general approach to the description of noise in quantum systems.

8.1 Classical noise and Markov processes

To understand noise in quantum systems it is helpful to build some intuition by understanding noise in classical systems. How should we model noise in a classical system? Let's look at some simple examples to understand how this is done, and what it can teach us about noise in quantum systems.

Imagine a bit is being stored on a hard disk drive attached to an ordinary classical computer. The bit starts out in the state 0 or 1, but after a long time it becomes likely that stray magnetic fields will cause the bit to become scrambled, possibly flipping its state. We can model this by a probability p for the bit to flip, and a probability $1 - p$ for the bit to remain the same. This process is illustrated in Figure 8.1.

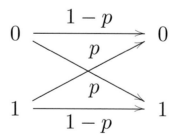

Figure 8.1. After a long time a bit on a hard disk drive may flip with probability p.

What is really going on, of course, is that the environment contains magnetic fields which can cause the bit to flip. To figure out the probability p for the bit to flip we need to understand two things. First, we need a model for the distribution of magnetic fields in the environment. Assuming that the user of the hard disk drive isn't doing anything silly like running a strong magnet near the disk drive, we can construct a realistic model by sampling the magnetic field in environments similar to the one in which the drive will be running. Second, we need a model for how magnetic fields in the environment will interact with bits on the disk. Fortunately, such a model is already well known to physicists, and goes by the name 'Maxwell's equations'. With these two elements in hand, we can in principle calculate the probability p that a bit flip will occur on the drive over some prescribed period of time.

This basic procedure – finding a model for the environment and for the system–environment interaction – is one we follow repeatedly in our study of noise, both classical and quantum. Interactions with the environment are the fundamental source of noise in both classical and quantum systems. It is often not easy to find exact models for the environment or the system–environment interaction; however, by being conservative in our modeling and closely studying the observed properties of a system to see if it obeys our model, it is possible to attain a high degree of accuracy in the modeling of noise in real physical systems.

The behavior of the bit on the hard disk can be succinctly summarized in a single equation. Suppose p_0 and p_1 are the initial probabilities that the bit is in the states 0 and 1, respectively. Let q_0 and q_1 be the corresponding probabilities after the noise has occurred. Let X be the initial state of the bit, and Y the final state of the bit. Then the law of total probability (Appendix 1) states that

$$p(Y = y) = \sum_x p(Y = y | X = x) p(X = x). \tag{8.1}$$

The conditional probabilities $p(Y = y | X = x)$ are called *transition probabilities*, since they summarize the changes that may occur in the system. Writing these equations out explicitly for the bit on a hard disk we have

$$\begin{bmatrix} q_0 \\ q_1 \end{bmatrix} = \begin{bmatrix} 1-p & p \\ p & 1-p \end{bmatrix} \begin{bmatrix} p_0 \\ p_1 \end{bmatrix}. \tag{8.2}$$

Let's look at a slightly more complicated example of noise in a classical system. Imagine that we are trying to build a classical circuit to perform some computational task. Unfortunately, we've been given faulty components to build the circuit. Our rather artificial circuit consists of a single input bit, X, to which are applied two consecutive (faulty) NOT gates, producing an intermediate bit Y, and a final bit Z. It seems reasonable to assume that whether the second NOT gate works correctly is *independent* of whether the first NOT gate worked correctly. This assumption – that the consecutive noise processes act independently – is a physically reasonable assumption to make in many situations. It results in a stochastic process $X \to Y \to Z$ of a special type known as a *Markov process*. Physically, this assumption of Markovicity corresponds to assuming that the environment causing the noise in the first NOT gate acts *independently* of the environment causing the noise in the second NOT gate, a reasonable assumption given that the gates are likely to be located a considerable distance apart in space.

Summarizing, noise in classical systems can be described using the theory of stochastic processes. Often, in the analysis of multi-stage processes it is a good assumption to use Markov processes. For a single stage process the output probabilities $\vec{q}$ are related to the input probabilities $\vec{p}$ by the equation

$$\vec{q} = E\vec{p}, \tag{8.3}$$

where E is a matrix of transition probabilities which we shall refer to as the *evolution matrix*. Thus, the final state of the system is linearly related to the starting state. This feature of linearity is echoed in the description of quantum noise, with density matrices replacing probability distributions.

What properties must the evolution matrix E possess? We require that if $\vec{p}$ is a valid probability distribution, then $E\vec{p}$ must also be a valid probability distribution. Satisfying this condition turns out to be equivalent to two conditions on E. First, all the entries of E must be non-negative, a condition known as the *positivity* requirement. If they weren't, then it would be possible to obtain negative probabilities in $E\vec{p}$. Second, all the columns of E must sum to one, a condition known as the *completeness* requirement. Suppose this weren't true. Imagine, for example, that the first column didn't sum to one. Letting $\vec{p}$ contain a one in the first entry and zeroes everywhere else, we see that $E\vec{p}$ would not be a valid probability distribution in this case.

Summarizing, the key features of classical noise are as follows: there is a linear relationship between input and output probabilities, described by a transition matrix with non-negative entries (*positivity*) and columns summing to one (*completeness*). Classical noise processes involving multiple stages are described as Markov processes, provided the noise is caused by independent environments. Each of these key features has important analogues in the theory of quantum noise. Of course, there are also some surprising new features of quantum noise!

8.2 Quantum operations

8.2.1 Overview

The quantum operations formalism is a general tool for describing the evolution of quantum systems in a wide variety of circumstances, including stochastic changes to quantum states, much as Markov processes describe stochastic changes to classical states. Just as a classical state is described by a vector of probabilities, we shall describe quantum states in terms of the density operator (density matrix) ρ, whose properties you may wish to review by rereading Section 2.4, beginning on page 98, before continuing to read this chapter. And similar to how classical states transform as described by (8.3), quantum states transform as

$$\rho' = \mathcal{E}(\rho). \tag{8.4}$$

The map $\mathcal{E}$ in this equation is a *quantum operation*. Two simple examples of quantum operations which we have encountered previously, in Chapter 2, are unitary transformations and measurements, for which $\mathcal{E}(\rho) = U\rho U^\dagger$, and $\mathcal{E}_m(\rho) = M_m\rho M_m^\dagger$, respectively (see Exercises 8.1 and 8.2, below). The quantum operation captures the dynamic change to a state which occurs as the result of some physical process; ρ is the initial state before the process, and $\mathcal{E}(\rho)$ is the final state after the process occurs, possibly up to some normalization factor.

Over the next few sections, we develop a general theory of quantum operations incorporating unitary evolution, measurement, and even more general processes! We shall develop *three* separate ways of understanding quantum operations, illustrated in Figure 8.2, all of which turn out to be equivalent. The first method is based on the idea of studying dynamics as the result of an interaction between a system and an environment, much as classical noise was described in Section 8.1. This method is concrete and easy to relate to the real world. Unfortunately, it suffers from the drawback of not being mathematically convenient. Our second method of understanding quantum operations, completely equivalent to the first, overcomes this mathematical inconvenience by providing a powerful mathematical representation for quantum operations known as the *operator-sum representation*. This method is rather abstract, but is very useful for calculations and theoretical work. Our third approach to quantum operations, equivalent to the other two, is via a set of physically motivated axioms that we would expect a dynamical map in quantum mechanics to satisfy. The advantage of this approach is that it is exceedingly *general* – it shows that quantum dynamics will be described by quantum operations under an amazingly wide range of circumstances. However, it does not offer the calculational convenience of the second approach, nor the concrete nature of the first. Taken together, these three approaches to quantum operations offer a powerful tool with which we can understand quantum noise and its effects.

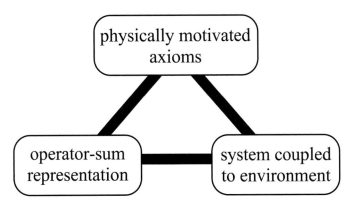

Figure 8.2. Three approaches to quantum operations which are equivalent, but offer different advantages depending upon the intended application.

Exercise 8.1: (Unitary evolution as a quantum operation)   Pure states evolve under unitary transforms as $|\psi\rangle \rightarrow U|\psi\rangle$. Show that, equivalently, we may write $\rho \rightarrow \mathcal{E}(\rho) \equiv U\rho U^\dagger$, for $\rho = |\psi\rangle\langle\psi|$.

Exercise 8.2: (Measurement as a quantum operation)   Recall from Section 2.2.3 (on page 84) that a quantum measurement with outcomes labeled by m is described by a set of measurement operators M_m such that $\sum_m M_m^\dagger M_m = I$. Let the state of the system immediately before the measurement be ρ. Show that for $\mathcal{E}_m(\rho) \equiv M_m\rho M_m^\dagger$, the state of the system immediately after the measurement is

$$\frac{\mathcal{E}_m(\rho)}{\mathrm{tr}(\mathcal{E}_m(\rho))}. \tag{8.5}$$

Also show that the probability of obtaining this measurement result is $p(m) = \mathrm{tr}(\mathcal{E}_m(\rho))$.

8.2.2 Environments and quantum operations

The dynamics of a closed quantum system are described by a unitary transform. Conceptually, we can think of the unitary transform as a box into which the input state enters and from which the output exits, as illustrated on the left hand side of Figure 8.3.

For our purposes, the interior workings of the box are not of concern to us; it could be implemented by a quantum circuit, or by some Hamiltonian system, or anything else.

A natural way to describe the dynamics of an *open* quantum system is to regard it as arising from an interaction between the system of interest, which we shall call the *principal system*, and an *environment*, which together form a closed quantum system, as illustrated on the right hand side of Figure 8.3. In other words, suppose we have a system in state ρ, which is sent into a box which is coupled to an environment. In general the final state of the system, $\mathcal{E}(\rho)$, may *not* be related by a unitary transformation to the initial state ρ. We *assume* (for now) that the system–environment input state is a product state, $\rho \otimes \rho_{\text{env}}$. After the box's transformation U the system no longer interacts with the environment, and thus we perform a partial trace over the environment to obtain the reduced state of the system alone:

$$\mathcal{E}(\rho) = \mathrm{tr}_{\text{env}} \left[U \left(\rho \otimes \rho_{\text{env}} \right) U^\dagger \right] . \tag{8.6}$$

Of course, if U does not involve any interaction with the environment, then $\mathcal{E}(\rho) = \tilde{U} \rho \tilde{U}^\dagger$, where $\tilde{U}$ is the part of U which acts on the system alone. Equation (8.6) is our first of three equivalent *definitions* of a quantum operation.

Figure 8.3. Models of closed (left) and open (right) quantum systems. An open quantum system consists of two parts, the principal system and an environment.

An important assumption is made in this definition – we assume that the system and the environment start in a product state. In general, of course, this is not true. Quantum systems interact constantly with their environments, building up correlations. One way this expresses itself is via the exchange of heat between the system and its environment. Left to itself a quantum system will relax to the same temperature as its environment, which causes correlations to exist between the two. However, in many cases of practical interest it *is* reasonable to assume that the system and its environment start out in a product state. When an experimentalist *prepares* a quantum system in a specified state they undo all the correlations between that system and the environment. Ideally, the correlations will be completely destroyed, leaving the system in a pure state. Even if this is not the case, we shall see later that the quantum operations formalism can even describe quantum dynamics when the system and environment do *not* start out in a product state.

Another issue one might raise is: how can U be specified if the environment has nearly infinite degrees of freedom? It turns out, very interestingly, that in order for this model to properly describe any possible transformation $\rho \to \mathcal{E}(\rho)$, if the principal system has a Hilbert space of d dimensions, then it suffices to model the environment as being in a Hilbert space of no more than d^2 dimensions. It also turns out not to be necessary for the environment to start out in a mixed state; a pure state will do. We shall return to these points in Section 8.2.3.

As an explicit example of the use of Equation (8.6), consider the two qubit quantum circuit shown in Figure 8.4, in which U is a controlled-NOT gate, with the principal system the control qubit, and the environment initially in the state $\rho_{\text{env}} = |0\rangle\langle 0|$ as the target qubit. Inserting into Equation (8.6), it is easily seen that

$$\mathcal{E}(\rho) = P_0 \rho P_0 + P_1 \rho P_1 , \qquad (8.7)$$

where $P_0 = |0\rangle\langle 0|$ and $P_1 = |1\rangle\langle 1|$ are projection operators. Intuitively, this dynamics occurs because the environment stays in the $|0\rangle$ state only when the system is $|0\rangle$; otherwise the environment is flipped to the state $|1\rangle$. In the next section we give a derivation of this equation as an example of the operator-sum representation.

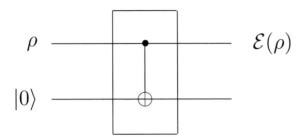

Figure 8.4. Controlled-NOT gate as an elementary example of a single qubit quantum operation.

We have described quantum operations as arising from the interaction of a principal system with an environment; however, it is convenient to generalize the definition somewhat to allow different input and output spaces. For example, imagine that a single qubit, which we label A, is prepared in an unknown state ρ. A three-level quantum system ('qutrit') labelled B is prepared in some standard state $|0\rangle$, and then interacts with system A via a unitary interaction U, causing the joint system to evolve into the state $U(\rho \otimes |0\rangle\langle 0|)U^\dagger$. We then discard system A, leaving system B in some final state ρ'. By definition, the quantum operation $\mathcal{E}$ describing this process is

$$\mathcal{E}(\rho) = \rho' = \text{tr}_A(U(\rho \otimes |0\rangle\langle 0|)U^\dagger). \qquad (8.8)$$

Notice that $\mathcal{E}$ maps density operators of the input system, A, to density operators of the output system, B. Most of our discussion of quantum operations below is concerned with quantum operations 'on' some system A, that is, they map density operators of system A to density operators of system A. However it is occasionally useful in applications to allow a more general definition. Such a definition is provided by defining quantum operations as the class of maps which arise as a result of the following processes: some initial system is prepared in an unknown quantum state ρ, and then brought into contact with other systems prepared in standard states, allowed to interact according to some unitary interaction, and then some part of the combined system is discarded, leaving just a final system in some state ρ'. The quantum operation $\mathcal{E}$ defining this process simply maps ρ to ρ'. We will see that this extension to allow different input and output spaces gels naturally with our treatment of quantum operations via the operator-sum representation, and also our axiomatic study. Nevertheless, for the most part it simplifies discussion if we assume that the input and output spaces of a quantum operation are the same, using the convenient distinction between 'principal system' and 'environment' which disappears in the general case, and giving occasional exercises to indicate the necessary extensions when the input and output spaces are different.

8.2.3 Operator-sum representation

Quantum operations can be represented in an elegant form known as the *operator-sum representation*, which is essentially a re-statement of Equation (8.6) explicitly in terms of operators on the principal system's Hilbert space alone. The main result is motivated by the following simple calculation. Let $|e_k\rangle$ be an orthonormal basis for the (finite dimensional) state space of the environment, and let $\rho_{\text{env}} = |e_0\rangle\langle e_0|$ be the initial state of the environment. There is no loss of generality in assuming that the environment starts in a pure state, since if it starts in a mixed state we are free to introduce an extra system purifying the environment (Section 2.5). Although this extra system is 'fictitious', it makes no difference to the dynamics experienced by the principal system, and thus can be used as an intermediate step in calculations. Equation (8.6) can thus be rewritten as

$$\mathcal{E}(\rho) = \sum_k \langle e_k | U \left[\rho \otimes |e_0\rangle\langle e_0| \right] U^\dagger | e_k \rangle \tag{8.9}$$

$$= \sum_k E_k \rho E_k^\dagger , \tag{8.10}$$

where $E_k \equiv \langle e_k | U | e_0 \rangle$ is an operator on the state space of the principal system. Equation (8.10) is known as the operator-sum representation of $\mathcal{E}$. The operators $\{E_k\}$ are known as *operation elements* for the quantum operation $\mathcal{E}$. The operator-sum representation is important; it will be used repeatedly for the remainder of the book.

The operation elements satisfy an important constraint known as the *completeness relation*, analogous to the completeness relation for evolution matrices in the description of classical noise. In the classical case, the completeness relation arose from the requirement that probability distributions be normalized to one. In the quantum case the completeness relation arises from the analogous requirement that the trace of $\mathcal{E}(\rho)$ be equal to one,

$$1 = \text{tr}(\mathcal{E}(\rho)) \tag{8.11}$$

$$= \text{tr} \left(\sum_k E_k \rho E_k^\dagger \right) \tag{8.12}$$

$$= \text{tr} \left(\sum_k E_k^\dagger E_k \rho \right) . \tag{8.13}$$

Since this relationship is true for *all* ρ it follows that we must have

$$\sum_k E_k^\dagger E_k = I. \tag{8.14}$$

This equation is satisfied by quantum operations which are *trace-preserving*. There are also non-trace-preserving quantum operations, for which $\sum_k E_k^\dagger E_k \leq I$, but they describe processes in which extra information about what occurred in the process is obtained by measurement, as we explain in more detail shortly. Maps $\mathcal{E}$ of the form of (8.10) for which $\sum_k E_k^\dagger E_k \leq I$ provide our second *definition* of a quantum operation. We show below that this definition is essentially equivalent to the first, Equation (8.6), and in fact is slightly more general, since it allows for non-trace-preserving operations. We will often have occasion to move backwards and forwards between these two definitions; it should be clear from context which definition we are working from at any given moment.

Exercise 8.3: Our derivation of the operator-sum representation implicitly assumed that the input and output spaces for the operation were the same. Suppose a composite system AB initially in an unknown quantum state ρ is brought into contact with a composite system CD initially in some standard state $|0\rangle$, and the two systems interact according to a unitary interaction U. After the interaction we discard systems A and D, leaving a state ρ' of system BC. Show that the map $\mathcal{E}(\rho) = \rho'$ satisfies

$$\mathcal{E}(\rho) = \sum_k E_k \rho E_k^\dagger, \tag{8.15}$$

for some set of linear operators E_k from the state space of system AB to the state space of system BC, and such that $\sum_k E_k^\dagger E_k = I$.

The operator-sum representation is important because it gives us an *intrinsic* means of characterizing the dynamics of the principal system. The operator-sum representation describes the dynamics of the principal system without having to explicitly consider properties of the environment; all that we need to know is bundled up into the operators E_k, which act on the principal system alone. This simplifies calculations and often provides considerable theoretical insight. Furthermore, many different environmental interactions may give rise to the same dynamics on the principal system. If it is only the dynamics of the principal system which are of interest then it makes sense to choose a representation of the dynamics which does not include unimportant information about other systems.

In the remainder of this section, we explore the properties of the operator-sum representation, and in particular, three features. First, we give it a physical interpretation, in terms of the operation elements E_k. A natural question which arises from this is how an operator-sum representation can be determined for any open quantum system (given, for example, the system–environment interaction or other specification). This is answered in the second topic addressed below, and the converse, how to construct a model open quantum system for any operator-sum representation, concludes.

Exercise 8.4: (Measurement) Suppose we have a single qubit principal system, interacting with a single qubit environment through the transform

$$U = P_0 \otimes I + P_1 \otimes X, \tag{8.16}$$

where X is the usual Pauli matrix (acting on the environment), and $P_0 \equiv |0\rangle\langle 0|, P_1 \equiv |1\rangle\langle 1|$ are projectors (acting on the system). Give the quantum operation for this process, in the operator-sum representation, assuming the environment starts in the state $|0\rangle$.

Exercise 8.5: (Spin flips) Just as in the previous exercise, but now let

$$U = \frac{X}{\sqrt{2}} \otimes I + \frac{Y}{\sqrt{2}} \otimes X, \tag{8.17}$$

Give the quantum operation for this process, in the operator-sum representation.

Exercise 8.6: (Composition of quantum operations) Suppose $\mathcal{E}$ and $\mathcal{F}$ are quantum operations on the same quantum system. Show that the composition $\mathcal{F} \circ \mathcal{E}$ is a quantum operation, in the sense that it has an operator-sum representation. State and prove an extension of this result to the case where $\mathcal{E}$ and $\mathcal{F}$ do not necessarily have the same input and output spaces.

Physical interpretation of the operator-sum representation

There is a nice interpretation that can be given to the operator-sum representation. Imagine that a measurement of the environment is performed in the basis $|e_k\rangle$ after the unitary transformation U has been applied. Applying the principle of implicit measurement, we see that such a measurement affects only the state of the environment, and does not change the state of the principal system. Let ρ_k be the state of the principal system given that outcome k occurs, so

$$\rho_k \propto \mathrm{tr}_E(|e_k\rangle\langle e_k|U(\rho \otimes |e_0\rangle\langle e_0|)U^\dagger|e_k\rangle\langle e_k|) = \langle e_k|U(\rho \otimes |e_0\rangle\langle e_0|)U^\dagger|e_k\rangle \quad (8.18)$$
$$= E_k \rho E_k^\dagger . \quad (8.19)$$

Normalizing ρ_k,

$$\rho_k = \frac{E_k \rho E_k^\dagger}{\mathrm{tr}(E_k \rho E_k^\dagger)}, \quad (8.20)$$

we find the probability of outcome k is given by

$$p(k) = \mathrm{tr}(|e_k\rangle\langle e_k|U(\rho \otimes |e_0\rangle\langle e_0|)U^\dagger|e_k\rangle\langle e_k|) \quad (8.21)$$
$$= \mathrm{tr}(E_k \rho E_k^\dagger). \quad (8.22)$$

Thus

$$\mathcal{E}(\rho) = \sum_k p(k)\rho_k = \sum_k E_k \rho E_k^\dagger . \quad (8.23)$$

This gives us a beautiful physical *interpretation* of what is going on in a quantum operation with operation elements $\{E_k\}$. The action of the quantum operation is equivalent to taking the state ρ and randomly replacing it by $E_k \rho E_k^\dagger/\mathrm{tr}(E_k \rho E_k^\dagger)$, with probability $\mathrm{tr}(E_k \rho E_k^\dagger)$. In this sense, it is very similar to the concept of noisy communication channels used in classical information theory; in this vein, we shall sometimes refer to certain quantum operations which describe quantum noise processes as being noisy quantum channels.

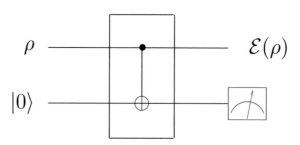

Figure 8.5. Controlled-NOT gate as an elementary model of single qubit measurement.

A simple example, based on Figure 8.4, illustrates this interpretation of the operator-sum representation. Suppose we choose the states $|e_k\rangle = |0_E\rangle$ and $|1_E\rangle$, where we include the E subscript to make it clear that the state is a state of the environment. This can be interpreted as doing a measurement in the computational basis of the environment qubit, as shown in Figure 8.5. Doing such a measurement does not, of course, change the state of the principal system. Using subscripts P to denote the principal system, the controlled-NOT may be expanded as

$$U = |0_P0_E\rangle\langle0_P0_E| + |0_P1_E\rangle\langle0_P1_E| + |1_P1_E\rangle\langle1_P0_E| + |1_P0_E\rangle\langle1_P1_E|. \quad (8.24)$$

Thus

$$E_0 = \langle0_E|U|0_E\rangle = |0_P\rangle\langle0_P| \quad (8.25)$$
$$E_1 = \langle1_E|U|0_E\rangle = |1_P\rangle\langle1_P|, \quad (8.26)$$

and therefore

$$\mathcal{E}(\rho) = E_0\rho E_0 + E_1\rho E_1, \quad (8.27)$$

in agreement with Equation (8.7).

Measurements and the operator-sum representation

Given a description of an open quantum system, how do we determine an operator-sum representation for its dynamics? We have already found one answer: given the unitary system–environment transformation operation U, and a basis of states $|e_k\rangle$ for the environment, the operation elements are

$$E_k \equiv \langle e_k|U|e_0\rangle. \quad (8.28)$$

It is possible to extend this result even further by allowing the possibility that a *measurement* is performed on the combined system–environment after the unitary interaction, allowing the acquisition of information about the quantum state. It turns out that this physical possibility is naturally connected to non-trace-preserving quantum operations, that is, maps $\mathcal{E}(\rho) = \sum_k E_k\rho E_k^\dagger$ such that $\sum_k E_k^\dagger E_k \leq I$.

Suppose the principal system is initially in a state ρ. For convenience we denote the principal system by the letter Q. Adjoined to Q is an environment system E. We suppose that Q and E are initially independent systems, and that E starts in some standard state, σ. The joint state of the system is thus initially

$$\rho^{QE} = \rho \otimes \sigma. \quad (8.29)$$

We suppose that the systems interact according to some unitary interaction U. After the unitary interaction a projective measurement is performed on the joint system, described by projectors P_m. The case where no measurement is made corresponds to the special case where there is only a single measurement outcome, $m = 0$, which corresponds to the projector $P_0 \equiv I$.

The situation is summarized in Figure 8.6. Our aim is to determine the final state of Q as a function of the initial state, ρ. The final state of QE is given by

$$\frac{P_mU(\rho \otimes \sigma)U^\dagger P_m}{\text{tr}(P_mU(\rho \otimes \sigma)U^\dagger P_m)}, \quad (8.30)$$

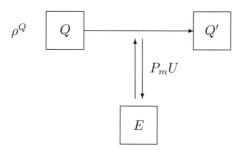

Figure 8.6. Environmental model for a quantum operation.

given that measurement outcome m occurred. Tracing out E we see that the final state of Q alone is

$$\frac{\mathrm{tr}_E(P_m U(\rho \otimes \sigma)U^\dagger P_m)}{\mathrm{tr}(P_m U(\rho \otimes \sigma)U^\dagger P_m)} . \tag{8.31}$$

This representation of the final state involves the initial state σ of the environment, the interaction U and the measurement operators P_m. Define a map

$$\mathcal{E}_m(\rho) \equiv \mathrm{tr}_E(P_m U(\rho \otimes \sigma)U^\dagger P_m), \tag{8.32}$$

so the final state of Q alone is $\mathcal{E}_m(\rho)/\mathrm{tr}(\mathcal{E}_m(\rho))$. Note that $\mathrm{tr}[\mathcal{E}_m(\rho)]$ is the probability of outcome m of the measurement occurring. Let $\sigma = \sum_j q_j |j\rangle\langle j|$ be an ensemble decomposition for σ. Introduce an orthonormal basis $|e_k\rangle$ for the system E. Note that

$$\mathcal{E}_m(\rho) = \sum_{jk} q_j \mathrm{tr}_E(|e_k\rangle\langle e_k| P_m U(\rho \otimes |j\rangle\langle j|)U^\dagger P_m |e_k\rangle\langle e_k|) \tag{8.33}$$

$$= \sum_{jk} E_{jk} \rho E_{jk}^\dagger , \tag{8.34}$$

where

$$E_{jk} \equiv \sqrt{q_j}\langle e_k| P_m U|j\rangle . \tag{8.35}$$

This equation is a generalization of Equation (8.10), and gives an explicit means for calculating the operators appearing in an operator-sum representation for $\mathcal{E}_m$, given that the initial state σ of E is known, and the dynamics between Q and E are known. The quantum operations $\mathcal{E}_m$ can be thought of as defining a kind of measurement process generalizing the description of measurements given in Chapter 2.

Exercise 8.7: Suppose that instead of doing a projective measurement on the combined principal system and environment we had performed a general measurement described by measurement operators $\{M_m\}$. Find operator-sum representations for the corresponding quantum operations $\mathcal{E}_m$ on the principal system, and show that the respective measurement probabilities are $\mathrm{tr}[\mathcal{E}(\rho)]$.

System–environment models for any operator-sum representation

We have shown that interacting quantum systems give rise in a natural way to an operator-sum representation for quantum operations. What about the converse problem? Given a set of operators $\{E_k\}$ is there some reasonable *model environmental system and dynamics* which give rise to a quantum operation with those operation elements? By 'reasonable' we mean that the dynamics must be either a unitary evolution or a projective measurement. Here, we show how to construct such a model. We will only show how to do this for quantum operations mapping the input space to the same output space, although it is mainly a matter of notation to generalize the construction to the more general case. In particular, we show that for any trace-preserving or non-trace-preserving quantum operation, $\mathcal{E}$, with operation elements $\{E_k\}$, there exists a model environment, E, starting in a pure state $|e_0\rangle$, and model dynamics specified by a unitary operator U and projector P onto E such that

$$\mathcal{E}(\rho) = \text{tr}_E(PU(\rho \otimes |e_0\rangle\langle e_0|)U^\dagger P). \tag{8.36}$$

To see this, suppose first that $\mathcal{E}$ is a trace-preserving quantum operation, with operator-sum representation generated by operation elements $\{E_k\}$ satisfying the completeness relation $\sum_k E_k^\dagger E_k = I$, so we are only attempting to find an appropriate unitary operator U to model the dynamics. Let $|e_k\rangle$ be an orthonormal basis set for E, in one-to-one correspondence with the index k for the operators E_k. Note that *by definition* E has such a basis; we are trying to find a *model* environment giving rise to a dynamics described by the operation elements $\{E_k\}$. Define an operator U which has the following action on states of the form $|\psi\rangle|e_0\rangle$,

$$U|\psi\rangle|e_0\rangle \equiv \sum_k E_k|\psi\rangle|e_k\rangle, \tag{8.37}$$

where $|e_0\rangle$ is just some standard state of the model environment. Note that for arbitrary states $|\psi\rangle$ and $|\varphi\rangle$ of the principal system,

$$\langle\psi|\langle e_0|U^\dagger U|\varphi\rangle|e_0\rangle = \sum_k \langle\psi|E_k^\dagger E_k|\varphi\rangle = \langle\psi|\varphi\rangle, \tag{8.38}$$

by the completeness relation. Thus the operator U can be extended to a unitary operator acting on the entire state space of the joint system. It is easy to verify that

$$\text{tr}_E(U(\rho \otimes |e_0\rangle\langle e_0|)U^\dagger) = \sum_k E_k\rho E_k^\dagger, \tag{8.39}$$

so this model provides a realization of the quantum operation $\mathcal{E}$ with operation elements $\{E_k\}$. This result is illustrated in Box 8.1.

Non-trace-preserving quantum operations can easily be modeled using a construction along the same lines (Exercise 8.8). A more interesting generalization of this construction is the case of a set of quantum operations $\{\mathcal{E}_m\}$ corresponding to possible outcomes from a measurement, so the quantum operation $\sum_m \mathcal{E}_m$ is trace-preserving, since the probabilities of the distinct outcomes sum to one, $1 = \sum_m p(m) = \text{tr}\left[\left(\sum_m \mathcal{E}_m\right)(\rho)\right]$ for all possible inputs ρ. See Exercise 8.9, below.

Exercise 8.8: (**Non-trace-preserving quantum operations**)   Explain how to construct a unitary operator for a system–environment model of a non-trace-preserving quantum operation, by introducing an extra operator, E_∞, into the set of operation elements E_k, chosen so that when summing over the complete set of k, including $k = \infty$, one obtains $\sum_k E_k^\dagger E_k = I$.

Exercise 8.9: (Measurement model) If we are given a set of quantum operations $\{\mathcal{E}_m\}$ such that $\sum_m \mathcal{E}_m$ is trace-preserving, then it is possible to construct a *measurement model* giving rise to this set of quantum operations. For each m, let E_{mk} be a set of operation elements for $\mathcal{E}_m$. Introduce an environmental system, E, with an orthonormal basis $|m, k\rangle$ in one-to-one correspondence with the set of indices for the operation elements. Analogously to the earlier construction, define an operator U such that

$$U|\psi\rangle|e_0\rangle = \sum_{mk} E_{mk}|\psi\rangle|m, k\rangle. \tag{8.40}$$

Next, define projectors $P_m \equiv \sum_k |m, k\rangle\langle m, k|$ on the environmental system, E. Show that performing U on $\rho \otimes |e_0\rangle\langle e_0|$, then measuring P_m gives m with probability $\mathrm{tr}(\mathcal{E}_m(\rho))$, and the corresponding post-measurement state of the principal system is $\mathcal{E}_m(\rho)/\mathrm{tr}(\mathcal{E}_m(\rho))$.

Box 8.1: Mocking up a quantum operation

Given a trace-preserving quantum operation expressed in the operator-sum representation, $\mathcal{E}(\rho) = \sum_k E_k \rho E_k^\dagger$, we can construct a physical model for it in the following way. From (8.10), we want U to satisfy

$$E_k = \langle e_k|U|e_0\rangle, \tag{8.41}$$

where U is some unitary operator, and $|e_k\rangle$ are orthonormal basis vectors for the environment system. Such a U is conveniently represented as the block matrix

$$U = \begin{bmatrix} [E_1] & \cdot & \cdot & \cdot & \cdots \\ [E_2] & \cdot & \cdot & \cdot & \cdots \\ [E_3] & \cdot & \cdot & \cdot & \cdots \\ [E_4] & \cdot & \cdot & \cdot & \cdots \\ \vdots & \vdots & \vdots & \vdots & \end{bmatrix} \tag{8.42}$$

in the basis $|e_k\rangle$. Note that the operation elements E_k only determine the first block column of this matrix (unlike elsewhere, here it is convenient to have the first label of the states be the environment, and the second, the principal system). Determination of the rest of the matrix is left up to us; we simply choose the entries such that U is unitary. Note that by the results of Chapter 4, U can be implemented by a quantum circuit.

8.2.4 Axiomatic approach to quantum operations

Until now the main motivation for our study of quantum operations has been that they provide an elegant way of studying systems which are interacting with an environment. We're going to switch to a different viewpoint now, where we try to write down physically motivated axioms which we expect quantum operations to obey. This viewpoint is more abstract than our earlier approach, based on explicit system–environment models, but it is also extremely powerful because of that abstraction.

The way we're going to proceed is as follows. First, we're going to *forget* everything we've learned about quantum operations, and start over by *defining* quantum operations according to a set of axioms, which we'll justify on physical grounds. That done, we'll *prove* that a map $\mathcal{E}$ satisfies these axioms if and only if it has an operator-sum representation, thus providing the missing link between the abstract axiomatic formulation, and our earlier discussion.

We *define* a quantum operation $\mathcal{E}$ as a map from the set of density operators of the input space Q_1 to the set of density operators for the output space Q_2, with the following three axiomatic properties: (note that for notational simplicity in the proofs we take $Q_1 = Q_2 = Q$)

- **A1**: First, $\mathrm{tr}[\mathcal{E}(\rho)]$ is the probability that the process represented by $\mathcal{E}$ occurs, when ρ is the initial state. Thus, $0 \le \mathrm{tr}[\mathcal{E}(\rho)] \le 1$ for any state ρ.
- **A2**: Second, $\mathcal{E}$ is a *convex-linear map* on the set of density matrices, that is, for probabilities $\{p_i\}$,

$$\mathcal{E}\left(\sum_i p_i \rho_i\right) = \sum_i p_i \mathcal{E}(\rho_i). \tag{8.43}$$

- **A3**: Third, $\mathcal{E}$ is a *completely positive* map. That is, if $\mathcal{E}$ maps density operators of system Q_1 to density operators of system Q_2, then $\mathcal{E}(A)$ must be positive for any positive operator A. Furthermore, if we introduce an extra system R of arbitrary dimensionality, it must be true that $(\mathcal{I} \otimes \mathcal{E})(A)$ is positive for any positive operator A on the combined system RQ_1, where $\mathcal{I}$ denotes the identity map on system R.

The first property is one of mathematical convenience. To cope with the case of measurements, it turns out that it is extremely convenient to make the convention that $\mathcal{E}$ does not necessarily preserve the trace property of density matrices, that $\mathrm{tr}(\rho) = 1$. Rather, we make the convention that $\mathcal{E}$ is to be defined in such a way that $\mathrm{tr}[\mathcal{E}(\rho)]$ is equal to the probability of the measurement outcome described by $\mathcal{E}$ occurring. For example, suppose that we are doing a projective measurement in the computational basis of a single qubit. Then two quantum operations are used to describe this process, defined by $\mathcal{E}_0(\rho) \equiv |0\rangle\langle0|\rho|0\rangle\langle0|$ and $\mathcal{E}_1(\rho) \equiv |1\rangle\langle1|\rho|1\rangle\langle1|$. Notice that the probabilities of the respective outcomes are correctly given by $\mathrm{tr}[\mathcal{E}_0(\rho)]$ and $\mathrm{tr}[\mathcal{E}_1(\rho)]$. With this convention the correctly normalized final quantum state is therefore

$$\frac{\mathcal{E}(\rho)}{\mathrm{tr}[\mathcal{E}(\rho)]}. \tag{8.44}$$

In the case where the process is deterministic, that is, no measurement is taking place, this reduces to the requirement that $\mathrm{tr}[\mathcal{E}(\rho)] = 1 = \mathrm{tr}(\rho)$, for all ρ. As previously discussed, in this case, we say that the quantum operation is a trace-preserving quantum operation, since on its own $\mathcal{E}$ provides a complete description of the quantum process. On the other hand, if there is a ρ such that $\mathrm{tr}[\mathcal{E}(\rho)] < 1$, then the quantum operation is non-trace-preserving, since on its own $\mathcal{E}$ does not provide a complete description of the processes that may occur in the system. (That is, other measurement outcomes may occur, with some probability.) A *physical* quantum operation is one that satisfies the requirement that probabilities never exceed 1, $\mathrm{tr}[\mathcal{E}(\rho)] \leq 1$.

The second property stems from a physical requirement on quantum operations. Suppose the input ρ to the quantum operation is obtained by randomly selecting the state from an ensemble $\{p_i, \rho_i\}$ of quantum states, that is, $\rho = \sum_i p_i \rho_i$. Then we would expect that the resulting state, $\mathcal{E}(\rho)/\mathrm{tr}[\mathcal{E}(\rho)] = \mathcal{E}(\rho)/p(\mathcal{E})$ corresponds to a random selection from the ensemble $\{p(i|\mathcal{E}), \mathcal{E}(\rho_i)/\mathrm{tr}[\mathcal{E}(\rho_i)]\}$, where $p(i|\mathcal{E})$ is the probability that the state prepared was ρ_i, given that the process represented by $\mathcal{E}$ occurred. Thus, we demand that

$$\mathcal{E}(\rho) = p(\mathcal{E}) \sum_i p(i|\mathcal{E}) \frac{\mathcal{E}(\rho_i)}{\mathrm{tr}[\mathcal{E}(\rho_i)]}, \tag{8.45}$$

where $p(\mathcal{E}) = \mathrm{tr}[\mathcal{E}(\rho)]$ is the probability that the process described by $\mathcal{E}$ occurs on input of ρ. By Bayes' rule (Appendix 1),

$$p(i|\mathcal{E}) = p(\mathcal{E}|i) \frac{p_i}{p(\mathcal{E})} = \frac{\mathrm{tr}[\mathcal{E}(\rho_i)] p_i}{p(\mathcal{E})} \tag{8.46}$$

so (8.45) reduces to (8.43).

The third property also originates from an important physical requirement, that not only must $\mathcal{E}(\rho)$ be a valid density matrix (up to normalization) so long as ρ is valid, but furthermore, if $\rho = \rho_{RQ}$ is the density matrix of some joint system of R and Q, if $\mathcal{E}$ acts only on Q, then $\mathcal{E}(\rho_{RQ})$ must still result in a valid density matrix (up to normalization) of the joint system. An example is given in Box 8.2. Formally, suppose we introduce a second (finite dimensional) system R. Let $\mathcal{I}$ denote the identity map on system R. Then the map $\mathcal{I} \otimes \mathcal{E}$ must take positive operators to positive operators.

It is perhaps surprising that these three axioms are sufficient to define quantum operations. However, the following theorem shows that they are equivalent to the earlier system-environment models and the definition in terms of an operator-sum representation:

Theorem 8.1: The map $\mathcal{E}$ satisfies axioms **A1**, **A2** and **A3** if and only if

$$\mathcal{E}(\rho) = \sum_i E_i \rho E_i^\dagger, \tag{8.50}$$

for some set of operators $\{E_i\}$ which map the input Hilbert space to the output Hilbert space, and $\sum_i E_i^\dagger E_i \leq I$.

Proof
Suppose $\mathcal{E}(\rho) = \sum_i E_i \rho E_i^\dagger$. $\mathcal{E}$ is obviously linear, so to check that $\mathcal{E}$ is a quantum operation we need only prove that it is completely positive. Let A be any positive operator acting on the state space of an extended system, RQ, and let $|\psi\rangle$ be some state of RQ. Defining $|\varphi_i\rangle \equiv (I_R \otimes E_i^\dagger)|\psi\rangle$, we have

$$\langle\psi|(I_R \otimes E_i)A(I_R \otimes E_i^\dagger)|\psi\rangle = \langle\varphi_i|A|\varphi_i\rangle \tag{8.51}$$
$$\geq 0, \tag{8.52}$$

by the positivity of the operator A. It follows that

$$\langle\psi|(\mathcal{I} \otimes \mathcal{E})(A)|\psi\rangle = \sum_i \langle\varphi_i|A|\varphi_i\rangle \geq 0, \tag{8.53}$$

Box 8.2: Complete positivity versus positivity

The transpose operation on a single qubit provides an example of why complete positivity is an important requirement for quantum operations. By definition, this map transposes the density operator in the computational basis:

$$\begin{bmatrix} a & b \\ c & d \end{bmatrix} \xrightarrow{T} \begin{bmatrix} a & c \\ b & d \end{bmatrix}. \tag{8.47}$$

This map preserves positivity of a single qubit. However, suppose that qubit is part of a two qubit system initially in the entangled state

$$\frac{|00\rangle + |11\rangle}{\sqrt{2}}, \tag{8.48}$$

and the transpose operation is applied to the first of these two qubits, while the second qubit is subject to trivial dynamics. Then the density operator of the system after the dynamics has been applied is

$$\frac{1}{2} \begin{bmatrix} 1 & 0 & 0 & 0 \\ 0 & 0 & 1 & 0 \\ 0 & 1 & 0 & 0 \\ 0 & 0 & 0 & 1 \end{bmatrix}. \tag{8.49}$$

A calculation shows that this operator has eigenvalues $1/2, 1/2, 1/2$ and $-1/2$, so this is not a valid density operator. Thus, the transpose operation is an example of a positive map which is not completely positive, that is, it preserves the positivity of operators on the principal system, but does not continue to preserve positivity when applied to systems which contain the principal system as a subsystem.

and thus for any positive operator A, the operator $(\mathcal{I} \otimes \mathcal{E})(A)$ is also positive, as required. The requirement $\sum_i E_i^\dagger E_i \leq I$ ensures that probabilities are less than or equal to 1. This completes the first part of the proof.

Suppose next that $\mathcal{E}$ satisfies axioms A1, A2 and A3. Our aim will be to find an operator-sum representation for $\mathcal{E}$. Suppose we introduce a system, R, with the same dimension as the original quantum system, Q. Let $|i_R\rangle$ and $|i_Q\rangle$ be orthonormal bases for R and Q. It will be convenient to use the same index, i, for these two bases, and this can certainly be done as R and Q have the same dimensionality. Define a joint state $|\alpha\rangle$ of RQ by

$$|\alpha\rangle \equiv \sum_i |i_R\rangle|i_Q\rangle. \tag{8.54}$$

The state $|\alpha\rangle$ is, up to a normalization factor, a maximally entangled state of the systems R and Q. This interpretation of $|\alpha\rangle$ as a maximally entangled state may help in understanding the following construction. Next, we define an operator σ on the state space of RQ by

$$\sigma \equiv (\mathcal{I}_R \otimes \mathcal{E})(|\alpha\rangle\langle\alpha|). \tag{8.55}$$

We may think of this as the result of applying the quantum operation $\mathcal{E}$ to one half of a maximally entangled state of the system RQ. It is a truly remarkable fact, which we will now demonstrate, that the operator σ completely specifies the quantum operation $\mathcal{E}$. That is, to know how $\mathcal{E}$ acts on an arbitrary state of Q, it is sufficient to know how it acts on a single maximally entangled state of Q with another system!

The trick which allows us to recover $\mathcal{E}$ from σ is as follows. Let $|\psi\rangle = \sum_j \psi_j |j_Q\rangle$ be any state of system Q. Define a corresponding state $|\tilde{\psi}\rangle$ of system R by the equation

$$|\tilde{\psi}\rangle \equiv \sum_j \psi_j^* |j_R\rangle. \tag{8.56}$$

Notice that

$$\langle\tilde{\psi}|\sigma|\tilde{\psi}\rangle = \langle\tilde{\psi}| \left(\sum_{ij} |i_R\rangle\langle j_R| \otimes \mathcal{E}(|i_Q\rangle\langle j_Q|) \right) |\tilde{\psi}\rangle \tag{8.57}$$

$$= \sum_{ij} \psi_i \psi_j^* \mathcal{E}(|i_Q\rangle\langle j_Q|) \tag{8.58}$$

$$= \mathcal{E}(|\psi\rangle\langle\psi|). \tag{8.59}$$

Let $\sigma = \sum_i |s_i\rangle\langle s_i|$ be some decomposition of σ, where the vectors $|s_i\rangle$ need not be normalized. Define a map

$$E_i(|\psi\rangle) \equiv \langle\tilde{\psi}|s_i\rangle. \tag{8.60}$$

A little thought shows that this map is a linear map, so E_i is a linear operator on the state space of Q. Furthermore, we have

$$\sum_i E_i|\psi\rangle\langle\psi|E_i^\dagger = \sum_i \langle\tilde{\psi}|s_i\rangle\langle s_i|\tilde{\psi}\rangle \tag{8.61}$$

$$= \langle\tilde{\psi}|\sigma|\tilde{\psi}\rangle \tag{8.62}$$

$$= \mathcal{E}(|\psi\rangle\langle\psi|). \tag{8.63}$$

Thus

$$\mathcal{E}(|\psi\rangle\langle\psi|) = \sum_i E_i|\psi\rangle\langle\psi|E_i^\dagger, \tag{8.64}$$

for all pure states, $|\psi\rangle$, of Q. By convex-linearity it follows that

$$\mathcal{E}(\rho) = \sum_i E_i \rho E_i^\dagger \tag{8.65}$$

in general. The condition $\sum_i E_i^\dagger E_i \leq I$ follows immediately from axiom **A1** identifying the trace of $\mathcal{E}(\rho)$ with a probability.   □

Freedom in the operator-sum representation

We have seen that the operator-sum representation provides a very general description of the dynamics of an open quantum system. Is it a *unique* description?

Consider quantum operations $\mathcal{E}$ and $\mathcal{F}$ acting on a single qubit with the operator-sum representations $\mathcal{E}(\rho) = \sum_k E_k \rho E_k^\dagger$ and $\mathcal{F}(\rho) = \sum_k F_k \rho F_k^\dagger$, where the operation elements for $\mathcal{E}$ and $\mathcal{F}$ are defined by

$$E_1 = \frac{I}{\sqrt{2}} = \frac{1}{\sqrt{2}}\begin{bmatrix} 1 & 0 \\ 0 & 1 \end{bmatrix} \qquad E_2 = \frac{Z}{\sqrt{2}} = \frac{1}{\sqrt{2}}\begin{bmatrix} 1 & 0 \\ 0 & -1 \end{bmatrix} \tag{8.66}$$

and

$$F_1 = |0\rangle\langle0| = \begin{bmatrix} 1 & 0 \\ 0 & 0 \end{bmatrix} \qquad F_2 = |1\rangle\langle1| = \begin{bmatrix} 0 & 0 \\ 0 & 1 \end{bmatrix}. \tag{8.67}$$

These appear to be very different quantum operations. What is interesting is that $\mathcal{E}$ and $\mathcal{F}$ are *actually the same quantum operation*. To see this, note that $F_1 = (E_1 + E_2)/\sqrt{2}$ and $F_2 = (E_1 - E_2)/\sqrt{2}$. Thus,

$$\mathcal{F}(\rho) = \frac{(E_1 + E_2)\rho(E_1^\dagger + E_2^\dagger) + (E_1 - E_2)\rho(E_1^\dagger - E_2^\dagger)}{2} \tag{8.68}$$

$$= E_1\rho E_1^\dagger + E_2\rho E_2^\dagger \tag{8.69}$$

$$= \mathcal{E}(\rho). \tag{8.70}$$

This example shows that the operation elements appearing in an operator-sum representation for a quantum operation are *not* unique.

The freedom in the representation is very interesting. Suppose we flipped a fair coin, and, depending on the outcome of the coin toss, applied either the unitary operator I or Z to the quantum system. This process corresponds to the first operator-sum representation for $\mathcal{E}$. The second operator-sum representation for $\mathcal{E}$ (labeled $\mathcal{F}$ above) corresponds to performing a projective measurement in the $\{|0\rangle, |1\rangle\}$ basis, with the outcome of the measurement unknown. These two apparently very different physical processes give rise to exactly the same dynamics for the principal system.

When do two sets of operation elements give rise to the same quantum operation? Understanding this question is important for at least two reasons. First, from a physical point of view, understanding the freedom in the representation gives us more insight into how different physical processes can give rise to the same system dynamics. Second, understanding the freedom in operator-sum representation is crucial to a good understanding of quantum error-correction.

Intuitively, it is clear that there must be a great deal of freedom in an operator-sum representation. Consider a trace-preserving quantum operation $\mathcal{E}$ which describes the dynamics of some system such as that shown in Figure 8.3. We have shown that the operation elements $E_k = \langle e_k|U|e_0\rangle$ for $\mathcal{E}$ may be associated with an orthonormal basis $|e_k\rangle$ for the environment. Suppose that we supplement the interaction U with an additional unitary action U' on the environment alone, as shown in Figure 8.7. Clearly this does not change the state of the principal system. What are the corresponding operation elements to this new process, $(I \otimes U')U$? We obtain:

$$F_k = \langle e_k|(I \otimes U')U|e_0\rangle \tag{8.71}$$

$$= \sum_j \left[I \otimes \langle e_k|U'|e_j\rangle \right] \langle e_j|U|e_0\rangle \tag{8.72}$$

$$= \sum_j U'_{kj} E_j, \tag{8.73}$$

where we have used the fact that $\sum_j |e_j\rangle\langle e_j| = I$, and U'_{kj} are the matrix elements of U' with respect to the basis $|e_k\rangle$. It turns out that the freedom in the operator-sum representation yielded by this physically motivated picture captures the essence of the complete freedom available in the operator-sum representation, as proved in the following theorem.

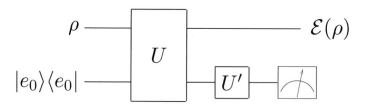

Figure 8.7. Origin of the unitary freedom in the operator-sum representation.

Theorem 8.2: (**Unitary freedom in the operator-sum representation**) Suppose $\{E_1, \ldots, E_m\}$ and $\{F_1, \ldots, F_n\}$ are operation elements giving rise to quantum operations $\mathcal{E}$ and $\mathcal{F}$, respectively. By appending zero operators to the shorter list of operation elements we may ensure that $m = n$. Then $\mathcal{E} = \mathcal{F}$ if and only if there exist complex numbers u_{ij} such that $E_i = \sum_j u_{ij} F_j$, and u_{ij} is an m by m unitary matrix.

Proof

The key to the proof is Theorem 2.6, on page 103. Recall that this result tells us that two sets of vectors $|\psi_i\rangle$ and $|\varphi_j\rangle$ generate the same operator if and only if

$$|\psi_i\rangle = \sum_j u_{ij} |\varphi_j\rangle, \tag{8.74}$$

where u_{ij} is a unitary matrix of complex numbers, and we 'pad' whichever set of states $|\psi_i\rangle$ or $|\varphi_j\rangle$ is smaller with additional states 0 so that the two sets have the same number of elements. This result allows us to characterize the freedom in operator-sum representations. Suppose $\{E_i\}$ and $\{F_j\}$ are two sets of operation elements for the same quantum operation, $\sum_i E_i \rho E_i^\dagger = \sum_j F_j \rho F_j^\dagger$ for all ρ. Define

$$|e_i\rangle \equiv \sum_k |k_R\rangle \left(E_i |k_Q\rangle\right) \tag{8.75}$$

$$|f_j\rangle \equiv \sum_k |k_R\rangle \left(F_j |k_Q\rangle\right). \tag{8.76}$$

Recall the definition of σ in Equation (8.55), from which it follows that $\sigma = \sum_i |e_i\rangle\langle e_i| = \sum_j |f_j\rangle\langle f_j|$, and thus there exists unitary u_{ij} such that

$$|e_i\rangle = \sum_j u_{ij} |f_j\rangle. \tag{8.77}$$

But for arbitrary $|\psi\rangle$ we have

$$E_i|\psi\rangle = \langle\tilde{\psi}|e_i\rangle \tag{8.78}$$

$$= \sum_j u_{ij} \langle\tilde{\psi}|f_j\rangle \tag{8.79}$$

$$= \sum_k u_{ij} F_j|\psi\rangle. \tag{8.80}$$

Thus

$$E_i = \sum_j u_{ij} F_j. \tag{8.81}$$

Conversely, supposing E_i and F_j are related by a unitary transformation of the form $E_i = \sum_{ij} u_{ij} F_j$, simple algebra shows that the quantum operation with operation elements $\{E_i\}$ is the same as the quantum operation with operation elements $\{F_j\}$. $\qquad\square$

Theorem 8.2 can be used to answer another interesting question: what is the maximum size of an environment that would be needed to mock up a given quantum operation?

Theorem 8.3: All quantum operations $\mathcal{E}$ on a system of Hilbert space dimension d can be generated by an operator-sum representation containing at most d^2 elements,

$$\mathcal{E}(\rho) = \sum_{k=1}^{M} E_k \rho E_k^\dagger, \qquad (8.82)$$

where $1 \leq M \leq d^2$.

The proof of this theorem is simple and is left as an exercise for you.

Exercise 8.10: Give a proof of Theorem 8.3 based on the freedom in the operator-sum representation, as follows. Let $\{E_j\}$ be a set of operation elements for $\mathcal{E}$. Define a matrix $W_{jk} \equiv \text{tr}(E_j^\dagger E_k)$. Show that the matrix W is Hermitian and of rank at most d^2, and thus there is unitary matrix u such that $uWu^\dagger$ is diagonal with at most d^2 non-zero entries. Use u to define a new set of at most d^2 non-zero operation elements $\{F_j\}$ for $\mathcal{E}$.

Exercise 8.11: Suppose $\mathcal{E}$ is a quantum operation mapping a d-dimensional input space to a d'-dimensional output space. Show that $\mathcal{E}$ can be described using a set of at most dd' operation elements $\{E_k\}$.

The freedom in the operator-sum representation is surprisingly useful. We use it, for example, in our study of quantum error-correction in Chapter 10. In that chapter we will see that certain sets of operators in the operator-sum representation give more useful information about the quantum error-correction process, and it behooves us to study quantum error-correction from that point of view. As usual, having multiple ways of understanding a process gives us much more insight into what is going on.

8.3 Examples of quantum noise and quantum operations

In this section we examine some concrete examples of quantum noise and quantum operations. These models illustrate the power of the quantum operations formalism we have been developing. They are also important in understanding the practical effects of noise on quantum systems, and how noise can be controlled by techniques such as error-correction.

We begin in Section 8.3.1 by considering how *measurement* can be described as a quantum operation, and in particular we consider the trace and partial trace operations. After that, we turn to noise processes, beginning in Section 8.3.2 with the presentation of a graphical method for understanding quantum operations on a single qubit. This method is used in the remainder of the section to illustrate elementary bit and phase flip error processes (in Section 8.3.3), the depolarizing channel (in Section 8.3.4), amplitude damping (in Section 8.3.5), and phase damping (in Section 8.3.6). Amplitude and phase damping are ideal models of noise that capture many of the most important features of the noise occurring in quantum mechanical systems, and we not only consider their abstract mathematical formulation, but also how the processes arise in real-world quantum systems.

8.3.1 Trace and partial trace

One of the main uses of the quantum operations formalism is to describe the effects of measurement. Quantum operations can be used to describe both the probability of getting a particular outcome from a measurement on a quantum system, and also the state change in the system effected by the measurement.

The simplest operation related to measurement is the trace map $\rho \rightarrow \text{tr}(\rho)$ – which we can show is indeed a quantum operation, in the following way. Let H_Q be any input Hilbert space, spanned by an orthonormal basis $|1\rangle \ldots |d\rangle$, and let H_Q' be a one-dimensional output space, spanned by the state $|0\rangle$. Define

$$\mathcal{E}(\rho) \equiv \sum_{i=1}^{d} |0\rangle\langle i|\rho|i\rangle\langle 0| , \tag{8.83}$$

so that $\mathcal{E}$ is a quantum operation, by Theorem 8.1. Note that $\mathcal{E}(\rho) = \text{tr}(\rho)|0\rangle\langle 0|$, so that, up to the unimportant $|0\rangle\langle 0|$ multiplier, this quantum operation is identical to the trace function.

An even more useful result is the observation that the partial trace is a quantum operation. Suppose we have a joint system QR, and wish to trace out system R. Let $|j\rangle$ be a basis for system R. Define a linear operator $E_i : H_{QR} \rightarrow H_Q$ by

$$E_i \left(\sum_j \lambda_j |q_j\rangle |j\rangle \right) \equiv \lambda_i |q_i\rangle , \tag{8.84}$$

where λ_j are complex numbers, and $|q_j\rangle$ are arbitrary states of system Q. Define $\mathcal{E}$ to be the quantum operation with operation elements $\{E_i\}$, that is,

$$\mathcal{E}(\rho) \equiv \sum_i E_i \rho E_i^\dagger . \tag{8.85}$$

By Theorem 8.1, this is a quantum operation from system QR to system Q. Notice that

$$\mathcal{E}(\rho \otimes |j\rangle\langle j'|) = \rho \delta_{j,j'} = \text{tr}_R(\rho \otimes |j\rangle\langle j'|) , \tag{8.86}$$

where ρ is any Hermitian operator on the state space of system Q, and $|j\rangle$ and $|j'\rangle$ are members of the orthonormal basis for system R. By linearity of $\mathcal{E}$ and tr_R, it follows that $\mathcal{E} = \text{tr}_R$.

8.3.2 Geometric picture of single qubit quantum operations

There is an elegant geometric method for picturing quantum operations on a single qubit. This method allows one to get an intuitive feel for the behavior of quantum operations in terms of their action on the Bloch sphere. Recall from Exercise 2.72 on page 105 that the state of a single qubit can always be written in the Bloch representation,

$$\rho = \frac{I + \vec{r} \cdot \vec{\sigma}}{2} , \tag{8.87}$$

where $\vec{r}$ is a three component real vector. Explicitly, this gives us

$$\rho = \frac{1}{2} \begin{bmatrix} 1 + r_z & r_x - ir_y \\ r_x + ir_y & 1 - r_z \end{bmatrix}. \tag{8.88}$$

In this representation, it turns out that an arbitrary trace-preserving quantum operation is equivalent to a map of the form

$$\vec{r} \xrightarrow{\mathcal{E}} \vec{r}' = M\vec{r} + \vec{c}, \tag{8.89}$$

where M is a 3×3 real matrix, and $\vec{c}$ is a constant vector. This is an *affine map*, mapping the Bloch sphere into itself. To see this, suppose the operators E_i generating the operator-sum representation for $\mathcal{E}$ are written in the form

$$E_i = \alpha_i I + \sum_{k=1}^{3} a_{ik} \sigma_k. \tag{8.90}$$

Then it is not difficult to check that

$$M_{jk} = \sum_l \left[a_{lj} a_{lk}^* + a_{lj}^* a_{lk} + \left(|\alpha_l|^2 - \sum_p a_{lp} a_{lp}^* \right) \delta_{jk} + i \sum_p \epsilon_{jkp} (\alpha_l a_{lp}^* - \alpha_l^* a_{lp}) \right] \tag{8.91}$$

$$c_k = 2i \sum_l \sum_{jp} \epsilon_{jpk} a_{lj} a_{lp}^*, \tag{8.92}$$

where we have made use of the completeness relation $\sum_i E_i^\dagger E_i = I$ to simplify the expression for $\vec{c}$.

The meaning of the affine map, Equation (8.89), is made clearer by considering the polar decomposition of the matrix M, $M = U|M|$, where U is unitary. Because M is real, it follows that $|M|$ is real and Hermitian, that is, $|M|$ is a symmetric matrix. Furthermore, because M is real we may assume that U is real, and is thus an orthogonal matrix, that is, $U^T U = I$, where T is the transpose operation. Thus we may write

$$M = OS, \tag{8.93}$$

where O is a real orthogonal matrix with determinant 1, representing a proper rotation, and S is a real symmetric matrix. Viewed this way, Equation (8.89) is just a deformation of the Bloch sphere along principal axes determined by S, followed by a proper rotation due to O, followed by a displacement due to $\vec{c}$.

Exercise 8.12: Why can we assume that O has determinant 1 in the decomposition (8.93)?

Exercise 8.13: Show that unitary transformations correspond to rotations of the Bloch sphere.

Exercise 8.14: Show that $\det(S)$ need not be positive.

8.3.3 Bit flip and phase flip channels

The geometric picture described above can be used to visualize some important quantum operations on single qubits, which will later be used in the theory of error-correction. The *bit flip* channel flips the state of a qubit from $|0\rangle$ to $|1\rangle$ (and vice versa) with probability $1 - p$. It has operation elements

$$E_0 = \sqrt{p}\, I = \sqrt{p} \begin{bmatrix} 1 & 0 \\ 0 & 1 \end{bmatrix} \qquad E_1 = \sqrt{1-p}\, X = \sqrt{1-p} \begin{bmatrix} 0 & 1 \\ 1 & 0 \end{bmatrix}. \qquad (8.94)$$

The effect of the bit flip channel is illustrated in Figure 8.8.

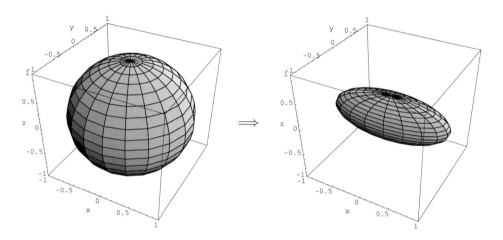

Figure 8.8. The effect of the bit flip channel on the Bloch sphere, for $p = 0.3$. The sphere on the left represents the set of all pure states, and the deformed sphere on the right represents the states after going through the channel. Note that the states on the $\hat{x}$ axis are left alone, while the $\hat{y}$-$\hat{z}$ plane is uniformly contracted by a factor of $1 - 2p$.

This geometric picture makes it very easy to verify certain facts about this quantum operation. For example, it is easy to verify that the quantity $\operatorname{tr}(\rho^2)$ for a single qubit is equal to $(1 + |r|^2)/2$, where $|r|$ is the norm of the Bloch vector. The contraction of the Bloch sphere illustrated in Figure 8.8 cannot increase the norm of the Bloch vector, and therefore we can immediately conclude that $\operatorname{tr}(\rho^2)$ can only ever decrease for the bit flip channel. This is but one example of the use of the geometric picture; once it becomes sufficiently familiar it becomes a great source of insight about the properties of quantum operations on a single qubit.

The *phase flip* channel has operation elements

$$E_0 = \sqrt{p}\, I = \sqrt{p} \begin{bmatrix} 1 & 0 \\ 0 & 1 \end{bmatrix} \qquad E_1 = \sqrt{1-p}\, Z = \sqrt{1-p} \begin{bmatrix} 1 & 0 \\ 0 & -1 \end{bmatrix}. \qquad (8.95)$$

The effect of the phase flip channel is illustrated in Figure 8.9. As a special case of the phase flip channel, consider the quantum operation which arises when we choose $p = 1/2$. Using the freedom in the operator-sum representation this operation may be written

$$\rho \to \mathcal{E}(\rho) = P_0 \rho P_0 + P_1 \rho P_1, \qquad (8.96)$$

where $P_0 = |0\rangle\langle 0|$, $P_1 = |1\rangle\langle 1|$, which corresponds to a measurement of the qubit in the $|0\rangle, |1\rangle$ basis, with the result of the measurement unknown. Using the above prescription it is easy to see that the corresponding map on the Bloch sphere is given by

$$(r_x, r_y, r_z) \rightarrow (0, 0, r_z). \tag{8.97}$$

Geometrically, the Bloch vector is projected along the z axis, and the x and y components of the Bloch vector are lost.

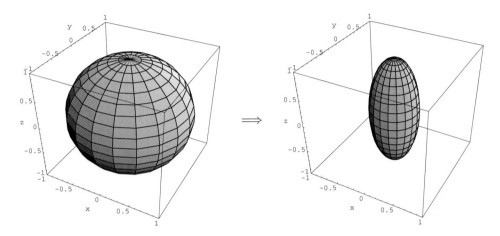

Figure 8.9. The effect of the phase flip channel on the Bloch sphere, for $p = 0.3$. Note that the states on the $\hat{z}$ axis are left alone, while the $\hat{x} - \hat{y}$ plane is uniformly contracted by a factor of $1 - 2p$.

The *bit–phase flip* channel has operation elements

$$E_0 = \sqrt{p}\, I = \sqrt{p} \begin{bmatrix} 1 & 0 \\ 0 & 1 \end{bmatrix} \qquad E_1 = \sqrt{1-p}\, Y = \sqrt{1-p} \begin{bmatrix} 0 & -i \\ i & 0 \end{bmatrix}. \tag{8.98}$$

As the name indicates, this is a combination of a phase flip and a bit flip, since $Y = iXZ$. The action of the bit–phase flip channel is illustrated in Figure 8.10.

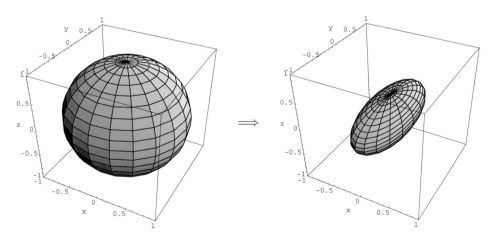

Figure 8.10. The effect of the bit–phase flip channel on the Bloch sphere, for $p = 0.3$. Note that the states on the $\hat{y}$ axis are left alone, while the $\hat{x}$-$\hat{z}$ plane is uniformly contracted by a factor of $1 - 2p$.

Exercise 8.15: Suppose a projective measurement is performed on a single qubit in the basis $|+\rangle, |-\rangle$, where $|\pm\rangle \equiv (|0\rangle \pm |1\rangle)/\sqrt{2}$. In the event that we are ignorant of the result of the measurement, the density matrix evolves according to the equation

$$\rho \rightarrow \mathcal{E}(\rho) = |+\rangle\langle+|\rho|+\rangle\langle+| \quad + \quad |-\rangle\langle-|\rho|-\rangle\langle-|. \tag{8.99}$$

Illustrate this transformation on the Bloch sphere.

Exercise 8.16: The graphical method for understanding single qubit quantum operations was derived for trace-preserving quantum operations. Find an explicit example of a non-trace-preserving quantum operation which cannot be described as a deformation of the Bloch sphere, followed by a rotation and a displacement.

8.3.4 Depolarizing channel

The *depolarizing channel* is an important type of quantum noise. Imagine we take a single qubit, and with probability p that qubit is *depolarized*. That is, it is replaced by the completely mixed state, $I/2$. With probability $1 - p$ the qubit is left untouched. The state of the quantum system after this noise is

$$\mathcal{E}(\rho) = \frac{pI}{2} + (1 - p)\rho. \tag{8.100}$$

The effect of the depolarizing channel on the Bloch sphere is illustrated in Figure 8.11.

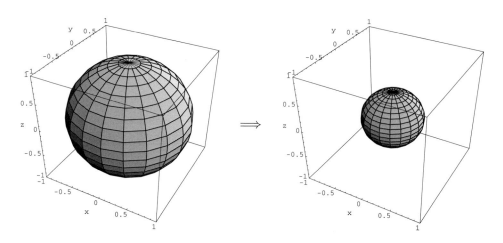

Figure 8.11. The effect of the depolarizing channel on the Bloch sphere, for $p = 0.5$. Note how the entire sphere contracts uniformly as a function of p.

A quantum circuit simulating the depolarizing channel is illustrated in Figure 8.12. The top line of the circuit is the input to the depolarizing channel, while the bottom two lines are an 'environment' to simulate the channel. We have used an environment with two mixed state inputs. The idea is that the third qubit, initially a mixture of the state $|0\rangle$ with probability $1 - p$ and state $|1\rangle$ with probability p acts as a control for whether or not the completely mixed state $I/2$ stored in the second qubit is swapped into the first qubit.

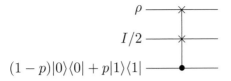

Figure 8.12. Circuit implementation of the depolarizing channel.

The form (8.100) is not in the operator-sum representation. However, if we observe that for arbitrary ρ,

$$\frac{I}{2} = \frac{\rho + X\rho X + Y\rho Y + Z\rho Z}{4} \tag{8.101}$$

and then substitute for $I/2$ into (8.100) we arrive at the equation

$$\mathcal{E}(\rho) = \left(1 - \frac{3p}{4}\right)\rho + \frac{p}{4}(X\rho X + Y\rho Y + Z\rho Z), \tag{8.102}$$

showing that the depolarizing channel has operation elements $\{\sqrt{1 - 3p/4}\,I, \sqrt{p}\,X/2, \sqrt{p}\,Y/2, \sqrt{p}\,Z/2\}$. Note, incidentally, that it is frequently convenient to parametrize the depolarizing channel in different ways, such as

$$\mathcal{E}(\rho) = (1 - p)\rho + \frac{p}{3}(X\rho X + Y\rho Y + Z\rho Z), \tag{8.103}$$

which has the interpretation that the state ρ is left alone with probability $1 - p$, and the operators X, Y and Z applied each with probability $p/3$.

Exercise 8.17: Verify (8.101) as follows. Define

$$\mathcal{E}(A) \equiv \frac{A + XAX + YAY + ZAZ}{4}, \tag{8.104}$$

and show that

$$\mathcal{E}(I) = I; \quad \mathcal{E}(X) = \mathcal{E}(Y) = \mathcal{E}(Z) = 0. \tag{8.105}$$

Now use the Bloch sphere representation for single qubit density matrices to verify (8.101).

The depolarizing channel can, of course, be generalized to quantum systems of dimension more than two. For a d-dimensional quantum system the depolarizing channel again replaces the quantum system with the completely mixed state I/d with probability p, and leaves the state untouched otherwise. The corresponding quantum operation is

$$\mathcal{E}(\rho) = \frac{pI}{d} + (1 - p)\rho. \tag{8.106}$$

Exercise 8.18: For $k \geq 1$ show that $\text{tr}(\rho^k)$ is never increased by the action of the depolarizing channel.

Exercise 8.19: Find an operator-sum representation for a generalized depolarizing channel acting on a d-dimensional Hilbert space.

8.3.5 Amplitude damping

An important application of quantum operations is the description of *energy dissipation* – effects due to loss of energy from a quantum system. What are the dynamics of an atom which is spontaneously emitting a photon? How does a spin system at high temperature approach equilibrium with its environment? What is the state of a photon in an interferometer or cavity when it is subject to scattering and attenuation?

Each of these processes has its own unique features, but the general behavior of all of them is well characterized by a quantum operation known as *amplitude damping*, which we can derive by considering the following scenario. Suppose we have a single optical mode containing the quantum state $a|0\rangle + b|1\rangle$, a superposition of zero or one photons. The scattering of a photon from this mode can be modeled by thinking of inserting a partially silvered mirror, a beamsplitter, in the path of the photon. As we saw in Section 7.4.2, this beamsplitter allows the photon to couple to another single optical mode (representing the environment), according to the unitary transformation $B = \exp\left[\theta\left(a^\dagger b - ab^\dagger\right)\right]$, where a, $a^\dagger$ and b, $b^\dagger$ are annihilation and creation operators for photons in the two modes. The output after the beamsplitter, assuming the environment starts out with no photons, is simply $B|0\rangle(a|0\rangle + b|1\rangle) = a|00\rangle + b(\cos\theta|01\rangle + \sin\theta|10\rangle)$, using Equation (7.34). Tracing over the environment gives us the quantum operation

$$\mathcal{E}_{AD}(\rho) = E_0\rho E_0^\dagger + E_1\rho E_1^\dagger, \tag{8.107}$$

where $E_k = \langle k|B|0\rangle$ are

$$E_0 = \begin{bmatrix} 1 & 0 \\ 0 & \sqrt{1-\gamma} \end{bmatrix}$$

$$E_1 = \begin{bmatrix} 0 & \sqrt{\gamma} \\ 0 & 0 \end{bmatrix}, \tag{8.108}$$

the operation elements for amplitude damping. $\gamma = \sin^2\theta$ can be thought of as the probability of losing a photon.

Observe that no linear combination can be made of E_0 and E_1 to give an operation element proportional to the identity (though compare with Exercise 8.23). The E_1 operation changes a $|1\rangle$ state into a $|0\rangle$ state, corresponding to the physical process of losing a quantum of energy to the environment. E_0 leaves $|0\rangle$ unchanged, but reduces the amplitude of a $|1\rangle$ state; physically, this happens because a quantum of energy was not lost to the environment, and thus the environment now perceives it to be more likely that the system is in the $|0\rangle$ state, rather than the $|1\rangle$ state.

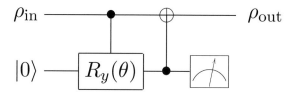

Figure 8.13. Circuit model for amplitude damping

Exercise 8.20: (Circuit model for amplitude damping) Show that the circuit in

Figure 8.13 models the amplitude damping quantum operation, with $\sin^2(\theta/2) = \gamma$.

Exercise 8.21: (Amplitude damping of a harmonic oscillator) Suppose that our principal system, a harmonic oscillator, interacts with an environment, modeled as another harmonic oscillator, through the Hamiltonian

$$H = \chi(a^\dagger b + b^\dagger a) \qquad (8.109)$$

where a and b are the annihilation operators for the respective harmonic oscillators, as defined in Section 7.3.

(1) Using $U = \exp(-iH\Delta t)$, denoting the eigenstates of $b^\dagger b$ as $|k_b\rangle$, and selecting the vacuum state $|0_b\rangle$ as the initial state of the environment, show that the operation elements $E_k = \langle k_b | U | 0_b \rangle$ are found to be

$$E_k = \sum_n \sqrt{\binom{n}{k}} \sqrt{(1-\gamma)^{n-k}\gamma^k} \, |n-k\rangle\langle n| , \qquad (8.110)$$

where $\gamma = 1 - \cos^2(\chi\Delta t)$ is the probability of loosing a single quantum of energy, and states such as $|n\rangle$ are eigenstates of $a^\dagger a$.

(2) Show that the operation elements E_k define a trace-preserving quantum operation.

Exercise 8.22: (Amplitude damping of single qubit density matrix) For the general single qubit state

$$\rho = \begin{bmatrix} a & b \\ b^* & c \end{bmatrix} \qquad (8.111)$$

show that amplitude damping leads to

$$\mathcal{E}_{AD}(\rho) = \begin{bmatrix} 1 - (1-\gamma)(1-a) & b\sqrt{1-\gamma} \\ b^*\sqrt{1-\gamma} & c(1-\gamma) \end{bmatrix} . \qquad (8.112)$$

Exercise 8.23: (Amplitude damping of dual-rail qubits) Suppose that a single qubit state is represented by using two qubits, as

$$|\psi\rangle = a\,|01\rangle + b\,|10\rangle . \qquad (8.113)$$

Show that $\mathcal{E}_{AD} \otimes \mathcal{E}_{AD}$ applied to this state gives a process which can be described by the operation elements

$$E_0^{dr} = \sqrt{1-\gamma}\, I \qquad (8.114)$$

$$E_1^{dr} = \sqrt{\gamma} \left[|00\rangle\langle 01| + |00\rangle\langle 10| \right] , \qquad (8.115)$$

that is, either nothing (E_0^{dr}) happens to the qubit, or the qubit is transformed (E_1^{dr}) into the state $|00\rangle$, which is orthogonal to $|\psi\rangle$. This is a simple error-detection code, and is also the basis for the robustness of the 'dual-rail' qubit discussed in Section 7.4.

Exercise 8.24: (Spontaneous emission is amplitude damping) A single atom coupled to a single mode of electromagnetic radiation undergoes spontaneous emission, as was described in Section 7.6.1. To see that this process is just

amplitude damping, take the unitary operation resulting from the
Jaynes–Cummings interaction, Equation (7.77), with detuning $\delta = 0$, and give
the quantum operation resulting from tracing over the field.

A general characteristic of a quantum operation is the set of states that are left invariant
under the operation. For example, we have seen how the phase flip channel leaves the $\hat{z}$
axis of the Bloch sphere unchanged; this corresponds to states of the form $p|0\rangle\langle 0| + (1 -
p)|1\rangle\langle 1|$ for arbitrary probability p. In the case of amplitude damping, only the ground
state $|0\rangle$ is left invariant. That is a natural consequence of our modeling the environment
as starting in the $|0\rangle$ state, as if it were at zero temperature.

What quantum operation describes the effect of dissipation to an environment at finite
temperature? This process, $\mathcal{E}_{\text{GAD}}$, called *generalized amplitude damping*, is defined for
single qubits by the operation elements

$$E_0 = \sqrt{p} \begin{bmatrix} 1 & 0 \\ 0 & \sqrt{1-\gamma} \end{bmatrix} \tag{8.116}$$

$$E_1 = \sqrt{p} \begin{bmatrix} 0 & \sqrt{\gamma} \\ 0 & 0 \end{bmatrix} \tag{8.117}$$

$$E_2 = \sqrt{1-p} \begin{bmatrix} \sqrt{1-\gamma} & 0 \\ 0 & 1 \end{bmatrix} \tag{8.118}$$

$$E_3 = \sqrt{1-p} \begin{bmatrix} 0 & 0 \\ \sqrt{\gamma} & 0 \end{bmatrix}, \tag{8.119}$$

where the stationary state

$$\rho_\infty = \begin{bmatrix} p & 0 \\ 0 & 1-p \end{bmatrix}, \tag{8.120}$$

satisfies $\mathcal{E}_{\text{GAD}}(\rho_\infty) = \rho_\infty$. Generalized amplitude damping describes the 'T_1' relaxation
processes due to coupling of spins to their surrounding lattice, a large system which is
in thermal equilibrium at a temperature often much higher than the spin temperature.
This is the case relevant to NMR quantum computation, where some of the properties
of $\mathcal{E}_{\text{GAD}}$ described in Box 8.3 become important.

Exercise 8.25: If we define the temperature T of a qubit by assuming that in
equilibrium the probabilities of being in the $|0\rangle$ and $|1\rangle$ states satisfy a
Boltzmann distribution, that is $p_0 = e^{-E_0/k_B T}/\mathcal{Z}$ and $p_1 = e^{-E_1/k_B T}/\mathcal{Z}$, where
E_0 is the energy of the state $|0\rangle$, E_1 the energy of the state $|1\rangle$, and
$\mathcal{Z} = e^{-E_0/k_B T} + e^{-E_1/k_B T}$, what temperature describes the state ρ_∞?

We can visualize the effect of amplitude damping in the Bloch representation as the
Bloch vector transformation

$$(r_x, r_y, r_z) \rightarrow \left(r_x\sqrt{1-\gamma}, r_y\sqrt{1-\gamma}, \gamma + r_z(1-\gamma) \right). \tag{8.122}$$

When γ is replaced with a time-varying function like $1 - e^{-t/T_1}$ (t is time, and T_1
just some constant characterizing the speed of the process), as is often the case for real
physical processes, we can visualize the effects of amplitude damping as a *flow* on the
Bloch sphere, which moves every point in the unit ball towards a fixed point at the north
pole, where $|0\rangle$ is located. This is shown in Figure 8.14.

Box 8.3: Generalized amplitude damping and effective pure states

The notion of 'effective pure states' introduced in Section 7.7 was found to be useful in NMR implementations of quantum computers. These states behave like pure states under unitary evolution and measurement of traceless observables. How do they behave under quantum operations? In general, non-unitary quantum operations ruin the effectiveness of these states, but surprisingly, they can behave properly under generalized amplitude damping.

Consider a single qubit effective pure state $\rho = (1 - p)I + (2p - 1)|0\rangle\langle0|$. Clearly, traceless measurement observables acting on $U\rho U^\dagger$ produce results which are proportional to those on the pure state $U|0\rangle\langle0|U^\dagger$. Suppose ρ is the stationary state of $\mathcal{E}_{\text{GAD}}$. Interestingly, in this case,

$$\mathcal{E}_{\text{GAD}}(U\rho U^\dagger) = (1 - p)I + (2p - 1)\mathcal{E}_{\text{AD}}(U\rho U^\dagger). \qquad (8.121)$$

That is, under generalized amplitude damping, an effective pure state can remain such, and moreover, the 'pure' component of ρ behaves as if it were undergoing amplitude damping to a reservoir at zero temperature!

Similarly, generalized amplitude damping performs the transformation

$$(r_x, r_y, r_z) \rightarrow \left(r_x\sqrt{1 - \gamma}, r_y\sqrt{1 - \gamma}, \gamma(2p - 1) + r_z(1 - \gamma)\right). \qquad (8.123)$$

Comparing (8.122) and (8.123), it is clear that amplitude damping and generalized amplitude damping differ only in the location of the fixed point of the flow; the final state is along the $\hat{z}$ axis, at the point $(2p - 1)$, which is a mixed state.

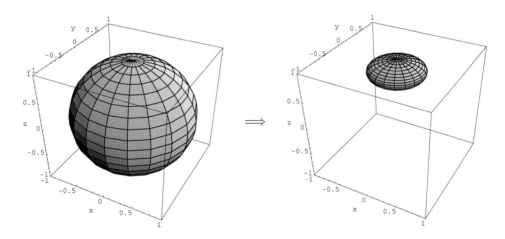

Figure 8.14. The effect of the amplitude damping channel on the Bloch sphere, for $p = 0.8$. Note how the entire sphere shrinks towards the north pole, the $|0\rangle$ state.

8.3.6 Phase damping

A noise process that is uniquely quantum mechanical, which describes the loss of quantum information without loss of energy, is *phase damping*. Physically it describes, for example, what happens when a photon scatters randomly as it travels through a waveguide, or how electronic states in an atom are perturbed upon interacting with distant electrical charges. The energy eigenstates of a quantum system do not change as a function of time, but do accumulate a phase which is proportional to the eigenvalue. When a system evolves for an amount of time which is not precisely known, partial information about this quantum phase – the *relative* phases between the energy eigenstates – is lost.

A very simple model for this kind of quantum noise is the following. Suppose that we have a qubit $|\psi\rangle = a|0\rangle + b|1\rangle$ upon which the rotation operation $R_z(\theta)$ is applied, where the angle of rotation θ is random. The randomness could originate, for example, from a deterministic interaction with an environment, which never again interacts with the system and thus is implicitly measured (see Section 4.4). We shall call this random R_z operation a *phase kick*. Let us assume that the phase kick angle θ is well represented as a random variable which has a Gaussian distribution with mean 0 and variance 2λ.

The output state from this process is given by the density matrix obtained from averaging over θ,

$$\rho = \frac{1}{\sqrt{4\pi\lambda}} \int_{-\infty}^{\infty} R_z(\theta)|\psi\rangle\langle\psi|R_z^{\dagger}(\theta)e^{-\theta^2/4\lambda} \, d\theta \qquad (8.124)$$

$$= \begin{bmatrix} |a|^2 & ab^* e^{-\lambda} \\ a^*b e^{-\lambda} & |b|^2 \end{bmatrix}. \qquad (8.125)$$

The random phase kicking causes the expected value of the off-diagonal elements of the density matrix to decay exponentially to zero with time. That is a characteristic result of phase damping.

Another way to derive the phase damping quantum operation is to consider an interaction between two harmonic oscillators, in a manner similar to how amplitude damping was derived in the last section, but this time with the interaction Hamiltonian

$$H = \chi \, a^{\dagger}a(b + b^{\dagger}), \qquad (8.126)$$

Letting $U = \exp(-iH\Delta t)$, considering only the $|0\rangle$ and $|1\rangle$ states of the a oscillator as our system, and taking the environment oscillator to initially be $|0\rangle$, we find that tracing over the environment gives the operation elements $E_k = \langle k_b|U|0_b\rangle$, which are

$$E_0 = \begin{bmatrix} 1 & 0 \\ 0 & \sqrt{1-\lambda} \end{bmatrix} \qquad (8.127)$$

$$E_1 = \begin{bmatrix} 0 & 0 \\ 0 & \sqrt{\lambda} \end{bmatrix}, \qquad (8.128)$$

where $\lambda = 1 - \cos^2(\chi\Delta t)$ can be interpreted as the probability that a photon from the system has been scattered (without loss of energy). As was the case for amplitude damping, E_0 leaves $|0\rangle$ unchanged, but reduces the amplitude of a $|1\rangle$ state; unlike amplitude damping, however, the E_1 operation destroys $|0\rangle$ and reduces the amplitude of the $|1\rangle$ state, and does not change it into a $|0\rangle$.

By applying Theorem 8.2, the unitary freedom of quantum operations, we find that a unitary recombination of E_0 and E_1 gives a new set of operation elements for phase damping,

$$\tilde{E}_0 = \sqrt{\alpha} \begin{bmatrix} 1 & 0 \\ 0 & 1 \end{bmatrix} \qquad (8.129)$$

$$\tilde{E}_1 = \sqrt{1-\alpha} \begin{bmatrix} 1 & 0 \\ 0 & -1 \end{bmatrix}, \tag{8.130}$$

where $\alpha = (1 + \sqrt{1-\lambda})/2$. Thus the phase damping quantum operation is *exactly* the same as the phase flip channel which we encountered in Section 8.3.3!

Since phase damping is the same as the phase flip channel, we have already seen how it is visualized on the Bloch sphere, in Figure 8.9. This corresponds to the Bloch vector transformation

$$(r_x, r_y, r_z) \rightarrow \left(r_x\sqrt{1-\lambda}, r_y\sqrt{1-\lambda}, r_z \right), \tag{8.131}$$

which has the effect of shrinking the sphere into ellipsoids. Phase damping is often referred to as a 'T_2' (or 'spin-spin') relaxation process, for historical reasons, where $e^{-t/2T_2} = \sqrt{1-\lambda}$. As a function of time, the amount of damping increases, corresponding to an inwards flow of all points in the unit ball towards the $\hat{z}$ axis. Note that states along the $\hat{z}$ axis remain invariant.

Historically, phase damping was a process that was almost always thought of, physically, as resulting from a random phase kick or scattering process. It was not until the connection to the phase flip channel was discovered that quantum error-correction was developed, since it was thought that phase errors were *continuous* and couldn't be described as a discrete process! In fact, single qubit phase errors can *always* be thought of as resulting from a process in which either nothing happens to a qubit, with probability α, or with probability $1 - \alpha$, the qubit is flipped by the Z Pauli operation. Although this might not be the actual microscopic physical process happening, from the standpoint of the transformation occurring to a qubit over a discrete time interval large compared to the underlying random process, there is no difference at all.

Phase damping is one of the most subtle and important processes in the study of quantum computation and quantum information. It has been the subject of an immense amount of study and speculation, particularly with regard to why the world around us appears to be so classical, with superposition states not a part of our everyday experience! Perhaps it is phase damping that is responsible for this absence of superposition states from the everyday (Exercise 8.31)? The pioneering quantum physicist Schrödinger was perhaps the first to pose this problem, and he did this in a particularly stark form, as discussed in Box 8.4.

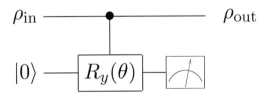

Figure 8.15. Circuit model for phase damping. The upper wire carries the input qubit with an unknown state, and the lower wire is an ancilla qubit used to model the environment.

Exercise 8.26: (Circuit model for phase damping) Show that the circuit in Figure 8.15 can be used to model the phase damping quantum operation, provided θ is chosen appropriately.

Exercise 8.27: (Phase damping = phase flip channel) Give the unitary transformation which relates the operation elements of (8.127)–(8.128) to those of (8.129)–(8.130); that is, find u such that $\tilde{E}_k = \sum_j u_{kj} E_j$.

Exercise 8.28: (One CNOT phase damping model circuit) Show that a single controlled-NOT gate can be used as a model for phase damping, if we let the initial state of the environment be a mixed state, where the amount of damping is determined by the probability of the states in the mixture.

Exercise 8.29: (Unitality) A quantum process $\mathcal{E}$ is *unital* if $\mathcal{E}(I) = I$. Show that the depolarizing and phase damping channels are unital, while amplitude damping is not.

Exercise 8.30: $(T_2 \leq T_1/2)$ The T_2 phase coherence relaxation rate is just the exponential decay rate of the off-diagonal elements in the qubit density matrix, while T_1 is the decay rate of the diagonal elements (see Equation (7.144)). Amplitude damping has *both* nonzero T_1 and T_2 rates; show that for amplitude damping $T_2 = T_1/2$. Also show that if amplitude and phase damping are *both* applied then $T_2 \leq T_1/2$.

Exercise 8.31: (Exponential sensitivity to phase damping) Using (8.126), show that the element $\rho_{nm} = \langle n|\rho|m \rangle$ in the density matrix of a harmonic oscillator decays exponentially as $e^{-\lambda(n-m)^2}$ under the effect of phase damping, for some constant λ.

8.4 Applications of quantum operations

As befits a powerful tool, the quantum operations formalism has numerous applications. In this section we describe two of these applications. Section 8.4.1 describes the theory of *master equations*, a picture of quantum noise complementary to the quantum operations formalism. The master equation approach describes quantum noise in *continuous time* using differential equations, and is the approach to quantum noise most often used by physicists. In Section 8.4.2 we describe *quantum process tomography*, a procedure to experimentally determine the dynamics of a quantum system.

8.4.1 Master equations

Open quantum systems occur in a wide range of disciplines, and many tools other than quantum operations can be employed in their study. In this section, we briefly describe one such tool, the approach of *master equations*.

The dynamics of open quantum systems have been studied extensively in the field of quantum optics. The main objective in this context is to describe the time evolution of an open system with a differential equation which properly describes non-unitary behavior. This description is provided by the master equation, which can be written most generally in the *Lindblad form* as

$$\frac{d\rho}{dt} = -\frac{i}{\hbar}[H, \rho] + \sum_j \left[2L_j \rho L_j^\dagger - \{L_j^\dagger L_j, \rho\} \right], \tag{8.134}$$

where $\{x, y\} = xy + yx$ denotes an anticommutator, H is the system Hamiltonian, a Hermitian operator representing the coherent part of the dynamics, and L_j are the *Lindblad*

Box 8.4: Schrödinger's cat

When I hear about Schrödinger's cat, I reach for my gun. – Stephen Hawking

Schrödinger's infamous cat faces life or death contingent upon an automatic device which breaks a vial of poison and kills the cat if an excited atomic state is observed to decay, as illustrated here:

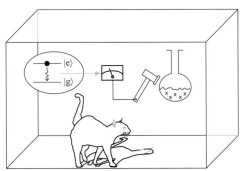

Schrödinger asked what happens when the atom is in a superposition state? Is the cat alive or dead? Why do superposition states such as this apparently not occur in the everyday world? This conundrum is resolved by realizing that it is very unlikely to occur in real life, because of extreme sensitivity of macroscopic superposition states to decoherence. Let the atom represent a single qubit. The joint system has the initial state $|\text{alive}\rangle|1\rangle$. Suppose that after one half-life of the atom, the state is the equal superposition $|\text{alive}\rangle(|0\rangle + |1\rangle)/\sqrt{2}$ (this represents a simplification of the actual physics, which are too involved to go into here). The apparatus kills the cat if the atom is in the $|0\rangle$ state; otherwise, the cat lives. This gives the state $|\psi\rangle = \left[|\text{dead}\rangle|0\rangle + |\text{alive}\rangle|1\rangle\right]/\sqrt{2}$, in which the cat's state has become entangled with that of the atom. This would seem to indicate the cat is simultaneously alive and dead, but suppose we consider the density matrix of this state,

$$\rho = |\psi\rangle\langle\psi| \tag{8.132}$$

$$= \frac{1}{2}\Big[|\text{alive}, 1\rangle\langle\text{alive}, 1| + |\text{dead}, 0\rangle\langle\text{dead}, 0|$$

$$+ |\text{alive}, 1\rangle\langle\text{dead}, 0| + |\text{dead}, 0\rangle\langle\text{alive}, 1|\Big] . \tag{8.133}$$

Now, in practice it is impossible to perfectly isolate the cat and the atom in their box, and thus information about this superposition state will leak into the external world. For example, heat from the cat's body could permeate the wall and give some indication of its state to the outside. Such effects may be modeled as phase damping, which exponentially damps out the final two (off-diagonal) terms in ρ. To a first approximation, we may model the cat–atom system as a simple harmonic oscillator. An important result about the decoherence of such a system is that coherence between states of high energy difference decays faster than between states with a lower energy difference (Exercise 8.31). Thus ρ will quickly be transformed into a nearly diagonal state, which represents an ensemble of cat–atom states which correspond to either dead or alive, and are not in a superposition of the two states.

operators, representing the coupling of the system to its environment. The differential equation takes on the above form in order that the process be completely positive in a sense similar to that described earlier for quantum operations. It is also generally assumed that the system and environment begin in a product state. Furthermore, in order to derive a master equation for a process, one usually begins with a system–environment model Hamiltonian, and then makes the Born and Markov approximations in order to determine L_j. Note that in the master equation approach, $\text{tr}[\rho(t)] = 1$ at all times.

As an example of a Lindblad equation, consider a two-level atom coupled to the vacuum, undergoing spontaneous emission. The coherent part of the atom's evolution is described by the Hamiltonian $H = -\hbar\omega\sigma_z/2$. $\hbar\omega$ is the energy difference of the atomic levels. Spontaneous emission causes an atom in the excited ($|1\rangle$) state to drop down into the ground ($|0\rangle$) state, emitting a photon in the process. This emission is described by the Lindblad operator $\sqrt{\gamma}\sigma_-$, where $\sigma_- \equiv |0\rangle\langle 1|$ is the atomic lowering operator, and γ is the rate of spontaneous emission. The master equation describing this process is

$$\frac{d\rho}{dt} = -\frac{i}{\hbar}[H, \rho] + \gamma\left[2\sigma_-\rho\sigma_+ - \sigma_+\sigma_-\rho - \rho\sigma_+\sigma_-\right], \tag{8.135}$$

where $\sigma_+ \equiv \sigma_-^\dagger$ is the atomic raising operator.

To solve the equation it is helpful to move to the interaction picture, that is, make the change of variables

$$\tilde{\rho}(t) \equiv e^{iHt}\rho(t)e^{-iHt}. \tag{8.136}$$

The equation of motion for $\tilde{\rho}$ is easily found to be

$$\frac{d\tilde{\rho}}{dt} = \gamma\left[2\tilde{\sigma}_-\tilde{\rho}\tilde{\sigma}_+ - \tilde{\sigma}_+\tilde{\sigma}_-\tilde{\rho} - \tilde{\rho}\tilde{\sigma}_+\tilde{\sigma}_-\right] \tag{8.137}$$

where

$$\tilde{\sigma}_- \equiv e^{iHt}\sigma_-e^{-iHt} = e^{-i\omega t}\sigma_- \tag{8.138}$$

$$\tilde{\sigma}_+ \equiv e^{iHt}\sigma_+e^{-iHt} = e^{i\omega t}\sigma_+. \tag{8.139}$$

Our final equation of motion is thus

$$\frac{d\tilde{\rho}}{dt} = \gamma\left[2\sigma_-\tilde{\rho}\sigma_+ - \sigma_+\sigma_-\tilde{\rho} - \tilde{\rho}\sigma_+\sigma_-\right]. \tag{8.140}$$

This equation of motion is easily solved using a Bloch vector representation for $\tilde{\rho}$. The solution is

$$\lambda_x = \lambda_x(0)e^{-\gamma t} \tag{8.141}$$

$$\lambda_y = \lambda_y(0)e^{-\gamma t} \tag{8.142}$$

$$\lambda_z = \lambda_z(0)e^{-2\gamma t} + 1 - e^{-2\gamma t}. \tag{8.143}$$

Defining $\gamma' = 1 - \exp(-2t\gamma)$ we can easily check that this evolution is equivalent to

$$\tilde{\rho}(t) = \mathcal{E}(\tilde{\rho}(0)) \equiv E_0\tilde{\rho}(0)E_0^\dagger + E_1\tilde{\rho}(0)E_1^\dagger, \tag{8.144}$$

where

$$E_0 \equiv \begin{bmatrix} 1 & 0 \\ 0 & \sqrt{1-\gamma'} \end{bmatrix} \tag{8.145}$$

$$E_1 \equiv \begin{bmatrix} 0 & \sqrt{\gamma'} \\ 0 & 0 \end{bmatrix} \tag{8.146}$$

are the operation elements defining the quantum operation $\mathcal{E}$. Note that the effect of $\mathcal{E}$ is amplitude damping; compare with Equation (8.108). The example we have considered is an instance of the spin-boson model, in which a small, finite dimensional quantum system interacts with a bath of simple harmonic oscillators. Physically, it is important in describing the interaction of atoms with electromagnetic radiation, as in cavity QED, or atom and ion traps.

The master equation approach is less general than the quantum operations formalism. Solving a master equation allows one to determine the time dependence of a density matrix. Knowing this, in turn, means that the result can be expressed as a quantum operation in the operator-sum representation,

$$\rho(t) = \sum_k E_k(t)\rho(0)E_k^\dagger(t)\,, \tag{8.147}$$

where $E_k(t)$ are time dependent operation elements, determined from the solution to the master equation. However, a quantum process described in terms of an operator-sum representation cannot necessarily be written down as a master equation. For example, quantum operations can describe non-Markovian dynamics, simply because they describe only state changes, not continuous time evolution. Nevertheless, each approach has its own place. In fact, even quantum operations do not provide the *most* general description; we consider in Section 8.5 some processes which are not described by quantum operations.

8.4.2 Quantum process tomography

Quantum operations provide a wonderful mathematical model for open quantum systems, and are conveniently visualized (at least for qubits) – but how do they relate to experimentally measurable quantities? What measurements should an experimentalist do if they wish to characterize the dynamics of a quantum system? For classical systems, this elementary task is known as *system identification*. Here, we show how its analogue, known as *quantum process tomography*, can be performed for finite dimensional quantum systems.

To understand process tomography we first need to understand another procedure called *quantum state tomography*. State tomography is the procedure of experimentally determining an unknown quantum state. Suppose we are given an unknown state, ρ, of a single qubit. How can we experimentally determine what the state of ρ is?

If we are given just a single copy of ρ then it turns out to be impossible to characterize ρ. The basic problem is that there is no quantum measurement which can distinguish non-orthogonal quantum states like $|0\rangle$ and $(|0\rangle + |1\rangle)/\sqrt{2}$ with certainty. However, it is possible to estimate ρ if we have a large number of copies of ρ. For instance, if ρ is the quantum state produced by some experiment, then we simply repeat the experiment many times to produce many copies of the state ρ.

Suppose we have many copies of a single qubit density matrix, ρ. The set $I/\sqrt{2}$, $X/\sqrt{2}$, $Y/\sqrt{2}$, $Z/\sqrt{2}$ forms an orthonormal set of matrices with respect to the Hilbert–Schmidt inner product, so ρ may be expanded as

$$\rho = \frac{\operatorname{tr}(\rho)I + \operatorname{tr}(X\rho)X + \operatorname{tr}(Y\rho)Y + \operatorname{tr}(Z\rho)Z}{2}\,. \tag{8.148}$$

Recall, however, that expressions like $\operatorname{tr}(A\rho)$ have an interpretation as the average value of *observables*. For example, to estimate $\operatorname{tr}(Z\rho)$ we measure the observable Z a large number of times, m, obtaining outcomes $z_1, z_2, \ldots, z_m$, all equal to $+1$ or -1. The empirical average of these quantities, $\sum_i z_i/m$, is an estimate for the true value of $\mathrm{tr}(Z\rho)$. We can use the central limit theorem to determine how well this estimate behaves for large m, where it becomes approximately Gaussian with mean equal to $\mathrm{tr}(Z\rho)$ and with standard deviation $\Delta(Z)/\sqrt{m}$, where $\Delta(Z)$ is the standard deviation for a single measurement of Z, which is upper bounded by 1, so the standard deviation in our estimate $\sum_i z_i/m$ is at most $1/\sqrt{m}$.

In a similar way we can estimate the quantities $\mathrm{tr}(X\rho)$ and $\mathrm{tr}(Y\rho)$ with a high degree of confidence in the limit of a large sample size, and thus obtain a good estimate for ρ. Generalizing this procedure to the case of more than one qubit is not difficult, at least in principle! Similar to the single qubit case, an arbitrary density matrix on n qubits can be expanded as

$$\rho = \sum_{\vec{v}} \frac{\mathrm{tr}\left(\sigma_{v_1} \otimes \sigma_{v_2} \otimes \cdots \otimes \sigma_{v_n}\, \rho\right) \sigma_{v_1} \otimes \sigma_{v_2} \otimes \cdots \otimes \sigma_{v_n}}{2^n}, \tag{8.149}$$

where the sum is over vectors $\vec{v} = (v_1, \ldots, v_n)$ with entries v_i chosen from the set $0, 1, 2, 3$. By performing measurements of observables which are products of Pauli matrices we can estimate each term in this sum, and thus obtain an estimate for ρ.

We've described how to do state tomography for systems comprised of qubits. What if non-qubit systems are involved? Not surprisingly, it is easy to generalize the above prescription to such systems. We won't explicitly do so here, but instead refer you to the end of chapter 'History and further reading' for references.

Now that we know how to do quantum state tomography, how can we use it to do quantum process tomography? The experimental procedure may be outlined as follows. Suppose the state space of the system has d dimensions; for example, $d = 2$ for a single qubit. We choose d^2 pure quantum states $|\psi_1\rangle, \ldots, |\psi_{d^2}\rangle$, chosen so that the corresponding density matrices $|\psi_1\rangle\langle\psi_1|, \ldots, |\psi_{d^2}\rangle\langle\psi_{d^2}|$ form a *basis set* for the space of matrices. We explain in more detail how to choose such a set below. For each state $|\psi_j\rangle$ we prepare the quantum system in that state and then subject it to the process which we wish to characterize. After the process has run to completion we use quantum state tomography to determine the state $\mathcal{E}(|\psi_j\rangle\langle\psi_j|)$ output from the process. From a purist's point of view we are now done, since in principle the quantum operation $\mathcal{E}$ is now determined by a linear extension of $\mathcal{E}$ to all states.

In practice, of course, we would like to have a way of determining a useful representation of $\mathcal{E}$ from experimentally available data. We will explain a general procedure for doing so, worked out explicitly for the case of a single qubit. Our goal is to determine a set of operation elements $\{E_i\}$ for $\mathcal{E}$,

$$\mathcal{E}(\rho) = \sum_i E_i \rho E_i^\dagger. \tag{8.150}$$

However, experimental results involve numbers, not operators, which are a theoretical concept. To determine the E_i from measurable parameters, it is convenient to consider an equivalent description of $\mathcal{E}$ using a *fixed* set of operators $\tilde{E}_i$, which form a basis for the set of operators on the state space, so that

$$E_i = \sum_m e_{im} \tilde{E}_m \tag{8.151}$$

for some set of complex numbers e_{im}. Equation (8.150) may thus be rewritten as

$$\mathcal{E}(\rho) = \sum_{mn} \tilde{E}_m \rho \tilde{E}_n^\dagger \chi_{mn}, \tag{8.152}$$

where $\chi_{mn} \equiv \sum_i e_{im} e_{in}^*$ are the entries of a matrix which is positive Hermitian by definition. This expression, known as the *chi matrix representation*, shows that $\mathcal{E}$ can be completely described by a complex number matrix, χ, once the set of operators $\tilde{E}_i$ has been fixed.

In general, χ will contain $d^4 - d^2$ independent real parameters, because a general linear map of d by d complex matrices to d by d matrices is described by d^4 independent parameters, but there are d^2 additional constraints due to the fact that ρ remains Hermitian with trace one; that is, the completeness relation

$$\sum_i E_i^\dagger E_i = I, \tag{8.153}$$

is satisfied, giving d^2 real constraints. We will show how to determine χ experimentally, and then show how an operator-sum representation of the form (8.150) can be recovered once the χ matrix is known.

Let ρ_j, $1 \leq j \leq d^2$ be a fixed, linearly independent basis for the space of $d \times d$ matrices; that is, any $d \times d$ matrix can be written as a unique linear combination of the ρ_j. A convenient choice is the set of operators $|n\rangle\langle m|$. Experimentally, the output state $\mathcal{E}(|n\rangle\langle m|)$ may be obtained by preparing the input states $|n\rangle$, $|m\rangle$, $|+\rangle = (|n\rangle + |m\rangle)/\sqrt{2}$, and $|-\rangle = (|n\rangle + i|m\rangle)/\sqrt{2}$ and forming linear combinations of $\mathcal{E}(|n\rangle\langle n|)$, $\mathcal{E}(|m\rangle\langle m|)$, $\mathcal{E}(|+\rangle\langle +|)$, and $\mathcal{E}(|-\rangle\langle -|)$, as follows:

$$\mathcal{E}(|n\rangle\langle m|) = \mathcal{E}(|+\rangle\langle +|) + i\mathcal{E}(|-\rangle\langle -|) - \frac{1+i}{2}\mathcal{E}(|n\rangle\langle n|) - \frac{1+i}{2}\mathcal{E}(|m\rangle\langle m|). \tag{8.154}$$

Thus, it is possible to determine $\mathcal{E}(\rho_j)$ by state tomography, for each ρ_j.

Furthermore, each $\mathcal{E}(\rho_j)$ may be expressed as a linear combination of the basis states,

$$\mathcal{E}(\rho_j) = \sum_k \lambda_{jk} \rho_k, \tag{8.155}$$

and since $\mathcal{E}(\rho_j)$ is known from the state tomography, λ_{jk} can be determined by standard linear algebraic algorithms. To proceed, we may write

$$\tilde{E}_m \rho_j \tilde{E}_n^\dagger = \sum_k \beta_{jk}^{mn} \rho_k, \tag{8.156}$$

where β_{jk}^{mn} are complex numbers which can be determined by standard algorithms from linear algebra given the $\tilde{E}_m$ operators and the ρ_j operators. Combining the last two expressions and (8.152) we have

$$\sum_k \sum_{mn} \chi_{mn} \beta_{jk}^{mn} \rho_k = \sum_k \lambda_{jk} \rho_k. \tag{8.157}$$

From the linear independence of the ρ_k it follows that for each k,

$$\sum_{mn} \beta_{jk}^{mn} \chi_{mn} = \lambda_{jk}. \tag{8.158}$$

This relation is a necessary and sufficient condition for the matrix χ to give the correct quantum operation $\mathcal{E}$. One may think of χ and λ as vectors, and β as a $d^4 \times d^4$ matrix with columns indexed by mn, and rows by jk. To show how χ may be obtained, let κ be the generalized inverse for the matrix β, satisfying the relation

$$\beta_{jk}^{mn} = \sum_{st,xy} \beta_{jk}^{st} \kappa_{st}^{xy} \beta_{xy}^{mn} . \tag{8.159}$$

Most computer packages for matrix manipulation are capable of finding such generalized inverses. We now prove that χ defined by

$$\chi_{mn} \equiv \sum_{jk} \kappa_{jk}^{mn} \lambda_{jk} \tag{8.160}$$

satisfies the relation (8.158).

The difficulty in verifying that χ defined by (8.160) satisfies (8.158) is that, in general, χ is not uniquely determined by Equation (8.158). For convenience we rewrite these equations in matrix form as

$$\beta\vec{\chi} = \vec{\lambda} \tag{8.161}$$

$$\vec{\chi} \equiv \kappa\vec{\lambda}. \tag{8.162}$$

From the construction that led to Equation (8.152) we know there exists at least one solution to Equation (8.161), which we shall call $\vec{\chi}'$. Thus $\vec{\lambda} = \beta\vec{\chi}'$. The generalized inverse satisfies $\beta\kappa\beta = \beta$. Premultiplying the definition of $\vec{\chi}$ by β gives

$$\beta\vec{\chi} = \beta\kappa\vec{\lambda} \tag{8.163}$$

$$= \beta\kappa\beta\vec{\chi}' \tag{8.164}$$

$$= \beta\vec{\chi}' \tag{8.165}$$

$$= \vec{\lambda}. \tag{8.166}$$

Thus χ defined by (8.162) satisfies the Equation (8.161), as we wanted to show.

Having determined χ one immediately obtains the operator-sum representation for $\mathcal{E}$ in the following manner. Let the unitary matrix $U^\dagger$ diagonalize χ,

$$\chi_{mn} = \sum_{xy} U_{mx} d_x \delta_{xy} U_{ny}^* . \tag{8.167}$$

From this it can easily be verified that

$$E_i = \sqrt{d_i} \sum_j U_{ji} \tilde{E}_j \tag{8.168}$$

are operation elements for $\mathcal{E}$. Our algorithm may thus be summarized as follows: λ is experimentally determined using state tomography, which in turn determines χ via the equation $\vec{\chi} = \kappa\lambda$, which gives us a complete description of $\mathcal{E}$, including a set of operation elements E_i.

In the case of a single qubit quantum process, only 12 parameters must be determined (Box 8.5). The dynamics of a two qubit quantum black box $\mathcal{E}_2$ pose an even greater challenge for our understanding. In this case there are 240 parameters which need to be determined in order to completely specify the quantum operation acting on the quantum system! Determining these would obviously be quite a considerable undertaking. However, as for the single qubit case, it is relatively straightforward to implement a numerical routine which will automate the calculation, provided experimental state tomography and state preparation procedures are available in the laboratory.

Box 8.5: Process tomography for a single qubit

The general method of process tomography can be simplified in the case of a one qubit operation to provide explicit formulas which may be useful in experimental contexts. This simplification is made possible by choosing the fixed operators $\tilde{E}_i$ to have commutation properties which conveniently allow the χ matrix to be determined by straightforward matrix multiplication. In the one qubit case, we use:

$$\tilde{E}_0 = I \tag{8.169}$$
$$\tilde{E}_1 = X \tag{8.170}$$
$$\tilde{E}_2 = -iY \tag{8.171}$$
$$\tilde{E}_3 = Z. \tag{8.172}$$

There are 12 parameters, specified by χ, which determine an arbitrary single qubit quantum operation $\mathcal{E}$.

These parameters may be measured using four sets of experiments. As a specific example, suppose the input states $|0\rangle$, $|1\rangle$, $|+\rangle = (|0\rangle + |1\rangle)/\sqrt{2}$ and $|-\rangle = (|0\rangle + i|1\rangle)/\sqrt{2}$ are prepared, and the four matrices

$$\rho_1' = \mathcal{E}(|0\rangle\langle 0|) \tag{8.173}$$
$$\rho_4' = \mathcal{E}(|1\rangle\langle 1|) \tag{8.174}$$
$$\rho_2' = \mathcal{E}(|+\rangle\langle +|) - i\mathcal{E}(|-\rangle\langle -|) - (1-i)(\rho_1' + \rho_4')/2 \tag{8.175}$$
$$\rho_3' = \mathcal{E}(|+\rangle\langle +|) + i\mathcal{E}(|-\rangle\langle -|) - (1+i)(\rho_1' + \rho_4')/2 \tag{8.176}$$

are determined using state tomography. These correspond to $\rho_j' = \mathcal{E}(\rho_j)$, where

$$\rho_1 = \begin{bmatrix} 1 & 0 \\ 0 & 0 \end{bmatrix}, \tag{8.177}$$

$\rho_2 = \rho_1 X$, $\rho_3 = X\rho_1$, and $\rho_4 = X\rho_1 X$. From (8.156) and Equations (8.169)–(8.172) we may determine β, and similarly ρ_j' determines λ. However, due to the particular choice of basis, and the Pauli matrix representation of $\tilde{E}_i$, we may express the β matrix as the Kronecker product $\beta = \Lambda \otimes \Lambda$, where

$$\Lambda = \frac{1}{2}\begin{bmatrix} I & X \\ X & -I \end{bmatrix}, \tag{8.178}$$

so that χ may be expressed conveniently as

$$\chi = \Lambda \begin{bmatrix} \rho_1' & \rho_2' \\ \rho_3' & \rho_4' \end{bmatrix} \Lambda, \tag{8.179}$$

in terms of block matrices.

We have shown how a useful representation for the dynamics of a quantum system may be experimentally determined using a systematic procedure. This procedure of quantum process tomography is analogous to the system identification step performed in classical control theory, and plays a similar role in understanding and controlling noisy quantum systems.

Exercise 8.32: Explain how to extend quantum process tomography to the case of non-trace-preserving quantum operations, such as arise in the study of measurement.

Exercise 8.33: (Specifying a quantum process) Suppose that one wished to completely specify an arbitrary single qubit operation $\mathcal{E}$ by describing how a set of points on the Bloch sphere $\{\vec{r}_k\}$ transform under $\mathcal{E}$. Prove that the set must contain at least four points.

Exercise 8.34: (Process tomography for two qubits) Show that the χ_2 describing the black box operations on two qubits can be expressed as

$$\chi_2 = \Lambda_2 \bar{\rho}' \Lambda_2 , \tag{8.180}$$

where $\Lambda_2 = \Lambda \otimes \Lambda$, Λ is as defined in Box 8.5, and $\bar{\rho}'$ is a block matrix of 16 measured density matrices,

$$\bar{\rho}' = P^T \begin{bmatrix} \rho'_{11} & \rho'_{12} & \rho'_{13} & \rho'_{14} \\ \rho'_{21} & \rho'_{22} & \rho'_{23} & \rho'_{24} \\ \rho'_{31} & \rho'_{32} & \rho'_{33} & \rho'_{34} \\ \rho'_{41} & \rho'_{42} & \rho'_{43} & \rho'_{44} \end{bmatrix} P , \tag{8.181}$$

where $\rho'_{nm} = \mathcal{E}(\rho_{nm})$, $\rho_{nm} = T_n |00\rangle\langle 00| T_m$, $T_1 = I \otimes I$, $T_2 = I \otimes X$, $T_3 = X \otimes I$, $T_4 = X \otimes X$, and $P = I \otimes [(\rho_{00} + \rho_{12} + \rho_{21} + \rho_{33}) \otimes I]$ is a permutation matrix.

Exercise 8.35: (Process tomography example) Consider a one qubit black box of unknown dynamics $\mathcal{E}_1$. Suppose that the following four density matrices are obtained from experimental measurements, performed according to Equations (8.173)–(8.176):

$$\rho'_1 = \begin{bmatrix} 1 & 0 \\ 0 & 0 \end{bmatrix} \tag{8.182}$$

$$\rho'_2 = \begin{bmatrix} 0 & \sqrt{1-\gamma} \\ 0 & 0 \end{bmatrix} \tag{8.183}$$

$$\rho'_3 = \begin{bmatrix} 0 & 0 \\ \sqrt{1-\gamma} & 0 \end{bmatrix} \tag{8.184}$$

$$\rho'_4 = \begin{bmatrix} \gamma & 0 \\ 0 & 1-\gamma \end{bmatrix} , \tag{8.185}$$

where γ is a numerical parameter. From an independent study of each of these input–output relations, one could make several important observations: the ground state $|0\rangle$ is left invariant by $\mathcal{E}_1$, the excited state $|1\rangle$ partially decays to the ground state, and superposition states are damped. Determine the χ matrix for this process.

8.5 Limitations of the quantum operations formalism

Are there interesting quantum systems whose dynamics are not described by quantum operations? In this section we will construct an artificial example of a system whose evolution is not described by a quantum operation, and try to understand the circumstances under which this is likely to occur.

Suppose a single qubit is prepared in some unknown quantum state, which we denote ρ. The preparation of this qubit involves certain procedures to be carried out in the laboratory in which the qubit is prepared. Suppose that among the laboratory degrees of freedom is a single qubit which, as a side effect of the state preparation procedure, is left in the state $|0\rangle$ if ρ is a state on the bottom half of the Bloch sphere, and is left in the state $|1\rangle$ if ρ is a state on the top half of the Bloch sphere. That is, the state of the system after preparation is

$$\rho \otimes |0\rangle\langle 0| \otimes \text{other degrees of freedom} \tag{8.186}$$

if ρ is a state on the bottom half of the Bloch sphere, and

$$\rho \otimes |1\rangle\langle 1| \otimes \text{other degrees of freedom} \tag{8.187}$$

if ρ is a state on the top half of the Bloch sphere.

Once the state preparation is done, the system begins to interact with the environment, in this case all the laboratory degrees of freedom. Suppose the interaction is such that a controlled-NOT is performed between the principal system and the extra qubit in the laboratory system. Thus, if the system's Bloch vector was initially in the bottom half of the Bloch sphere it is left invariant by the process, while if it was initially in the top half of the Bloch sphere it is rotated into the bottom half of the Bloch sphere.

Obviously, this process is not an affine map acting on the Bloch sphere, and therefore, by the results of Section 8.3.2, it *cannot be a quantum operation*. The lesson to be learned from this discussion is that *a quantum system which interacts with the degrees of freedom used to prepare that system after the preparation is complete will in general suffer a dynamics which is not adequately described within the quantum operations formalism*. This is an important conclusion to have reached, as it indicates that there are physically reasonable circumstances under which the quantum operations formalism may not adequately describe the processes taking place in a quantum system. This should be kept in mind, for example, in applications of the quantum process tomography procedure discussed in the previous section.

For the remainder of this book we will, however, work within the quantum operations formalism. It provides a powerful, and reasonably general tool for describing the dynamics experienced by quantum systems. Most of all, it provides a means by which concrete progress can be made on problems related to quantum information processing. It is an interesting problem for further research to study quantum information processing beyond the quantum operations formalism.

Problem 8.1: (Lindblad form to quantum operation) In the notation of Section 8.4.1, explicitly work through the steps to solve the differential equation

$$\dot{\rho} = -\frac{\lambda}{2}(\sigma_+\sigma_-\rho + \rho\sigma_+\sigma_- - 2\sigma_-\rho\sigma_+) \tag{8.188}$$

for $\rho(t)$. Express the map $\rho(0) \to \rho(t)$ as $\rho(t) = \sum_k E_k(t)\rho(0)E_k^\dagger(t)$.

Problem 8.2: (Teleportation as a quantum operation) Suppose Alice is in possession of a single qubit, denoted as system 1, which she wishes to teleport to

Bob. Unfortunately, she and Bob only share an imperfectly entangled pair of qubits. Alice's half of this pair is denoted system 2, and Bob's half is denoted system 3. Suppose Alice performs a measurement described by a set of quantum operations $\mathcal{E}_m$ with result m on systems 1 and 2. Show that this induces an operation $\tilde{\mathcal{E}}_m$ relating the initial state of system 1 to the final state of system 3, and that teleportation is accomplished if Bob can reverse this operation using a trace-preserving quantum operation $\mathcal{R}_m$, to obtain

$$\mathcal{R}_m \left(\frac{\tilde{\mathcal{E}}_m(\rho)}{\text{tr}[\tilde{\mathcal{E}}_m(\rho)]} \right) = \rho, \tag{8.189}$$

where ρ is the initial state of system 1.

Problem 8.3: (Random unitary channels) It is tempting to believe that all unital channels, that is, those for which $\mathcal{E}(I) = I$, result from averaging over random unitary operations, that is, $\mathcal{E}(\rho) = \sum_k p_k U_k \rho U_k^\dagger$, where U_k are unitary operators and the p_k form a probability distribution. Show that while this is true for single qubits, it is untrue for larger systems.

Summary of Chapter 8: Quantum noise and quantum operations

- **The operator-sum representation**: The behavior of an open quantum system can be modeled as

$$\mathcal{E}(\rho) = \sum_k E_k \rho E_k^\dagger, \qquad (8.190)$$

where E_k are operation elements, satisfying $\sum_k E_k^\dagger E_k = I$ if the quantum operation is trace-preserving.

- **Environmental models for quantum operations**: A trace-preserving quantum operation can always be regarded as arising from the unitary interaction of a system with an initially uncorrelated environment, and vice versa. Non-trace-preserving quantum operations may be treated similarly, except an additional projective measurement is performed on the composite of system and environment, with the different outcomes corresponding to different non-trace-preserving quantum operations.

- **Quantum process tomography**: A quantum operation on a d-dimensional quantum system can be completely determined by experimentally measuring the output density matrices produced from d^2 pure state inputs.

- **Operation elements for important single qubit quantum operations**:

depolarizing channel	$\sqrt{1 - \dfrac{3p}{4}} \begin{bmatrix} 1 & 0 \\ 0 & 1 \end{bmatrix},$ $\sqrt{\dfrac{p}{4}} \begin{bmatrix} 0 & -i \\ i & 0 \end{bmatrix},$	$\sqrt{\dfrac{p}{4}} \begin{bmatrix} 0 & 1 \\ 1 & 0 \end{bmatrix},$ $\sqrt{\dfrac{p}{4}} \begin{bmatrix} 1 & 0 \\ 0 & -1 \end{bmatrix}$
amplitude damping	$\begin{bmatrix} 1 & 0 \\ 0 & \sqrt{1-\gamma} \end{bmatrix},$	$\begin{bmatrix} 0 & \sqrt{\gamma} \\ 0 & 0 \end{bmatrix}$
phase damping	$\begin{bmatrix} 1 & 0 \\ 0 & \sqrt{1-\gamma} \end{bmatrix},$	$\begin{bmatrix} 0 & 0 \\ 0 & \sqrt{\gamma} \end{bmatrix}$
phase flip	$\sqrt{p} \begin{bmatrix} 1 & 0 \\ 0 & 1 \end{bmatrix},$	$\sqrt{1-p} \begin{bmatrix} 1 & 0 \\ 0 & -1 \end{bmatrix}$
bit flip	$\sqrt{p} \begin{bmatrix} 1 & 0 \\ 0 & 1 \end{bmatrix},$	$\sqrt{1-p} \begin{bmatrix} 0 & 1 \\ 1 & 0 \end{bmatrix}$
bit–phase flip	$\sqrt{p} \begin{bmatrix} 1 & 0 \\ 0 & 1 \end{bmatrix},$	$\sqrt{1-p} \begin{bmatrix} 0 & -i \\ i & 0 \end{bmatrix}$

History and further reading

Quantum noise is an important topic in several fields, and there is an enormous literature on the subject. We will necessarily be restricted to citing only a small sample of the resources available on the topic. An early treatise on quantum noise from a rather mathematical perspective is due to Davies[Dav76]. Caldeira and Leggett[CL83] did some of the first and most complete studies of an important model known as the *spin-boson* model, using an approach based upon the Feynman path integral. Gardiner[Gar91] studied quantum noise from the perspective of quantum optics. More recently, the quantum optics community has developed what is known as the *quantum trajectories* approach to quantum noise. Reviews of this subject may be found in the articles by Zoller and Gardiner[ZG97], and Plenio and Knight[PK98].

A large literature exists on the subject of quantum operations. We mention just a few key references, primarily the book by Kraus[Kra83], which contains references to much earlier work on the subject. Influential early papers on the subject include those by Hellwig and Kraus[HK69, HK70], and by Choi[Cho75]. Lindblad[Lin76] connected the quantum operations formalism to the theory of continuous time quantum evolution, introducing what is now known as the Lindblad form. Schumacher[Sch96b] and Caves[Cav99] have written excellent summaries of the quantum operations formalism from the point of view of quantum error-correction.

Quantum state tomography was suggested by Vogel and Risken[VR89]. Leonhardt[Leo97] has written a recent review containing references to other work. The need for quantum process tomography was pointed out in a paper by Turchette, Hood, Lange, Mabuchi, and Kimble[THL+95]. The theory was developed independently by Chuang and Nielsen[CN97], and by Poyatos, Cirac and Zoller[PCZ97]. Jones[Jon94] had earlier sketched out the main ideas of quantum process tomography.

An unfortunate confusion of terms has arisen with the word 'decoherence'. Historically, it has been used to refer just to a phase damping process, particularly by Zurek[Zur91]. Zurek and other researchers recognized that phase damping has a unique role in the transition from quantum to classical physics; for certain environmental couplings, it occurs on a time scale which is much faster than any amplitude damping process, and can therefore be much more important in determining the loss of quantum coherence. The major point of these studies has been this emergence of classicality due to environmental interactions. However, by and large, the usage of decoherence in quantum computation and quantum information is to refer to *any noise process* in quantum processing. In this book, we prefer the more generic term 'quantum noise' and tend towards its usage, although occasionally *decoherence* finds a proper place in the context.

A more detailed discussion of some of the limitations of the quantum operations formalism (and in particular, the assumption of a system and environment initially in a product state) is provided by Royer[Roy96].

Problem 8.2 is due to Nielsen and Caves[NC97]. Problem 8.3 is due to Landau and Streater[LS93] as part of an in-depth study of the extremal points of the convex set of doubly stochastic quantum operations.

9 Distance measures for quantum information

What does it mean to say that two items of information are similar? What does it mean to say that information is preserved by some process? These questions are central to a theory of quantum information processing, and the purpose of this chapter is the development of *distance measures* giving quantitative answers to these questions. Motivated by our two questions we will be concerned with two broad classes of distance measures, *static measures* and *dynamic measures*. Static measures quantify how close two quantum states are, while dynamic measures quantify how well information has been preserved during a dynamic process. The strategy we take is to begin by developing good static measures of distance, and then to use those static measures as the basis for the development of dynamic measures of distance.

There is a certain arbitrariness in the way distance measures are defined, both classically and quantum mechanically, and the community of people studying quantum computation and quantum information has found it convenient to use a variety of distance measures over the years. Two of those measures, the *trace distance* and the *fidelity*, have particularly wide currency today, and we discuss both these measures in detail in this chapter. For the most part the properties of both are quite similar, however for certain applications one may be easier to deal with than the other. It is for this reason and because both are widely used within the quantum computation and quantum information community that we discuss both measures.

9.1 Distance measures for classical information

> *The idea of distinguishing probability distributions is slippery business.*
> – Christopher Fuchs

Let's start out in an arena where we can easily apply our intuition – distance measures for classical information. What are the objects to be compared in classical information theory? We might consider comparing strings of bits like 00010 and 10011. One way of quantifying the distance between these is the *Hamming distance*, defined to be the number of places at which two bit strings are not equal. For example, the bit strings 00010 and 10011 differ in the first and last place, so the Hamming distance between them is two. Unfortunately, the Hamming distance between two objects is simply a matter of labeling, and *a priori* there aren't any labels in the Hilbert space arena of quantum mechanics!

A much better place to launch the study of distance measures for quantum information is with the comparison of classical probability distributions. In fact, in classical information theory an information source is usually modeled as a random variable, that is, as a probability distribution over some source alphabet. For example, an unknown source of English text may be modeled as a sequence of random variables over the Roman alphabet.

Before the text is read we can make a fair guess at the relative frequency of the letters that appear in the text, and certain correlations among them, such as the fact that occurrences of the pair of letters 'th' are much more common than the pair 'zx' in English text. This characterization of information sources as probability distributions over some alphabet encourages us to concentrate on the comparison of probability distributions in our search for measures of distance.

What does it mean to say that two probability distributions $\{p_x\}$ and $\{q_x\}$ over the same index set, x, are similar to one another? It is difficult to give an answer to this question which is obviously the unique 'correct' answer, so instead we propose two different answers, each of which is widely used by the quantum computation and quantum information community. The first measure is the *trace distance*, defined by the equation:

$$D(p_x, q_x) \equiv \frac{1}{2} \sum_x |p_x - q_x|. \tag{9.1}$$

This quantity is sometimes known as the L_1 *distance* or *Kolmogorov distance*. We prefer the term trace distance because it anticipates the later quantum mechanical analogue of this quantity, which is defined using the trace function. The trace distance turns out to be a metric on probability distributions, (a metric $D(x, y)$ must be symmetric, $D(x, y) = D(y, x)$, and satisfy the triangle inequality, $D(x, z) \leq D(x, y) + D(y, z)$) so the use of the term 'distance' is justified.

Exercise 9.1: What is the trace distance between the probability distribution $(1, 0)$ and the probability distribution $(1/2, 1/2)$? Between $(1/2, 1/3, 1/6)$ and $(3/4, 1/8, 1/8)$?

Exercise 9.2: Show that the trace distance between probability distributions $(p, 1 - p)$ and $(q, 1 - q)$ is $|p - q|$.

A second measure of distance between probability distributions, the *fidelity* of the probability distributions $\{p_x\}$ and $\{q_x\}$, is defined by

$$F(p_x, q_x) \equiv \sum_x \sqrt{p_x q_x}. \tag{9.2}$$

The fidelity is a very different way of measuring distance between probability distributions than is the trace distance. To begin with, it is not a metric, although later we discuss a metric derived from the fidelity. To see that the fidelity is not a metric note that when the distributions $\{p_x\}$ and $\{q_x\}$ are identical, $F(p_x, q_x) = \sum_x p_x = 1$. A better geometric understanding of the fidelity is illustrated in Figure 9.1; the fidelity is just the inner product between vectors with components $\sqrt{p_x}$ and $\sqrt{q_x}$, which lie on a unit sphere.

Exercise 9.3: What is the fidelity of the probability distributions $(1, 0)$ and $(1/2, 1/2)$? Of $(1/2, 1/3, 1/6)$ and $(3/4, 1/8, 1/8)$?

The trace distance and fidelity are mathematically useful means of defining the notion of a distance between two probability distributions. Do these measures have physically motivated operational meanings? In the case of the trace distance, the answer to this

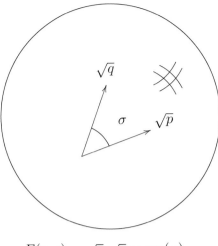

$$F(p,q) = \sqrt{p}.\sqrt{q} = \cos(\sigma)$$

Figure 9.1. Geometric interpretation of the fidelity as the inner product between vectors $\sqrt{p_x}$ and $\sqrt{q_x}$ lying on a unit sphere. (Since $1 = \sum_x (\sqrt{p_x})^2 = \sum_x (\sqrt{q_x})^2$.)

question is yes. In particular, it is simple to prove that

$$D(p_x, q_x) = \max_S |p(S) - q(S)| = \max_S \left| \sum_{x \in S} p_x - \sum_{x \in S} q_x \right|, \tag{9.3}$$

where the maximization is over all subsets S of the index set $\{x\}$. The quantity being maximized is the difference between the probability that the event S occurs, according to the distribution $\{p_x\}$, and the probability that the event S occurs, according to the distribution $\{q_x\}$. The event S is thus in some sense the optimal event to examine when trying to distinguish the distributions $\{p_x\}$ and $\{q_x\}$, with the trace distance governing how well it is possible to make this distinction.

Unfortunately, a similarly clear interpretation for the fidelity is not known. However, in the next section we show that the fidelity is a sufficiently useful quantity for mathematical purposes to justify its study, even without a clear physical interpretation. Moreover, we cannot rule out the possibility that a clear interpretation of the fidelity will be discovered in the future. Finally, it turns out that there are close connections between the fidelity and the trace distance, so properties of one quantity can often be used to deduce properties of the other, a fact which is useful surprisingly often.

Exercise 9.4: Prove (9.3).

Exercise 9.5: Show that the absolute value signs may be removed from Equation (9.3), that is,

$$D(p_x, q_x) = \max_S (p(S) - q(S)) = \max_S \left(\sum_{x \in S} p_x - \sum_{x \in S} q_x \right). \tag{9.4}$$

The trace distance and fidelity are static measures of distance for comparing two fixed probability distributions. There is a third notion of distance which is a *dynamic measure* of distance in the sense that it measures how well information is preserved by some physical process. Suppose a random variable X is sent through a noisy channel, giving as output another random variable Y, to form a Markov process $X \rightarrow Y$. For convenience we assume both X and Y have the same range of values, denoted by x. Then the probability that Y is not equal to X, $p(X \neq Y)$, is an obvious and important measure of the degree to which information has been preserved by the process.

Surprisingly, this dynamic measure of distance can be understood as a special case of the static trace distance! Imagine that the random variable X is given to you, and you make a copy of X, creating a new random variable $\tilde{X} = X$. The random variable X now passes through the noisy channel, leaving as output the random variable Y, as illustrated in Figure 9.2. How close is the initial perfectly correlated pair, $(\tilde{X}, X)$, to the final pair, $(\tilde{X}, Y)$? Using the trace distance as our measure of 'closeness', we obtain with some simple algebra,

$$D((\tilde{X}, X), (X, Y)) = \frac{1}{2} \sum_{xx'} |\delta_{xx'} p(X = x) - p(\tilde{X} = x, Y = x')| \tag{9.5}$$

$$= \frac{1}{2} \sum_{x \neq x'} p(\tilde{X} = x, Y = x') + \frac{1}{2} \sum_{x} |p(X = x) - p(\tilde{X} = x, Y = x)| \tag{9.6}$$

$$= \frac{1}{2} \sum_{x \neq x'} p(\tilde{X} = x, Y = x') + \frac{1}{2} \sum_{x} (p(X = x) - p(\tilde{X} = x, Y = x)) \tag{9.7}$$

$$= \frac{p(\tilde{X} \neq Y) + 1 - p(\tilde{X} = Y)}{2} \tag{9.8}$$

$$= \frac{p(X \neq Y) + p(\tilde{X} \neq Y)}{2} \tag{9.9}$$

$$= p(X \neq Y). \tag{9.10}$$

Thus, as illustrated in Figure 9.3, the probability of an error in the channel is equal to the trace distance between the probability distribution for $(\tilde{X}, X)$ and $(\tilde{X}, Y)$. This is an important construction, since it will be the basis for analogous quantum constructions. This is necessary because there is no *direct* quantum analogue of the probability $p(X \neq Y)$, since there is no notion in quantum mechanics analogous to the joint probability distribution for variables X and Y that *exist at different times*. Instead, to define dynamic measures of quantum distance we use an approach similar to the construction just given, based on the idea that it is quantum entanglement, rather than classical correlation, which is the important thing to preserve during a quantum channel's dynamics.

Figure 9.2. Given a Markov process $X \rightarrow Y$ we may first make a copy of X, $\tilde{X}$, before subjecting X to the noise which turns it into Y.

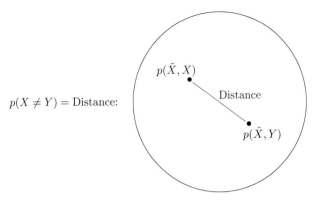

Figure 9.3. The probability of an error in the channel is equal to the trace distance between the probability distributions for $(\tilde{X}, X)$ and $(\tilde{X}, Y)$.

9.2 How close are two quantum states?

How close are two quantum states? Over the next few sections we describe quantum generalizations of the classical notions of trace distance and fidelity, and discuss in detail the properties of these quantities.

9.2.1 Trace distance

We begin by defining the *trace distance* between quantum states ρ and σ,

$$D(\rho, \sigma) \equiv \frac{1}{2}\mathrm{tr}|\rho - \sigma|. \tag{9.11}$$

where as per usual we define $|A| \equiv \sqrt{A^\dagger A}$ to be the positive square root of $A^\dagger A$. Notice that the quantum trace distance generalizes the classical trace distance in the sense that if ρ and σ commute then the (quantum) trace distance between ρ and σ is equal to the classical trace distance between the eigenvalues of ρ and σ. More explicitly, if ρ and σ commute they are diagonal in the same basis,

$$\rho = \sum_i r_i|i\rangle\langle i|; \quad \sigma = \sum_i s_i|i\rangle\langle i|, \tag{9.12}$$

for some orthonormal basis $|i\rangle$. Thus

$$D(\rho, \sigma) = \frac{1}{2}\mathrm{tr}\left|\sum_i (r_i - s_i)|i\rangle\langle i|\right| \tag{9.13}$$

$$= D(r_i, s_i). \tag{9.14}$$

Exercise 9.6: What is the trace distance between the density operators

$$\frac{3}{4}|0\rangle\langle 0| + \frac{1}{4}|1\rangle\langle 1|; \quad \frac{2}{3}|0\rangle\langle 0| + \frac{1}{3}|1\rangle\langle 1|? \tag{9.15}$$

Between:

$$\frac{3}{4}|0\rangle\langle 0| + \frac{1}{4}|1\rangle\langle 1|; \quad \frac{2}{3}|+\rangle\langle +| + \frac{1}{3}|-\rangle\langle -|? \tag{9.16}$$

(Recall that $|\pm\rangle \equiv (|0\rangle \pm |1\rangle)/\sqrt{2}$.)

A good way of getting a feel for the trace distance is to understand it for the special case of a qubit, in the Bloch sphere representation. Suppose ρ and σ have respective Bloch vectors $\vec{r}$ and $\vec{s}$,

$$\rho = \frac{I + \vec{r} \cdot \vec{\sigma}}{2} \; ; \quad \sigma = \frac{I + \vec{s} \cdot \vec{\sigma}}{2}. \tag{9.17}$$

(Recall that $\vec{\sigma}$ denotes a vector of Pauli matrices; it should not be confused with the state σ.) The trace distance between ρ and σ is easily calculated:

$$D(\rho, \sigma) = \frac{1}{2} \mathrm{tr} |\rho - \sigma| \tag{9.18}$$

$$= \frac{1}{4} \mathrm{tr} \left| (\vec{r} - \vec{s}) \cdot \vec{\sigma} \right|. \tag{9.19}$$

$(\vec{r} - \vec{s}) \cdot \vec{\sigma}$ has eigenvalues $\pm |\vec{r} - \vec{s}|$, so the trace of $|(\vec{r} - \vec{s}) \cdot \sigma|$ is $2|\vec{r} - \vec{s}|$, and we see that

$$D(\rho, \sigma) = \frac{|\vec{r} - \vec{s}|}{2}. \tag{9.20}$$

That is, the distance between two single qubit states is equal to one half the ordinary Euclidean distance between them on the Bloch sphere!

This intuitive geometric picture of the trace distance for qubits is often useful when trying to understand general properties of the trace distance. Conjectured properties can be suggested, refuted, or gain plausibility by looking at simple examples on the Bloch sphere. For example, rotations of the Bloch sphere leave the Euclidean distance invariant. This suggests that the trace distance might be preserved under unitary transformations in general,

$$D(U\rho U^\dagger, U\sigma U^\dagger) = D(\rho, \sigma), \tag{9.21}$$

a conjecture which you can easily verify with a moment's thought. We will come back to the Bloch sphere picture often in our investigation of distance measures.

To understand the properties of the trace distance a good starting point is to prove a formula for the trace distance generalizing Equation (9.3) for the classical trace distance:

$$D(\rho, \sigma) = \max_P \mathrm{tr}(P(\rho - \sigma)), \tag{9.22}$$

where the maximization may be taken alternately over all projectors, P, or over all positive operators $P \leq I$; the formula is valid in either case. This formula gives rise to an appealing interpretation of the trace distance. Recalling that POVM elements are positive operators $P \leq I$, the trace distance is therefore equal to the difference in probabilities that a measurement outcome with POVM element P may occur, depending on whether the state is ρ or σ, maximized over all possible POVM elements P.

We prove Equation (9.22) for the case where the maximization is over projectors; the case of positive operators $P \leq I$ follows the same reasoning. The proof is based on the fact that $\rho - \sigma$ can be expressed as $\rho - \sigma = Q - S$, where Q and S are *positive* operators with orthogonal support (see Exercise 9.7). It implies that $|\rho - \sigma| = Q + S$, so $D(\rho, \sigma) = (\mathrm{tr}(Q) + \mathrm{tr}(S))/2$. But $\mathrm{tr}(Q - S) = \mathrm{tr}(\rho - \sigma) = 0$, so $\mathrm{tr}(Q) = \mathrm{tr}(S)$, and therefore $D(\rho, \sigma) = \mathrm{tr}(Q)$. Let P be the projector onto the support of Q. Then $\mathrm{tr}(P(\rho - \sigma)) = \mathrm{tr}(P(Q - S)) = \mathrm{tr}(Q) = D(\rho, \sigma)$. Conversely, let P be any projector.

Then $\text{tr}(P(\rho - \sigma)) = \text{tr}(P(Q - S)) \leq \text{tr}(PQ) \leq \text{tr}(Q) = D(\rho, \sigma)$. This completes the proof.

Exercise 9.7: Show that for any states ρ and σ, one may write $\rho - \sigma = Q - S$, where Q and S are positive operators with support on orthogonal vector spaces. (*Hint:* use the spectral decomposition $\rho - \sigma = UDU^\dagger$, and split the diagonal matrix D into positive and negative parts. This fact will continue to be useful later.)

There is a closely related way of viewing the quantum trace distance which relates it more closely to the classical trace distance:

Theorem 9.1: Let $\{E_m\}$ be a POVM, with $p_m \equiv \text{tr}(\rho E_m)$ and $q_m \equiv \text{tr}(\sigma E_m)$ as the probabilities of obtaining a measurement outcome labeled by m. Then

$$D(\rho, \sigma) = \max_{\{E_m\}} D(p_m, q_m), \tag{9.23}$$

where the maximization is over all POVMs $\{E_m\}$.

Proof
Note that

$$D(p_m, q_m) = \frac{1}{2} \sum_m |\text{tr}(E_m(\rho - \sigma))|. \tag{9.24}$$

Using the spectral decomposition we may write $\rho - \sigma = Q - S$, where Q and S are positive operators with orthogonal support. Thus $|\rho - \sigma| = Q + S$, and

$$|\text{tr}(E_m(\rho - \sigma))| = |\text{tr}(E_m(Q - S))| \tag{9.25}$$
$$\leq \text{tr}(E_m(Q + S)) \tag{9.26}$$
$$\leq \text{tr}(E_m|\rho - \sigma|). \tag{9.27}$$

Thus

$$D(p_m, q_m) \leq \frac{1}{2} \sum_m \text{tr}(E_m|\rho - \sigma|) \tag{9.28}$$

$$= \frac{1}{2}\text{tr}(|\rho - \sigma|) \tag{9.29}$$
$$= D(\rho, \sigma), \tag{9.30}$$

where we have applied the completeness relation for POVM elements, $\sum_m E_m = I$.

Conversely, by choosing a measurement whose POVM elements include projectors onto the support of Q and S, we see that there exist measurements which give rise to probability distributions such that $D(p_m, q_m) = D(\rho, \sigma)$. $\qquad\square$

Thus, if two density operators are close in trace distance, then any measurement performed on those quantum states will give rise to probability distributions which are close together in the classical sense of trace distance, giving a second interpretation of the trace distance between two quantum states as an achievable upper bound on the trace distance between probability distributions arising from measurements performed on those quantum states.

We call the trace distance a 'distance', so we should check to see whether it has the property of being a metric on the space of density operators. From our geometric picture for a single qubit this is obviously true for a single qubit; is it true more generally? It is clear that $D(\rho, \sigma) = 0$ if and only if $\rho = \sigma$, and that $D(\cdot, \cdot)$ is a symmetric function of its inputs. All that remains to check is that the triangle inequality holds,

$$D(\rho, \tau) \leq D(\rho, \sigma) + D(\sigma, \tau). \tag{9.31}$$

To see this, note from Equation (9.22) that there exists a projector P such that

$$D(\rho, \tau) = \mathrm{tr}(P(\rho - \tau)) \tag{9.32}$$
$$= \mathrm{tr}(P(\rho - \sigma)) + \mathrm{tr}(P(\sigma - \tau)) \tag{9.33}$$
$$\leq D(\rho, \sigma) + D(\sigma, \tau), \tag{9.34}$$

establishing that the trace distance is a metric.

At this stage, we don't know a whole lot about the trace distance. However, we're in a good position to prove some genuinely spectacular results, useful in a wide variety of contexts. The most interesting result is that no physical process ever increases the distance between two quantum states, a result illustrated in Figure 9.4. We state this more formally as a theorem:

Theorem 9.2: (**Trace-preserving quantum operations are contractive**) Suppose $\mathcal{E}$ is a trace-preserving quantum operation. Let ρ and σ be density operators. Then

$$D(\mathcal{E}(\rho), \mathcal{E}(\sigma)) \leq D(\rho, \sigma). \tag{9.35}$$

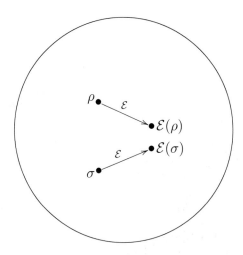

Figure 9.4. Trace-preserving quantum operations cause a *contraction* on the space of density operators.

Proof

Use the spectral decomposition to write $\rho - \sigma = Q - S$, where Q and S are positive matrices with orthogonal support, and let P be a projector such that $D(\mathcal{E}(\rho), \mathcal{E}(\sigma)) = \mathrm{tr}[P(\mathcal{E}(\rho) - \mathcal{E}(\sigma))]$. Observe that $\mathrm{tr}(Q) - \mathrm{tr}(S) = \mathrm{tr}(\rho) - \mathrm{tr}(\sigma) = 0$, so $\mathrm{tr}(Q) = \mathrm{tr}(S)$ and thus $\mathrm{tr}(\mathcal{E}(Q)) = \mathrm{tr}(\mathcal{E}(S))$. Using this observation we see that

$$D(\rho, \sigma) = \frac{1}{2}\mathrm{tr}|\rho - \sigma| \tag{9.36}$$

$$= \frac{1}{2}\mathrm{tr}|Q - S| \tag{9.37}$$

$$= \frac{1}{2}\mathrm{tr}(Q) + \frac{1}{2}\mathrm{tr}(S) \tag{9.38}$$

$$= \frac{1}{2}\mathrm{tr}(\mathcal{E}(Q)) + \frac{1}{2}\mathrm{tr}(\mathcal{E}(S)) \tag{9.39}$$

$$= \mathrm{tr}(\mathcal{E}(Q)) \tag{9.40}$$

$$\geq \mathrm{tr}(P\mathcal{E}(Q)) \tag{9.41}$$

$$\geq \mathrm{tr}(P(\mathcal{E}(Q) - \mathcal{E}(S))) \tag{9.42}$$

$$= \mathrm{tr}(P(\mathcal{E}(\rho) - \mathcal{E}(\sigma))) \tag{9.43}$$

$$= D(\mathcal{E}(\rho), \mathcal{E}(\sigma)), \tag{9.44}$$

which completes the proof.   □

There is an important special case of this result which can be understood by the following analogy. Imagine somebody shows you two different paintings in a gallery. Provided you have reasonably good vision, you shouldn't have any difficulty telling them apart. On the other hand, if somebody covers up most of the two paintings then you might have more difficulty telling the two apart, as illustrated in Figure 9.5. Similarly, if we 'cover up' parts of two quantum states then we can show that the distance between those two states is never increased. To prove this, recall from page 374 that the partial trace is a trace-preserving quantum operation. By Theorem 9.2, if we take quantum states ρ^{AB} and σ^{AB} of a composite quantum system AB then the distance between $\rho^A = \mathrm{tr}_B(\rho^{AB})$ and $\sigma^A = \mathrm{tr}_B(\sigma^{AB})$ is never more than the distance between ρ^{AB} and σ^{AB},

$$D(\rho^A, \sigma^A) \leq D(\rho^{AB}, \sigma^{AB}). \tag{9.45}$$

Figure 9.5. Objects become less distinguishable when only partial information is available.

In many applications we want to estimate the trace distance for a mixture of inputs. Such estimates are greatly aided by the following theorem:

Theorem 9.3: (**Strong convexity of the trace distance**) Let $\{p_i\}$ and $\{q_i\}$ be probability distributions over the same index set, and ρ_i and σ_i be density operators, also with indices from the same index set. Then

$$D\left(\sum_i p_i \rho_i, \sum_i q_i \sigma_i\right) \leq D(p_i, q_i) + \sum_i p_i D(\rho_i, \sigma_i), \tag{9.46}$$

where $D(p_i, q_i)$ is the classical trace distance between the probability distributions $\{p_i\}$ and $\{q_i\}$.

This result can be used to prove convexity results for the trace distance so we refer to this property as the *strong convexity* property for trace distance.

Proof

By Equation (9.22) there exists a projector P such that

$$D\left(\sum_i p_i \rho_i, \sum_i q_i \sigma_i\right) = \sum_i p_i \text{tr}(P\rho_i) - \sum_i q_i \text{tr}(P\sigma_i) \tag{9.47}$$

$$= \sum_i p_i \text{tr}(P(\rho_i - \sigma_i)) + \sum_i (p_i - q_i) \text{tr}(P\sigma_i) \tag{9.48}$$

$$\leq \sum_i p_i D(\rho_i, \sigma_i) + D(p_i, q_i), \tag{9.49}$$

where $D(p_i, q_i)$ is the trace distance between the probability distributions $\{p_i\}$ and $\{q_i\}$, and we used Equation (9.22) in the last line. □

As a special case of this result, we see that the trace distance is *jointly convex* in its inputs,

$$D\left(\sum_i p_i \rho_i, \sum_i p_i \sigma_i\right) \leq \sum_i p_i D(\rho_i, \sigma_i). \tag{9.50}$$

Exercise 9.8: (Convexity of the trace distance) Show that the trace distance is convex in its first input,

$$D\left(\sum_i p_i \rho_i, \sigma\right) \leq \sum_i p_i D(\rho_i, \sigma). \tag{9.51}$$

By symmetry convexity in the second entry follows from convexity in the first.

Exercise 9.9: (Existence of fixed points) *Schauder's fixed point theorem* is a classic result from mathematics that implies that any continuous map on a convex, compact subset of a Hilbert space has a fixed point. Use Schauder's fixed point theorem to prove that any trace-preserving quantum operation $\mathcal{E}$ has a fixed point, that is, ρ such that $\mathcal{E}(\rho) = \rho$.

Exercise 9.10: Suppose $\mathcal{E}$ is a *strictly contractive* trace-preserving quantum operation, that is, for any ρ and σ, $D(\mathcal{E}(\rho), \mathcal{E}(\sigma)) < D(\rho, \sigma)$. Show that $\mathcal{E}$ has a unique fixed point.

Exercise 9.11: Suppose $\mathcal{E}$ is a trace-preserving quantum operation for which there exists a density operator ρ_0 and a trace-preserving quantum operation $\mathcal{E}'$ such that

$$\mathcal{E}(\rho) = p\rho_0 + (1 - p)\mathcal{E}'(\rho), \tag{9.52}$$

for some p, $0 < p \leq 1$. Physically, this means that with probability p the input state is thrown out and replaced with the fixed state ρ_0, while with probability

$1 - p$ the operation $\mathcal{E}'$ occurs. Use joint convexity to show that $\mathcal{E}$ is a strictly contractive quantum operation, and thus has a unique fixed point.

Exercise 9.12: Consider the depolarizing channel introduced in Section 8.3.4 on page 378, $\mathcal{E}(\rho) = pI/2 + (1 - p)\rho$. For arbitrary ρ and σ find $D(\mathcal{E}(\rho), \mathcal{E}(\sigma))$ using the Bloch representation, and prove explicitly that the map $\mathcal{E}$ is strictly contractive, that is, $D(\mathcal{E}(\rho), \mathcal{E}(\sigma)) < D(\rho, \sigma)$.

Exercise 9.13: Show that the bit flip channel (Section 8.3.3) is contractive but not strictly contractive. Find the set of fixed points for the bit flip channel.

9.2.2 Fidelity

A second measure of distance between quantum states is the *fidelity*. The fidelity is not a metric on density operators, but we will see that it does give rise to a useful metric. This section reviews the definition and basic properties of the fidelity. The fidelity of states ρ and σ is defined to be

$$F(\rho, \sigma) \equiv \mathrm{tr}\sqrt{\rho^{1/2}\sigma\rho^{1/2}}. \tag{9.53}$$

It is certainly not immediately obvious that this is a useful measure of distance between ρ and σ. It doesn't even look symmetric! Yet we will see that the fidelity is symmetric in its inputs, and has many of the other properties we expect of a good distance measure.

There are two important special cases where it is possible to give more explicit formulae for the fidelity. The first is when ρ and σ commute, that is, are diagonal in the same basis,

$$\rho = \sum_i r_i |i\rangle\langle i|; \quad \sigma = \sum_i s_i |i\rangle\langle i|, \tag{9.54}$$

for some orthonormal basis $|i\rangle$. In this case we see that

$$F(\rho, \sigma) = \mathrm{tr}\sqrt{\sum_i r_i s_i |i\rangle\langle i|} \tag{9.55}$$

$$= \mathrm{tr}\left(\sum_i \sqrt{r_i s_i}|i\rangle\langle i|\right) \tag{9.56}$$

$$= \sum_i \sqrt{r_i s_i} \tag{9.57}$$

$$= F(r_i, s_i). \tag{9.58}$$

That is, when ρ and σ commute the quantum fidelity $F(\rho, \sigma)$ reduces to the classical fidelity $F(r_i, s_i)$ between the eigenvalue distributions r_i and s_i of ρ and σ.

Our second example is to calculate the fidelity between a pure state $|\psi\rangle$ and an arbitrary state, ρ. From Equation (9.53) we see that

$$F(|\psi\rangle, \rho) = \mathrm{tr}\sqrt{\langle\psi|\rho|\psi\rangle\,|\psi\rangle\langle\psi|} \tag{9.59}$$

$$= \sqrt{\langle\psi|\rho|\psi\rangle}. \tag{9.60}$$

That is, the fidelity is equal to the square root of the overlap between $|\psi\rangle$ and ρ. This is an important result which we will make use of often.

For the case of a qubit we were able to explicitly evaluate the trace distance between two states, and give it a simple geometric interpretation as half the Euclidean distance between points on the Bloch sphere. Unfortunately, no similarly clear geometric interpretation is known for the fidelity between two states of a qubit.

However, the fidelity does satisfy many of the same properties as the trace distance. For example, it is invariant under unitary transformations:

$$F(U\rho U^\dagger, U\sigma U^\dagger) = F(\rho, \sigma). \tag{9.61}$$

Exercise 9.14: (Invariance of fidelity under unitary transforms) Prove (9.61) by using the fact that for any positive operator A, $\sqrt{UAU^\dagger} = U\sqrt{A}U^\dagger$.

There is also a useful characterization of the fidelity analogous to the characterization (9.22) for the trace distance.

Theorem 9.4: **(Uhlmann's theorem)** Suppose ρ and σ are states of a quantum system Q. Introduce a second quantum system R which is a copy of Q. Then

$$F(\rho, \sigma) = \max_{|\psi\rangle, |\varphi\rangle} |\langle\psi|\varphi\rangle|, \tag{9.62}$$

where the maximization is over all purifications $|\psi\rangle$ of ρ and $|\varphi\rangle$ of σ into RQ.

Before giving the proof of Uhlmann's theorem we need an easily proved lemma.

Lemma 9.5: Let A be any operator, and U unitary. Then

$$|\text{tr}(AU)| \leq \text{tr}|A|, \tag{9.63}$$

with equality being attained by choosing $U = V^\dagger$, where $A = |A|V$ is the polar decomposition of A.

Proof

Equality is clearly attained under the conditions stated. Observe that

$$|\text{tr}(AU)| = |\text{tr}(|A|VU)| = \left|\text{tr}(|A|^{1/2}|A|^{1/2}VU)\right|. \tag{9.64}$$

The Cauchy–Schwarz inequality for the Hilbert–Schmidt inner product gives:

$$|\text{tr}(AU)| \leq \sqrt{\text{tr}|A|\ \text{tr}\left(U^\dagger V^\dagger|A|VU\right)} = \text{tr}|A|, \tag{9.65}$$

which completes the proof. $\square$

Proof

(Uhlmann's theorem)

Fix orthonormal bases $|i_R\rangle$ and $|i_Q\rangle$ in systems R and Q. Because R and Q are of the same dimension, the index i may be assumed to run over the same set of values. Define $|m\rangle \equiv \sum_i |i_R\rangle|i_Q\rangle$. Let $|\psi\rangle$ be any purification of ρ. Then the Schmidt decomposition and a moment's thought should convince you that

$$|\psi\rangle = (U_R \otimes \sqrt{\rho}U_Q)\,|m\rangle, \tag{9.66}$$

for some unitary operators U_R and U_Q on systems R and Q. Similarly, if $|\varphi\rangle$ is any purification of σ then there exist unitary operators V_R and V_Q such that

$$|\varphi\rangle = (V_R \otimes \sqrt{\sigma} V_Q)\,|m\rangle. \tag{9.67}$$

Taking the inner product gives

$$|\langle\psi|\varphi\rangle| = \left|\langle m|\left(U_R^\dagger V_R \otimes U_Q^\dagger \sqrt{\rho}\sqrt{\sigma} V_Q\right)|m\rangle\right|. \tag{9.68}$$

Using Exercise 9.16 on this page we see that

$$|\langle\psi|\varphi\rangle| = \left|\mathrm{tr}\left(V_R^\dagger U_R U_Q^\dagger \sqrt{\rho}\sqrt{\sigma} V_Q\right)\right|. \tag{9.69}$$

Setting $U \equiv V_Q V_R^\dagger U_R U_Q^\dagger$ we see that

$$|\langle\psi|\varphi\rangle| = \left|\mathrm{tr}\left(\sqrt{\rho}\sqrt{\sigma} U\right)\right|. \tag{9.70}$$

By Lemma 9.5,

$$|\langle\psi|\varphi\rangle| \le \mathrm{tr}\left|\sqrt{\rho}\sqrt{\sigma}\right| = \mathrm{tr}\sqrt{\rho^{1/2}\sigma\rho^{1/2}}. \tag{9.71}$$

To see that equality may be attained, suppose $\sqrt{\rho}\sqrt{\sigma} = |\sqrt{\rho}\sqrt{\sigma}|V$ is the polar decomposition of $\sqrt{\rho}\sqrt{\sigma}$. Choosing $U_Q = U_R = V_R = I$ and $V_Q = V^\dagger$ we see that equality is attained.   □

Exercise 9.15: Show that

$$F(\rho,\sigma) = \max_{|\varphi\rangle} |\langle\psi|\varphi\rangle|, \tag{9.72}$$

where $|\psi\rangle$ is any *fixed* purification of ρ, and the maximization is over all purifications of σ.

Exercise 9.16: (**The Hilbert–Schmidt inner product and entanglement**)
Suppose R and Q are two quantum systems with the same Hilbert space. Let $|i_R\rangle$ and $|i_Q\rangle$ be orthonormal basis sets for R and Q. Let A be an operator on R and B an operator on Q. Define $|m\rangle \equiv \sum_i |i_R\rangle|i_Q\rangle$. Show that

$$\mathrm{tr}(A^\dagger B) = \langle m|(A \otimes B)|m\rangle, \tag{9.73}$$

where the multiplication on the left hand side is of *matrices*, and it is understood that the matrix elements of A are taken with respect to the basis $|i_R\rangle$ and those for B with respect to the basis $|i_Q\rangle$.

Uhlmann's formula (9.62) does not provide a calculational tool for evaluating the fidelity, as does Equation (9.53). However, in many instances, properties of the fidelity are more easily proved using Uhlmann's formula than Equation (9.53). For example, Uhlmann's formula makes it clear that the fidelity is symmetric in its inputs, $F(\rho,\sigma) = F(\sigma,\rho)$, and that the fidelity is bounded between 0 and 1, $0 \le F(\rho,\sigma) \le 1$. If $\rho = \sigma$ then it is clear that $F(\rho,\sigma) = 1$, from Uhlmann's formula. If $\rho \ne \sigma$ then $|\psi\rangle \ne |\varphi\rangle$ for any purifications $|\psi\rangle$ and $|\varphi\rangle$ of ρ and σ, respectively, so $F(\rho,\sigma) < 1$. On the other hand, Equation (9.53) is sometimes useful as a means for understanding properties of the fidelity. For instance, we see that $F(\rho,\sigma) = 0$ if and only if ρ and σ have support on orthogonal subspaces. Intuitively, when ρ and σ are supported on orthogonal subspaces they are perfectly distinguishable, so we should expect the fidelity to be minimized at this point. Summarizing, the fidelity is symmetric in its inputs, $0 \leq F(\rho, \sigma) \leq 1$, with equality in the first inequality if and only if ρ and σ have orthogonal support, and equality in the second inequality if and only if $\rho = \sigma$.

We saw that the quantum trace distance could be related to the classical trace distance by considering the probability distributions induced by a measurement. In a similar way, we can show that

$$F(\rho, \sigma) = \min_{\{E_m\}} F(p_m, q_m), \tag{9.74}$$

where the minimum is over all POVMs $\{E_m\}$, and $p_m \equiv \mathrm{tr}(\rho E_m), q_m \equiv \mathrm{tr}(\sigma E_m)$ are the probability distributions for ρ and σ corresponding to the POVM $\{E_m\}$. To see that this is true, apply the polar decomposition $\sqrt{\rho^{1/2} \sigma \rho^{1/2}} = \sqrt{\rho}\sqrt{\sigma} U$, and note that

$$F(\rho, \sigma) = \mathrm{tr}(\sqrt{\rho}\sqrt{\sigma} U) \tag{9.75}$$

$$= \sum_m \mathrm{tr}(\sqrt{\rho}\sqrt{E_m}\sqrt{E_m}\sqrt{\sigma} U). \tag{9.76}$$

The Cauchy–Schwarz inequality and some simple algebra gives

$$F(\rho, \sigma) \leq \sum_m \sqrt{\mathrm{tr}(\rho E_m)\mathrm{tr}(\sigma E_m)} \tag{9.77}$$

$$= F(p_m, q_m), \tag{9.78}$$

which establishes that

$$F(\rho, \sigma) \leq \min_{\{E_m\}} F(p_m, q_m). \tag{9.79}$$

To see that equality can be attained in this inequality, we need to find a POVM $\{E_m\}$ such that the Cauchy–Schwarz inequality is satisfied with equality for each term in the sum, that is, $\sqrt{E_m}\sqrt{\rho} = \alpha_m\sqrt{E_m}\sqrt{\sigma} U$ for some set of complex numbers α_m. But $\sqrt{\rho}\sqrt{\sigma} U = \sqrt{\rho^{1/2}\sigma\rho^{1/2}}$, so for invertible ρ,

$$\sqrt{\sigma} U = \rho^{-1/2}\sqrt{\rho^{1/2}\sigma\rho^{1/2}}. \tag{9.80}$$

Substituting, we find that the equality conditions are that

$$\sqrt{E_m}(I - \alpha_m M) = 0, \tag{9.81}$$

where $M \equiv \rho^{-1/2}\sqrt{\rho^{1/2}\sigma\rho^{1/2}}\rho^{-1/2}$. If $M = \sum_m \beta_m |m\rangle\langle m|$ is a spectral decomposition for M then we choose $E_m = |m\rangle\langle m|$ and $\alpha_m = 1/\beta_m$. The case of non-invertible ρ follows from continuity.

We proved three important properties of the trace distance – the metric property, contractivity, and strong convexity. Rather remarkably, analogous properties all hold for the fidelity. What is more, the proof techniques used for fidelity differ considerably from those used for the trace distance. For that reason it's worth looking at these results in some detail.

The fidelity is not a metric; however, there is a simple way of turning the fidelity into a metric. The basic idea can be gleaned from Figure 9.6, by noticing that the *angle* between two points on the sphere is a metric. For the quantum case, Uhlmann's theorem tells us that the fidelity between two states is equal to the maximum inner product between purifications of those states. This suggests that we define the *angle* between states ρ and σ by

$$A(\rho, \sigma) \equiv \arccos F(\rho, \sigma). \tag{9.82}$$

Obviously the angle is non–negative, symmetric in its inputs, and is equal to zero if and only if $\rho = \sigma$. If we can show that the angle obeys the triangle inequality then we will have established that the angle is a metric.

We prove the triangle inequality using Uhlmann's theorem and some obvious facts about vectors in three dimensions. Let $|\varphi\rangle$ be a purification of σ, and choose purifications $|\psi\rangle$ of ρ and $|\gamma\rangle$ of τ such that

$$F(\rho, \sigma) = \langle \psi | \varphi \rangle \tag{9.83}$$
$$F(\sigma, \tau) = \langle \varphi | \gamma \rangle, \tag{9.84}$$

and $\langle \psi | \gamma \rangle$ is real and positive. (Note that this can always be done, by multiplying $|\psi\rangle$, $|\varphi\rangle$ and $|\gamma\rangle$ by appropriate phase factors, if necessary.) It is clear from Figure 9.6 that

$$\arccos(\langle \psi | \gamma \rangle) \le A(\rho, \sigma) + A(\sigma, \tau). \tag{9.85}$$

But by Uhlmann's theorem, $F(\rho, \tau) \ge \langle \psi | \gamma \rangle$, so $A(\rho, \tau) \le \arccos(\langle \psi | \gamma \rangle)$. Combining with the previous inequality gives the triangle inequality,

$$A(\rho, \tau) \le A(\rho, \sigma) + A(\sigma, \tau). \tag{9.86}$$

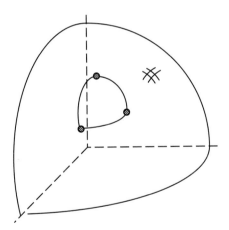

Figure 9.6. The angle between points on the surface of the unit sphere is a metric.

Exercise 9.17: Show that $0 \le A(\rho, \sigma) \le \pi/2$, with equality in the first inequality if and only if $\rho = \sigma$.

Qualitatively, the fidelity behaves like an 'upside-down' version of the trace distance, decreasing as two states become more distinguishable, and increasing as they become less distinguishable. Therefore, we should not expect the *contractivity* or *non-increasing* property of the trace distance to hold for fidelity. Instead, the analogous property of being *non-decreasing* does hold for fidelity. We will refer to this as the *monotonicity* of the fidelity under quantum operations.

Theorem 9.6: (**Monotonicity of the fidelity**) Suppose $\mathcal{E}$ is a trace-preserving quantum operation. Let ρ and σ be density operators. Show that

$$F(\mathcal{E}(\rho), \mathcal{E}(\sigma)) \geq F(\rho, \sigma). \tag{9.87}$$

Proof

Let $|\psi\rangle$ and $|\varphi\rangle$ be purifications of ρ and σ into a joint system RQ such that $F(\rho, \sigma) = |\langle\psi|\varphi\rangle|$. Introduce a model environment E for the quantum operation, $\mathcal{E}$, which starts in a pure state $|0\rangle$, and interacts with the quantum system Q via a unitary interaction U. Note that $U|\psi\rangle|0\rangle$ is a purification of $\mathcal{E}(\rho)$, and $U|\varphi\rangle|0\rangle$ is a purification of $\mathcal{E}(\sigma)$. By Uhlmann's theorem it follows that

$$F(\mathcal{E}(\rho), \mathcal{E}(\sigma)) \geq |\langle\psi|\langle0|U^{\dagger}U|\varphi\rangle|0\rangle| \tag{9.88}$$
$$= |\langle\psi|\varphi\rangle| \tag{9.89}$$
$$= F(\rho, \sigma), \tag{9.90}$$

establishing the property we set out to prove.   □

Exercise 9.18: (**Contractivity of the angle**) Let $\mathcal{E}$ be a trace-preserving quantum operation. Show that

$$A(\mathcal{E}(\rho), \mathcal{E}(\sigma)) \leq A(\rho, \sigma). \tag{9.91}$$

We finish off our study of the elementary properties of fidelity by proving a result for the fidelity analogous to the strong convexity of the trace distance, using Uhlmann's theorem.

Theorem 9.7: (**Strong concavity of the fidelity**) Let p_i and q_i be probability distributions over the same index set, and ρ_i and σ_i density operators also indexed by the same index set. Then

$$F\left(\sum_i p_i\rho_i, \sum_i q_i\sigma_i\right) \geq \sum_i \sqrt{p_iq_i}F(\rho_i, \sigma_i). \tag{9.92}$$

Not surprisingly, this result may be used to prove concavity results for the fidelity, for which reason we dub it the *strong concavity property* for the fidelity. This property is not strictly analogous to the strong convexity of the trace distance; however, the similarity in spirit leads us to use similar nomenclature.

Proof

Let $|\psi_i\rangle$ and $|\varphi_i\rangle$ be purifications of ρ_i and σ_i chosen such that $F(\rho_i, \sigma_i) = \langle\psi_i|\varphi_i\rangle$. Introduce an ancillary system which has orthonormal basis states $|i\rangle$ corresponding to the index set i for the probability distributions. Define

$$|\psi\rangle \equiv \sum_i \sqrt{p_i}|\psi_i\rangle|i\rangle; \quad |\varphi\rangle \equiv \sum_i \sqrt{q_i}|\varphi_i\rangle|i\rangle. \tag{9.93}$$

Note that $|\psi\rangle$ is a purification of $\sum_i p_i\rho_i$ and $|\varphi\rangle$ is a purification of $\sum_i q_i\sigma_i$, so by Uhlmann's formula,

$$F\left(\sum_i p_i\rho_i, \sum_i q_i\sigma_i\right) \geq |\langle\psi|\varphi\rangle| = \sum_i \sqrt{p_iq_i}\langle\psi_i|\varphi_i\rangle = \sum_i \sqrt{p_iq_i}F(\rho_i, \sigma_i), \tag{9.94}$$

which establishes the result we set out to prove.   □

Exercise 9.19: (Joint concavity of fidelity) Prove that the fidelity is *jointly concave*,

$$F\left(\sum_i p_i \rho_i, \sum_i p_i \sigma_i\right) \geq \sum_i p_i F(\rho_i, \sigma_i). \tag{9.95}$$

Exercise 9.20: (Concavity of fidelity) Prove that the fidelity is concave in the first entry,

$$F\left(\sum_i p_i \rho_i, \sigma\right) \geq \sum_i p_i F(\rho_i, \sigma). \tag{9.96}$$

By symmetry the fidelity is also concave in the second entry.

9.2.3 Relationships between distance measures

The trace distance and the fidelity are closely related, despite their very different forms. Qualitatively, they may be considered to be *equivalent* measures of distance for many applications. In this section we quantify more precisely the relationship between trace distance and fidelity.

In the case of pure states, the trace distance and the fidelity are completely equivalent to one another. To see this, consider the trace distance between two pure states, $|a\rangle$ and $|b\rangle$. Using the Gram–Schmidt procedure we may find orthonormal states $|0\rangle$ and $|1\rangle$ such that $|a\rangle = |0\rangle$ and $|b\rangle = \cos\theta|0\rangle + \sin\theta|1\rangle$. Note that $F(|a\rangle, |b\rangle) = |\cos\theta|$. Furthermore,

$$D(|a\rangle, |b\rangle) = \frac{1}{2}\mathrm{tr}\left|\begin{bmatrix} 1 - \cos^2\theta & -\cos\theta\sin\theta \\ -\cos\theta\sin\theta & -\sin^2\theta \end{bmatrix}\right| \tag{9.97}$$

$$= |\sin\theta| \tag{9.98}$$

$$= \sqrt{1 - F(|a\rangle, |b\rangle)^2} . \tag{9.99}$$

Thus the trace distance between two pure states is a function of the fidelity of those states, and vice versa. This relationship at the level of pure states can be used to deduce a relationship at the level of mixed states. Let ρ and σ be any two quantum states, and let $|\psi\rangle$ and $|\varphi\rangle$ be purifications chosen such that $F(\rho, \sigma) = |\langle\psi|\varphi\rangle| = F(|\psi\rangle, |\varphi\rangle)$. Recalling that trace distance is non-increasing under the partial trace, we see that

$$D(\rho, \sigma) \leq D(|\psi\rangle, |\varphi\rangle) \tag{9.100}$$

$$= \sqrt{1 - F(\rho, \sigma)^2}. \tag{9.101}$$

Thus, if the fidelity between two states is close to one, it follows that the states are also close in trace distance. The converse is also true. To see this, let $\{E_m\}$ be a POVM such that

$$F(\rho, \sigma) = \sum_m \sqrt{p_m q_m}, \tag{9.102}$$

where $p_m \equiv \mathrm{tr}(\rho E_m), q_m \equiv \mathrm{tr}(\sigma E_m)$ are the probabilities for obtaining outcome m for the states ρ and σ, respectively. Observe first that

$$\sum_m (\sqrt{p_m} - \sqrt{q_m})^2 = \sum_m p_m + \sum_m q_m - 2F(\rho, \sigma) \tag{9.103}$$

$$= 2(1 - F(\rho, \sigma)). \tag{9.104}$$

However, it is also true that $\left|\sqrt{p_m} - \sqrt{q_m}\right| \le \left|\sqrt{p_m} + \sqrt{q_m}\right|$, so

$$\sum_m (\sqrt{p_m} - \sqrt{q_m})^2 \le \sum_m \left|\sqrt{p_m} - \sqrt{q_m}\right| \left|\sqrt{p_m} + \sqrt{q_m}\right| \tag{9.105}$$

$$= \sum_m |p_m - q_m| \tag{9.106}$$

$$= 2D(p_m, q_m) \tag{9.107}$$

$$\le 2D(\rho, \sigma). \tag{9.108}$$

Comparing (9.104) and (9.108) we see that

$$1 - F(\rho, \sigma) \le D(\rho, \sigma). \tag{9.109}$$

Summarizing, we have

$$1 - F(\rho, \sigma) \le D(\rho, \sigma) \le \sqrt{1 - F(\rho, \sigma)^2}. \tag{9.110}$$

The implication is that the trace distance and the fidelity are qualitatively equivalent measures of closeness for quantum states. Indeed, for many purposes it does not matter whether the trace distance or the fidelity is used to quantify distance, since results about one may be used to deduce equivalent results about the other.

Exercise 9.21: When comparing pure states and mixed states it is possible to make a stronger statement than (9.110) about the relationship between trace distance and fidelity. Prove that

$$1 - F(|\psi\rangle, \sigma)^2 \le D(|\psi\rangle, \sigma). \tag{9.111}$$

9.3 How well does a quantum channel preserve information?

> *Friends come and go, but enemies accumulate*
> – Jones' Law, attributed to Thomas Jones

How well does a quantum channel preserve information? More precisely, suppose a quantum system is in the state $|\psi\rangle$ and some physical process occurs, changing the quantum system to the state $\mathcal{E}(|\psi\rangle\langle\psi|)$. How well has the channel $\mathcal{E}$ preserved the state $|\psi\rangle$ of the quantum system? The static measures of distance discussed in previous sections will be used in this section to develop measures of how well a quantum channel preserves information.

This type of scenario occurs often in quantum computation and quantum information. For example, in the memory of a quantum computer, $|\psi\rangle$ is the initial state of the memory, and $\mathcal{E}$ represents the dynamics that the memory undergoes, including noise processes arising from interaction with the environment. A second example is provided by a quantum communication channel for transmitting the state $|\psi\rangle$ from one location to another. No channel is ever perfect, so the action of the channel is described by a quantum operation $\mathcal{E}$.

An obvious way of quantifying how well the state $|\psi\rangle$ is preserved by the channel is to make use of the static distance measures introduced in the previous section. For example, we can compute the fidelity between the starting state $|\psi\rangle$ and the ending state $\mathcal{E}(|\psi\rangle\langle\psi|)$. For the case of the depolarizing channel, we obtain

$$F(|\psi\rangle, \mathcal{E}(|\psi\rangle\langle\psi|)) = \sqrt{\langle\psi| \left(p\frac{I}{2} + (1-p)|\psi\rangle\langle\psi|\right)|\psi\rangle} \tag{9.112}$$

$$= \sqrt{1 - \frac{p}{2}}. \tag{9.113}$$

This result agrees well with our intuition – the higher the probability p of depolarizing, the lower the fidelity of the final state with the initial state. Provided p is very small the fidelity is close to one, and the state $\mathcal{E}(\rho)$ is practically indistinguishable from the initial state $|\psi\rangle$.

There is nothing special about the use of the fidelity in the above expression. We could equally well have used the trace distance. However, for the remainder of this chapter we are going to restrict ourselves to measures of distance based upon the fidelity and derived quantities. Using the properties of the trace distance established in the last section it is not difficult, for the most part, to give a parallel development based upon the trace distance. However, it turns out that the fidelity is an easier tool to calculate with, and for that reason we restrict ourselves to considerations based upon the fidelity.

Our prototype measure for information preservation, the fidelity $F(|\psi\rangle, \mathcal{E}(|\psi\rangle\langle\psi|))$, has some drawbacks which need to be remedied. In a real quantum memory or quantum communications channel, we don't know in advance what the initial state $|\psi\rangle$ of the system will be. However, we can quantify the worst-case behavior of the system by *minimizing* over all possible initial states,

$$F_{\min}(\mathcal{E}) \equiv \min_{|\psi\rangle} F(|\psi\rangle, \mathcal{E}(|\psi\rangle\langle\psi|)). \tag{9.114}$$

For example, for the p-depolarizing channel $F_{\min} = \sqrt{1 - p/2}$, as the fidelity of the channel is the same for all input states $|\psi\rangle$. A more interesting example is the phase damping channel,

$$\mathcal{E}(\rho) = p\rho + (1-p)Z\rho Z. \tag{9.115}$$

For the phase damping channel the fidelity is given by

$$F(|\psi\rangle, \mathcal{E}(|\psi\rangle\langle\psi|)) = \sqrt{\langle\psi| \left(p|\psi\rangle\langle\psi| + (1-p)Z|\psi\rangle\langle\psi|Z\right)|\psi\rangle} \tag{9.116}$$

$$= \sqrt{p + (1-p)\langle\psi|Z|\psi\rangle^2}. \tag{9.117}$$

The second term under the square root sign is non-negative, and equal to zero when $|\psi\rangle = (|0\rangle + |1\rangle)/\sqrt{2}$. Thus for the phase damping channel the minimum fidelity is

$$F_{\min}(\mathcal{E}) = \sqrt{p}. \tag{9.118}$$

You might wonder why we minimized over *pure states* in the definition of $F_{\min}$. After all, might not the quantum system of interest start in a mixed state ρ? For example, a quantum memory might be entangled with the rest of the quantum computer, and therefore would start out in a mixed state. Fortunately, the joint concavity of the fidelity can be used to show that allowing mixed states does not change $F_{\min}$. To see this, suppose that $\rho = \sum_i \lambda_i |i\rangle\langle i|$ is the initial state of the quantum system. Then we have

$$F(\rho, \mathcal{E}(\rho)) = F\left(\sum_i \lambda_i |i\rangle\langle i|, \sum_i \lambda_i \mathcal{E}(|i\rangle\langle i|)\right) \tag{9.119}$$

$$\geq \sum_i \lambda_i F(|i\rangle, \mathcal{E}(|i\rangle\langle i|)). \tag{9.120}$$

It follows that

$$F(\rho, \mathcal{E}(\rho)) \geq F(|i\rangle, \mathcal{E}(|i\rangle\langle i|)) \tag{9.121}$$

for at least one of the states $|i\rangle$, and thus $F(\rho, \mathcal{E}(\rho)) \geq F_{\min}$.

Of course, we are interested not only in protecting quantum states as they are transmitted through a quantum communications channel, but also as they dynamically undergo computation. Suppose, for example, that as part of a quantum computation we attempt to implement a quantum gate described by the unitary operator U. As described in the last chapter, any such attempt will inevitably encounter some (hopefully not too severe) noise, so the correct description of the gate is using a trace-preserving quantum operation $\mathcal{E}$. A natural measure of how successful our gate has been is the *gate fidelity*,

$$F(U, \mathcal{E}) \equiv \min_{|\psi\rangle} F(U|\psi\rangle, \mathcal{E}(|\psi\rangle\langle\psi|)). \tag{9.122}$$

Suppose, for example, that we try to implement a NOT gate on a single qubit, but instead implement the noisy operation $\mathcal{E}(\rho) = (1 - p)X\rho X + pZ\rho Z$, for some small noise parameter p. Then the gate fidelity for this operation is given by

$$F(X, \mathcal{E}) = \min_{|\psi\rangle} \sqrt{\langle\psi|X\left((1-p)X|\psi\rangle\langle\psi|X + pZ|\psi\rangle\langle\psi|Z\right)X|\psi\rangle} \tag{9.123}$$

$$= \min_{|\psi\rangle} \sqrt{(1-p) + p\langle\psi|Y|\psi\rangle^2} \tag{9.124}$$

$$= \sqrt{(1-p)}. \tag{9.125}$$

In Exercise 9.22 you will show that performing a sequence of gates each with high fidelity is sufficient to ensure that the total operation has high fidelity, and thus for the purposes of quantum computation it is sufficient to perform each gate in the computation with high fidelity. (Compare also the similar but less general arguments of Chapter 4 on approximating quantum circuits.)

Exercise 9.22: (Chaining property for fidelity measures) Suppose U and V are unitary operators, and $\mathcal{E}$ and $\mathcal{F}$ are trace-preserving quantum operations meant to approximate U and V. Letting $d(\cdot, \cdot)$ be any metric on the space of density matrices satisfying $d(U\rho U^\dagger, U\sigma U^\dagger) = d(\rho, \sigma)$ for all density matrices ρ and σ and unitary U (such as the angle $\arccos(F(\rho, \sigma))$), define the corresponding *error* $E(U, \mathcal{E})$ by

$$E(U, \mathcal{E}) \equiv \max_\rho d(U\rho U^\dagger, \mathcal{E}(\rho)), \tag{9.126}$$

and show that $E(VU, \mathcal{F}\circ\mathcal{E}) \leq E(U, \mathcal{E}) + E(V, \mathcal{F})$. Thus, to perform a quantum computation with high fidelity it suffices to complete each step of the computation with high fidelity.

Quantum sources of information and the entanglement fidelity

We've been talking about dynamic measures of information preservation, without defining exactly what we mean by a quantum source of information. We'll now explain two possible definitions for this notion, and use these definitions to motivate the introduction of some dynamic measures of information preservation. *A priori*, it is not entirely clear how best to go about defining the notion of a quantum source of information. Classically, the best solution to this definitional problem is not at all obvious, and it is possible to come up with inequivalent definitions, each yielding a rich and useful theory of information. Since quantum information contains classical information as a subfield, it should not be surprising if there are even more ways of defining the notion of an information source quantum mechanically! To conclude this chapter we introduce two possible quantum definitions for the notion of an information source, explain how they motivate corresponding distance measures for the preservation of information, and prove some elementary properties of these distance measures. Further discussion of quantum sources of information is deferred until Chapter 12.

One attractive definition for quantum sources is to imagine a stream of identical quantum systems (say, qubits) being produced by some physical process, where the states of the respective systems are given by $\rho_{X_1}, \rho_{X_2}, \ldots$, the X_j are independent and identically distributed random variables, and ρ_j is some fixed set of density operators. For example, one might imagine a stream of qubits, each of which is prepared in the state $|0\rangle$ with probability one half, or in the state $(|0\rangle + |1\rangle)/\sqrt{2}$ with probability one half.

This *ensemble* notion of a quantum source leads naturally to a notion of *ensemble average fidelity* which captures the idea that the source is well-preserved under the action of a noisy channel described by a trace-preserving quantum operation $\mathcal{E}$, namely

$$\bar{F} = \sum_j p_j F(\rho_j, \mathcal{E}(\rho_j))^2, \tag{9.127}$$

where p_j are the respective probabilities for the different possible preparations of the system ρ_j. Obviously, $0 \leq \bar{F} \leq 1$, and provided $\bar{F} \approx 1$ we can be confident that, on average, the channel $\mathcal{E}$ preserves the states emitted by the source with a high degree of accuracy. You may wonder why the fidelity appearing on the right hand side of the definition is squared. There are two answers to this question, one simple, and one complex. The simple answer is that including this square term makes the ensemble fidelity more naturally related to the entanglement fidelity, as defined below. The more complex answer is that quantum information is, at present, in a state of infancy and it is not entirely clear what the 'correct' definitions for notions such as information preservation are! Nevertheless, as we shall see in Chapter 12, the ensemble average fidelity and the entanglement fidelity give rise to a rich theory of quantum information, which leads us to believe that these measures are on the right track, even though a complete theory of quantum information has not yet been developed.

Exercise 9.23: Show that $\bar{F} = 1$ if and only if $\mathcal{E}(\rho_j) = \rho_j$ for all j such that $p_j > 0$.

There is a second notion of a quantum source we may consider, motivated by the idea that a channel which preserves information well is a channel that preserves *entanglement* well. The basic idea comes from the discussion of the classical probability of error in Section 9.1. As noted there, a direct analogue of the probability of error $p(X \neq Y)$

cannot be defined for quantum processes, for there is no direct quantum analogue of a probability distribution defined at two times. Instead, we will use a quantum analogue of the idea illustrated in Figure 9.7, which is that a dynamic measure of distance can be defined by first copying the random variable X to $\tilde{X}$, then subjecting X to noise to obtain Y, and using as our measure of distance some metric quantity $D[(\tilde{X}, X), (\tilde{X}', Y)]$ between the joint distributions for $(\tilde{X}, X)$ and $(\tilde{X}', Y)$.

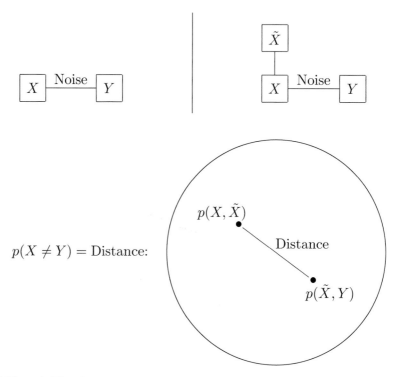

Figure 9.7. The probability of an error in the channel is equal to the trace distance between the probability distributions for $(\tilde{X}, X)$ and $(\tilde{X}, Y)$.

A quantum analogue of this model is as follows. A quantum system, Q, is prepared in a state ρ. The state of Q is assumed to be *entangled* in some way with the external world. This entanglement replaces the correlation between X and $\tilde{X}$ in the classical model. We represent the entanglement by introducing a fictitious system R, such that the joint state of RQ is a pure state. It turns out that all results that we prove do not depend in any way on how this purification is performed, so we may as well suppose that this is an arbitrary entanglement with the outside world. The system Q is then subjected to a dynamics described by a quantum operation, $\mathcal{E}$. The basic situation is illustrated in Figure 9.8.

How well is the entanglement between R and Q preserved by the quantum operation $\mathcal{E}$? We quantify this by the *entanglement fidelity* $F(\rho, \mathcal{E})$ which is a function of ρ and $\mathcal{E}$ defined for trace-preserving quantum operations $\mathcal{E}$ by

$$F(\rho, \mathcal{E}) \equiv F(RQ, R'Q')^2 \tag{9.128}$$

$$= \langle RQ| \left[(\mathcal{I}_R \otimes \mathcal{E})(|RQ\rangle\langle RQ|) \right] |RQ\rangle, \tag{9.129}$$

where the use of a prime indicates the state of a system after the quantum operation has been applied, and the absence of a prime indicates the state of a system before the

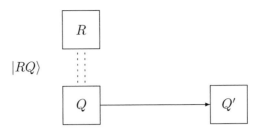

Figure 9.8. The RQ picture of a quantum channel. The initial state of RQ is a pure state.

quantum operation has been applied. The quantity appearing on the right hand side of this definition is the *square* of the static fidelity between the initial and final states of RQ. The use of the square of the static fidelity is purely a convenience, since it simplifies certain properties of the entanglement fidelity. Note that the entanglement fidelity depends only upon ρ and $\mathcal{E}$, and not (as it may appear) upon the details of the purification $|RQ\rangle$. To see this, we use the fact, proved in Exercise 2.81, that any two purifications $|R_1Q_1\rangle$ and $|R_2Q_2\rangle$ of ρ are related by a unitary operation, U, *that acts upon R alone*, $|R_2Q_2\rangle = U|R_1Q_1\rangle$. Thus

$$F(|R_2Q_2\rangle, \rho^{R_2'Q_2'}) = F(|R_1Q_1\rangle, \rho^{R_1'Q_2'}), \tag{9.130}$$

which establishes the result. The entanglement fidelity provides a measure of how well the entanglement between R and Q is preserved by the process $\mathcal{E}$, with values close to 1 indicating that the entanglement has been well preserved, and values close to 0 indicating that most of the entanglement has been destroyed. The choice of whether to use the static fidelity squared or the static fidelity is essentially arbitrary; the present definition results in slightly more attractive mathematical properties.

One of the attractive properties of the entanglement fidelity is that there is a very simple formula which enables it to be calculated exactly. Suppose E_i is a set of operation elements for a quantum operation $\mathcal{E}$. Then

$$F(\rho, \mathcal{E}) = \langle RQ|\rho^{R'Q'}|RQ\rangle = \sum_i |\langle RQ|E_i|RQ\rangle|^2. \tag{9.131}$$

Suppose we write $|RQ\rangle = \sum_j \sqrt{p_j}|j\rangle|j\rangle$, where $\rho = \sum_j p_j|j\rangle\langle j|$. Then

$$\langle RQ|E_i|RQ\rangle = \sum_{jk} \sqrt{p_jp_k}\langle j|k\rangle\langle j|E_i|k\rangle \tag{9.132}$$

$$= \sum_j p_j\langle j|E_i|j\rangle \tag{9.133}$$

$$= \mathrm{tr}(E_i\rho). \tag{9.134}$$

Substituting this expression into Equation (9.131) we obtain the useful computational formula

$$F(\rho, \mathcal{E}) = \sum_i |\mathrm{tr}(\rho E_i)|^2. \tag{9.135}$$

Thus, for example, the entanglement fidelity for the phase damping channel $\mathcal{E}(\rho) = p\rho + (1 - p)Z\rho Z$ is given by

$$F(\rho, \mathcal{E}) = p\left|\text{tr}(\rho)\right|^2 + (1 - p)\left|\text{tr}(\rho Z)\right|^2 = p + (1 - p)\text{tr}(\rho Z)^2, \qquad (9.136)$$

so we see that as p decreases, the entanglement fidelity decreases, as we intuitively expect.

We've defined two notions of a quantum information source and associated distance measures, one based on the idea that we want to preserve an ensemble of quantum states with high average fidelity, the other based on the idea that it is entanglement between the source and some reference system that we wish to preserve. Perhaps surprisingly, these two definitions turn out to be closely related! The reason for this lies in two useful properties of the entanglement fidelity. First, the entanglement fidelity is a lower bound on the square of the static fidelity between the input and output to the process,

$$F(\rho, \mathcal{E}) \leq \left[F(\rho, \mathcal{E}(\rho))\right]^2 . \qquad (9.137)$$

Intuitively, this result states that it is harder to preserve a state plus entanglement with the outside world than it is to merely preserve the state alone. The proof is an elementary application of the monotonicity of the static fidelity under partial trace, $F(\rho, \mathcal{E}) = F(|RQ\rangle, \rho^{R'Q'})^2 \leq F(\rho^Q, \rho^{Q'})^2$.

The second property of entanglement fidelity we need to relate it to the ensemble average definition is that it is a *convex* function of ρ. To see this, define $f(x) \equiv F(x\rho_1 + (1 - x)\rho_2, \mathcal{E})$, and using Equation (9.135) and elementary calculus we find

$$f''(x) = \sum_i \left|\text{tr}((\rho_1 - \rho_2)E_i)\right|^2 , \qquad (9.138)$$

so that $f''(x) \geq 0$, which implies the convexity of the entanglement fidelity, as required. Combining these two results we see that

$$F\left(\sum_j p_j\rho_j, \mathcal{E}\right) \leq \sum_j p_j F(\rho_j, \mathcal{E}) \qquad (9.139)$$

$$\leq \sum_j p_j F(\rho_j, \mathcal{E}(\rho_j))^2, \qquad (9.140)$$

and thus

$$F\left(\sum_j p_j\rho_j, \mathcal{E}\right) \leq \bar{F} \quad ! \qquad (9.141)$$

Thus, any quantum channel $\mathcal{E}$ which does a good job of preserving the entanglement between a source described by a density operator ρ and a reference system will automatically do a good job of preserving an ensemble source described by probabilities p_j and states ρ_j such that $\rho = \sum_j p_j\rho_j$. In this sense the notion of a quantum source based on entanglement fidelity is a more stringent notion than the ensemble definition, and for this reason we will prefer the entanglement fidelity based definition in our study of quantum information theory, in Chapter 12.

We conclude the chapter with a short list of some easily proved properties of the entanglement fidelity that will be useful in later chapters:

(1) $0 \leq F(\rho, \mathcal{E}) \leq 1$. Immediate from properties of the static fidelity.

(2) The entanglement fidelity is linear in the quantum operation input. This is immediate from the definition of the entanglement fidelity.

(3) For pure state inputs, the entanglement fidelity is equal to the static fidelity squared between input and output,

$$F(|\psi\rangle, \mathcal{E}) = F(|\psi\rangle, \mathcal{E}(|\psi\rangle\langle\psi|))^2. \tag{9.142}$$

This is immediate from the observation that the state $|\psi\rangle$ is a purification of itself, and the definition of the entanglement fidelity.

(4) $F(\rho, \mathcal{E}) = 1$ if and only if for all pure states $|\psi\rangle$ in the support of ρ,

$$\mathcal{E}(|\psi\rangle\langle\psi|) = |\psi\rangle\langle\psi|. \tag{9.143}$$

To prove this, suppose $F(\rho, \mathcal{E}) = 1$, and $|\psi\rangle$ is a pure state in the support of ρ. Define $p \equiv 1/\langle\psi|\rho^{-1}|\psi\rangle > 0$ (compare Exercise 2.73 on page 105) and σ to be a density operator such that $(1 - p)\sigma = \rho - p|\psi\rangle\langle\psi|$. Then by convexity,

$$1 = F(\rho, \mathcal{E}) \le p\sqrt{F(|\psi\rangle, \mathcal{E})} + (1 - p), \tag{9.144}$$

and thus $F(|\psi\rangle, \mathcal{E}) = 1$, establishing the result one way. The converse is a straightforward application of the definition of the entanglement fidelity.

(5) Suppose that $\langle\psi|\mathcal{E}(|\psi\rangle\langle\psi|)|\psi\rangle \ge 1 - \eta$ for all $|\psi\rangle$ in the support of ρ, for some η. Then $F(\rho, \mathcal{E}) \ge 1 - (3\eta/2)$. (See Problem 9.3.)

Problem 9.1: (Alternate characterization of the fidelity) Show that

$$F(\rho, \sigma) = \inf_P \mathrm{tr}(\rho P)\mathrm{tr}(\sigma P^{-1}), \tag{9.145}$$

where the infimum is taken over all invertible positive matrices P.

Problem 9.2: Let $\mathcal{E}$ be a trace-preserving quantum operation. Show that for each ρ there is a set of operation elements $\{E_i\}$ for $\mathcal{E}$ such that

$$F(\rho, \mathcal{E}) = |\mathrm{tr}(\rho E_1)|^2 . \tag{9.146}$$

Problem 9.3: Prove fact (5) on this page.

Summary of Chapter 9: Distance measures for quantum information

- **Trace distance:** $D(\rho, \sigma) \equiv \frac{1}{2}\mathrm{tr}|\rho - \sigma|$. Doubly convex metric on density operators, contractive under quantum operations.

- **Fidelity:**

$$F(\rho, \sigma) \equiv \mathrm{tr}\sqrt{\rho^{1/2}\sigma\rho^{1/2}} = \max_{|\psi\rangle, |\varphi\rangle} |\langle\psi|\varphi\rangle|.$$

Strongly concave, $F(\sum_i p_i\rho_i, \sum_i q_i\sigma_i) \ge \sum_i \sqrt{p_iq_i}F(\rho_i, \sigma_i)$.

- **Entanglement fidelity:** $F(\rho, \mathcal{E})$. Measure of how well entanglement is preserved during a quantum mechanical process, starting with the state ρ of a system Q, which is assumed to be entangled with another quantum system, R, and applying the quantum operation $\mathcal{E}$ to system Q.

History and further reading

Readers wishing to learn more about distance measures for quantum information would be well advised to start with Fuchs' 1996 Ph.D. Dissertation[Fuc96]. It contains a wealth of material on distance measures for quantum information, including a list of 528 references on related topics, organized into subject areas. Notably, the proof of Equation (9.74) may be found there, as well as much else of interest. Also see the article by Fuchs and van de Graaf[FvdG99]; this paper is the origin of the inequality (9.110), and is also a good overview of distance measures for quantum information, especially in the context of quantum cryptography. The contractivity of the trace distance was proved by Ruskai[Rus94]. The monotonicity of the fidelity was proved by Barnum, Caves, Fuchs, Jozsa and Schumacher[BCF+96]. In the literature both the quantity we call fidelity and its square have been referred to as the fidelity. Uhlmann's paper[Uhl76] in which he proves the eponymous theorem also contains an extensive discussion of the elementary properties of fidelity. The proof of Uhlmann's theorem given here is due to Jozsa[Joz94]. Chaining properties for fidelity measures and the relation to noisy quantum computation are discussed in more detail by Aharonov, Kitaev and Nisan[AKN98]. Schumacher[Sch96b] introduced the *entanglement fidelity* and proved many elementary properties. Knill and Laflamme[KL97] established the connection between subspace fidelity and entanglement fidelity, Problem 9.3. A more detailed proof of this fact appeared in Barnum, Knill and Nielsen[BKN98]. Problem 9.1 is due to Alberti[Alb83].

10 Quantum error-correction

We have learned that it is possible to fight entanglement with entanglement.
– John Preskill

To be an Error and to be Cast out is part of God's Design
– William Blake

This chapter explains how to do quantum information processing reliably in the presence of noise. The chapter covers three broad topics: the basic theory of *quantum error-correcting codes*, *fault-tolerant quantum computation*, and the *threshold theorem*. We begin by developing the basic theory of quantum error-correcting codes, which protect quantum information against noise. These codes work by *encoding* quantum states in a special way that make them resilient against the effects of noise, and then *decoding* when it is wished to recover the original state. Section 10.1 explains the basic ideas of classical error-correction, and some of the conceptual challenges that must be overcome to make quantum error-correction possible. Section 10.2 explains a simple example of a quantum error-correcting code, which we then generalize into a theory of quantum error-correcting codes in Section 10.3. Section 10.4 explains some ideas from the classical theory of linear codes, and how they give rise to an interesting class of quantum codes known as *Calderbank–Shor–Steane* (CSS) codes. Section 10.5 concludes our introductory survey of quantum error-correcting codes with a discussion of *stabilizer* codes, a richly structured class of codes with a close connection to classical error-correcting codes.

Our discussion of quantum error-correction assumes that encoding and decoding of quantum states can be done perfectly, without error. This is useful, for example, if we wish to send quantum states over a noisy communication channel and can use almost-noiseless quantum computers to perform pretty good encoding and decoding of the states at each end of the channel. However, this assumption cannot be made if the quantum gates used to do the encoding and decoding are themselves noisy. Fortunately, the theory of *fault-tolerant quantum computation*, developed in Section 10.6, allows us to remove the assumption of perfect encoding and decoding. Even more impressively, fault-tolerance allow us to perform logical operations on *encoded* quantum states, in a manner which tolerates faults in the underlying gate operations. The chapter culminates in Section 10.6.4 with the *threshold theorem* for quantum computation: *provided the noise in individual quantum gates is below a certain constant threshold it is possible to efficiently perform an arbitrarily large quantum computation.* Of course there are caveats to this result, which we spend some time discussing. Nevertheless, the threshold theorem is a remarkable result indicating that noise likely poses no fundamental barrier to the performance of large-scale quantum computations.

10.1 Introduction

Noise is a great bane of information processing systems. Whenever possible we build our systems to avoid noise completely, and where that is not possible, we try to protect against the effects of noise. For example, components in modern computers are extremely reliable, with a failure rate typically below one error in 10^{17} operations. For most practical purposes we can act as if computer components are completely noiseless. On the other hand, many systems in widespread use do suffer from a substantial noise problem. Modems and CD players both make use of error-correcting codes to protect against the effects of noise. The details of the techniques used to protect against noise in practice are sometimes rather complicated, but the basic principles are easily understood. The key idea is that if we wish to protect a message against the effects of noise, then we should *encode* the message by adding some redundant information to the message. That way, even if some of the information in the encoded message is corrupted by noise, there will be enough redundancy in the encoded message that it is possible to recover or *decode* the message so that all the information in the original message is recovered.

For example, suppose we wish to send a bit from one location to another through a noisy classical communications channel. The effect of the noise in the channel is to flip the bit being transmitted with probability $p > 0$, while with probability $1 - p$ the bit is transmitted without error. Such a channel is known as a *binary symmetric channel*, and is illustrated in Figure 10.1. A simple means of protecting the bit against the effects of noise in the binary symmetric channel is to replace the bit we wish to protect with three copies of itself:

$$0 \to 000 \tag{10.1}$$

$$1 \to 111. \tag{10.2}$$

The bit strings 000 and 111 are sometimes referred to as the *logical* 0 and *logical* 1, since they play the role of 0 and 1, respectively. We now send all three bits through the channel. At the receiver's end of the channel three bits are output, and the receiver has to decide what the value of the original bit was. Suppose 001 were output from the channel. Provided the probability p of a bit flip is not too high, it is very likely that the third bit was flipped by the channel, and that 0 was the bit that was sent.

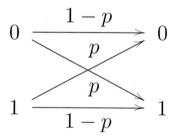

Figure 10.1. Binary symmetric channel.

This type of decoding is called *majority voting*, since the decoded output from the channel is whatever value, 0 or 1, appears more times in the actual channel output. Majority voting fails if two or more of the bits sent through the channel were flipped, and succeeds otherwise. The probability that two or more of the bits are flipped is

$3p^2(1-p) + p^3$, so the probability of error is $p_e = 3p^2 - 2p^3$. Without encoding, the probability of an error was p, so the code makes the transmission more reliable provided $p_e < p$, which occurs whenever $p < 1/2$.

The type of code just described is called a *repetition code*, since we encode the message to be sent by repeating it a number of times. A similar technique has been used for millennia as a part of everyday conversation: if we're having difficulty understanding someone's spoken language, perhaps because they have a foreign accent, we ask them to repeat what they're saying. We may not catch all the words either time, but we can put the iterations together to comprehend a coherent message. Many clever techniques have been developed in the theory of classical error-correcting codes; however, the key idea is always to encode messages by adding enough redundancy that the original message is recoverable after noise has acted on the encoded message, with the amount of redundancy needing to be added depending on the severity of noise in the channel.

10.1.1 The three qubit bit flip code

To protect quantum states against the effects of noise we would like to develop *quantum error-correcting codes* based upon similar principles. There are some important differences between classical information and quantum information that require new ideas to be introduced to make such quantum error-correcting codes possible. In particular, at a first glance we have three rather formidable difficulties to deal with:

- *No cloning*: One might try to implement the repetition code quantum mechanically by duplicating the quantum state three or more times. This is forbidden by the no-cloning theorem discussed in Box 12.1 on page 532. Even if cloning were possible, it would not be possible to measure and compare the three quantum states output from the channel.
- *Errors are continuous*: A continuum of different errors may occur on a single qubit. Determining which error occurred in order to correct it would appear to require infinite precision, and therefore infinite resources.
- *Measurement destroys quantum information*: In classical error-correction we observe the output from the channel, and decide what decoding procedure to adopt. Observation in quantum mechanics generally destroys the quantum state under observation, and makes recovery impossible.

Fortunately, none of these problems is fatal, as we shall demonstrate. Suppose we send qubits through a channel which leaves the qubits untouched with probability $1 - p$, and flips the qubits with probability p. That is, with probability p the state $|\psi\rangle$ is taken to the state $X|\psi\rangle$, where X is the usual Pauli sigma x operator, or *bit flip operator*. This channel is called the *bit flip channel*, and we now explain the *bit flip code*, which may be used to protect qubits against the effects of noise from this channel.

Suppose we encode the single qubit state $a|0\rangle + b|1\rangle$ in three qubits as $a|000\rangle + b|111\rangle$. A convenient way to write this encoding is

$$|0\rangle \rightarrow |0_L\rangle \equiv |000\rangle \tag{10.3}$$

$$|1\rangle \rightarrow |1_L\rangle \equiv |111\rangle, \tag{10.4}$$

where it is understood that superpositions of basis states are taken to corresponding superpositions of encoded states. The notation $|0_L\rangle$ and $|1_L\rangle$ indicates that these are the *logical* $|0\rangle$ and *logical* $|1\rangle$ states, not the *physical* zero and one states. A circuit performing this encoding is illustrated in Figure 10.2.

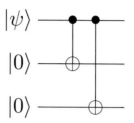

Figure 10.2. Encoding circuit for the three qubit bit flip code. The data to be encoded enters the circuit on the top line.

Exercise 10.1: Verify that the encoding circuit in Figure 10.2 works as claimed.

Suppose the initial state $a|0\rangle + b|1\rangle$ has been perfectly encoded as $a|000\rangle + b|111\rangle$. Each of the three qubits is passed through an independent copy of the bit flip channel. Suppose a bit flip occurred on one or fewer of the qubits. There is a simple two stage *error-correction* procedure which can be used to recover the correct quantum state in this case:

(1) *Error-detection* or *syndrome diagnosis*: We perform a measurement which tells us what error, if any, occurred on the quantum state. The measurement result is called the *error syndrome*. For the bit flip channel there are four error syndromes, corresponding to the four projection operators:

$$P_0 \equiv |000\rangle\langle 000| + |111\rangle\langle 111| \text{ no error} \tag{10.5}$$
$$P_1 \equiv |100\rangle\langle 100| + |011\rangle\langle 011| \text{ bit flip on qubit one} \tag{10.6}$$
$$P_2 \equiv |010\rangle\langle 010| + |101\rangle\langle 101| \text{ bit flip on qubit two} \tag{10.7}$$
$$P_3 \equiv |001\rangle\langle 001| + |110\rangle\langle 110| \text{ bit flip on qubit three.} \tag{10.8}$$

Suppose for example that a bit flip occurs on qubit one, so the corrupted state is $a|100\rangle + b|011\rangle$. Notice that $\langle\psi|P_1|\psi\rangle = 1$ in this case, so the outcome of the measurement result (the error syndrome) is certainly 1. Furthermore, the syndrome measurement does not cause any change to the state: it is $a|100\rangle + b|011\rangle$ both before and after syndrome measurement. Note that the syndrome contains only information about what error has occurred, and does not allow us to infer anything about the value of a or b, that is, it contains no information about the state being protected. This is a generic feature of syndrome measurements, since to obtain information about the identity of a quantum state it is in general necessary to perturb that state.

(2) *Recovery*: We use the value of the error syndrome to tell us what procedure to use to recover the initial state. For example, if the error syndrome was 1, indicating a bit flip on the first qubit, then we flip that qubit again, recovering the original state $a|000\rangle + b|111\rangle$ with perfect accuracy. The four possible error syndromes and the recovery procedure in each case are: 0 (no error) – do nothing; 1 (bit flip on first qubit) – flip the first qubit again; 2 (bit flip on second qubit) – flip the second qubit

again; 3 (bit flip on third qubit) – flip the third qubit again. For each value of the error syndrome it is easy to see that the original state is recovered with perfect accuracy, given that the corresponding error occurred.

This error-correction procedure works perfectly, provided bit flips occur on one or fewer of the three qubits. This occurs with probability $(1 - p)^3 + 3p(1 - p)^2 = 1 - 3p^2 + 2p^3$. The probability of an error remaining uncorrected is therefore $3p^2 - 2p^3$, just as for the classical repetition code we studied earlier. Once again, provided $p < 1/2$ the encoding and decoding improve the reliability of storage of the quantum state.

Improving the error analysis

This error analysis is not completely adequate. The problem is that not all errors and states in quantum mechanics are created equal: quantum states live in a continuous space, so it is possible for some errors to corrupt a state by a tiny amount, while others mess it up completely. An extreme example is provided by the bit flip 'error' X, which does not affect the state $(|0\rangle + |1\rangle)/\sqrt{2}$ at all, but flips the $|0\rangle$ state so it becomes a $|1\rangle$. In the former case we would not be worried about a bit flip error occurring, while in the latter case we would obviously be very worried.

To address this problem we make use of the *fidelity* quantity introduced in Chapter 9. Recall that the fidelity between a pure and a mixed state is given by $F(|\psi\rangle, \rho) = \sqrt{\langle\psi|\rho|\psi\rangle}$. The object of quantum error-correction is to increase the fidelity with which quantum information is stored (or communicated) up near the maximum possible fidelity of one. Let's compare the *minimum* fidelity achieved by the three qubit bit flip code with the fidelity when no error-correction is performed. Suppose the quantum state of interest is $|\psi\rangle$. Without using the error-correcting code the state of the qubit after being sent through the channel is

$$\rho = (1 - p)|\psi\rangle\langle\psi| + pX|\psi\rangle\langle\psi|X. \tag{10.9}$$

The fidelity is given by

$$F = \sqrt{\langle\psi|\rho|\psi\rangle} = \sqrt{(1 - p) + p\langle\psi|X|\psi\rangle\langle\psi|X|\psi\rangle}. \tag{10.10}$$

The second term under the square root is non-negative, and equal to zero when $|\psi\rangle = |0\rangle$, so we see that the minimum fidelity is $F = \sqrt{1 - p}$. Suppose the three qubit error-correcting code is used to protect the state $|\psi\rangle = a|0_L\rangle + b|1_L\rangle$. The quantum state after both the noise and error-correction is:

$$\rho = \left[(1 - p)^3 + 3p(1 - p)^2\right]|\psi\rangle\langle\psi| + \cdots. \tag{10.11}$$

The omitted terms represent contributions from bit flips on two or three qubits. All the omitted terms are positive operators, so the fidelity we calculate will be a *lower bound* on the true fidelity. We see that $F = \sqrt{\langle\psi|\rho|\psi\rangle} \geq \sqrt{(1 - p)^3 + 3p(1 - p)^2}$. That is, the fidelity is at least $\sqrt{1 - 3p^2 + 2p^3}$, so the fidelity of storage for the quantum state is improved provided $p < 1/2$, which is the same conclusion we came to earlier based on a much cruder analysis.

Exercise 10.2: The action of the bit flip channel can be described by the quantum operation $\mathcal{E}(\rho) = (1 - p)\rho + pX\rho X$. Show that this may be given an alternate operator-sum representation, as $\mathcal{E}(\rho) = (1 - 2p)\rho + 2pP_+\rho P_+ + 2pP_-\rho P_-$ where

P_+ and P_- are projectors onto the $+1$ and -1 eigenstates of X, $(|0\rangle + |1\rangle)/\sqrt{2}$ and $(|0\rangle - |1\rangle)/\sqrt{2}$, respectively. This latter representation can be understood as a model in which the qubit is left alone with probability $1 - 2p$, and is 'measured' by the environment in the $|+\rangle, |-\rangle$ basis with probability $2p$.

There is a different way of understanding the syndrome measurement that is useful in generalizing the three qubit code. Suppose that instead of measuring the four projectors P_0, P_1, P_2, P_3 we performed two measurements, the first of the observable Z_1Z_2 (that is, $Z \otimes Z \otimes I$), and the second of the observable Z_2Z_3. Each of these observables has eigenvalues ± 1, so each measurement provides a single bit of information, for a total of two bits of information – four possible syndromes, just as in the earlier description. The first measurement, of Z_1Z_2, can be thought of as *comparing* the first and second qubits to see if they are the same. To see why this is so, note that Z_1Z_2 has spectral decomposition

$$Z_1Z_2 = (|00\rangle\langle00| + |11\rangle\langle11|) \otimes I - (|01\rangle\langle01| + |10\rangle\langle10|) \otimes I, \tag{10.12}$$

which corresponds to a projective measurement with projectors $(|00\rangle\langle00| + |11\rangle\langle11|) \otimes I$ and $(|01\rangle\langle01| + |10\rangle\langle10|) \otimes I$. Thus, measuring Z_1Z_2 can be thought of as comparing the values of the first and second qubits, giving $+1$ if they are the same, and -1 if they are different. Similarly, measuring Z_2Z_3 compares the values of the second and third qubits, giving $+1$ if they are the same, and -1 if they are different. Combining these two measurement results we can determine whether a bit flip occurred on one of the qubits or not, and if so, which one: if both measurement results give $+1$ then with high probability no bit flip has occurred; if measuring Z_1Z_2 gives $+1$ and measuring Z_2Z_3 gives -1 then with high probability just the third qubit flipped; if measuring Z_1Z_2 gives -1 and measuring Z_2Z_3 gives $+1$ then with high probability just the first qubit flipped; and finally if both measurements give -1 then with high probability just the second qubit flipped. What is crucial to the success of these measurements is that neither measurement gives any information about the amplitudes a and b of the encoded quantum state, and thus neither measurement destroys the superpositions of quantum states that we wish to preserve using the code.

Exercise 10.3: Show by explicit calculation that measuring Z_1Z_2 followed by Z_2Z_3 is equivalent, up to labeling of the measurement outcomes, to measuring the four projectors defined by (10.5)–(10.8), in the sense that both procedures result in the same measurement statistics and post-measurement states.

10.1.2 Three qubit phase flip code

The bit flip code is interesting, but it does not appear to be that significant an innovation over classical error-correcting codes, and leaves many problems open (for example, many kinds of errors other than bit flips can happen to qubits). A more interesting noisy quantum channel is the *phase flip* error model for a single qubit. In this error model the qubit is left alone with probability $1 - p$, and with probability p the relative phase of the $|0\rangle$ and $|1\rangle$ states is flipped. More precisely, the phase flip operator Z is applied to the qubit with probability $p > 0$, so the state $a|0\rangle + b|1\rangle$ is taken to the state $a|0\rangle - b|1\rangle$ under the phase flip. There is no classical equivalent to the phase flip channel, since classical channels don't have any property equivalent to phase. However, there is an easy way to turn the phase flip channel into a bit flip channel. Suppose we work in the qubit basis

$|+\rangle \equiv (|0\rangle + |1\rangle)/\sqrt{2}$, $|-\rangle \equiv (|0\rangle - |1\rangle)/\sqrt{2}$. With respect to this basis the operator Z takes $|+\rangle$ to $|-\rangle$ and vice versa, that is, it acts just like a bit flip with respect to the labels $+$ and $-$! This suggests using the states $|0_L\rangle \equiv |+ + +\rangle$ and $|1_L\rangle \equiv |- - -\rangle$ as logical zero and one states for protection against phase flip errors. All the operations needed for error-correction – encoding, error-detection, and recovery – are performed just as for the bit flip channel, but with respect to the $|+\rangle, |-\rangle$ basis instead of the $|0\rangle, |1\rangle$ basis. To accomplish this basis change we simply apply the Hadamard gate and its inverse (also the Hadamard gate) at appropriate points in the procedure, since the Hadamard gate accomplishes the change back and forth between the $|0\rangle, |1\rangle$ basis and the $|+\rangle, |-\rangle$ basis.

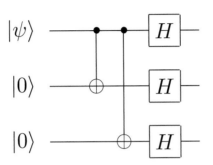

Figure 10.3. Encoding circuits for the phase flip code.

More explicitly, the encoding for the phase flip channel is performed in two steps: first, we encode in three qubits exactly as for the bit flip channel; second, we apply a Hadamard gate to each qubit (Figure 10.3). Error-detection is achieved by applying the same projective measurements as before, but conjugated by Hadamard gates: $P_j \rightarrow P'_j \equiv H^{\otimes 3} P_j H^{\otimes 3}$. Equivalently, syndrome measurement may be performed by measuring the observables $H^{\otimes 3} Z_1 Z_2 H^{\otimes 3} = X_1 X_2$ and $H^{\otimes 3} Z_2 Z_3 H^{\otimes 3} = X_2 X_3$. It is interesting to interpret these measurements along similar lines to the measurement of $Z_1 Z_2$ and $Z_2 Z_3$ for the bit flip code. Measurement of the observables $X_1 X_2$ and $X_2 X_3$ corresponds to comparing the *sign* of qubits one and two, and two and three, respectively, in the sense that measurement of $X_1 X_2$, for example, gives $+1$ for states like $|+\rangle|+\rangle \otimes (\cdot)$ or $|-\rangle|-\rangle \otimes (\cdot)$, and -1 for states like $|+\rangle|-\rangle \otimes (\cdot)$ or $|-\rangle|+\rangle \otimes (\cdot)$. Finally, error-correction is completed with the recovery operation, which is the Hadamard-conjugated recovery operation from the bit flip code. For example, suppose we detected a flip in the sign of the first qubit from $|+\rangle$ to $|-\rangle$. Then we recover by applying $H X_1 H = Z_1$ to the first qubit. Similar procedures apply for other error syndromes.

Obviously this code for the phase flip channel has the same characteristics as the code for the bit flip channel. In particular, the minimum fidelity for the phase flip code is the same as that for the bit flip code, and we have the same criteria for the code producing an improvement over the case with no error-correction. We say that these two channels are *unitarily equivalent*, since there is a unitary operator U (in this case the Hadamard gate) such that the action of one channel is the same as the other, provided the first channel is preceded by U and followed by $U^\dagger$. These operations may be trivially incorporated into the encoding and error-correction operations. For general unitary operators these ideas are worked out in Problem 10.1 on page 495.

Exercise 10.4: Consider the three qubit bit flip code. Suppose we had performed the error syndrome measurement by measuring the eight orthogonal projectors corresponding to projections onto the eight computational basis states.

(1) Write out the projectors corresponding to this measurement, and explain how the measurement result can be used to diagnose the error syndrome: either *no bits flipped* or *bit number j flipped*, where j is in the range one to three.
(2) Show that the recovery procedure works only for computational basis states.
(3) What is the minimum fidelity for the error-correction procedure?

10.2 The Shor code

There is a simple quantum code which can protect against the effects of an *arbitrary* error on a single qubit! The code is known as the *Shor code*, after its inventor. The code is a combination of the three qubit phase flip and bit flip codes. We first encode the qubit using the phase flip code: $|0\rangle \rightarrow |+++\rangle, |1\rangle \rightarrow |---\rangle$. Next, we encode each of these qubits using the three qubit bit flip code: $|+\rangle$ is encoded as $(|000\rangle + |111\rangle)\sqrt{2}$ and $|-\rangle$ is encoded as $(|000\rangle - |111\rangle)\sqrt{2}$. The result is a nine qubit code, with codewords given by:

$$|0\rangle \rightarrow |0_L\rangle \equiv \frac{(|000\rangle + |111\rangle)(|000\rangle + |111\rangle)(|000\rangle + |111\rangle)}{2\sqrt{2}}$$

$$|1\rangle \rightarrow |1_L\rangle \equiv \frac{(|000\rangle - |111\rangle)(|000\rangle - |111\rangle)(|000\rangle - |111\rangle)}{2\sqrt{2}}. \qquad (10.13)$$

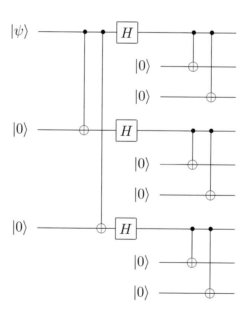

Figure 10.4. Encoding circuit for the Shor nine qubit code. Some of the $|0\rangle$ states appear indented, simply to emphasize the concatenated nature of the encoding.

The quantum circuit encoding the Shor code is shown in Figure 10.4. As described above, the first part of the circuit encodes the qubit using the three qubit phase flip code;

comparison with Figure 10.3 shows that the circuits are identical. The second part of the circuit encodes each of these three qubits using the bit flip code, using three copies of the bit flip code encoding circuit, Figure 10.2. This method of encoding using a hierarchy of levels is known as *concatenation*. It's a great trick for obtaining new codes from old, and we use it again later to prove some important results about quantum error-correction.

The Shor code is able to protect against phase flip and bit flip errors on any qubit. To see this, suppose a bit flip occurs on the first qubit. As for the bit flip code, we perform a measurement of $Z_1 Z_2$ comparing the first two qubits, and find that they are different. We conclude that a bit flip error occurred on the first or second qubit. Next we compare the second and third qubit by performing a measurement of $Z_2 Z_3$. We find that they are the same, so it could not have been the second qubit which flipped. We conclude that the first qubit must have flipped, and recover from the error by flipping the first qubit again, back to its original state. In a similar way we can detect and recover from the effects of bit flip errors on any of the nine qubits in the code.

We cope in a similar manner with phase flips on the qubits. Suppose a phase flip occurs on the first qubit. Such a phase flip flips the sign of the first block of qubits, changing $|000\rangle + |111\rangle$ to $|000\rangle - |111\rangle$, and vice versa. Indeed, a phase flip on *any* of the first three qubits has this effect, and the error-correction procedure we describe works for any of these three possible errors. Syndrome measurement begins by comparing the sign of the first and second blocks of three qubits, just as syndrome measurement for the phase flip code began by comparing the sign of the first and second qubits. For example, $(|000\rangle - |111\rangle)(|000\rangle - |111\rangle)$ has the same sign $(-)$ in both blocks of qubits, while $(|000\rangle - |111\rangle)(|000\rangle + |111\rangle)$ has different signs. When a phase flip occurs on any of the first three qubits we find that the signs of the first and second blocks are different. The second and final stage of syndrome measurement is to compare the sign of the second and third blocks of qubits. We find that these are the same, and conclude that the phase must have flipped in the first block of three qubits. We recover from this by flipping the sign in the first block of three qubits back to its original value. We can recover from a phase flip on any of the nine qubits in a similar manner.

Exercise 10.5: Show that the syndrome measurement for detecting phase flip errors in the Shor code corresponds to measuring the observables $X_1 X_2 X_3 X_4 X_5 X_6$ and $X_4 X_5 X_6 X_7 X_8 X_9$.

Exercise 10.6: Show that recovery from a phase flip on any of the first three qubits may be accomplished by applying the operator $Z_1 Z_2 Z_3$.

Suppose both bit and phase flip errors occur on the first qubit, that is, the operator $Z_1 X_1$ is applied to that qubit. Then it is easy to see that the procedure for detecting a bit flip error will detect a bit flip on the first qubit, and correct it, and the procedure for detecting a phase flip error will detect a phase flip on the first block of three qubits, and correct it. Thus, the Shor code also enables the correction of combined bit and phase flip errors on a single qubit.

Indeed, the Shor code protects against much more than just bit and phase flip errors on a single qubit – we now show that it protects against completely *arbitrary* errors, provided they only affect a single qubit! The error can be tiny – a rotation about the z axis of the Bloch sphere by $\pi/263$ radians, say – or it can be an apparently disastrous error like removing the qubit entirely and replacing it with garbage! The interesting thing is, no additional work needs to be done in order to protect against arbitrary errors – the procedure already described works just fine. This is an example of the extraordinary fact that the apparent *continuum* of errors that may occur on a single qubit can all be corrected by correcting only a *discrete* subset of those errors; all other possible errors being corrected automatically by this procedure! This discretization of the errors is central to why quantum error-correction works, and should be regarded in contrast to classical error-correction for analog systems, where no such discretization of errors is possible.

To simplify the analysis, suppose noise of an arbitrary type is occurring on the first qubit only; we'll come back to what happens when noise is affecting other qubits as well. Following Chapter 8 we describe the noise by a trace-preserving quantum operation $\mathcal{E}$. It is most convenient to analyze error-correction by expanding $\mathcal{E}$ in an operator-sum representation with operation elements $\{E_i\}$. Supposing the state of the encoded qubit is $|\psi\rangle = \alpha|0_L\rangle + \beta|1_L\rangle$ before the noise acts, then after the noise has acted the state is $\mathcal{E}(|\psi\rangle\langle\psi|) = \sum_i E_i|\psi\rangle\langle\psi|E_i^\dagger$. To analyze the effects of error-correction it's easiest to focus on the effect error-correction has on a single term in this sum, say $E_i|\psi\rangle\langle\psi|E_i^\dagger$. As an operator on the first qubit alone E_i may be expanded as a linear combination of the identity, I, the bit flip, X_1, the phase flip, Z_1, and the combined bit and phase flip, X_1Z_1:

$$E_i = e_{i0}I + e_{i1}X_1 + e_{i2}Z_1 + e_{i3}X_1Z_1 . \tag{10.14}$$

The (un-normalized) quantum state $E_i|\psi\rangle$ can thus be written as a superposition of four terms, $|\psi\rangle, X_1|\psi\rangle, Z_1|\psi\rangle, X_1Z_1|\psi\rangle$. Measuring the error syndrome collapses this superposition into one of the four states $|\psi\rangle, X_1|\psi\rangle, Z_1|\psi\rangle$ or $X_1Z_1|\psi\rangle$ from which recovery may then be performed by applying the appropriate inversion operation, resulting in the final state $|\psi\rangle$. The same is true for all the other operation elements E_i. Thus, error-correction results in the original state $|\psi\rangle$ being recovered, despite the fact that the error on the first qubit was arbitrary. This is a fundamental and deep fact about quantum error-correction, that by correcting just a discrete set of errors – the bit flip, phase flip, and combined bit–phase flip, in this example – a quantum error-correcting code is able to automatically correct an apparently much larger (continuous!) class of errors.

What happens when noise is affecting more than just the first qubit? Two basic ideas are used to cope with this. First, in many situations it is a good approximation to assume that the noise acts on qubits independently. Provided the effect of the noise on one qubit is fairly small, we can expand the total effect of the noise as a sum over terms involving errors on no qubits, on a single qubit, on two qubits, and so on, with the terms with errors on no qubits and on a single qubit dominating the higher order terms. Performing error-correction results in the zeroth and first order terms being corrected properly, and leaves only the much smaller second and higher order errors, achieving a net suppression of error. A more detailed analysis of this idea will be given later. Sometimes, of course, it is not reasonable to assume that the noise acts on qubits independently. When this occurs we use a different idea – error-correcting codes which can correct errors on more than a single qubit. Such codes may be constructed along similar lines to the Shor code, and we explain the basic ideas behind how this can be done later in this chapter.

10.3 Theory of quantum error-correction

Can we construct a general theory of quantum error-correcting codes? This section develops a general framework for studying quantum error-correction, including the *quantum error-correction conditions*, a set of equations which must be satisfied if quantum error-correction is to be possible. Of course, possessing such a framework doesn't guarantee us that good quantum error-correcting codes exist – that topic is taken up in Section 10.4 on page 445! But it does provide background to enable us to find good quantum error-correcting codes.

The basic ideas of the theory of quantum error-correction generalize in a natural way the ideas introduced by the Shor code. Quantum states are encoded by a unitary operation into a *quantum error-correcting code*, formally defined as a subspace C of some larger Hilbert space. It is useful to have a notation for the projector onto the code space C, so we use the notation P; for the three qubit bit flip code $P = |000\rangle\langle000| + |111\rangle\langle111|$. After encoding the code is subjected to noise, following which a syndrome measurement is performed to diagnose the type of error which occurred, that is, the *error syndrome*. Once this has been determined, a *recovery* operation is performed, to return the quantum system to the original state of the code. The basic picture is illustrated in Figure 10.5: different error syndromes correspond to undeformed and orthogonal subspaces of the total Hilbert space. The subspaces must be orthogonal, otherwise they couldn't be reliably distinguished by the syndrome measurement. Furthermore, the different subspaces must be undeformed versions of the original code space, in the sense that the errors mapping to the different subspaces must take the (orthogonal) codewords to orthogonal states, in order to be able to recover from the error. This intuitive picture is essentially the content of the quantum error-correction conditions discussed below.

To develop a general theory of quantum error-correction it behooves us to make as few assumptions as possible about the nature of noise and about the procedure used to do the error-correction. That is, we won't necessarily assume that the error-correction is done via a two-stage detection–recovery method, and we won't make any assumptions about the noise occurring for qubit systems, or being weak. Instead, we just make two very broad assumptions: the noise is described by a quantum operation $\mathcal{E}$, and the complete error-correction procedure is effected by a trace-preserving quantum operation $\mathcal{R}$, which we call the *error-correction* operation. This error-correction operation bundles into one piece the two steps that we have called error-detection and recovery above. In order for error-correction to be deemed successful, we require that for any state ρ whose support lies in the code C,

$$(\mathcal{R} \circ \mathcal{E})(\rho) \propto \rho. \qquad (10.15)$$

You may be wondering why we wrote $\propto$ rather than $=$ in the last equation. If $\mathcal{E}$ were a trace-preserving quantum operation then by taking traces of both sides of the equation we see that $\propto$ would be $=$. However, sometimes we may be interested in error-correcting non-trace-preserving operations $\mathcal{E}$, such as measurements, for which $\propto$ is appropriate. Of course, the error-correction step $\mathcal{R}$ must succeed with probability 1, which is why we required $\mathcal{R}$ to be trace-preserving.

The *quantum error-correction conditions* are a simple set of equations which can be checked to determine whether a quantum error-correcting code protects against a particular type of noise $\mathcal{E}$. We will use these conditions to construct a plethora of quantum

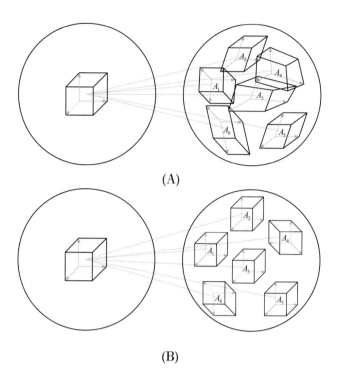

(A)

(B)

Figure 10.5. The packing of Hilbert spaces in quantum coding: (A) bad code, with non-orthogonal, deformed resultant spaces, and (B) good code, with orthogonal (distinguishable), undeformed spaces.

codes, and also to investigate some of the general properties of quantum error-correcting codes.

Theorem 10.1: (**Quantum error-correction conditions**) Let C be a quantum code, and let P be the projector onto C. Suppose $\mathcal{E}$ is a quantum operation with operation elements $\{E_i\}$. A necessary and sufficient condition for the existence of an error-correction operation $\mathcal{R}$ correcting $\mathcal{E}$ on C is that

$$PE_i^\dagger E_j P = \alpha_{ij} P, \qquad (10.16)$$

for some Hermitian matrix α of complex numbers.

We call the operation elements $\{E_i\}$ for the noise $\mathcal{E}$ *errors*, and if such an $\mathcal{R}$ exists we say that $\{E_i\}$ constitutes a *correctable set of errors*.

Proof

We prove sufficiency first, by constructing an explicit error-correction operation $\mathcal{R}$ whenever (10.16) is satisfied. The construction we use is of the two-part form used for the Shor code – error-detection and then recovery – so the proof also shows that error-correction can always be accomplished using such a two-part procedure. Suppose $\{E_i\}$ is a set of operation elements satisfying the quantum error-correction conditions, (10.16). By assumption α is a Hermitian matrix, and thus can be diagonalized, $d = u^\dagger \alpha u$, where u is a unitary matrix and d is diagonal. Define operators $F_k \equiv \sum_i u_{ik} E_i$. Recalling Theorem 8.2, we see that $\{F_k\}$ is also a set of operation elements for $\mathcal{E}$. By direct substitution,

$$PF_k^\dagger F_l P = \sum_{ij} u_{ki}^* u_{jl} PE_i^\dagger E_j P. \tag{10.17}$$

Substituting (10.16) simplifies this to $PF_k^\dagger F_l P = \sum_{ij} u_{ki}^* \alpha_{ij} u_{jl} P$ and since $d = u^\dagger \alpha u$ we obtain

$$PF_k^\dagger F_l P = d_{kl} P, \tag{10.18}$$

which can be thought of as a simplification of the quantum error-correction conditions (10.16), because d_{kl} is diagonal.

We use the simplified conditions (10.18) and the polar decomposition (Section 2.1.10 on page 78) to define the syndrome measurement. From the polar decomposition, we see that $F_k P = U_k \sqrt{PF_k^\dagger F_k P} = \sqrt{d_{kk}} U_k P$ for some unitary U_k. The effect of F_k is therefore to rotate the coding subspace into the subspace defined by the projector $P_k \equiv U_k PU_k^\dagger = F_k PU_k^\dagger / \sqrt{d_{kk}}$. Equation (10.18) implies that these subspaces are orthogonal, since when $k \neq l$,

$$P_l P_k = P_l^\dagger P_k = \frac{U_l PF_l^\dagger F_k PU_k^\dagger}{\sqrt{d_{ll} d_{kk}}} = 0. \tag{10.19}$$

The syndrome measurement is a projective measurement defined by the projectors P_k, augmented by an additional projector if necessary to satisfy the completeness relation $\sum_k P_k = I$. Recovery is accomplished simply by applying $U_k^\dagger$. To see that this error-correction procedure works, note that the combined detection–recovery step corresponds to the quantum operation $\mathcal{R}(\sigma) = \sum_k U_k^\dagger P_k \sigma P_k U_k$. For states ρ in the code, simple algebra and the definitions show that:

$$U_k^\dagger P_k F_l \sqrt{\rho} = U_k^\dagger P_k^\dagger F_l P \sqrt{\rho} \tag{10.20}$$

$$= \frac{U_k^\dagger U_k PF_k^\dagger F_l P \sqrt{\rho}}{\sqrt{d_{kk}}} \tag{10.21}$$

$$= \delta_{kl} \sqrt{d_{kk}} P \sqrt{\rho} \tag{10.22}$$

$$= \delta_{kl} \sqrt{d_{kk}} \sqrt{\rho}. \tag{10.23}$$

Thus

$$\mathcal{R}(\mathcal{E}(\rho)) = \sum_{kl} U_k^\dagger P_k F_l \rho F_l^\dagger P_k U_k \tag{10.24}$$

$$= \sum_{kl} \delta_{kl} d_{kk} \rho \tag{10.25}$$

$$\propto \rho, \tag{10.26}$$

as required.

To prove necessity of the quantum error-correction conditions (10.16), suppose $\{E_i\}$ is a set of errors which is perfectly correctable by an error-correction operation $\mathcal{R}$ with operation elements $\{R_j\}$. Define a quantum operation $\mathcal{E}_C$ by $\mathcal{E}_C(\rho) \equiv \mathcal{E}(P\rho P)$. Since $P\rho P$ is in the code space for all ρ, it follows that

$$\mathcal{R}(\mathcal{E}_C(\rho)) \propto P\rho P, \tag{10.27}$$

for all states ρ. Moreover, the proportionality factor must be a constant c, not depending on ρ, if both right and left hand sides are to be linear. Rewriting the last equation explicitly in terms of the operation elements gives

$$\sum_{ij} R_j E_i P \rho P E_i^\dagger R_j^\dagger = cP\rho P. \tag{10.28}$$

This equation holds for all ρ. It follows that the quantum operation with operation elements $\{R_j E_i\}$ is identical to the quantum operation with a single operation element $\sqrt{c}P$. Theorem 8.2 implies that there exist complex numbers c_{ki} such that

$$R_k E_i P = c_{ki} P. \tag{10.29}$$

Taking the adjoint of this equation gives $PE_i^\dagger R_k^\dagger = c_{ki}^* P$ and therefore $PE_i^\dagger R_k^\dagger R_k E_j P = c_{ki}^* c_{kj} P$. But R is a trace-preserving operation, so $\sum_k R_k^\dagger R_k = I$. Summing the equation $PE_i^\dagger R_k^\dagger R_k E_j P = c_{ki}^* c_{kj} P$ over k we deduce that

$$PE_i^\dagger E_j P = \alpha_{ij} P, \tag{10.30}$$

where $\alpha_{ij} \equiv \sum_k c_{ki}^* c_{kj}$ is just a Hermitian matrix of complex numbers. These are the quantum error-correction conditions. □

Direct verification of the quantum error-correction conditions is an easy but time-consuming task. In Sections 10.4 and 10.5 we describe a theoretical formalism which uses the quantum error-correction conditions as a launching point for the construction of many interesting classes of codes, and which circumvents much of the difficulty associated with verifying the quantum error-correction conditions directly. For now, you should work through the following example, which demonstrates the quantum error-correction conditions in action:

Exercise 10.7: Consider the three qubit bit flip code of Section 10.1.1, with corresponding projector $P = |000\rangle\langle000| + |111\rangle\langle111|$. The noise process this code protects against has operation elements $\{\sqrt{(1-p)^3}I, \sqrt{p(1-p)^2}X_1, \sqrt{p(1-p)^2}X_2, \sqrt{p(1-p)^2}X_3\}$, where p is the probability that a bit flips. Note that this quantum operation is not trace-preserving, since we have omitted operation elements corresponding to bit flips on two and three qubits. Verify the quantum error-correction conditions for this code and noise process.

10.3.1 Discretization of the errors

We have discussed the protection of quantum information against a specific noise process $\mathcal{E}$. In general, however, we don't know exactly what noise is afflicting a quantum system. It would be useful if a specific code C and error-correction operation $\mathcal{R}$ could be used to protect against an entire class of noise processes. Fortunately, the quantum error-correction conditions are easily adapted to provide exactly this sort of protection.

Theorem 10.2: Suppose C is a quantum code and $\mathcal{R}$ is the error-correction operation constructed in the proof of Theorem 10.1 to recover from a noise process $\mathcal{E}$ with operation elements $\{E_i\}$. Suppose $\mathcal{F}$ is a quantum operation with operation elements $\{F_j\}$ which are *linear combinations* of the E_i, that is $F_j = \sum_i m_{ji} E_i$ for some matrix m_{ji} of complex numbers. Then the error-correction operation $\mathcal{R}$ also corrects for the effects of the noise process $\mathcal{F}$ on the code C.

Box 10.1: Quantum error-correction without measurement

In the main text we have described quantum error-correction as a two stage process: an error-detection step effected using quantum measurement, followed by a recovery step effected using conditioned unitary operations. It is possible to perform quantum error-correction using only unitary operations and ancilla systems prepared in standard states. The advantage of knowing how to do this is that for some real-world quantum systems it is very difficult to perform the quantum measurements needed for quantum error-correction, so an alternate procedure is needed. The techniques we use to do that are essentially the same as those described in Chapter 8 for mocking up an arbitrary quantum operation; we now recap the basic idea in the context of quantum error-correction.

Suppose the syndrome measurement on the principal system – the one being error-corrected – is described by measurement operators M_i, and the corresponding conditional unitary operation is U_i. Introduce an ancilla system with basis states $|i\rangle$ corresponding to the possible error syndromes. The ancilla starts in a standard pure state $|0\rangle$ before error-correction. Define a unitary operator U on the principal system plus ancilla by

$$U|\psi\rangle|0\rangle \equiv \sum_i (U_i M_i |\psi\rangle)|i\rangle. \tag{10.31}$$

This can be extended to a unitary operator acting on the whole space since

$$\langle\varphi|\langle 0|U^\dagger U|\psi\rangle|0\rangle = \sum_{ij} \langle\varphi|M_i^\dagger M_j|\psi\rangle\delta_{ij} \tag{10.32}$$

$$= \sum_i \langle\varphi|M_i^\dagger M_i|\psi\rangle \tag{10.33}$$

$$= \langle\varphi|\psi\rangle. \tag{10.34}$$

That is, U preserves inner products, and can thus be extended to a unitary operator on the entire state space. The effect of U is to effect the transformation $\mathcal{R}(\sigma) = \sum_i U_i M_i \sigma M_i^\dagger U_i^\dagger$ on the system being error-corrected, exactly the same quantum operation as described in the main text for the performance of quantum error-correction. Note that in order for this error-correction procedure to work it is necessary to use a fresh ancilla each time error-correction is performed.

Proof

By Theorem 10.1 the operation elements $\{E_i\}$ must satisfy the quantum error-correction conditions, $PE_iE_j^\dagger P = \alpha_{ij}P$. As shown in the proof of Theorem 10.1, without loss of generality we may assume that the operation elements for $\mathcal{E}$ have been chosen such that $\alpha_{ij} = d_{ij}$ is diagonal with real entries. The error-correction operation $\mathcal{R}$ has operation elements $U_k^\dagger P_k$, where by Equation (10.23) the U_k and P_k are chosen such that for any ρ in the code space:

$$U_k^\dagger P_k E_i \sqrt{\rho} = \delta_{ki}\sqrt{d_{kk}}\sqrt{\rho}. \tag{10.35}$$

Substituting $F_j = \sum_i m_{ji} E_i$ gives

$$U_k^\dagger P_k F_j \sqrt{\rho} = \sum_i m_{ji} \delta_{ki} \sqrt{d_{kk}} \sqrt{\rho} \tag{10.36}$$

$$= m_{jk} \sqrt{d_{kk}} \sqrt{\rho}, \tag{10.37}$$

and thus

$$\mathcal{R}(\mathcal{F}(\rho)) = \sum_{kj} U_k^\dagger P_k F_j \rho F_j^\dagger P_k U_k \tag{10.38}$$

$$= \sum_{kj} |m_{jk}|^2 d_{kk} \rho \tag{10.39}$$

$$\propto \rho, \tag{10.40}$$

as required.   □

This result enables the introduction of a more powerful language to describe quantum error-correcting codes. Instead of talking about the class of error processes $\mathcal{E}$ correctable by a code C and error-correction operation $\mathcal{R}$ we can talk about a set of *error operators* (or simply *errors*) $\{E_i\}$ which are correctable. By this, we mean that the quantum error-correction conditions hold for these operators,

$$P E_i E_j^\dagger P = \alpha_{ij} P. \tag{10.41}$$

Theorems 10.1 and 10.2 together imply that any noise process $\mathcal{E}$ whose operation elements are built from linear combinations of these error operators $\{E_i\}$ will be corrected by the recovery operation $\mathcal{R}$!

Let's look at an example of this powerful new viewpoint in action. Suppose $\mathcal{E}$ is a quantum operation acting on a single qubit. Then its operation elements $\{E_i\}$ can each be written as a linear combination of the Pauli matrices $\sigma_0, \sigma_1, \sigma_2, \sigma_3$. Therefore, to check that the Shor code corrects against *arbitrary* single qubit errors on the first qubit it is sufficient to verify that the equations

$$P \sigma_i^1 \sigma_j^1 P = \alpha_{ij} P, \tag{10.42}$$

are satisfied, where σ_i^1 are the Pauli matrices $(I, X, Y$ and $Z)$ acting on the first qubit. Once this is done it is assured that any error process whatsoever on the first qubit may be corrected! (The actual calculation is quite simple, and is done as part of Exercise 10.10.) Indeed this example explains a point that can be somewhat mysterious when first confronted with the literature on quantum error-correction: many authors have what appears to be a suspicious fondness for the depolarizing channel, $\mathcal{E}(\rho) = (1-p)\rho + \frac{p}{3}(X\rho X + Y\rho Y + Z\rho Z)$. It is tempting to assume that this greatly limits the validity of their models of error-correction, but this is not so, for as the discussion just now implies, the ability to error-correct the depolarizing channel automatically implies the ability to error-correct an *arbitrary* single qubit quantum operation.

Summarizing, we have learnt that it is possible to *discretize* quantum errors, that to fight the continuum of errors possible on a single qubit it is sufficient merely to win the war against a finite set of errors, the four Pauli matrices. Similar results hold for higher-dimensional quantum systems. This stands in remarkable contrast to the theory of error-correction for *analog* classical systems. Error-correction in such systems is extremely difficult because in principle there are an infinite number of different error syndromes.

Digital error-correction for classical information processing is much more successful because it involves only a finite number of error syndromes. The surprising thing we have learnt is that quantum error-correction seems much more similar to classical digital error-correction than it is to classical analog error-correction.

Exercise 10.8: Verify that the three qubit phase flip code
$|0_L\rangle = |+++\rangle, |1_L\rangle = |---\rangle$ satisfies the quantum error-correction conditions for the set of error operators $\{I, Z_1, Z_2, Z_3\}$.

Exercise 10.9: Again, consider the three qubit phase flip code. Let P_i and Q_i be the projectors onto the $|0\rangle$ and $|1\rangle$ states, respectively, of the ith qubit. Prove that the three qubit phase flip code protects against the error set $\{I, P_1, Q_1, P_2, Q_2, P_3, Q_3\}$.

Exercise 10.10: Explicitly verify the quantum error-correction conditions for the Shor code, for the error set containing I and the error operators X_j, Y_j, Z_j for $j = 1$ through 9.

Exercise 10.11: Construct operation elements for a single qubit quantum operation $\mathcal{E}$ that upon input of any state ρ replaces it with the completely randomized state $I/2$. It is amazing that even such noise models as this may be corrected by codes such as the Shor code!

10.3.2 Independent error models

How can we make the connection between quantum error-correction and the criteria for doing reliable quantum information processing introduced in Chapter 9? In this section we explain the basic idea of how this may be done using the assumption of *independent errors* on different qubits. Intuitively, if a noise process acts independently on the different qubits in the code, then provided the noise is sufficiently weak error-correction should improve the storage fidelity of the encoded state over the unencoded state. To illustrate this, we begin with the example of the depolarizing channel, which provides an especially simple demonstration of the basic ideas, and then broaden the ideas to include other important channels.

Recall that the depolarizing channel may be described by a single parameter, a probability p. The action of the depolarizing channel on a single qubit is defined by the equation $\mathcal{E}(\rho) = (1-p)\rho + p/3[X\rho X + Y\rho Y + Z\rho Z]$, and can be interpreted as saying that nothing happens to the qubit with probability $1-p$, and each of the operators X, Y and Z is applied to the qubit with probability $p/3$. The depolarizing channel is especially easy to analyze in the context of quantum error-correction because it has such a nice interpretation in terms of the four basic errors I, X, Y and Z which are most commonly used in the analysis of quantum codes. We'll explain how this analysis is performed, and then return to the question of what happens when we consider a process which doesn't have such a simple interpretation in terms of the I, X, Y and Z operations. A simple calculation shows that the minimum fidelity for states sent through a depolarizing channel is given by $F = \sqrt{1 - 2p/3} = 1 - p/3 + O(p^2)$.

Exercise 10.12: Show that the fidelity between the state $|0\rangle$ and $\mathcal{E}(|0\rangle\langle 0|)$ is

$\sqrt{1 - 2p/3}$, and use this to argue that the minimum fidelity for the depolarizing channel is $\sqrt{1 - 2p/3}$.

Suppose we encode a single qubit of information in an n qubit quantum code which corrects errors on any single qubit. Suppose the depolarizing channel with parameter p acts independently on each of the qubits, giving rise to a joint action on all n qubits of

$$\mathcal{E}^{\otimes n}(\rho) = (1-p)^n \rho + \sum_{j=1}^{n}\sum_{k=1}^{3}(1-p)^{n-1}\frac{p}{3}\sigma_k^j \rho \sigma_k^j + \cdots, \tag{10.43}$$

where the '...' indicates higher-order terms which are positive and will drop out of the analysis. After error-correction has been performed all terms appearing in this sum will be returned to the state ρ, provided ρ was in the code originally,

$$\left(\mathcal{R} \otimes \mathcal{E}^{\otimes n}\right)(\rho) = \left[(1-p)^n + n(1-p)^{n-1}p\right]\rho + \cdots, \tag{10.44}$$

so the fidelity satisfies

$$F \geq \sqrt{(1-p)^{n-1}(1-p+np)} = 1 - \frac{\binom{n}{2}}{2}p^2 + O(p^3). \tag{10.45}$$

Thus, provided the probability of error p is sufficiently small, using the quantum error-correcting code leads to an improvement in the fidelity of the quantum states being protected by the code.

Not all noisy channels can be interpreted so easily as a random combination of no error, bit flips, phase flips and combinations of the two. Many naturally occurring quantum channels have no such interpretation. Consider the example of amplitude damping (Section 8.3.5), which has operation elements E_0 and E_1:

$$E_0 = \begin{bmatrix} 1 & 0 \\ 0 & \sqrt{1-\gamma} \end{bmatrix}; \quad E_1 = \begin{bmatrix} 0 & \sqrt{\gamma} \\ 0 & 0 \end{bmatrix}. \tag{10.46}$$

The parameter γ is a small positive parameter characterizing the strength of the amplitude damping process – as γ gets closer to zero, the strength decreases, until ultimately we end up with an essentially noise-free quantum channel. We might reasonably guess that the amplitude damping channel has an equivalent description in terms of a set of operation elements including a term proportional to the identity, $\{f(\gamma)I, E_1', E_2', \ldots\}$, where $f(\gamma) \to 1$ as $\gamma \to 0$. If this were the case then an analysis of error-correction for the amplitude damping channel acting independently on multiple qubits could be done that was similar to the analysis of error-correction performed for the depolarizing channel. Surprisingly, it turns out that no such description is possible! This follows from Theorem 8.2 simply because for $\gamma > 0$ no linear combination of E_0 and E_1 can ever be proportional to the identity, and thus no set of operation elements for the amplitude damping channel can ever include a term proportional to the identity.

Similarly, many other noise processes in quantum mechanics are close to the identity in a physical sense, yet no operator-sum representation for the process contains a large identity component. Intuitively it seems reasonable that in such a circumstance error-correction should result in a net gain in the storage fidelity for quantum information, provided the noise is sufficiently weak. We will now show that this is in fact the case, using the specific example of the amplitude damping channel for concreteness. A simple calculation shows that the minimum fidelity for the amplitude damping channel applied to a single qubit is $\sqrt{1-\gamma}$. Suppose the qubit is encoded in an n qubit quantum code capable of correcting arbitrary errors on a single qubit, and that amplitude damping channels of parameter γ act *independently* on each qubit. We will sketch the basic idea showing that the effect of quantum error-correction is to change the fidelity of storage to $1 - O(\gamma^2)$, so for small γ encoding the qubit in the quantum code results in a net suppression of error.

Exercise 10.13: Show that the minimum fidelity $F(|\psi\rangle, \mathcal{E}(|\psi\rangle\langle\psi|))$ when $\mathcal{E}$ is the amplitude damping channel with parameter γ, is $\sqrt{1-\gamma}$.

Using $E_{j,k}$ to denote the action of E_j on the kth qubit, the effect of the noise on the encoded qubits may be written:

$$\mathcal{E}^{\otimes n}(\rho) = \left(E_{0,1} \otimes E_{0,2} \otimes \cdots \otimes E_{0,n}\right) \rho \left(E_{0,1}^\dagger \otimes E_{0,2}^\dagger \otimes \cdots \otimes E_{0,n}^\dagger\right)$$

$$+ \sum_{j=1}^n \left[E_{1,j} \otimes \left(\bigotimes_{k \neq j} E_{0,k}\right)\right] \rho \left[E_{1,j}^\dagger \otimes \left(\bigotimes_{k \neq j} E_{0,k}^\dagger\right)\right]$$

$$+ O(\gamma^2). \tag{10.47}$$

Suppose we write $E_0 = (1-\gamma/4)I + \gamma Z/4 + O(\gamma^2)$, and $E_1 = \sqrt{\gamma}(X+iY)/2$. Substituting these expressions in (10.47) gives

$$\mathcal{E}^{\otimes n}(\rho) = \left(1 - \frac{\gamma}{4}\right)^{2n} \rho + \frac{\gamma}{4}\left(1 - \frac{\gamma}{4}\right)^{2n-1} \sum_{j=1}^n (Z_j\rho + \rho Z_j)$$

$$+ \frac{\gamma}{4}\left(1 - \frac{\gamma}{4}\right)^{2n-2} \sum_{j=1}^n (X_j + iY_j)\,\rho\,(X_j - iY_j) + O(\gamma^2). \tag{10.48}$$

Suppose ρ is a state of the code. Obviously, the effect of error-correction on ρ is to leave it invariant! The effect of error-correction on terms like $Z_j\rho$ and ρZ_j is most easily understood by considering the effect on $Z_j|\psi\rangle\langle\psi|$, where $|\psi\rangle$ is a state of the code. We suppose the code is such that the error Z_j takes $|\psi\rangle$ to a subspace which is orthogonal to the code, so that when the syndrome measurement is performed terms like $Z_j|\psi\rangle\langle\psi|$ disappear. (Note that even if this orthogonality assumption is not made a similar analysis can still be made, by working in terms of error operators which do take the code to orthogonal subspaces.) Thus terms like $Z_j\rho$ vanish after error-correction, as do terms like ρZ_j, $X_j\rho Y_j$ and $Y_j\rho X_j$. Furthermore, error-correction takes $X_j\rho X_j$ and $Y_j\rho Y_j$ back to ρ, since the code can correct errors on one qubit. Thus, after error-correction the state of the system is

$$\left(1 - \frac{\gamma}{4}\right)^{2n} \rho + 2n\frac{\gamma}{4}\left(1 - \frac{\gamma}{4}\right)^{2n-2} \rho + O(\gamma^2) = \rho + O(\gamma^2). \tag{10.49}$$

Thus, to order γ^2 error-correction returns the quantum system to its original state ρ, and for weak noise (small γ) error-correction results in a net suppression of errors, just as for the depolarizing channel. Our analysis here was for the amplitude damping noise model, but it is not difficult to generalize this argument to obtain similar conclusions for other noise models. In general, however, for the remainder of this chapter we work mainly with noise models which can be understood as stochastic application of errors corresponding to the Pauli matrices, similar to the depolarizing channel, allowing us to do the analysis using familiar concepts from classical probability theory. Keep in mind that the ideas we describe can be extended beyond this simple error model to apply to a much wider range of error models using principles similar to those we have just outlined.

10.3.3 Degenerate codes

In many respects quantum error-correcting codes are quite similar to classical codes – an error is identified by measuring the error syndrome, and then corrected as appropriate, just as in the classical case. There is, however, an interesting class of quantum codes known as *degenerate* codes possessing a striking property unknown in classical codes. The idea is most easily illustrated for the case of the Shor code. Consider the effect of the errors Z_1 and Z_2 on the codewords for the Shor code. As we have already noted, the effect of these errors is the *same* on both codewords. For classical error-correcting codes errors on different bits necessarily lead to different corrupted codewords. The phenomenon of degenerate quantum codes is a sort of good news–bad news situation for quantum codes. The bad news is that some of the proof techniques used classically to prove bounds on error-correction fall down because they can't be applied to degenerate codes. We'll see an example of this in the next section and the quantum Hamming bound. The good news is that degenerate quantum codes seem to be among the most interesting quantum codes! In some sense they are able to 'pack more information in' than are classical codes, because distinct errors do not necessarily have to take the code space to orthogonal spaces, and it is possible (though has not yet been shown) that this extra ability may lead to degenerate codes that can store quantum information more efficiently than any non-degenerate code.

10.3.4 The quantum Hamming bound

For applications we would like to make use of the 'best' possible quantum codes. What 'best' means in any given circumstance depends on the application. For this reason we would like to have criteria for determining whether or not a code with particular characteristics exists or not. In this section we develop the *quantum Hamming bound*, a simple bound which gives some insights into the general properties of quantum codes. Unfortunately the quantum Hamming bound only applies to non-degenerate codes, but it gives us an idea of what more general bounds may look like. Suppose a non-degenerate code is used to encode k qubits in n qubits in such a way that it can correct errors on any subset of t or fewer qubits. Suppose j errors occur, where $j \leq t$. There are $\binom{n}{j}$ sets of locations where errors may occur. With each such set of locations there are three possible errors – the three Pauli matrices X, Y, Z – that may occur in each qubit, for a total of 3^j possible errors. The total number of errors that may occur on t or fewer qubits is therefore

$$\sum_{j=0}^{t} \binom{n}{j} 3^j. \tag{10.50}$$

(Note that $j = 0$ corresponds to the case of no errors on any qubit, the 'error' I.) In order to encode k qubits in a non-degenerate way each of these errors must correspond to an orthogonal 2^k-dimensional subspace. All of these subspaces must be fitted into the total 2^n-dimensional space available to n qubits, leading to the inequality

$$\sum_{j=0}^{t} \binom{n}{j} 3^j 2^k \leq 2^n,$$
(10.51)

which is the quantum Hamming bound. Consider, for example, the case where we wish to encode one qubit in n qubits in such a way that errors on one qubit are tolerated. In this case the quantum Hamming bound reads:

$$2(1 + 3n) \leq 2^n.$$
(10.52)

Substitution shows that this inequality is not satisfied for $n \leq 4$, while it is for values of $n \geq 5$. Therefore, there is no non-degenerate code encoding one qubit in fewer than five qubits in such a way as to protect from all possible errors on a single qubit.

Of course, not all quantum codes are non-degenerate, so the quantum Hamming bound is more useful as a rule of thumb than as a hard and fast bound on the existence of quantum codes. (In particular, at the time of writing no codes are known that violate the quantum Hamming bound, even allowing degenerate codes.) Later, we will have occasion to look at some bounds on quantum codes that apply to all quantum codes, not just non-degenerate codes. For example, in Section 12.4.3 we prove the quantum Singleton bound, which implies that any quantum code encoding k qubits in n qubits and able to correct errors on any t qubits must satisfy $n \geq 4t + k$. It follows that the smallest code encoding a single qubit and able to correct an arbitrary error on a single qubit must satisfy $n \geq 4 + 1 = 5$, and indeed, we will soon exhibit such a five qubit code.

10.4 Constructing quantum codes

We now have a theoretical framework for the study of quantum error-correcting codes, but we don't as yet have many examples of such codes! We begin to remedy this defect by taking a brief tour into the theory of classical linear codes in Section 10.4.1, and then in Section 10.4.2 explain how ideas from classical linear codes can be used to construct a large class of quantum codes known as Calderbank–Shor–Steane (CSS) codes. Our job is completed in Section 10.5 with the development of the stabilizer codes, a class of codes even more general than the CSS codes which offers a powerful means for constructing a wide variety of quantum codes.

10.4.1 Classical linear codes

Classical error-correcting codes have many varied and important technological applications, so it is unsurprising that a powerful theory of such codes has been developed. Our interest in the techniques of classical error-correction is that many of these techniques have important implications for quantum error-correction, especially the theory of *classical linear codes*, which can be used to develop a wide variety of good quantum error-correcting codes. In this section we review classical linear codes, emphasizing especially those ideas important to quantum error-correction.

A *linear code* C encoding k bits of information into an n bit code space is specified by an n by k *generator matrix* G whose entries are all elements of $\mathbf{Z}_2$, that is, zeroes and ones. The matrix G maps messages to their encoded equivalent. Thus the k bit message x is encoded as Gx, where the message x is treated as a column vector in the obvious way. Furthermore, the multiplication operation, and all our other arithmetic operations in this section, are done modulo 2. As a simple example, the repetition code mapping a single bit to three repetitions is specified by the generator matrix

$$G = \begin{bmatrix} 1 \\ 1 \\ 1 \end{bmatrix}, \tag{10.53}$$

since G maps the possible messages, 0 and 1, to their encoded form, $G[0] = (0,0,0)$ and $G[1] = (1,1,1)$. (Recall that $(a, b, \ldots, z)$ is our shorthand notation for column vectors.) We say that a code using n bits to encode k bits of information is an $[n, k]$ code; this example is therefore a $[3, 1]$ code. A slightly more complicated example is to encode two bits using three repetitions of each bit – a $[6, 2]$ code. This has generator matrix

$$G = \begin{bmatrix} 1 & 0 \\ 1 & 0 \\ 1 & 0 \\ 0 & 1 \\ 0 & 1 \\ 0 & 1 \end{bmatrix}, \tag{10.54}$$

from which we see that

$$G(0, 0) = (0, 0, 0, 0, 0, 0); \quad G(0, 1) = (0, 0, 0, 1, 1, 1); \tag{10.55}$$
$$G(1, 0) = (1, 1, 1, 0, 0, 0); \quad G(1, 1) = (1, 1, 1, 1, 1, 1), \tag{10.56}$$

just as we expect. The set of possible codewords for the code corresponds to the vector space spanned by the columns of G, so in order that all messages be uniquely encoded we require that the columns of G be linearly independent, but otherwise place no constraints on G.

Exercise 10.14: Write an expression for a generator matrix encoding k bits using r repetitions for each bit. This is an $[rk, k]$ linear code, and should have an $rk \times k$ generator matrix.

Exercise 10.15: Show that adding one column of G to another results in a generator matrix generating the same code.

A great advantage of linear codes over general error-correcting codes is their compact specification. A general code encoding k bits in n bits requires 2^k codewords each of length n to specify the encoding, a total of $n2^k$ bits to specify a description of the code. With a linear code we need only specify the kn bits of the generator matrix, an exponential saving in the amount of memory required! This compact description is mirrored in the ability to do efficient encoding and decoding, important features which classical linear codes share with their quantum cousins, the stabilizer codes. We can already see how to perform efficient encoding of a classical linear code: one simply multiplies the k bit message by the n by k generator matrix G to obtain the n bit encoded message, a procedure which can be done using $O(nk)$ operations.

One of the attractive features of the generator matrix definition of linear codes is the transparent connection between the messages we wish to encode and how they are encoded. It is not so clear how to perform error-correction. Error-correction for linear codes is most easily understood by introducing an alternative (but equivalent) formulation of linear codes in terms of *parity check* matrices. In this definition an $[n, k]$ code is defined to consist of all n-element vectors x over $\mathbf{Z}_2$ such that

$$Hx = 0, \tag{10.57}$$

where H is an $n - k$ by n matrix known as the *parity check matrix*, with entries all zeroes and ones. Equivalently, but more succinctly, the code is defined to be the kernel of H. A code encoding k bits has 2^k possible codewords so the kernel of H must be k-dimensional, and therefore we require that H have linearly independent rows.

Exercise 10.16:   Show that adding one row of the parity check matrix to another does not change the code. Using Gaussian elimination and swapping of bits it is therefore possible to assume that the parity check matrix has the *standard form* $[A|I_{n-k}]$, where A is an $(n - k) \times k$ matrix.

To connect the parity check picture of linear codes with the generator matrix picture we need to develop a procedure that enables us to convert back and forth between the parity check matrix H and the generator matrix G. To go from the parity check matrix to the generator matrix, pick k linearly independent vectors $y_1, \ldots, y_k$ spanning the kernel of H, and set G to have columns y_1 through y_k. To go from the generator matrix to the parity check matrix, pick $n - k$ linearly independent vectors $y_1, \ldots, y_{n-k}$ orthogonal to the columns of G, and set the rows of H to be $y_1^T, \ldots, y_{n-k}^T$. (By orthogonal, we mean that the inner product modulo 2 must be zero.) As an example, consider the $[3, 1]$ repetition code defined by the generator matrix (10.53). To construct H we pick out $3 - 1 = 2$ linearly independent vectors orthogonal to the columns of G, say $(1, 1, 0)$ and $(0, 1, 1)$, and define the parity check matrix as:

$$H \equiv \begin{bmatrix} 1 & 1 & 0 \\ 0 & 1 & 1 \end{bmatrix}. \tag{10.58}$$

It is easy to check that $Hx = 0$ only for the codewords $x = (0, 0, 0)$ and $x = (1, 1, 1)$.

Exercise 10.17:   Find a parity check matrix for the $[6, 2]$ repetition code defined by the generator matrix in (10.54).

Exercise 10.18:   Show that the parity check matrix H and generator matrix G for the same linear code satisfy $HG = 0$.

Exercise 10.19:   Suppose an $[n, k]$ linear code C has a parity check matrix of the form $H = [A|I_{n-k}]$, for some $(n - k) \times k$ matrix A. Show that the corresponding generator matrix is

$$G = \begin{bmatrix} I_k \\ \hline -A \end{bmatrix}. \tag{10.59}$$

(Note that $-A = A$ since we are working modulo 2; however, this equation also holds for linear codes over more general fields than $\mathbf{Z}_2$.)

The parity check matrix makes error-detection and recovery quite transparent. Suppose that we encode the message x as $y = Gx$, but an error e due to noise corrupts y giving the corrupted codeword $y' = y + e$. (Note that $+$ here denotes bitwise addition modulo 2.) Because $Hy = 0$ for all codewords, it follows that $Hy' = He$. We call Hy' the *error syndrome*, and it plays a role similar to the role played by the error syndrome in quantum error-correction; it is a function Hy' of the corrupted state y', just as the quantum error syndrome is determined by measuring the corrupted quantum state, and because of the relation $Hy' = He$, the error syndrome contains information about the error that occurred that hopefully will enable recovery to the original codeword y. To see how this is possible, suppose no errors or just one error occurred. Then the error syndrome Hy' is equal to 0 in the no error case and is equal to He_j when an error occurs on the jth bit, where e_j is the unit vector with component 1 in the jth component. If we assume that errors occur on at most one bit, it is therefore possible to perform error-correction by computing the error syndrome Hy' and comparing it to the different possible values of He_j to determine which (if any) bit needs to be corrected.

More generally, insight into how error-correction may be performed with a linear code can be attained using the concept of *distance*. Suppose x and y are words of n bits each. The *(Hamming) distance* $d(x, y)$ between x and y is defined to be the number of places at which x and y differ. Thus $d((1, 1, 0, 0), (0, 1, 0, 1)) = 2$, for example. The *(Hamming) weight* of a word x is defined to be the distance from the string of all zeroes, $\text{wt}(x) \equiv d(x, 0)$, that is, the number of places at which x is non-zero. Note that $d(x, y) = \text{wt}(x + y)$. To understand the connection with error-correction suppose we encode x as $y = Gx$ using a linear error-correcting code. Noise corrupts the encoded bit string producing $y' = y + e$. Provided the probability of a bit flip is less than $1/2$, the most likely codeword to have been encoded is the codeword y which minimizes the number of bit flips needed to get from y to y', that is, which minimizes $\text{wt}(e) = d(y, y')$. In principle, error-correction with a linear code may be accomplished by simply replacing y' by such a y. In practice this may be rather inefficient, since determining the minimal distance $d(y, y')$ in general requires searching all 2^k possible codewords y. A great deal of effort in classical coding theory has gone into constructing codes with special structure that enable error-correction to be performed more efficiently, however these constructions are beyond the scope of this book.

The global properties of the code can also be understood using the Hamming distance. We define the *distance* of a code to be the minimum distance between any two codewords,

$$d(C) \equiv \min_{x, y \in C, x \neq y} d(x, y). \tag{10.60}$$

But $d(x, y) = \text{wt}(x + y)$. Since the code is linear, $x + y$ is a codeword if x and y are, so we see that

$$d(C) = \min_{x \in C, x \neq 0} \text{wt}(x). \tag{10.61}$$

Setting $d \equiv d(C)$, we say that C is an $[n, k, d]$ code. The importance of the distance is that a code with distance at least $2t + 1$ for some integer t is able to correct errors on up to t bits, simply by decoding the corrupted encoded message y' as the unique codeword y satisfying $d(y, y') \leq t$.

Exercise 10.20: Let H be a parity check matrix such that any $d - 1$ columns are linearly independent, but there exists a set of d linearly dependent columns. Show that the code defined by H has distance d.

Exercise 10.21: (Singleton bound) Show that an $[n, k, d]$ code must satisfy
$$n - k \geq d - 1.$$

A good illustrative class of linear error-correcting codes are the Hamming codes. Suppose $r \geq 2$ is an integer and let H be the matrix whose columns are all $2^r - 1$ bit strings of length r which are not identically 0. This parity check matrix defines a $[2^r - 1, 2^r - r - 1]$ linear code known as a *Hamming code*. An especially important example for quantum error-correction is the case $r = 3$, which is a $[7, 4]$ code having parity check matrix:

$$H = \begin{bmatrix} 0 & 0 & 0 & 1 & 1 & 1 & 1 \\ 0 & 1 & 1 & 0 & 0 & 1 & 1 \\ 1 & 0 & 1 & 0 & 1 & 0 & 1 \end{bmatrix}. \tag{10.62}$$

Any two columns of H are different, and therefore linearly independent; the first three columns are linearly dependent, so by Exercise 10.20 the distance of the code is 3. It follows that this code is able to correct an error on any single bit. Indeed, the error-correction method is very simple. Suppose an error occurs on the jth bit. Inspection of (10.62) reveals that the syndrome He_j is just a binary representation for j, telling us which bit to flip to correct the error.

Exercise 10.22: Show that all Hamming codes have distance 3, and thus can correct an error on a single bit. The Hamming codes are therefore $[2^r - 1, 2^r - r - 1, 3]$ codes.

What can we say more generally about the properties of linear codes? In particular, we would like conditions telling us whether or not codes with particular code parameters exist. Not surprisingly, many techniques for proving such conditions exist. One such set of conditions is known as the *Gilbert–Varshamov bound*, which states that for large n there exists an $[n, k]$ error-correcting code protecting against errors on t bits for some k such that

$$\frac{k}{n} \geq 1 - H\left(\frac{2t}{n}\right), \tag{10.63}$$

where $H(x) \equiv -x \log(x) - (1 - x) \log(1 - x)$ is the binary Shannon entropy, studied in detail in Chapter 11. The importance of the Gilbert–Varshamov bound is that it guarantees the existence of good codes, provided one doesn't try to encode too many bits (k) into too small a number of bits (n). The proof of the Gilbert–Varshamov bound is quite simple, and is left as an exercise.

Exercise 10.23: Prove the Gilbert–Varshamov bound.

We conclude our survey of classical error-correction by explaining an important construction for codes known as the *dual* construction. Suppose C is an $[n, k]$ code with generator matrix G and parity check matrix H. Then we can define another code, the *dual* of C, denoted $C^{\perp}$, to be the code with generator matrix H^T and parity check matrix G^T. Equivalently, the dual of C consists of all codewords y such that y is orthogonal to all the codewords in C. A code is said to be *weakly self-dual* if $C \subseteq C^{\perp}$, and *strictly self-dual* if $C = C^{\perp}$. Rather remarkably, the dual construction for classical linear codes arises naturally in the study of quantum error-correction, and is the key to the construction of an important class of quantum codes known as CSS codes.

Exercise 10.24: Show that a code with generator matrix G is weakly self-dual if and only if $G^T G = 0$.

Exercise 10.25: Let C be a linear code. Show that if $x \in C^\perp$ then $\sum_{y \in C} (-1)^{x \cdot y} = |C|$, while if $x \notin C^\perp$ then $\sum_{y \in C} (-1)^{x \cdot y} = 0$.

10.4.2 Calderbank–Shor–Steane codes

Our first example of a large class of quantum error-correcting codes is the *Calderbank–Shor–Steane* codes, more usually known as CSS codes, after the initials of the inventors of this class of codes. CSS codes are an important subclass of the more general class of stabilizer codes.

Suppose C_1 and C_2 are $[n, k_1]$ and $[n, k_2]$ classical linear codes such that $C_2 \subset C_1$ and C_1 and $C_2^\perp$ both correct t errors. We will define an $[n, k_1 - k_2]$ quantum code CSS(C_1, C_2) capable of correcting errors on t qubits, the *CSS code of C_1 over C_2*, via the following construction. Suppose $x \in C_1$ is any codeword in the code C_1. Then we define the quantum state $|x + C_2\rangle$ by

$$|x + C_2\rangle \equiv \frac{1}{\sqrt{|C_2|}} \sum_{y \in C_2} |x + y\rangle, \tag{10.64}$$

where $+$ is bitwise addition modulo 2. Suppose x' is an element of C_1 such that $x - x' \in C_2$. Then it is easy to see that $|x + C_2\rangle = |x' + C_2\rangle$, and thus the state $|x + C_2\rangle$ depends only upon the coset of C_1/C_2 which x is in, explaining the coset notation we have used for $|x + C_2\rangle$. Furthermore, if x and x' belong to different cosets of C_2, then for no $y, y' \in C_2$ does $x + y = x' + y'$, and therefore $|x + C_2\rangle$ and $|x' + C_2\rangle$ are orthonormal states. The quantum code CSS(C_1, C_2) is defined to be the vector space spanned by the states $|x + C_2\rangle$ for all $x \in C_1$. The number of cosets of C_2 in C_1 is $|C_1|/|C_2|$ so the dimension of CSS(C_1, C_2) is $|C_1|/|C_2| = 2^{k_1 - k_2}$, and therefore CSS$(C_1, C_2)$ is an $[n, k_1 - k_2]$ quantum code.

We can exploit the classical error-correcting properties of C_1 and $C_2^\perp$ to detect and correct quantum errors! In fact, it is possible to error-correct up to t bit and phase flip errors on CSS(C_1, C_2) by making use of the error-correcting properties of C_1 and $C_2^\perp$, respectively. Suppose the bit flip errors are described by an n bit vector e_1 with 1s where bit flips occurred, and 0s elsewhere, and the phase flip errors are described by an n bit vector e_2 with 1s where phase flips occurred, and 0s elsewhere. If $|x + C_2\rangle$ was the original state then the corrupted state is:

$$\frac{1}{\sqrt{|C_2|}} \sum_{y \in C_2} (-1)^{(x+y) \cdot e_2} |x + y + e_1\rangle. \tag{10.65}$$

To detect where bit flips occurred it is convenient to introduce an ancilla containing sufficient qubits to store the syndrome for the code C_1, and initially in the all zero state $|0\rangle$. We use reversible computation to apply the parity matrix H_1 for the code C_1, taking $|x + y + e_1\rangle|0\rangle$ to $|x + y + e_1\rangle|H_1(x + y + e_1)\rangle = |x + y + e\rangle|H_1 e_1\rangle$, since $(x + y) \in C_1$ is annihilated by the parity check matrix. The effect of this operation is to produce the state:

$$\frac{1}{\sqrt{|C_2|}} \sum_{y \in C_2} (-1)^{(x+y) \cdot e_2} |x + y + e_1\rangle|H_1 e_1\rangle. \tag{10.66}$$

Exercise 10.26: Suppose H is a parity check matrix. Explain how to compute the transformation $|x\rangle|0\rangle \rightarrow |x\rangle|Hx\rangle$ using a circuit composed entirely of controlled-NOTs.

Error-detection for the bit flip errors is completed by measuring the ancilla to obtain the result $H_1 e_1$ and discarding the ancilla, giving the state

$$\frac{1}{\sqrt{|C_2|}} \sum_{y \in C_2} (-1)^{(x+y)\cdot e_2} |x + y + e_1\rangle. \tag{10.67}$$

Knowing the error syndrome $H_1 e_1$ we can infer the error e_1 since C_1 can correct up to t errors, which completes the error-detection. Recovery is performed simply by applying NOT gates to the qubits at whichever positions in the error e_1 a bit flip occurred, removing all the bit flip errors and giving the state

$$\frac{1}{\sqrt{|C_2|}} \sum_{y \in C_2} (-1)^{(x+y)\cdot e_2} |x + y\rangle. \tag{10.68}$$

To detect phase flip errors we apply Hadamard gates to each qubit, taking the state to

$$\frac{1}{\sqrt{|C_2|2^n}} \sum_z \sum_{y \in C_2} (-1)^{(x+y)\cdot(e_2+z)} |z\rangle, \tag{10.69}$$

where the sum is over all possible values for n bit z. Setting $z' \equiv z + e_2$, this state may be rewritten:

$$\frac{1}{\sqrt{|C_2|2^n}} \sum_{z'} \sum_{y \in C_2} (-1)^{(x+y)\cdot z'} |z' + e_2\rangle. \tag{10.70}$$

(The next step appeared as Exercise 10.25 on page 450.) Supposing $z' \in C_2^\perp$ it is easy to see that $\sum_{y \in C_2} (-1)^{y \cdot z'} = |C_2|$, while if $z' \notin C_2^\perp$ then $\sum_{y \in C_2} (-1)^{y \cdot z'} = 0$. Thus the state may be rewritten:

$$\frac{1}{\sqrt{2^n/|C_2|}} \sum_{z' \in C_2^\perp} (-1)^{x \cdot z'} |z' + e_2\rangle, \tag{10.71}$$

which looks just like a bit flip error described by the vector e_2! As for the error-detection for bit flips we introduce an ancilla and reversibly apply the parity check matrix H_2 for $C_2^\perp$ to obtain $H_2 e_2$, and correct the 'bit flip error' e_2, obtaining the state

$$\frac{1}{\sqrt{2^n/|C_2|}} \sum_{z' \in C_2^\perp} (-1)^{x \cdot z'} |z'\rangle. \tag{10.72}$$

The error-correction is completed by again applying Hadamard gates to each qubit; we can either compute the result of these gates directly, or note that the effect is to apply Hadamard gates to the state in (10.71) with $e_2 = 0$; since the Hadamard gate is self-inverse this takes us back to the state in (10.68) with $e_2 = 0$:

$$\frac{1}{\sqrt{|C_2|}} \sum_{y \in C_2} |x + y\rangle, \tag{10.73}$$

which is the original encoded state!

One important application of CSS codes is to prove a quantum version of the Gilbert–Varshamov bound, guaranteeing the existence of good quantum codes. This states that in the limit as n becomes large, an $[n, k]$ quantum code protecting against errors on up to t qubits exists for some k such that

$$\frac{k}{n} \geq 1 - 2H\left(\frac{2t}{n}\right). \tag{10.74}$$

Thus, good quantum error-correcting codes exist, provided one doesn't try to pack too many qubits k into an n qubit code. The proof of the Gilbert–Varshamov bound for CSS codes is rather more complex than the proof of the classical Gilbert–Varshamov bound, due to the constraints on the classical codes C_1 and C_2, and is left as an end of chapter problem.

Summarizing, suppose C_1 and C_2 are $[n, k_1]$ and $[n, k_2]$ classical linear codes, respectively, such that $C_2 \subset C_1$, and both C_1 and $C_2^{\perp}$ can correct errors on up to t bits. Then $CSS(C_1, C_2)$ is an $[n, k_1 - k_2]$ quantum error-correcting code which can correct arbitrary errors on up to t qubits. Furthermore, the error-detection and correction steps require only the application of Hadamard and controlled-NOT gates, in each case a number linear in the size of the code. Encoding and decoding can also be performed using a number of gates linear in the size of the code, but we won't discuss that here; it's discussed later in more generality in Section 10.5.8.

Exercise 10.27: Show that the codes defined by

$$|x + C_2\rangle \equiv \frac{1}{\sqrt{|C_2|}} \sum_{y \in C_2} (-1)^{u \cdot y} |x + y + v\rangle \tag{10.75}$$

and parameterized by u and v are equivalent to $CSS(C_1, C_2)$ in the sense that they have the same error-correcting properties. These codes, which we'll refer to as $CSS_{u,v}(C_1, C_2)$, will be useful later in our study of quantum key distribution, in Section 12.6.5.

The Steane code

An important example of a CSS code may be constructed using the $[7, 4, 3]$ Hamming code whose parity check matrix we recall here:

$$H = \begin{bmatrix} 0 & 0 & 0 & 1 & 1 & 1 & 1 \\ 0 & 1 & 1 & 0 & 0 & 1 & 1 \\ 1 & 0 & 1 & 0 & 1 & 0 & 1 \end{bmatrix}. \tag{10.76}$$

Suppose we label this code C and define $C_1 \equiv C$ and $C_2 \equiv C^{\perp}$. To use these codes to define a CSS code we need first to check that $C_2 \subset C_1$. By definition the parity check matrix of $C_2 = C^{\perp}$ is equal to the transposed generator matrix of $C_1 = C$:

$$H[C_2] = G[C_1]^T = \begin{bmatrix} 1 & 0 & 0 & 0 & 0 & 1 & 1 \\ 0 & 1 & 0 & 0 & 1 & 0 & 1 \\ 0 & 0 & 1 & 0 & 1 & 1 & 0 \\ 0 & 0 & 0 & 1 & 1 & 1 & 1 \end{bmatrix}. \tag{10.77}$$

Exercise 10.28: Verify that the transpose of the matrix in (10.77) is the generator of the $[7, 4, 3]$ Hamming code.

Comparing with (10.76) we see that the span of the rows of $H[C_2]$ strictly contains the span of the rows of $H[C_1]$, and since the corresponding codes are the kernels of $H[C_2]$ and $H[C_1]$ we conclude that $C_2 \subset C_1$. Furthermore, $C_2^\perp = (C^\perp)^\perp = C$, so both C_1 and $C_2^\perp$ are distance 3 codes which can correct errors on 1 bit. Since C_1 is a $[7, 4]$ code and C_2 is a $[7, 3]$ code it follows that $\mathrm{CSS}(C_1, C_2)$ is a $[7, 1]$ quantum code which can correct errors on a single qubit.

This $[7, 1]$ quantum code has nice properties that make it easy to work with, and will be used in many of the examples for the remainder of this chapter. It is known as the *Steane code*, after its inventor. The codewords of C_2 are easily determined from (10.77) and Exercise 10.28. Rather than write them out explicitly, we write them out implicitly as the entries in the logical $|0_L\rangle$ for the Steane code, $|0 + C_2\rangle$:

$$|0_L\rangle = \frac{1}{\sqrt{8}} \Big[|0000000\rangle + |1010101\rangle + |0110011\rangle + |1100110\rangle$$

$$+ |0001111\rangle + |1011010\rangle + |0111100\rangle + |1101001\rangle \Big] . \quad (10.78)$$

To determine the other logical codeword we need to find an element of C_1 that is not in C_2. An example of such an element is $(1, 1, 1, 1, 1, 1, 1)$, giving:

$$|1_L\rangle = \frac{1}{\sqrt{8}} \Big[|1111111\rangle + |0101010\rangle + |1001100\rangle + |0011001\rangle$$

$$+ |1110000\rangle + |0100101\rangle + |1000011\rangle + |0010110\rangle \Big] . \quad (10.79)$$

10.5 Stabilizer codes

We cannot clone, perforce; instead, we split
Coherence to protect it from that wrong
That would destroy our valued quantum bit
And make our computation take too long.

Correct a flip and phase – that will suffice.
If in our code another error's bred,
We simply measure it, then God plays dice,
Collapsing it to X or Y or zed.

We start with noisy seven, nine, or five
And end with perfect one. To better spot
Those flaws we must avoid, we first must strive
To find which ones commute and which do not.

With group and eigenstate, we've learned to fix
Your quantum errors with our quantum tricks.
– 'Quantum Error Correction Sonnet', by Daniel Gottesman

Stabilizer codes, sometimes known as *additive* quantum codes, are an important class of quantum codes whose construction is analogous to classical linear codes. In order to understand stabilizer codes it is useful to first develop the *stabilizer formalism*, a powerful method for understanding a wide class of operations in quantum mechanics. The applications of the stabilizer formalism extend far beyond quantum error-correction; however, in this book our main concern is with this specific application. After defining the stabilizer formalism, we explain how unitary gates and measurements may be described using it, and an important theorem which quantifies the limitations of stabilizer operations. We then present stabilizer constructions for stabilizer codes, along with explicit examples, a useful standard form, and circuits for encoding, decoding, and correction.

10.5.1 The stabilizer formalism

The central insight of the stabilizer formalism is easily illustrated by an example. Consider the EPR state of two qubits

$$|\psi\rangle = \frac{|00\rangle + |11\rangle}{\sqrt{2}}. \tag{10.80}$$

It is easy to verify that this state satisfies the identities $X_1 X_2|\psi\rangle = |\psi\rangle$ and $Z_1 Z_2|\psi\rangle = |\psi\rangle$; we say that the state $|\psi\rangle$ is *stabilized* by the operators $X_1 X_2$ and $Z_1 Z_2$. A little less obviously, the state $|\psi\rangle$ is the *unique* quantum state (up to a global phase) which is stabilized by these operators $X_1 X_2$ and $Z_1 Z_2$. The basic idea of the stabilizer formalism is that many quantum states can be more easily described by working with the operators that stabilize them than by working explicitly with the state itself. This claim is perhaps rather surprising at first sight; nevertheless it is true. It turns out that many quantum codes (including CSS codes and the Shor code) can be much more compactly described using stabilizers than in the state vector description. Even more importantly, errors on the qubits and operations such as the Hadamard gate, phase gate, and even the controlled-NOT gate and measurements in the computational basis are all easily described using the stabilizer formalism!

The key to the power of the stabilizer formalism lies in the clever use of *group theory*, whose basic elements are reviewed in Appendix 2. The group of principal interest is the *Pauli group* G_n on n qubits. For a single qubit, the Pauli group is defined to consist of all the Pauli matrices, together with multiplicative factors $\pm 1, \pm i$:

$$G_1 \equiv \{\pm I, \pm iI, \pm X, \pm iX, \pm Y, \pm iY, \pm Z, \pm iZ\}. \tag{10.81}$$

This set of matrices forms a group under the operation of matrix multiplication. You might wonder why we don't omit the multiplicative factors ± 1 and $\pm i$; the reason these are included is to ensure that G_1 is closed under multiplication, and thus forms a legitimate group. The general Pauli group on n qubits is defined to consist of all n-fold tensor products of Pauli matrices, and again we allow multiplicative factors $\pm 1, \pm i$.

We can now define stabilizers a little more precisely. Suppose S is a subgroup of G_n and define V_S to be the set of n qubit states which are fixed by every element of S. V_S is the *vector space stabilized by* S, and S is said to be the *stabilizer* of the space V_S, since every element of V_S is stable under the action of elements in S. You should convince yourself of the truth of the following simple exercise:

Exercise 10.29: Show that an arbitrary linear combination of any two elements of V_S is also in V_S. Therefore, V_S is a subspace of the n qubit state space. Show that V_S is the intersection of the subspaces fixed by each operator in S (that is, the eigenvalue one eigenspaces of elements of S).

Let's look at a simple example of the stabilizer formalism in action, a case with $n = 3$

qubits and $S \equiv \{I, Z_1 Z_2, Z_2 Z_3, Z_1 Z_3\}$. The subspace fixed by $Z_1 Z_2$ is spanned by $|000\rangle$, $|001\rangle$, $|110\rangle$ and $|111\rangle$, and the subspace fixed by $Z_2 Z_3$ is spanned by $|000\rangle$, $|100\rangle$, $|011\rangle$ and $|111\rangle$. Note that the elements $|000\rangle$ and $|111\rangle$ are common to both these lists. With these observations and a little thought one realizes that V_S must be the subspace spanned by the states $|000\rangle$ and $|111\rangle$.

In this example we determined V_S simply by looking at the subspaces stabilized by two of the operators in S. This is a manifestation of an important general phenomenon – the description of a group by its *generators*. As explained in Appendix 2 a set of elements $g_1, \ldots, g_l$ in a group G is said to *generate* the group G if every element of G can be written as a product of elements from the list $g_1, \ldots, g_l$, and we write $G = \langle g_1, \ldots, g_l \rangle$. In the example $S = \langle Z_1 Z_2, Z_2 Z_3 \rangle$ as $Z_1 Z_3 = (Z_1 Z_2)(Z_2 Z_3)$ and $I = (Z_1 Z_2)^2$. The great advantage of using generators to describe groups is that they provide a *compact* means of describing the group. Indeed, in Appendix 2 we show that a group G with size $|G|$ has a set of at most $\log(|G|)$ generators. Furthermore, to see that a particular vector is stabilized by a group S we need only check that the vector is stabilized by the generators, since it is then automatically stabilized by products of the generators, making this a most convenient representation. (The notation $\langle \cdots \rangle$ which we use for group generators may potentially be confused with the notation for observable averages introduced in Section 2.2.5 beginning on page 87; however, in practice, it is always clear from context how the notation is being used.)

Not just any subgroup S of the Pauli group can be used as the stabilizer for a non-trivial vector space. For example, consider the subgroup of G_1 consisting of $\{\pm I, \pm X\}$. Obviously the only solution to $(-I)|\psi\rangle = |\psi\rangle$ is $|\psi\rangle = 0$, and thus $\{\pm I, \pm X\}$ is the stabilizer for the trivial vector space. What conditions must be satisfied by S in order that it stabilize a non-trivial vector space, V_S? Two conditions easily seen to be necessary are that (a) the elements of S commute, and (b) $-I$ is not an element of S. We don't yet have all the tools to prove it, but will show later that these two conditions are also sufficient for V_S to be non-trivial.

Exercise 10.30: Show that $-I \notin S$ implies $\pm iI \notin S$.

To see that these two conditions are necessary, suppose V_S is non-trivial, so it contains a non-zero vector $|\psi\rangle$. Let M and N be elements of S. Then M and N are tensor products of Pauli matrices, possibly with an overall multiplicative factor; because the Pauli matrices all commute or anti-commute with one another, it follows that M and N must either commute or anti-commute. To establish condition (a), that they commute, we suppose that M and N anti-commute and show that this leads to a contradiction. By assumption $-NM = MN$ so we have $-|\psi\rangle = -NM|\psi\rangle = MN|\psi\rangle = |\psi\rangle$, where the first and last equalities follow from the fact that M and N stabilize $|\psi\rangle$. So we have $-|\psi\rangle = |\psi\rangle$, which implies that $|\psi\rangle$ is the zero vector, which is the desired contradiction. To establish condition (b), that $-I \notin S$ just note that if $-I$ is an element of S then we have $-I|\psi\rangle = |\psi\rangle$, which again leads to a contradiction.

Exercise 10.31: Suppose S is a subgroup of G_n generated by elements $g_1, \ldots, g_l$. Show that all the elements of S commute if and only if g_i and g_j commute for each pair i, j.

A beautiful example of the stabilizer formalism is provided by the seven qubit Steane

Name	Operator
g_1	$I\,I\,I\,X\,X\,X\,X$
g_2	$I\,X\,X\,I\,I\,X\,X$
g_3	$X\,I\,X\,I\,X\,I\,X$
g_4	$I\,I\,I\,Z\,Z\,Z\,Z$
g_5	$I\,Z\,Z\,I\,I\,Z\,Z$
g_6	$Z\,I\,Z\,I\,Z\,I\,Z$

Figure 10.6. Stabilizer generators for the Steane seven qubit code. The entries represent tensor products on the respective qubits; for example, $ZIZIZIZ = Z \otimes I \otimes Z \otimes I \otimes Z \otimes I \otimes Z = Z_1 Z_3 Z_5 Z_7$.

code. It turns out that the six generators g_1 through g_6 listed in Figure 10.6 generate a stabilizer for the code space of the Steane code. Observe how clean and compact this description is when compared with the rather messy specification in terms of state vectors, (10.78) and (10.79); even further advantages will manifest when we examine quantum error-correction from this viewpoint. Note also the similarity in structure between the generators in Figure 10.6 and the parity check matrices for the classical linear codes C_1 and $C_2^\perp$ used in the construction of the Steane code. (Recall that for the Steane code $C_1 = C_2^\perp$ is the Hamming [7, 4, 3] code with parity check matrix given by (10.76).) The first three generators of the stabilizer have Xs in locations corresponding to the locations of the 1s in the parity check matrix for C_1, while the final three generators g_4 through g_6 have Zs in locations corresponding to the locations of the 1s in the parity check matrix for $C_2^\perp$. With these observations in hand the solution to the following exercise becomes almost self-evident.

Exercise 10.32: Verify that the generators in Figure 10.6 stabilize the codewords for the Steane code, as described in Section 10.4.2.

This use of the stabilizer formalism to describe a quantum code foreshadows our later use of stabilizers to describe a wide class of quantum codes, but for now it is important to appreciate that there is nothing special about the Steane code's status as a quantum code – it is merely a subspace of a vector space which happens to have a description using stabilizers.

In practice, we want our generators $g_1, \dots, g_l$ to be *independent* in the sense that removing any generator g_i makes the group generated smaller,

$$\langle g_1, \dots, g_{i-1}, g_{i+1}, \dots, g_l \rangle \neq \langle g_1, \dots, g_l \rangle. \tag{10.82}$$

Determining whether a particular set of generators is independent or not is rather time-consuming with our current understanding; fortunately, there is a simple way this can be done based on an idea known as the check matrix, so-named because it plays a role in the theory of stabilizer codes analogous to the parity check matrix in classical linear codes.

Suppose $S = \langle g_1, \dots, g_l \rangle$. There is an extremely useful way of presenting the generators $g_1, \dots, g_l$ of S using the *check matrix*. This is an $l \times 2n$ matrix whose rows correspond to the generators g_1 through g_l; the left hand side of the matrix contains 1s to indicate which generators contain Xs, and the right hand side contains 1s to indicate which generators contain Zs; the presence of a 1 on both sides indicates a Y in the generator. More explicitly, the ith row is constructed as follows. If g_i contains an I on the jth qubit then the jth and $n + j$th column elements are 0; if it contains an X on the jth qubit then the jth column element is a 1 and the $n + j$th column element is a 0; if it contains a Z on the jth qubit then the jth column element is 0 and the $n + j$th column element is 1; if it contains a Y on the jth qubit then both the jth and $n + j$th columns are 1. In the case of the Steane seven qubit code we can read the check matrix off from Figure 10.6:

$$
\left[
\begin{array}{ccccccc|ccccccc}
0 & 0 & 0 & 1 & 1 & 1 & 1 & 0 & 0 & 0 & 0 & 0 & 0 & 0 \\
0 & 1 & 1 & 0 & 0 & 1 & 1 & 0 & 0 & 0 & 0 & 0 & 0 & 0 \\
1 & 0 & 1 & 0 & 1 & 0 & 1 & 0 & 0 & 0 & 0 & 0 & 0 & 0 \\
0 & 0 & 0 & 0 & 0 & 0 & 0 & 0 & 0 & 0 & 1 & 1 & 1 & 1 \\
0 & 0 & 0 & 0 & 0 & 0 & 0 & 0 & 1 & 1 & 0 & 0 & 1 & 1 \\
0 & 0 & 0 & 0 & 0 & 0 & 0 & 1 & 0 & 1 & 0 & 1 & 0 & 1
\end{array}
\right] .
\tag{10.83}
$$

The check matrix doesn't contain any information about the multiplicative factors out the front of the generators, but it does contain much other useful information, so much so that we use $r(g)$ to denote the $2n$-dimensional row vector representation of an element g of the Pauli group. Suppose we define a $2n \times 2n$ matrix Λ by

$$
\Lambda = \begin{bmatrix} 0 & I \\ I & 0 \end{bmatrix}
\tag{10.84}
$$

where the I matrices on the off-diagonals are $n \times n$. Elements g and g' of the Pauli group are easily seen to commute if and only if $r(g)\Lambda r(g')^T = 0$; the formula $x \Lambda y^T$ defines a sort of 'twisted' inner product between row matrices x and y expressing whether the elements of the Pauli group corresponding to x and y commute or not.

Exercise 10.33: Show that g and g' commute if and only if $r(g)\Lambda r(g')^T = 0$. (In the check matrix representation, arithmetic is done modulo two.)

Exercise 10.34: Let $S = \langle g_1, \ldots, g_l \rangle$. Show that $-I$ is not an element of S if and only if $g_j^2 = I$ for all j, and $g_j \neq -I$ for all j.

Exercise 10.35: Let S be a subgroup of G_n such that $-I$ is not an element of S. Show that $g^2 = I$ for all $g \in S$, and thus $g^\dagger = g$.

A useful connection between independence of the generators and the check matrix is established by means of the following proposition:

Proposition 10.3: Let $S = \langle g_1, \ldots, g_l \rangle$ be such that $-I$ is not an element of S. The generators g_1 through g_l are independent if and only if the rows of the corresponding check matrix are linearly independent.

Proof
We prove the contrapositive. Note first that g_i^2 must equal I for all i, by Exercise 10.35. Observe that $r(g) + r(g') = r(gg')$, so addition in the row vector representation corresponds to multiplication of group elements. Thus the rows of the check matrix are linearly dependent, $\sum_i a_i r(g_i) = 0$, and $a_j \neq 0$ for some j, if and only if $\prod_i g_i^{a_i}$ is equal to the

identity, up to an overall multiplicative factor. But $-I \notin S$ so the multiplicative factor must be 1, and the last condition corresponds to the condition $g_j = g_j^{-1} = \prod_{i \neq j} g_i^{a_i}$, and thus $g_1, \ldots, g_l$ are not independent generators. $\square$

The following innocuous looking proposition is surprisingly useful, and can be immediately leveraged into a proof that V_S is of dimension 2^k when S is generated by $l = n - k$ independent commuting generators, and $-I \notin S$. We will use the proposition repeatedly throughout the remainder of this chapter. Once again the tool of choice in the proof is the check matrix representation.

Proposition 10.4: Let $S = \langle g_1, \ldots, g_l \rangle$ be generated by l independent generators and satisfy $-I \notin S$. Fix i in the range $1, \ldots, l$. Then there exists $g \in G_n$ such that $g g_i g^\dagger = -g_i$ and $g g_j g^\dagger = g_j$ for all $j \neq i$.

Proof

Let G be the check matrix associated to $g_1, \ldots, g_l$. The rows of G are linearly independent by Proposition 10.3, so there exists a $2n$-dimensional vector x such that $G \Lambda x = e_i$, where e_i is the l-dimensional vector with a 1 in the ith position and 0s elsewhere. Let g be such that $r(g) = x^T$. Then by definition of x we have $r(g_j) \Lambda r(g)^T = 0$ for $j \neq i$ and $r(g_i) \Lambda r(g)^T = 1$, and thus $g g_i g^\dagger = -g_i$ and $g g_j g^\dagger = g_j$ for $j \neq i$. $\square$

We conclude our look at the most basic elements of the stabilizer formalism by fulfilling our earlier promise that V_S is non-trivial provided S is generated by independent commuting generators and $-I \notin S$. Indeed, if there are $l = n - k$ generators, then it is at least plausible (and we will prove) that V_S is 2^k-dimensional, based on the intuitive argument that each additional generator for the stabilizer cuts the dimension of V_S by a factor of $1/2$, as we might naively expect because the $+1$ and -1 eigenspaces for a tensor product of Pauli matrices divide the total Hilbert space into two subspaces of equal dimension.

Proposition 10.5: Let $S = \langle g_1 \ldots, g_{n-k} \rangle$ be generated by $n - k$ independent and commuting elements from G_n, and such that $-I \notin S$. Then V_S is a 2^k-dimensional vector space.

In all our later discussion of the stabilizer formalism we use the convention that stabilizers are always described in terms of independent commuting generators such that $-I \notin S$.

Proof

Let $x = (x_1, \ldots, x_{n-k})$ be a vector of $n - k$ elements of $\mathbf{Z}_2$. Define

$$P_S^x \equiv \frac{\prod_{j=1}^{n-k} (I + (-1)^{x_j} g_j)}{2^{n-k}}. \tag{10.85}$$

Because $(I + g_j)/2$ is the projector onto the $+1$ eigenspace of g_j, it is easy to see that $P_S^{(0,\ldots,0)}$ must be the projector onto V_S. By Proposition 10.4 for each x there exists g_x in G_n such that $g_x P_S^{(0,\ldots,0)}(g_x)^\dagger = P_S^x$, and therefore the dimension of P_S^x is the same as the dimension of V_S. Furthermore, for distinct x the P_S^x are easily seen to be orthogonal.

The proof is concluded with the algebraic observation that

$$I = \sum_x P_S^x.$$ (10.86)

The left hand side is a projector onto a 2^n-dimensional space, while the right hand side is a sum over 2^{n-k} orthogonal projectors of the same dimension as V_S, and thus the dimension of V_S must be 2^k. □

10.5.2 Unitary gates and the stabilizer formalism

We have been discussing the use of the stabilizer formalism to describe vector spaces. The formalism can also be used to describe the *dynamics* of those vector spaces in the larger state space, under a variety of interesting quantum operations. Aside from the intrinsic interest of understanding quantum dynamical operations, this goal is especially relevant because we will describe quantum error-correcting codes using the stabilizer formalism, and would like an elegant means for understanding the effects of noise and other dynamical processes on those codes. Suppose we apply a unitary operation U to a vector space V_S stabilized by the group S. Let $|\psi\rangle$ be any element of V_S. Then for any element g of S,

$$U|\psi\rangle = Ug|\psi\rangle = UgU^\dagger U|\psi\rangle,$$ (10.87)

and thus the state $U|\psi\rangle$ is stabilized by $UgU^\dagger$, from which we deduce that the vector space UV_S is stabilized by the group $USU^\dagger \equiv \{UgU^\dagger | g \in S\}$. Furthermore, if $g_1, \ldots, g_l$ generate S, then $Ug_1U^\dagger, \ldots, Ug_lU^\dagger$ generate $USU^\dagger$, so to compute the change in the stabilizer we need only compute how it affects the generators of the stabilizer.

The great advantage of this approach to dynamics is that for certain special unitary operations U this transformation of the generators takes on a particularly appealing form. Suppose, for example, that we apply a Hadamard gate to a single qubit. Note that

$$HXH^\dagger = Z; \quad HYH^\dagger = -Y; \quad HZH^\dagger = X.$$ (10.88)

As a consequence we correctly deduce that after a Hadamard gate is applied to the quantum state stabilized by Z ($|0\rangle$), the resulting state will be stabilized by X ($|+\rangle$).

Not very impressive, you may think! Imagine though that we had n qubits in a state whose stabilizer is $\langle Z_1, Z_2, \ldots, Z_n \rangle$. It is easy to see that this is the state $|0\rangle^{\otimes n}$. Applying the Hadamard gate to each of the n qubits we see that the state afterwards has a stabilizer $\langle X_1, X_2, \ldots, X_n \rangle$; again it is easy to see that this is just the familiar state which is an equal superposition of all computational basis states. What is remarkable about this example is that the usual (state vector) description of the final state requires 2^n amplitudes to be specified, compared with the description provided by the generators: $\langle X_1, \ldots, X_n \rangle$, which is linear in n! Still, you might say, after applying the Hadamard gate to each of the n qubits there is no entanglement in the quantum computer, so it is not so surprising that a compact description can be obtained. But much more is possible within the stabilizer formalism, including an efficient description of the controlled-NOT, which together with the Hadamard gate *is* capable of generating entanglement. To understand how this works, consider how the operators X_1, X_2, Z_1 and Z_2 behave under conjugation by the controlled-NOT. Denoting by U the controlled-NOT gate with qubit 1 as control

Operation	Input	Output
controlled-NOT	X_1	X_1X_2
	X_2	X_2
	Z_1	Z_1
	Z_2	Z_1Z_2
H	X	Z
	Z	X
S	X	Y
	Z	Z
X	X	X
	Z	$-Z$
Y	X	$-X$
	Z	$-Z$
Z	X	$-X$
	Z	Z

Figure 10.7. Transformation properties of elements of the Pauli group under conjugation by various common operations. The controlled-NOT has qubit 1 as the control and qubit 2 as the target.

and qubit 2 as the target, we have

$$
UX_1U^\dagger = \begin{bmatrix} 1 & 0 & 0 & 0 \\ 0 & 1 & 0 & 0 \\ 0 & 0 & 0 & 1 \\ 0 & 0 & 1 & 0 \end{bmatrix} \begin{bmatrix} 0 & 0 & 1 & 0 \\ 0 & 0 & 0 & 1 \\ 1 & 0 & 0 & 0 \\ 0 & 1 & 0 & 0 \end{bmatrix} \begin{bmatrix} 1 & 0 & 0 & 0 \\ 0 & 1 & 0 & 0 \\ 0 & 0 & 0 & 1 \\ 0 & 0 & 1 & 0 \end{bmatrix}
$$

$$
= \begin{bmatrix} 0 & 0 & 0 & 1 \\ 0 & 0 & 1 & 0 \\ 0 & 1 & 0 & 0 \\ 1 & 0 & 0 & 0 \end{bmatrix}
$$

$$
= X_1X_2 . \tag{10.89}
$$

Similar calculations show that $UX_2U^\dagger = X_2$, $UZ_1U^\dagger = Z_1$ and $UZ_2U^\dagger = Z_1Z_2$. To see how U conjugates other operators in the two qubit Pauli group we need only take products of the facts we already know. For example, to calculate $UX_1X_2U^\dagger$ we observe that $UX_1X_2U^\dagger = UX_1U^\dagger UX_2U^\dagger = (X_1X_2)X_2 = X_1$. The Y Pauli matrices may be dealt with similarly, as for example $UY_2U^\dagger = iUX_2Z_2U^\dagger = iUX_2U^\dagger UZ_2U^\dagger = iX_2(Z_1Z_2) = Z_1Y_2$.

Exercise 10.36: Explicitly verify that $UX_1U^\dagger = X_1X_2$, $UX_2U^\dagger = X_2$, $UZ_1U^\dagger = Z_1$, and $UZ_2U^\dagger = Z_1Z_2$. These and other useful conjugation relations for the Hadamard, phase, and Pauli gates are summarized in Figure 10.7.

Exercise 10.37: What is $UY_1U^\dagger$?

As an example of using the stabilizer formalism to understand unitary dynamics, consider the swap circuit introduced in Section 1.3.4 on page 23; for convenience, the circuit is illustrated in Figure 10.8. Consider the way the operators Z_1 and Z_2 transform by conjugation by the gates in the circuit. The operator Z_1 transforms through the sequence $Z_1 \rightarrow Z_1 \rightarrow Z_1 Z_2 \rightarrow Z_2$ and the operator Z_2 transforms through the sequence $Z_2 \rightarrow Z_1 Z_2 \rightarrow Z_1 \rightarrow Z_1$. Similarly, $X_1 \rightarrow X_2$ and $X_2 \rightarrow X_1$ under the circuit. Of course, if we take U to be the swap operator then it is obvious that $U Z_1 U^\dagger = Z_2$ and $U Z_2 U^\dagger = Z_1$, and similarly for X_1 and X_2, just as for the circuit in Figure 10.8. Proving that this implies that the circuit implements U is left as an exercise:

Exercise 10.38: Suppose U and V are unitary operators on two qubits which transform Z_1, Z_2, X_1, and X_2 by conjugation in the same way. Show this implies that $U = V$.

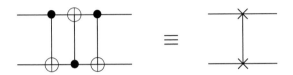

Figure 10.8. Circuit swapping two qubits.

The example of the swap circuit is interesting but doesn't do justice to the feature of the stabilizer formalism which makes it truly useful – the ability to describe certain types of quantum entanglement. We've already seen that the stabilizer formalism can be used to describe Hadamard gates and controlled-NOT gates, and of course these gates together can be used to create entangled states (compare Section 1.3.6). We will see that the stabilizer formalism can in fact be used to describe a wide class of entangled states, including many quantum error-correcting codes.

What gates other than the Hadamard and controlled-NOT gates can be described within the stabilizer formalism? The most important addition to this set is the phase gate, a single qubit gate whose definition we now recall,

$$ S = \begin{bmatrix} 1 & 0 \\ 0 & i \end{bmatrix} . \tag{10.90} $$

The action of the phase gate by conjugation on the Pauli matrices is easily computed:

$$ SXS^\dagger = Y; \qquad SZS^\dagger = Z . \tag{10.91} $$

Exercise 10.39: Verify (10.91).

Indeed, it turns out that *any* unitary operation taking elements of G_n to elements of G_n under conjugation can be composed from the Hadamard, phase and controlled-NOT gates. By definition, we say the set of U such that $U G_n U^\dagger = G_n$ is the *normalizer* of G_n, and denote it by $N(G_n)$, so we are claiming that the normalizer of G_n is generated by the Hadamard, phase and controlled-NOT gates, in view of which the Hadamard, phase and controlled-NOT gates are sometimes referred to simply as the normalizer gates. The proof of this result is simple but instructive, and you will work through it in Exercise 10.40 on page 462.

Theorem 10.6: Suppose U is any unitary operator on n qubits with the property that if $g \in G_n$ then $UgU^\dagger \in G_n$. Then up to a global phase U may be composed from $O(n^2)$ Hadamard, phase and controlled-NOT gates.

Exercise 10.40: Provide an inductive proof of Theorem 10.6 as follows.

(1) Prove that the Hadamard and phase gates can be used to perform any normalizer operation on a single qubit.

(2) Suppose U is an $n + 1$ qubit gate in $N(G_{n+1})$ such that $UZ_1U^\dagger = X_1 \otimes g$ and $UX_1U^\dagger = Z_1 \otimes g'$ for some $g, g' \in G_n$. Define U' on n qubits by $U'|\psi\rangle \equiv \sqrt{2}\langle 0|U(|0\rangle \otimes |\psi\rangle)$. Use the inductive hypothesis to show that the construction for U in Figure 10.9 may be implemented using $O(n^2)$ Hadamard, phase and controlled-NOT gates.

(3) Show that any gate $U \in N(G_{n+1})$ may be implemented using $O(n^2)$ Hadamard, phase and controlled-NOT gates.

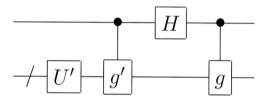

Figure 10.9. Construction used to prove that the Hadamard, phase and controlled-NOT gates generate the normalizer $N(G_n)$.

We've seen that many interesting quantum gates are in the normalizer $N(G_n)$; are there any gates which aren't? It turns out that most quantum gates are outside the normalizer. Two gates of particular interest which are not in the normalizer are the $\pi/8$ and Toffoli gates. Letting U denote the Toffoli gate with qubits 1 and 2 as controls and qubit 3 as the target, and recalling that T denotes the $\pi/8$ gate, we can easily calculate the action by conjugation of the $\pi/8$ and Toffoli gates on Pauli matrices as

$$TZT^\dagger = Z \qquad\qquad TXT^\dagger = \frac{X + Y}{\sqrt{2}} \qquad\qquad (10.92)$$

and

$$UZ_1U^\dagger = Z_1 \qquad\qquad UX_1U^\dagger = X_1 \otimes \frac{I + Z_2 + X_3 - Z_2X_3}{2} \qquad\qquad (10.93)$$

$$UZ_2U^\dagger = Z_2 \qquad\qquad UX_2U^\dagger = X_2 \otimes \frac{I + Z_1 + X_3 - Z_1X_3}{2} \qquad\qquad (10.94)$$

$$UX_3U^\dagger = X_3 \qquad\qquad UZ_3U^\dagger = Z_3 \otimes \frac{I + Z_1 + Z_2 - Z_1Z_2}{2}. \qquad\qquad (10.95)$$

Unfortunately, this makes analyzing quantum circuits including $\pi/8$ and Toffoli gates via the stabilizer formalism much less convenient than circuits which only contain Hadamard, phase and controlled-NOT gates. Fortunately, encoding, decoding, error-detection and recovery for stabilizer quantum codes can all be accomplished using only such normalizer gates, so the stabilizer formalism is very convenient for the analysis of such codes!

Exercise 10.41: Verify Equations (10.92) through (10.95).

10.5.3 Measurement in the stabilizer formalism

We have explained how a limited class of unitary gates may be conveniently described within the stabilizer formalism, but even more is true! Measurements in the computational basis may also be easily described within the stabilizer formalism. To understand how this works, imagine we make a measurement of $g \in G_n$ (recall that g is a Hermitian operator, and can thus be regarded as an observable in the sense of Section 2.2.5). For convenience we assume without loss of generality that g is a product of Pauli matrices with no multiplicative factor of -1 or $\pm i$ out the front. The system is assumed to be in a state $|\psi\rangle$ with stabilizer $\langle g_1, \ldots, g_n \rangle$. How does the stabilizer of the state transform under this measurement? There are two possibilities:

- g commutes with all the generators of the stabilizer.
- g anti-commutes with one or more of the generators of the stabilizer. Suppose the stabilizer has generators $g_1, \ldots, g_n$, and that g anti-commutes with g_1. Without loss of generality we may assume that g commutes with $g_2, \ldots, g_n$, since if it does not commute with one of these elements (say g_2) then it is easy to verify that g *does* commute with $g_1 g_2$, and we simply replace the generator g_2 by $g_1 g_2$ in our list of generators for the stabilizer.

In the first instance it follows that either g or $-g$ is an element of the stabilizer by the following argument. Since $g_j g |\psi\rangle = g g_j |\psi\rangle = g |\psi\rangle$ for each stabilizer generator, $g |\psi\rangle$ is in V_S and is thus a multiple of $|\psi\rangle$. Because $g^2 = I$ it follows that $g |\psi\rangle = \pm |\psi\rangle$, whence either g or $-g$ must be in the stabilizer. We assume that g is in the stabilizer, with the discussion for $-g$ proceeding analogously. In this instance $g |\psi\rangle = |\psi\rangle$ and thus a measurement of g yields $+1$ with probability one, and the measurement does not disturb the state of the system, and thus leaves the stabilizer invariant.

What about the second instance, when g anti-commutes with g_1 and commutes with all the other generators of the stabilizer? Note that g has eigenvalues ± 1 and so the projectors for the measurement outcomes ± 1 are given by $(I \pm g)/2$, respectively, and thus the measurement probabilities are given by

$$p(+1) = \text{tr} \left(\frac{I + g}{2} |\psi\rangle \langle\psi| \right) \tag{10.96}$$

$$p(-1) = \text{tr} \left(\frac{I - g}{2} |\psi\rangle \langle\psi| \right). \tag{10.97}$$

Using the facts that $g_1 |\psi\rangle = |\psi\rangle$ and $g g_1 = -g_1 g$ gives

$$p(+1) = \text{tr} \left(\frac{I + g}{2} g_1 |\psi\rangle \langle\psi| \right) \tag{10.98}$$

$$= \text{tr} \left(g_1 \frac{I - g}{2} |\psi\rangle \langle\psi| \right). \tag{10.99}$$

Applying the cyclic property of trace we may take g_1 to the right hand end of the trace and absorb it into $\langle\psi|$ using $g_1 = g_1^\dagger$ (Exercise 10.35 on page 457), giving

$$p(+1) = \text{tr} \left(\frac{I - g}{2} |\psi\rangle \langle\psi| \right) = p(-1). \tag{10.100}$$

Since $p(+1) + p(-1) = 1$ we deduce that $p(+1) = p(-1) = 1/2$. Suppose the result $+1$ occurs. In this instance the new state of the system is $|\psi^+\rangle \equiv (I + g)|\psi\rangle / \sqrt{2}$, which has stabilizer $\langle g, g_2, \ldots, g_n \rangle$. Similarly, if the result -1 occurs the posterior state is stabilized by $\langle -g, g_2, \ldots, g_n \rangle$.

10.5.4 The Gottesman–Knill theorem

The results about using stabilizers to describe unitary dynamics and measurements may be summarized in the remarkable *Gottesman–Knill* theorem:

Theorem 10.7: (**Gottesman–Knill theorem**) Suppose a quantum computation is performed which involves only the following elements: state preparations in the computational basis, Hadamard gates, phase gates, controlled-NOT gates, Pauli gates, and measurements of observables in the Pauli group (which includes measurement in the computational basis as a special case), together with the possibility of classical control conditioned on the outcome of such measurements. Such a computation may be efficiently simulated on a classical computer.

We have already implicitly proved the Gottesman–Knill theorem. The way the classical computer performs the simulation is simply to keep track of the generators of the stabilizer as the various operations are being performed in the computation. For example, to simulate a Hadamard gate we simply update each of the n generators describing the quantum state. Similarly, simulation of the state preparation, phase gate, controlled-NOT gate, Pauli gates, and measurements of observables in the Pauli group may all be done using $O(n^2)$ steps on a classical computer, so that a quantum computation involving m operations from this set can be simulated using $O(n^2 m)$ operations on a classical computer.

The Gottesman–Knill theorem highlights how subtle is the power of quantum computation. It shows that some quantum computations involving highly entangled states may be simulated efficiently on classical computers. Of course, not all quantum computations (and therefore, not all types of entanglement) can be described efficiently within the stabilizer formalism, but an impressive variety can be. Consider that interesting quantum information processing tasks such as quantum teleportation (Section 1.3.7) and super-dense coding (Section 2.3) can be performed using only the Hadamard gate, controlled-NOT gate, and measurements in the computational basis, and can therefore be efficiently simulated on a classical computer, by the Gottesman–Knill theorem. Moreover, we will shortly see that a wide variety of quantum error-correcting codes can be described within the stabilizer formalism. There is much more to quantum computation than just the power bestowed by quantum entanglement!

Exercise 10.42: Use the stabilizer formalism to verify that the circuit of Figure 1.13 on page 27 teleports qubits, as claimed. Note that the stabilizer formalism restricts the class of states being teleported, so in some sense this is not a complete description of teleportation, nevertheless it does allow an understanding of the dynamics of teleportation.

10.5.5 Stabilizer code constructions

The stabilizer formalism is ideally suited to the description of quantum codes. In this section we describe how this may be done, and use it to illustrate several important codes, such as Shor's nine qubit code, CSS codes, and a five qubit code which is the smallest code that can be used to protect against the effects of arbitrary errors on a single qubit. The basic idea is very simple: an $[n, k]$ *stabilizer code* is defined to be the vector space V_S stabilized by a subgroup S of G_n such that $-I \notin S$ and S has $n - k$ independent and commuting generators, $S = \langle g_1, \ldots, g_{n-k} \rangle$. We denote this code $C(S)$.

What are the logical basis states for the code $C(S)$? In principle, given $n - k$ generators for the stabilizer S we can choose any 2^k orthonormal vectors in the code $C(S)$ to act as our logical computational basis states. In practice it makes a great deal more sense to choose the states in a more systematic way. One method is as follows. First, we choose operators $\bar{Z}_1, \ldots, \bar{Z}_k \in G_n$ such that $g_1, \ldots, g_{n-k}, \bar{Z}_1, \ldots, \bar{Z}_k$ forms an independent and commuting set. (We explain in detail how this may be done a little later.) The $\bar{Z}_j$ operator plays the role of a logical Pauli sigma z operator on logical qubit number j, so the logical computational basis state $|x_1, \ldots, x_k\rangle_L$ is therefore defined to be the state with stabilizer

$$\langle g_1, \ldots, g_{n-k}, (-1)^{x_1} \bar{Z}_1, \ldots, (-1)^{x_k} \bar{Z}_k \rangle. \tag{10.101}$$

Similarly, we define $\bar{X}_j$ to be that product of Pauli matrices which takes $\bar{Z}_j$ to $-\bar{Z}_j$ under conjugation, and leaves all other $\bar{Z}_i$ and g_i alone when acting by conjugation. Clearly $\bar{X}_j$ has the effect of a quantum NOT gate acting on the jth encoded qubit. The operator $\bar{X}_j$ satisfies $\bar{X}_j g_k \bar{X}_j^\dagger = g_k$, and thus commutes with all the generators of the stabilizer. It is also easy to check that $\bar{X}_j$ commutes with all the $\bar{Z}_i$ except for $\bar{Z}_j$, with which it anti-commutes.

How are the error-correcting properties of a stabilizer code related to the generators of its stabilizer? Suppose we encode a state using an $[n, k]$ stabilizer code $C(S)$ with stabilizer $S = \langle g_1, \ldots, g_{n-k} \rangle$, and an error E occurs corrupting the data. In three stages of analysis we are going to determine what types of errors can be detected using $C(S)$ and when recovery can be performed. First, we'll take a look at the effect different types of errors have on the code space, simply to gain some intuition about what types of errors can be detected and corrected; there will be no proofs, as this stage is simply to build intuition. The second stage is the statement and proof of a general theorem telling us what kinds of errors can be detected and corrected by a stabilizer code, based upon the quantum error-correction conditions. The third stage of our analysis is to give a practical prescription for performing error-detection and recovery, using notions such as error syndrome.

Suppose $C(S)$ is a stabilizer code corrupted by an error $E \in G_n$. What happens to the code space when E anti-commutes with an element of the stabilizer? In this case E takes $C(S)$ to an orthogonal subspace, and the error can in principle be detected (and perhaps corrected after detection) by performing an appropriate projective measurement. If $E \in S$ we don't need to worry since the 'error' E doesn't corrupt the space at all. The real danger comes when E commutes with all the elements of S but is not actually in S, that is, $Eg = gE$ for all $g \in S$. The set of $E \in G_n$ such that $Eg = gE$ for all $g \in S$ is known as the *centralizer* of S in G_n and is denoted $Z(S)$. In fact, for the stabilizer groups S of concern to us, the centralizer is identical to a more familiar group, the normalizer of S, denoted $N(S)$, which is defined to consist of all elements E of G_n such that $EgE^\dagger \in S$ for all $g \in S$.

Exercise 10.43: Show that $S \subseteq N(S)$ for any subgroup S of G_n.

Exercise 10.44: Show that $N(S) = Z(S)$ for any subgroup S of G_n not containing $-I$.

These observations about the effect of various types of error operator E motivate the statement and proof of the following theorem, which is essentially just a translation of the quantum error-correction conditions (Theorem 10.1) into the terms of stabilizer codes.

Theorem 10.8: (**Error-correction conditions for stabilizer codes**) Let S be the stabilizer for a stabilizer code $C(S)$. Suppose $\{E_j\}$ is a set of operators in G_n such that $E_j^\dagger E_k \notin N(S) - S$ for all j and k. Then $\{E_j\}$ is a correctable set of errors for the code $C(S)$.

Without loss of generality we can restrict ourselves to considering errors E_j in G_n such that $E_j^\dagger = E_j$, which reduces the error-correction conditions for stabilizer codes to having $E_j E_k \notin N(S) - S$ for all j and k.

Proof
Let P be the projector onto the code space $C(S)$. For given j and k there are two possibilities: either $E_j^\dagger E_k$ in S or $E_j^\dagger E_k$ in $G_n - N(S)$. Consider the first case. Then $P E_j^\dagger E_k P = P$ since P is invariant under multiplication by elements of S. Suppose $E_j^\dagger E_k \in G_n - N(S)$ so that $E_j^\dagger E_k$ must anticommute with some element g_1 of S. Let $g_1, \ldots, g_{n-k}$ be a set of generators of S, so that

$$P = \frac{\prod_{l=1}^{n-k}(I + g_l)}{2^{n-k}}. \tag{10.102}$$

Using the anti-commutativity gives

$$E_j^\dagger E_k P = (I - g_1)E_j^\dagger E_k \frac{\prod_{l=2}^{n-k}(I + g_l)}{2^{n-k}}. \tag{10.103}$$

But $P(I - g_1) = 0$ since $(I + g_1)(I - g_1) = 0$ and therefore $P E_j^\dagger E_k P = 0$ whenever $E_j^\dagger E_k \in G_n - N(S)$. It follows that the set of errors $\{E_j\}$ satisfies the quantum error-correction conditions, and thus forms a correctable set of errors.   □

The statement and proof of Theorem 10.8 are wonderful theoretical results, but they don't explicitly tell us how to perform the error-correction operation when it is in fact possible! To understand how this is achieved, suppose $g_1, \ldots, g_{n-k}$ is a set of generators for the stabilizer of an $[n, k]$ stabilizer code, and that $\{E_j\}$ is a set of correctable errors for the code. Error-detection is performed by measuring the generators of the stabilizer g_1 through g_{n-k} in turn, to obtain the error syndrome, which consists of the results of the measurements, β_1 through β_{n-k}. If the error E_j occurred then the error syndrome is given by β_l such that $E_j g_l E_j^\dagger = \beta_l g_l$. In the case when E_j is the unique error operator having this syndrome recovery may be achieved simply by applying $E_j^\dagger$. In the case when there are two distinct errors E_j and $E_{j'}$ giving rise to the same error syndrome, it follows that $E_j P E_j^\dagger = E_{j'} P E_{j'}^\dagger$, where P is the projector onto the code space, so $E_j^\dagger E_{j'} P E_{j'}^\dagger E_j = P$, whence $E_j^\dagger E_{j'} \in S$, and thus applying $E_j^\dagger$ after the error $E_{j'}$ has occurred results in a successful recovery. Thus, for each possible error syndrome we simply pick out a single error E_j with that syndrome, and apply $E_j^\dagger$ to achieve recovery when that syndrome is observed.

Theorem 10.8 motivates the definition of a notion of *distance* for a quantum code analogous to the distance for a classical code. We define the *weight* of an error $E \in G_n$ to be the number of terms in the tensor product which are not equal to the identity. For example, the weight of $X_1 Z_4 Y_8$ is three. The distance of a stabilizer code $C(S)$ is defined to be the minimum weight of an element of $N(S) - S$, and if $C(S)$ is an $[n, k]$ code with distance d then we say that $C(S)$ is an $[n, k, d]$ stabilizer code. By Theorem 10.8 a code with distance at least $2t + 1$ is able to correct arbitrary errors on any t qubits, just as in the classical case.

Exercise 10.45: (Correcting located errors) Suppose $C(S)$ is an $[n, k, d]$ stabilizer code. Suppose k qubits are encoded in n qubits using this code, which is then subjected to noise. Fortunately, however, we are told that only $d - 1$ of the qubits are affected by the noise, and moreover, we are told precisely which $d - 1$ qubits have been affected. Show that it is possible to correct the effects of such *located* errors.

10.5.6 Examples

We now give a few simple examples of stabilizer codes, including already familiar codes such as the Shor nine qubit code and CSS codes, but from the new point of view of the stabilizer formalism. In each case, the properties of the codes follow easily by applying Theorem 10.8 to the generators of the stabilizer. With the examples under our belt we will turn our attention to finding quantum circuits for performing encoding and decoding.

The three qubit bit flip code

Consider the familiar three qubit bit flip code spanned by the states $|000\rangle$ and $|111\rangle$, with stabilizer generated by $Z_1 Z_2$ and $Z_2 Z_3$. By inspection we see that every possible product of two elements from the error set $\{I, X_1, X_2, X_3\} - I, X_1, X_2, X_3, X_1 X_2, X_1 X_3, X_2 X_3$ – anti-commutes with at least one of the generators of the stabilizer (except for I, which is in S), and thus by Theorem 10.8 the set $\{I, X_1, X_2, X_3\}$ forms a correctable set of errors for the three qubit bit flip code with stabilizer $\langle Z_1 Z_2, Z_2 Z_3 \rangle$.

Error-detection for the bit flip code is effected by measuring the stabilizer generators, $Z_1 Z_2$ and $Z_2 Z_3$. If, for example, the error X_1 occurred, then the stabilizer is transformed to $\langle -Z_1 Z_2, Z_2 Z_3 \rangle$, so the syndrome measurement gives the results -1 and $+1$. Similarly, the error X_2 gives error syndrome -1 and -1, the error X_3 gives error syndrome $+1$ and -1, and the trivial error I gives error syndrome $+1$ and $+1$. In each instance recovery is effected in the obvious way simply by applying the inverse operation to the error indicated by the error syndrome. The error-correction operation for the bit flip code is summarized in Figure 10.10.

Of course, the procedure we have outlined is exactly the same as that described earlier for the three qubit bit flip code! All this group-theoretic analysis would hardly be worthwhile if this was all the insight we gained. The real utility of the stabilizer formalism only starts to become apparent as we move to more complex examples.

Exercise 10.46: Show that the stabilizer for the three qubit phase flip code is generated by $X_1 X_2$ and $X_2 X_3$.

Z_1Z_2	Z_2Z_3	Error type	Action
+1	+1	no error	no action
+1	−1	bit 3 flipped	flip bit 3
−1	+1	bit 1 flipped	flip bit 1
−1	−1	bit 2 flipped	flip bit 2

Figure 10.10. Error-correction for the three qubit bit flip code in the language of stabilizer codes.

Name	Operator
g_1	$Z\,Z\,I\,I\,I\,I\,I\,I\,I$
g_2	$I\,Z\,Z\,I\,I\,I\,I\,I\,I$
g_3	$I\,I\,I\,Z\,Z\,I\,I\,I\,I$
g_4	$I\,I\,I\,I\,Z\,Z\,I\,I\,I$
g_5	$I\,I\,I\,I\,I\,I\,Z\,Z\,I$
g_6	$I\,I\,I\,I\,I\,I\,I\,Z\,Z$
g_7	$X\,X\,X\,X\,X\,X\,I\,I\,I$
g_8	$I\,I\,I\,X\,X\,X\,X\,X\,X$
$\bar{Z}$	$X\,X\,X\,X\,X\,X\,X\,X\,X$
$\bar{X}$	$Z\,Z\,Z\,Z\,Z\,Z\,Z\,Z\,Z$

Figure 10.11. The eight generators for the Shor nine qubit code, and the logical Z and logical X operations. (Yes, they really are the reverse of what one might naively expect!)

The nine qubit Shor code

The stabilizer for the Shor code has eight generators, as illustrated in Figure 10.11. It is easy to check the conditions of Theorem 10.8 for the error set containing I and all single qubit errors. Consider, for example, single qubit errors like X_1 and Y_4. The product X_1Y_4 anti-commutes with Z_1Z_2, and thus is not in $N(S)$. Similarly, all other products of two errors from this error set are either in S or else anti-commute with at least one element of S and thus are not in $N(S)$, implying that the Shor code can be used to correct an arbitrary single qubit error.

Exercise 10.47: Verify that the generators of Figure 10.11 generate the two codewords of Equation (10.13).

Exercise 10.48: Show that the operations $\bar{Z} = X_1X_2X_3X_4X_5X_6X_7X_8X_9$ and $\bar{X} = Z_1Z_2Z_3Z_4Z_5Z_6Z_7Z_8Z_9$ act as logical Z and X operations on a Shor-code encoded qubit. That is, show that this $\bar{Z}$ is independent of and commutes with the generators of the Shor code, and that $\bar{X}$ is independent of and commutes with the generators of the Shor code, and anti-commutes with $\bar{Z}$.

The five qubit code

What is the minimum size for a quantum code which encodes a single qubit so that any error on a single qubit in the encoded state can be detected and corrected? It turns out that the answer to this question is five qubits. (See Section 12.4.3). The stabilizer for

Name	Operator
g_1	$XZZXI$
g_2	$IXZZX$
g_3	$XIXZZ$
g_4	$ZXIXZ$
$\bar{Z}$	$ZZZZZ$
$\bar{X}$	XXXXX

Figure 10.12. The four generators for the five qubit code, and the logical Z and logical X operations. Note that the last three generators may be obtained by shifting the first right.

the five qubit code has the generators given in Figure 10.12. Because the five qubit code is the smallest capable of protecting against a single error it might be thought that it is the most useful code; however, for many applications it is more transparent to use the Steane seven qubit code.

Exercise 10.49: Use Theorem 10.8 to verify that the five qubit code can protect against an arbitrary single qubit error.

The logical codewords for the five qubit code are

$$
\begin{aligned}
|0_L\rangle = \frac{1}{4} \big[& |00000\rangle + |10010\rangle + |01001\rangle + |10100\rangle \\
& +|01010\rangle - |11011\rangle - |00110\rangle - |11000\rangle \\
& - |11101\rangle - |00011\rangle - |11110\rangle - |01111\rangle \\
& - |10001\rangle - |01100\rangle - |10111\rangle + |00101\rangle \big]
\end{aligned}
\tag{10.104}
$$

$$
\begin{aligned}
|1_L\rangle = \frac{1}{4} \big[& |11111\rangle + |01101\rangle + |10110\rangle + |01011\rangle \\
& +|10101\rangle - |00100\rangle - |11001\rangle - |00111\rangle \\
& - |00010\rangle - |11100\rangle - |00001\rangle - |10000\rangle \\
& - |01110\rangle - |10011\rangle - |01000\rangle + |11010\rangle \big]
\end{aligned}
\tag{10.105}
$$

Exercise 10.50: Show that the five qubit code saturates the quantum Hamming bound, that is, it satisfies the inequality of (10.51) with equality.

CSS codes and the seven qubit code

The CSS codes are an excellent example of a class of stabilizer codes, demonstrating beautifully how easy the stabilizer formalism makes it to understand quantum code construction. Suppose C_1 and C_2 are $[n, k_1]$ and $[n, k_2]$ classical linear codes such that $C_2 \subset C_1$ and C_1 and $C_2^{\perp}$ both correct t errors. Define a check matrix with the form

$$
\left[\begin{array}{c|c} H(C_2^{\perp}) & 0 \\ \hline 0 & H(C_1) \end{array} \right].
\tag{10.106}
$$

To see that this defines a stabilizer code, we need the check matrix to satisfy the commutativity condition $H(C_2^{\perp})H(C_1)^T = 0$. But we have $H(C_2^{\perp})H(C_1)^T = [H(C_1)G(C_2)]^T = 0$

because of the assumption $C_2 \subset C_1$. Indeed, it's an easy exercise to see that this code is exactly $\mathrm{CSS}(C_1, C_2)$, and that it is capable of correcting arbitrary errors on any t qubits.

The seven qubit Steane code is an example of a CSS code, whose check matrix we have already seen, in Equation (10.83). Encoded Z and X operators may be defined for the Steane code as

$$\bar{Z} \equiv Z_1 Z_2 Z_3 Z_4 Z_5 Z_6 Z_7; \quad \bar{X} \equiv X_1 X_2 X_3 X_4 X_5 X_6 X_7 . \tag{10.107}$$

Exercise 10.51: Verify that the check matrix defined in (10.106) corresponds to the stabilizer of the CSS code $\mathrm{CSS}(C_1, C_2)$, and use Theorem 10.8 to show that arbitrary errors on up to t qubits may be corrected by this code.

Exercise 10.52: Verify by direct operation on the codewords that the operators of (10.107) act appropriately, as logical Z and X.

10.5.7 Standard form for a stabilizer code

The construction of the logical Z and X operators for a stabilizer code is made much easier to understand if we put the code into *standard form*. To understand what the standard form is, consider the check matrix for an $[n, k]$ stabilizer code C:

$$G = \left[G_1 | G_2 \right] . \tag{10.108}$$

This matrix has $n - k$ rows. Swapping rows of this matrix corresponds to relabeling generators, swapping corresponding columns on both sides of the matrix to relabeling qubits, and adding two rows corresponds to multiplying generators; it is easy to see that we may always replace a generator g_i by $g_i g_j$ when $i \neq j$. Thus there is an equivalent code with a different set of generators whose corresponding check matrix corresponds to the matrix G where Gaussian elimination has been done on G_1, swapping qubits as necessary:

$$\begin{array}{c} r\{ \\ n-k-r\{ \end{array} \left[\begin{array}{cc|cc} \overbrace{I}^{r} & \overbrace{A}^{n-r} & \overbrace{B}^{r} & \overbrace{C}^{n-r} \\ 0 & 0 & D & E \end{array} \right], \tag{10.109}$$

where r is the rank of G_1. Next, swapping qubits as necessary we perform a Gaussian elimination on E to obtain

$$\begin{array}{c} r\{ \\ n-k-r-s\{ \\ s\{ \end{array} \left[\begin{array}{ccc|ccc} \overbrace{I}^{r} & \overbrace{A_1}^{n-k-r-s} & \overbrace{A_2}^{k+s} & \overbrace{B}^{r} & \overbrace{C_1}^{n-k-r-s} & \overbrace{C_2}^{k+s} \\ 0 & 0 & 0 & D_1 & I & E_2 \\ 0 & 0 & 0 & D_2 & 0 & 0 \end{array} \right]. \tag{10.110}$$

The last s generators cannot commute with the first r generators unless $D_2 = 0$, and thus we may assume that $s = 0$. Furthermore, we may also set $C_1 = 0$ by taking appropriate linear combinations of rows, so our check matrix has the form:

$$\begin{array}{c} r\{ \\ n-k-r\{ \end{array} \left[\begin{array}{ccc|ccc} \overbrace{I}^{r} & \overbrace{A_1}^{n-k-r} & \overbrace{A_2}^{k} & \overbrace{B}^{r} & \overbrace{0}^{n-k-r} & \overbrace{C}^{k} \\ 0 & 0 & 0 & D & I & E \end{array} \right], \tag{10.111}$$

where we have relabeled E_2 as E and D_1 as D. It is not difficult to see that this procedure is not unique; however, we will say that any code with check matrix in the form (10.111) is in standard form.

Given the standard form for a quantum code it is easy to define encoded Z operations for the code. That is, we have to pick k operators independent of the generators of the stabilizer and of one another, yet commuting with one another, and also with the generators of the stabilizer. Suppose we write the check matrix for these k encoded Z operators in the form $G_z = [F_1 F_2 F_3 | F_4 F_5 F_6]$, where all the matrices have k rows, and the respective column sizes are $r, n - k - r, k, r, n - k - r$ and k. We choose these matrices such that $G_z = [000 | A_2^T 0 I]$. The commutativity of these encoded Z operations with the elements of the stabilizer follows from the equation $I \times (A_2^T)^T + A_2 = 0$, and it is clear that the encoded Z operations commute with one another since they only contain products of Z operators. The independence of the encoded Z operators from the first r generators of the stabilizer follows from the fact that no X terms appear in the definition of the encoded Z operators, the independence from the set of $n - k - r$ generators follows from the $(n - k - r) \times (n - k - r)$ identity matrix appearing in the check matrix for those generators, and the lack of any corresponding terms in the check matrix for the encoded Z operators. In a similar way we may pick the encoded X operators, with $k \times 2n$ check matrix $[0 E^T I | C^T 00]$.

Exercise 10.53: Prove that the encoded Z operators are independent of one another.

Exercise 10.54: Prove that with the check matrix for the encoded X operators defined as above, the encoded X operators are independent of one another and of the generators, commute with the generators of the stabilizer, with each other, and $\bar{X}_j$ commutes with all the $\bar{Z}_k$ except $\bar{Z}_j$, with which it anti-commutes.

As an example we bring the check matrix for the Steane code (Equation 10.83) into standard form. We have $n = 7$ and $k = 1$ for this code, and inspection of the check matrix shows that the rank of the σ_x part is $r = 3$. The matrix may be brought into standard form by swapping qubits 1 and 4, 3 and 4, and 6 and 7, then by adding row 6 to row 4, then row 6 to row 5, and finally adding rows 4 and 5 to row 6. The resulting standard form is:

$$\begin{bmatrix} 1 & 0 & 0 & 0 & 1 & 1 & 1 & 0 & 0 & 0 & 0 & 0 & 0 & 0 \\ 0 & 1 & 0 & 1 & 0 & 1 & 1 & 0 & 0 & 0 & 0 & 0 & 0 & 0 \\ 0 & 0 & 1 & 1 & 1 & 1 & 0 & 0 & 0 & 0 & 0 & 0 & 0 & 0 \\ 0 & 0 & 0 & 0 & 0 & 0 & 0 & 1 & 0 & 1 & 1 & 0 & 0 & 1 \\ 0 & 0 & 0 & 0 & 0 & 0 & 0 & 0 & 1 & 1 & 0 & 1 & 0 & 1 \\ 0 & 0 & 0 & 0 & 0 & 0 & 0 & 1 & 1 & 1 & 0 & 0 & 1 & 0 \end{bmatrix}. \qquad (10.112)$$

We read off $A_2 = (1, 1, 0)$ and thus the encoded Z has check matrix $[0000000 | 1100001]$, which corresponds to $\bar{Z} = Z_1 Z_2 Z_7$. Recalling that qubits 1 and 4, 3 and 4, 6 and 7 were swapped, this corresponds to an encoded Z of $\bar{Z} = Z_2 Z_4 Z_6$ in the original code. This may appear rather puzzling in the light of Equation (10.107), which states that the encoded Z is $\bar{Z} = Z_1 Z_2 Z_3 Z_4 Z_5 Z_6 Z_7$; however, the puzzle is resolved by noting that the two 'different' encoded Z operations differ only by a factor $Z_1 Z_3 Z_5 Z_7$, which is an element of the stabilizer of the Steane code, and thus both have the same effect on Steane code states.

Exercise 10.55: Find the $\bar{X}$ operator for the standard form of the Steane code.

Exercise 10.56: Show that replacing an encoded X or Z operator by g times that operator, where g is an element of the stabilizer, does not change the action of the operator on the code.

Exercise 10.57:   Give the check matrices for the five and nine qubit codes in standard form.

10.5.8 Quantum circuits for encoding, decoding, and correction

One of the features of stabilizer codes is that their structure enables systematic construction of procedures for encoding, decoding, and error-correction. We describe the general method first, then present some explicit circuit constructions as examples. Let us begin with a general case, an $[n, k]$ stabilizer code with generators $g_1, \ldots, g_{n-k}$, and logical Z operators $\bar{Z}_1, \ldots, \bar{Z}_k$.

Preparing an encoded $|0\rangle^{\otimes k}$ state, which is the standard state for beginning a quantum computation, is quite simple. To do this we can start with any easily-prepared state – say, the state $|0\rangle^{\otimes n}$ – and measure each of the observables $g_1, \ldots, g_{n-k}, \bar{Z}_1, \ldots, \bar{Z}_k$ in turn. Depending on the measurement outcomes the resulting quantum state will have stabilizer $\langle \pm g_1, \ldots, \pm g_{n-k}, \pm \bar{Z}_1, \ldots, \pm \bar{Z}_k \rangle$, with the various signs being determined by the respective measurement outcomes. The signs of all the stabilizer generators and the $\bar{Z}_j$ can then be fixed up by applying products of Pauli operators, as described in the proof of Proposition 10.4, resulting in a state with stabilizer $\langle g_1, \ldots, g_{n-k}, \bar{Z}_1, \ldots, \bar{Z}_k \rangle$, that is, to the encoded $|0\rangle^{\otimes k}$. Once this state is prepared it is possible to change it to an arbitrary encoded computational basis state $|x_1, \ldots, x_k\rangle$ by applying the appropriate operators from the set $\bar{X}_1, \ldots, \bar{X}_k$. Of course, this approach to encoding has the disadvantage that it is not unitary. To obtain fully unitary encoding a different approach based upon the standard form of the check matrix may be used; this approach is outlined in Problem 10.3. Also, if you wish to encode an *unknown* state, this can also be done systematically, starting from an encoded $|0\rangle^{\otimes k}$ state, as explained in Problem 10.4. For our purposes it will be sufficient to prepare encoded $|0\rangle^{\otimes k}$ states.

Decoding quantum codes is also quite simple, however it is worth explaining why, for many purposes, a full decoding is not necessary. It turns out that the techniques of fault-tolerant quantum computation can be used to perform logical operations directly on encoded data, without the need to decode the data. Furthermore, the output of a computation performed in this way can be directly determined simply by measuring the logical Z operators, without the need to decode and measure in the computational basis. Thus, doing a fully unitary decoding which preserves the encoded quantum information is not so important for our purposes. If such a decoding procedure is desired for some reason – perhaps one is using quantum error-correcting codes to transmit information over a noisy communication channel – then it may be achieved by running the unitary encoding circuit found in Problem 10.3 backwards.

The error-correction procedure for a stabilizer code has already been described in Section 10.5.5, and is much like the encoding procedure: simply measure each of the generators $g_1, \ldots g_{n-k}$ in turn, obtaining the error syndrome $\beta_1, \ldots, \beta_{n-k}$. Classical computation is then used to determine from β_j the required recovery operations $E_j^\dagger$.

The key to constructing encoding, decoding, and error-correction circuits in each of the above descriptions is understanding how to measure operators. Recall that this is a generalization of the normal projective measurements we have widely used, in which the objective is to project a state into an eigenstate of the operator and to obtain both the projected state, and an indicator of the eigenvalue. If this reminds you of the phase estimation algorithm of Chapter 5, it's no coincidence! Recall from that chapter, and from Exercise 4.34 on page 188, that the circuit shown in Figure 10.13 can be used to measure the single qubit operator M (with eigenvalues ± 1), given a gate which performs a controlled-M operation. Two useful versions of this circuit, which measure X and Z, are given in Figures 10.14 and 10.15.

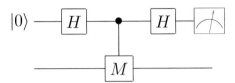

Figure 10.13. Quantum circuit for measuring a single qubit operator M with eigenvalues ± 1. The top qubit is the ancilla used for the measurement, and the bottom qubit is being measured.

Figure 10.14. Quantum circuit for measuring the X operator. Two equivalent circuits are given; the one on the left is the usual construction (as in Figure 10.13), and the one on the right is a useful equivalent circuit.

Figure 10.15. Quantum circuit for measuring the Z operator. Two equivalent circuits are given; the one on the left is the usual construction (as in Figure 10.13), and the one on the right is a useful simplification.

Of course, there is nothing special about the fact that M is a single qubit operator: the circuit in Figure 10.13 works just as well if we replace the second qubit with a bundle of qubits, and M is an arbitrary Hermitian operator with eigenvalues ± 1. Such operators include, for example, the products of Pauli operators we need to measure during the encoding, decoding and error-correction procedures for stabilizer codes.

As a concrete example, consider the syndrome measurement and encoding procedures for the seven qubit Steane code. A convenient starting point is the standard form of the check matrix for the code, Equation (10.112), because we can immediately read off the generators we need to measure directly from this matrix. Specifically, recall that the left block corresponds to X generators, and the right, Z, so the quantum circuit shown in Figure 10.16 immediately results. Note how the location of the zeroes and ones in the matrix corresponds to the location of the targets for the gates in the left half (which measure X), and the targets in the right half (which measure Z). This circuit can be used to perform error-correction by following the measurement results with products of Pauli operators on the code qubits to correct the errors. Or, by adding an additional measurement of $\bar{Z}$ and fixing the signs in the generators of the stabilizer, as described earlier, the circuit can be used to prepare the encoded logical state $|0_L\rangle$.

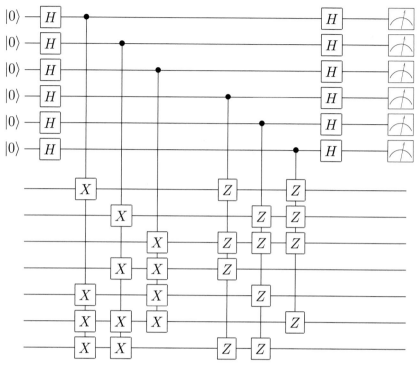

Figure 10.16. Quantum circuit for measuring the generators of the Steane code, to give the error syndrome. The top six qubits are the ancilla used for the measurement, and the bottom seven are the code qubits.

Exercise 10.58: Verify that the circuits in Figures 10.13–10.15 work as described, and check the claimed circuit equivalences.

Exercise 10.59: Show that by using the identities of Figures 10.14 and 10.15, the syndrome circuit of Figure 10.16 can be replaced with the circuit of Figure 10.17

Exercise 10.60: Construct a syndrome measuring circuit analogous to that in Figure 10.16, but for the nine and five qubit codes.

Exercise 10.61: Describe explicit recovery operations $E_j^\dagger$ corresponding to the different possible error syndromes that may be measured using the circuit in Figure 10.16.

10.6 Fault-tolerant quantum computation

One of the most powerful applications of quantum error-correction is not merely the protection of stored or transmitted quantum information, but the protection of quantum information as it *dynamically* undergoes *computation*. Remarkably, it turns out that arbitrarily good quantum computation can be achieved even with faulty logic gates, provided only that the error probability per gate is below a certain constant *threshold*. Over

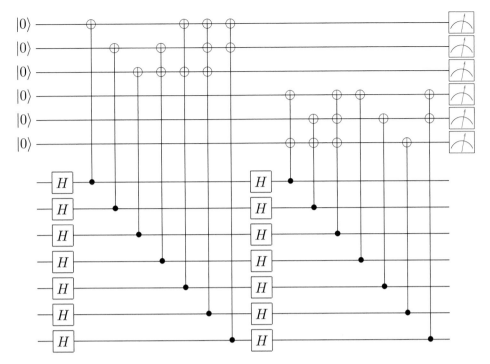

Figure 10.17. Quantum circuit equivalent to the one in Figure 10.16.

the next few sections we explain the techniques of *fault-tolerant quantum computation* which are used to achieve this remarkable result. We look first at the big picture in Section 10.6.1, before examining in detail the various elements of fault-tolerant quantum computation in Sections 10.6.2 and 10.6.3, and conclude in Section 10.6.4 with a discussion of some of the limitations and possible extensions of the fault-tolerant constructions. Note that a rigorous discussion of the many subtleties of fault-tolerant quantum computation lies somewhat beyond our scope; the interested reader is referred to the end of chapter 'History and further reading'.

10.6.1 Fault-tolerance: the big picture

The theory of fault-tolerant quantum computation integrates many different ideas *en route* to the threshold condition. We now describe each of these ideas in turn. We begin with the notion of computing on encoded data, and explain how the fundamental problems of error propagation and error accumulation require that our circuits for computing on encoded data satisfy certain fault-tolerance criteria. We then introduce our fundamental noise model for quantum circuits, which allows us to give more precise definitions for the notion of a fault-tolerant operation. We work through a specific example of a fault-tolerant operation in action – the fault-tolerant controlled-NOT – explaining how it can be used to prevent the propagation and accumulation of errors. We conclude by explaining how fault-tolerant operations can be combined with a procedure known as *concatenation* to obtain the threshold theorem for quantum computation, and give a simple estimate for the threshold.

Fundamental issues

The basic idea of fault-tolerant quantum computation is to compute directly on *encoded quantum states* in such a manner that decoding is never required. Suppose we are given a simple quantum circuit such as that shown in Figure 10.18. Unfortunately, noise afflicts each of the elements used to build this circuit – the state preparation procedures, quantum logic gates, measurement of the output, and even the simple transmission of quantum information along the quantum wires. To combat the effect of this noise we replace each qubit in the original circuit with an *encoded block* of qubits, using an error-correcting code such as the seven qubit Steane code, and replace each gate in the original circuit with a *procedure* for performing an *encoded gate* acting on the encoded state, as shown in Figure 10.19. By performing error-correction periodically on the encoded state we prevent accumulation of errors in the state. Of course, merely performing error-correction periodically is not sufficient to prevent the build-up of errors, even if it is applied after every encoded gate. The reasons for this are two-fold. First, and most importantly, the encoded gates can cause errors to *propagate*. For example, the encoded controlled-NOT illustrated in Figure 10.20 may cause an error on the encoded control qubit to propagate to the encoded target qubit. Thus errors in the qubits forming the encoded control qubit can propagate to become errors in the encoded target qubit. Encoded gates should therefore be designed very carefully so that a failure anywhere during the procedure for performing the encoded gate can only propagate to a small number of qubits in each block of the encoded data, in order that error-correction will be effective at removing the errors. Such procedures for performing encoded gates are referred to as *fault-tolerant* procedures, and we will show that it is possible to perform a universal set of logical operations – the Hadamard, phase, controlled-NOT and $\pi/8$ gates – using fault-tolerant procedures. The second issue that must be addressed is that error-correction itself can introduce errors on the encoded qubits, so we must be careful to design error-correction procedures that do not introduce too many errors into the encoded data. This can be achieved using techniques similar to those used to prevent propagation of errors by encoded gates, by taking care to ensure that failures during the procedure for error-correction do not propagate to cause too many errors in the encoded data.

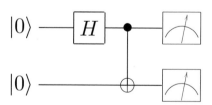

Figure 10.18. A simple quantum circuit. If each component in the circuit fails with probability p then the probability of an error at the output is $O(p)$.

Fault-tolerant operations: definitions

Let's pin down a little more precisely what it means for a particular procedure implementing an encoded quantum gate to be fault-tolerant. We *define* the fault-tolerance of a procedure to be the property that if only one component in the procedure fails then the failure causes at most one error in each encoded block of qubits output from the procedure. For example, the failure of a single component in a fault-tolerant recovery

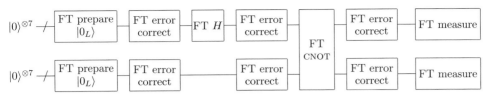

Figure 10.19. A simulation of the circuit in Figure 10.18, using encoded qubits and encoded logical operations. If fault-tolerant procedures are used to perform all the operations then the probability of error at the output is $O(p^2)$, where p is the probability for any individual component in the circuit to fail. An interesting feature is the second error-correction step performed on the second qubit. The 'operation' being error-corrected by this step appears to be trivial: nothing happens to the qubit at all! Nevertheless, simply *storing* qubits for a period of time introduces errors into the qubits, and should be periodically error-corrected in order to prevent the accumulation of errors.

Figure 10.20. A controlled-NOT gate can cause an error to propagate so that instead of affecting one qubit, it affects two. This is also true when encoded qubits are used, and an encoded controlled-NOT is implemented, as discussed in the text.

procedure for quantum error-correction results in the recovery procedure being performed correctly, up to an error on a single qubit of the output. By 'component' we mean any of the elementary operations used in the encoded gate, which might include noisy gates, noisy measurements, noisy quantum wires, and noisy state preparations. This definition of a fault-tolerant procedure for a quantum gate is sometimes generalized in the literature to cope with some of the more subtle issues that arise in the theory of fault-tolerant computation, but at our level of detail it is sufficient.

Of course, encoded quantum gates aren't all we wish to perform during our quantum computations. It is also useful to define the notions of a fault-tolerant measurement procedure, and fault-tolerant state preparation. A procedure for measuring an observable on a set of encoded qubits is said to be fault-tolerant if the failure of any single component in the procedure results in an error in at most one qubit in each encoded block of qubits at the output of the procedure. Furthermore, we require that if only one component fails then the measurement result reported must have probability of error of $O(p^2)$, where p is the (maximum) probability of a failure in any one of the components used to implement the measurement procedure. A procedure for preparing a fixed encoded state is said to be fault-tolerant if, given that a single component failed during the procedure, there is at most a single qubit in error in each encoded block of qubits output from the procedure.

To make these notions of fault-tolerance more precise, we need to be a little more specific about our error model. One of the major simplifications we are going to make in our analysis is to describe errors on qubits as being of one of four types: I, X, Y or Z, occurring stochastically with appropriate probabilities. We allow correlated errors to occur on two qubits, when performing gates such as the controlled-NOT, but again will assume that they are of the form of a tensor product of Pauli matrices, occurring with some probability. This probabilistic analysis enables us to use familiar concepts from classical probability theory to determine the total probability that the output from a circuit is correct or not. In more sophisticated presentations of fault-tolerance (see

'History and further reading') much more general error models can be considered, for example, allowing correlated errors of arbitrary type across several qubits. Nevertheless, the techniques used in those more sophisticated analyses are essentially generalizations of those we describe, combined with the deep insight we obtained earlier in the chapter that to perform error-correction of a continuum of possible errors it suffices to correct a discrete set of errors.

With our noise model at hand, we can be more precise about what we mean when we say that an error 'propagates' through the circuit. Consider, for example, the CNOT gate in Figure 10.20. Imagine that an X error on the first qubit occurs just before the CNOT is applied. If the unitary operator for the CNOT gate is denoted U, then the effective action of the circuit is $UX_1 = UX_1U^\dagger U = X_1X_2U$, that is, it is as though the controlled-NOT was applied correctly, but an X error occurred on both qubits after the CNOT. Over the remainder of this chapter we repeatedly use this trick of conjugating errors through gates to study how errors propagate through our circuits. A slightly more challenging example of error propagation is to suppose it is the CNOT gate itself which fails. What happens then? Suppose our noisy CNOT gate implements the quantum operation $\mathcal{E}$. Then this may be rewritten as $\mathcal{E} = \mathcal{E} \circ \mathcal{U}^{-1} \circ \mathcal{U}$, where $\mathcal{U}$ is the quantum operation implementing a perfect CNOT gate. Thus, the noisy CNOT gate is equivalent to applying a perfect CNOT followed by the operation $\mathcal{E} \circ \mathcal{U}^{-1}$, which is approximately the identity if our noisy CNOT is reasonably good, and can be understood in our usual error model of tensor products such as $X \otimes Z$ occurring on the two qubits with some small probability p.

Over the next few sections we explain in detail procedures for performing each class of fault-tolerant operations we have described – fault-tolerant quantum logic with a universal set of gates, fault-tolerant measurement, and fault-tolerant state preparation. The actual constructions we describe are for the Steane code, but they generalize fairly easily to the case of more general stabilizer codes. For now, however, we imagine that we have all of these procedures at our disposal. How can we put them together to perform quantum computation?

Example: fault-tolerant controlled-NOT

Let's examine a procedure for implementing a fault-tolerant controlled-NOT gate, followed by a fault-tolerant error-correction step, as illustrated in Figure 10.21. Our analysis of this circuit proceeds in 4 steps. Step 1 is the point of entry into the circuit, step 2 is after the encoded CNOT has been performed, step 3 is after the syndrome measurement, and step 4 is after the recovery operation has been applied. Our goal is to show that the probability that this circuit introduces two or more errors in the first encoded block behaves as $O(p^2)$, where p is the probability of failure of individual components in the circuit. Because a (hypothetical) perfect decoding of the first block of qubits only fails if there are two or more errors in the block, it follows that the probability that a perfectly decoded state contains errors is at most $O(p^2)$ larger after the action of this circuit than it was before.

To show that this procedure introduces two errors into the first encoded block with probability $O(p^2)$, let's identify all the possible ways this circuit can introduce two or more errors into the first encoded block of qubits at the output:

(1) There is a single pre-existing error entering the circuit at step 1 in each encoded block of qubits. This can cause two errors in the output from the first block because,

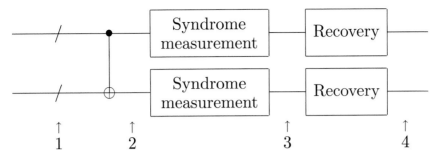

Figure 10.21. Block diagram of the fault-tolerant procedure construction, including error-correction.

for example, the error on the second block may propagate through the encoded CNOT circuit to cause an error on the first block of qubits. Provided the operations up to this stage have been done fault-tolerantly, we can argue that the probability of such an error entering on the first block is at most $c_0 p$ for some constant c_0, since such an error must have occurred during either the syndrome measurement or recovery stages in the previous stage of the quantum circuit. c_0 is the total number of places at which a failure may occur during syndrome measurement or recovery in the previous stage of the circuit. If we assume for simplicity that the probability of a single pre-existing error entering at step 1 on the second block is also $c_0 p$, and that these two errors occur independently, then the probability of this event is at most $c_0^2 p^2$. For the Steane code construction described below, there are contributions to c_0 from the six separate syndrome measurements, each of which has approximately 10^1 locations at which a failure may occur, together with a recovery operation involving seven components, for a total of approximately $c_0 \approx 70$.

(2) A single pre-existing error enters in step 1 on either the first or second block of qubits, and a single failure occurs during the fault-tolerant controlled-NOT. The probability of this is $c_1 p^2$, where c_1 is a constant defined to be the number of pairs of points at which a failure may occur. For the Steane code construction described below, we argued previously that there are roughly 70 locations times two blocks where a failure may have occurred causing an error to enter the circuit, for a total of 140 locations. There are a further 7 locations at which a failure may occur during the circuit, for a total of $c_1 \approx 7 \times 140 \approx 10^3$ locations at which a pair of failures may occur.

(3) Two failures occur during the fault-tolerant CNOT. This happens with probability at most $c_2 p^2$, where c_2 is the number of pairs of points at which a failure may occur. For the Steane code, $c_2 \approx 10^2$.

(4) A failure occurs during the CNOT and during the syndrome measurement. The only way two or more errors can occur at the output is if the syndrome measurement gives the incorrect result, which occurs with probability $c_3 p^2$ for some constant c_3 (for the Steane code, $c_3 \approx 10^2$). Another case which appears to be of interest, but which does not actually matter, is when the syndrome measurement gives the correct result, in which case the error introduced by the CNOT is correctly diagnosed and corrected by recovery, leaving only a single error at the output, introduced during syndrome measurement.

(5) Two or more failures occur during the syndrome measurement. This happens with

probability at most $c_4 p^2$, where c_4 is the number of pairs of points at which a failure may occur. For the Steane code, $c_4 \approx 70^2 \approx 5 \times 10^3$.

(6) A failure occurs during syndrome measurement and a failure during recovery. This happens with probability at most $c_5 p^2$, where c_5 is the number of pairs of points at which a failure may occur. For the Steane code, $c_5 \approx 70 \times 7 \approx 500$.

(7) Two or more failures occur during recovery. This happens with probability at most $c_6 p^2$, where c_6 is the number of pairs of points at which a failure may occur. For the Steane code, $c_6 \approx 7^2 \approx 50$.

Thus, the probability that this circuit introduces two or more errors into the encoded first block of qubits is at most cp^2 for some constant $c = c_0^2 + c_1 + c_2 + c_3 + c_4 + c_5 + c_6$ which is approximately equal to 10^4 for the Steane code. If a perfect decoding were to be performed at the end of the circuit, the probability of an error would therefore be at most cp^2. This is a truly remarkable result: we have managed to find an implementation for the CNOT with the property that individual components may fail with probability p, but the encoded procedure succeeds with probability $1 - cp^2$, and thus, provided p is small enough, in the example, $p < 10^{-4}$, there is a net gain to be had by using the encoded procedure! Similar conclusions can be drawn for all the other operations one might wish to do during a quantum computation, so that by doing any of our operations fault-tolerantly we can reduce our probability of a failure from p to cp^2, for some constant c. We have estimated c for the CNOT, however estimates for other fault-tolerant operations do not differ all that greatly, and we will continue to use $c \approx 10^4$ in our numerical estimates.

Concatenated codes and the threshold theorem

There is a beautiful construction based on *concatenated codes* which can be used to reduce the effective error rate achieved by the computation even further! The idea is to recursively apply the scheme described above for simulating a circuit using an encoded circuit, constructing a hierarchy of quantum circuits C_0 (the original circuit we wish to simulate), $C_1, C_2, \ldots$. In the first stage of this construction, each qubit in the original circuit is encoded in a quantum code whose qubits are themselves encoded in a quantum code, whose own qubits are encoded yet again, and so forth *ad infinitum*, as illustrated in Figure 10.22. In the second stage of this construction, any given gate in the original circuit C_0, such as a Hadamard gate, is replaced in the circuit C_1 by a fault-tolerant procedure implementing an encoded Hadamard gate and error-correction. Each of the components used in the circuit C_1 is then replaced in the circuit C_2 by a fault-tolerant procedure implementing an encoded version of the component and error-correction, and so on, *ad infinitum*. Suppose we do two levels of concatenation, as illustrated. If the failure probability of components at the lowest level of the code – the actual physical qubits – is p, then the failure probability at the middle level (one level of encoding) is at most cp^2, and at the highest level (two levels of encoding) – the level at which the circuit must function correctly if the computation is to produce the correct output – is $c(cp^2)^2$. Thus, if we concatenate k times, the failure probability for a procedure at the highest level is $(cp)^{2^k}/c$, while the size of the simulating circuit goes as d^k times the size of the original circuit, where d is a constant representing the maximum number of operations used in a fault-tolerant procedure to do an encoded gate and error-correction.

Suppose then that we wish to simulate a circuit containing $p(n)$ gates, where n specifies the size of some problem, and $p(n)$ is a polynomial function in n. This might be, for

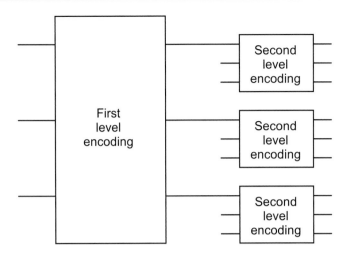

Figure 10.22. A two level concatenated code, encoding a single qubit in nine qubits. We use a three qubit code merely to keep the figure simple; in practice a code such as the Steane code which can correct arbitrary errors on one or more qubits would be used.

example, the circuit for the quantum factoring algorithm. Suppose we wish to achieve a final accuracy of ϵ in our simulation of this algorithm. To do so our simulation of each gate in the algorithm must be accurate to $\epsilon/p(n)$, so we must concatenate a number of times k such that

$$\frac{(cp)^{2^k}}{c} \le \frac{\epsilon}{p(n)}. \tag{10.113}$$

Provided $p < p_{\text{th}} \equiv 1/c$ such a k can be found. This condition – that $p < p_{\text{th}}$ – is known as the *threshold condition* for quantum computation, since provided it is satisfied we can achieve arbitrary accuracy in our quantum computations. How large a simulating circuit is required to achieve this level of accuracy? Note that we have

$$d^k = \left(\frac{\log(p(n)/c\epsilon)}{\log(1/pc)}\right)^{\log d} = O\left(\text{poly}(\log p(n)/\epsilon)\right), \tag{10.114}$$

where poly indicates a polynomial of fixed degree, and thus the simulating circuit contains

$$O(\text{poly}(\log p(n)/\epsilon)p(n)) \tag{10.115}$$

gates, which is only polylogarithmically larger than the size of the original circuit. Summarizing, we have the *threshold theorem for quantum computation*:

> **Threshold theorem for quantum computation:** A quantum circuit containing $p(n)$ gates may be simulated with probability of error at most ϵ using
>
> $$O(\text{poly}(\log p(n)/\epsilon)p(n)) \tag{10.116}$$
>
> gates on hardware whose components fail with probability at most p, provided p is below some constant *threshold*, $p < p_{\text{th}}$, and given reasonable assumptions about the noise in the underlying hardware.

What is the value of p_{th}? For the Steane code, $c \approx 10^4$ according to our counting, so a very rough estimate gives $p_{th} \approx 10^{-4}$. It needs to be emphasized that our estimates are (very) far from rigorous, however much more sophisticated calculations for the threshold have typically yielded values in the range 10^{-5}–10^{-6}. Note that the precise value of the threshold depends greatly on the assumptions made about the computational capabilities! For example, if parallel operations are not possible, then the threshold condition is impossible to achieve, because errors accumulate in the circuit too quickly for error-correction to cope with. Classical computation is also required in addition to the quantum operations, to process the measured syndromes and determine what quantum gates to apply to correct errors. Some discussion of limitations on estimates of the threshold are given in Section 10.6.4.

Exercise 10.62: Show by explicit construction of generators for the stabilizer that concatenating an $[n_1, 1]$ stabilizer code with an $[n_2, 1]$ stabilizer code gives an $[n_1 n_2, 1]$ stabilizer code.

10.6.2 Fault-tolerant quantum logic

A key technique in the construction of fault-tolerant quantum circuits is the method of constructing *fault-tolerant operations* to do logic on encoded states. In Section 4.5.3 of Chapter 4 we learned that the Hadamard, phase, controlled-NOT and $\pi/8$ gates form a universal set in terms of which any quantum computation may be expressed. We now explain how each of these gates can be implemented fault-tolerantly.

Normalizer operations

We begin with fault-tolerant constructions for the normalizer operations – the Hadamard, phase and controlled-NOT gates – for the specific case of the Steane code. By understanding the basic principles underlying the constructions for this concrete example it is easy to generalize them to any stabilizer code. Recall from Equation (10.107) that for the Steane code the Pauli $\bar{Z}$ and $\bar{X}$ operators on the encoded states can be written in terms of operators on the unencoded qubits, as

$$\bar{Z} = Z_1 Z_2 Z_3 Z_4 Z_5 Z_6 Z_7; \quad \bar{X} = X_1 X_2 X_3 X_4 X_5 X_6 X_7. \qquad (10.117)$$

An encoded Hadamard gate $\bar{H}$ should interchange $\bar{Z}$ and $\bar{X}$ under conjugation, just as the Hadamard gate interchanges Z and X under conjugation. $\bar{H} = H_1 H_2 H_3 H_4 H_5 H_6 H_7$ accomplishes this task, so that a Hadamard on the encoded qubit can be implemented as shown in Figure 10.23.

Exercise 10.63: Suppose U is any unitary operation mapping the Steane code into itself, and such that $U\bar{Z}U^\dagger = \bar{X}$ and $U\bar{X}U^\dagger = \bar{Z}$. Prove that up to a global phase the action of U on the encoded states $|0_L\rangle$ and $|1_L\rangle$ is $|0_L\rangle \rightarrow (|0_L\rangle + |1_L\rangle)/\sqrt{2}$ and $|1_L\rangle \rightarrow (|0_L\rangle - |1_L\rangle)/\sqrt{2}$.

This is a good first step, but just doing logic on the encoded state is not sufficient to make this operation fault-tolerant! We also need to understand how errors propagate. Because the circuit implementing $\bar{H} = H^{\otimes 7}$ does not involve more than one qubit in the encoded block in interaction, it seems physically reasonable to assume that the failure of a single component in the circuit can cause at most one error in the block of qubits output

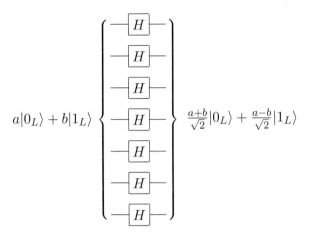

$$a|0_L\rangle + b|1_L\rangle \qquad \frac{a+b}{\sqrt{2}}|0_L\rangle + \frac{a-b}{\sqrt{2}}|1_L\rangle$$

Figure 10.23. Transversal Hadamard gate on a qubit encoded in the Steane code.

from the procedure. To see that this is true, imagine an error on the first qubit occurred just before the encoded H gate was applied. For the sake of definiteness, suppose that the error is a Z error, so the combined operation on the qubit is HZ. As in our earlier analysis of error propagation for the CNOT gate, inserting the identity $H^\dagger H = I$ gives $HZ = HZH^\dagger H = XH$, so such an error is equivalent to first applying H then the error X occurring. Similarly, a failure during the gate operation itself is equivalent to a perfect gate, followed by a small amount of noise acting on the qubit, which we can think of in terms of our usual model of X, Y and Z all occurring with some small probability. The circuit in Figure 10.23 thus really is a fault-tolerant operation, because a single failure occurring anywhere in the procedure doesn't propagate to affect other qubits, and thus causes at most one error in the block of qubits output from the procedure.

Are there any general principles we can distill from the circuit in Figure 10.23? One useful observation is that encoded gates are automatically fault-tolerant if they can be implemented in a bitwise fashion, since that property ensures that a single failure anywhere in the encoded gate introduces at most one error per block of the code, and thus error probabilities do not grow out of the control of the error-correcting code. This property, that an encoded gate can be implemented in a bitwise fashion, is known as the *transversality* property of an encoded quantum gate. Transversality is interesting because it offers a general design principle for finding fault-tolerant quantum circuits, and we see below that many gates other than the Hadamard gate can be given transversal implementations. Keep in mind, though, that it is possible to find fault-tolerant constructions which *aren't* transversal, as we'll see below with the example of the fault-tolerant $\pi/8$ gate.

Using the Steane code, many gates other than the Hadamard gate are easily given transversal (and thus fault-tolerant) implementations. Three of the most interesting, in addition to the Hadamard, are the phase gate and the Pauli X and Z gates. Suppose we apply the X gates bitwise to each of the seven qubits of the Steane code. This transforms each Z operator to $-Z$ under conjugation, so $\bar{Z} \to (-1)^7\bar{Z} = -\bar{Z}$ and $\bar{X} \to \bar{X}$ under conjugation by the bitwise application of X, and thus this circuit effects an encoded X operation on the states of the Steane code. This circuit is transversal, and thus is automatically fault-tolerant. In a similar way, applying Z bitwise to the states of the Steane code gives a fault-tolerant implementation of an encoded Z. The transversal implementation of the phase gate is a little more challenging. Under conjugation, $\bar{S}$ must take $\bar{Z}$ to $\bar{Z}$ and $\bar{X}$ to $\bar{Y} = i\bar{X}\bar{Z}$. However, applying the obvious guess $\bar{S} = S_1 S_2 S_3 S_4 S_5 S_6 S_7$ takes $\bar{Z}$ to $\bar{Z}$ under conjugation, and $\bar{X}$ to $-\bar{Y}$. The minus sign in front of the $-\bar{Y}$ may be fixed up by applying $\bar{Z}$. Thus, applying the operation ZS to each qubit in the code effects an encoded phase gate, which is transversal and thus fault-tolerant.

In contrast to the Hadamard, Pauli, and phase gates, implementing the controlled-NOT fault-tolerantly appears at first to be a challenge, because it involves two separate code blocks of seven qubits. How can we realize a CNOT which does not introduce more than one error per block of the code? Fortunately, this turns out to be very simple when using the Steane code, as illustrated in Figure 10.24: it is easily seen to be effected by seven CNOT gates applied pairwise between the seven qubits in the two blocks! You might worry that this transversal construction violates our own rules; after all, can't the controlled-NOT gates we are doing cause errors to propagate beyond a single qubit? This is correct, but there is no problem, because the error propagation only ever affects at most one qubit *in another block*; it does not adversely affect qubits within the same block. Remember that affecting qubits in other blocks is okay because each block can handle errors on single qubits!

More precisely, suppose that an X error on the first qubit occurs just before the controlled-NOT between the first qubit of each block, which we'll label qubits 1 and 8. If this controlled-NOT gate is denoted U, then the effective action is $UX_1 = UX_1U^\dagger U = X_1 X_8 U$, that is, it is as though the controlled-NOT was applied correctly, but an X error occurred on the first qubit of both blocks of encoded qubits! Slightly more challenging, suppose one of the CNOT gates fails. What happens then? Suppose our noisy CNOT gate implements the quantum operation $\mathcal{E}$. Then this may be rewritten as $\mathcal{E} = \mathcal{E} \circ \mathcal{U}^{-1} \circ \mathcal{U}$, where $\mathcal{U}$ is the quantum operation implementing a perfect CNOT gate. Thus, the noisy CNOT gate is equivalent to applying a perfect CNOT followed by the operation $\mathcal{E} \circ \mathcal{U}^{-1}$, which is approximately the identity if our noisy CNOT is reasonably good, and can be understood in our usual error model of tensor products such as $X \otimes Z$ occurring on the two qubits with some small probability. Fortunately, while such errors involve two qubits, they only involve a single qubit in each block of encoded qubits. Similar conclusions about error propagation apply to errors at other locations. It follows that the failure of a single component anywhere within the procedure we have described propagates to cause no more than one error in each block of encoded qubits, and thus this implementation of the encoded controlled-NOT is fault-tolerant.

Having found fault-tolerant implementations of the Hadamard, phase and controlled-NOT gates, it follows from Theorem 10.6 that any operation in the normalizer can be realized fault-tolerantly. Of course, normalizer operations do not exhaust the complete set of unitary gates required to do quantum computation, but this is a promising start!

Exercise 10.64: (Back propagation of errors) It is clear that an X error on the control qubit of a CNOT gate propagates to the target qubit. In addition, it turns out that a Z error on the target propagates back to the control! Show this using the stabilizer formalism, and also directly using quantum circuit identities. You may find Exercise 4.20 on page 179 useful.

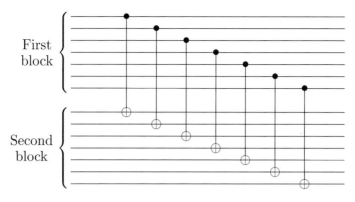

Figure 10.24. Transversal controlled-NOT between two qubits encoded in separate blocks with the Steane code.

Fault-tolerant $\pi/8$ gate

The one remaining gate we require to complete the standard set of gates for universal quantum computation is the $\pi/8$ gate. Alternatively, as was noted in Section 4.5.3, adding a fault-tolerant Toffoli gate to our current set of fault-tolerant Hadamard, phase and controlled-NOT gates would also give us a universal set, allowing us to perform all the gates required by a quantum computer in a fault-tolerant manner. It turns out that the fault-tolerant $\pi/8$ gate is very simple to realize, and using a similar but more elaborate construction a fault-tolerant Toffoli gate can be realized.

Our basic strategy in constructing the fault-tolerant $\pi/8$ gate is to split the construction into three parts. The first part of the construction is a simple circuit to simulate the $\pi/8$ gate using elements we already know how to do fault-tolerantly, such as the controlled-NOT, phase and X gates. There are, however, two parts of this circuit which we don't yet know how to make fault-tolerant. The first is the preparation of an *ancilla state* for input into the circuit. In order that this ancilla be adequate, we require that the failure of any component during the ancilla preparation should lead to at most a single error in the block of qubits making up the ancilla. We explain how such fault-tolerant ancilla preparation can be done later in this section. The second operation we need is measurement. In order to make the measurement fault-tolerant, we require that the failure of a single component during the procedure for measurement should not affect the measurement outcome. If it did, then the error would propagate to cause errors on many qubits in the first block, since whether the encoded SX operation is performed or not is determined by the measurement result. How to do such a fault-tolerant measurement is described in the next section. (Strictly speaking, for the fault-tolerant measurement procedure we describe the measurement outcome may actually be incorrect with probability $O(p^2)$, where p is the probability of failure of a single component. We will ignore this for the purposes of the present discussion; it is easily dealt with by a slightly more sophisticated analysis along similar lines.)

Figure 10.25 shows a circuit implementing a $\pi/8$ gate. All elements in the circuit can be performed fault-tolerantly, except perhaps those in the dashed box and the measurement. The circuit starts with two encoded qubits, one of which is the qubit $|\psi\rangle = a|0\rangle + b|1\rangle$ we wish to operate on (let $|0\rangle$ and $|1\rangle$ denote logical states here). The other qubit is prepared

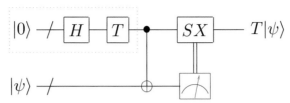

Figure 10.25. Quantum circuit which fault-tolerantly implements a $\pi/8$ gate. The dashed box represents a (non-fault-tolerant) preparation procedure for the ancilla state $(|0\rangle + \exp(i\pi/4)|1\rangle)/\sqrt{2}$; how to do this preparation fault-tolerantly is explained in the text. The slash on the wire denotes a bundle of seven qubits, and the double-line wire represents the classical bit resulting from the measurement. Note that the final SX operation is controlled by the measurement result.

in the state

$$|\Theta\rangle = \frac{|0\rangle + \exp(i\pi/4)|1\rangle}{\sqrt{2}}, \tag{10.118}$$

which is the state generated by the circuit in the dotted box in the figure. We explain how this ancilla preparation step may be done fault-tolerantly in a moment. Next, perform a fault-tolerant controlled-NOT operation, giving

$$\frac{1}{\sqrt{2}}\left[|0\rangle\left(a|0\rangle + b|1\rangle\right) + \exp(i\pi/4)|1\rangle\left(a|1\rangle + b|0\rangle\right)\right]$$

$$= \frac{1}{\sqrt{2}}\left[\left(a|0\rangle + b\exp(i\pi/4)|1\rangle\right)|0\rangle + \left(b|0\rangle + a\exp(i\pi/4)|1\rangle\right)|1\rangle\right]. \tag{10.119}$$

Finally, measure the second qubit, and if it is 0 then we are done. Otherwise, perform the operation

$$SX = \begin{bmatrix} 1 & 0 \\ 0 & i \end{bmatrix}\begin{bmatrix} 0 & 1 \\ 1 & 0 \end{bmatrix} \tag{10.120}$$

to the remaining qubit. Either way, we are left with the state $a|0\rangle + b\exp(i\pi/4)|1\rangle$ up to an irrelevant global phase, as required for a $\pi/8$ gate. This wonderful result may seem to have come out of nowhere, but in fact it is the result of a systematic construction which is explained in the exercises below. The same construction is used to realize a fault-tolerant Toffoli gate, as shown in Exercise 10.68.

The construction of the fault-tolerant $\pi/8$ gate requires a fault-tolerant method to produce the ancilla state $|\Theta\rangle$. This preparation can be achieved using the techniques for fault-tolerant measurements, as explained in detail in the next section. For now we explain the connection to fault-tolerant measurement. As shown in Figure 10.25, $|\Theta\rangle$ may be produced by applying a Hadamard gate and then a $\pi/8$ gate to the state $|0\rangle$. The state $|0\rangle$ is a $+1$ eigenstate of Z, so it follows that $|\Theta\rangle$ is a $+1$ eigenstate of $THZHT^\dagger = TXT^\dagger = e^{-i\pi/4}SX$. $|\Theta\rangle$ can therefore be prepared by first preparing an encoded $|0\rangle$, and then fault-tolerantly measuring $e^{-i\pi/4}SX$. If the result $+1$ is obtained we conclude that the state has been correctly prepared. If the result -1 is obtained, we have one of two options. We can either start over, repeating the procedure until the fault-tolerant measurement of $e^{-i\pi/4}SX$ gives the result $+1$, or we can use the more elegant and efficient observation that since $ZSXZ = -SX$, applying a fault-tolerant Z operation changes the state from the -1 eigenstate of $e^{-i\pi/4}SX$ to the $+1$ eigenstate of

$e^{-i\pi/4}SX, |\Theta\rangle$. Whichever procedure is used, a single failure anywhere in the procedure produces an error in at most one qubit in the ancilla state $|\Theta\rangle$.

It is not difficult to see that the procedure we have described is fault-tolerant as a whole; however, it may be useful to look at an explicit example to see this. Suppose a single component failure occurred during ancilla construction, leading to an error on a single qubit in the ancilla. This propagates through the encoded controlled-NOT gate to cause one error in each of the first and second blocks of qubits. Fortunately, an error on a single qubit in the second encoded qubit doesn't affect the result of our fault-tolerant measurement procedure, so SX is applied or not applied as appropriate, and thus the error on the first block of qubits propagates through to cause a single error in the output from the encoded gate. Similarly, it is not difficult to convince yourself that a single failure anywhere else in this procedure for an encoded $\pi/8$ gate leads to an error on only a single qubit in the output block of encoded qubits.

Exercise 10.65: An unknown qubit in the state $|\psi\rangle$ can be swapped with a second qubit which is prepared in the state $|0\rangle$ using only two controlled-NOT gates, with the circuit

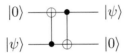

Show that the two circuits below, which use only a single CNOT gate, with measurement and a classically controlled single qubit operation, also accomplish the same task:

Exercise 10.66: (Fault-tolerant $\pi/8$ gate construction) One way to implement a $\pi/8$ gate is to first swap the qubit state $|\psi\rangle$ you wish to transform with some known state $|0\rangle$, then to apply a $\pi/8$ gate to the resulting qubit. Here is a quantum circuit which does that:

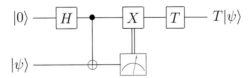

Doing this does not seem particularly useful, but actually it leads to something which is! Show that by using the relations $TX = \exp(-i\pi/4)SX$ and $TU = UT$ (U is the controlled-NOT gate, and T acts on the control qubit) we may obtain the circuit of Figure 10.25.

Exercise 10.67: Show that the following circuit identities hold:

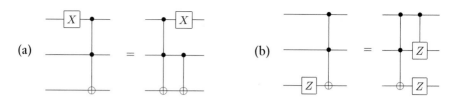

Exercise 10.68: (Fault-tolerant Toffoli gate construction) A procedure similar
to the above sequence of exercises for the $\pi/8$ gate gives a fault-tolerant Toffoli
gate.

(1) First, swap the three qubit state $|xyz\rangle$ you wish to transform with some
known state $|000\rangle$, then apply a Toffoli gate to the resulting qubits. Show
that the following circuit accomplishes this task:

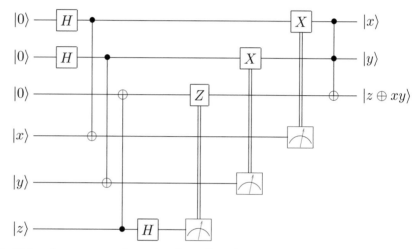

(2) Using the commutation rules from Exercise 10.67, show that moving the
final Toffoli gate all the way back to the left side gives the circuit

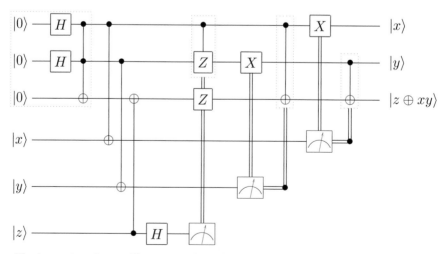

(3) Assuming the ancilla preparation shown in the leftmost dotted box can be
done fault-tolerantly, show that this circuit can be used to give a
fault-tolerant implementation of the Toffoli gate using the Steane code.

10.6.3 Fault-tolerant measurement

An extremely useful and important tool in the construction of fault-tolerant circuits is the ability to *measure an operator* M. Measurements are used to do encoding, read out the result of a quantum computation, to diagnose the syndrome in error-correction, and to do ancilla state preparation in the construction of the fault-tolerant $\pi/8$ and Toffoli gates, and thus are absolutely crucial to fault-tolerant quantum computation. In order that a procedure for performing an encoded measurement be considered to be fault-tolerant, we require that two things be true in order to prevent the propagation of errors. First, a single failure anywhere in the procedure should lead to at most one error in any block of qubits at the end of the procedure. Second, even if a single failure occurs during the procedure, we require that the measurement result be correct with probability $1 - O(p^2)$. This latter requirement is extremely important, since the measurement result may be used to control other operations in the quantum computer, and if it is incorrect then it may propagate to affect many qubits in other blocks of encoded qubits.

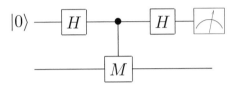

Figure 10.26. Quantum circuit for measuring a single qubit operator M with eigenvalues ± 1. The top qubit is the ancilla used for the measurement, and the bottom qubit is being measured.

Recall that measurement of a single qubit observable M may be performed using the circuit shown in Figure 10.26. Suppose M can be given a transversal encoded implementation on a quantum code as bitwise application of a gate M' to each qubit of the code. For example, for the Steane code, $M = H$ can be given a transversal implementation as bitwise application of $M' = H$, while a transversal implementation of $M = S$ uses bitwise application of $M' = ZS$. This suggests a possible circuit for measuring the encoded M on the encoded data, as shown schematically in Figure 10.27. Note that a real quantum code, such as the Steane code, would require more qubits. Unfortunately, the circuit in Figure 10.27 is not fault-tolerant. To see this, imagine a single failure occurs at the very beginning of the circuit, on the ancilla qubit. This will propagate forward to affect all the encoded qubits, so the circuit is not fault-tolerant.

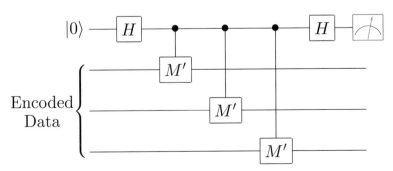

Figure 10.27. Schematic procedure for performing a measurement of an encoded observable M with a transversal implementation as bitwise application of M'. The circuit is not fault-tolerant. Note that a real code would require more than three qubits.

A nice way to make the measurement circuit fault-tolerant is schematically illustrated in Figure 10.28. For simplicity, this figure shows the data to be measured encoded in only three qubits; in practice more qubits will be used, such as the seven qubit Steane code, and for concreteness we imagine that it is the Steane code which is being used here. In addition to the encoded data, the circuit introduces *one* ancilla qubit for each data qubit, initially each in the state $|0\rangle$. The first step is to prepare the ancilla in a 'cat' state, $|00\ldots0\rangle + |11\ldots1\rangle$. (Note that the cat state is *not* encoded in any code.) The circuit used to do this preparation is not itself fault-tolerant, because a single failure during the circuit can cause errors on multiple qubits in the cat state. Nevertheless, this does not affect the fault-tolerance of the entire procedure, because we follow the ancilla preparation by several *verification* steps (only a single verification step is shown in the Figure).

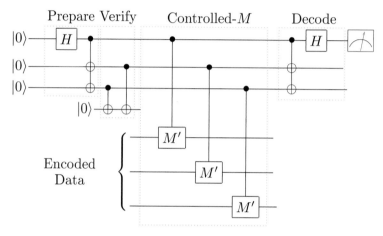

Figure 10.28. Schematic procedure for fault-tolerant measurement of an observable M, which is performed on encoded data. This procedure is repeated three times and a majority vote of the measurement outcomes taken, with the result that the measurement result is wrong with probability $O(p^2)$, where p is the failure probability of any individual component, and a single error anywhere in the circuit produces at most one error in the data.

The verification works as follows. The basic idea is that to verify that the state is a cat state it is sufficient to show that measurement of $Z_i Z_j$ for all pairs of qubits i and j in the cat state will give 1; that is, the parity of any pair of qubits in the cat state is even. To verify this for a particular pair $Z_i Z_j$ ($Z_2 Z_3$ in the example) we introduce an extra qubit, initially in the state $|0\rangle$. We compute the parity of two of the qubits in the ancilla by implementing two controlled-NOTs with the ancilla qubits as controls, and the extra qubit as the target, before measuring the extra qubit. If the measured parity is 1 then we know that the ancilla is not in the cat state, discard it, and start again. Suppose a single component failure occurs somewhere during this sequence of parity checks. This procedure is not fault-tolerant, because it is easy to show that there are single component failures which lead to more than one phase flip in the ancilla state. For example, if there is a Z error on the extra qubit between the CNOT gates then this can propagate forward to cause Z errors on two of the ancilla qubits. Fortunately, it is easy to show that multiple Z errors in the ancilla qubits do *not* propagate to the encoded data, although they may cause the final measurement result to be incorrect. To cope with this problem, and as described in more detail below, we repeat this procedure for measurement three times and take a majority vote, so the probability that the measurement will be wrong two or more times in this way is at most $O(p^2)$, where p is the probability of failure for a single component. What about X or Y errors? Well, these *can* propagate to cause errors in the encoded data, but it is a fortunate fact that a single failure during the cat state preparation and verification can cause at most one X or Y error in the ancilla after verification, and thus at most one error in the encoded data, ensuring fault-tolerance!

Exercise 10.69: Show that a single failure anywhere in the ancilla preparation and verification can lead to at most one X or Y error in the ancilla output.

Exercise 10.70: Show that Z errors in the ancilla do not propagate to affect the encoded data, but result in an incorrect measurement result being observed.

After the cat state has been verified, controlled-M' gates are performed between pairs of ancilla and data qubits, with no ancilla qubit being used more than once. Thus, if the ancilla is in the state $|00...0\rangle$ this results in nothing being done to the encoded data, while if the ancilla is in the state $|11...1\rangle$ the encoded M operation is applied to the data. The value of the cat state is that it ensures errors do not propagate from one controlled-M' gate to another, so a single error in the verification stage or the sequence of controlled-M' gates results in at most a single error in the encoded data. Finally, the measurement result is obtained by decoding the cat state with a series of CNOT gates and a Hadamard; the resulting qubit is 0 or 1 depending on the eigenvalue of the state of the data. These final gates do not involve the data, and thus an error in these gates does not propagate to affect the data at all. But what if an error in this final sequence of gates results in an incorrect measurement result? By repeating the measurement procedure three times and taking a majority vote of the results, we can ensure that the probability of an error in the measurement result is $O(p^2)$, where p is the probability of failure in an individual component.

We have described a method for performing fault-tolerant measurements such that the measurement gives an erroneous result with probability $O(p^2)$, where p is the failure probability for the individual components, and a single failure anywhere in the procedure results in an error on at most one qubit in the encoded data. The construction can be applied for any single qubit observable M which can be implemented in a transversal fashion. For the Steane code, this includes the Hadamard, phase and Pauli gates, and with a slight modification, the observable $M = e^{-i\pi/4}SX$. To perform the controlled-M operation on the Steane code for this choice of M we apply controlled-ZSX gates transversally for each pair of qubits in the ancilla and the code, followed by seven T gates applied transversally to the ancilla qubits. As described in Section 10.6.2, a fault-tolerant measurement of this observable can be used to create the ancilla used in the fault-tolerant circuit for the $\pi/8$ gate.

Exercise 10.71: Verify that when $M = e^{-i\pi/4}SX$ the procedure we have described gives a fault-tolerant method for measuring M.

Exercise 10.72: (Fault-tolerant Toffoli ancilla state construction) Show how to fault-tolerantly prepare the state created by the circuit in the dotted box of

Exercise 10.68, that is,

$$\frac{|000\rangle + |010\rangle + |100\rangle + |111\rangle}{2}. \tag{10.121}$$

You may find it helpful to first give the stabilizer generators for this state.

Measurement of stabilizer generators

We have described the fault-tolerant measurement procedure when M is an encoded observable for a single qubit, however the techniques immediately generalize to other cases. For our purposes it is sufficient to be able to measure stabilizer generators, which take the form of a tensor product of Pauli matrices. Such measurements allow us to perform fault-tolerant error-correction, the initial encoding for the quantum computer, and to measure encoded Z operators for the final readout stage of the computation.

As a simple example, suppose we wished to measure an operator like $X_1 Z_2 X_3$ on the first three qubits of a block of seven qubits encoded using the Steane code. An obvious generalization of Figure 10.28 can be used to perform this measurement, as shown in Figure 10.29. Once again, we perform verified cat state preparations before applying transversal controlled operations on the encoded data, in order to achieve a fault-tolerant measurement procedure for the operator $X_1 Z_2 X_3$. With the ability to fault-tolerantly measure such observables we automatically obtain the ability to perform the steps of encoding, syndrome measurement and measurement in the (logical) computational basis that are required to perform quantum computation. For the purposes of encoding, it suffices for quantum computation to prepare an encoded $|0\rangle$ state. For a stabilizer code such as the Steane code, such a preparation can be achieved by fault-tolerantly measuring all the stabilizer generators and the encoded $\bar{Z}$ operator, and then fixing the signs of the stabilizer generators and the encoded Z by applying appropriate fault-tolerant operations, according to the prescription in the proof of Proposition 10.4 in Section 10.5.1. An example illustrating how the Steane code encoded $|0\rangle$ state may be prepared fault-tolerantly is explained in Exercise 10.73. Syndrome measurement for error-correction and final read-out in the encoded computational basis of the quantum computer are realized fault-tolerantly along similar lines.

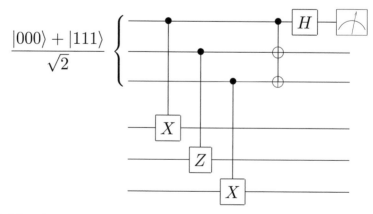

Figure 10.29. Schematic procedure for performing a fault-tolerant measurement of the operator XZX on three qubits.

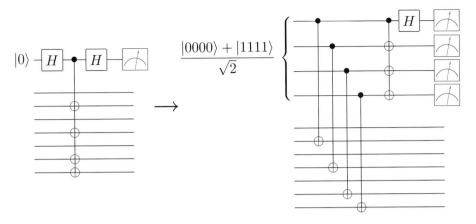

Figure 10.30. One step in fault-tolerantly producing the Steane code encoded $|0\rangle$ state.

Exercise 10.73: (Fault-tolerant encoded state construction) Show that the
Steane code encoded $|0\rangle$ state can be constructed fault-tolerantly in the following
manner.

(1) Begin with the circuit of Figure 10.16, and replace the measurement of each
generator, as shown in Figure 10.30, with each ancilla qubit becoming a cat
state $|00\ldots0\rangle + |11\ldots1\rangle$, and the operations rearranged to have their
controls on different qubits, so that errors do not propagate within the code
block.

(2) Add a stage to fault-tolerantly measure $\bar{Z}$.

(3) Calculate the error probability of this circuit, and of the circuit when the
generator measurements are repeated three times and majority voting is done.

(4) Enumerate the operations which should be performed conditioned on the
measurement results and show that they can be done fault-tolerantly.

Exercise 10.74: Construct a quantum circuit to fault-tolerantly generate the encoded
$|0\rangle$ state for the five qubit code (Section 10.5.6).

10.6.4 Elements of resilient quantum computation

The most spectacular success of quantum error-correcting codes – the *threshold for
quantum computation* – is that *provided the noise in individual quantum gates is below a certain constant threshold it is possible to efficiently perform an arbitrarily large
quantum computation*. Stated another way, noise is not a serious problem in principle
for quantum computation. The basic idea in the proof of the threshold, as was outlined
in Section 10.6.1, is to perform fault-tolerant operations directly on encoded states, in-
terleaved with error-correction steps, resulting in a net reduction in error probability
from p to $O(p^2)$. By concatenating our codes multiple times and creating hierarchical
fault-tolerant procedures, the error probability can be reduced further to $O(p^4)$, then
$O(p^8)$ and so on, ultimately being reduced to as low a level as desired, so long as the
the original error p is less than some threshold value p_{th}. Using the procedures we have
described, we estimate a threshold of approximately $p_{\text{th}} \sim 10^{-5}\text{-}10^{-6}$.

A bold claim such as the threshold theorem obviously needs qualifiers. It is *not* the
case that it is possible to protect a quantum computation against the effect of completely

arbitrary noise. The threshold theorem relies for its functioning on a small number of *physically reasonable* assumptions about the type of noise occurring in the quantum computer, and the quantum computer architecture available in order to reach its powerful conclusion. The error model we have considered is rather simplistic, and it is certainly the case that real quantum computers will experience more varied types of noise than we have considered here. Nevertheless, it seems plausible that the techniques introduced here when coupled with more sophisticated quantum error-correcting codes and with more sophisticated tools for analysis, can result in a threshold for quantum computation applicable in a much wider variety of circumstances than those we have considered.

We have not the space here to dive into a more sophisticated analysis, but several observations are in order. First, it is interesting to note that the threshold result *requires* a high degree of parallelism in our circuits. Even if all we wish to do is store quantum information in a quantum memory this operation will require periodic error-correction that demands a high degree of parallelism. Thus, a desirable goal for would-be designers of quantum computers is to develop architectures which are parallelizable, in order that the techniques of fault-tolerant quantum computation may be applied. Second, we note that our presentation of the threshold has completely neglected the cost of the *classical* computations and communication that are done during state preparation, syndrome measurement, and recovery. The cost of these could potentially be quite high; for example, to do recovery at the highest levels of the concatenated code requires communication between all parts of the quantum system. If this communication cannot be accomplished much faster than the time scale over which errors occur in the system then errors will begin to creep back in, negating the effect of the error-correction. More sophisticated analyses can deal with this problem; however, as with other complications there is a con-comitant cost in the form of a more stringent threshold for quantum computation. Third, our fault-tolerant constructions for measurement and the $\pi/8$ gate made use of ancilla qubits in the state $|0\rangle$, perhaps with some slight additional noise. It can be shown that in fact a constant supply of such fresh ancilla qubits is necessary for the threshold theorem to apply, and thus quantum computer designers must provide architectures which are not only parallelizable, but which also allow fresh ancilla qubits to be brought up on a regular basis.

Our presentation has focused on basic principles, not on optimizing the methods used, and it is likely that in practice much more streamlined versions of our constructions would be used. A simple but important guiding principle is to *choose your codes well*. We have focused on the Steane code because it is easy to work with and demonstrates all the fundamental principles; however, in practice other codes may work much better. For example, it may pay handsome dividends at the first level of concatenation to use a code optimized to protect against the type of noise known to occur in the particular physical system being used for implementation.

Although the theoretical ideas behind the threshold theorem can be adapted in a variety of different ways, depending upon the noise prevalent in a specific implementation of quantum computation, a skeptic might still claim that all such noise models for which a threshold may be proved are overly restrictive, and will not be realized within any real physical system. Such skepticism is, finally, only answerable in the laboratory with a demonstration of large-scale fault-tolerant quantum computation. The marvel of the present result is that it proves that, to the best of our current knowledge, no principle of physics will limit quantum computers from being realized someday.

Summarizing, in this chapter we have outlined the basic principles by which quantum information processing may be performed in a resilient manner, focusing on the specific example of quantum computation. The same basic techniques apply also to any other system in which quantum information processing may be performed, such as quantum communications channels for performing tasks such as quantum cryptography. The extreme fragility of quantum information in all known systems makes it likely that some form of quantum error-correction will need to be used in any practical quantum information processing system, but surprisingly, these techniques work so well that arbitrarily reliable quantum computations can be performed using noisy components, provided the error probability in those components is less than some constant threshold.

Problem 10.1:   Channels $\mathcal{E}_1$ and $\mathcal{E}_2$ are said to be *equivalent* if there exist unitary channels $\mathcal{U}$ and $\mathcal{V}$ such that $\mathcal{E}_2 = \mathcal{U} \circ \mathcal{E}_1 \circ \mathcal{V}$.

(1) Show that the relation of channel equivalence is an equivalence relation.
(2) Show how to turn an error-correcting code for $\mathcal{E}_1$ into an error-correcting code for $\mathcal{E}_2$. Assume that the error-correction procedure for $\mathcal{E}_1$ is performed as a projective measurement followed by a conditional unitary operation, and explain how the error-correction procedure for $\mathcal{E}_2$ can be performed in the same fashion.

Problem 10.2: (Gilbert–Varshamov bound)   Prove the Gilbert–Varshamov bound for CSS codes, namely, that an $[n, k]$ CSS code correcting t errors exists for some k such that

$$\frac{k}{n} \geq 1 - 2H\left(\frac{2t}{n}\right). \tag{10.122}$$

As a challenge, you may like to try proving the Gilbert–Varshamov bound for a general stabilizer code, namely, that there exists an $[n, k]$ stabilizer code correcting errors on t qubits, with

$$\frac{k}{n} \geq 1 - \frac{2\log(3)t}{n} - H\left(\frac{2t}{n}\right). \tag{10.123}$$

Problem 10.3: (Encoding stabilizer codes)   Suppose we assume that the generators for the code are in standard form, and that the encoded Z and X operators have been constructed in standard form. Find a circuit taking the $n \times 2n$ check matrix corresponding to a listing of all the generators for the code together with the encoded Z operations from

$$G = \begin{bmatrix} 0 & 0 & 0 & I & 0 & 0 \\ 0 & 0 & 0 & 0 & I & 0 \\ 0 & 0 & 0 & 0 & 0 & I \end{bmatrix} \tag{10.124}$$

to the standard form

$$\begin{bmatrix} I & A_1 & A_2 & B & 0 & C_2 \\ 0 & 0 & 0 & D & I & E \\ 0 & 0 & 0 & A_2^T & 0 & I \end{bmatrix}. \tag{10.125}$$

Problem 10.4: (Encoding by teleportation) Suppose you are given a qubit $|\psi\rangle$ to encode in a stabilizer code, but you are not told anything about how $|\psi\rangle$ was constructed: it is an unknown state. Construct a circuit to perform the encoding in the following manner:

(1) Explain how to fault-tolerantly construct the partially encoded state

$$\frac{|0\rangle|0_L\rangle + |1\rangle|1_L\rangle}{\sqrt{2}}, \tag{10.126}$$

by writing this as a stabilizer state, so it can be prepared by measuring stabilizer generators.

(2) Show how to fault-tolerantly perform a Bell basis measurement with $|\psi\rangle$ and the unencoded qubit from this state.

(3) Give the Pauli operations which you need to fix up the remaining encoded qubit after this measurement, so that it becomes $|\psi\rangle$, as in the usual quantum teleportation scheme.

Compute the probability of error of this circuit. Also show how to modify the circuit to perform fault-tolerant decoding.

Problem 10.5: Suppose $C(S)$ is an $[n, 1]$ stabilizer code capable of correcting errors on a single qubit. Explain how a fault-tolerant implementation of the controlled-NOT gate may be implemented between two logical qubits encoded using this code, using only fault-tolerant stabilizer state preparation, fault-tolerant measurement of elements of the stabilizer, and normalizer gates applied transversally.

Summary of Chapter 10: Quantum error-correction

- **Quantum error-correcting code**: An $[n, k, d]$ quantum error-correcting code uses n qubits to encode k qubits, with distance d.

- **Quantum error-correction conditions**: Let C be a quantum error-correcting code and P be the projector onto C. Then the code can correct the error set $\{E_i\}$ if and only if

$$PE_i^\dagger E_j P = \alpha_{ij} P, \tag{10.127}$$

  for some Hermitian matrix α of complex numbers.

- **Stabilizer codes**: Let S be the stabilizer for a stabilizer code $C(S)$ and suppose $\{E_j\}$ is a set of errors in the Pauli group such that $E_j^\dagger E_k \notin N(S) - S$ for all j and k. Then $\{E_j\}$ is a correctable set of errors for the code $C(S)$.

- **Fault-tolerant quantum computation**: A universal set of logical operations on *encoded* quantum states can be performed in such a way that the effective failure probability in the encoded states scales like $O(p^2)$, where p is the failure probability in the underlying gates.

- **The threshold theorem**: Provided the noise in individual quantum gates is below a certain constant threshold and obeys certain physically reasonable assumptions, it is possible to reliably perform an arbitrarily long quantum computation, with only a small overhead in the size of the circuit necessary to ensure reliability.

History and further reading

There are many excellent texts on error-correcting codes in classical information theory. We especially recommend the wonderful text of MacWilliams and Sloane[MS77]. This begins at a very elementary level, but quickly and smoothly moves into more advanced topics, covering an enormous range of material. A more recent introduction, also very good, is the text by Welsh[Wel88].

Quantum error-correction was independently discovered by Shor[Sho95], who found the nine qubit code presented in Section 10.2, and by Steane[Ste96a], who used a different approach, in which he studied the interference properties of multiple particle entangled states. The quantum error-correction conditions were proved independently by Bennett, DiVincenzo, Smolin and Wootters[BDSW96], and by Knill and Laflamme[KL97], building upon earlier work by Ekert and Macchiavello[EM96]. The five qubit code was discovered by Bennett, DiVincenzo, Smolin and Wootters[BDSW96], and independently by Laflamme, Miquel, Paz and Zurek[LMPZ96].

Calderbank and Shor[CS96], and Steane[Ste96b] used ideas from classical error-correction to develop the CSS (Calderbank–Shor–Steane) codes. Calderbank and Shor also stated and proved the Gilbert–Varshamov bound for CSS codes. Gottesman[Got96] invented the stabilizer formalism, and used it to define stabilizer codes, and investigated some of the properties of some specific codes. Independently, Calderbank, Rains, Shor and Sloane[CRSS97] invented an essentially equivalent approach to quantum error-correction based on ideas from classical coding theory. They were able to classify almost all known quantum codes using a $GF(4)$ orthogonal geometry approach[CRSS98], and also provided the first proof of the quantum Gilbert–Varshamov bound for general stabilizer codes, which had earlier been stated by Ekert and Macchiavello[EM96]. The Gottesman–Knill theorem seems to have first been stated by Gottesman in [Got97], where he attributes the result to Knill, with the proof based upon the stabilizer formalism Gottesman had introduced. Gottesman has applied the stabilizer formalism to a wide variety of problems with considerable success; see for example [Got97] for a sample and further references. Our presentation of the stabilizer formalism is based primarily upon [Got97], wherein may be found most of the results we describe, including the result that the Hadamard, phase and controlled-NOT generate the normalizer $N(G_n)$.

Many constructions for specific classes of quantum codes are known; we point to just a few here. Rains, Hardin, Shor and Sloane[RHSS97] have constructed interesting examples of quantum codes lying outside the stabilizer codes we have considered. Many people have considered quantum codes based on systems other than qubits; we mention especially the work of Gottesman[Got98a] and Rains[Rai99b] which construct non-binary codes and consider fault-tolerant computation with such codes. Aharonov and Ben-Or[ABO99] construct non-binary codes using interesting techniques based on polynomials over finite fields, and also investigate fault-tolerant computation with such codes. *Approximate* quantum error-correction is another topic we have not touched; that approximate quantum error-correction can lead to improved codes was shown by Leung, Nielsen, Chuang and Yamamoto[LNCY97].

A large and interesting class of quantum error-correcting codes (but beyond the scope of this chapter) are known variously by the names *noiseless quantum codes* and *decoherence free subspaces*. A substantial body of work exists on these subjects (and clarifying the connections between them). An entry into the literature may be found through the work of Zanardi and Rasetti[ZR98, Zan99], Lidar, Chuang, and Whaley[LCW98], Bacon, Kempe, Lidar and Whaley[BKLW99, LBW99], and of Knill, Laflamme and Viola[KLV99].

Many bounds on quantum error-correcting codes are known, often adapted from similar classical bounds. Ekert and Macchiavello[EM96] pointed out the possibility of proving a quantum analogue to the Hamming bound; this construction and the role of 'degenerate' quantum codes was subsequently clarified by Gottesman[Got96]. Shor and Laflamme[SL97] proved a quantum analogue of a result in classical coding theory, the MacWilliams identities, which touched off a great deal of work studying the properties of certain polynomials related to quantum codes (the weight enumerators) as well as more general work on the problem of bounds for quantum codes, by Ashikhmin[Ash97], Ashikhmin and Lytsin[AL99], and several papers on the topic by Rains[Rai98, Rai99c, Rai99a].

The theory of fault-tolerant computation for classical computers was worked out by von Neumann[von56], and is discussed in the monograph by Winograd and Cowan[WC67]. Shor[Sho96] introduced the idea of fault-tolerance into quantum computation, and showed how to perform all the basic fault tolerant steps (state preparation, quantum logic, error-correction, and measurement). Kitaev[Kit97b, Kit97c] independently developed many similar ideas, including fault-tolerant constructions for many basic quantum logic gates. Cirac, Pellizzari and Zoller[CPZ96], and Zurek and Laflamme[ZL96], also took early steps toward fault-tolerant quantum computation. DiVincenzo and Shor generalized Shor's original construction to show how fault-tolerant measurement of syndromes for any stabilizer code could be performed[DS96], and Gottesman[Got98b] generalized all the fault-tolerant constructions, showing how to perform fault-tolerant computation with any stabilizer code. A general review of this work as well as much other survey material may be found in [Got97]; this includes a construction to solve Problem 10.5. The fault-tolerant $\pi/8$ and Toffoli gate constructions are based on a line of thought developed by Gottesman and Chuang[GC99], and Zhou and Chuang[ZLC00]; the circuit for the fault-tolerant Toffoli described in Exercise 10.68 is actually Shor's original construction[Sho96]. Steane[Ste99] has developed many ingenious constructions for fault-tolerant procedures.

Kitaev[Kit97a, Kit97b] has introduced a beautiful set of ideas for implementing fault-tolerance, using *topological* methods to assist in the performance of quantum error-correction. The basic idea is that if information is stored in the topology of a system, then that information will naturally be very robust against the effects of noise. These and many other elegant ideas have been developed in further papers by Bravyi and Kitaev[BK98b] and by Freedman and Meyer[FM98]. Preskill[Pre97] is an excellent review of the field of quantum error-correction as a whole, and contains an especially beautiful description of topological quantum error-correction, as well as a provocative discussion of whether topological error-correction can be used to gain insight into fundamental questions about black holes and quantum gravity!

Many different groups proved threshold results for quantum computation. These results hold for a wide variety of assumptions, giving essentially different threshold theorems. Aharonov and Ben-Or[ABO97, ABO99] and Kitaev's[Kit97c, Kit97b] threshold proofs do not require fast or reliable classical computation. Aharonov and Ben-Or also showed that in order for a threshold result to hold, there must be constant parallelism in the quantum computer at each timestep[ABO97]. In their threshold proofs, Gottesman[Got97] and Preskill[Pre98c, GP10] have an especially detailed optimization of the value of the threshold. Knill, Laflamme and Zurek[KLZ98a, KLZ98b]'s results concentrate on proving the threshold theorem for a wide class of error models. Aharonov, Ben-Or, Impagliazzo, and Nisan have also shown that a supply of fresh qubits is necessary for the threshold[ABOIN96]. Further references and historical material may be found within the cited works. In particular, each group built on Shor's pioneering work[Sho96] on fault-tolerant quantum computation.

Numerous excellent reviews of fault-tolerant quantum computation have been written, developing the basic ideas in much greater detail than we have here, from a variety of different points of view. Aharonov's thesis[Aha99a] develops the threshold theorem and much material of related interest in a self-contained way. Gottesman's thesis[Got97] also provides a review of fault-tolerant quantum computation, with more emphasis on properties of quantum codes, and developing fault-tolerant constructions for a wide variety of different codes. Knill, Laflamme and Zurek have written a semi-popular overview of the threshold result[KLZ98a]. Finally, Preskill has written two superb articles[Pre98c, Pre98a] explaining quantum error-correction and fault-tolerant quantum computation.

11 Entropy and information

Entropy is a key concept of quantum information theory. It measures how much uncertainty there is in the state of a physical system. In this chapter we review the basic definitions and properties of entropy in both classical and quantum information theory. In places the chapter contains rather detailed and lengthy mathematical arguments. On a first reading these sections may be read lightly and returned to later for reference purposes.

11.1 Shannon entropy

The key concept of classical information theory is the *Shannon entropy*. Suppose we learn the value of a random variable X. The Shannon entropy of X quantifies how much information we gain, on average, when we learn the value of X. An alternative view is that the entropy of X measures the amount of *uncertainty* about X before we learn its value. These two views are complementary; we can view the entropy either as a measure of our uncertainty *before* we learn the value of X, or as a measure of how much information we have gained *after* we learn the value of X.

Intuitively, the information content of a random variable should not depend on the labels attached to the different values that may be taken by the random variable. For example, we expect that a random variable taking the values 'heads' and 'tails' with respective probabilities $1/4$ and $3/4$ contains the same amount of information as a random variable that takes the values 0 and 1 with respective probabilities $1/4$ and $3/4$. For this reason, the entropy of a random variable is defined to be a function of the probabilities of the different possible values the random variable takes, and is not influenced by the labels used for those values. We often write the entropy as a function of a probability distribution, $p_1, \ldots, p_n$. The *Shannon entropy* associated with this probability distribution is defined by

$$H(X) \equiv H(p_1, \ldots, p_n) \equiv -\sum_x p_x \log p_x. \tag{11.1}$$

We justify this definition shortly. Note that in the definition – and throughout this book – logarithms indicated by 'log' are taken to base two, while 'ln' indicates a natural logarithm. It is conventional to say that entropies are measured in 'bits' with this convention for the logarithm. You may wonder what happens when $p_x = 0$, since $\log 0$ is undefined. Intuitively, an event which can never occur should not contribute to the entropy, so by convention we agree that $0 \log 0 \equiv 0$. More formally, note that $\lim_{x \to 0} x \log x = 0$, which provides further support for our convention.

Why is the entropy defined in this way? Later in this section, Exercise 11.2 gives an intuitive justification for this definition of the entropy, based on certain 'reasonable' axioms which might be expected of a measure of information. This intuitive justification is reassuring but it is not the whole story. The best reason for this definition of entropy is that it can be used to *quantify the resources needed to store information*. More concretely, suppose there is some *source* (perhaps a radio antenna) which is producing information of some sort, say in the form of a bit string. Let's consider a very simple model for a source: we model it as producing a string $X_1, X_2, \ldots$ of independent, identically distributed random variables. Most real information sources don't behave quite this way, but it's often a good approximation to reality. Shannon asked what minimal physical resources are required to store the information being produced by the source, in such a way that at a later time the information can be reconstructed? The answer to this question turns out to be the entropy, that is, $H(X)$ bits are required per source symbol, where $H(X) \equiv H(X_1) = H(X_2) = \ldots$ is the entropy of each random variable modeling the source. This result is known as *Shannon's noiseless coding theorem*, and we prove both classical and quantum versions of it in Chapter 12.

For a concrete example of Shannon's noiseless channel coding theorem, suppose an information source produces one of four symbols, 1, 2, 3 or 4. Without compression two bits of storage space corresponding to the four possible outputs are consumed for each use of the source. Suppose however that the symbol 1 is produced by the source with probability 1/2, the symbol 2 with probability 1/4, and the symbols 3 and 4 both with probability 1/8. We can make use of the bias between the source outputs to compress the source, using fewer bits to store commonly occurring symbols such as 1, and more bits to store rarely occurring symbols like 3 and 4. One possible compression scheme is to encode 1 as the bit string 0, 2 as the bit string 10, 3 as the bit string 110, and 4 as the bit string 111. Notice that the average length of the compressed string is $\frac{1}{2} \cdot 1 + \frac{1}{4} \cdot 2 + \frac{1}{8} \cdot 3 + \frac{1}{8} \cdot 3 = 7/4$ bits of information per use of the source. This is less than is required in the naive approach to storing this source! Amazingly, this matches the entropy of the source, $H(X) = -1/2 \log(1/2) - 1/4 \log(1/4) - 1/8 \log(1/8) - 1/8 \log(1/8) = 7/4$! Moreover, it turns out that any attempt to compress the source further results in data being irretrievably lost; the entropy quantifies the optimal compression that may be achieved.

This *operational* motivation for the definition of entropy in terms of data compression expresses a key philosophy of information theory, both quantum and classical: *fundamental measures of information arise as the answers to fundamental questions about the physical resources required to solve some information processing problem.*

Exercise 11.1: (Simple calculations of entropy) What is the entropy associated with the toss of a fair coin? With the roll of a fair die? How would the entropy behave if the coin or die were unfair?

Exercise 11.2: (Intuitive justification for the definition of entropy) Suppose we are trying to quantify how much information is provided by an event E which may occur in a probabilistic experiment. We do this using an 'information function' $I(E)$ whose value is determined by the event E. Suppose we make the following assumptions about this function:

(1) $I(E)$ is a function only of the probability of the event E, so we may write $I = I(p)$, where p is a probability in the range 0 to 1.

(2) I is a smooth function of probability.

(3) $I(pq) = I(p) + I(q)$ when $p, q > 0$. (*Interpretation:* The information gained

when two independent events occur with individual probabilities p and q is the sum of the information gained from each event alone.)

Show that $I(p) = k \log p$, for some constant, k. It follows that the average information gain when one of a mutually exclusive set of events with probabilities $p_1, \ldots, p_n$ occurs is $k \sum_i p_i \log p_i$, which is just the Shannon entropy, up to a constant factor.

11.2 Basic properties of entropy

11.2.1 The binary entropy

The entropy of a two-outcome random variable is so useful that we give it a special name, the *binary entropy*, defined as

$$H_{\text{bin}}(p) \equiv -p \log p - (1-p) \log(1-p), \tag{11.8}$$

where p and $1-p$ are the probabilities of the two outcomes. Where context makes the meaning clear we write $H(p)$ rather than $H_{\text{bin}}(p)$. The binary entropy function is plotted in Figure 11.1. Notice that $H(p) = H(1-p)$ and that $H(p)$ attains its maximum value of 1 at $p = 1/2$.

The binary entropy is an excellent testing ground for the understanding of more general properties of entropy. One property of particular interest is how the entropy behaves when we mix two or more probability distributions. Imagine, for example, that Alice has in her possession two coins, one a quarter from the US, the other a dollar coin from Australia. Both coins have been altered to exhibit bias, with the probability of heads on the US coin being p_U, and the probability of heads on the Australian coin being p_A. Suppose Alice flips the US coin with probability q and the Australian coin with probability $1-q$, telling Bob whether the result was heads or tails. How much information does Bob gain on average? Intuitively it is clear that Bob should gain at least as much information as the average of the information he would have gained from a US coin flip or an Australian coin flip. As an equation this intuition may be expressed as:

$$H(qp_U + (1-q)p_A) \geq qH(p_U) + (1-q)H(p_A). \tag{11.9}$$

Sometimes the inequality can be strict, because Bob gains information not only about the value (heads or tails) of the coin, but also some *additional* information about the identity of the coin. For instance, if $p_U = 1/3$ and $p_A = 5/6$, and heads comes up, then Bob has received a pretty good indicator that the coin was Australian.

Equation (11.9) is easily shown to be correct. It is an example of a broader concept, that of *concavity*, which we met in Chapter 9 when discussing distance measures. Recall that a real-valued function f is said to be concave if for any p in the range 0 to 1,

$$f(px + (1-p)y) \geq pf(x) + (1-p)f(y). \tag{11.10}$$

The binary entropy is easily seen to be concave, as can be grasped visually by examining Figure 11.1 and observing that the graph of the binary entropy goes *above* any line cutting the graph. We will be much concerned with concavity properties of the entropy, both classical and quantum. Don't let the simplicity of the above intuitive argument beguile you into a false complacency: many of the deepest results in quantum information have their roots in skilful application of concavity properties of classical or quantum entropies.

Box 11.1: Entropic quantum uncertainty principle

There is an elegant entropic way of reformulating the uncertainty principle of quantum mechanics. Recall the Heisenberg uncertainty principle from Box 2.4 on page 89. This states that the standard deviations $\Delta(C)$ and $\Delta(D)$ for observables C and D must satisfy the relation

$$\Delta(C)\Delta(D) \geq \frac{|\langle\psi|[C,D]|\psi\rangle|}{2}, \tag{11.2}$$

for a quantum system in the state $|\psi\rangle$.

Let $C = \sum_c c|c\rangle\langle c|$ and $D = \sum_d d|d\rangle\langle d|$ be spectral decompositions for C and D. Define $f(C,D) \equiv \max_{c,d} |\langle c|d\rangle|$ to be the maximum fidelity between any two eigenvectors of $|c\rangle$ and $|d\rangle$. For example, $f(X,Z) = 1/\sqrt{2}$ for the Pauli matrices X and Z.

Suppose the quantum system is prepared in a state $|\psi\rangle$, and let $p(c)$ be the probability distribution associated with a measurement of C, with associated entropy $H(C)$, and $q(d)$ the probability distribution associated with a measurement of D, with associated entropy $H(D)$. The entropic uncertainty principle states that

$$H(C) + H(D) \geq 2\log\left(\frac{1}{f(C,D)}\right). \tag{11.3}$$

A full proof of this result would take us too far afield (see 'History and further reading' for references); however, we can give a simple proof of the weaker result:

$$H(C) + H(D) \geq -2\log\frac{1 + f(C,D)}{2}. \tag{11.4}$$

To prove this, note that

$$H(C) + H(D) = -\sum_{cd} p(c)q(d)\log\left(p(c)q(d)\right). \tag{11.5}$$

We aim to bound $p(c)q(d) = |\langle c|\psi\rangle\langle\psi|d\rangle|^2$ from above. To do this, let $|\tilde\psi\rangle$ be the projection of $|\psi\rangle$ into the plane spanned by $|c\rangle$ and $|d\rangle$, so $|\tilde\psi\rangle$ has norm λ less than or equal to one. If θ is the angle $|d\rangle$ makes with $|c\rangle$ in this plane, and φ is the angle $|\tilde\psi\rangle$ makes with $|d\rangle$, then we see $p(c)q(d) = |\langle c|\tilde\psi\rangle\langle\tilde\psi|d\rangle|^2 = \lambda^2\cos^2(\theta - \varphi)\cos^2(\varphi)$. Calculus implies that the maximum is reached when $\lambda = 1$ and $\varphi = \theta/2$, at which point $p(c)q(d) = \cos^4(\theta/2)$, which can be put in the form

$$p(c)q(d) = \left(\frac{1 + |\langle c|d\rangle|}{2}\right)^2. \tag{11.6}$$

Combining this with Equation (11.5) gives

$$H(C) + H(D) \geq -2\log\frac{1 + f(C,D)}{2}, \tag{11.7}$$

as claimed.

Moreover, for quantum entropies it is sometimes rather difficult to justify our intuitive beliefs about what concavity properties entropy ought to have.

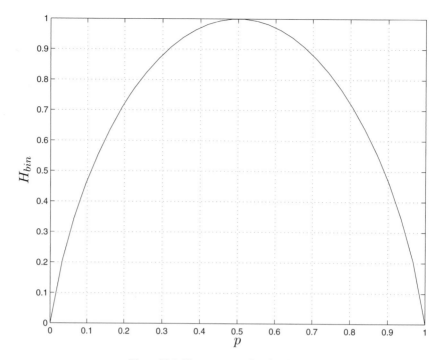

Figure 11.1. Binary entropy function $H(p)$.

Exercise 11.3: Prove that the binary entropy $H_{\text{bin}}(p)$ attains its maximum value of one at $p = 1/2$.

Exercise 11.4: (Concavity of the binary entropy) From Figure 11.1 it appears that the binary entropy is a concave function. Prove that this is so, that is:

$$H_{\text{bin}}(px_1 + (1-p)x_2) \geq pH_{\text{bin}}(x_1) + (1-p)H_{\text{bin}}(x_2), \qquad (11.11)$$

where $0 \leq p, x_1, x_2 \leq 1$. Prove in addition that the binary entropy is *strictly concave*, that is, the above inequality is an equality only for the trivial cases $x_1 = x_2$, or $p = 0$, or $p = 1$.

11.2.2 The relative entropy

The *relative entropy* is a very useful entropy-like measure of the closeness of two probability distributions, $p(x)$ and $q(x)$, over the same index set, x. Suppose $p(x)$ and $q(x)$ are two probability distributions on the same index set, x. Define the *relative entropy* of $p(x)$ to $q(x)$ by

$$H(p(x)\|q(x)) \equiv \sum_x p(x) \log \frac{p(x)}{q(x)} \equiv -H(X) - \sum_x p(x) \log q(x). \qquad (11.12)$$

We define $-0 \log 0 \equiv 0$ and $-p(x) \log 0 \equiv +\infty$ if $p(x) > 0$.

It is not immediately obvious what the relative entropy is good for, or even why it is a good measure of distance between two distributions. The following theorem gives some motivation as to why it is regarded as being like a distance measure.

Theorem 11.1: (**Non-negativity of the relative entropy**) The relative entropy is non-negative, $H(p(x)\|q(x)) \geq 0$, with equality if and only if $p(x) = q(x)$ for all x.

Proof
A very useful inequality in information theory is $\log x \ln 2 = \ln x \leq x - 1$, for all positive x, with equality if and only if $x = 1$. Here we need to rearrange the result slightly, to $-\log x \geq (1 - x)/\ln 2$, and then note that

$$H(p(x)\|q(x)) = -\sum_x p(x) \log \frac{q(x)}{p(x)} \tag{11.13}$$

$$\geq \frac{1}{\ln 2} \sum_x p(x) \left(1 - \frac{q(x)}{p(x)}\right) \tag{11.14}$$

$$= \frac{1}{\ln 2} \sum_x (p(x) - q(x)) \tag{11.15}$$

$$= \frac{1}{\ln 2}(1 - 1) = 0, \tag{11.16}$$

which is the desired inequality. The equality conditions are easily deduced by noting that equality occurs in the second line if and only if $q(x)/p(x) = 1$ for all x, that is, the distributions are identical. $\qquad\square$

The relative entropy is often useful, not in itself, but because other entropic quantities can be regarded as special cases of the relative entropy. Results about the relative entropy then give as special cases results about other entropic quantities. For example, we can use the non-negativity of the relative entropy to prove the following fundamental fact about entropies. Suppose $p(x)$ is a probability distribution for X, over d outcomes. Let $q(x) \equiv 1/d$ be the uniform probability distribution over those outcomes. Then

$$H(p(x)\|q(x)) = H(p(x)\|1/d) = -H(X) - \sum_x p(x)\log(1/d) = \log d - H(X). \tag{11.17}$$

From the non-negativity of the relative entropy, Theorem 11.1, we see that $\log d - H(X) \geq 0$, with equality if and only if X is uniformly distributed. This is an elementary fact, but so important that we restate it formally as a theorem.

Theorem 11.2: Suppose X is a random variable with d outcomes. Then $H(X) \leq \log d$, with equality if and only if X is uniformly distributed over those d outcomes.

We use this technique – finding expressions for entropic quantities in terms of the relative entropy – often in the study of both classical and quantum entropies.

Exercise 11.5: (**Subadditivity of the Shannon entropy**) Show that $H(p(x, y)\|p(x)p(y)) = H(p(x)) + H(p(y)) - H(p(x, y))$. Deduce that $H(X, Y) \leq H(X) + H(Y)$, with equality if and only if X and Y are independent random variables.

11.2.3 Conditional entropy and mutual information
Suppose X and Y are two random variables. How is the information content of X related to the information content of Y? In this section we introduce two concepts – the

conditional entropy and the *mutual information* – which help answer this question. The definitions we give for these concepts are rather formal, and at times you may be confused as to why a particular quantity – say, the conditional entropy – is to be interpreted in the way we indicate. Keep in mind that the ultimate justification for these definitions is that they answer resource questions, which we investigate in more detail in Chapter 12, and that the interpretations given to the quantities depend on the nature of the resource question being answered.

We already met the *joint entropy* of a pair of random variables implicitly in the previous section. For clarity, we now make this definition explicit. The *joint entropy* of X and Y is defined in the obvious way,

$$H(X, Y) \equiv - \sum_{x,y} p(x, y) \log p(x, y), \tag{11.18}$$

and may be extended in the obvious way to any vector of random variables. The joint entropy measures our total uncertainty about the pair (X, Y). Suppose we know the value of Y, so we have acquired $H(Y)$ bits of information about the pair, (X, Y). The remaining uncertainty about the pair (X, Y), is associated with our remaining lack of knowledge about X, even given that we know Y. The *entropy of X conditional on knowing Y* is therefore defined by

$$H(X|Y) \equiv H(X, Y) - H(Y). \tag{11.19}$$

The conditional entropy is a measure of how uncertain we are, on average, about the value of X, given that we know the value of Y.

A second quantity, the *mutual information content of X and Y*, measures how much information X and Y have in common. Suppose we add the information content of X, $H(X)$, to the information content of Y. Information which is common to X and Y will have been counted twice in this sum, while information which is not common will have been counted exactly once. Subtracting off the joint information of (X, Y), $H(X, Y)$, we therefore obtain the common or *mutual information* of X and Y:

$$H(X : Y) \equiv H(X) + H(Y) - H(X, Y). \tag{11.20}$$

Note the useful equality $H(X : Y) = H(X) - H(X|Y)$ relating the conditional entropy and mutual information.

To get some feeling for how the Shannon entropy behaves, we now give some simple relationships between the different entropies.

Theorem 11.3: (**Basic properties of Shannon entropy**)
  (1) $H(X, Y) = H(Y, X)$, $H(X : Y) = H(Y : X)$.
  (2) $H(Y|X) \geq 0$ and thus $H(X : Y) \leq H(Y)$, with equality if and only if Y is a function of X, $Y = f(X)$.
  (3) $H(X) \leq H(X, Y)$, with equality if and only if Y is a function of X.
  (4) **Subadditivity:** $H(X, Y) \leq H(X) + H(Y)$ with equality if and only if X and Y are independent random variables.
  (5) $H(Y|X) \leq H(Y)$ and thus $H(X : Y) \geq 0$, with equality in each if and only if X and Y are independent random variables.
  (6) **Strong subadditivity:** $H(X, Y, Z) + H(Y) \leq H(X, Y) + H(Y, Z)$, with equality if and only if $Z \to Y \to X$ forms a Markov chain.

(7) **Conditioning reduces entropy:** $H(X|Y,Z) \leq H(X|Y)$.

Most of the proofs are either obvious or easy exercises. We give a few hints below.

Proof

(1) Obvious from the relevant definitions.

(2) Since $p(x,y) = p(x)p(y|x)$ we have

$$H(X,Y) = -\sum_{xy} p(x,y) \log p(x)p(y|x) \tag{11.21}$$

$$= -\sum_{x} p(x) \log p(x) - \sum_{xy} p(x,y) \log p(y|x) \tag{11.22}$$

$$= H(X) - \sum_{xy} p(x,y) \log p(y|x). \tag{11.23}$$

Thus $H(Y|X) = -\sum_{xy} p(x,y) \log p(y|x)$. But $-\log p(y|x) \geq 0$, so $H(Y|X) \geq 0$ with equality if and only if Y is a deterministic function of X.

(3) Follows from the previous result.

(4) To prove subadditivity and, later, strong subadditivity we again use the fact that $\log x \leq (x-1)/\ln 2$ for all positive x, with equality if and only if $x = 1$. We find that

$$\sum_{x,y} p(x,y) \log \frac{p(x)p(y)}{p(x,y)} \leq \frac{1}{\ln 2} \sum_{x,y} p(x,y) \left(\frac{p(x)p(y)}{p(x,y)} - 1 \right) \tag{11.24}$$

$$= \frac{1}{\ln 2} \sum_{x,y} p(x)p(y) - p(x,y) = \frac{1-1}{\ln 2} = 0. \tag{11.25}$$

Subadditivity follows. Notice that equality is achieved if and only if $p(x,y) = p(x)p(y)$ for all x and y. That is, the subadditivity inequality is saturated if and only if X and Y are independent.

(5) Follows from subadditivity and the relevant definitions.

(6) Strong subadditivity of Shannon entropy follows from the same technique used to prove subadditivity; the difficulty level is only slightly higher than that proof. You will be asked to supply this proof as Exercise 11.6.

(7) Intuitively, we expect that the uncertainty about X, given that we know the value of Y and Z, is less than our uncertainty about X, given that we only know Y. More formally, inserting the relevant definitions, the result that conditioning reduces entropy is equivalent to

$$H(X,Y,Z) - H(Y,Z) \leq H(X,Y) - H(Y), \tag{11.26}$$

which is a rearranged version of the strong subadditivity inequality. □

Exercise 11.6: (Proof of classical strong subadditivity) Prove that $H(X,Y,Z) + H(Y) \leq H(X,Y) + H(Y,Z)$, with equality if and only if $Z \to Y \to X$ forms a Markov chain.

Exercise 11.7: In Exercise 11.5 you implicitly showed that the mutual information $H(X:Y)$ could be expressed as the relative entropy of two probability distributions, $H(X:Y) = H(p(x,y)\|p(x)p(y))$. Find an expression for the conditional entropy $H(Y|X)$ as a relative entropy between two probability distributions. Use this expression to deduce that $H(Y|X) \geq 0$, and to find the equality conditions.

The various relationships between entropies may mostly be deduced from the 'entropy Venn diagram' shown in Figure 11.2. Such figures are not completely reliable as a guide to the properties of entropy, but they provide a useful mnemonic for remembering the various definitions and properties of entropy.

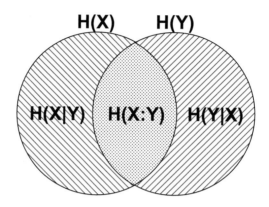

Figure 11.2. Relationships between different entropies.

We conclude our study of the elementary properties of conditional entropy and mutual information with a simple and useful chaining rule for conditional entropies.

Theorem 11.4: (**Chaining rule for conditional entropies**) Let $X_1, \ldots, X_n$ and Y be any set of random variables. Then

$$H(X_1, \ldots, X_n | Y) = \sum_{i=1}^{n} H(X_i | Y, X_1, \ldots, X_{i-1}). \qquad (11.27)$$

Proof
We prove the result for $n = 2$, and then induct on n. Using only the definitions and some simple algebra we have

$$H(X_1, X_2 | Y) = H(X_1, X_2, Y) - H(Y) \qquad (11.28)$$
$$= H(X_1, X_2, Y) - H(X_1, Y) + H(X_1, Y) - H(Y) \qquad (11.29)$$
$$= H(X_2 | Y, X_1) + H(X_1 | Y), \qquad (11.30)$$

which establishes the result for $n = 2$. Now we assume the result for general n, and show the result holds for $n + 1$. Using the already established $n = 2$ case, we have

$$H(X_1, \ldots, X_{n+1} | Y) = H(X_2, \ldots, X_{n+1} | Y, X_1) + H(X_1 | Y). \qquad (11.31)$$

Applying the inductive hypothesis to the first term on the right hand side gives

$$H(X_1, \ldots, X_{n+1} | Y) = \sum_{i=2}^{n+1} H(X_i | Y, X_1, \ldots, X_{i-1}) + H(X_1 | Y) \qquad (11.32)$$

$$= \sum_{i=1}^{n+1} H(X_i | Y, X_1, \ldots, X_{i-1}),$$ (11.33)

and the induction goes through.   □

Exercise 11.8: (Mutual information is not always subadditive) Let X and Y be
   independent identically distributed random variables taking the values 0 or 1
   with probability $1/2$. Let $Z \equiv X \oplus Y$, where $\oplus$ denotes addition modulo 2.
   Show that the mutual information in this case is not subadditive,

$$H(X, Y : Z) \not\leq H(X : Z) + H(Y : Z).$$ (11.34)

Exercise 11.9: (Mutual information is not always superadditive) Let X_1 be a
   random variable taking values 0 or 1 with respective probabilities of $1/2$ and
   $X_2 \equiv Y_1 \equiv Y_2 \equiv X_1$. Show that the mutual information in this case is not
   superadditive,

$$H(X_1 : Y_1) + H(X_2 : Y_2) \not\leq H(X_1, X_2 : Y_1, Y_2)$$ (11.35)

11.2.4 The data processing inequality

In many applications of interest we perform computations on the information we have
available, but that information is imperfect, as it has been subjected to noise before it
becomes available to us. A basic inequality of information theory, the *data processing
inequality*, states that information about the output of a source can only *decrease with
time*: once information has been lost, it is gone forever. Making this statement more
precise is the goal of this section.

   The intuitive notion of *information processing* is captured in the idea of a *Markov
chain* of random variables. A Markov chain is a sequence $X_1 \to X_2 \to \cdots$ of random
variables such that X_{n+1} is independent of $X_1, \ldots, X_{n-1}$, given X_n. More formally,

$$p(X_{n+1} = x_{n+1} | X_n = x_n, \ldots, X_1 = x_1) = p(X_{n+1} = x_{n+1} | X_n = x_n).$$ (11.36)

Under what conditions does a Markov chain lose information about its early values, as
time progresses? The following *data processing inequality* gives an information-theoretic
way of answering this question.

Theorem 11.5: **(Data processing inequality)** Suppose $X \to Y \to Z$ is a Markov
   chain. Then

$$H(X) \geq H(X : Y) \geq H(X : Z).$$ (11.37)

   Moreover, the first inequality is saturated if and only if, given Y, it is possible to
   reconstruct X.

   This result is intuitively plausible: it tells us that if a random variable X is subject to
noise, producing Y, then further actions on our part ('data processing') cannot be used
to increase the amount of mutual information between the output of the process and the
original information X.

Proof

The first inequality was proved in Theorem 11.3 on page 507. From the definitions we see that $H(X\!:\!Z) \le H(X\!:\!Y)$ is equivalent to $H(X|Y) \le H(X|Z)$. From the fact that $X \to Y \to Z$ is a Markov chain it is easy to prove (Exercise 11.10) that $Z \to Y \to X$ is also a Markov chain, and thus $H(X|Y) = H(X|Y, Z)$. The problem is thus reduced to proving that $H(X, Y, Z) - H(Y, Z) = H(X|Y, Z) \le H(X|Z) = H(X, Z) - H(Z)$. This is just the already proved strong subadditivity inequality.

Suppose $H(X\!:\!Y) < H(X)$. Then it is not possible to reconstruct X from Y, since if Z is the attempted reconstruction based only on knowledge of Y, then $X \to Y \to Z$ must be a Markov chain, and thus $H(X) > H(X\!:\!Z)$ by the data processing inequality. Thus $Z \ne X$. On the other hand, if $H(X\!:\!Y) = H(X)$, then we have $H(X|Y) = 0$ and thus whenever $p(X = x, Y = y) > 0$ we have $p(X = x|Y = y) = 1$. That is, if $Y = y$ then we can infer with certainty that X was equal to x, allowing us to reconstruct X.

$\square$

As noted above, if $X \to Y \to Z$ is a Markov chain, then so is $Z \to Y \to X$. Thus, as a corollary to the data processing inequality we see that if $X \to Y \to Z$ is a Markov chain then

$$H(Z\!:\!Y) \ge H(Z\!:\!X). \tag{11.38}$$

We refer to this result as the *data pipelining inequality*. Intuitively, it says that any information Z shares with X must be information which Z also shares with Y; the information is 'pipelined' from X through Y to Z.

Exercise 11.10: Show that if $X \to Y \to Z$ is a Markov chain then so is $Z \to Y \to X$.

11.3 Von Neumann entropy

The Shannon entropy measures the uncertainty associated with a classical probability distribution. Quantum states are described in a similar fashion, with density operators replacing probability distributions. In this section we generalize the definition of the Shannon entropy to quantum states.

Von Neumann defined the *entropy* of a quantum state ρ by the formula

$$S(\rho) \equiv -\mathrm{tr}(\rho \log \rho). \tag{11.39}$$

In this formula logarithms are taken to base two, as usual. If λ_x are the eigenvalues of ρ then von Neumann's definition can be re-expressed

$$S(\rho) = -\sum_x \lambda_x \log \lambda_x, \tag{11.40}$$

where we define $0 \log 0 \equiv 0$, as for the Shannon entropy. For calculations it is usually this last formula which is most useful. For instance, the completely mixed density operator in a d-dimensional space, I/d, has entropy $\log d$.

From now on, when we refer to entropy, it will usually be clear from context whether we mean the Shannon or von Neumann entropy.

Exercise 11.11: (Example calculations of entropy) Calculate $S(\rho)$ for

$$\rho = \begin{bmatrix} 1 & 0 \\ 0 & 0 \end{bmatrix} \tag{11.41}$$

$$\rho = \frac{1}{2}\begin{bmatrix} 1 & 1 \\ 1 & 1 \end{bmatrix} \tag{11.42}$$

$$\rho = \frac{1}{3}\begin{bmatrix} 2 & 1 \\ 1 & 1 \end{bmatrix}. \tag{11.43}$$

Exercise 11.12: (Comparison of quantum and classical entropies) Suppose $\rho = p|0\rangle\langle 0| + (1-p)\frac{(|0\rangle+|1\rangle)(\langle 0|+\langle 1|)}{2}$. Evaluate $S(\rho)$. Compare the value of $S(\rho)$ to $H(p, 1-p)$.

11.3.1 Quantum relative entropy

As for the Shannon entropy, it is extremely useful to define a quantum version of the relative entropy. Suppose ρ and σ are density operators. The *relative entropy* of ρ to σ is defined by

$$S(\rho||\sigma) \equiv \mathrm{tr}(\rho \log \rho) - \mathrm{tr}(\rho \log \sigma). \tag{11.50}$$

As with the classical relative entropy, the quantum relative entropy can sometimes be infinite. In particular, the relative entropy is defined to be $+\infty$ if the kernel of σ (the vector space spanned by the eigenvectors of σ with eigenvalue 0) has non-trivial intersection with the support of ρ (the vector space spanned by the eigenvectors of ρ with non-zero eigenvalue), and is finite otherwise. The quantum relative entropy is non-negative, a result sometimes known as *Klein's inequality*:

Theorem 11.7: **(Klein's inequality)** The quantum relative entropy is non-negative,

$$S(\rho||\sigma) \geq 0, \tag{11.51}$$

with equality if and only if $\rho = \sigma$.

Proof
Let $\rho = \sum_i p_i |i\rangle\langle i|$ and $\sigma = \sum_j q_j |j\rangle\langle j|$ be orthonormal decompositions for ρ and σ. From the definition of the relative entropy we have

$$S(\rho||\sigma) = \sum_i p_i \log p_i - \sum_i \langle i|\rho \log \sigma|i\rangle. \tag{11.52}$$

We substitute into this equation the equations $\langle i|\rho = p_i\langle i|$ and

$$\langle i| \log \sigma|i\rangle = \langle i| \left(\sum_j \log(q_j)|j\rangle\langle j| \right) |i\rangle = \sum_j \log(q_j)P_{ij}, \tag{11.53}$$

where $P_{ij} \equiv \langle i|j\rangle\langle j|i\rangle \geq 0$, to give

$$S(\rho||\sigma) = \sum_i p_i \left(\log p_i - \sum_j P_{ij} \log(q_j) \right). \tag{11.54}$$

Note that P_{ij} satisfies the equations $P_{ij} \geq 0, \sum_i P_{ij} = 1$ and $\sum_j P_{ij} = 1$. (Regarding P_{ij}

Box 11.2: Continuity of the entropy

Suppose we vary ρ by a small amount. How much does $S(\rho)$ change? *Fannes'
inequality* tells us that the answer is 'not much', and even provides a bound on
how small the change is.

Theorem 11.6: (**Fannes' inequality**) Suppose ρ and σ are density matrices such
that the trace distance between them satisfies $T(\rho, \sigma) \leq 1/e$. Then

$$|S(\rho) - S(\sigma)| \leq T(\rho, \sigma) \log d + \eta(T(\rho, \sigma)), \qquad (11.44)$$

where d is the dimension of the Hilbert space, and $\eta(x) \equiv -x \log x$.
Removing the restriction that $T(\rho, \sigma) \leq 1/e$ we can prove the weaker
inequality

$$|S(\rho) - S(\sigma)| \leq T(\rho, \sigma) \log d + \frac{1}{e}. \qquad (11.45)$$

Proof

To prove Fannes' inequality we need a simple result relating the trace distance
between two operators to their eigenvalues. Let $r_1 \geq r_2 \geq \cdots \geq r_d$ be the eigenvalues of ρ, in descending order, and $s_1 \geq s_2 \geq \cdots \geq s_d$ be the eigenvalues
of σ, again in descending order. By the spectral decomposition we may decompose $\rho - \sigma = Q - R$, where Q and R are positive operators with orthogonal
support, so $T(\rho, \sigma) = \text{tr}(R) + \text{tr}(Q)$. Defining $V \equiv R + \rho = Q + \sigma$, we have
$T(\rho, \sigma) = \text{tr}(R) + \text{tr}(Q) = \text{tr}(2V) - \text{tr}(\rho) - \text{tr}(\sigma)$. Let $t_1 \geq t_2 \geq \cdots \geq t_d$ be the
eigenvalues of T. Note that $t_i \geq \max(r_i, s_i)$, so $2t_i \geq r_i + s_i + |r_i - s_i|$, and it
follows that

$$T(\rho, \sigma) \geq \sum_i |r_i - s_i|. \qquad (11.46)$$

By calculus whenever $|r - s| \leq 1/2$ it follows that $|\eta(r) - \eta(s)| \leq \eta(|r - s|)$. A
moment's thought shows that $|r_i - s_i| \leq 1/2$ for all i, so

$$|S(\rho) - S(\sigma)| = \left| \sum_i (\eta(r_i) - \eta(s_i)) \right| \leq \sum_i \eta(|r_i - s_i|). \qquad (11.47)$$

Setting $\Delta \equiv \sum_i |r_i - s_i|$ and observing that $\eta(|r_i - s_i|) = \Delta \eta(|r_i - s_i|/\Delta) - |r_i -
s_i| \log(\Delta)$, we see that

$$|S(\rho) - S(\sigma)| \leq \Delta \sum \eta(|r_i - s_i|/\Delta) + \eta(\Delta) \leq \Delta \log d + \eta(\Delta), \qquad (11.48)$$

where we applied Theorem 11.2 to obtain the second inequality. But $\Delta \leq T(\rho, \sigma)$
by (11.46), so by the monotonicity of $\eta(\cdot)$ on the interval $[0, 1/e]$,

$$|S(\rho) - S(\sigma)| \leq T(\rho, \sigma) \log d + \eta(T(\rho, \sigma)), \qquad (11.49)$$

whenever $T(\rho, \sigma) \leq 1/e$, which is Fannes' inequality. The weaker form of Fannes'
inequality for general $T(\rho, \sigma)$ follows with minor modifications.   □

as a matrix, this property is known as *double stochasticity*.) Because $\log(\cdot)$ is a strictly concave function it follows that $\sum_j P_{ij} \log q_j \leq \log r_i$, where $r_i \equiv \sum_j P_{ij} q_j$, with equality if and only if there exists a value of j for which $P_{ij} = 1$. Thus

$$S(\rho||\sigma) \geq \sum_i p_i \log \frac{p_i}{r_i}, \tag{11.55}$$

with equality if and only if for each i there exists a value of j such that $P_{ij} = 1$, that is, if and only if P_{ij} is a permutation matrix. This has the form of the classical relative entropy. From the non-negativity of the classical relative entropy, Theorem 11.1, we deduce that

$$S(\rho||\sigma) \geq 0, \tag{11.56}$$

with equality if and only if $p_i = r_i$ for all i, and P_{ij} is a permutation matrix. To simplify the equality conditions further, note that by relabeling the eigenstates of σ if necessary, we can assume that P_{ij} is the identity matrix, and thus that ρ and σ are diagonal in the same basis. The condition $p_i = r_i$ tells us that the corresponding eigenvalues of ρ and σ are identical, and thus $\rho = \sigma$ are the equality conditions. □

11.3.2 Basic properties of entropy

The von Neumann entropy has many interesting and useful properties:

Theorem 11.8: (**Basic properties of von Neumann entropy**)

(1) The entropy is non-negative. The entropy is zero if and only if the state is pure.
(2) In a d-dimensional Hilbert space the entropy is at most $\log d$. The entropy is equal to $\log d$ if and only if the system is in the completely mixed state I/d.
(3) Suppose a composite system AB is in a pure state. Then $S(A) = S(B)$.
(4) Suppose p_i are probabilities, and the states ρ_i have support on orthogonal subspaces. Then

$$S\left(\sum_i p_i \rho_i\right) = H(p_i) + \sum_i p_i S(\rho_i). \tag{11.57}$$

(5) **Joint entropy theorem:** Suppose p_i are probabilities, $|i\rangle$ are orthogonal states for a system A, and ρ_i is any set of density operators for another system, B. Then

$$S\left(\sum_i p_i |i\rangle\langle i| \otimes \rho_i\right) = H(p_i) + \sum_i p_i S(\rho_i). \tag{11.58}$$

Proof
(1) Clear from the definition.
(2) The result follows from the non-negativity of the relative entropy,
$$0 \leq S(\rho||I/d) = -S(\rho) + \log d.$$
(3) From the Schmidt decomposition we know that the eigenvalues of the density operators of systems A and B are the same. (Recall the discussion following Theorem 2.7 on page 109.) The entropy is determined completely by the eigenvalues, so $S(A) = S(B)$.

(4) Let λ_i^j and $|e_i^j\rangle$ be the eigenvalues and corresponding eigenvectors of ρ_i. Observe that $p_i\lambda_i^j$ and $|e_i^j\rangle$ are the eigenvalues and eigenvectors of $\sum_i p_i\rho_i$, and thus

$$S\left(\sum_i p_i\rho_i\right) = -\sum_{ij} p_i\lambda_i^j \log p_i\lambda_i^j \tag{11.59}$$

$$= -\sum_i p_i \log p_i - \sum_i p_i \sum_j \lambda_i^j \log \lambda_i^j \tag{11.60}$$

$$= H(p_i) + \sum_i p_i S(\rho_i), \tag{11.61}$$

as required.

(5) Immediate from the preceding result. □

Exercise 11.13: (Entropy of a tensor product) Use the joint entropy theorem to show that $S(\rho \otimes \sigma) = S(\rho) + S(\sigma)$. Prove this result directly from the definition of the entropy.

By analogy with the Shannon entropies it is possible to define quantum joint and conditional entropies and quantum mutual information, for composite quantum systems. The joint entropy $S(A, B)$ for a composite system with two components A and B is defined in the obvious way, $S(A, B) \equiv -\mathrm{tr}(\rho^{AB} \log(\rho^{AB}))$, where ρ^{AB} is the density matrix of the system AB. We define the conditional entropy and mutual information by:

$$S(A|B) \equiv S(A, B) - S(B) \tag{11.62}$$

$$S(A:B) \equiv S(A) + S(B) - S(A, B) \tag{11.63}$$

$$= S(A) - S(A|B) = S(B) - S(B|A). \tag{11.64}$$

Some properties of the Shannon entropy fail to hold for the von Neumann entropy, and this has many interesting consequences for quantum information theory. For instance, for random variables X and Y, the inequality $H(X) \leq H(X, Y)$ holds. This makes intuitive sense: surely we cannot be more uncertain about the state of X than we are about the joint state of X and Y. This intuition fails for quantum states. Consider a system AB of two qubits in the entangled state $(|00\rangle + |11\rangle)/\sqrt{2}$. This is a pure state, so $S(A, B) = 0$. On the other hand, system A has density operator $I/2$, and thus has entropy equal to one. Another way of stating this result is that, for this system, the quantity $S(B|A) = S(A, B) - S(A)$ is negative.

Exercise 11.14: (Entanglement and negative conditional entropy) Suppose $|AB\rangle$ is a pure state of a composite system belonging to Alice and Bob. Show that $|AB\rangle$ is entangled if and only if $S(B|A) < 0$.

11.3.3 Measurements and entropy

How does the entropy of a quantum system behave when we perform a measurement on that system? Not surprisingly, the answer to this question depends on the type of measurement which we perform. Nevertheless, there are some surprisingly general assertions we can make about how the entropy behaves.

Suppose, for example, that a projective measurement described by projectors P_i is performed on a quantum system, but we never learn the result of the measurement. If the state of the system before the measurement was ρ then the state after is given by

$$\rho' = \sum_i P_i \rho P_i. \qquad (11.65)$$

The following result shows that the entropy is never decreased by this procedure, and remains constant only if the state is not changed by the measurement:

Theorem 11.9: (**Projective measurements increase entropy**) Suppose P_i is a complete set of orthogonal projectors and ρ is a density operator. Then the entropy of the state $\rho' \equiv \sum_i P_i \rho P_i$ of the system after the measurement is at least as great as the original entropy,

$$S(\rho') \geq S(\rho), \qquad (11.66)$$

with equality if and only if $\rho = \rho'$.

Proof
The proof is to apply Klein's inequality to ρ and ρ',

$$0 \leq S(\rho'||\rho) = -S(\rho) - \mathrm{tr}(\rho \log \rho'). \qquad (11.67)$$

The result will follow if we can prove that $-\mathrm{tr}(\rho \log \rho') = S(\rho')$. To do this, we apply the completeness relation $\sum_i P_i = I$, the relation $P_i^2 = P_i$, and the cyclic property of the trace, to obtain

$$-\mathrm{tr}(\rho \log \rho') = -\mathrm{tr}\left(\sum_i P_i \rho \log \rho'\right) \qquad (11.68)$$

$$= -\mathrm{tr}\left(\sum_i P_i \rho \log \rho' P_i\right). \qquad (11.69)$$

Note that $\rho' P_i = P_i \rho P_i = P_i \rho'$. That is, P_i commutes with ρ' and thus with $\log \rho'$, so

$$-\mathrm{tr}(\rho \log \rho') = -\mathrm{tr}\left(\sum_i P_i \rho P_i \log \rho'\right) \qquad (11.70)$$

$$= -\mathrm{tr}(\rho' \log \rho') = S(\rho'). \qquad (11.71)$$

This completes the proof. ☐

Exercise 11.15: (**Generalized measurements can decrease entropy**) Suppose a qubit in the state ρ is measured using the measurement operators $M_1 = |0\rangle\langle0|$ and $M_2 = |0\rangle\langle1|$. If the result of the measurement is unknown to us then the state of the system afterwards is $M_1 \rho M_1^\dagger + M_2 \rho M_2^\dagger$. Show that this procedure can *decrease* the entropy of the qubit.

11.3.4 Subadditivity
Suppose distinct quantum systems A and B have a joint state ρ^{AB}. Then the joint entropy for the two systems satisfies the inequalities

$$S(A, B) \leq S(A) + S(B) \qquad (11.72)$$

$$S(A, B) \geq |S(A) - S(B)|. \qquad (11.73)$$

The first of these inequalities is known as the *subadditivity* inequality for Von Neumann entropy, and holds with equality if and only if systems A and B are uncorrelated, that is, $\rho^{AB} = \rho^A \otimes \rho^B$. The second is called the *triangle* inequality, or sometimes the *Araki-Lieb* inequality; it is the quantum analogue of the inequality $H(X, Y) \geq H(X)$ for Shannon entropies.

The proof of subadditivity is a simple application of Klein's inequality, $S(\rho) \leq -\text{tr}(\rho \log \sigma)$. Setting $\rho \equiv \rho^{AB}$ and $\sigma \equiv \rho^A \otimes \rho^B$, note that

$$-\text{tr}(\rho \log \sigma) = -\text{tr}(\rho^{AB}(\log \rho^A + \log \rho^B)) \tag{11.74}$$
$$= -\text{tr}(\rho^A \log \rho^A) - \text{tr}(\rho^B \log \rho^B) \tag{11.75}$$
$$= S(A) + S(B). \tag{11.76}$$

Klein's inequality therefore gives $S(A, B) \leq S(A) + S(B)$, as desired. The equality conditions $\rho = \sigma$ for Klein's inequality give equality conditions $\rho^{AB} = \rho^A \otimes \rho^B$ for subadditivity.

To prove the triangle inequality, introduce a system R which purifies systems A and B, as in Section 2.5. Applying subadditivity we have

$$S(R) + S(A) \geq S(A, R). \tag{11.77}$$

Since ABR is in a pure state, $S(A, R) = S(B)$ and $S(R) = S(A, B)$. The previous inequality may be rearranged to give

$$S(A, B) \geq S(B) - S(A). \tag{11.78}$$

The equality conditions for this inequality are not quite as easily stated as those for subadditivity. Formally, the equality conditions are that $\rho^{AR} = \rho^A \otimes \rho^R$. Intuitively, what this means is that A is already as entangled as it can possibly be with the outside world, given its existing correlations with system B. A more detailed mathematical statement of the equality conditions is given in Exercise 11.16 on this page.

By symmetry between the systems A and B we also have $S(A, B) \geq S(A) - S(B)$. Combining this with $S(A, B) \geq S(B) - S(A)$ gives the triangle inequality.

Exercise 11.16: (Equality conditions for $S(A, B) \geq S(B) - S(A)$) Let $\rho^{AB} = \sum_i \lambda_i |i\rangle\langle i|$ is a spectral decomposition for ρ^{AB}. Show that $S(A, B) = S(B) - S(A)$ if and only if the operators $\rho_i^A \equiv \text{tr}_B(|i\rangle\langle i|)$ have a common eigenbasis, and the $\rho_i^B \equiv \text{tr}_A(|i\rangle\langle i|)$ have orthogonal support.

Exercise 11.17: Find an explicit non-trivial example of a mixed state ρ for AB such that $S(A, B) = S(B) - S(A)$.

11.3.5 Concavity of the entropy

The entropy is a *concave* function of its inputs. That is, given probabilities p_i – nonnegative real numbers such that $\sum_i p_i = 1$ – and corresponding density operators ρ_i, the entropy satisfies the inequality:

$$S\left(\sum_i p_i \rho_i\right) \geq \sum_i p_i S(\rho_i). \tag{11.79}$$

The intuition is that the $\sum_i p_i \rho_i$ expresses the state of a quantum system which is in an unknown state ρ_i with probability p_i, and our uncertainty about this mixture of states should be higher than the average uncertainty of the states ρ_i, since the state $\sum_i p_i \rho_i$ expresses ignorance not only due to the states ρ_i, but also a contribution due to our ignorance of the index i.

Suppose the ρ_i are states of a system A. Introduce an *auxiliary system B* whose state space has an orthonormal basis $|i\rangle$ corresponding to the index i on the density operators ρ_i. Define a joint state of AB by:

$$\rho^{AB} \equiv \sum_i p_i \rho_i \otimes |i\rangle\langle i|. \tag{11.80}$$

To prove concavity we use the subadditivity of the entropy. Note that for the density matrix ρ^{AB} we have:

$$S(A) = S\left(\sum_i p_i \rho_i\right) \tag{11.81}$$

$$S(B) = S\left(\sum_i p_i |i\rangle\langle i|\right) = H(p_i) \tag{11.82}$$

$$S(A, B) = H(p_i) + \sum_i p_i S(\rho_i). \tag{11.83}$$

Applying the subadditivity inequality $S(A, B) \le S(A) + S(B)$ we obtain

$$\sum_i p_i S(\rho_i) \le S\left(\sum_i p_i \rho_i\right), \tag{11.84}$$

which is concavity! Note that equality holds if and only if all the states ρ_i for which $p_i > 0$ are identical; that is, the entropy is a strictly concave function of its inputs.

It's worth pausing to think about the strategy we've employed in the proof of concavity, and the similar strategy used to prove the triangle inequality. We introduced an *auxiliary system B* in order to prove a result about the system A. Introducing auxiliary systems is something done often in quantum information theory, and we'll see this trick again and again. The intuition behind the introduction of B in this particular instance is as follows: we want to find a system of which part is in the state $\sum_i p_i \rho_i$, where the value of i is not known. System B effectively stores the 'true' value of i; if A were 'truly' in state ρ_i, the system B would be in state $|i\rangle\langle i|$, and observing system B in the $|i\rangle$ basis would reveal this fact. Using auxiliary systems to encode our intuition in a rigorous way is an art, but it is also an essential part of many proofs in quantum information theory.

Exercise 11.18: Prove that equality holds in the concavity inequality (11.79) if and only if all the ρ_is are the same.

Exercise 11.19: Show that there exists a set of unitary matrices U_j and a probability distribution p_j such that for any matrix A,

$$\sum_i p_i U_i A U_i^\dagger = \text{tr}(A)\frac{I}{d}, \tag{11.85}$$

where d is the dimension of the Hilbert space A lives in. Use this observation and the strict concavity of the entropy to give an alternate proof that the completely mixed state I/d on a space of d dimensions is the unique state of maximal entropy.

Exercise 11.20: Let P be a projector and $Q = I - P$ the complementary projector. Prove that there are unitary operators U_1 and U_2 and a probability p such that for all ρ, $P\rho P + Q\rho Q = pU_1\rho U_1^\dagger + (1 - p)U_2\rho U_2^\dagger$. Use this observation to give an alternate proof of Theorem 11.9 based on concavity.

Exercise 11.21: (Concavity of the Shannon entropy) Use the concavity of the von Neumann entropy to deduce that the Shannon entropy is concave in probability distributions.

Exercise 11.22: (Alternate proof of concavity) Define $f(p) \equiv S(p\rho + (1 - p)\sigma)$. Argue that to show concavity it is sufficient to prove that $f''(p) \leq 0$. Prove that $f''(p) \leq 0$, first for the case where ρ and σ are invertible, and then for the case where they are not.

11.3.6 The entropy of a mixture of quantum states

The flip side of concavity is the following useful theorem which provides an upper bound on the entropy of a mixture of quantum states. Taken together the two results imply that for a mixture $\sum_i p_i\rho_i$ of quantum states ρ_i the following inequality holds:

$$\sum_i p_i S(\rho_i) \leq S\left(\sum_i p_i\rho_i\right) \leq \sum_i p_i S(\rho_i) + H(p_i). \tag{11.86}$$

The intuition behind the upper bound on the right hand side is that our uncertainty about the state $\sum_i p_i\rho_i$ is never more than our average uncertainty about the state ρ_i, plus an additional contribution of $H(p_i)$ which represents the maximum possible contribution our uncertainty about the index i contributes to our total uncertainty. We now prove this upper bound.

Theorem 11.10: Suppose $\rho = \sum_i p_i\rho_i$, where p_i are some set of probabilities, and the ρ_i are density operators. Then

$$S(\rho) \leq \sum_i p_i S(\rho_i) + H(p_i), \tag{11.87}$$

with equality if and only if the states ρ_i have support on orthogonal subspaces.

Proof
We begin with the pure state case, $\rho_i = |\psi_i\rangle\langle\psi_i|$. Suppose the ρ_i are states of a system A, and introduce an auxiliary system B with an orthonormal basis $|i\rangle$ corresponding to the index i on the probabilities p_i. Define

$$|AB\rangle \equiv \sum_i \sqrt{p_i}|\psi_i\rangle|i\rangle. \tag{11.88}$$

Since $|AB\rangle$ is a pure state we have

$$S(B) = S(A) = S\left(\sum_i p_i|\psi_i\rangle\langle\psi_i|\right) = S(\rho). \tag{11.89}$$

Suppose we perform a projective measurement on the system B in the $|i\rangle$ basis. After the measurement the state of system B is

$$\rho^{B'} = \sum_i p_i|i\rangle\langle i|. \tag{11.90}$$

But by Theorem 11.9 projective measurements never decrease entropy, so $S(\rho) = S(B) \leq S(B') = H(p_i)$. Observing that $S(\rho_i) = 0$ for the pure state case, we have proved that

$$S(\rho) \leq H(p_i) + \sum_i p_i S(\rho_i), \tag{11.91}$$

when the states ρ_i are pure states. Furthermore, equality holds if and only if $B = B'$, which is easily seen to occur if and only if the states $|\psi_i\rangle$ are orthogonal.

The mixed state case is now easy. Let $\rho_i = \sum_j p_j^i |e_j^i\rangle\langle e_j^i|$ be orthonormal decompositions for the states ρ_i, so $\rho = \sum_{ij} p_i p_j^i |e_j^i\rangle\langle e_j^i|$. Applying the pure state result and the observation that $\sum_j p_j^i = 1$ for each i, we have

$$S(\rho) \leq -\sum_{ij} p_i p_j^i \log(p_i p_j^i) \tag{11.92}$$

$$= -\sum_i p_i \log p_i - \sum_i p_i \sum_j p_j^i \log p_j^i \tag{11.93}$$

$$= H(p_i) + \sum_i p_i S(\rho_i), \tag{11.94}$$

which is the desired result. The equality conditions for the mixed state case follow immediately from the equality conditions for the pure state case. □

11.4 Strong subadditivity

The subadditivity and triangle inequalities for two quantum systems can be extended to three systems. The basic result is known as the *strong subadditivity* inequality, and it is one of the most important and useful results in quantum information theory. The inequality states that for a trio of quantum systems, A, B, C,

$$S(A, B, C) + S(B) \leq S(A, B) + S(B, C). \tag{11.95}$$

Unfortunately, unlike the classical case, all known proofs of the quantum strong subadditivity inequality are quite difficult. However, it is so useful in quantum information theory that we give a full proof of this result. The basic structure of the proof is presented in Section 11.4.1, with some of the details of the proof left to Appendix 6.

11.4.1 Proof of strong subadditivity

The proof of the strong subadditivity inequality which we shall give is based upon a deep mathematical result known as *Lieb's theorem*. We begin with a definition which allows us to state Lieb's theorem.

Suppose $f(A, B)$ is a real-valued function of two matrices, A and B. Then f is said to be *jointly concave* in A and B if for all $0 \leq \lambda \leq 1$,

$$f(\lambda A_1 + (1 - \lambda)A_2, \lambda B_1 + (1 - \lambda)B_2) \geq \lambda f(A_1, B_1) + (1 - \lambda)f(A_2, B_2). \tag{11.96}$$

Exercise 11.23: (Joint concavity implies concavity in each input) Let $f(A, B)$ be a jointly concave function. Show that $f(A, B)$ is concave in A, with B held fixed. Find a function of two variables that is concave in each of its inputs, but is not jointly concave.

Theorem 11.11: (**Lieb's theorem**) Let X be a matrix, and $0 \le t \le 1$. Then the function

$$f(A, B) \equiv \text{tr}(X^\dagger A^t X B^{1-t}) \tag{11.97}$$

is jointly concave in positive matrices A and B.

Proof
See Appendix 6 for the proof of Lieb's theorem.   □

Lieb's theorem implies a sequence of results, each interesting in its own right, culminating in the proof of strong subadditivity. We begin with the convexity of the relative entropy.

Theorem 11.12: (**Convexity of the relative entropy**) The relative entropy $S(\rho\|\sigma)$ is jointly convex in its arguments.

Proof
For arbitrary matrices A and X acting on the same space define

$$I_t(A, X) \equiv \text{tr}(X^\dagger A^t X A^{1-t}) - \text{tr}(X^\dagger X A). \tag{11.98}$$

The first term in this expression is concave in A, by Lieb's theorem, and the second term is linear in A. Thus, $I_t(A, X)$ is concave in A. Define

$$I(A, X) \equiv \left.\frac{d}{dt}\right|_{t=0} I_t(A, X) = \text{tr}(X^\dagger(\log A)XA) - \text{tr}(X^\dagger X(\log A)A). \tag{11.99}$$

Noting that $I_0(A, X) = 0$ and using the concavity of $I_t(A, X)$ in A we have

$$I(\lambda A_1 + (1 - \lambda)A_2, X) = \lim_{\Delta \to 0} \frac{I_\Delta(\lambda A_1 + (1 - \lambda)A_2, X)}{\Delta} \tag{11.100}$$

$$\ge \lambda \lim_{\Delta \to 0} \frac{I_\Delta(A_1, X)}{\Delta} + (1 - \lambda) \lim_{\Delta \to 0} \frac{I_\Delta(A_2, X)}{\Delta} \tag{11.101}$$

$$= \lambda I(A_1, X) + (1 - \lambda)I(A_2, X). \tag{11.102}$$

That is, $I(A, X)$ is a concave function of A. Defining the block matrices

$$A \equiv \begin{bmatrix} \rho & 0 \\ 0 & \sigma \end{bmatrix}, \quad X \equiv \begin{bmatrix} 0 & 0 \\ I & 0 \end{bmatrix} \tag{11.103}$$

we can easily verify that $I(A, X) = -S(\rho\|\sigma)$. The joint convexity of $S(\rho\|\sigma)$ follows from the concavity of $I(A, X)$ in A.   □

Corollary 11.13: (**Concavity of the quantum conditional entropy**) Let AB be a composite quantum system with components A and B. Then the conditional entropy $S(A|B)$ is concave in the state ρ^{AB} of AB.

Proof

Let d be the dimension of system A. Note that

$$S\left(\rho^{AB} \left\| \frac{I}{d} \otimes \rho^B \right.\right) = -S(A,B) - \text{tr}\left(\rho_{AB} \log\left(\frac{I}{d} \otimes \rho^B\right)\right) \quad (11.104)$$

$$= -S(A,B) - \text{tr}(\rho^B \log \rho^B) + \log d \quad (11.105)$$

$$= -S(A|B) + \log d. \quad (11.106)$$

Thus $S(A|B) = \log d - S(\rho^{AB}\|I/d \otimes \rho^B)$. The concavity of $S(A|B)$ follows from the joint convexity of the relative entropy. □

Theorem 11.14: (**Strong subadditivity**) For any trio of quantum systems, A, B, C, the inequalities

$$S(A) + S(B) \le S(A,C) + S(B,C) \quad (11.107)$$

$$S(A,B,C) + S(B) \le S(A,B) + S(B,C) \quad (11.108)$$

hold.

Proof

The two inequalities are in fact equivalent. We will use concavity of the conditional entropy to prove the first, and then show that the second follows. Define a function $T(\rho^{ABC})$ of density operators on the system ABC,

$$T(\rho^{ABC}) \equiv S(A) + S(B) - S(A,C) - S(B,C) = -S(C|A) - S(C|B). (11.109)$$

From the concavity of the conditional entropy we see that $T(\rho^{ABC})$ is a convex function of ρ^{ABC}. Let $\rho^{ABC} = \sum_i p_i |i\rangle\langle i|$ be a spectral decomposition of ρ^{ABC}. From the convexity of T, $T(\rho^{ABC}) \le \sum_i p_i T(|i\rangle\langle i|)$. But $T(|i\rangle\langle i|) = 0$ as for a pure state $S(A,C) = S(B)$ and $S(B,C) = S(A)$. It follows that $T(\rho^{ABC}) \le 0$, and thus

$$S(A) + S(B) - S(A,C) - S(B,C) \le 0, \quad (11.110)$$

which is the first inequality we set out to prove.

To obtain the second inequality, introduce an auxiliary system R purifying the system ABC. Then using the just-proved inequality we have

$$S(R) + S(B) \le S(R,C) + S(B,C). \quad (11.111)$$

Since $ABCR$ is a pure state, $S(R) = S(A,B,C)$ and $S(R,C) = S(A,B)$, so (11.111) becomes

$$S(A,B,C) + S(B) \le S(A,B) + S(B,C), \quad (11.112)$$

as we set out to show. □

Exercise 11.24: We obtained strong subadditivity as a consequence of the inequality $S(A) + S(B) \le S(A,C) + S(B,C)$. Show that this inequality can be obtained as a consequence of strong subadditivity.

Exercise 11.25: We obtained strong subadditivity as a consequence of the concavity of the conditional information, $S(A|B)$. Show that the concavity of the conditional entropy may be deduced from strong subadditivity. (*Hint:* You may need to introduce an auxiliary system into the problem.)

11.4.2 Strong subadditivity: elementary applications

Strong subadditivity and related results have many useful implications for quantum information theory. Let's take a look at a few of the elementary consequences of these results.

First, it is worth emphasizing how remarkable it is that the inequality $S(A) + S(B) \leq S(A,C) + S(B,C)$ holds. The corresponding inequality holds also for Shannon entropies, but for different reasons. For Shannon entropies it is true that $H(A) \leq H(A,C)$ and $H(B) \leq H(B,C)$, so the sum of the two inequalities must necessarily be true. In the quantum case, it is possible to have either $S(A) > S(A,C)$ or $S(B) > S(B,C)$, yet somehow Nature conspires in such a way that both of these possibilities are not true simultaneously, in order to ensure that the condition $S(A) + S(B) \leq S(A,C) + S(B,C)$ is always satisfied. Other ways of rephrasing this are in terms of conditional entropies and mutual informations,

$$0 \leq S(C|A) + S(C|B) \tag{11.113}$$

$$S(A:B) + S(A:C) \leq 2S(A), \tag{11.114}$$

both of which are also remarkable inequalities, for similar reasons. Note, however, that the inequality $0 \leq S(A|C) + S(B|C)$, which one might hope to be true based upon (11.114), is not, as can be seen by choosing ABC to be a product of a pure state for A with an EPR state for BC.

Exercise 11.26: Prove that $S(A:B) + S(A:C) \leq 2S(A)$. Note that the corresponding inequality for Shannon entropies holds since $H(A:B) \leq H(A)$. Find an example where $S(A:B) > S(A)$.

For practical applications strong subadditivity is often most easily applied by using a rephrasing in terms of the conditional or mutual informations. The following theorem lists three very simple reformulations of strong subadditivity that provide a powerful intuitive guide to the properties of quantum entropy.

Theorem 11.15:

(1) **Conditioning reduces entropy**: Suppose ABC is a composite quantum system. Then $S(A|B,C) \leq S(A|B)$.

(2) **Discarding quantum systems never increases mutual information**: Suppose ABC is a composite quantum system. Then $S(A:B) \leq S(A:B,C)$.

(3) **Quantum operations never increase mutual information**: Suppose AB is a composite quantum system and $\mathcal{E}$ is a trace-preserving quantum operation on system B. Let $S(A:B)$ denote the mutual information between systems A and B before $\mathcal{E}$ is applied to system B, and $S(A':B')$ the mutual information after $\mathcal{E}$ is applied to system B. Then $S(A':B') \leq S(A:B)$.

Proof

(1) The proof is the same as the classical proof (part of Theorem 11.3 on page 506), which we reproduce for convenience: $S(A|B,C) \leq S(A|B)$ is equivalent to $S(A,B,C) - S(B,C) \leq S(A,B) - S(B)$, which is equivalent to $S(A,B,C) + S(B) \leq S(A,B) + S(B,C)$, which is strong subadditivity.

(2) $S(A\!:\!B) \leq S(A\!:\!B,C)$ is equivalent to
$S(A) + S(B) - S(A,B) \leq S(A) + S(B,C) - S(A,B,C)$, which is equivalent to $S(A,B,C) + S(B) \leq S(A,B) + S(B,C)$, which is strong subadditivity.

(3) By the constructions of Chapter 8 the action of $\mathcal{E}$ on B may be simulated by introducing a third system C, initially in a pure state $|0\rangle$, and a unitary interaction U between B and C. The action of $\mathcal{E}$ on B is equivalent to the action of U followed by discarding system C. Letting primes denote the state of the systems after U has acted we have $S(A\!:\!B) = S(A\!:\!B,C)$ because C starts out in a product state with AB, and clearly $S(A\!:\!B,C) = S(A'\!:\!B',C')$. Discarding systems cannot increase mutual information, so $S(A'\!:\!B') \leq S(A'\!:\!B',C')$. Putting it all together gives $S(A'\!:\!B') \leq S(A\!:\!B)$, as required.   □

There is an interesting set of questions related to the subadditivity properties of quantum conditional entropies. We saw earlier that the Shannon mutual information is not subadditive, and thus the quantum mutual information is not subadditive either. What about the subadditivity of the conditional entropy? That is, is it true that

$$S(A_1, A_2|B_1, B_2) \leq S(A_1|B_1) + S(A_2|B_2), \tag{11.115}$$

for any four quantum systems A_1, A_2, B_1 and B_2? It turns out that this inequality is correct. What's more, the conditional entropy is also subadditive in the first and second entries. Proving these facts is an instructive exercise in the application of strong subadditivity.

Theorem 11.16: (**Subadditivity of the conditional entropy**) Let $ABCD$ be a composite of four quantum systems. Then the conditional entropy is jointly subadditive in the first and second entries:

$$S(A, B|C, D) \leq S(A|C) + S(B|D). \tag{11.116}$$

Let ABC be a composite of three quantum systems. Then the conditional entropy is subadditive in each of the first and second entries:

$$S(A, B|C) \leq S(A|C) + S(B|C) \tag{11.117}$$
$$S(A|B, C) \leq S(A|B) + S(A|C). \tag{11.118}$$

Proof

To prove joint subadditivity in both entries, note that by strong subadditivity

$$S(A, B, C, D) + S(C) \leq S(A, C) + S(B, C, D). \tag{11.119}$$

Adding $S(D)$ to each side of this inequality, we obtain

$$S(A, B, C, D) + S(C) + S(D) \leq S(A, C) + S(B, C, D) + S(D). \tag{11.120}$$

Applying strong subadditivity to the last two terms of the right hand side gives

$$S(A, B, C, D) + S(C) + S(D) \leq S(A, C) + S(B, D) + S(C, D). \quad (11.121)$$

Rearranging this inequality gives

$$S(A, B|C, D) \leq S(A|C) + S(B|D), \quad (11.122)$$

which is the joint subadditivity of the conditional entropy.

Subadditivity of the conditional entropy in the first entry, $S(A, B|C) \leq S(A|C) + S(B|C)$, is trivially seen to be equivalent to strong subadditivity. Subadditivity in the second entry is slightly more challenging. We wish to show that $S(A|B, C) \leq S(A|B) + S(A|C)$. Note that this is equivalent to demonstrating the inequality

$$S(A, B, C) + S(B) + S(C) \leq S(A, B) + S(B, C) + S(A, C). \quad (11.123)$$

To prove this, note that at least one of the inequalities $S(C) \leq S(A, C)$ or $S(B) \leq S(A, B)$ must be true, as $S(A|B) + S(A|C) \geq 0$ by the first inequality in Theorem 11.14. Suppose $S(C) \leq S(A, C)$. Adding to this inequality the strong subadditivity inequality, $S(A, B, C) + S(B) \leq S(A, B) + S(B, C)$ gives the result. A similar proof holds in the case when $S(B) \leq S(A, B)$. □

When we introduced the relative entropy it was described as being rather like a measure of distance between probability distributions or between density operators. Imagine that a quantum system consists of two parts, labeled A and B, and that we are given two density operators ρ^{AB} and σ^{AB}. In order to fulfil its distance-like promise, a very desirable property of $S(\cdot\|\cdot)$ is that it decrease when part of the system is ignored, that is:

$$S(\rho^A\|\sigma^A) \leq S(\rho^{AB}\|\sigma^{AB}). \quad (11.124)$$

This result is known as the *monotonicity* of the relative entropy. Intuitively this is a very reasonable property for a measure of distance to have; we expect that ignoring part of a physical system makes it harder to distinguish two states of that system (compare Section 9.2.1), and thus decrease any reasonable measure of distance between them.

Theorem 11.17: (**Monotonicity of the relative entropy**) Let ρ^{AB} and σ^{AB} be any two density matrices of a composite system AB. Then

$$S(\rho^A\|\sigma^A) \leq S(\rho^{AB}\|\sigma^{AB}). \quad (11.125)$$

Proof
Exercise 11.19 on page 517 implies that there exist unitary transformations U_j on the space B and probabilities p_j such that

$$\rho^A \otimes \frac{I}{d} = \sum_j p_j U_j \rho^{AB} U_j^\dagger \quad (11.126)$$

for all ρ^{AB}. From the convexity of the relative entropy we obtain

$$S\left(\rho^A \otimes \frac{I}{d} \middle\| \sigma^A \otimes \frac{I}{d}\right) \leq \sum_j p_j S\left(U_j \rho^{AB} U_j^\dagger \middle\| U_j \sigma^{AB} U_j^\dagger\right). \quad (11.127)$$

But the relative entropy is invariant under unitary conjugation, so this gives

$$S\left(\rho^A \otimes \frac{I}{d} \middle\| \sigma^A \otimes \frac{I}{d}\right) \le \sum_j p_j S\left(\rho^{AB} \middle\| \sigma^{AB}\right) = S\left(\rho^{AB} \middle\| \sigma^{AB}\right). \quad (11.128)$$

Combining this with the easily verified observation that

$$S\left(\rho^A \otimes \frac{I}{d} \middle\| \sigma^A \otimes \frac{I}{d}\right) = S(\rho^A \| \sigma^A), \quad (11.129)$$

gives the monotonicity of the relative entropy. $\qquad\square$

Problem 11.1: (Generalized Klein's inequality) Suppose $f(\cdot)$ is a convex function from real numbers to real numbers. Then f induces a natural function $f(\cdot)$ on Hermitian operators, as described in Section 2.1.8 on page 75. Prove that

$$\mathrm{tr}(f(A) - f(B)) \ge \mathrm{tr}((A - B)f'(B)). \quad (11.130)$$

Use this result to show that the relative entropy is non-negative.

Problem 11.2: (Generalized relative entropy) The definition of the relative entropy may be extended to apply to any two positive operators r and s,

$$S(r\|s) \equiv \mathrm{tr}(r \log r) - \mathrm{tr}(r \log s). \quad (11.131)$$

The earlier argument proving joint convexity of the relative entropy goes directly through for this generalized definition:

(1) For $\alpha, \beta > 0$ show that

$$S(\alpha r \| \beta s) = \alpha S(r\|s) + \alpha\,\mathrm{tr}(r) \log(\alpha/\beta). \quad (11.132)$$

(2) Prove that the joint convexity of the relative entropy implies the subadditivity of the relative entropy,

$$S(r_1 + r_2 \| s_1 + s_2) \le S(r_1 \| s_1) + S(r_2 \| s_2). \quad (11.133)$$

(3) Prove that subadditivity of the relative entropy implies joint convexity of the relative entropy.

(4) Let p_i and q_i be probability distributions over the same set of indices. Show that

$$S\left(\sum_i p_i r_i \middle\| \sum_i q_i s_i\right) \le \sum_i p_i S(r_i \| s_i) + \sum_i p_i \mathrm{tr}(r_i) \log(p_i/q_i). \quad (11.134)$$

In the case where the r_i are density operators so $\mathrm{tr}(r_i) = 1$, this reduces to the pretty formula

$$S\left(\sum_i p_i r_i \middle\| \sum_i q_i s_i\right) \le \sum_i p_i S(r_i \| s_i) + H(p_i \| q_i), \quad (11.135)$$

where $H(\cdot\|\cdot)$ is the Shannon relative entropy.

Problem 11.3: (Analogue of the triangle inequality for conditional entropy)

(1) Show that $H(X, Y|Z) \ge H(X|Z)$.

(2) Show that it is not always true that $S(A, B|C) \ge S(A|C)$.

(3) Prove the conditional version of the triangle inequality,

$$S(A, B|C) \geq S(A|C) - S(B|C). \tag{11.136}$$

Problem 11.4: (Conditional forms of strong subadditivity)

(1) Prove that $S(A, B, C|D) + S(B|D) \leq S(A, B|D) + S(B, C|D)$.
(2) Show by explicit example that it is not always true that
$H(D|A, B, C) + H(D|B) \leq H(D|A, B) + H(D|B, C)$.

Problem 11.5: (Strong subadditivity – Research) Find a simple proof of the strong subadditivity inequality for quantum entropies.

Summary of Chapter 11: Entropy and information

- **Fundamental measures of information arise as the answers to questions about the quantity of physical resources required to solve some information processing problem.**

- **Basic definitions:**

(entropy)	$S(A) = -\text{tr}(\rho^A \log \rho^A)$
(relative entropy)	$S(\rho\|\sigma) = -S(\rho) - \text{tr}(\rho \log \sigma)$
(conditional entropy)	$S(A|B) = S(A, B) - S(B)$
(mutual information)	$S(A:B) = S(A) + S(B) - S(A, B)$

- **Strong subadditivity:** $S(A, B, C) + S(B) \leq S(A, B) + S(B, C)$. The other entropy inequalities we discussed are corollaries of this or the joint convexity of the relative entropy.

- **The relative entropy is jointly convex in its arguments.**

- **The relative entropy is monotonic:** $S(\rho^A\|\sigma^A) \leq S(\rho^{AB}\|\sigma^{AB})$.

History and further reading

Historically, the concept of entropy first arose in the study of thermodynamics and statistical mechanics. But the modern information-theoretic foundation for the study of entropy came in Shannon's wonderful papers on information theory [Sha48]. A good general reference on properties of the Shannon entropy (and much else in information theory) are Chapters 2 and 16 of Cover and Thomas[CT91]. General references on the von Neumann entropy are the review article by Wehrl[Weh78], and the book by Ohya and Petz[OP93].

The entropic uncertainty principle we prove is due to Deutsch[Deu83]. Many other people have worked on entropic uncertainty relations, and we just mention two other papers of interest. Kraus[Kra87] conjectured an entropic uncertainty relation strengthening Deutsch's for a particular class of measurements, and Maassen and Uffink[MU88] proved Kraus' conjecture. The relative entropy was introduced by Kullback and Leibler[KL51], and its quantum generalization is due to Umegaki[Ume62]. Fannes' inequality appeared in [Fan73]. Klein's inequality was proved in [Kle31]. The triangle inequality is due to Araki and Lieb[AL70]. Strong subadditivity has an interesting history. Robinson and Ruelle[RR67] first noted the importance of classical strong subadditivity for *statistical physics*. The quantum version was then conjectured in 1968 by Lanford and Robinson[LR68]. Obtaining a proof of the result was rather difficult however. Finally, in 1973 the theorem was proved in two papers: Lieb's eponymous theorem in [Lie73], while the surprising connection to strong subadditivity was developed in Lieb and Ruskai[LR73b]; see also[LR73a]. Lieb's theorem is a generalization of the *Wigner–Yanase–Dyson* conjecture made in 1963 by Wigner and Yanase[WY63] and subsequently extended by Dyson (unpublished); prior to 1973 it was not known that the Wigner–Yanase–Dyson conjecture and strong subadditivity were related! See Wehrl[Weh78] for a discussion of the Wigner–Yanase–Dyson conjecture. The proof of Lieb's theorem we have given is due to Simon[Sim79], and is a variant of a proof by Uhlmann[Uhl77]. Other proofs of Lieb's theorem are also known. See for example, Epstein[Eps73], Ando[And79], and Petz[Pet86]. Subadditivity of the relative entropy in the first and second entries was proved by Lieb[Lie75]. Joint subadditivity of the quantum conditional entropy was proved by Nielsen[Nie98]. The monotonicity of the relative entropy was first noted by Lindblad[Lin75]. Problem 11.2 is due to Ruskai[Rus94].

12 Quantum information theory

Classical information theory is mostly concerned with the problem of sending classical information – letters in an alphabet, speech, strings of bits – over *communications channels* which operate in accordance with the laws of classical physics. How does the picture change if we can build quantum-mechanical communications channels? Can we transmit information more efficiently? Can we make use of quantum mechanics to transmit secret information without being eavesdropped on? These are just two of the questions we may ask when communication channels are allowed to be quantum mechanical. This redefinition of what a channel is causes us to go back and re-examine the fundamental questions motivating classical information theory, in the search for new answers. This chapter surveys what is known about quantum information theory, including some surprising and intriguing possibilities made possible by quantum communication channels.

Quantum information theory is motivated by the study of communications channels, but it has a much wider domain of application, and it is a thought-provoking challenge to capture in a verbal nutshell the goals of the field. As described in Section 1.6, we can identify three fundamental goals uniting work on quantum information theory: to identify elementary classes of static resources in quantum mechanics (which we identify as types of 'information'); to identify elementary classes of dynamical processes in quantum mechanics (identified as types of 'information processing'); and to quantify resource tradeoffs incurred performing elementary dynamical processes. Quantum information theory is fundamentally richer than classical information theory, because quantum mechanics includes so many more elementary classes of static and dynamic resources – not only does it support all the familiar classical types, but there are entirely new classes such as the static resource of entanglement to make life even more interesting than it is classically.

The title of the chapter is 'Quantum information theory', and you may be forgiven for wondering how it is that we can hope to cover all aspects of quantum information theory in a single chapter. In fact, quantum information theory contains many facets other than those described here, including the study of quantum operations, the definition and study of fidelity measures, quantum error-correcting codes, and various notions of entropy – all topics which we have described in detail in earlier chapters. The purpose of the present chapter is to describe quantum information theory in its 'purest' form; those other chapters are focused on developing specific tools, while we are concerned here with the grand sweep of things, with the most general statements one can make about the properties of quantum information.

We begin in Section 12.1 with a discussion of some of the unique properties of quantum states when compared to classical states, in the language of information theory. Not only are quantum states impossible to copy, generally, but also they cannot be perfectly distinguished! This is quantified by the *Holevo bound*. We then consider, in Section 12.2, an elementary information-theoretic task, *data compression*, and show how quantum states can be compressed much as classical ones are. This is done by paralleling the theorem of typical sequences with the typical subspace theorem, to prove Schumacher's quantum *noiseless channel coding theorem*, the analogue of Shannon's noiseless channel coding theorem. A natural generalization of this problem is the capacity of a noisy channel for classical information, and in Section 12.3 we define and prove the analogue to Shannon's noisy channel coding theorem, known as the *Holevo–Schumacher–Westmoreland theorem*. The most difficult challenge is the capacity of a *noisy* quantum channel for *quantum* information. This is the subject of Section 12.4, in which the entropy exchange, the quantum Fano inequality, and the quantum data processing inequality are defined, but the open question of the capacity is left unresolved. Two applications of the noisy channel relations, the quantum Singleton bound, and the exorcism of Maxwell's demon, are presented, and the first half of this chapter is summarized. Two recurring themes throughout this exploration of quantum information are entanglement and non-orthogonality, and these subjects are the focus of the last two sections of the chapter. Section 12.5 describes how entanglement can be thought of as a *physical resource*, and explains how it can be transformed, distilled, and diluted. And finally, in Section 12.6, we present *quantum cryptography*, a provably secure means of communication whose success arises from the many properties of quantum information considered in this chapter.

12.1 Distinguishing quantum states and the accessible information

There is a simple game we can play to illustrate the remarkable differences between quantum and classical information. We will describe this game using two fictitious protagonists, Alice and Bob; of course the results can be rephrased in more abstract language, but the anthropocentric language makes the results easier to think (and write!) about.

Suppose Alice has a classical information source producing symbols $X = 0, \ldots, n$ according to a probability distribution $p_0, \ldots, p_n$. The aim of the game is for Bob to determine the value of X as best he can. To achieve this goal, Alice prepares a quantum state ρ_X chosen from some fixed set $\rho_0, \ldots, \rho_n$, and gives the state to Bob, who makes a quantum measurement on the state he has been given, and then tries to make the best guess he can as to the identity of X, based on his measurement result Y.

A good measure of how much information Bob has gained about X through the measurement is the mutual information $H(X : Y)$ between X and the measurement outcome Y, as defined in Chapter 11. By the data processing inequality we know that Bob can infer X from Y if and only if $H(X : Y) = H(X)$, and that in general $H(X : Y) \leq H(X)$; we will see later that the closeness of $H(X : Y)$ to $H(X)$ actually provides a quantitative measure of how well Bob can determine X. Bob's goal is to choose a measurement which maximizes $H(X : Y)$, bringing it as close as possible to $H(X)$. To this end, we define Bob's *accessible information* to be the maximum of the mutual information $H(X : Y)$ over all possible measurement schemes. The accessible information is a measure of how well Bob can do at inferring the state Alice prepared.

In classical information theory, the accessible information is not so interesting; while in practice it may be difficult to distinguish two classical states – consider the troubles we have reading bad handwriting – in principle it is always possible. In contrast, in quantum mechanics it is not always possible to distinguish distinct states, even in principle. For example, we saw in Box 2.3 on page 87 that there is no quantum mechanical procedure to reliably distinguish two non-orthogonal quantum states. Stated in terms of accessible information, if Alice prepares the state $|\psi\rangle$ with probability p, and another, non-orthogonal state $|\varphi\rangle$ with probability $1 - p$, the accessible information of this preparation is strictly less than $H(p)$, as it is not possible for Bob to determine the identity of the state with full reliability. Classically, if Alice prepares one of two classical states – say a bit in the state 0 with probability p, or in the state 1 with probability $1 - p$ – then there is no fundamental reason of principle why Bob cannot distinguish between these two states, and thus the accessible information is the same as the entropy of preparation, $H(p)$.

There's an important caveat to this discussion, a context in which the concept of accessible information does make sense classically. The context is that of distinguishing probability distributions. Imagine that Alice prepares the state 0 or 1 according to one of two probability distributions, either $(p, 1 - p)$ or $(q, 1 - q)$. Given the state, Bob's goal is to identify which probability distribution Alice used to prepare the state. Clearly, it is not always possible for Bob to perform this identification with perfect reliability! Nevertheless, this example (analogous to the accessible information for a quantum system prepared in one of a set of mixed states) is of subsidiary importance. What is most important and remarkable is that the fundamental objects in quantum mechanics – the pure quantum states – enjoy distinguishability properties that are markedly different and much richer than the corresponding properties for the fundamental objects of classical information theory, such as '0' or '1'.

The *no-cloning theorem* provides another perspective on the lack of accessibility suffered by quantum information in comparison to classical information. Classical information can, of course, be *copied*. This can be done exactly with digital information, like the multiply backed-up LaTeX file being used to generate this book, or approximately, as with the analog images appearing on each page of this book, which have been copied by printing press prior to distribution. Surprisingly, the no-cloning theorem states that quantum mechanics does not allow unknown quantum states to be copied exactly, and places severe limitations on our ability to make approximate copies. The no-cloning theorem is proved in Box 12.1.

At first sight the no-cloning theorem appears rather peculiar. After all, isn't classical physics a special case of quantum mechanics? How is it possible that we can copy classical information if we can't copy quantum states? The answer is that the no-cloning theorem does not prevent *all* quantum states from being copied, it simply says that non-orthogonal quantum states cannot be copied. More precisely, suppose $|\psi\rangle$ and $|\varphi\rangle$ are two non-orthogonal quantum states. Then the no-cloning theorem implies that it is not possible to build a quantum device that, when input with $|\psi\rangle$ or $|\varphi\rangle$, will output two copies of the input state, $|\psi\rangle|\psi\rangle$ or $|\varphi\rangle|\varphi\rangle$. On the other hand, if $|\psi\rangle$ and $|\varphi\rangle$ are orthogonal, then the no-cloning theorem doesn't prohibit their cloning. Indeed, it is rather easy to design quantum circuits which copy such states! This observation resolves the apparent contradiction between the no-cloning theorem and the ability to copy classical information, for the different states of classical information can be thought of merely as orthogonal quantum states.

Exercise 12.1: Suppose $|\psi\rangle$ and $|\varphi\rangle$ are two orthogonal quantum states of a single qubit. Design a quantum circuit with two input qubits (the 'data' and the 'target' qubits), with the data qubit in either the state $|\psi\rangle$ or $|\varphi\rangle$, and the target qubit prepared in the standard state $|0\rangle$, which produces as output $|\psi\rangle|\psi\rangle$ or $|\varphi\rangle|\varphi\rangle$, depending on whether $|\psi\rangle$ or $|\varphi\rangle$ was input to the data qubit.

What is the connection between cloning and accessible information? Suppose Alice prepares one of two non-orthogonal quantum states $|\psi\rangle$ and $|\varphi\rangle$ with respective probabilities p and $1-p$. Suppose it were the case that Bob's accessible information about these states was $H(p)$, that is, the laws of quantum mechanics allowed Bob to obtain enough information by measurement to identify which of the two states $|\psi\rangle$ and $|\varphi\rangle$ Alice had prepared. Then Bob could clone the states in a very simple manner: he would perform a measurement to determine which of $|\psi\rangle$ and $|\varphi\rangle$ had been prepared by Alice, and once he had made the identification, could prepare at will multiple copies of whichever state $|\psi\rangle$ or $|\varphi\rangle$ Alice had given him. Thus, the no-cloning theorem can be seen as a consequence of the fact that the accessible information for these states is strictly less than $H(p)$. It's also possible to turn this perspective around, and view the fact that the accessible information is less than $H(p)$ as a consequence of the no-cloning theorem! This is done as follows. Imagine that it is possible to clone non-orthogonal states. After receiving the state $|\psi\rangle$ or $|\varphi\rangle$ from Alice, Bob could repeatedly apply such a cloning device to obtain the state $|\psi\rangle^{\otimes n}$ or $|\varphi\rangle^{\otimes n}$. In the limit of large n these two states become very nearly orthogonal and it is possible to distinguish them with arbitrarily high reliability by a projective measurement. That is, if it were possible to clone then Bob could identify with arbitrarily high probability of success whether the state $|\psi\rangle$ or $|\varphi\rangle$ had been prepared, and thus the accessible information would be $H(p)$. We can therefore view the no-cloning theorem as being equivalent to the statement than in quantum mechanics the accessible information for non-orthogonal states is in general less than the entropy of preparation.

As we have emphasized throughout the book, the hidden nature of quantum information lies at the heart of the power of quantum computation and quantum information, and the accessible information captures in a quantitative way this hidden nature of quantum information. Unfortunately, no general method for calculating the accessible information is known; however, a variety of important bounds can be proved, the most important of which is the Holevo bound.

12.1.1 The Holevo bound

The Holevo bound is an exceedingly useful upper bound on the accessible information that plays an important role in many applications of quantum information theory.

Theorem 12.1: (**The Holevo bound**) Suppose Alice prepares a state ρ_X where $X = 0, \ldots, n$ with probabilities $p_0, \ldots, p_n$. Bob performs a measurement described by POVM elements $\{E_y\} = \{E_0, \ldots, E_m\}$ on that state, with measurement outcome Y. The Holevo bound states that for any such measurement Bob may do:

$$H(X:Y) \le S(\rho) - \sum_x p_x S(\rho_x), \tag{12.6}$$

where $\rho = \sum_x p_x \rho_x$.

The Holevo bound is thus an upper bound on the accessible information. The quantity appearing on the right hand side of the Holevo bound is so useful in quantum information theory that it is given a name, the *Holevo χ quantity*, and is sometimes denoted χ.

Box 12.1: The no-cloning theorem

Is it possible to make a copy of an unknown quantum state? Surprisingly, it turns out that the answer to this question is no. In this box we describe an elementary proof of this fact that captures the essential reason this is not possible.

Suppose we have a quantum machine with two slots labeled A and B. Slot A, the *data slot*, starts out in an unknown but pure quantum state, $|\psi\rangle$. This is the state which is to be copied into slot B, the *target slot*. We assume that the target slot starts out in some standard pure state, $|s\rangle$. Thus the initial state of the copying machine is

$$|\psi\rangle \otimes |s\rangle. \tag{12.1}$$

Some unitary evolution U now effects the copying procedure, ideally,

$$|\psi\rangle \otimes |s\rangle \xrightarrow{U} U\left(|\psi\rangle \otimes |s\rangle\right) = |\psi\rangle \otimes |\psi\rangle. \tag{12.2}$$

Suppose this copying procedure works for two particular pure states, $|\psi\rangle$ and $|\varphi\rangle$. Then we have

$$U\left(|\psi\rangle \otimes |s\rangle\right) = |\psi\rangle \otimes |\psi\rangle \tag{12.3}$$
$$U\left(|\varphi\rangle \otimes |s\rangle\right) = |\varphi\rangle \otimes |\varphi\rangle. \tag{12.4}$$

Taking the inner product of these two equations gives

$$\langle\psi|\varphi\rangle = (\langle\psi|\varphi\rangle)^2. \tag{12.5}$$

But $x = x^2$ has only two solutions, $x = 0$ and $x = 1$, so either $|\psi\rangle = |\varphi\rangle$ or $|\psi\rangle$ and $|\varphi\rangle$ are orthogonal. Thus a cloning device can only clone states which are orthogonal to one another, and therefore a general quantum cloning device is impossible. A potential quantum cloner cannot, for example, clone the qubit states $|\psi\rangle = |0\rangle$ and $|\varphi\rangle = (|0\rangle + |1\rangle)/\sqrt{2}$, since these states are not orthogonal.

What we have shown is that it is impossible to perfectly clone an unknown quantum state using unitary evolution. Several questions naturally arise: what if we try to copy a mixed state? What if we allow cloning devices that are not unitary? What if we are willing to allow imperfect copies that nevertheless are 'good' according to some interesting measure of fidelity? These are all good questions which have been the subject of much investigation, as can be seen from the end of chapter 'History and further reading.' The short summary of this work is that even if one allows non-unitary cloning devices, the cloning of non-orthogonal pure states remains impossible unless one is willing to tolerate a finite loss of fidelity in the copied states. Similar conclusions hold also for mixed states, although a somewhat more sophisticated approach is necessary to even define what is meant by the notion of cloning a mixed state.

Proof

The proof of the Holevo bound is via a simple and beautiful construction involving three quantum systems, which we label P, Q and M. The system Q is the quantum system Alice gives to Bob; P and M are fictitious auxiliary systems introduced to ease the proof, just as was done in proving many of the entropic inequalities in Chapter 11. Intuitively, P may be thought of as the 'preparation' system. By definition it has an orthonormal basis $|x\rangle$ whose elements correspond to the labels $0, \ldots, n$ on the possible preparations for the quantum system, Q. M can be thought of intuitively as Bob's 'measuring device', and it has a basis $|y\rangle$ whose elements correspond to the possible outcomes $1, \ldots, n$ of Bob's measurement. The initial state of the total system is assumed to be

$$\rho^{PQM} = \sum_x p_x |x\rangle\langle x| \otimes \rho_x \otimes |0\rangle\langle 0|, \tag{12.7}$$

where we write the tensor product decomposition in the order PQM. Intuitively, this state represents the situation that Alice has chosen a value of x with probability p_x, prepared a corresponding ρ_x and given it to Bob, who is about to use his measuring apparatus, initially in the standard state $|0\rangle$, to perform the measurement. To describe the measurement we introduce a quantum operation $\mathcal{E}$ acting on the systems Q and M only (not P), whose action is to perform a measurement with POVM elements $\{E_y\}$ on the system Q, and to store the result of the measurement in system M:

$$\mathcal{E}(\sigma \otimes |0\rangle\langle 0|) \equiv \sum_y \sqrt{E_y} \sigma \sqrt{E_y} \otimes |y\rangle\langle y|, \tag{12.8}$$

where σ is any state of system Q, and $|0\rangle$ is the initial state of the measuring apparatus. In the following exercise you show that $\mathcal{E}$ is a trace-preserving quantum operation.

Exercise 12.2:   Define U_y to be the unitary operator acting on system M whose action on a basis is $U_y|y'\rangle \equiv |y' + y\rangle$, where the addition is done modulo $n + 1$. Show that $\{\sqrt{E_y} \otimes U_y\}$ is a set of operation elements defining a trace-preserving quantum operation $\mathcal{E}$ whose action on states of the form $\sigma \otimes |0\rangle\langle 0|$ agrees with (12.8).

The proof of Holevo's bound now proceeds as follows. Using primes to denote states of PQM after application of $\mathcal{E}$, and unprimed states to indicate the time prior to application of $\mathcal{E}$ we have $S(P:Q) = S(P:Q, M)$, since M is initially uncorrelated with P and Q, and $S(P:Q, M) \geq S(P':Q', M')$, since applying the quantum operation $\mathcal{E}$ to QM can't increase the mutual information between P and QM (Theorem 11.15 on page 522), and finally $S(P':Q', M') \geq S(P':M')$, since discarding systems can't increase mutual information (also Theorem 11.15). Putting these results together gives

$$S(P':M') \leq S(P:Q). \tag{12.9}$$

This result, with a little simple algebra, is easily understood to be the Holevo bound! Let's first calculate the quantity on the right hand side. Note that

$$\rho^{PQ} = \sum_x p_x |x\rangle\langle x| \otimes \rho_x, \tag{12.10}$$

from which it follows that $S(P) = H(p_x)$, $S(Q) = S(\rho)$, and $S(P, Q) = H(p_x) + \sum_x p_x S(\rho_x)$ (by Theorem 11.10 on page 518), whence

$$S(P:Q) = S(P) + S(Q) - S(P, Q) = S(\rho) - \sum_x p_x S(\rho_x), \tag{12.11}$$

which is exactly what we want on the right hand side of the Holevo bound! To calculate the quantity on the left hand side of (12.9), note that

$$\rho^{P'Q'M'} = \sum_{xy} p_x |x\rangle\langle x| \otimes \sqrt{E_y} \rho_x \sqrt{E_y} \otimes |y\rangle\langle y|. \tag{12.12}$$

Tracing out system Q' and using the observation that the joint distribution $p(x, y)$ for the pair (X, Y) satisfies $p(x, y) = p_x p(y|x) = p_x \text{tr}(\rho_x E_y) = p_x \text{tr}(\sqrt{E_y} \rho_x \sqrt{E_y})$, gives

$$\rho^{P'M'} = \sum_{xy} p(x, y) |x\rangle\langle x| \otimes |y\rangle\langle y|, \tag{12.13}$$

whence $S(P' : M') = H(X : Y)$, which is exactly what we want on the left hand side of the Holevo bound! This completes the proof of the Holevo bound. □

12.1.2 Example applications of the Holevo bound

The Holevo bound is a keystone in the proof of many results in quantum information theory. For now, we'll give just a taste of how this important result may be applied. Recall Theorem 11.10, which implies that

$$S(\rho) - \sum_x p_x S(\rho_x) \le H(X), \tag{12.14}$$

with equality if and only if the states ρ_x have orthogonal support. Suppose that the states ρ_x do not have orthogonal support, so the inequality in (12.14) is strict. Then the Holevo bound implies that $H(X : Y)$ is strictly less than $H(X)$, and thus it is impossible for Bob to determine X with perfect reliability based on his measurement result Y, which generalizes our existing understanding that if the states prepared by Alice are not orthogonal then it is not possible for Bob to determine with certainty which state Alice prepared.

A concrete example involves Alice preparing a single qubit in one of two quantum states according to the outcome of a fair coin toss. If the coin toss yields heads, then Alice prepares the state $|0\rangle$, and if the coin toss yields tails, then Alice prepares the state $\cos\theta|0\rangle + \sin\theta|1\rangle$, where θ is some real parameter. In the $|0\rangle, |1\rangle$ basis it follows that ρ may be written

$$\rho = \frac{1}{2}\begin{bmatrix} 1 & 0 \\ 0 & 0 \end{bmatrix} + \frac{1}{2}\begin{bmatrix} \cos^2\theta & \cos\theta\sin\theta \\ \cos\theta\sin\theta & \sin^2\theta \end{bmatrix}. \tag{12.15}$$

A simple calculation shows that the eigenvalues of ρ are $(1 \pm \cos\theta)/2$, and the Holevo bound is therefore given by the binary entropy $H((1 + \cos\theta)/2)$, as illustrated in Figure 12.1. Notice that the Holevo bound is maximized when $\theta = \pi/2$, attaining a value of 1 bit, corresponding to the case of Alice preparing states chosen from an orthogonal set, at which point it is possible for Bob to determine with surety which state Alice prepared. For other values of θ the Holevo bound is strictly less than 1 bit, and it is impossible for Bob to determine with surety which state Alice prepared.

The Holevo bound may be given more operational meaning by making use of the *Fano inequality* (see Box 12.2 on page 536 for a derivation). Suppose Bob makes a guess $\tilde{X} = f(Y)$ as to which state Alice prepared, based on the outcome of his measurement Y and some rule for making a guess, encapsulated in the function $f(\cdot)$. Then according to the Fano inequality and the Holevo bound,

$$H(p(\tilde{X} \ne X)) + p(\tilde{X} \ne X)\log(|X| - 1) \ge H(X|Y)$$

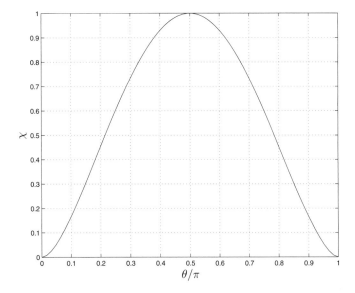

Figure 12.1. Plot of the Holevo bound χ as a function of θ when the states $|0\rangle$ and $\cos\theta|0\rangle + \sin\theta|1\rangle$ are prepared with equal probability. Notice that the Holevo bound reaches a maximum when $\theta = \pi/2$, corresponding to orthogonal states. It is only at this point that it is possible for Bob to determine with certainty which state Alice prepared.

$$= H(X) - H(X:Y)$$
$$\geq H(X) - \chi, \qquad (12.19)$$

which allows us to place bounds on how well Bob may infer the value of X. Heuristically, the smaller χ is, the harder it is for Bob to determine which state Alice prepared. This is illustrated in Figure 12.2 for the case where Alice prepares $|0\rangle$ with probability one-half and $\cos\theta|0\rangle + \sin\theta|1\rangle$ with probability one-half, for which the bound reduces to $H(p(\tilde{X} \neq X)) \geq 1 - \chi$ and $\chi = H((1 + \cos(\theta))/2)$ as we noted before. Notice that when $\theta \neq \pi/2$, Bob has some finite probability of making an error in his guess. This error gets larger as θ gets closer to zero. Finally, when $\theta = 0$ and the two states are indistinguishable, the lower bound tells us that Bob's probability of error is at least one-half – he can do no better than chance in guessing which state Alice prepared, as we would expect.

Exercise 12.3: Use the Holevo bound to argue that n qubits can not be used to transmit more than n bits of classical information.

Exercise 12.4: Suppose Alice sends Bob an equal mixture of the four pure states

$$|X_1\rangle = |0\rangle \qquad (12.20)$$

$$|X_2\rangle = \sqrt{\frac{1}{3}}\left[|0\rangle + \sqrt{2}|1\rangle\right] \qquad (12.21)$$

$$|X_3\rangle = \sqrt{\frac{1}{3}}\left[|0\rangle + \sqrt{2}e^{2\pi i/3}|1\rangle\right] \qquad (12.22)$$

$$|X_4\rangle = \sqrt{\frac{1}{3}}\left[|0\rangle + \sqrt{2}e^{4\pi i/3}|1\rangle\right]. \qquad (12.23)$$

Box 12.2: Fano's inequality

Suppose we wish to infer the value of a random variable X based on knowledge of another random variable Y. Intuitively, we expect that the conditional entropy $H(X|Y)$ limits how well we may perform this inference. The *Fano inequality* makes this intuition rigorous, and provides a useful bound on how well we may infer X, given Y.

Suppose $\tilde{X} \equiv f(Y)$ is some function of Y which we are using as our best guess for X. Let $p_e \equiv p(X \neq \tilde{X})$ be the probability that this guess is incorrect. Then the Fano inequality states that

$$H(p_e) + p_e \log(|X| - 1) \geq H(X|Y), \tag{12.16}$$

where $H(\cdot)$ is the binary entropy and $|X|$ is the number of values X may assume. Qualitatively, what the inequality tells us is that if $H(X|Y)$ is large (that is, comparable in size to $\log(|X| - 1)$) then the probability p_e of making an error in inference must also be large.

To prove the Fano inequality, define an 'error' random variable, $E \equiv 1$ if $X \neq \tilde{X}$, and $E \equiv 0$ if $X = \tilde{X}$. Notice that $H(E) = H(p_e)$. Using the chaining rule for conditional entropies (page 508), we have $H(E, X|Y) = H(E|X, Y) + H(X|Y)$. But E is completely determined once X and Y are known, so $H(E|X, Y) = 0$ and thus $H(E, X|Y) = H(X|Y)$. Applying the chain rule for entropies again but to different variables, we obtain $H(E, X|Y) = H(X|E, Y) + H(E|Y)$. Conditioning reduces entropy, so $H(E|Y) \leq H(E) = H(p_e)$, whence $H(X|Y) = H(E, X|Y) \leq H(X|E, Y) + H(p_e)$. The proof of the Fano inequality is concluded by bounding $H(X|E, Y)$ as follows (we have omitted a few simple algebraic details, which you can easily fill in):

$$H(X|E, Y) = p(E = 0)H(X|E = 0, Y) + p(E = 1)H(X|E = 1, Y) \tag{12.17}$$
$$\leq p(E = 0) \times 0 + p_e \log(|X| - 1) = p_e \log(|X| - 1), \tag{12.18}$$

where $H(X|E = 1, Y) \leq \log(|X| - 1)$ follows from the fact that when $E = 1$, $X \neq Y$, and X can assume at most $|X| - 1$ values, bounding its entropy, and thus its conditional entropy by $\log(|X| - 1)$. Substituting $H(X|E, Y) \leq p_e \log(|X| - 1)$ into $H(X|Y) \leq H(X|E, Y) + H(p_e)$ gives the Fano inequality $H(X|Y) \leq H(p_e) + p_e \log(|X| - 1)$.

Show that the maximum mutual information between Bob's measurement and Alice's transmission is less than one bit. A POVM which achieves ≈ 0.415 bits is known. Can you construct this or, better yet, one which achieves the Holevo bound?

12.2 Data compression

Let's switch tacks now and investigate an elementary dynamical process – data compression – which arises in both classical and quantum information theory. In its most general form the problem of data compression is to determine *what are the minimal*

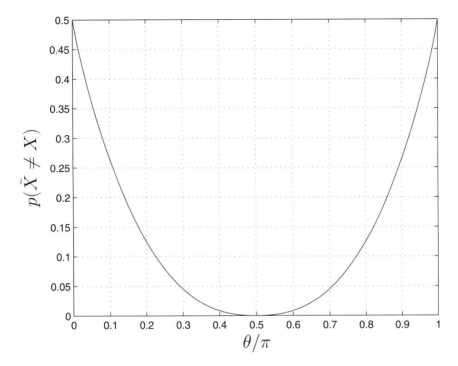

Figure 12.2. A lower bound on the probability of Bob making an error in inferring whether Alice prepared the state $|0\rangle$ or $\cos\theta|0\rangle + \sin\theta|1\rangle$. This lower bound is obtained by combining Fano's inequality with the Holevo bound. Observe that the bound decreases to zero as θ gets close to $\pi/2$, where the states may be reliably distinguished.

physical requirements needed to store an information source? It is one of the fundamental problems of information theory, with implications far beyond its immediate scope. In both classical and quantum information theory the techniques utilized in solving this problem turn out to have a far wider range of applicability than mere data compression, yet receive perhaps their simplest and most elegant expression in the solution of the data compression problem. In this section we examine in detail both quantum and classical data compression.

12.2.1 Shannon's noiseless channel coding theorem

Shannon's noiseless channel coding theorem quantifies the extent to which we can compress the information being produced by a classical information source. What do we mean by a classical information source? Many models of such a source are possible. A simple and very fruitful model is that a source consists of a sequence of random variables $X_1, X_2, \ldots$ whose values represent the output of the source. We will find it convenient to assume that the random variables take values from a finite alphabet of symbols, although extensions to infinite alphabets also hold. Furthermore, we assume that the different uses of the source are *independent* and *identically distributed*; that is, the source is what is known as an *i.i.d* information source. In the real world sources often don't behave in this way. It is easy to see, for example, that the letters in the English text in front of you don't occur in an independent fashion; strong correlations exist between the letters. To take a simple example the letter 't' is followed by the letter 'h' more frequently than one would expect based upon the overall frequency of the letter 'h' in normal English text; we say that occurrences of 't' and 'h' do not occur independently, but are correlated. Nevertheless, for a wide variety of information sources (including English text) the assumption of an i.i.d. source works pretty well in practice, and the ideas introduced to deal with the special case of an i.i.d. source can be generalized to more sophisticated sources.

Before we get into the technical details of Shannon's theorem, let's use a simple example to understand the intuition behind the result. Suppose an i.i.d. information source is producing bits $X_1, X_2, X_3, \ldots$, each being equal to zero with probability p, and equal to one with probability $1 - p$. The key idea behind Shannon's theorem is to divide the possible sequences of values $x_1, \ldots, x_n$ for the random variables $X_1, \ldots, X_n$ up into two types – sequences which are highly likely to occur, known as *typical sequences*, and sequences which occur rarely, known as *atypical sequences*. How is this done? As n gets large, we expect that with high probability a fraction p of the symbols output from the source will be equal to zero, and that a fraction $1 - p$ will be equal to one. The sequences $x_1, \ldots, x_n$ for which this assumption is correct are known as *typical sequences*. Combining this definition with the independence assumption for the source gives

$$p(x_1, \ldots, x_n) = p(x_1)p(x_2) \ldots p(x_n) \approx p^{np}(1 - p)^{(1-p)n} \qquad (12.24)$$

for typical sequences. Taking logarithms on both sides gives

$$-\log p(x_1, \ldots, x_n) \approx -np \log p - n(1 - p) \log(1 - p) = nH(X), \qquad (12.25)$$

where X is a random variable distributed according to the source distribution and $H(X) = -p \log(p) - (1 - p) \log(1 - p)$ is the entropy of the source distribution, also known as the *entropy rate* of the source. Thus $p(x_1, \ldots, x_n) \approx 2^{-nH(X)}$, from which we see that there can be at most $2^{nH(X)}$ typical sequences, since the total probability of all typical sequences cannot be greater than one.

We now have the tools to understand a simple scheme for data compression. Suppose the output from the source is $x_1, \ldots, x_n$. To compress this output, we check to see whether $x_1, \ldots, x_n$ is a typical sequence. If it's not, we give up – declare an error. Fortunately, as n becomes large this happens very rarely, since nearly all sequences are typical in the limit of large n. If the output is a typical sequence, we record that fact. Since there are at most $2^{nH(X)}$ typical sequences, it only requires $nH(X)$ bits to uniquely identify a particular typical sequence. We choose some such scheme for identification, and compress the output from the source to the corresponding string of $nH(X)$ bits describing which typical sequence occurred, which can later be decompressed. As n becomes large this scheme succeeds with probability approaching one.

Several criticisms can be made of this scheme: (a) It has a small but finite chance of failing. Slightly more sophisticated schemes make use of similar ideas to remove the possibility of an error occurring. (b) To do the compression we have to wait until the source has emitted a large number, n, of symbols. Again, adaptations exist which allow the processing to be done as the symbols are emitted by the source. (c) No explicit scheme mapping outputs from the source to the compressed sequences has been given. Once again, slightly more sophisticated schemes can be developed which address this issue. (d) The exact procedure being used to do the data compression depends on the output distribution of the source. What if this is not known? Clever *universal compression* algorithms exist which can cope with this possibility. The reader who is interested in these and other issues is referred to the book of Cover and Thomas listed in the end of chapter 'History and further reading'.

Let's generalize the notion of typical sequences beyond the binary case. Suppose $X_1, X_2, \ldots$ is an i.i.d. information source. Typically, the frequency of occurrence of any given letter x in the sequence output from the source will be close to the probability $p(x)$ of that letter occurring on any given use of the source. With this intuitive understanding in mind we make the following rigorous definition of the notion of a typical sequence. Given $\epsilon > 0$ we say that a string of source symbols $x_1 x_2 \ldots x_n$ is ϵ-typical if

$$2^{-n(H(X)+\epsilon)} \leq p(x_1, \ldots, x_n) \leq 2^{-n(H(X)-\epsilon)}, \tag{12.26}$$

and denote the set of all such ϵ-typical sequences of length n by $T(n, \epsilon)$. A useful equivalent reformulation of the definition is as

$$\left| \frac{1}{n} \log \frac{1}{p(x_1, \ldots, x_n)} - H(X) \right| \leq \epsilon. \tag{12.27}$$

Using the law of large numbers (stated and proved in Box 12.3 on page 541) we can prove the *theorem of typical sequences*, which makes rigorous the idea that in the limit of large n most sequences output by an information source are typical.

Theorem 12.2: (**Theorem of typical sequences**)

(1) Fix $\epsilon > 0$. Then for any $\delta > 0$, for sufficiently large n, the probability that a sequence is ϵ-typical is at least $1 - \delta$.

(2) For any fixed $\epsilon > 0$ and $\delta > 0$, for sufficiently large n, the number $|T(n, \epsilon)|$ of ϵ-typical sequences satisfies

$$(1 - \delta)2^{n(H(X)-\epsilon)} \leq |T(n, \epsilon)| \leq 2^{n(H(X)+\epsilon)}. \tag{12.28}$$

(3) Let $S(n)$ be a collection of size at most 2^{nR}, of length n sequences from the source, where $R < H(X)$ is fixed. Then for any $\delta > 0$ and for sufficiently large n,

$$\sum_{x \in S(n)} p(x) \leq \delta. \tag{12.29}$$

Proof

Part 1: A direct application of the law of large numbers. Notice that $-\log p(X_i)$ are independent, identically distributed random variables. By the law of large numbers for any $\epsilon > 0$ and $\delta > 0$ for sufficiently large n we have

$$p\left(\left| \sum_{i=1}^{n} \frac{-\log p(X_i)}{n} - \mathbf{E}(-\log p(X)) \right| \leq \epsilon \right) \geq 1 - \delta. \tag{12.30}$$

But $\mathbf{E}(\log p(X)) = -H(X)$ and $\sum_{i=1}^{n} \log p(X_i) = \log(p(X_1, \ldots, X_n))$. Thus

$$p\left(|-\log(p(X_1, \ldots, X_n))/n - H(X)| \leq \epsilon \right) \geq 1 - \delta. \tag{12.31}$$

That is, the probability that a sequence is ϵ-typical is at least $1 - \delta$.

Part 2: Follows from the definition of typicality, and the observation that the sum of the probabilities of the typical sequences must lie in the range $1 - \delta$ (by part 1) to 1 (as probabilities cannot sum to more than 1). Thus

$$1 \geq \sum_{x \in T(n,\epsilon)} p(x) \geq \sum_{x \in T(n,\epsilon)} 2^{-n(H(X)+\epsilon)} = |T(n,\epsilon)| 2^{-n(H(X)+\epsilon)}, \qquad (12.32)$$

from which we deduce $|T(n,\epsilon)| \leq 2^{n(H(X)+\epsilon)}$, and

$$1 - \delta \leq \sum_{x \in T(n,\epsilon)} p(x) \leq \sum_{x \in T(n,\epsilon)} 2^{-n(H(X)-\epsilon)} = |T(n,\epsilon)| 2^{-n(H(X)-\epsilon)}, \qquad (12.33)$$

from which we deduce $|T(n,\epsilon)| \geq (1-\delta) 2^{n(H(X)-\epsilon)}$.

Part 3: The idea is to split the sequences in $S(n)$ up into typical and atypical sequences. The atypical sequences have small probability in the large n limit. The number of typical sequences in $S(n)$ is obviously no larger than the total number of sequences in $S(n)$, which is at most 2^{nR}, and each typical sequence has probability about $2^{-nH(X)}$ so the total probability of the typical sequences in $S(n)$ scales like $2^{n(R-H(X))}$, which goes to zero when $R < H(X)$.

More rigorously, choose ϵ so $R < H(X) - \delta$ and $0 < \epsilon < \delta/2$. Split the sequences in $S(n)$ up into the ϵ-typical sequences, and the ϵ-atypical sequences. By part 1, for n sufficiently large the total probability of the atypical sequences can be made less than $\delta/2$. There are at most 2^{nR} typical sequences in $S(n)$, each with probability at most $2^{-n(H(X)-\epsilon)}$, so the probability of the typical sequences is at most $2^{-n(H(X)-\epsilon-R)}$, which goes to zero as n goes to infinity. Thus the total probability of the sequences in $S(n)$ is less than δ for n sufficiently large. $\qquad\square$

Shannon's noiseless channel coding theorem is an easy application of the theorem of typical sequences. We give here a very simple version of the noiseless channel coding theorem; more sophisticated versions are left to the exercises and the end of chapter 'History and further reading'. The basic setting is to suppose that $X_1, X_2, \ldots$ is an i.i.d. classical information source over some finite alphabet containing d symbols. A *compression scheme of rate R* maps possible sequences $x = (x_1, \ldots, x_n)$ to a bit string of length nR which we denote by $C^n(x) = C^n(x_1, \ldots, x_n)$. (Note that nR may not be an integer; our notation is simplified by agreeing that in this case $nR = \lfloor nR \rfloor$.) The matching *decompression scheme* takes the nR compressed bits and maps them back to a string of n letters from the alphabet, $D^n(C^n(x))$. A compression–decompression scheme (C^n, D^n) is said to be *reliable* if the probability that $D^n(C^n(x)) = x$ approaches one as n approaches ∞. Shannon's noiseless channel coding theorem specifies for what values of the rate R a reliable compression scheme exists, revealing a remarkable operational interpretation for the entropy rate $H(X)$: it is just the minimal physical resources necessary and sufficient to reliably store the output from the source.

Theorem 12.4: (Shannon's noiseless channel coding theorem) Suppose $\{X_i\}$ is an i.i.d. information source with entropy rate $H(X)$. Suppose $R > H(X)$. Then there exists a reliable compression scheme of rate R for the source. Conversely, if $R < H(X)$ then any compression scheme will not be reliable.

Proof

Suppose $R > H(X)$. Choose $\epsilon > 0$ such that $H(X) + \epsilon < R$. Consider the set $T(n, \epsilon)$

Box 12.3: The law of large numbers

Suppose we repeat an experiment a large number of times, each time measuring the value of some parameter, X. We label the results of the experiments $X_1, X_2, \ldots$. Assuming that the results of the experiments are independent, we intuitively expect that the value of the estimator $S_n \equiv \sum_{i=1}^{n} X_i/n$ of the average $\mathbf{E}(X)$, should approach $\mathbf{E}(X)$ as $n \to \infty$. The *law of large numbers* is a rigorous statement of this intuition.

Theorem 12.3: (**Law of large numbers**) Suppose $X_1, X_2, \ldots$ are independent random variables all having the same distribution as a random variable X with finite first and second moments, $|\mathbf{E}(X)| < \infty$ and $\mathbf{E}(X^2) < \infty$. Then for any $\epsilon > 0$, $p(|S_n - \mathbf{E}(X)| > \epsilon) \to 0$ as $n \to \infty$.

Proof

To begin we assume that $\mathbf{E}(X) = 0$ and discuss what happens when $\mathbf{E}(X) \neq 0$ upon completion of the proof. Since the random variables are independent with mean zero, it follows that $\mathbf{E}(X_i X_j) = \mathbf{E}(X_i)\mathbf{E}(X_j) = 0$ when $i \neq j$, and thus

$$\mathbf{E}(S_n^2) = \frac{\sum_{i,j=1}^{n} \mathbf{E}(X_i X_j)}{n^2} = \frac{\sum_{i=1}^{n} \mathbf{E}(X_i^2)}{n^2} = \frac{\mathbf{E}(X^2)}{n}, \tag{12.34}$$

where the final equality follows from the fact that $X_1, \ldots, X_n$ are identically distributed to X. By the same token, from the definition of the expectation we have

$$\mathbf{E}(S_n^2) = \int dP\, S_n^2, \tag{12.35}$$

where dP is the underlying probability measure. It is clear that either $|S_n| \leq \epsilon$ or $|S_n| > \epsilon$, so we can split this integral into two pieces, and then drop one of these pieces with the justification that it is non-negative,

$$\mathbf{E}(S_n^2) = \int_{|S_n| \leq \epsilon} dP\, S_n^2 + \int_{|S_n| > \epsilon} dP\, S_n^2 \geq \int_{|S_n| > \epsilon} dP\, S_n^2. \tag{12.36}$$

In the region of integration $S_n^2 > \epsilon^2$, and thus

$$\mathbf{E}(S_n^2) \geq \epsilon^2 \int_{|S_n| > \epsilon} dP = \epsilon^2 p(|S_n| > \epsilon). \tag{12.37}$$

Comparing this inequality with (12.34) we see that

$$p(|S_n| > \epsilon) \leq \frac{\mathbf{E}(X^2)}{n\epsilon^2}. \tag{12.38}$$

Letting $n \to \infty$ completes the proof. In the case when $\mathbf{E}(X) \neq 0$, it is easy to obtain the result, by defining

$$Y_i \equiv X_i - \mathbf{E}(X), \qquad Y \equiv X - \mathbf{E}(X). \tag{12.39}$$

Y and $Y_1, Y_2, \ldots$ are a sequence of independent, identically distributed random variables with $\mathbf{E}(Y) = 0$ and $\mathbf{E}(Y^2) < \infty$. The result follows from the earlier reasoning. $\qquad \square$

of ϵ-typical sequences. For any $\delta > 0$ and for sufficiently large n, there are at most $2^{n(H(x)+\epsilon)} < 2^{nR}$ such sequences, and the probability of the source producing such a sequence is at least $1 - \delta$. The method of compression therefore is simply to examine the output of the source to see if it is ϵ-typical. If it is not, then compress to some fixed nR bit string which indicates failure; the decompression operation simply outputs a random sequence $x_1, \ldots, x_n$ as a guess to the information produced by the source; effectively we give up on compression in this case. If the output of the source is typical then we compress the output simply by storing an index for the particular sequence using nR bits in the obvious way, allowing later recovery.

Suppose $R < H(X)$. The combined compression–decompression operation has at most 2^{nR} possible outputs, so at most 2^{nR} of the sequences output from the source can be compressed and decompressed without an error occurring. By the theorem of typical sequences, for sufficiently large n the probability of a sequence output from the source lying in a subset of 2^{nR} sequences goes to zero, for $R < H(X)$. Thus any such compression scheme cannot be reliable.   □

Exercise 12.5: (Variable-length zero error data compression) Consider the
following heuristic for a variable length data compression scheme. Let $x_1, \ldots, x_n$
be the output from n uses of an i.i.d. source with entropy rate $H(X)$. If
$x_1, \ldots, x_n$ is typical, then send a $H(X)$ bit index indicating which typical
sequence it is. If $x_1, \ldots, x_n$ is atypical, send an uncompressed $\log d^n$ bit index
for the sequence (recall that d is the alphabet size). Turn this heuristic into a
rigorous argument that the source can be compressed to an average of R bits per
source symbol, for any $R > H(X)$, with zero probability of error.

12.2.2 Schumacher's quantum noiseless channel coding theorem

A great conceptual breakthrough of quantum information theory is to realize that we can treat quantum states *as if they were information*, and ask information-theoretic questions about those quantum states. In this section we define the notion of a quantum source of information, and study the question: to what extent can the 'information' – quantum states – being produced by that source be compressed?

How might we define the notion of a quantum information source? As with the definition of a classical information source it is by no means obvious what the best means of making this definition is, and it is possible to come up with several different definitions, not all of which are equivalent. The definition we are going to use is based on the idea that *entanglement* is what we are trying to compress and decompress. More formally, an (i.i.d) quantum source will be described by a Hilbert space H, and a density matrix ρ on that Hilbert space. We imagine that the state ρ of the system is merely part of a larger system which is in a pure state, and the mixed nature of ρ is due to entanglement between H and the remainder of the system. A *compression scheme of rate R* for this source consists of two families of quantum operations C^n and D^n, analogous to the compression and decompression schemes used in the classical case. C^n is the compression operation, taking states in $H^{\otimes n}$ to states in a 2^{nR}-dimensional state space, the *compressed space*. We may regard the compressed space as representing nR qubits. The operation D^n is a decompression operation, which takes states in the compressed space to states in the original state space. The combined compression–decompression operation is therefore $D^n \circ C^n$. Our criterion for reliability is that in the limit of large n the entanglement fidelity

$F(\rho^{\otimes n}, \mathcal{D}^n \circ \mathcal{C}^n)$ should tend towards one. The basic idea of quantum data compression is illustrated in Figure 12.3.

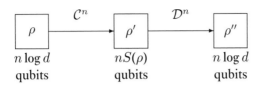

Figure 12.3. Quantum data compression. The compression operation $\mathcal{C}^n$ compresses a quantum source ρ stored in $n \log d$ qubits into $nS(\rho)$ qubits. The source is accurately recovered via the decompression operation $\mathcal{D}^n$.

The key technical idea making the quantum noiseless channel coding theorem possible is a quantum version of the idea of typical sequences. Suppose the density operator ρ associated with a quantum source has orthonormal decomposition

$$\rho = \sum_x p(x)|x\rangle\langle x|, \tag{12.40}$$

where $|x\rangle$ is an orthonormal set, and $p(x)$ are the eigenvalues of ρ. The eigenvalues $p(x)$ of ρ obey the same rules as a probability distribution: they are non–negative and sum to one. Furthermore, $H(p(x)) = S(\rho)$. Therefore, it makes sense to talk of an ϵ-typical sequence, $x_1, \ldots, x_n$ for which

$$\left| \frac{1}{n} \log \left(\frac{1}{p(x_1)p(x_2)\ldots p(x_n)} \right) - S(\rho) \right| \le \epsilon, \tag{12.41}$$

in exactly the same fashion as for the classical definition. An ϵ-typical state is a state $|x_1\rangle|x_2\rangle \ldots |x_n\rangle$ for which the sequence $x_1, x_2, \ldots, x_n$ is ϵ-typical. Define the ϵ-*typical subspace* to be the subspace spanned by all ϵ-typical states, $|x_1\rangle \ldots |x_n\rangle$. We'll denote the ϵ-typical subspace by $T(n, \epsilon)$, and the projector onto the ϵ-typical subspace by $P(n, \epsilon)$. Notice that

$$P(n, \epsilon) = \sum_{x \ \epsilon-\text{typical}} |x_1\rangle\langle x_1| \otimes |x_2\rangle\langle x_2| \otimes \ldots |x_n\rangle\langle x_n|. \tag{12.42}$$

The theorem of typical sequences may now be translated into an equivalent quantum form, the typical subspace theorem.

Theorem 12.5: (**Typical subspace theorem**)

    (1) Fix $\epsilon > 0$. Then for any $\delta > 0$, for sufficiently large n,

$$\text{tr}(P(n, \epsilon)\rho^{\otimes n}) \ge 1 - \delta . \tag{12.43}$$

    (2) For any fixed $\epsilon > 0$ and $\delta > 0$, for sufficiently large n, the dimension $|T(n, \epsilon)| = \text{tr}(P(n, \epsilon))$ of $T(n, \epsilon)$ satisfies:

$$(1 - \delta)2^{n(S(\rho)-\epsilon)} \le |T(n, \epsilon)| \le 2^{n(S(\rho)+\epsilon)} . \tag{12.44}$$

    (3) Let $S(n)$ be a projector onto any subspace of $H^{\otimes n}$ of dimension at most 2^{nR}, where $R < S(\rho)$ is fixed. Then for any $\delta > 0$, and for sufficiently large n,

$$\text{tr}(S(n)\rho^{\otimes n}) \le \delta . \tag{12.45}$$

In each case the result can be obtained directly using the law of large numbers, but we prefer to use the theorem of typical sequences to emphasize the close connection to the techniques used in the proof of Shannon's noiseless channel coding theorem.

Proof

Part 1: Observe that

$$\text{tr}(P(n, \epsilon)\rho^{\otimes n}) = \sum_{x \ \epsilon-\text{typical}} p(x_1)p(x_2)\dots p(x_n). \tag{12.46}$$

The result follows immediately from part 1 of the theorem of typical sequences.

Part 2: Follows immediately from part 2 of the theorem of typical sequences.

Part 3: We split the trace up into a trace over the typical subspace and the atypical subspace,

$$\text{tr}(S(n)\rho^{\otimes n}) = \text{tr}(S(n)\rho^{\otimes n}P(n, \epsilon)) + \text{tr}(S(n)\rho^{\otimes n}(I - P(n, \epsilon))), \tag{12.47}$$

and bound each term separately. For the first term observe that

$$\rho^{\otimes n}P(n, \epsilon) = P(n, \epsilon)\rho^{\otimes n}P(n, \epsilon), \tag{12.48}$$

since $P(n, \epsilon)$ is a projector which commutes with $\rho^{\otimes n}$. But

$$\text{tr}(S(n)P(n, \epsilon)\rho^{\otimes n}P(n, \epsilon)) \leq 2^{nR}2^{-n(S(\rho)-\epsilon)}, \tag{12.49}$$

since the eigenvalues of $P(n, \epsilon)\rho^{\otimes n}P(n, \epsilon)$ are bounded above by $2^{-n(S(\rho)-\epsilon)}$. Letting $n \to \infty$ we see the first term tends to zero. For the second term note that $S(n) \leq I$. Since $S(n)$ and $\rho^{\otimes}(I - P(n, \epsilon))$ are both positive operators it follows that $0 \leq \text{tr}(S(n)\rho^{\otimes n}(I - P(n, \epsilon)) \leq \text{tr}(\rho^{\otimes n}(I - P(n, \epsilon)) \to 0$ as $n \to \infty$, so the second term also tends to zero as n becomes large, giving the result. $\qquad\square$

With the typical subspace theorem under our belts it is not difficult to prove a quantum analogue of Shannon's noiseless channel coding theorem. The main ideas of the proof are analogous, but the technical analysis is made a little more difficult by the appearance of non-commuting operators in the proof, which have no classical analogue.

Theorem 12.6: (**Schumacher's noiseless channel coding theorem**) Let $\{H, \rho\}$ be an i.i.d. quantum source. If $R > S(\rho)$ then there exists a reliable compression scheme of rate R for the source $\{H, \rho\}$. If $R < S(\rho)$ then any compression scheme of rate R is not reliable.

Proof

Suppose $R > S(\rho)$ and let $\epsilon > 0$ be such that $S(\rho) + \epsilon \leq R$. By the typical subspace theorem, for any $\delta > 0$ and for all n sufficiently large, $\text{tr}(\rho^{\otimes n}P(n, \epsilon)) \geq 1 - \delta$, and $\dim(T(n, \epsilon)) \leq 2^{nR}$. Let H_c^n be any 2^{nR}-dimensional Hilbert space containing $T(n, \epsilon)$. The encoding is done in the following fashion. First a measurement is made, described by the complete set of orthogonal projectors $P(n, \epsilon), I - P(n, \epsilon)$, with corresponding outcomes we denote 0 and 1. If outcome 0 occurs nothing more is done and the state is left in the typical subspace. If outcome 1 occurs then we replace the state of the system with some standard state '$|0\rangle$' chosen from the typical subspace; it doesn't matter what state is used. It follows that the encoding is a map $C^n : H^{\otimes n} \to H^n_c$ into the 2^{nR}-dimensional subspace H^n_c, with operator-sum representation

$$C^n(\sigma) \equiv P(n, \epsilon)\sigma P(n, \epsilon) + \sum_i A_i \sigma A_i^\dagger , \tag{12.50}$$

where $A_i \equiv |0\rangle\langle i|$ and $|i\rangle$ is an orthonormal basis for the orthocomplement of the typical subspace.

The decoding operation $D^n : H^n_c \to H^{\otimes n}$ is defined to be the identity on H^n_c, $D^n(\sigma) = \sigma$. With these definitions for the encoding and decoding it follows that

$$F(\rho^{\otimes n}, D^n \circ C^n) = |\mathrm{tr}(\rho^{\otimes n} P(n, \epsilon))|^2 + \sum_i |\mathrm{tr}(\rho^{\otimes n} A_i)|^2 \tag{12.51}$$

$$\geq |\mathrm{tr}(\rho^{\otimes n} P(n, \epsilon))|^2 \tag{12.52}$$

$$\geq |1 - \delta|^2 \geq 1 - 2\delta , \tag{12.53}$$

where the last line follows from the typical subspace theorem. But δ can be made arbitrarily small for sufficiently large n, and thus it follows that there exists a reliable compression scheme $\{C^n, D^n\}$ of rate R whenever $S(\rho) < R$.

To prove the converse, suppose $R < S(\rho)$. Without loss of generality we suppose that the compression operation maps from $H^{\otimes n}$ to a 2^{nR}-dimensional subspace with corresponding projector $S(n)$. Let C_j be operation elements for the compression operation C^n, and D_k operation elements for the decompression operation D^n. Then we have

$$F(\rho^{\otimes n}, D^n \circ C^n) = \sum_{jk} \left|\mathrm{tr}(D_k C_j \rho^{\otimes n})\right|^2 . \tag{12.54}$$

Each of the operators C_j maps to within the subspace with projector $S(n)$ so $C_j = S(n)C_j$. Let $S^k(n)$ be the projector onto the subspace to which the subspace $S(n)$ is mapped by D_k, so we have $S^k(n)D_k S(n) = D_k S(n)$ and thus $D_k C_j = D_k S(n)C_j = S^k(n)D_k S(n)C_j = S^k(n)D_k C_j$, whence

$$F(\rho^{\otimes n}, D^n \circ C^n) = \sum_{jk} \left|\mathrm{tr}(D_k C_j \rho^{\otimes n} S^k(n))\right|^2 . \tag{12.55}$$

Applying the Cauchy–Schwarz inequality gives

$$F(\rho^{\otimes n}, D^n \circ C^n) \leq \sum_{jk} \mathrm{tr}(D_k C_j \rho^{\otimes n} C_j^\dagger D_k^\dagger)\mathrm{tr}(S^k(n)\rho^{\otimes n}) . \tag{12.56}$$

Applying part 3 of the typical subspace theorem we see that for any $\delta > 0$ and for sufficiently large n, $\mathrm{tr}(S^k(n)\rho^{\otimes n}) \leq \delta$. Moreover, the proof of the typical subspace theorem implies that the size n needs to be for this to hold does not depend on k. Thus

$$F(\rho^{\otimes n}, D^n \circ C^n) \leq \delta \sum_{jk} \mathrm{tr}(D_k C_j \rho^{\otimes n} C_j^\dagger D_k^\dagger) \tag{12.57}$$

$$= \delta , \tag{12.58}$$

since C^n and D^n are trace-preserving. Since δ was arbitrary it follows that $F(\rho^{\otimes n}, D^n \circ C^n) \to 0$ as $n \to \infty$, and thus the compression scheme is not reliable. $\qquad\square$

Schumacher's theorem not only discusses the *existence* of a reliable compression scheme, but it also gives clues as to how to actually construct one. The key is to be able to efficiently perform the mapping $C^n : H^{\otimes n} \to H_c^n$ into the 2^{nR}-dimensional typical subspace H_c^n. Classical compression techniques such as enumerative coding, Huffman coding, and arithmetic coding can be applied but with one strong restriction: the encoding circuit must be completely *reversible*, and also entirely erase the original state in the process of creating the compressed one! After all, by the no-cloning theorem, it cannot copy the original state, so it cannot leave it behind as normal classical compression schemes typically do. A simple example illustrating how quantum compression works is given in Box 12.4.

Exercise 12.6: In the notation of Box 12.4, give an explicit expression for C_X in terms of X. Also, describe how to construct a quantum circuit to perform U_n for arbitrary n. How many elementary operations do you require, as a function of n?

Exercise 12.7: (Data compression circuit) Outline the construction of a circuit to reliably compress a qubit source with $\rho = p|0\rangle\langle 0| + (1-p)|1\rangle\langle 1|$ into nR qubits for any $R > S(\rho) = H(p)$.

Exercise 12.8: (Compression of an ensemble of quantum states) Suppose that instead of adopting the definition of a quantum source based on a single density matrix ρ and the entanglement fidelity, we instead adopted the following *ensemble* definition, that an (i.i.d.) quantum source is specified by an ensemble $\{p_j, |\psi_j\rangle\}$ of quantum states, and that consecutive uses of the source are independent and produce a state $|\psi_j\rangle$ with probability p_j. A compression–decompression scheme $(\mathcal{C}^n, \mathcal{D}^n)$ is said to be reliable in this definition if the *ensemble average fidelity* approaches 1 as $n \to \infty$:

$$\bar{F} \equiv \sum_J p_{j_1} \cdots p_{j_n} F(\rho_J, (\mathcal{D}^n \circ \mathcal{C}^n)(\rho_J))^2, \tag{12.61}$$

where $J = (j_1, \ldots, j_n)$ and $\rho_J \equiv |\psi_{j_1}\rangle\langle\psi_{j_1}| \otimes \cdots \otimes |\psi_{j_n}\rangle\langle\psi_{j_n}|$. Define $\rho \equiv \sum_j p_j |\psi_j\rangle\langle\psi_j|$ and show that provided $R > S(\rho)$ there exists a reliable compression scheme of rate R with respect to this definition of fidelity.

12.3 Classical information over noisy quantum channels

Anything that can go wrong, will
– Attributed to Edward A. Murphy, Jr.

We all have difficulty talking on the telephone from time to time. We say we have a 'bad line' when we have exceptional difficulty understanding the person on the other end of the line. This is an example of the general phenomenon of *noise* which is present to some extent in all information processing systems. As described in Chapter 10 error-correcting codes can be used to combat the effects of noise, allowing reliable communication and computation to take place even in the presence of quite severe noise. Given a particular noisy communications channel $\mathcal{N}$ an interesting question is how much information can be transmitted reliably through that channel. For example, it might be possible that 1000 uses of the channel can be used to transmit 500 bits of information using an appropriate error-correcting code, with high probability of recovery from any errors the channel introduces.

Box 12.4: Schumacher compression

Consider an i.i.d. quantum source characterized by the single qubit density matrix

$$\rho = \frac{1}{4}\begin{bmatrix} 3 & 1 \\ 1 & 1 \end{bmatrix}. \qquad (12.59)$$

This could originate, for example, as a small part of a of much larger entangled system. An alternate way of viewing this source (compare Section 9.3) is that it is producing the state $|\psi_0\rangle = |0\rangle$ or $|\psi_1\rangle = (|0\rangle + |1\rangle)/\sqrt{2}$ with equal probabilities one half each (see Exercise 12.8). ρ has orthonormal decomposition $p|\bar{0}\rangle\langle\bar{0}| + (1 - p)|\bar{1}\rangle\langle\bar{1}|$, where $|\bar{0}\rangle = \cos\frac{\pi}{8}|0\rangle + \sin\frac{\pi}{8}|1\rangle$, $|\bar{1}\rangle = -\sin\frac{\pi}{8}|0\rangle + \cos\frac{\pi}{8}|1\rangle$, and $p = [3 + \tan(\pi/8)]/4$. In this basis, a block of n qubits can be written as the state

$$\sum_{X=\{\bar{0}\bar{0}...\bar{0},\, \bar{0}...\bar{0}\bar{1},\, ...,\, \bar{1}\bar{1}...\bar{1}\}} C_X |X\rangle. \qquad (12.60)$$

By Theorem 12.6, only $|X\rangle$ for which the Hamming weight is approximately equal to np (that is, a basis for the typical subspace) need be transmitted in order to enable reconstruction of the original state with high fidelity. This is easy to appreciate, because $|\langle\bar{0}|\psi_k\rangle| = \cos(\pi/8)$ (for $k = \{0, 1\}$) is much larger than $|\langle\bar{1}|\psi_k\rangle| = \sin(\pi/8)$, and for X with large Hamming weight, the coefficients C_X are very small.

How do we realize such a compression scheme? One approximate way is the following. Suppose we have quantum circuit U_n, which permutes basis states $|X\rangle$ such that states are re-ordered lexicographically according to Hamming weight. For example, for $n = 4$ it does

0000 → 0000	1000 → 0100	1001 → 1000	1011 → 1100
0001 → 0001	0011 → 0101	1010 → 1001	1101 → 1101
0010 → 0010	0101 → 0110	1100 → 1010	1110 → 1110
0100 → 0011	0110 → 0111	0111 → 1011	1111 → 1111

Such a transform, which can be realized using just controlled-NOT and Toffoli gates, reversibly packs the typical subspace into the first $\approx nH(p)$ qubits (from left to right). To complete the scheme, we also need a quantum gate V, which rotates single qubits into the $|\bar{0}\rangle, |\bar{1}\rangle$ basis. The desired compression scheme is then $C^n = (V^\dagger)^{\otimes n} U_n V^{\otimes n}$, and we need send only the first $nH(p)$ qubits output from C^n, to enable a sequence of states from the source to be reconstructed with high fidelity, using a decoder which is the inverse of this circuit. A more efficient coding scheme would pack just the states with Hamming weight $\approx np$ into the first $nH(p)$ qubit space; this can be done using a quantum version of arithmetic coding, for example.

We say such a code has rate $500/1000 = 1/2$. A fundamental problem of information theory is to determine the *maximum* rate for reliable communication through the channel $\mathcal{N}$, a number known as the *capacity* of the channel.

For noisy classical communications channels the capacity of the channel may be calculated using a beautiful result known as *Shannon's noisy channel coding theorem*. We begin our investigation of the communication of classical information in the presence of noise in Section 12.3.1 with a discussion of some of the main ideas behind Shannon's noisy channel coding theorem. We don't get too detailed, however, because in Section 12.3.2 we move on to take a detailed look at a generalization of the problem whereby two parties attempt to communicate classical information by the use of a noisy quantum channel!

12.3.1 Communication over noisy classical channels

Many of the main ideas about noisy channel coding, both quantum and classical, can be understood by examining the binary symmetric channel. Recall from Section 10.1 that the binary symmetric channel is a noisy communications channel for a single bit of information, whose effect is to flip the bit being transmitted with probability $p > 0$, while with probability $1 - p$ the bit is transmitted without error, as illustrated in Figure 12.4.

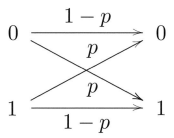

Figure 12.4. Binary symmetric channel.

How much information can we reliably transmit per use of a binary symmetric channel? Using error-correcting codes it is possible to transmit information through the channel, but at an overhead in the number of bits used to accomplish the communication. We will argue that the maximum rate at which information can be reliably transmitted is $1 - H(p)$, where $H(\cdot)$ is the Shannon entropy.

What does it mean that the transmission be reliably accomplished? This is a good question, since different answers give rise to different possible rates. We are going to use the following definition for reliability: we assume that inputs to the channel may be encoded in large *blocks* all at once, and require that the probability for an error in transmission using the code goes to zero as the blocksize is made large. Another possible definition of reliability is to suppose again that the encoding may be performed in blocks, but that as the blocksize becomes large the probability of error becomes *exactly* zero. Unfortunately, this definition turns out to be too optimistic about what can be achieved with error-correction, and leads to zero capacity for the binary symmetric channel! Similarly, if we don't allow encoding to be performed in large blocks the capacity turns out to be zero. Indeed, it is rather amazing (and not at all obvious) that even with our less ambitious definition of reliability a non-zero rate of information transmission can be achieved. To show that this is possible several clever ideas are needed.

Random coding for the binary symmetric channel

Suppose we want to transmit nR bits of information using n uses of our binary symmetric channel; that is, we want to transmit information at a rate R through the channel. We are going to outline a proof that an error-correcting code exists that accomplishes this with low probability of error in the limit of large n, and provided $R < 1 - H(p)$. The first idea we need is a *random coding* method for constructing an error-correcting code. Suppose $(q, 1 - q)$ is any fixed probability distribution over the possible inputs to the channel (0 and 1). (This distribution is often called the *a priori* distribution of the code – the introduction of this distribution is just a technical device to enable the random coding method to work, and the randomness present in the distribution should not be confused with the randomness in the channel.) Then we pick out a codeword $x = (x_1, \ldots, x_n)$ for our code simply by choosing $x_j = 0$ with probability q and $x_j = 1$ with probability $1 - q$, independently for each $j = 1, \ldots, n$. We repeat this procedure 2^{nR} times, creating a *codebook* C of 2^{nR} entries; we denote a generic entry in the codebook by x^j.

Obviously it's possible to construct some pretty lousy error-correcting codes using this procedure! We might get really unlucky and construct a code all of whose codewords consist of the string of n zeroes, which obviously is not much use for the transmission of information. Nevertheless, it turns out that on average this random coding procedure gives a pretty good error-correcting code. To understand why this is so, let's look at what the channel does to a *single* codeword in the code. Since all the codewords are constructed in the same way, we may as well look at the first, x^1.

What is the effect of the binary symmetric channel on x^1? On a codeword of length n we expect roughly np of the bits to be flipped, so with high probability the output from the channel will be have a Hamming distance of about np from the codeword x^1, as illustrated in Figure 12.5; we say that such an output is on the *Hamming sphere* of radius np around x^1. How many elements are there in this Hamming sphere? The answer to this question is roughly $2^{nH(p)}$, since the Hamming sphere consists of all the typically occurring outputs $y = x^1 \oplus e$ from the channel, where e is the error that occurs in the channel, $\oplus$ denotes bitwise addition modulo 2, and by the theorem of typical sequences the number of such typical errors e is about $2^{nH(p)}$.

We've focused on a single codeword, but of course this same type of corruption occurs for all the codewords. We can imagine the space of all the codewords and their surrounding Hamming spheres, as depicted in Figure 12.6. If, as we've shown there, the Hamming spheres don't overlap, then there's an easy way for Bob to decode the output from the channel. He simply checks to see if the output is in one of the Hamming spheres, outputs the corresponding codeword if so, and outputs 'error' if not. Since we've assumed the spheres are non–overlapping, given any codeword as input it's highly likely that this will result in a successful decoding. Indeed, even if the spheres overlap slightly, it is still possible for Bob to perform the decoding with a good chance of success, provided the overlap is small – with high probability the output from the channel will belong to one (not zero or two or more) of the Hamming spheres, and will result in a successful decoding.

When does this small overlap condition occur? To understand this we need to better understand the structure of the possible outputs from the channel. We obtained the codewords for our code by sampling 2^{nR} times from a set $(X_1, \ldots, X_n)$ of random variables which are independent and identically distributed with $X_j = 0$ with probability q and $X_j = 1$ with probability $1 - q$. Suppose we let Y_j be the result of sending X_j through the binary symmetric channel. The theorem of typical sequences implies that the set of typical values for $(Y_1, \ldots, Y_n)$ is of size roughly $2^{nH(Y)}$, where Y is distributed as each of the Y_j. What's more, each of these typical output values has roughly equal probability.

Now, if we sample one hundred times uniformly from a population of size one million,

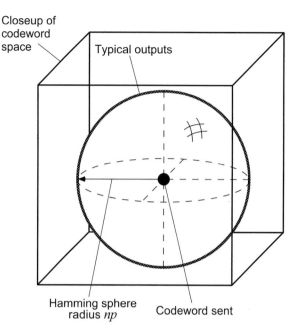

Figure 12.5. Suppose the codeword x^1 is sent through n uses of the binary symmetric channel. Then a typical output from the channel is an element of the Hamming sphere of radius np around the sequence which has been sent. (This figure is a closeup of Figure 12.6.)

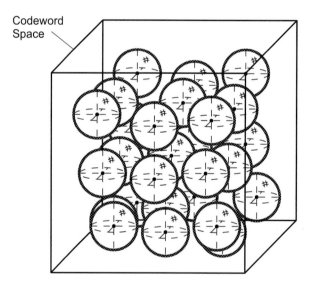

Figure 12.6. Randomly chosen codewords for the binary symmetric channel, surrounded by their Hamming spheres of 'typical' outputs. A closeup of an individual codeword may be found in Figure 12.5.

it's not too likely that we're going to get any repeats. In fact, even if we sample one hundred thousand times the number of repeats is going to be pretty small. It's not until we get out to about a million samples that the number of repeats is going to start getting large relative to the size of the sample. In a similar fashion, the amount of overlap between our 2^{nR} Hamming spheres of radius np is not going to start getting large until the combined number of elements in all the spheres approaches the size of the space $- 2^{nH(Y)}$ – that we are effectively sampling from. Since each sphere contains roughly $2^{nH(p)}$ elements, this means that we are very likely to have a good error-correcting code provided

$$2^{nR} \times 2^{nH(p)} < 2^{nH(Y)}, \tag{12.62}$$

which corresponds to the condition

$$R < H(Y) - H(p). \tag{12.63}$$

Now the entropy $H(Y)$ depends on the *a priori* distribution $(q, 1 - q)$ chosen for the X_j. To make the rate as high as possible we try to maximize $H(Y)$. A simple calculation shows that this is achieved by using the uniform *a priori* distribution corresponding to $q = 1/2$, for which $H(Y) = 1$, and therefore it is possible to achieve the rate R for any R less than $1 - H(p)$.

We've just outlined a proof that it is possible to reliably transmit information through a binary symmetric channel at any rate up to $1 - H(p)$. The proof is rather sketchy, but in fact contains many of the key ideas needed for a rigorous treatment, even in the quantum case. It turns out that the rates we have shown how to achieve are also the fastest it's possible to transmit the information through the binary symmetric channel; any faster than a rate $1 - H(p)$ and the Hamming spheres start to overlap too much to determine what codeword was sent, no matter how the codewords were chosen! Thus $1 - H(p)$ is the capacity of the binary symmetric channel.

How practical is random coding as a method to achieve high rate codes for the binary symmetric channel? It is true that if we use a random code then with high probability we can operate at a rate near the capacity. Unfortunately, there is a major difficulty with this procedure. In order to do the encoding and decoding, the sender and receiver ('Alice' and 'Bob') must first agree on a strategy for doing these tasks. In the case of random codes, this means that Alice must send Bob a list of all her random codewords. Doing this takes as much or more communication than Alice and Bob will be able to extract from the noisy channel. Clearly, this is undesirable for many applications! The random coding method is merely a method of demonstrating the *existence* of high rate codes, it is not a practical method for their construction. For wide practical application, what we would like is a method for achieving rates near the channel capacity which does not incur an unacceptable communication overhead for Alice and Bob. It is quite remarkable that methods for constructing such codes have only recently been discovered even for noisy classical channels, despite many decades of intense effort, and it remains an interesting open problem to find similar constructions for noisy quantum channels.

Shannon's noisy channel coding theorem

Shannon's noisy channel coding theorem generalizes the capacity result for the binary symmetric channel to the case of a *discrete memoryless* channel. Such a channel has a finite *input alphabet* $\mathcal{I}$, and a finite *output alphabet* $\mathcal{O}$. For the binary symmetric channel, $\mathcal{I} = \mathcal{O} = \{0, 1\}$. The action of the channel is described by a set of *conditional probabilities*, $p(y|x)$, where $x \in \mathcal{I}$ and $y \in \mathcal{O}$. These represent the probabilities of the different outputs y from the channel, given that the input was x, and satisfy the rules

$$p(y|x) \geq 0 \tag{12.64}$$

$$\sum_y p(y|x) = 1 \text{ for all } x. \tag{12.65}$$

The channel is memoryless in the sense that the channel acts the same way each time it is used, and different uses are independent of one another. We shall use the symbol $\mathcal{N}$ to denote a classical noisy channel.

Of course, there are many interesting communications channels which aren't discrete memoryless channels, such as the telephone line example given earlier, which has a continuous set of inputs and outputs. More general channels may be technically more difficult to understand than discrete memoryless channels, but many of the underlying ideas are the same and we refer you to the end of chapter 'History and further reading' for books containing information on this subject.

Let's look at the actual statement of Shannon's noisy channel coding theorem. We won't give the details of the proof, since we prove a more general result for quantum channels in the next section, but it is instructive to look at the statement of the classical result. First, we need to make our notion of reliable information transmission a little more precise. The basic idea is illustrated in Figure 12.7. In the first stage one of 2^{nR} possible messages M is produced by Alice and is encoded using a map $C^n : \{1, \ldots, 2^{nR}\} \to \mathcal{I}^n$ which assigns to each of Alice's possible messages an input string which is sent through n uses of the channel to Bob, who decodes the channel output using a map $D^n : \mathcal{O}^n \to \{1, \ldots, 2^{nR}\}$ which assigns a message to each string for each possible output from the channel. For a given encoding–decoding pair, the *probability of error* is defined to be the maximum probability over all messages M that the decoded output of the channel $D(Y)$ is not equal to the message M:

$$p(C^n, D^n) \equiv \max_M p(D^n(Y) \neq M | X = C^n(M)). \tag{12.66}$$

We say a rate R is *achievable* if there exists such a sequence of encoding–decoding pairs (C^n, D^n), and require in addition that $p(C^n, D^n) \to 0$ as $n \to \infty$. The *capacity* $C(\mathcal{N})$ of a given noisy channel $\mathcal{N}$ is defined to be the supremum over all achievable rates for the channel.

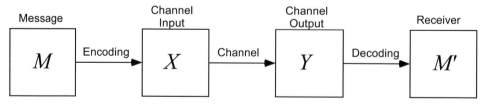

Figure 12.7. The noisy coding problem for classical messages. We require that every one of the 2^{nR} possible messages should be sent uncorrupted through the channel with high probability.

A priori it is not at all obvious how to calculate the capacity of a channel – a bare-hands calculation would involve taking a supremum over a very large (infinite!) class of possible encoding and decoding methods, and does not appear to be a particularly promising approach. Shannon's noisy channel coding theorem enormously simplifies the calculation of capacity, reducing it to a simple and well-defined optimization problem that can be solved exactly in many cases, and which is computationally quite tractable even when an exact solution is not feasible.

Theorem 12.7: (**Shannon's noisy channel coding theorem**) For a noisy channel $\mathcal{N}$ the capacity is given by

$$C(\mathcal{N}) = \max_{p(x)} H(X:Y), \tag{12.67}$$

where the maximum is taken over all input distributions $p(x)$ for X, for one use of the channel, and Y is the corresponding induced random variable at the output of the channel.

As an example of the noisy channel coding theorem, consider the case of a binary symmetric channel flipping bits with probability p, and with input distribution $p(0) = q, p(1) = 1 - q$. We have

$$H(X:Y) = H(Y) - H(Y|X) \tag{12.68}$$
$$= H(Y) - \sum_x p(x) H(Y|X = x). \tag{12.69}$$

But for each x, $H(Y|X = x) = H(p)$ so $H(X:Y) = H(Y) - H(p)$, which is maximized by choosing $q = 1/2$, so $H(Y) = 1$ and therefore $C(\mathcal{N}) = 1 - H(p)$ by Shannon's noisy channel coding theorem, just as our earlier intuitive calculation of the channel capacity of the binary symmetric channel suggested.

Exercise 12.9: The *erasure channel* has two inputs, 0 and 1, and three outputs, 0, 1 and e. With probability $1 - p$ the input is left alone. With probability p the input is 'erased', and replaced by e.

(1) Show that the capacity of the erasure channel is $1 - p$.
(2) Prove that the capacity of the erasure channel is greater than the capacity of the binary symmetric channel. Why is this result intuitively plausible?

Exercise 12.10: Suppose $\mathcal{N}_1$ and $\mathcal{N}_2$ are two discrete memoryless channels such that the input alphabet of $\mathcal{N}_2$ is equal to the output alphabet of $\mathcal{N}_1$. Show that

$$C(\mathcal{N}_2 \circ \mathcal{N}_1) \leq \min(C(\mathcal{N}_1), C(\mathcal{N}_2)). \tag{12.70}$$

Find an example where the inequality is strict.

A slight peculiarity of the noisy channel coding theorem we have presented is that nowhere does the notion of a classical information source appear! Recall that earlier we defined a classical information source as a sequence of independent and identically distributed random variables. We can combine this notion of an information source in an interesting way with the noisy channel coding theorem to obtain what is known as a *source–channel* coding theorem. The basic idea is illustrated in Figure 12.8. An information source with entropy rate $H(X)$ is producing information. By Shannon's noiseless channel coding theorem it is possible to compress the information from the source so that it only requires $nH(X)$ bits to describe; this step is sometimes known as source coding. The compressed output of the source is now used as the input message for the noisy channel. Transmitting at a rate R less than capacity, it requires $nH(X)/R$ uses of the channel to reliably transmit the compressed data to the receiver, who can then decompress it to recover the original output from the source.

You might wonder whether a better scheme for transmitting an information source over the noisy channel is possible. Perhaps it is possible to do something more efficient

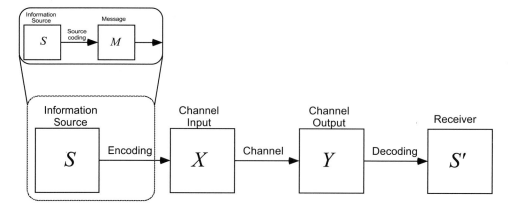

Figure 12.8. The noisy coding problem for a classical information source, sometimes known as the *source-coding* model.

than this two stage compress–encode and decode–decompress method? In fact, this turns out not to be the case, and the method of source–channel coding described is in fact optimal, but a proof of this fact is beyond our scope; see the end of chapter 'History and further reading' for more details.

12.3.2 Communication over noisy quantum channels

Suppose that instead of using a noisy classical communications channel to communicate Alice and Bob make use of a noisy quantum communications channel. More precisely, Alice has some message M that she wants to send to Bob. She encodes that message, just as she did in the classical case, but now the message is encoded as a *quantum state*, which is sent over the noisy quantum channel. By performing the encoding in just the right way, we hope that Bob will be able to determine what Alice's message was, with low probability of failure. Moreover, we'd like the *rate* at which Alice can send information to Bob to be as high as possible. What we want, in other words, is a procedure for computing the *capacity for classical information of a noisy quantum channel*. This problem has not yet been completely solved, but a great deal of progress has been made, and in this section we examine this progress.

What is known is how to calculate the capacity for a channel $\mathcal{E}$ assuming that Alice encodes her messages using *product states* of the form $\rho_1 \otimes \rho_2 \otimes \ldots$, where each of the $\rho_1, \rho_2, \ldots$ are potential inputs for one use of the channel $\mathcal{E}$. We call the capacity with this restriction the *product state capacity*, and denote it $C^{(1)}(\mathcal{E})$ to indicate that input states cannot be entangled across two or more uses of the channel. Note that this restricted model of communication between Alice and Bob *does* allow Bob to decode using measurements entangled across multiple uses of the channel; in fact, it turns out that this is essential. The only restriction (and an unfortunate restriction it is) is that Alice can only prepare product state inputs. It is believed by many researchers, but has not been proved, that allowing entangled signals doesn't increase the capacity. The result which allows us to calculate the product state capacity is known as the Holevo–Schumacher–Westmoreland (HSW) theorem, after its discoverers. As does Shannon's noisy channel coding theorem for classical noisy channels, the HSW theorem provides an effective means for computing the product state capacity for a specified noisy channel $\mathcal{E}$, and in some instances may even allow the derivation of an exact expression.

Theorem 12.8: **(Holevo–Schumacher–Westmoreland (HSW) theorem)** Let $\mathcal{E}$ be a trace-preserving quantum operation. Define

$$\chi(\mathcal{E}) \equiv \max_{\{p_j, \rho_j\}} \left[S\left(\mathcal{E}\left(\sum_j p_j \rho_j \right) \right) - \sum_j p_j S(\mathcal{E}(\rho_j)) \right], \qquad (12.71)$$

where the maximum is over all ensembles $\{p_j, \rho_j\}$ of possible input states ρ_j to the channel. Then $\chi(\mathcal{E})$ is the product state capacity for the channel $\mathcal{E}$, that is, $\chi(\mathcal{E}) = C^{(1)}(\mathcal{E})$.

The maximum in (12.71) is potentially over an unbounded set. In practice, we use the results of the following exercise to restrict the maximization to pure state ensembles containing at most d^2 elements, where d is the dimension of the input to the channel.

Exercise 12.11: Show that the maximum in the expression (12.71) may be achieved using an ensemble of pure states. Show further that it suffices to consider only ensembles of at most d^2 pure states, where d is the dimension of the input to the channel.

The proof of the HSW theorem involves several different ideas and it is easiest to understand the proof by breaking the discussion up into smaller pieces, and then putting the pieces together to obtain the HSW theorem.

Random coding

Suppose ρ_j is a set of possible inputs to the channel $\mathcal{E}$ and let $\sigma_j \equiv \mathcal{E}(\rho_j)$ be the corresponding outputs. We are going to develop a random coding technique similar to that described earlier for the binary symmetric channel, allowing Alice and Bob to communicate using codewords which are products of the states ρ_j. We let p_j be a probability distribution over the indices j, the *a priori distribution*. Alice wants to send a message M chosen from the set $\{1, \ldots, 2^{nR}\}$ to Bob. To each possible message M she associates a codeword $\rho_{M_1} \otimes \rho_{M_2} \otimes \ldots \otimes \rho_{M_n}$, where $M_1, \ldots, M_n$ are chosen from the index set $\{j\}$. (The $M_1, \ldots, M_n$ are not meant to be a decimal representation of M or anything of that sort!) For each message M Alice chooses M_1 by sampling from the distribution $\{p_j\}$. She chooses M_2 similarly and so on through to M_n, which completes the specification of the codeword. Abusing notation slightly we write $\rho_M \equiv \rho_{M_1} \otimes \cdots \otimes \rho_{M_n}$. The corresponding output states are simply denoted with a σ instead of a ρ, so for example we have $\sigma_{M_1} = \mathcal{E}(\rho_{M_1})$ and $\sigma_M = \mathcal{E}^{\otimes n}(\rho_M)$.

When Bob receives a particular state σ_M (corresponding to Alice trying to communicate the message M) he performs a measurement in an attempt to determine what the message was. Because we are only interested in the measurement statistics and not in the post-measurement state of Bob's system, it is sufficient to describe this measurement using the POVM formalism. We suppose that for each possible message M Bob has a corresponding POVM element E_M. It is possible that Bob might have one (or more) POVM elements that don't correspond to any specific message sent by Alice; obviously these can all be summed together into a single POVM element E_0 satisfying $E_0 = I - \sum_{M \neq 0} E_M$. The probability of Bob successfully identifying M is $\text{tr}(\sigma_M E_M)$, and therefore the probability of an error being made for the message M is $p_M^e \equiv 1 - \text{tr}(\sigma_M E_M)$.

What we want to do is prove the existence of high rate codes such that the probability of error p_M^e is small for *all* messages M. To do this we use a counter-intuitive and rather clever trick introduced by Shannon for the classical problem. We imagine that Alice is producing the messages M by choosing uniformly from the set $\{1, \ldots, 2^{nR}\}$, and analyze the *average* probability of error

$$p_{av} \equiv \frac{\sum_M p_M^e}{2^{nR}} = \frac{\sum_M (1 - \text{tr}(\sigma_M E_M))}{2^{nR}}. \tag{12.72}$$

The first step of the proof is to show that high rate codes exist with p_{av} tending to zero as n becomes large. After this has been done we will use Shannon's trick to show that this implies the existence of codes with essentially the same rate for which p_M^e is close to zero for *all* M. We begin by constructing a POVM $\{E_M\}$ which represents a pretty good (though perhaps not optimal) method for Bob to decode the outputs σ_M from the channel. The key idea in the construction, as for the classical binary symmetric channel, is the idea of typicality.

Let $\epsilon > 0$. Suppose we define $\bar{\sigma} \equiv \sum_j p_j \sigma_j$, and let P be the projector onto the ϵ-typical subspace of $\bar{\sigma}^{\otimes n}$. By the theorem of typical sequences it follows that for any $\delta > 0$ and for sufficiently large n,

$$\text{tr}\left(\bar{\sigma}^{\otimes n}(I - P)\right) \leq \delta. \tag{12.73}$$

For a given message M we are also going to define a notion of an ϵ-typical subspace for σ_M, based on the idea that typically σ_M is a tensor product of about np_1 copies of ρ_1, np_2 copies of ρ_2, and so on. Define $\bar{S} \equiv \sum_j p_j S(\sigma_j)$. Suppose σ_j has spectral decomposition $\sum_k \lambda_k^j |e_k^j\rangle\langle e_k^j|$, so

$$\sigma_M = \sum_K \lambda_K^M |E_K^M\rangle\langle E_K^M|, \tag{12.74}$$

where $K = (K_1, \ldots, K_n)$, and for convenience we define $\lambda_K^M \equiv \lambda_{K_1}^{M_1} \lambda_{K_2}^{M_2} \ldots \lambda_{K_n}^{M_n}$ and $|E_K^M\rangle \equiv |e_{K_1}^{M_1}\rangle|e_{K_2}^{M_2}\rangle \ldots |e_{K_n}^{M_n}\rangle$. Define P_M to be the projector onto the space spanned by all $|E_K^M\rangle$ such that

$$\left|\frac{1}{n}\log\frac{1}{\lambda_K^M} - \bar{S}\right| \leq \epsilon. \tag{12.75}$$

(It will be useful to denote by T_M the set of all K such that this condition is satisfied.) In a similar manner to the proof of the theorem of typical sequences, the law of large numbers implies that for any $\delta > 0$ and for sufficiently large n we have $\mathbf{E}(\text{tr}(\sigma_M P_M)) \geq 1 - \delta$, where the expectation is taken with respect to the distribution over codewords ρ_M (for a fixed message M) induced by random coding, and thus for each M,

$$\mathbf{E}\left[\text{tr}\left(\sigma_M(I - P_M)\right)\right] \leq \delta. \tag{12.76}$$

Also note that by the definition (12.75) the dimension of the subspace onto which P_M projects can be at most $2^{n(\bar{S}+\epsilon)}$, and thus

$$\mathbf{E}(\text{tr}(P_M)) \leq 2^{n(\bar{S}+\epsilon)}. \tag{12.77}$$

We now use the typicality notions to define Bob's decoding POVM. We define

$$E_M \equiv \left(\sum_{M'} P P_{M'} P\right)^{-1/2} P P_M P \left(\sum_{M'} P P_{M'} P\right)^{-1/2}, \tag{12.78}$$

where $A^{-1/2}$ denotes the generalized inverse of $A^{1/2}$, that is, the operator which is inverse to $A^{1/2}$ on the support of A and is otherwise zero. It follows that $\sum_M E_M \leq I$, and we can define one more positive operator $E_0 \equiv I - \sum_M E_M$ to complete the POVM. The intuition behind this construction is similar to the decoding method described for the binary symmetric channel. In particular, up to small corrections E_M is equal to the projector P_M, and Bob's measurement of $\{E_M\}$ corresponds essentially to checking to see if the output from the channel falls into the space on which P_M projects; the space onto which this projector projects can be thought of as analogous to the Hamming sphere of radius np around the codewords used for the binary symmetric channel.

The main technical part of the proof that random coding works is to obtain an upper bound on the average probability of error p_{av}. The details of how this is done are given in Box 12.5. The result is

$$ p_{av} \leq \frac{1}{2^{nR}} \sum_M \left[3\mathrm{tr}\,(\sigma_M(I - P)) + \sum_{M' \neq M} \mathrm{tr}(P\sigma_M P P_{M'}) + \mathrm{tr}\,(\sigma_M(I - P_M)) \right]. $$

$$ (12.79) $$

The quantity p_{av} is defined with respect to a specific choice of codewords. We are going to calculate the *expectation* of this quantity over all random codes. By construction $\mathbf{E}(\sigma_M) = \bar{\sigma}^{\otimes n}$, and σ_M and $P_{M'}$ are independent when $M' \neq M$, so we obtain

$$ \mathbf{E}(p_{av}) \leq 3\mathrm{tr}(\bar{\sigma}^{\otimes n}(I - P)) + (2^{nR} - 1)\mathrm{tr}(P\bar{\sigma}^{\otimes n} P\mathbf{E}(P_1)) + \mathbf{E}\,(\mathrm{tr}(\sigma_1(I - P_1))) \,(12.80) $$

Substituting (12.73) and (12.76) we obtain

$$ \mathbf{E}(p_{av}) \leq 4\delta + (2^{nR} - 1)\mathrm{tr}(P\bar{\sigma}^{\otimes n} P\mathbf{E}(P_1)). \qquad (12.81) $$

But $P\bar{\sigma}^{\otimes n} P \leq 2^{-n(S(\bar{\sigma})-\epsilon)}I$ and by (12.77) we have $\mathbf{E}(\mathrm{tr}(P_1)) \leq 2^{n(\bar{S}+\epsilon)}$ whence

$$ \mathbf{E}(p_{av}) \leq 4\delta + (2^{nR} - 1)2^{-n(S(\bar{\sigma})-\bar{S}-2\epsilon)}. \qquad (12.82) $$

Provided $R < S(\bar{\sigma}) - \bar{S}$ it follows that $\mathbf{E}(p_{av}) \to 0$ as $n \to \infty$. Indeed, by choosing the ensemble $\{p_j, \rho_j\}$ to achieve the maximum in (12.71) we see that this must be true whenever $R < \chi(\mathcal{E})$. Thus there must exist a sequence of codes of rate R such that $p_{av} \to 0$ as the block-size n of the code is increased. It follows that for any fixed $\epsilon > 0$ (note that this is a new meaning of ϵ to replace the old, which is no longer needed!) for sufficiently large n

$$ p_{av} = \frac{\sum_M p_M^e}{2^{nR}} < \epsilon. \qquad (12.83) $$

Obviously in order for this to be true at least half the messages M must satisfy $p_M^e < 2\epsilon$. So we construct a new code by deleting half the codewords (the codewords with high p_M^e) from the code with rate R and $p_{av} < \epsilon$, obtaining a new code with $2^{nR}/2 = 2^{n(R-1/n)}$ codewords, and with $p_M^e < 2\epsilon$ for all messages M. Obviously this code also has asymptotic rate R, and the probability of an error can be made arbitrarily small for *all* codewords, not just on average, as n becomes large.

Summing up, we have shown that for any rate R less than $\chi(\mathcal{E})$ as defined in (12.71), there exists a code using product state inputs enabling transmission through the channel $\mathcal{E}$ at rate R. Our proof suffers the same flaw as do random coding proofs of Shannon's classical noisy channel coding theorem, namely, it does not provide a constructive procedure for performing the coding, but it does at least demonstrate the existence of codes at rates up to capacity.

Proof of the upper bound

Suppose R is greater than $\chi(\mathcal{E})$ as defined in (12.71). We will show that it is impossible for Alice to reliably send information to Bob at this rate through the channel $\mathcal{E}$. Our general strategy is to imagine that Alice is producing messages M uniformly at random from the set $\{1, \ldots, 2^{nR}\}$ and to show that her *average* error probability must be bounded away from zero, and therefore the maximum error probability must also be bounded away from zero.

Suppose Alice encodes message M as $\rho_M = \rho_1^M \otimes \cdots \otimes \rho_n^M$ with corresponding outputs denoted using σ instead of ρ, and Bob decodes using a POVM $\{E_M\}$ which, without loss of generality, we may suppose contains an element E_M for each message, and possibly an extra element E_0 to ensure that the completeness relation $\sum_M E_M = I$ is satisfied. This gives an average error probability:

$$p_{\mathrm{av}} = \frac{\sum_M (1 - \mathrm{tr}(\sigma_M E_M))}{2^{nR}}. \tag{12.96}$$

From Exercise 12.3 we know that $R \le \log(d)$, where d is the dimension of the input to the channel, and thus the POVM $\{E_M\}$ contains at most $d^n + 1$ elements. By Fano's inequality it follows that

$$H(p_{\mathrm{av}}) + p_{\mathrm{av}} \log(d^n) \ge H(M|Y), \tag{12.97}$$

where Y is the measurement outcome from Bob's decoding, and thus

$$n p_{\mathrm{av}} \log d \ge H(M) - H(M:Y) - H(p_{\mathrm{av}}) = nR - H(M:Y) - H(p_{\mathrm{av}}). \tag{12.98}$$

Applying first the Holevo bound and then the subadditivity of entropy gives

$$H(M:Y) \le S(\bar{\sigma}) - \sum_M \frac{S(\sigma_1^M \otimes \cdots \otimes \sigma_n^M)}{2^{nR}} \tag{12.99}$$

$$\le \sum_{j=1}^n \left(S(\bar{\sigma}^j) - \sum_M \frac{S(\sigma_j^M)}{2^{nR}} \right), \tag{12.100}$$

where $\bar{\sigma}^j \equiv \sum_M \sigma_j^M / 2^{nR}$. Each of the n terms in the sum on the right hand side is no greater than $\chi(\mathcal{E})$ as defined in (12.71), so

$$H(M:Y) \le n\chi(\mathcal{E}). \tag{12.101}$$

Substituting into (12.98) gives $n p_{\mathrm{av}} \log d \ge n(R - \chi(\mathcal{E})) - H(p_{\mathrm{av}})$, and thus in the limit as n becomes large we obtain

$$p_{\mathrm{av}} \ge \frac{(R - \chi(\mathcal{E}))}{\log(d)}, \tag{12.102}$$

which is bounded away from zero when $R > \chi(\mathcal{E})$, which completes the proof that $\chi(\mathcal{E})$ is an upper bound on the product state capacity.

Box 12.5: HSW theorem: the error estimate

The most technically complicated part of the proof of the HSW theorem is obtaining an estimate for p_{av}. We outline the details of how this is done here; missing steps should be regarded as exercises to be filled in. Suppose we define $|\tilde{E}_K^M\rangle \equiv P|E_K^M\rangle$. Then

$$E_M = \left(\sum_{M'} \sum_{K \in T_{M'}} |\tilde{E}_K^{M'}\rangle\langle\tilde{E}_K^{M'}| \right)^{-1/2} \sum_{K \in T_M} |\tilde{E}_K^M\rangle\langle\tilde{E}_K^M| \left(\sum_{M'} \sum_{K \in T_{M'}} |\tilde{E}_K^{M'}\rangle\langle\tilde{E}_K^{M'}| \right)^{-1/2}$$
(12.84)

Defining

$$\alpha_{(M,K),(M',K')} \equiv \langle\tilde{E}_K^M| \left(\sum_{M''} \sum_{K'' \in T_{M''}} |\tilde{E}_{K''}^{M''}\rangle\langle\tilde{E}_{K''}^{M''}| \right)^{-1/2} |\tilde{E}_{K'}^{M'}\rangle, \quad (12.85)$$

the average probability of error can be written

$$p_{av} = \frac{1}{2^{nR}} \sum_M \left[1 - \sum_K \sum_{K' \in T_M} \lambda_K^M |\alpha_{(M,K),(M,K')}|^2 \right]. \quad (12.86)$$

Using $\sum_K \lambda_K^M = 1$ and omitting non-positive terms, we see that

$$p_{av} \leq \frac{1}{2^{nR}} \sum_M \left[\sum_{K \in T_M} \lambda_K^M (1 - \alpha_{(M,K),(M,K)}^2) + \sum_{K \notin T_M} \lambda_K^M \right]. \quad (12.87)$$

Define a matrix Γ with entries $\gamma_{(M,K),(M',K')} \equiv \langle\tilde{E}_K^M|\tilde{E}_{K'}^{M'}\rangle$, where the indices are such that $K \in T_M$ and $K' \in T_{M'}$. It is convenient to work in the matrix space defined by these index conventions, to let E denote the unit matrix with respect to these indices, and to use sp (for *spur*) to denote the trace operations with respect to these indices. A calculation shows that $\Gamma^{1/2} = [\alpha_{(M,K),(M',K')}]$ and it follows that $\alpha_{(M,K),(M,K)}^2 \leq \gamma_{(M,K),(M,K)} \leq 1$. Using the observation that $1 - x^2 = (1+x)(1-x) \leq 2(1-x)$ when $0 \leq x \leq 1$ together with (12.87) gives

$$p_{av} \leq \frac{1}{2^{nR}} \sum_M \left[2 \sum_{K \in T_M} \lambda_K^M (1 - \alpha_{(M,K),(M,K)}) + \sum_{K \notin T_M} \lambda_K^M \right]. \quad (12.88)$$

Define the diagonal matrix $\Lambda \equiv \text{diag}(\lambda_K^M)$ and observe that

$$2(E - \Gamma^{1/2}) = (E - \Gamma^{1/2})^2 + (E - \Gamma) \quad (12.89)$$
$$= (E - \Gamma)^2(E + \Gamma^{1/2})^{-2} + (E - \Gamma) \quad (12.90)$$
$$\leq (E - \Gamma)^2 + (E - \Gamma). \quad (12.91)$$

Thus,

$$2 \sum_M \sum_{K \in T_M} \lambda_K^M (1 - \alpha_{(M,K),(M,K)}) = 2\text{sp}(\Lambda(E - \Gamma^{1/2})) \quad (12.92)$$

$$\leq \text{sp}(\Lambda(E - \Gamma)^2) + \text{sp}(\Lambda(E - \Gamma)). \quad (12.93)$$

(continued)

Box 12.5 (continued):

Calculating the spurs on the right hand side, substituting into (12.88), and doing some simple algebra gives

$$p_{av} \leq \frac{1}{2^{nR}} \sum_M \left[\sum_K \lambda_K^M \left(2 - 2\gamma_{(M,K),(M,K)} + \sum_{K' \neq K} |\gamma_{(M,K),(M,K')}|^2 \right. \right.$$

$$\left. \left. + \sum_{M' \neq M, K' \in T_{M'}} |\gamma_{(M,K),(M',K')}|^2 \right) + \sum_{K \notin T_M} \lambda_K^M \right]. \quad (12.94)$$

Substituting definitions and doing some simple algebra gives

$$p_{av} \leq \frac{1}{2^{nR}} \sum_M \left[2\mathrm{tr}\left(\sigma_M(I - P)\right) + \mathrm{tr}(\sigma_M(I - P)P_M(I - P)) \right.$$

$$\left. + \sum_{M' \neq M} \mathrm{tr}(P\sigma_M P P_{M'}) + \mathrm{tr}\left(\sigma_M(I - P_M)\right) \right]. \quad (12.95)$$

The second term is less than $\mathrm{tr}(\sigma_M(I - P))$, which gives the desired error estimate, (12.79).

Examples

An interesting implication of the HSW theorem is that *any* quantum channel $\mathcal{E}$ whatsoever can be used to transmit classical information, provided the channel is not simply a constant. For if the channel is not a constant then there exist pure states $|\psi\rangle$ and $|\varphi\rangle$ such that $\mathcal{E}(|\psi\rangle\langle\psi|) \neq \mathcal{E}(|\varphi\rangle\langle\varphi|)$. Substituting the ensemble made up of these two states with equal probabilities $1/2$ into the expression (12.71) for the product state capacity we see that

$$C^{(1)}(\mathcal{E}) \geq S\left(\frac{\mathcal{E}(|\psi\rangle\langle\psi|) + \mathcal{E}(|\varphi\rangle\langle\varphi|)}{2}\right) - \frac{1}{2}\mathcal{E}(|\psi\rangle\langle\psi|) - \frac{1}{2}\mathcal{E}(|\varphi\rangle\langle\varphi|) > 0, \quad (12.103)$$

where the second inequality follows from the strict concavity of the entropy established in Section 11.3.5.

Let's look at a simple example where the product state capacity can be calculated exactly, the case of the depolarizing channel with parameter p. Let $\{p_j, |\psi_j\rangle\}$ be an ensemble of quantum states. Then we have

$$\mathcal{E}(|\psi_j\rangle\langle\psi_j|) = p|\psi_j\rangle\langle\psi_j| + (1 - p)\frac{I}{2}, \quad (12.104)$$

a quantum state which has eigenvalues $(1+p)/2$ and $(1-p)/2$ from which it follows that

$$S(\mathcal{E}(|\psi_j\rangle\langle\psi_j|)) = H\left(\frac{1+p}{2}\right), \quad (12.105)$$

which does not depend on $|\psi_j\rangle$ at all. Thus the maximum in (12.71) is achieved by maximizing the entropy $S(\sum_j \mathcal{E}(|\psi_j\rangle\langle\psi_j|))$, which may be done by simply choosing the $|\psi_j\rangle$ to form an orthonormal basis (say $|0\rangle$ and $|1\rangle$) for the state space of a single qubit, giving a value for the entropy of one bit, and a product state capacity of

$$C(\mathcal{E}) = 1 - H\left(\frac{1+p}{2}\right) \tag{12.106}$$

for the depolarizing channel with parameter p.

Exercise 12.12: Adapt the proof of the HSW theorem to find a proof of Shannon's noisy channel coding theorem, simplifying the proof wherever possible.

12.4 Quantum information over noisy quantum channels

How much quantum information can be reliably transmitted over a noisy quantum channel? This problem of determining the *quantum channel capacity* is less well understood than is the problem of determining the capacity for sending classical information through a noisy quantum channel. We now present some of the information-theoretic tools that have been developed to understand the capacity of a quantum channel for quantum information, most notably quantum information-theoretic analogues to the Fano inequality (Box 12.2 on page 536), data processing inequality (Section 11.2.4 on page 509) and Singleton bound (Exercise 10.21 on page 449).

As for quantum data compression, our point of view in studying these problems is to regard a quantum source as being a quantum system in a mixed state ρ which is entangled with another quantum system, and the measure of reliability for transmission of quantum information by the quantum operation $\mathcal{E}$ is the entanglement fidelity $F(\rho, \mathcal{E})$. It is useful to introduce, as in Chapter 9, labels Q for the system ρ lives in and R, the reference system, which initially purifies Q. In this picture the entanglement fidelity is a measure of how well the entanglement between Q and R was preserved by the action of $\mathcal{E}$ on system Q.

12.4.1 Entropy exchange and the quantum Fano inequality

How much noise does a quantum operation cause when applied to the state ρ of a quantum system Q? One measure of this is the extent to which the state of RQ, initially pure, becomes mixed as a result of the quantum operation. We define the *entropy exchange* of the operation $\mathcal{E}$ upon input of ρ by

$$S(\rho, \mathcal{E}) \equiv S(R', Q'). \tag{12.107}$$

Suppose that the action of the quantum operation $\mathcal{E}$ is mocked up by introducing an environment E, initially in a pure state, and then causing a unitary interaction between Q and E, as described in Chapter 8. Then the state of RQE after the interaction is a pure state, whence $S(R', Q') = S(E')$, so the entropy exchange may also be identified with the amount of entropy introduced by the operation $\mathcal{E}$ into an initially pure environment E.

Note that the entropy exchange does not depend upon the way in which the initial state of Q, ρ, is purified into RQ. The reason is because any two purifications of Q into RQ are related by a unitary operation on the system R, as shown in Exercise 2.81 on page 111. This unitary operation on R obviously commutes with the action of the quantum operation on Q, and thus the final states of $R'Q'$ induced by the two different purifications are related by a unitary transformation on R, and thus give rise to the same value for the entropy exchange. Furthermore, it follows from these results that $S(E')$ does not depend upon the particular environmental model for $\mathcal{E}$ which is used, provided the model starts with E in a pure state.

A useful explicit formula for the entropy exchange can be given based upon the operator-sum representation for quantum operations. Suppose a trace-preserving quantum operation $\mathcal{E}$ has operation elements $\{E_i\}$. Then, as shown in Section 8.2.3, a unitary model for this quantum operation is given by defining a unitary operator U on QE such that

$$U|\psi\rangle|0\rangle = \sum_i E_i|\psi\rangle|i\rangle, \qquad (12.108)$$

where $|0\rangle$ is the initial state of the environment, and $|i\rangle$ is an orthonormal basis for the environment. Note that the state of E' after application of $\mathcal{E}$ is:

$$\rho^{E'} = \sum_{i,j} \text{tr}(E_i\rho E_j^\dagger)|i\rangle\langle j|. \qquad (12.109)$$

That is, $\text{tr}(E_i\rho E_j^\dagger)$ are the matrix elements of E' in the $|i\rangle$ basis. Given a quantum operation with operation elements $\{E_i\}$ it is therefore natural to define a matrix W (the *w-matrix*) with matrix elements $W_{ij} \equiv \text{tr}(E_i\rho E_j^\dagger)$, that is, W is the matrix of E' in an appropriate basis. This representation for $\rho^{E'}$ gives rise to a formula for the entropy exchange which is useful in making explicit calculations,

$$S(\rho, \mathcal{E}) = S(W) \equiv -\text{tr}(W \log W). \qquad (12.110)$$

Given a quantum operation $\mathcal{E}$ and a state ρ, it is always possible to choose operation elements $\{F_j\}$ for $\mathcal{E}$ such that W is diagonal; we say W is in *canonical form*. To see that such a set of operation elements exists, recall from Chapter 8 that a quantum operation may have many different sets of operation elements. In particular, two sets of operators $\{E_i\}$ and $\{F_j\}$ are operation elements for the same quantum operation if and only if $F_j = \sum_j u_{ji} E_i$, where u is a unitary matrix of complex numbers, and it may be necessary to append 0 operators to the sets E_i or F_j so that the matrix u is a square matrix. Let W be the w-matrix associated to a particular choice of operation elements $\{E_i\}$ for $\mathcal{E}$. W is a matrix representation of the environmental density operator, and thus is a positive matrix which may be diagonalized by a unitary matrix v, $D = vWv^\dagger$, where D is a diagonal matrix with non-negative entries. Define operators F_j by the equation $F_j \equiv \sum_i v_{ji} E_i$, so the F_j are also a set of operation elements for $\mathcal{E}$, giving rise to a new w-matrix $\widetilde{W}$ with matrix elements

$$\widetilde{W}_{kl} = \text{tr}(F_k\rho F_l^\dagger) = \sum_{mn} v_{km}v_{ln}^* W_{mn} = D_{kl}. \qquad (12.111)$$

Thus, the w-matrix is diagonal if calculated with respect to the operation elements $\{F_j\}$. Any such set of operation elements $\{F_j\}$ for $\mathcal{E}$ for which the corresponding w-matrix is diagonal is said to be a *canonical representation* for $\mathcal{E}$ with respect to the input ρ. We see later that canonical representations turn out to have a special significance for quantum error-correction.

Many properties of the entropy exchange follow easily from properties of the entropy discussed in Chapter 11. For example, working in a canonical representation for a trace-preserving quantum operation $\mathcal{E}$ on a d-dimensional space, we see immediately that $S(I/d, \mathcal{E}) = 0$ if and only if $\mathcal{E}$ is a unitary quantum operation. Therefore, $S(I/d, \mathcal{E})$ can be thought of as quantifying the extent to which incoherent quantum noise occurs on the system as a whole. A second example is that the matrix W is linear in ρ, and by the concavity of the entropy it follows that $S(\rho, \mathcal{E})$ is concave in ρ. Since the system RQ can always be chosen to be at most d^2-dimensional, where d is the dimension of Q, it follows that the entropy exchange is bounded above by $2 \log d$.

Exercise 12.13: Show that the entropy exchange is concave in the quantum operation $\mathcal{E}$.

Intuitively, if the quantum source Q is subject to noise which results in the entanglement RQ becoming mixed, then the fidelity of the final state $R'Q'$ with the initial state RQ cannot be perfect. Moreover, the greater the noise the worse the fidelity. In Section 12.1.1 an analogous situation arose in the study of classical channels, where the uncertainty $H(X|Y)$ about the input of a channel, X, given the output, Y, was related to the probability of being able to recover the state of X from Y by the Fano inequality. There is a very useful quantum analogue of this result, relating the entropy exchange $S(\rho, \mathcal{E})$ to the entanglement fidelity $F(\rho, \mathcal{E})$.

Theorem 12.9: (**Quantum Fano inequality**) Let ρ be a quantum state and $\mathcal{E}$ a trace-preserving quantum operation. Then

$$S(\rho, \mathcal{E}) \leq H(F(\rho, \mathcal{E})) + (1 - F(\rho, \mathcal{E})) \log(d^2 - 1), \qquad (12.112)$$

where $H(\cdot)$ is the binary Shannon entropy.

Inspection of the quantum Fano inequality reveals an attractive intuitive meaning: if the entropy exchange for a process is large, then the entanglement fidelity for the process must necessarily be small, indicating that the entanglement between R and Q has not been well preserved. Moreover, we note that in the quantum Fano inequality the entropy exchange $S(\rho, \mathcal{E})$ plays a role analogous to the role played by the conditional entropy $H(X|Y)$ in classical information theory.

Proof
To prove the quantum Fano inequality, let $|i\rangle$ be an orthonormal basis for the system RQ chosen so the first state in the set $|1\rangle = |RQ\rangle$. If we form the quantities $p_i \equiv \langle i|\rho^{R'Q'}|i\rangle$, then from the results of Section 11.3.3 it follows that

$$S(R', Q') \leq H(p_1, \ldots, p_{d^2}), \qquad (12.113)$$

where $H(p_i)$ is the Shannon information of the set $\{p_i\}$. Elementary algebra shows that

$$H(p_1, \ldots, p_{d^2}) = H(p_1) + (1 - p_1) H\left(\frac{p_2}{1 - p_1}, \ldots, \frac{p_{d^2}}{1 - p_1}\right). \qquad (12.114)$$

Combining this with the observation that $H(\frac{p_2}{1-p_1}, \ldots, \frac{p_{d^2}}{1-p_1}) \leq \log(d^2 - 1)$ and $p_1 = F(\rho, \mathcal{E})$ by definition gives,

$$S(\rho, \mathcal{E}) \leq H(F(\rho, \mathcal{E})) + (1 - F(\rho, \mathcal{E})) \log(d^2 - 1), \qquad (12.115)$$

which is the quantum Fano inequality.   □

12.4.2 The quantum data processing inequality

In Section 11.2.4 we discussed the classical *data processing inequality*. Recall that the data processing inequality states that for a Markov process $X \rightarrow Y \rightarrow Z$,

$$H(X) \geq H(X:Y) \geq H(X:Z), \tag{12.116}$$

with equality in the first stage if and only if the random variable X can be recovered from Y with probability one. Thus the data processing inequality provides information-theoretic necessary and sufficient conditions for error-correction to be possible.

There is a quantum analogue to the data processing inequality applicable to a two stage quantum process described by quantum operations $\mathcal{E}_1$ and $\mathcal{E}_2$,

$$\rho \xrightarrow{\mathcal{E}_1} \rho' \xrightarrow{\mathcal{E}_2} \rho''. \tag{12.117}$$

We define the quantum *coherent information* by

$$I(\rho, \mathcal{E}) \equiv S(\mathcal{E}(\rho)) - S(\rho, \mathcal{E}). \tag{12.118}$$

This quantity, coherent information, is suspected (but not known) to play a role in quantum information theory analogous to the role played by the mutual information $H(X:Y)$ in classical information theory. One reason for this belief is that the coherent information satisfies a quantum data processing inequality analogous to the classical data processing inequality.

Theorem 12.10: (**Quantum data processing inequality**) Let ρ be a quantum state and $\mathcal{E}_1$ and $\mathcal{E}_2$ trace-preserving quantum operations. Then

$$S(\rho) \geq I(\rho, \mathcal{E}_1) \geq I(\rho, \mathcal{E}_2 \circ \mathcal{E}_1), \tag{12.119}$$

with equality in the first inequality if and only if it is possible to *perfectly reverse* the operation $\mathcal{E}_1$, in the sense that there exists a trace-preserving reversal operation $\mathcal{R}$ such that $F(\rho, \mathcal{R} \circ \mathcal{E}) = 1$.

Comparison with the classical data processing inequality shows that the coherent information plays a role in the quantum data processing inequality identical to the role played by the mutual information in the classical data processing inequality. Of course, such a heuristic argument cannot be regarded as any sort of a rigorous justification for the view that the coherent information is the correct quantum analogue of the classical mutual information. In order to obtain such a justification, the coherent information ought to be related to the quantum channel capacity in a similar way to the relation between classical mutual information and classical channel capacity, and such a relationship has not yet been established. (See the end of chapter 'History and further reading' for some partial progress.)

How is the notion of perfect reversibility defined in the quantum data processing inequality connected with more familiar notions such as that arising in the context of quantum error-correction? By definition, we say a trace-preserving quantum operation $\mathcal{E}$ is *perfectly reversible* upon input of ρ if there exists a trace-preserving quantum operation $\mathcal{R}$ such that

$$F(\rho, \mathcal{R} \circ \mathcal{E}) = 1. \tag{12.120}$$

But from item (4) on page 423, it follows that a quantum operation is perfectly reversible if and only if for every state $|\psi\rangle$ in the support of ρ,

$$(\mathcal{R} \circ \mathcal{E})(|\psi\rangle\langle\psi|) = |\psi\rangle\langle\psi|. \tag{12.121}$$

This observation connects the notion of perfect reversibility to quantum error-correcting codes. Recall that a quantum error-correcting code is a subspace of some larger Hilbert space spanned by logical codewords. To be resilient against the noise induced by a quantum operation $\mathcal{E}$ it is necessary that the quantum operation $\mathcal{E}$ be reversible by a trace-preserving reversal operation $\mathcal{R}$ in the sense that for all states $|\psi\rangle$ in the code, $(\mathcal{R} \circ \mathcal{E})(|\psi\rangle\langle\psi|) = |\psi\rangle\langle\psi|$. This condition is equivalent to the criterion of perfect reversibility in the statement of the data processing inequality, that $F(\rho, \mathcal{R} \circ \mathcal{E}) = 1$, for some ρ whose support is the code space.

Proof

The quantum data processing inequality is proved using a four system construction: R and Q appear in their familiar roles, while E_1 and E_2 are systems initially in pure states, chosen such that a unitary interaction between Q and E_1 generates the dynamics $\mathcal{E}_1$, and a unitary interaction between Q and E_2 generates the dynamics $\mathcal{E}_2$. The proof of the first stage of the quantum data processing inequality is to apply the subadditivity inequality $S(R', E_1') \leq S(R') + S(E_1')$ to obtain

$$\begin{align} I(\rho, \mathcal{E}_1) &= S(\mathcal{E}_1(\rho)) - S(\rho, \mathcal{E}_1) \tag{12.122}\\ &= S(Q') - S(E_1') \tag{12.123}\\ &= S(R', E_1') - S(E_1') \tag{12.124}\\ &\leq S(R') + S(E_1') - S(E_1') = S(R') \tag{12.125}\\ &= S(R) = S(Q) = S(\rho). \tag{12.126} \end{align}$$

The proof of the second part of the data processing inequality is to apply the strong subadditivity inequality,

$$S(R'', E_1'', E_2'') + S(E_1'') \leq S(R'', E_1'') + S(E_1'', E_2''). \tag{12.127}$$

From the purity of the total state of $R''Q''E_1''E_2''$ it follows that

$$S(R'', E_1'', E_2'') = S(Q''). \tag{12.128}$$

Neither of the systems R or E_1 are involved in the second stage of the dynamics in which Q and E_2 interact unitarily. Thus, their state does not change during this stage: $\rho^{R''E_1''} = \rho^{R'E_1'}$. But from the purity of RQE_1 after the first stage of the dynamics,

$$S(R'', E_1'') = S(R', E_1') = S(Q'). \tag{12.129}$$

The remaining two terms in the strong subadditivity inequality (12.127) are now recognized as entropy exchanges,

$$S(E_1'') = S(E_1') = S(\rho, \mathcal{E}_1); \quad S(E_1'', E_2'') = S(\rho, \mathcal{E}_2 \circ \mathcal{E}_1). \tag{12.130}$$

Making these substitutions into (12.127) yields

$$S(Q'') + S(\rho, \mathcal{E}_1) \leq S(Q') + S(\rho, \mathcal{E}_2 \circ \mathcal{E}_1), \tag{12.131}$$

which can be rewritten as the second stage of the data processing inequality, $I(\rho, \mathcal{E}_1) \geq I(\rho, \mathcal{E}_2 \circ \mathcal{E}_1)$.

To complete the proof we need to show that $\mathcal{E}$ is perfectly reversible upon input of ρ if and only if the first inequality in the quantum data processing inequality is satisfied with equality,

$$S(\rho) = I(\rho, \mathcal{E}) = S(\rho') - S(\rho, \mathcal{E}). \tag{12.132}$$

To prove the necessity of this condition for reversal, suppose that $\mathcal{E}$ is perfectly reversible upon input of ρ, with reversal operation $\mathcal{R}$. From the second stage of the quantum data processing inequality it follows that

$$S(\rho') - S(\rho, \mathcal{E}) \geq S(\rho'') - S(\rho, \mathcal{R} \circ \mathcal{E}). \tag{12.133}$$

From the reversibility requirement it follows that $\rho'' = \rho$. Furthermore, from the quantum Fano inequality (12.112) and the perfect reversibility requirement $F(\rho, \mathcal{R} \circ \mathcal{E}) = 1$ it follows that $S(\rho, \mathcal{R} \circ \mathcal{E}) = 0$. Thus the second stage of the quantum data processing inequality when applied to $\rho \to \mathcal{E}(\rho) \to (\mathcal{R} \circ \mathcal{E})(\rho)$ may be rewritten

$$S(\rho') - S(\rho, \mathcal{E}) \geq S(\rho). \tag{12.134}$$

Combining this with the first part of the quantum data processing inequality, $S(\rho) \geq S(\rho') - S(\rho, \mathcal{E})$, we deduce that

$$S(\rho') = S(\rho) - S(\rho, \mathcal{E}), \tag{12.135}$$

for any $\mathcal{E}$ which is perfectly reversible upon input of ρ.

Next, we give a constructive proof that satisfaction of the condition

$$S(\rho) = S(\rho') - S(\rho, \mathcal{E}) \tag{12.136}$$

implies that the quantum operation $\mathcal{E}$ is reversible upon input of ρ. Noting that $S(\rho) = S(Q) = S(R) = S(R')$, $S(\rho') = S(Q') = S(R', E')$ and $S(\rho, \mathcal{E}) = S(E')$, we see that $S(R') + S(E') = S(R', E')$, which we saw in Section 11.3.4 is equivalent to the condition that $\rho^{R'E'} = \rho^{R'} \otimes \rho^{E'}$. Suppose the initial state of Q is $\rho = \sum_i p_i |i\rangle\langle i|$, and that we purify this state into RQ as $|RQ\rangle = \sum_i \sqrt{p_i}|i\rangle|i\rangle$, where the first system is R and the second system is Q. Note that $\rho^{R'} = \rho^R = \sum_i p_i |i\rangle\langle i|$. Furthermore, suppose that $\rho^{E'} = \sum_j q_j |j\rangle\langle j|$ for some orthonormal set $|j\rangle$, so that

$$\rho^{R'E'} = \sum_{ij} p_i q_j |i\rangle\langle i| \otimes |j\rangle\langle j|. \tag{12.137}$$

This has eigenvectors $|i\rangle|j\rangle$ so by the Schmidt decomposition we may write the total state of $R'Q'E'$ after the quantum operation $\mathcal{E}$ has been applied as

$$|R'Q'E'\rangle = \sum_{ij} \sqrt{p_i q_j}|i\rangle|i,j\rangle|j\rangle, \tag{12.138}$$

where $|i, j\rangle$ is some orthonormal set of states for system Q. Define projectors P_j by $P_j \equiv \sum_i |i, j\rangle\langle i, j|$. The idea of the restoration operation is to first perform a measurement described by the projectors P_j, which reveals the state $|j\rangle$ of the environment, and then do a unitary rotation U_j conditional on j which restores the state $|i, j\rangle$ to $|i\rangle$: $U_j|i, j\rangle \equiv |i\rangle$.

That is, j is the measurement syndrome, and U_j the corresponding recovery operation. The complete recovery operation may be written

$$\mathcal{R}(\sigma) \equiv \sum_j U_j P_j \sigma P_j U_j^\dagger. \tag{12.139}$$

The projectors P_j are orthogonal, by the orthogonality of the states $|i, j\rangle$, but may not be complete. If this is the case, then to ensure that the quantum operation $\mathcal{R}$ is trace-preserving, it is necessary to add an extra projector $\tilde{P} \equiv I - \sum_j P_j$ to the set of projectors to make the operation trace-preserving.

The final state of the system RQE after the reversal operation is given by

$$\sum_j U_j P_j |R'Q'E'\rangle\langle R'Q'E'| P_j U_j^\dagger$$

$$= \sum_j \sum_{i_1 i_2} \sqrt{p_{i_1} p_{i_2}} q_j |i_1\rangle\langle i_2| \otimes (U_j |i_1, j\rangle\langle i_2, j| U_j^\dagger) \otimes |j\rangle\langle j| \tag{12.140}$$

$$= \sum_{i_1, i_2} \sqrt{p_{i_1} p_{i_2}} |i_1\rangle\langle i_2| \otimes |i_1\rangle\langle i_2| \otimes \rho^{E'}, \tag{12.141}$$

from which we see that $\rho^{R''Q''} = \rho^{RQ}$, and thus $F(\rho, \mathcal{R} \circ \mathcal{E}) = 1$, that is, the operation $\mathcal{E}$ is perfectly reversible upon input of the state ρ, as we desired to show. □

This completes the proof of the information-theoretic reversibility conditions for trace-preserving quantum operations. Intuition about the result may be obtained by imagining that Q is a memory element in a quantum computer, R is the remainder of the quantum computer, and E is an environment whose interaction with Q causes noise. The information-theoretic reversibility condition is most elegantly understood as the statement that the state of the environment E' after the noise has occurred should not be correlated with the state of the remainder of the quantum computer, R'. To state it in more anthropomorphic terms, error-correction is possible precisely when the environment does not learn anything about the remainder of the quantum computer through interacting with Q!

Even more concretely, suppose Q is an n qubit system and C is an $[n, k]$ quantum error-correcting code living in the system Q with orthonormal codewords $|x\rangle$ and projector P onto the codespace. Consider the density matrix $P/2^k$, which may be purified to a pure state of RQ:

$$\frac{1}{\sqrt{2^k}} \sum_x |x\rangle|x\rangle. \tag{12.142}$$

Imagine this code is able to correct arbitrary errors on some subset Q_1 of the qubits. Then, in particular, it must be able to correct that error which simply swaps those qubits out into the environment and replaces them with some standard state. The information-theoretic reversibility condition that $\rho^{R'E'} = \rho^{R'} \otimes \rho^{E'}$ can in this case be rephrased as the condition $\rho^{RQ_1} = \rho^R \otimes \rho^{Q_1}$. Thus the reference system R and subsystem Q_1 on which errors can be corrected must initially be uncorrelated if correction is to be possible!

Exercise 12.14: Show that the condition $\rho^{RQ_1} = \rho^R \otimes \rho^{Q_1}$ is also *sufficient* to be able to correct errors on the subsystem Q_1.

The reasoning used in the proof of the quantum data processing inequality can be adapted to prove a wide variety of other inequalities. For example, suppose we have a quantum system Q in a state ρ that is subjected to the quantum operation $\mathcal{E}$. The first stage of the data processing inequality follows by applying the subadditivity inequality for entropy to the systems $R'E'$. What if instead we applied the subadditivity inequality to the systems $Q'E'$, obtaining

$$S(\rho) = S(R) = S(R') = S(Q', E') \leq S(Q') + S(E') = S(\mathcal{E}(\rho)) + S(\rho, \mathcal{E}). \quad (12.143)$$

That is,

$$\Delta S + S(\rho, \mathcal{E}) \geq 0, \quad (12.144)$$

where $\Delta S \equiv S(\mathcal{E}(\rho)) - S(\rho)$ is the change in entropy caused by the process $\mathcal{E}$. Loosely speaking, this inequality says that the change in entropy of the system plus the change in entropy of the environment must be non-negative, an eminently reasonable statement in accord with the second law of thermodynamics, and one which will aid us in Section 12.4.4 in our thermodynamic analysis of quantum error-correction!

Exercise 12.15: Apply all possible combinations of the subadditivity and strong subadditivity inequalities to deduce other inequalities for the two stage quantum process $\rho \to \rho' = \mathcal{E}_1(\rho) \to \rho'' = (\mathcal{E}_2 \circ \mathcal{E}_1)(\rho)$, expressing the results whenever possible in terms of entropy exchanges and the entropies $S(\rho), S(\rho'), S(\rho'')$. When it is not possible to express a quantity appearing in such an inequality in these terms, give a prescription for calculating the quantity using only a knowledge of ρ and operation elements $\{E_j\}$ for $\mathcal{E}_1$ and $\{F_k\}$ for $\mathcal{E}_2$.

12.4.3 Quantum Singleton bound

The information-theoretic approach to quantum error-correction can be used to prove a beautiful bound on the ability of quantum error-correcting codes to correct errors, the *quantum Singleton bound*. Recall that an $[n, k, d]$ code uses n qubits to encode k qubits, and is able to correct located errors (Exercise 10.45) on up to $d - 1$ of the qubits. The quantum Singleton bound states that we must have $n - k \geq 2(d - 1)$. Contrast this with the classical Singleton bound, Exercise 10.21 on page 449, which states that for an $[n, k, d]$ classical code we must have $n - k \geq d - 1$. Because a quantum code to correct errors on up to t qubits must have distance at least $2t + 1$ it follows that $n - k \geq 4t$. Thus, for example, a code to encode $k = 1$ qubits and capable of correcting errors on $t = 1$ of the qubits must satisfy $n - 1 \geq 4$, that is, n must be at least 5, so the five qubit code described in Chapter 10 is the smallest possible code for this task.

The proof of the quantum Singleton bound is an extension of the information-theoretic techniques we have been using to analyze quantum error-correction. Suppose the code is a 2^k-dimensional subspace associated with the system Q, with orthonormal basis denoted by $|x\rangle$. Introduce a 2^k-dimensional reference system R also with 2^k orthonormal basis vectors denoted $|x\rangle$, and consider the entangled state of RQ,

$$|RQ\rangle = \frac{1}{\sqrt{2^k}} \sum_x |x\rangle|x\rangle. \quad (12.145)$$

We divide the n qubits of Q up into three disjoint blocks, the first and second, Q_1 and Q_2, consisting of $d - 1$ qubits each, and the third Q_3 consisting of the remaining $n - 2(d - 1)$

qubits. Because the code has distance d any set of $d-1$ located errors may be corrected, and thus it is possible to correct errors on either Q_1 or Q_2. It follows that R and Q_1 must be uncorrelated, as are R and Q_2. By this observation, the overall purity of the state of $RQ_1Q_2Q_3$, and the subadditivity of entropy we have:

$$S(R) + S(Q_1) = S(R, Q_1) = S(Q_2, Q_3) \leq S(Q_2) + S(Q_3) \qquad (12.146)$$

$$S(R) + S(Q_2) = S(R, Q_2) = S(Q_1, Q_3) \leq S(Q_1) + S(Q_3). \qquad (12.147)$$

Adding these two inequalities gives

$$2S(R) + S(Q_1) + S(Q_2) \leq S(Q_1) + S(Q_2) + 2S(Q_3). \qquad (12.148)$$

Canceling terms and substituting $S(R) = k$ gives $k \leq S(Q_3)$. But Q_3 is $n - 2(d-1)$ qubits in size, so $S(Q_3) \leq n - 2(d-1)$, giving $k \leq n - 2(d-1)$, whence $2(d-1) \leq n-k$, the quantum Singleton bound.

As an example of the quantum Singleton bound in action, consider the depolarizing channel $\mathcal{E}(\rho) = p\rho + (1-p)/3(X\rho X + Y\rho Y + Z\rho Z)$. Suppose the depolarizing channel acts independently on a large number n of qubits. If $p < 3/4$ then more than one quarter of those qubits will suffer errors, so any code capable of recovering from the errors must have $t > n/4$. But the quantum Singleton bound implies that $n - k \geq 4t > n$, and thus k must be negative, that is, it is not possible to encode *any* qubits at all in this case. Thus, when $p < 3/4$, the quantum Singleton bound implies that the capacity of the depolarizing channel for quantum information is zero!

12.4.4 Quantum error-correction, refrigeration and Maxwell's demon

Quantum error-correction may be thought of as a type of refrigeration process, capable of keeping a quantum system at a constant entropy, despite the influence of noise processes which tend to change the entropy of the system. Indeed, quantum error-correction may even appear rather puzzling from this point of view, as it appears to allow a reduction in entropy of the quantum system in apparent violation of the second law of thermodynamics! To understand why there is no violation of the second law we do an analysis of quantum error-correction similar to that used to analyze Maxwell's demon in Box 3.5 on page 162. Quantum error-correction is essentially a special type of Maxwell's demon – we can imagine a 'demon' performing syndrome measurements on the quantum system, and then correcting errors according to the result of the syndrome measurement. Just as in the analysis of the classical Maxwell's demon, the storage of the syndrome in the demon's memory carries with it a thermodynamic cost, in accordance with Landauer's principle. In particular, because any memory is finite, the demon must eventually begin erasing information from its memory, in order to have space for new measurement results. Landauer's principle states that erasing one bit of information from the memory increases the total entropy of the system – quantum system, demon, and environment – by at least one bit.

Rather more precisely, we can consider a four-stage error-correction 'cycle' as depicted in Figure 12.9:

(1) The system, starting in a state ρ, is subjected to a noisy quantum evolution that takes it to a state ρ'. In typical scenarios for error-correction, we are interested in cases where the entropy of the system increases, $S(\rho') > S(\rho)$, although this is not necessary.

(2) A demon performs a (syndrome) measurement on the state ρ' described by measurement operators $\{M_m\}$, obtaining result m with probability $p_m = \mathrm{tr}(M_m \rho' M_m^\dagger)$, resulting in the posterior state $\rho'_m = M_m \rho' M_m^\dagger / p_m$.

(3) The demon applies a unitary operation V_m (the recovery operation) that creates a final system state

$$\rho''_m = V_m \rho'_m V_m^\dagger = \frac{V_m M_m \rho' M_m^\dagger V_m^\dagger}{p_m}. \tag{12.149}$$

(4) The cycle is restarted. In order that this actually be a cycle and that it be a successful error-correction, we must have $\rho''_m = \rho$ for each measurement outcome m.

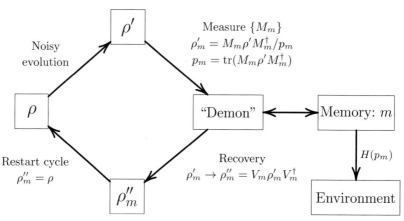

Figure 12.9. The quantum error-correction cycle.

We now show that any reduction in entropy during the second and third stages – the error-correction stages – comes at the expense of entropy production in the environment which is at least as large as the entropy reduction in the quantum system being error-corrected. After the third stage the only record of the measurement result m is the record kept in the demon's memory. To reset its memory for the next cycle, the demon must erase its record of the measurement result, causing a net increase in the entropy of the environment by Landauer's principle. The number of bits that must be erased is determined by the representation the demon uses to store the measurement result m; by Shannon's noiseless channel coding theorem, at least $H(p_m)$ bits are required, on average, to store the measurement result, and thus a single error-correction cycle on average involves the dissipation of $H(p_m)$ bits of entropy into the environment when the measurement record is erased.

Before error-correction the state of the quantum system is ρ'. After error-correction the state of the quantum system is ρ, so the net change in entropy of the system due to the error-correction is $\Delta S \equiv S(\rho) - S(\rho')$. There is an additional entropic cost of $H(p_m)$ (on average) associated with erasing the measurement record, for a total cost of $\Delta(S) + H(p_m)$. Our goal is to bound this thermodynamic cost, and in so doing demonstrate that the second law of thermodynamics is never violated. To do so, it helps to introduce two items of notation: let $\mathcal{E}$ represent the noise process occurring during stage 1 of the error-correction cycle, $\rho \to \rho' = \mathcal{E}(\rho)$, and let $\mathcal{R}$ be the quantum operation representing the error-correction operation,

$$R(\sigma) \equiv \sum_m V_m M_m \sigma M_m^\dagger V_m^\dagger. \tag{12.150}$$

With input ρ' the w-matrix for this process has elements $W_{mn} = \text{tr}(V_m M_m \rho' M_n^\dagger V_n^\dagger)$, and thus has diagonal elements $W_{mm} = \text{tr}(V_m M_m \rho' M_m^\dagger V_m^\dagger) = \text{tr}(M_m \rho' M_m^\dagger)$, which is just the probability p_m the demon obtains measurement outcome m when measuring the error syndrome. By Theorem 11.9 on page 515 the entropy of the diagonal elements of W is at least as great as the entropy of W, so

$$H(p_m) \geq S(W) = S(\rho', R), \tag{12.151}$$

with equality if and only if the operators $V_m M_m$ are a canonical decomposition of R with respect to ρ' so that the off-diagonal terms in W vanish. By Equation (12.144) on page 568 it follows that

$$\Delta S + S(\rho', R) = S(\rho) - S(\rho') + S(\rho', R) \geq 0. \tag{12.152}$$

Combining this result with (12.151) we deduce that $\Delta S + H(p_m) \geq 0$. But $\Delta S + H(p_m)$ was the total entropy change caused by the error-correction procedure. We conclude that error-correction can only ever result in a net *increase* in total entropy, with any decrease in system entropy due to error-correction being paid for with entropy production when the error syndrome produced during error-correction is erased.

Exercise 12.16: Show that in the case where R perfectly corrects $\mathcal{E}$ for the input ρ, the inequality

$$S(\rho) - S(\rho') + S(\rho', R) \geq 0 \tag{12.153}$$

must actually be satisfied with equality.

12.5 Entanglement as a physical resource

Thus far our study of quantum information has been focused on resources that are not too far distant from the resources considered in classical information theory. For your convenience Figure 12.10 summarizes many of these results in both their quantum and classical guises. One of the delights of quantum computation and quantum information is that quantum mechanics also contains essentially new types of resource that differ vastly from the sort of resource traditionally regarded as information in classical information theory. Perhaps the best understood of these is quantum entanglement, and it is to this resource that we now turn.

We say 'best understood', but that is not saying a whole lot! We are a long way from having a general theory of quantum entanglement. Nevertheless, some encouraging progress towards such a general theory has been made, revealing an intriguing structure to the entangled states, and some quite remarkable connections between the properties of noisy quantum channels and various types of entanglement transformation. We are just going to take a quick peek at what is known, focusing on the transformation properties of entanglement distributed between two systems ('bi-partite' entanglement), Alice and Bob. There is, of course, a great deal of interest in developing a general theory of entanglement for multi-partite systems, but how to do this is not well understood.

Information Theory

Classical

Shannon entropy

$$H(X) = -\sum_x p(x) \log p(x)$$

Quantum

von Neumann entropy

$$S(\rho) = -\text{tr}(\rho \log \rho)$$

Distinguishability and accessible information

Letters always distinguishable

$$N = |X|$$

Holevo bound

$$H(X:Y) \leq S(\rho) - \sum_x p_x S(\rho_x)$$

$$\rho = \sum_x p_x \rho_x$$

Noiseless channel coding

Shannon's theorem

$$n_{\text{bits}} = H(X)$$

Schumacher's theorem

$$n_{\text{qubits}} = S\left(\sum_x p_x \rho_x\right)$$

Capacity of noisy channels for classical information

Shannon's noisy coding
theorem

$$C(\mathcal{N}) = \max_{p(x)} H(X:Y)$$

Holevo–Schumacher–Westmoreland
theorem

$$C^{(1)}(\mathcal{E}) = \max_{\{p_x, \rho_x\}} \left[S(\rho') - \sum_x p_x S(\rho'_x) \right]$$

$$\rho'_x = \mathcal{E}(\rho_x), \quad \rho' = \sum_x p_x \rho'_x$$

Information-theoretic relations

Fano inequality

$$H(p_e) + p_e \log(|X| - 1) \geq H(X|Y)$$

Mutual information

$$H(X:Y) = H(Y) - H(Y|X)$$

Data processing inequality

$$X \to Y \to Z$$

$$H(X) \geq H(X:Y) \geq H(X:Z)$$

Quantum Fano inequality

$$H(F(\rho, \mathcal{E})) + (1 - F(\rho, \mathcal{E})) \log(d^2 - 1) \geq S(\rho, \mathcal{E})$$

Coherent information

$$I(\rho, \mathcal{E}) = S(\mathcal{E}(\rho)) - S(\rho, \mathcal{E})$$

Quantum data processing inequality

$$\rho \to \mathcal{E}_1(\rho) \to (\mathcal{E}_2 \circ \mathcal{E}_1)(\rho)$$

$$S(\rho) \geq I(\rho, \mathcal{E}_1) \geq I(\rho, \mathcal{E}_2 \circ \mathcal{E}_1)$$

Figure 12.10. Summary of some important classical information relations, and quantum analogues of those relations.

12.5.1 Transforming bi-partite pure state entanglement

The starting point for our investigation is the following simple question: given that Alice and Bob share an entangled pure state $|\psi\rangle$, into what other types of entanglement $|\varphi\rangle$ may they transform $|\psi\rangle$, given that they can each perform *arbitrary* operations on their local systems, including measurement, but can only communicate using classical communication? No quantum communication between Alice and Bob is allowed, constraining the class of transformations they may achieve.

As an example, imagine that Alice and Bob share an entangled pair of qubits in the Bell state $(|00\rangle + |11\rangle)/\sqrt{2}$. Alice performs a two outcome measurement described by measurement operators M_1 and M_2:

$$M_1 = \begin{bmatrix} \cos\theta & 0 \\ 0 & \sin\theta \end{bmatrix} ; \quad M_2 = \begin{bmatrix} \sin\theta & 0 \\ 0 & \cos\theta \end{bmatrix} . \tag{12.154}$$

After the measurement the state is either $\cos\theta|00\rangle + \sin\theta|11\rangle$ or $\cos\theta|11\rangle + \sin\theta|00\rangle$, depending on the measurement outcome, 1 or 2. In the latter case, Alice applies a NOT gate after the measurement, resulting in the state $\cos\theta|01\rangle + \sin\theta|10\rangle$. She then sends the measurement result (1 or 2) to Bob, who does nothing to the state if the measurement result is 1, and performs a NOT gate if the result is 2. The final state of the joint system is therefore $\cos\theta|11\rangle + \sin\theta|00\rangle$, regardless of the measurement outcome obtained by Alice. That is, Alice and Bob have transformed their initial entangled resource $(|00\rangle + |11\rangle)/\sqrt{2}$ into the state $\cos\theta|00\rangle + \sin\theta|11\rangle$ using only local operations on their individual systems, and classical communication.

It is perhaps not immediately obvious what is the significance of the problem of entanglement transformation. There is a certain intrinsic interest to the class of transformations we are allowing – local operations and classical communication (LOCC) – however, it is by no means clear *a priori* that this is truly an interesting problem. It turns out, however, that generalizations of this entanglement transformation problem exhibit deep and unexpected connections to quantum error-correction. Furthermore, the techniques introduced in the solution to the problem are of quite considerable interest, and give unexpected insight into the properties of entanglement. In particular, we will discover a close connection between entanglement and the theory of *majorization*, an area of mathematics that actually predates quantum mechanics!

Before jumping into the study of entanglement transformation, let's first acquaint ourselves with a few relevant facts about majorization. Majorization is an *ordering* on d-dimensional real vectors intended to capture the notion that one vector is more or less disordered than another. More precisely, suppose $x = (x_1, \ldots, x_d)$ and $y = (y_1, \ldots, y_d)$ are two d-dimensional vectors. We use the notation $x^\downarrow$ to mean x re-ordered so the components are in decreasing order so, for example, $x_1^\downarrow$ is the largest component of x. We say x is *majorized* by y, written $x \prec y$, if $\sum_{j=1}^{k} x_j^\downarrow \le \sum_{j=1}^{k} y_j^\downarrow$ for $k = 1, \ldots, d$, with equality instead of inequality when $k = d$. What this definition has to do with notions of disorder will become clear shortly!

The connection between majorization and entanglement transformation is easily stated yet rather surprising. Suppose $|\psi\rangle$ and $|\varphi\rangle$ are states of the joint Alice–Bob system. Define $\rho_\psi \equiv \mathrm{tr}_B(|\psi\rangle\langle\psi|), \rho_\varphi \equiv \mathrm{tr}_B(|\varphi\rangle\langle\varphi|)$ to be the corresponding reduced density matrices of Alice's system, and let λ_ψ and λ_φ be the vectors whose entries are the eigenvalues of ρ_ψ and ρ_φ. We will show that $|\psi\rangle$ may be transformed to $|\varphi\rangle$ by LOCC if and only if $\lambda_\psi \prec \lambda_\varphi$! To demonstrate this we need first a few simple facts about majorization.

Exercise 12.17: Show that $x \prec y$ if and only if for all real t,
$$\sum_{j=1}^{d} \max(x_j - t, 0) \leq \sum_{j=1}^{d} \max(y_j - t, 0), \text{ and } \sum_{j=1}^{d} x_j = \sum_{j=1}^{d} y_j.$$

Exercise 12.18: Use the previous exercise to show that the set of x such that $x \prec y$ is convex.

The following proposition gives a more intuitive meaning to the notion of majorization, showing that $x \prec y$ if and only if x can be written as a convex combination of permutations of y. Intuitively, therefore, $x \prec y$ if x is more disordered than y in the sense that x can be obtained by permuting the elements of y and mixing the resulting vectors. This representation theorem is one of the most useful results in the study of majorization.

Proposition 12.11: $x \prec y$ if and only if $x = \sum_j p_j P_j y$ for some probability distribution p_j and permutation matrices P_j.

Proof
Suppose $x \prec y$. Without loss of generality we may suppose $x = x^{\downarrow}$ and $y = y^{\downarrow}$. We will prove that $x = \sum_j p_j P_j y$ by induction on the dimension d. For $d = 1$ the result is clear. Suppose x and y are $d + 1$-dimensional vectors such that $x \prec y$. Then $x_1 \leq y_1$. Choose j such that $y_j \leq x_1 \leq y_{j-1}$, and define t in the range $[0, 1]$ such that $x_1 = ty_1 + (1 - t)y_j$. Define a convex combination of permutations $D \equiv tI + (1 - t)T$, where T is the permutation matrix which transposes the 1st and jth matrix elements. Then

$$Dy = (x_1, y_2, \ldots, y_{j-1}, (1 - t)y_1 + ty_j, y_{j+1}, \ldots, y_{d+1}). \tag{12.155}$$

Define $x' \equiv (x_2, \ldots, x_{d+1})$ and $y' \equiv (y_2, \ldots, y_{j-1}, (1 - t)y_1 + ty_j, y_{j+1}, \ldots, y_{d+1})$. In Exercise 12.19 on this page you show $x' \prec y'$, so by the inductive hypothesis $x' = \sum_j p'_j P'_j y'$ for probabilities p'_j and permutation matrices P'_j, whence $x = \left(\sum_j p'_j P'_j \right) Dy$, where the P'_j are extended to $d+1$ dimensions by acting trivially on the first entry. Since $D = (tI + (1 - t)T)$, and a product of permutation matrices is a permutation matrix, the result follows.

Exercise 12.19: Verify that $x' \prec y'$.

Conversely, suppose $x = \sum_j p_j P_j y$. It is clear that $P_j y \prec y$ and by Exercise 12.18 it follows that $x = \sum_j p_j P_j y \prec y$. □

Matrices which are convex combinations of permutation matrices have many interesting properties. Notice, for example, that the entries of such a matrix must be non-negative, and that each of the rows and columns must sum to one. A matrix with these properties is known as a *doubly stochastic* matrix, and there is a result known as *Birkhoff's theorem* which implies that the doubly stochastic matrices correspond exactly to the set of matrices which can be written as convex combinations of permutation matrices. We won't prove Birkhoff's theorem here (see the end of chapter 'History and further reading'), but merely state it:

Theorem 12.12: (**Birkhoff's theorem**) A d by d matrix D is doubly stochastic (that is, has non-negative entries and each row and column sums to 1) if and only if D can be written as a convex combination of permutation matrices, $D = \sum_j p_j P_j$.

From Birkhoff's theorem and Proposition 12.11 it follows that $x \prec y$ if and only if $x = Dy$ for some doubly stochastic D. This result allows us to prove a striking and useful operator generalization of Proposition 12.11. Suppose H and K are two Hermitian operators. Then we say $H \prec K$ if $\lambda(H) \prec \lambda(K)$, where we use $\lambda(H)$ to denote the vector of eigenvalues of a Hermitian operator H. Then we have:

Theorem 12.13: Let H and K be Hermitian operators. Then $H \prec K$ if and only if there is a probability distribution p_j and unitary matrices U_j such that

$$H = \sum_j p_j U_j K U_j^\dagger. \tag{12.156}$$

Proof

Suppose $H \prec K$. Then $\lambda(H) = \sum_j p_j P_j \lambda(K)$ by Proposition 12.11. Let $\Lambda(H)$ denote the diagonal matrix whose entries are the eigenvalues of H. Then the vector equation $\lambda(H) = \sum_j p_j P_j \lambda(K)$ may be re-expressed as

$$\Lambda(H) = \sum_j p_j P_j \Lambda(K) P_j^\dagger. \tag{12.157}$$

But $H = V \Lambda(H) V^\dagger$ and $\Lambda(K) = W K W^\dagger$ for some unitary matrices V and W, giving $H = \sum_j p_j U_j K U_j^\dagger$ where $U_j \equiv V P_j W$ is a unitary matrix, completing the proof in the forward direction.

Conversely, suppose $H = \sum_j p_j U_j K U_j^\dagger$. Similarly to before, this is equivalent to $\Lambda(H) = \sum_j p_j V_j \Lambda(K) V_j^\dagger$ for some unitary matrices V_j. Writing the matrix components of V_j as $V_{j,kl}$, we have:

$$\lambda(H)_k = \sum_{jl} p_j V_{j,kl} \lambda(K)_l V_{j,lk}^\dagger = \sum_{jl} p_j |V_{j,kl}|^2 \lambda(K)_l. \tag{12.158}$$

Define a matrix D with entries $D_{kl} \equiv \sum_j p_j |V_{j,kl}|^2$ so we have $\lambda(H) = D\lambda(K)$. The entries of D are non-negative by definition, and the rows and columns of D all sum to one because the rows and columns of the unitary matrices V_j are unit vectors, so D is doubly stochastic and thus $\lambda(H) \prec \lambda(K)$. $\qquad\square$

We now have in hand all the facts about majorization needed in the study of LOCC transformations of bipartite pure state entanglement. The first step of the argument is to reduce the problem from the study of general protocols, which may involve two way classical communication, to protocols involving only one-way classical communication.

Proposition 12.14: Suppose $|\psi\rangle$ can be transformed to $|\varphi\rangle$ by LOCC. Then this transformation can be achieved by a protocol involving just the following steps: Alice performs a single measurement described by measurement operators M_j, sends the result j to Bob, who performs a unitary operation U_j on his system.

Proof

Without loss of generality we may suppose the protocol consists of Alice performing a measurement, sending the result to Bob, who performs a measurement (whose nature may depend on the information received from Alice), and sends the result back to Alice, who performs a measurement... and so on. The idea of the proof is simply to show that the effect of any measurement Bob can do may be simulated by Alice (with one small caveat) so all Bob's actions can actually be replaced by actions by Alice! To see that this is the case, imagine Bob performs a measurement with measurement operators M_j on a pure state $|\psi\rangle$. Suppose this pure state has Schmidt decomposition $|\psi\rangle = \sum_l \sqrt{\lambda_l}|l_A\rangle|l_B\rangle$ and define operators N_j on Alice's system to have a matrix representation with respect to Alice's Schmidt basis which is the same as the matrix representation of Bob's operators M_j with respect to his Schmidt basis. That is, if $M_j = \sum_{kl} M_{j,kl}|k_B\rangle\langle l_B|$ then we define

$$N_j \equiv \sum_{kl} M_{j,kl}|k_A\rangle\langle l_A|. \tag{12.159}$$

Suppose Bob performs the measurement defined by the measurement operators M_j. Then the post-measurement state is $|\psi_j\rangle \propto M_j|\psi\rangle = \sum_{kl} M_{j,kl}\sqrt{\lambda_l}|l_A\rangle|k_B\rangle$, with probability $\sum_{kl} \lambda_l|M_{j,kl}|^2$. On the other hand, if Alice had measured N_j then the post-measurement state is $|\varphi_j\rangle \propto N_j|\psi\rangle = \sum_{kl} M_{j,kl}\sqrt{\lambda_l}|k_A\rangle|l_B\rangle$, also with probability $\sum_{kl} \lambda_l|M_{j,kl}|^2$. Note furthermore that $|\psi_j\rangle$ and $|\varphi_j\rangle$ are the same states up to interchange of Alice and Bob's systems via the map $|k_A\rangle \leftrightarrow |k_B\rangle$, and therefore must have the same Schmidt components. It follows from Exercise 2.80 on page 111 that there exists a unitary U_j on Alice's system and V_j on Bob's system such that $|\psi_j\rangle = (U_j \otimes V_j)|\varphi_j\rangle$. Therefore, Bob performing a measurement described by measurement operators M_j is equivalent to Alice performing the measurement described by measurement operators $U_j N_j$ followed by Bob performing the unitary transformation V_j. Summarizing, a measurement by Bob on a known pure state can be simulated by a measurement by Alice, up to a unitary transformation by Bob.

Imagine then that Alice and Bob engage in a multi-round protocol transforming $|\psi\rangle$ to $|\varphi\rangle$. Without loss of generality, we may suppose that the first round of the protocol consists of Alice performing a measurement and sending the result to Bob. The second round consists of Bob performing a measurement (perhaps with the type of measurement determined by the result of the first) and sending the result to Alice. Instead, however, we can suppose that this measurement is simulated by a measurement performed by Alice, up to a unitary transformation by Bob. Indeed, we can replace all the measurements by Bob and communication from Bob to Alice by measurements by Alice, with a unitary to be done by Bob conditional on Alice's measurement result. Finally, all the measurements performed by Alice can be combined into one single measurement (Exercise 2.57 on page 86), whose result determines a unitary transformation to be performed by Bob; the net effect of this protocol is exactly the same as the original protocol with two-way communication. □

Theorem 12.15: A bipartite pure state $|\psi\rangle$ may be transformed to another pure state $|\varphi\rangle$ by LOCC if and only if $\lambda_\psi \prec \lambda_\varphi$.

Proof
Suppose $|\psi\rangle$ may be transformed to $|\varphi\rangle$ by LOCC. By Proposition 12.14 we may assume that the transformation is effected by Alice performing a measurement with measurement operators M_j, then sending the result to Bob, who performs a unitary transformation U_j. From Alice's point of view she starts with the state ρ_ψ and ends with the state ρ_φ regardless of the measurement outcome, so we must have

$$M_j \rho_\psi M_j^\dagger = p_j \rho_\varphi, \tag{12.160}$$

where p_j is the probability of outcome j. Polar decomposing $M_j\sqrt{\rho_\psi}$ implies that there exists a unitary V_j such that

$$M_j\sqrt{\rho_\psi} = \sqrt{M_j\rho_\psi M_j^\dagger}V_j = \sqrt{p_j\rho_\varphi}V_j. \tag{12.161}$$

Premultiplying this equation by its adjoint gives

$$\sqrt{\rho_\psi}M_j^\dagger M_j\sqrt{\rho_\psi} = p_j V_j^\dagger \rho_\varphi V_j. \tag{12.162}$$

Summing on j and using the completeness relation $\sum_j M_j^\dagger M_j = I$ gives

$$\rho_\psi = \sum_j p_j V_j^\dagger \rho_\varphi V_j, \tag{12.163}$$

whence $\lambda_\psi \prec \lambda_\varphi$ by Theorem 12.13.

The proof of the converse is essentially to run the proof in the forwards direction backwards. Suppose $\lambda_\psi \prec \lambda_\varphi$, so $\rho_\psi \prec \rho_\varphi$ and by Theorem 12.13 there exist probabilities p_j and unitary operators U_j such that $\rho_\psi = \sum_j p_j U_j \rho_\varphi U_j^\dagger$. Assume for now that ρ_ψ is invertible (this assumption is easily removed; see Exercise 12.20) and define operators M_j for Alice's system by

$$M_j\sqrt{\rho_\psi} \equiv \sqrt{p_j\rho_\varphi}U_j^\dagger. \tag{12.164}$$

To see that these operators define a measurement we need to check the completeness relation. We have $M_j = \sqrt{p_j\rho_\varphi}U_j^\dagger\rho_\psi^{-1/2}$ and thus

$$\sum_j M_j^\dagger M_j = \rho_\psi^{-1/2}\left(\sum_j p_j U_j \rho_\varphi U_j^\dagger\right)\rho_\psi^{-1/2} = \rho_\psi^{-1/2}\rho_\psi\rho_\psi^{-1/2} = I, \tag{12.165}$$

which is the completeness relation. Suppose Alice performs the measurement described by the operators M_j, obtaining the outcome j and the corresponding state $|\psi_j\rangle \propto M_j|\psi\rangle$. Let ρ_j denote Alice's reduced density matrix corresponding to the state $|\psi_j\rangle$, so

$$\rho_j \propto M_j\rho_\psi M_j^\dagger = p_j\rho_\varphi, \tag{12.166}$$

by substitution of (12.164), and thus $\rho_j = \rho_\varphi$. It follows by Exercise 2.81 that Bob may convert $|\psi_j\rangle$ into $|\varphi\rangle$ by application of a suitable unitary transformation V_j. $\square$

Exercise 12.20: Show that the assumption that ρ_ψ is invertible may be removed from the proof of the converse part of Theorem 12.15.

Exercise 12.21: (Entanglement catalysis) Suppose Alice and Bob share a pair of four level systems in the state $|\psi\rangle = \sqrt{0.4}|00\rangle + \sqrt{0.4}|11\rangle + \sqrt{0.1}|22\rangle + \sqrt{0.1}|33\rangle$. Show that it is not possible for them to convert this state by LOCC to the state $|\varphi\rangle = \sqrt{0.5}|00\rangle + \sqrt{0.25}|11\rangle + \sqrt{0.25}|22\rangle$. Imagine, however, that a friendly bank is willing to offer them the loan of a *catalyst*, an entangled pair of qubits in the state $|c\rangle = \sqrt{0.6}|00\rangle + \sqrt{0.4}|11\rangle$. Show that it is possible for Alice and Bob to convert the state $|\psi\rangle|c\rangle$ to $|\varphi\rangle|c\rangle$ by local operations and classical communication, returning the catalyst $|c\rangle$ to the bank after the transformation is complete.

Exercise 12.22: (Entanglement conversion without communication) Suppose Alice and Bob are trying to convert a pure state $|\psi\rangle$ into a pure state $|\varphi\rangle$ using local operations only – no classical communication. Show that this is possible if and only if $\lambda_\psi \cong \lambda_\varphi \otimes x$, where x is some real vector with non-negative entries summing to 1, and '$\cong$' means that the vectors on the left and the right have identical non-zero entries.

12.5.2 Entanglement distillation and dilution

Suppose that instead of being supplied with a single copy of a state $|\psi\rangle$ Alice and Bob are supplied with a large number of copies. What types of entanglement transformation can they accomplish with all these copies? We are going to focus on two particular types of entanglement transformation, known as *entanglement distillation* and *entanglement dilution*. The idea of entanglement distillation is for Alice and Bob to convert some large number of copies of a known pure state $|\psi\rangle$ into as many copies of the Bell state $(|00\rangle + |11\rangle)/\sqrt{2}$ as possible using local operations and classical communication, requiring not that they succeed exactly, but only with high fidelity. Entanglement dilution is the reverse process of using LOCC to convert a large number of copies of the Bell state $(|00\rangle + |11\rangle)/\sqrt{2}$ into copies of $|\psi\rangle$, again with high fidelity in the limit where a large number of copies of the Bell state are initially available.

What motivates the study of entanglement distillation and dilution? Suppose we take seriously the idea that entanglement is a physical resource and that as such it should be possible to quantify entanglement, much as we quantify other physical resources such as energy or entropy. Suppose we decide to pick the Bell state $(|00\rangle + |11\rangle)/\sqrt{2}$ as our *standard unit* of entanglement – the basic measure, rather like the standard kilogram or the standard meter. We can associate a measure of entanglement to a quantum state $|\psi\rangle$ in a similar way to the way we associate a mass to an object. Suppose, for instance, that it takes 15 chocolate biscuits of a particular brand to attain a mass equivalent to the standard kilogram; we say that the chocolate biscuits have a mass of 1/15th of a kilogram. Now, strictly speaking, if the chocolate biscuits had a mass of 1/14.8 kilograms we'd be in a bit of trouble, because no integer number of chocolate biscuits is going to balance the standard kilogram, and it's not so obvious how to define a non-integer number of chocolate biscuits. Fortunately, what we do is notice that 148 chocolate biscuits exactly balances 10 standard kilograms, so the mass of the chocolate biscuits is 10/148 kilograms. But what if the real mass is not 1/14.8 kilograms, but is something even more esoteric, like 1/14.7982 ... kilograms? Well, we simply go to the limit of balancing a large number m of chocolate biscuits with another large number n of standard kilograms, and declare the mass of a chocolate biscuit to be the limiting ratio n/m as both m and n become very large.

In a similar way, a potential approach to defining the amount of entanglement present in a pure state $|\psi\rangle$ is to imagine that we are given a large number n of Bell states $(|00\rangle + |11\rangle)/\sqrt{2}$, and are asked to produce as many (high-fidelity) copies of $|\psi\rangle$ as possible using local operations and classical communication. If the number of copies of $|\psi\rangle$ that can be produced is m then we define the limiting ratio of n/m to be the *entanglement of formation* of the state $|\psi\rangle$. Alternately, we might imagine performing the process in reverse, going from m copies of $|\psi\rangle$ to n copies of $(|00\rangle + |11\rangle)/\sqrt{2}$ using LOCC, and defining the limiting ratio n/m to be the *distillable entanglement* of the state $|\psi\rangle$. It is by no means obvious that these two definitions give the same number for the quantity of entanglement; we will see that for pure states $|\psi\rangle$ the entanglement of formation and the distillable entanglement are in fact exactly the same!

Let's take a look at a simple protocol for entanglement dilution, and another for entanglement distillation. Suppose an entangled state $|\psi\rangle$ has Schmidt decomposition

$$|\psi\rangle = \sum_x \sqrt{p(x)}|x_A\rangle|x_B\rangle. \tag{12.167}$$

We write the squared Schmidt co-efficients $p(x)$ in the form we usually reserve for probability distributions, both because the co-efficients satisfy the usual rules of probability distributions (non-negative and sum to one), and because ideas from probability theory turn out to be useful in understanding entanglement distillation and dilution. The m-fold tensor product $|\psi\rangle^{\otimes m}$ may be written

$$|\psi\rangle^{\otimes m} = \sum_{x_1,x_2,\ldots,x_m} \sqrt{p(x_1)p(x_2)\ldots p(x_m)}|x_{1A}x_{2A}\ldots x_{mA}\rangle|x_{1B}x_{2B}\ldots x_{mB}\rangle. \tag{12.168}$$

Suppose that we defined a new quantum state $|\varphi_m\rangle$ by omitting all those terms $x_1, \ldots, x_m$ which are not ϵ-typical, in the sense defined in Section 12.2.1:

$$|\varphi_m\rangle \equiv \sum_{x\ \epsilon-\text{typical}} \sqrt{p(x_1)p(x_2)\ldots p(x_m)}|x_{1A}x_{2A}\ldots x_{mA}\rangle|x_{1B}x_{2B}\ldots x_{mB}\rangle. \tag{12.169}$$

The state $|\varphi_m\rangle$ is not quite a properly normalized quantum state; to normalize it we define $|\varphi'_m\rangle \equiv |\varphi_m\rangle/\sqrt{\langle\varphi_m|\varphi_m\rangle}$. By part 1 of the theorem of typical sequences the fidelity $F(|\psi\rangle^{\otimes m}, |\varphi'_m\rangle) \to 1$ as $m \to \infty$. Furthermore, by part 2 of the theorem of typical sequences the number of terms in the sum (12.169) is at most $2^{m(H(p(x))+\epsilon)} = 2^{m(S(\rho_\psi)+\epsilon)}$, where ρ_ψ is the result of tracing out Bob's part of $|\psi\rangle$.

Suppose then that Alice and Bob are in joint possession of $n = m(S(\rho_\psi) + \epsilon)$ Bell states. Alice prepares 'both parts' of $|\varphi'_m\rangle$ locally, and then uses the Bell states shared with Bob to teleport what should be Bob's half of the state $|\varphi'_m\rangle$ over to Bob. In this way, Alice and Bob can dilute their n Bell states to obtain $|\varphi'_m\rangle$, a pretty good approximation to $|\psi\rangle^{\otimes m}$. This entanglement dilution procedure has $n = m(S(\rho_\psi) + \epsilon)$, so the ratio n/m tends to $S(\rho_\psi) + \epsilon$. We may choose ϵ as small as we like, so we conclude that the entanglement of formation for the state $|\psi\rangle$ is no larger than $S(\rho_\psi)$, since we have just shown that (asymptotically) $S(\rho_\psi)$ Bell states may be converted into a single copy of $|\psi\rangle$.

An entanglement distillation protocol for converting copies of $|\psi\rangle$ into Bell states also follows along similar lines. Suppose that Alice and Bob are in possession of m copies of $|\psi\rangle$. By performing a measurement onto the ϵ-typical subspace of ρ_ψ Alice may, with high fidelity, convert the state $|\psi\rangle^{\otimes m}$ into the state $|\varphi'_m\rangle$. The largest Schmidt co-efficient appearing in $|\varphi_m\rangle$ is at most $2^{-m(S(\rho_\psi)-\epsilon)}$ by the definition of typical sequences. The renormalized state $|\varphi'_m\rangle$ has Schmidt co-efficients at most a factor $1/\sqrt{(1-\delta)}$ larger, since the theorem of typical sequences tells us that $1 - \delta$ is a lower bound on the probability that a sequence is ϵ-typical, and may be made arbitrarily close to 1 for sufficiently large m. Thus, the largest eigenvalue of the state $\rho_{\varphi'_m}$ is at most $2^{-m(S(\rho_\psi)-\epsilon)}/(1-\delta)$. Suppose we choose any n such that

$$\frac{2^{-m(S(\rho_\psi)-\epsilon)}}{1-\delta} \le 2^{-n}. \tag{12.170}$$

Then the vector of eigenvalues for $\rho_{\varphi'_m}$ is majorized by the vector $(2^{-n}, 2^{-n}, \ldots, 2^{-n})$, and thus by Theorem 12.15 the state $|\varphi'_m\rangle$ may be converted to n Bell states by local operations and classical communication. Examining (12.170) we see that this is possible provided $n \approx mS(\rho_\psi)$, and thus the entanglement of distillation is at least $S(\rho_\psi)$.

We have exhibited strategies for distilling $|\psi\rangle$ into $S(\rho_\psi)$ Bell states and for diluting

$S(\rho_\psi)$ Bell states into a copy of $|\psi\rangle$. In fact, it is not very difficult to see that the procedures we have described are really the optimal methods for doing entanglement dilution and distillation! Suppose, for example, that a more efficient protocol for entanglement dilution existed, that could dilute $|\psi\rangle$ into $S > S(\rho_\psi)$ Bell states. Then starting with $S(\rho_\psi)$ Bell states Alice and Bob could produce a copy of $|\psi\rangle$ using the protocol already described, and then use the hypothetical protocol to produce S Bell states. Thus, by local operations and classical communication, Alice and Bob have taken $S(\rho_\psi)$ Bell states and turned them into $S > S(\rho_\psi)$ Bell states! It is not very difficult to convince yourself (and see Exercise 12.24) that increasing the number of Bell states present using local operations and classical communication is not possible, so such a hypothetical dilution protocol can not exist. In a similar way we can see that the procedure for entanglement distillation is optimal. Thus, the entanglement of formation and entanglement of distillation for the state $|\psi\rangle$ are the same, and both equal $S(\rho_\psi)$!

Exercise 12.23: Prove that the procedure for entanglement distillation we have described is optimal.

Exercise 12.24: Recall that the Schmidt number of a bi-partite pure state is the number of non-zero Schmidt components. Prove that the Schmidt number of a pure quantum state cannot be increased by local operations and classical communication. Use this result to argue that the number of Bell states shared between Alice and Bob cannot be increased by local operations and classical communication.

We've learned how to transform Bell states of a bipartite quantum system into copies of another entangled state, $|\psi\rangle$, and back again, in an optimal fashion, motivating us to define the *amount* of entanglement present in that quantum state to be the number of Bell states into which copies of $|\psi\rangle$ may be inter-converted, that is, $S(\rho_\psi)$. What do we learn from this definition? Below we will see that it is possible to obtain some interesting insights into quantum error-correction by further generalizing the notion of distillable entanglement. Nevertheless, at the time of writing it seems fair to say that the study of entanglement is in its infancy, and it is not yet entirely clear what advances in our understanding of quantum computation and quantum information can be expected as a result of the study of quantitative measures of entanglement. We have a reasonable understanding of the properties of pure states of bi-partite quantum systems, but a very poor understanding of systems containing three or more components, or even of mixed states for bi-partite systems. Developing a better understanding of entanglement and connecting that understanding to topics such as quantum algorithms, quantum error-correction and quantum communication is a major outstanding task of quantum computation and quantum information!

12.5.3 Entanglement distillation and quantum error-correction

We defined entanglement distillation for pure states, but there is no reason that definition cannot be extended to mixed states. More precisely, suppose ρ is a general state of a bi-partite quantum system belonging to Alice and Bob. They are supplied with a large number m of copies of these states, and using local operations and classical communication attempt to convert these states to the largest possible number n of Bell states, with high fidelity. The distillable entanglement $D(\rho)$ of ρ is the limiting value of the ratio n/m

for the best possible distillation protocol; for pure states $|\psi\rangle$ we have already shown that $D(|\psi\rangle) = S(\rho_\psi)$, but we do not yet know how to evaluate $D(\rho)$ for a mixed state.

A considerable number of techniques for doing entanglement distillation have been developed, giving lower bounds on the value for $D(\rho)$ for specific classes of states ρ. We won't review these techniques here (see the end of chapter 'History and further reading'). What we will describe is a fascinating connection between distillable entanglement and quantum error-correction.

Imagine that Alice is attempting to send Bob quantum information through a noisy quantum channel $\mathcal{E}$. We suppose the channel is a qubit channel such as the depolarizing channel, although the same basic ideas adapt easily to non-qubit channels. One method for sending quantum information through the channel is as follows. Alice prepares a large number m of Bell states, and sends half of each Bell pair through the channel. Suppose the result of applying $\mathcal{E}$ to half a Bell pair is to create the state ρ, so Alice and Bob end up sharing m copies of ρ. Alice and Bob now perform entanglement distillation, producing $mD(\rho)$ Bell pairs. Alice can now prepare an $mD(\rho)$ qubit state and teleport it to Bob using the $mD(\rho)$ Bell pairs which they share.

Thus, entanglement distillation protocols can be used as a type of error-correction for quantum communications channels between two parties Alice and Bob, enabling Alice to reliably send $mD(\rho)$ qubits of information to Bob, where $D(\rho)$ is the distillable entanglement of ρ, the state that results when one half of a Bell pair is sent through the noisy channel $\mathcal{E}$ connecting Alice and Bob.

What is truly remarkable is that this method of communication using entanglement distillation may sometimes work even when conventional quantum error-correction techniques fail. For example, for the depolarizing channel with $p = 3/4$ we saw in Section 12.4.3 that no quantum information may be transmitted through the channel. However, entanglement distillation protocols are known which can produce a non-zero rate of transmission $D(\rho)$ even for this channel! The reason this is possible is because the entanglement distillation protocols allow classical communication back and forth between Alice and Bob, whereas conventional quantum error-correction does not allow any such classical communication.

This example allows us to explain the claim we made way back in Chapter 1, illustrated in Figure 12.11, that there are channels with zero capacity for quantum information which, when one such channel is connecting Alice to Bob, and another connects Bob to Alice, can be used to achieve a net flow of quantum information! The way this is done is very simple and is based on entanglement distillation. Now, in order that entanglement distillation be possible, we need Alice and Bob to be able to communicate classically, so we set aside half the forward uses of the channel, and all the backward uses of the channel to be used for the transmission of classical information used by the distillation protocol; these channels have non-zero rate for classical information transmission by the HSW theorem. The remaining forward uses of the channel are used to transmit halves of Bell pairs from Alice to Bob, with entanglement distillation being used to extract good Bell pairs out of the resulting states, and then teleportation with the good Bell pairs to achieve a net transmission of quantum information, providing yet another vivid demonstration of the remarkable properties of quantum information!

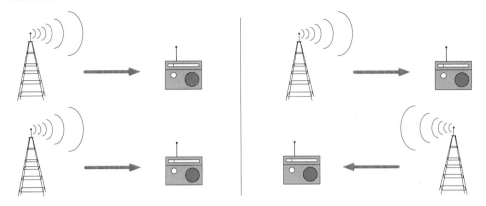

Figure 12.11. Classically, if we have two very noisy channels of zero capacity running side by side, then the combined channel has zero capacity to send information. Not surprisingly, if we reverse the direction of one of the channels, we still have zero capacity to send information. Quantum mechanically, reversing one of the zero capacity channels can actually allow us to send information!

12.6 Quantum cryptography

A fitting conclusion to this chapter is provided by a most remarkable application of quantum information. As we saw in Chapter 5, quantum computers can be used to break some of the best public key cryptosystems. Fortunately, however, what quantum mechanics takes away with one hand, it gives back with the other: a procedure known as *quantum cryptography* or *quantum key distribution* exploits the principles of quantum mechanics to enable provably secure distribution of private information. In this section we describe this procedure, and discuss its security. We begin by explaining the basic ideas of a classical technique, *private key cryptography*, in Section 12.6.1. Private key cryptography is a much older form of cryptography than the public key cryptosystems (mentioned in Chapter 5), and the principles of private key cryptography are used in quantum cryptosystems. Two other important classical techniques, privacy amplification and information reconciliation, which are also used in the quantum systems, are described in Section 12.6.2. Three different protocols for quantum key distribution are then presented in Section 12.6.3. How secure are these protocols? It turns out, as we see in Section 12.6.4, that the *coherent information*, a measure of quantum information we first encountered in Section 12.4.1, gives an information-theoretic lower bound on the in-principle ability to use a quantum communication channel to send private information! This suggests that ideas of quantum information may be useful in proving the security of specific quantum key distribution protocols, and indeed, they are: we conclude the chapter in Section 12.6.5 with a sketch of how the theory of quantum error-correction provides proof of the security of quantum cryptography.

12.6.1 Private key cryptography

Until the invention of public key cryptography in the 1970s, all cryptosystems operated on a different principle, now known as *private key cryptography*. In a private key cryptosystem, if Alice wishes to send messages to Bob then Alice must have an *encoding key*, which allows her to encrypt her message, and Bob must have a matching *decoding key*, which allows Bob to decrypt the encrypted message. A simple, yet highly effective private key cryptosystem is the *Vernam cipher*, sometimes called a *one time pad*. Alice and Bob begin with n-bit secret key strings, which are identical. Alice encodes her n-bit message by adding the message and key together, and Bob decodes by subtracting to invert the encoding, as illustrated in Figure 12.12.

The great feature of this system is that as long as the key strings are truly secret, it is provably secure. That is, when the protocol used by Alice and Bob succeeds, it does so with arbitrarily high probability (an eavesdropper Eve can always jam the communication channel, but Alice and Bob can detect this jamming and declare failure). And for any eavesdropping strategy employed by Eve, Alice and Bob can guarantee that Eve's mutual information with their unencoded message can be made as small as desired. In contrast, the security of public key cryptography (Appendix 5) relies on unproven mathematical assumptions about the difficulty of solving certain problems like factoring (with classical computers!), even though it is widely used and more convenient.

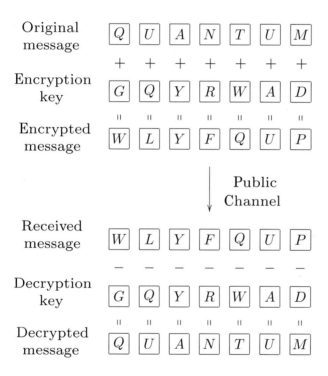

Figure 12.12. The Vernam cipher. Alice encrypts by adding the random key bits (or in this example, letters of the alphabet) to the original message. Bob decrypts by subtracting the key bits to recover the message.

The major difficulty of private key cryptosystems is secure distribution of key bits. In particular, the Vernam cipher is secure only as long as the number of key bits is at least as large as the size of the message being encoded, and key bits cannot be reused! Thus, the large amount of key bits needed makes such schemes impractical for general use. Furthermore, key bits must be delivered in advance, guarded assiduously until used, then destroyed afterwards; otherwise, in principle, such classical information can be copied without disturbing the originals, thus compromising the security of the whole protocol. Despite these drawbacks, private key cryptosystems such as the Vernam cipher continue to be used because of their provable security, with key material delivered by clandestine meetings, trusted couriers, or private secured communication links.

Exercise 12.25: Consider a system with n users, any pair of which would like to be able to communicate privately. Using public key cryptography how many keys are required? Using private key cryptography how many keys are required?

12.6.2 Privacy amplification and information reconciliation

The first step in private key cryptography is distribution of the key string. What if Alice and Bob start out with imperfect keys? Specifically, suppose that Alice and Bob share correlated random classical bit strings X and Y, and they also have an upper bound on Eve's mutual information with X and Y. From these imperfect keys, how can they obtain a good enough key to conduct a secure cryptographic protocol? We now show that by performing two steps, *information reconciliation*, followed by *privacy amplification*, they can systematically increase the correlation between their key strings, while reducing eavesdropper Eve's mutual information about the result, to any desired level of security. These classical steps will be used in the next section, in the quantum key distribution protocol.

Information reconciliation is nothing more than error-correction conducted over a public channel, which reconciles errors between X and Y to obtain a shared bit string W while divulging as little as possible to Eve. After this procedure, suppose Eve has obtained a random variable Z which is partially correlated with W. Privacy amplification is then used by Alice and Bob to distill from W a smaller set of bits S whose correlation with Z is below a desired threshold. Since this last step is conceptually new, let us consider it first.

A detailed proof of why privacy amplification succeeds is beyond the scope of this book, but we will describe the basic method and present the main theorem. One way to accomplish privacy amplification uses the class of *universal hash functions* G, which map the set of n-bit strings A to the set of m-bit strings B, such that for any distinct $a_1, a_2 \in A$, when g is chosen uniformly at random from G, then the probability that $g(a_1) = g(a_2)$ is at most $1/|B|$,

The *collision entropy* of the random variable X with probability distribution $p(x)$ is defined as

$$H_c(X) = - \log \left[\sum_x p(x)^2 \right] . \tag{12.171}$$

(This is sometimes known as the Rènyi entropy of order 2.) Using the concavity of the log function, it is not difficult to show that the Shannon entropy provides an upper bound on this quantity: $H(X) \geq H_c(X)$. H_c is important in the following theorem about universal hash functions:

Theorem 12.16: Let X be a random variable on alphabet $\mathcal{X}$ with probability distribution $p(x)$ and collision entropy $H_c(X)$, and let G be the random variable corresponding to the random choice (with uniform distribution) of a member of the universal class of hash functions from $\mathcal{X}$ to $\{0, 1\}^m$. Then

$$H(G(X)|G) \geq H_c(G(X)|G) \geq m - 2^{m-H_c(X)} . \tag{12.172}$$

Theorem 12.16 can be applied to privacy amplification in the following manner. Alice and Bob publicly select $g \in G$ and each apply it to W, giving a new bit string S, which they choose as their secret key. If Eve's uncertainty about W given her knowledge $Z = z$ (about a specific instance of the protocol) is known in terms of the collision entropy to be bounded below by some number, say $H_c(W|Z = z) > d$, then it follows from Theorem 12.16 that

$$H_c(S|G, Z = z) \geq m - 2^{m-d}. \tag{12.173}$$

In other words, m may be chosen small enough such that $H_c(S|G, Z = z)$ is nearly equal to m. This maximizes Eve's uncertainty about the key S, making it a secure secret.

Information reconciliation further reduces the number of bits that Alice and Bob can obtain, but this can be bounded as follows. By computing a series of parity checks on subsets of her bits X, Alice can compose a (classical) message u consisting of subset specifications and parities, which, when transmitted to Bob, allows him to correct the errors in his string Y, after which both have the same string W. Clearly, this will require sending $k > H(W|Y)$ bits of information in u. However, this procedure gives Eve additional knowledge $U = u$, thus increasing her collision entropy to $H_c(W|Z = z, U = u)$. On average (over possible reconciliation messages u), this increase is bounded below by $H_c(W|Z = z, U = u) \geq H_c(W|Z = z) - H(U)$, where $H(U)$ is the usual Shannon entropy of U, but this bound is too weak, because it implies that the probability that the leaked information $U = u$ decreases H_c by more than $mH(U)$ is only at most $1/m$. A stronger bound is provided by this theorem:

Theorem 12.17: Let X and U be random variables with alphabets $\mathcal{X}$ and $\mathcal{U}$, respectively, where X has probability distribution $p(x)$, and U is jointly distributed with X according to $p(x, u)$. Also let $s > 0$ be an arbitrary parameter. Then, with probability at least $1 - 2^{-s}$, U takes on a value u for which

$$H_c(X|U = u) \geq H_c(X) - 2\log|\mathcal{U}| - 2s. \tag{12.174}$$

Here, s is known as a *security parameter*. Applying this to the reconciliation protocol gives the conclusion that Alice and Bob can choose s such that Eve's collision entropy is bounded from below by $H_c(W|Z = z, U = u) \geq d - 2(k + s)$, with probability better than $1 - 2^{-s}$. Following this step with privacy amplification allows them to distill m secret key bits S, for which Eve's total information is less than $2^{m-d+2(k+s)}$ bits.

CSS code privacy amplification & information reconciliation

As we noted above, information reconciliation is nothing more than error-correction; it turns out that privacy amplification is also intimately related to error-correction, and both tasks can be implemented by using classical codes. This viewpoint provides a simple conceptual picture which will be useful in the proof of the security of quantum key distribution, in Section 12.6.5, since we have a well-developed theory of quantum error-correction codes. With this in mind, it is useful to observe the following.

Decoding from a randomly chosen CSS code (see Section 10.4.2) can be thought of as performing information reconciliation and privacy amplification. Although CSS codes are usually used to encode quantum information, for our present purpose we can just consider their classical properties. Consider two classical linear codes C_1 and C_2 which satisfy the conditions for a t error-correcting $[n, m]$ CSS code: $C_2 \subset C_1$ and C_1 and $C_2^\perp$

both correct t errors. Alice chooses a random n bit string X and transmits it to Bob, who receives Y.

Let us assume that *a priori* it is known that along a communication channel between Alice and Bob, the expected number of errors per code block caused by all noise sources *including eavesdropping* is less than t; in practice, this can be established by random testing of the channel. Furthermore, suppose that Eve knows nothing about the codes C_1 and C_2; this can be ensured by Alice choosing the code randomly. Finally, suppose that Alice and Bob have an upper bound on the mutual information between Eve's data Z, and their own data, X and Y.

Bob receives $Y = X + \epsilon$, where ϵ is some error. Since it is known that less than t errors occurred, if Alice and Bob both correct their states to the nearest codeword in C_1, their results $X', Y' \in C_1$ are identical, $W = X' = Y'$. This step is nothing more than information reconciliation. Of course, Eve's mutual information with W may still be unacceptably large. To reduce this, Alice and Bob identify which of the 2^m cosets of C_2 in C_1 their state W belongs to; that is, that they compute the coset of $W + C_2$ in C_1. The result is their m bit key string, S. By virtue of Eve's lack of knowledge about C_2, and the error-correcting properties of C_2, this procedure can reduce Eve's mutual information with S to an acceptable level, performing privacy amplification.

12.6.3 Quantum key distribution

Quantum key distribution (QKD) is a protocol which is *provably* secure, by which private key bits can be created between two parties over a *public* channel. The key bits can then be used to implement a classical private key cryptosystem, to enable the parties to communicate securely. The only requirement for the QKD protocol is that qubits can be communicated over the public channel with an error rate lower than a certain threshold. The security of the resulting key is guaranteed by the properties of quantum information, and thus is conditioned only on fundamental laws of physics being correct!

The basic idea behind QKD is the following fundamental observation: Eve cannot gain any information from the qubits transmitted from Alice to Bob without disturbing their state. First of all, by the no-cloning theorem (Box 12.1), Eve cannot clone Alice's qubit. Second, we have the following proposition:

Proposition 12.18: (**Information gain implies disturbance**) In any attempt to distinguish between two non-orthogonal quantum states, information gain is only possible at the expense of introducing disturbance to the signal.

Proof

Let $|\psi\rangle$ and $|\varphi\rangle$ be the non-orthogonal quantum states Eve is trying to obtain information about. By the results of Section 8.2, we may assume without loss of generality that the process she uses to obtain information is to unitarily interact the state ($|\psi\rangle$ or $|\varphi\rangle$) with an ancilla prepared in a standard state $|u\rangle$. Assuming that this process does not disturb the states, in the two cases one obtains

$$|\psi\rangle|u\rangle \rightarrow |\psi\rangle|v\rangle \tag{12.175}$$

$$|\varphi\rangle|u\rangle \rightarrow |\varphi\rangle|v'\rangle . \tag{12.176}$$

Eve would like $|v\rangle$ and $|v'\rangle$ to be different so that she can acquire information about the identity of the state. However, since inner products are preserved under unitary transformations, it must be that

$$\langle v|v'\rangle \langle \psi|\varphi\rangle = \langle u|u\rangle \langle \psi|\varphi\rangle \tag{12.177}$$

$$\langle v|v'\rangle = \langle u|u\rangle = 1, \tag{12.178}$$

which implies that $|v\rangle$ and $|v'\rangle$ must be identical. Thus, distinguishing between $|\psi\rangle$ and $|\varphi\rangle$ must inevitably disturb at least one of these states. □

We make use of this idea by transmitting non-orthogonal qubit states between Alice and Bob. By checking for disturbance in their transmitted states, they establish an upper bound on any noise or eavesdropping occurring in their communication channel. These 'check' qubits are interspersed randomly among data qubits (from which key bits are later extracted), so that the upper bound applies to the data qubits as well. Alice and Bob then perform information reconciliation and privacy amplification to distill a shared secret key string. The threshold for the maximum tolerable error rate is thus determined by the efficacy of the best information reconciliation and privacy amplification protocols. Three different QKD protocols which work in this way are presented below.

The BB84 protocol

Alice begins with a and b, two strings each of $(4 + \delta)n$ random classical bits. She then encodes these strings as a block of $(4 + \delta)n$ qubits,

$$|\psi\rangle = \bigotimes_{k=1}^{(4+\delta)n} |\psi_{a_k b_k}\rangle, \tag{12.179}$$

where a_k is the kth bit of a (and similarly for b), and each qubit is one of the four states

$$|\psi_{00}\rangle = |0\rangle \tag{12.180}$$

$$|\psi_{10}\rangle = |1\rangle \tag{12.181}$$

$$|\psi_{01}\rangle = |+\rangle = (|0\rangle + |1\rangle)/\sqrt{2} \tag{12.182}$$

$$|\psi_{11}\rangle = |-\rangle = (|0\rangle - |1\rangle)/\sqrt{2}. \tag{12.183}$$

The effect of this procedure is to encode a in the basis X or Z, as determined by b. Note that the four states are not all mutually orthogonal, and therefore no measurement can distinguish between (all of) them with certainty. Alice then sends $|\psi\rangle$ to Bob, over their public quantum communication channel.

Bob receives $\mathcal{E}(|\psi\rangle\langle\psi|)$, where $\mathcal{E}$ describes the quantum operation due to the combined effect of the channel and Eve's actions. He then publicly announces this fact. At this point, Alice, Bob, and Eve each have their own states described by separate density matrices. Note also that at this point, since Alice hasn't revealed b, Eve has no knowledge of what basis she should have measured in to eavesdrop on the communication; at best, she can only guess, and if her guess was wrong, then she would have disturbed the state received by Bob. Moreover, whereas in reality the noise $\mathcal{E}$ may be partially due to the environment (a poor channel) in addition to Eve's eavesdropping, it doesn't help Eve to have complete control over the channel, so that she is entirely responsible for $\mathcal{E}$.

Of course, Bob also finds $\mathcal{E}(|\psi\rangle\langle\psi|)$ uninformative at this point, because he does not know anything about b. Nevertheless, he goes ahead and measures each qubit in basis X or Z, as determined by a random $(4 + \delta)n$ bit string b' which he creates on his own. Let

Bob's measurement result be a'. After this, Alice publicly announces b, and by discussion over a public channel, Bob and Alice discard all bits in $\{a', a\}$ except those for which corresponding bits of b' and b are equal. Their remaining bits satisfy $a' = a$, since for these bits Bob measured in the same basis Alice prepared in. Note that b reveals nothing about either a, or the bits a' resulting from Bob's measurement, but it is important that Alice not publish b until after Bob announces reception of Alice's qubits. For simplicity in the following explanation, let Alice and Bob keep just $2n$ bits of their result; δ can be chosen sufficiently large so that this can be done with exponentially high probability.

Now Alice and Bob perform some tests to determine how much noise or eavesdropping happened during their communication. Alice selects n bits (of their $2n$ bits) at random, and publicly announces the selection. Bob and Alice then publish and compare the values of these check bits. If more than t bits disagree, then they abort and re-try the protocol from the start. t is selected such that if the test passes, then they can apply information reconciliation and privacy amplification algorithms to obtain m acceptably secret shared key bits from the remaining n bits.

This protocol, known as BB84 after its inventors (see the end of chapter 'History and further reading'), is summarized in Figure 12.13, and an experimental realization is described in Box 12.7. Related versions of this protocol, such as using fewer check bits, are also known by the same name.

The BB84 QKD protocol

1: Alice chooses $(4 + \delta)n$ random data bits.
2: Alice chooses a random $(4 + \delta)n$-bit string b. She encodes each data bit as $\{|0\rangle, |1\rangle\}$ if the corresponding bit of b is 0 or $\{|+\rangle, |-\rangle\}$ if b is 1.
3: Alice sends the resulting state to Bob.
4: Bob receives the $(4 + \delta)n$ qubits, announces this fact, and measures each qubit in the X or Z basis at random.
5: Alice announces b.
6: Alice and Bob discard any bits where Bob measured a different basis than Alice prepared. With high probability, there are at least $2n$ bits left (if not, abort the protocol). They keep $2n$ bits.
7: Alice selects a subset of n bits that will to serve as a check on Eve's interference, and tells Bob which bits she selected.
8: Alice and Bob announce and compare the values of the n check bits. If more than an acceptable number disagree, they abort the protocol.
9: Alice and Bob perform information reconciliation and privacy amplification on the remaining n bits to obtain m shared key bits.

Figure 12.13. The four state quantum key distribution protocol known as BB84.

Exercise 12.26: Let a'_k be Bob's measurement result of qubit $|\psi_{a_k b_k}\rangle$, assuming a noiseless channel with no eavesdropping. Show that when $b'_k \neq b_k$, a'_k is random and completely uncorrelated with a_k. But when $b'_k = b_k$, $a'_k = a_k$.

Exercise 12.27: (Random sampling tests) The random test of n of $2n$ check bits allows Alice and Bob to place an upper bound on the number of errors in their untested bits, with high probability. Specifically, for any $\delta > 0$, the probability of obtaining less than δn errors on the check bits, and more than $(\delta + \epsilon)n$ errors on the remaining n bits is asymptotically less than $\exp[-O(\epsilon^2 n)]$, for large n. We prove this claim here.

(1) Without loss of generality, you may assume that there are μn errors in the $2n$ bits, where $0 \le \mu \le 2$. Now, if there are δn errors on the check bits, and $(\delta + \epsilon)n$ errors on the rest, then $\delta = (\mu - \epsilon)/2$. The two conditional statements in the claim thus imply the following:

$$< \delta n \text{ errors on check bits} \quad \Rightarrow \quad < \delta n \text{ errors on check bits} \quad (12.184)$$

$$> (\delta + \epsilon)n \text{ errors on rest} \quad \Rightarrow \quad > (\mu - \delta)n \text{ errors on rest}, \quad (12.185)$$

and in fact, the top claim on the right implies the bottom one on the right. Using this, show that the probability p which we would like to bound satisfies

$$p < \binom{2n}{n}^{-1} \binom{\mu n}{\delta n} \binom{(2 - \mu)n}{(1 - \delta)n} \delta n . \quad (12.186)$$

(2) Show that for large n, you can bound

$$\frac{1}{an + 1} 2^{anH(b/a)} \le \binom{an}{bn} \le 2^{anH(b/a)}, \quad (12.187)$$

where $H(\cdot)$ is the binary entropy function, Equation (11.8). Apply this to the above bound for p.

(3) Apply the bound $H(x) < 1 - 2(x - 1/2)^2$ to obtain the final result, $p < \exp[-O(\epsilon^2 n)]$. You may replace μ by a constant which expresses the worst possible case.

(4) Compare the result with the Chernoff bound, Box 3.4. Can you come up with a different way to derive an upper bound on p?

The B92 protocol

The BB84 protocol can be generalized to use other states and bases, and similar conclusions hold. In fact, a particularly simple protocol exists in which only two states are used. For simplicity, it is sufficient to consider what happens to a single bit at a time; the description easily generalizes to block tests just as is done in BB84.

Suppose Alice prepares one random classical bit a, and, depending on the result, sends Bob

$$|\psi\rangle = \begin{cases} |0\rangle & \text{if } a = 0 \\ \dfrac{|0\rangle + |1\rangle}{\sqrt{2}} & \text{if } a = 1. \end{cases} \quad (12.188)$$

Depending on a random classical bit a' which he generates, Bob subsequently measures the qubit he receives from Alice in either the Z basis $|0\rangle, |1\rangle$ (if $a' = 0$), or in the X basis $|\pm\rangle = (|0\rangle \pm |1\rangle)/\sqrt{2}$ (if $a' = 1$). From his measurement, he obtains the result b, which is 0 or 1, corresponding to the -1 and $+1$ eigenstates of X and Z. Bob then publicly announces b (but keeps a' secret), and Alice and Bob conduct a public discussion keeping

---

Box 12.7: Experimental quantum cryptography

Quantum key distribution is particularly interesting and astonishing because it is easily experimentally realized. Here is a schematic diagram of one system employing commercial fiber-optic components to deliver key bits over a ten kilometer distance, which has been built at IBM:

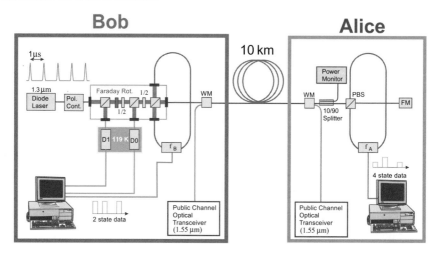

Bob initially generates strong coherent states $|\alpha\rangle$ using a diode laser emitting light at a wavelength of 1.3 μm, and transmits them to Alice, who attenuates them to (approximately) generate a single photon. She also polarizes the photon in one of the four states of the BB84 protocol, using as $|0\rangle$ and $|1\rangle$ states horizontal and vertical polarization. She then returns the photon to Bob, who measures it using a polarization analyzer, in a random basis. By using this special configuration in which the photon traverses the same path twice, the apparatus can be made to autocompensate for imperfections (such as slowly fluctuating path lengths and polarization shifts) along the fiber link. Alice and Bob then select the subset of results in which they used the same basis, reconcile their information, and perform privacy amplification, communicating over a public channel with photons (over the same fiber) of 1.55 μm wavelength. Key bits can be exchanged at the rate of a few hundred per second. Ultimately, improvements in the light source and detector should allow the rate to be improved by a few orders of magnitude. Quantum key distribution over distances exceeding 40 kilometers, and also in installed telecommunication fiber (under Lake Geneva) has been demonstrated.

---

only those pairs $\{a, a'\}$ for which $b = 1$. Note that when $a = a'$, then $b = 0$ always. Only if $a' = 1 - a$ will Bob obtain $b = 1$, and that occurs with probability $1/2$. The final key is a for Alice, and $1 - a'$ for Bob.

This protocol, known as B92 (see the end of chapter 'History and further reading'), highlights how the impossibility of perfect distinction between non-orthogonal states lies at the heart of quantum cryptography. As in BB84, because it is impossible for any eavesdropper to distinguish between Alice's states without disrupting the correlation between the bits Alice and Bob finally keep, this protocol allows Alice and Bob to created shared key bits while also placing an upper bound on the noise and eavesdropping during their communication. They can then apply information reconciliation and privacy amplification to extract secret bits from their resulting correlated random bit strings.

Exercise 12.28: Show that when $b = 1$, then a and a' are perfectly correlated with each other.

Exercise 12.29: Give a protocol using six states, the eigenstates of X, Y, and Z, and argue why it is also secure. Discuss the sensitivity of this protocol to noise and eavesdropping, in comparison with that of BB84 and B92.

The EPR protocol

The key bits generated in the BB84 and B92 protocols may appear to have been originated by Alice. However, it turns out that the key can be seen to arise from a fundamentally random process involving the properties of entanglement. This is illustrated by the following protocol.

Suppose Alice and Bob share a set of n entangled pairs of qubits in the state

$$\frac{|00\rangle + |11\rangle}{\sqrt{2}}. \tag{12.189}$$

These states are known as EPR pairs. Obtaining these states could have come about in many different ways; for example, Alice could prepare the pairs and then send half of each to Bob, or vice versa. Alternatively, a third party could prepare the pairs and send halves to Alice and Bob. Or they could have met a long time ago and shared them, storing them until the present. Alice and Bob then select a random subset of the EPR pairs, and test to see if they violate Bell's inequality (Equation (2.225), on page 115 in Section 2.6), or some other appropriate test of fidelity. Passing the test certifies that they continue to hold sufficiently pure, entangled quantum states, placing a lower bound on the fidelity of the remaining EPR pairs (and thus any noise or eavesdropping). And when they measure these in jointly determined random bases, Alice and Bob obtain correlated classical bit strings from which they can obtain secret key bits as in the B92 and BB84 protocols. Using an argument based on Holevo's bound, the fidelity of their EPR pairs can be used to establish an upper bound on Eve's accessible information about the key bits.

Where do the key bits come from in this EPR protocol? Since it is symmetric – Alice and Bob perform identical tasks on their qubits, even possibly simultaneously – it cannot be said that either Alice or Bob generates the key. Rather, the key is truly random. In fact the same applies to the BB84 protocol, since it can be reduced to an instance of a generalized version of the EPR protocol. Suppose that Alice prepares a random classical bit b, and according to it, measures her half of the EPR pair in either the $|0\rangle, |1\rangle$ basis, or in the basis $|\pm\rangle = (|0\rangle \pm |1\rangle)/\sqrt{2}$, obtaining a. Let Bob do identically, measuring in (his randomly chosen) basis b' and obtaining a'. Now they communicate b and b' over a public classical channel, and keep as their key only those $\{a, a'\}$ for which $b = b'$. Note that this key is *undetermined* until Alice or Bob performs a measurement on their EPR pair half. Similar observations can be made about the B92 protocol. For this reason, quantum cryptography is sometimes thought of not as secret key exchange or transfer, but rather as secret key *generation*, since fundamentally neither Alice nor Bob can pre-determine the key they will ultimately end up with upon completion of the protocol.

12.6.4 Privacy and coherent information

Thus far, we have described the basic protocol for QKD and argued that it is secure, but we have not provided quantitative bounds. How secure is it? It turns out there is an interesting and fundamental connection between the basic quantitative measures of quantum information discussed in this chapter, and the in-principle obtainable security of quantum cryptography, which we describe below.

The quantum coherent information $I(\rho, \mathcal{E})$ gives a lower bound on the ability of a quantum channel to send private information. In the most general circumstance, Alice prepares states ρ_k^A, where $k = 0, 1, \ldots$ denotes the different possible states she may send, each with some probability p_k. Bob receives states $\rho_k^B = \mathcal{E}(\rho_k^A)$ which may be different from ρ_k^A because of channel noise or an eavesdropper, Eve. The mutual information between the result of any measurement Bob may do, and Alice's value k, $H_{\text{bob:alice}}$, is bounded above by Holevo's bound, (12.6),

$$H_{\text{bob:alice}} \leq \chi^B = S(\rho^B) - \sum_k p_k S(\rho_k^B), \qquad (12.190)$$

where $\rho^B = \sum_k p_k \rho_k^B$. Similarly, Eve's mutual information is bounded above,

$$H_{\text{eve:alice}} \leq \chi^E = S(\rho^E) - \sum_k p_k S(\rho_k^E). \qquad (12.191)$$

Since any excess information Bob has relative to Eve (at least above a certain threshold) can in principle be exploited by Bob and Alice to distill a shared secret key using techniques such as privacy amplification, it makes sense to define the quantity

$$\mathcal{P} = \sup \left[H_{\text{bob:alice}} - H_{\text{eve:alice}} \right], \qquad (12.192)$$

as the guaranteed privacy of the channel, where the supremum is taken over all strategies Alice and Bob may employ to use the channel. This is the maximum excess classical information that Bob may obtain relative to Eve about Alice's quantum signal. By the HSW theorem, Alice and Bob can employ a strategy such that $H_{\text{bob:alice}} = \chi^B$, while for any strategy Eve may employ, $H_{\text{eve:alice}} \leq \chi^E$. Thus, $\mathcal{P} \geq \chi^B - \chi^E$, for a suitable choice of strategy.

From Exercise 12.11, it follows that a lower bound on the privacy $\mathcal{P}$ can be obtained by assuming that all of Alice's signal states $\rho_k^A = |\psi_k^A\rangle\langle\psi_k^A|$ are pure states, which are initially unentangled with Eve, who starts out in some state $|0^E\rangle$ (that may also be assumed to be pure, without loss of generality). In general, the channel from Alice to Bob will include interactions with some environment other than Eve, but to give Eve the greatest possible advantage, all such interactions may be attributed to her, such that the final joint state received by Eve and Bob after transmission is

$$|\psi^{EB}\rangle = U|\psi_k^A\rangle|0^E\rangle. \qquad (12.193)$$

Since this is a pure state, the reduced density matrices ρ_k^E and ρ_k^B will have the same non-zero eigenvalues, and thus the same entropies, $S(\rho_k^E) = S(\rho_k^B)$. Thus,

$$\mathcal{P} \geq \chi^B - \chi^E \qquad (12.194)$$

$$= S(\rho^B) - \sum_k p_k S(\rho_k^B) - S(\rho^E) + \sum_k p_k S(\rho_k^E) \qquad (12.195)$$

$$= S(\rho^B) - S(\rho^E) \qquad (12.196)$$

$$= I(\rho, \mathcal{E}). \tag{12.197}$$

That is, a lower bound on the guaranteed privacy of the channel $\mathcal{E}$ is given by the quantum coherent information $I(\rho, \mathcal{E})$, as defined in Equation (12.118). Note that this result is not specific to any protocol (which may have its own security flaws). Also, the protocol must perform tests, which are not considered in this calculation, to actually determine properties of the channel $\mathcal{E}$, after which this bound can be applied. So although the information-theoretic bound we have arrived at here is quite elegant, we still have a way to go before being able to quantify the security of QKD!

12.6.5 The security of quantum key distribution

How secure is quantum key distribution? Because of the inevitability of disturbance of the communicated state, upon information gain by an eavesdropper, we have good reason to believe in the security of QKD. What we need, however, to conclude that the protocol truly is secure, is a *quantifiable* definition of security which explicitly bounds Eve's knowledge about the final key, given some measure of Alice and Bob's effort. The following criterion is acceptable:

> A QKD protocol is defined as being *secure* if, for any security parameters $s > 0$ and $\ell > 0$ chosen by Alice and Bob, and for any eavesdropping strategy, either the scheme aborts, or it succeeds with probability at least $1 - O(2^{-s})$, and guarantees that Eve's mutual information with the final key is less than $2^{-\ell}$. The key string must also be essentially random.

In this final section, we give the main elements of a proof that BB84 is secure. This proof serves as a fitting conclusion to the chapter, because of its elegant use of many concepts of quantum information in providing an argument that is sufficiently simple and transparent so as to be incontestable. The origin of this proof, intriguingly, comes from the observation that after information reconciliation and privacy amplification are performed, the ultimately obtainable key rate turns out to coincide with the achievable qubit transmission rate for CSS codes (Section 10.4.2) over noisy communication channels!

Here is the main idea, in outline. It is relatively straightforward to exhaustively establish that the BB84, B92, and EPR protocols are secure if Eve can only attack the transmission one qubit at a time. The difficulty lies in dealing with the possibility of collective attacks, in which Eve manipulates and possibly stores large blocks of the transmitted qubits. To address this, we need a more general and powerful argument. Suppose we somehow know that Eve never introduces more than t qubit errors per block. Then Alice could encode her qubits in a t error-correcting quantum code, such that all of Eve's meddling could be removed by Bob decoding the code. Two things must be established to make this feasible: first, how can an upper bound be placed on t? This turns out to be possible by sampling the channel in the appropriate manner, leaving us with a protocol which is secure, even against collective attacks! Unfortunately, this protocol generally requires a fault-tolerant quantum computer in order to encode and decode qubits robustly. The second challenge is thus to choose a quantum code such that the full sequence of encoding, decoding, and measurement can be performed using no quantum computation or storage – just single qubit preparation and measurement. Using CSS codes does the trick (after some simplification), and in fact gives just the BB84 protocol. Below, we begin with a manifestly secure EPR pair based QKD protocol, and then apply solutions to the two challenges to systematically simplify the initial protocol into BB84.

Requirements for a secure QKD protocol

Suppose that Alice has n pairs of entangled qubits, each in the state

$$|\beta_{00}\rangle = \frac{|00\rangle + |11\rangle}{\sqrt{2}}. \tag{12.198}$$

Denote this state as $|\beta_{00}\rangle^{\otimes n}$. Alice then transmits half of each pair to Bob; because of noise and eavesdropping on the channel, the resulting state may be impure, and can be described as the density matrix ρ. Alice and Bob then perform local measurements to obtain a key, as described previously. The following Lemma can be used to show that the fidelity of ρ with respect to $|\beta_{00}\rangle^{\otimes n}$ places an upper bound on the mutual information Eve has with the key.

Lemma 12.19: (**High fidelity implies low entropy**) If $F(\rho, |\beta_{00}\rangle^{\otimes n})^2 > 1 - 2^{-s}$,
then $S(\rho) < (2n + s + 1/\ln 2)2^{-s} + O(2^{-2s})$.

Proof
If $F(\rho, |\beta_{00}\rangle^{\otimes n})^2 = {}^{\otimes n}\langle\beta_{00}|\rho|\beta_{00}\rangle^{\otimes n} > 1 - 2^{-s}$, then the largest eigenvalue of ρ must be larger than $1 - 2^{-s}$. Therefore, the entropy of ρ is bounded above by the entropy of a diagonal density matrix $\rho_{\max}$ with diagonal entries $1 - 2^{-s}, 2^{-s}/(2^{2n} - 1), 2^{-s}/(2^{2n} - 1), \ldots, 2^{-s}/(2^{2n} - 1)$, that is, $\rho_{\max}$ has a large entry $1 - 2^{-s}$, and the remaining probability is distributed equally among the remaining $2^{2n} - 1$ entries. Since

$$S(\rho_{\max}) = -(1 - 2^{-s})\log(1 - 2^{-s}) - 2^{-s}\log\frac{2^{-s}}{2^{2n} - 1}, \tag{12.199}$$

the desired result follows. $\square$

By Holevo's bound (12.6), $S(\rho)$ is an upper bound on the information accessible to Eve, resulting from Alice and Bob's measurements of ρ. This implies that if a QKD protocol can provide Alice and Bob with EPR pairs of fidelity at least $1 - 2^{-s}$ (with high probability), then it is secure.

Exercise 12.30: Simplify (12.199) to obtain the expression for $S(\rho)$ given in the statement of the Lemma.

Exercise 12.31: It may be unclear why $S(\rho)$ bounds Eve's mutual information with Alice and Bob's measurement results. Show that this follows from assuming the worst about Eve, giving her *all* the control over the channel.

Random sampling can upper-bound eavesdropping

How can a protocol place a lower bound on the fidelity of Alice and Bob's EPR pairs? The key idea is a classical argument, based on random sampling, which we encountered in the description of the BB84 protocol (Exercise 12.27). Arguments based on classical probability, however, need not apply when considering outcomes of quantum measurements. This is vividly demonstrated by Bell's inequality (Section 2.6). On the other hand, quantum experiments *do* allow a classical interpretation whenever measurement observables that refer to only one basis are considered. And fortunately, it happens that measurements in only one basis are required for Alice and Bob to bound the fidelity of their EPR pairs.

According to (10.14), a qubit transmitted through a noisy quantum channel can be described as having had one of four things happen to it: nothing (I), a bit flip (X), a phase flip (Z), or a combined bit and phase flip (Y). Recall that the Bell basis is defined by the four states

$$|\beta_{00}\rangle = \frac{|00\rangle + |11\rangle}{\sqrt{2}}, \quad |\beta_{10}\rangle = \frac{|00\rangle - |11\rangle}{\sqrt{2}}, \quad |\beta_{01}\rangle = \frac{|01\rangle + |10\rangle}{\sqrt{2}}, \quad |\beta_{11}\rangle = \frac{|01\rangle - |10\rangle}{\sqrt{2}}.$$
(12.200)

Let the second qubit in each pair be the one that Alice sends to Bob. If a bit flip error occurs to this qubit, then $|\beta_{00}\rangle$ is transformed into $|\beta_{01}\rangle$. Similarly, a phase flip gives $|\beta_{10}\rangle$, and a combination of the two errors gives $|\beta_{11}\rangle$ (up to irrelevant overall phases). A natural measurement which detects if a bit flip has occurred is given by projectors $\Pi_{bf} = |\beta_{01}\rangle\langle\beta_{01}| + |\beta_{11}\rangle\langle\beta_{11}|$ and $I - \Pi_{bf}$, and likewise, the measurement described by projectors $\Pi_{pf} = |\beta_{10}\rangle\langle\beta_{10}| + |\beta_{11}\rangle\langle\beta_{11}|$ and $I - \Pi_{pf}$ detects phase flips. Since both these measurements commute with the Bell basis, their outcomes obey classical probability arguments. In fact, any measurement which commutes with the Bell basis will also satisfy the same classical arguments.

More precisely, Alice and Bob and can bound the fidelity of their EPR pairs by randomly sampling a subset of them. Suppose Alice sends $2n$ EPR pair halves to Bob. They subsequently randomly select n of them and check these qubits by jointly measuring either Π_{bf} or Π_{pf} (again, chosen randomly). By the same classical arguments used in the random sampling tests in BB84 (Exercise 12.27), if δn bit or phase flip errors are detected, then the remaining n EPR pairs would be exponentially certain to have the same number of errors, were they also to be measured in the Bell basis.

Bell states are nonlocal, and generally measurements in the Bell basis require nonlocal operations, which can be difficult. Fortunately, however, they are not required in the present scheme, because $\Pi_{bf} = (I \otimes I - Z \otimes Z)/2$, and $\Pi_{pf} = (I \otimes I - X \otimes X)/2$. Thus, Alice and Bob can perform the desired checks with local measurements of Pauli operators, either by both measuring Z, or both measuring X.

Exercise 12.32: Note that the local measurements that Alice and Bob perform, such as $I \otimes X$ and $X \otimes I$, do *not* commute with the Bell basis. Show that despite this, the statistics which Alice and Bob compile from their measurements are the same as those which they would have obtained had they actually measured Π_{bf} and Π_{pf}.

The modified Lo–Chau protocol

Random sampling in the Bell basis thus provides Alice and Bob with EPR pairs ρ with known fidelity to the ideal state $|\beta_{00}\rangle^{\otimes n}$, and as previously discussed, this bounds Eve's mutual information with any measurements that might be performed on ρ. For ρ to be useful for key generation, however, Alice and Bob must reduce Eve's mutual information with their state until it is exponentially small. This task can be achieved by applying classical privacy amplification to their measurement results. Equivalently, Alice and Bob can first perform entanglement distillation, as introduced in Section 12.5.2, to obtain ρ' which is very close to $|\beta_{00}\rangle^{\otimes m}$ for some $m < n$, *then* measure the final state. This sort of 'quantum privacy amplification' will be useful to us.

A rough argument is as follows. Entanglement distillation can be accomplished by performing quantum error-correction. Since ρ is nearly certain to have δn errors, encoding these qubits in a δn correcting quantum error-correction code allows up to δn errors to be perfectly corrected. As we saw in Sections 10.5.5 and 10.5.8, if an $[n, m]$ stabilizer code is used, then encoding, syndrome measurement and error recovery can be performed by measurements of Pauli operators determined by the rows of the check matrix for the code. Alice and Bob simply perform identical measurements and recovery operations on their respective n qubit halves of ρ, producing an error-corrected state which has a fidelity relative to $|\beta_{00}\rangle^{\otimes m}$ that is on the order of one minus the the probability of more than δn errors occurring. By construction, the syndrome measurements turn out to commute with the Bell basis, since Alice and Bob perform identical tasks.

Putting the random sampling and entanglement distillation pieces together gives us the *modified Lo–Chau* protocol, presented in Figure 12.14. A few notes about this protocol are in order. The random Hadamard transforms performed in steps **3** and **7** create a symmetry in Eve's strategy between detecting information encoded in the X and Z bases (and thus causing X and Z errors). They also enable random selection of a measurement of Π_{bf} or Π_{pf} on the check qubits. The specific procedure employed in step **9** can be justified for the case of any stabilizer code, as in Exercise 12.34. The Gilbert–Varshamov bound for CSS codes, Equation (10.74), shows that good quantum codes exist for large block lengths, so that n can be chosen sufficiently large that for a δn error-correcting $[n, m]$ quantum code, the criteria for security can be satisfied.

QKD protocol: modified Lo–Chau

1: Alice creates $2n$ EPR pairs in the state $|\beta_{00}\rangle^{\otimes 2n}$.

2: Alice randomly selects n of the $2n$ EPR pairs to serve as checks to check for Eve's interference. She does not do anything with them yet.

3: Alice selects a random $2n$-bit string b, and performs a Hadamard transform on the second qubit of each pair for which b is 1.

4: Alice sends the second qubit of each pair to Bob.

5: Bob receives the qubits and publicly announces this fact.

6: Alice announces b and which n qubits are to provide check bits.

7: Bob performs Hadamards on the qubits where b is 1.

8: Alice and Bob each measure their n check qubits in the $|0\rangle$, $|1\rangle$ basis, and publicly share the results. If more than t of these disagree, they abort the protocol.

9: Alice and Bob measure their remaining n qubits according to the check matrix for a pre-determined $[n, m]$ quantum code correcting up to t errors. They share the results, compute the syndromes for the errors, and then correct their state, obtaining m nearly perfect EPR pairs.

10: Alice and Bob measure the m EPR pairs in the $|0\rangle$, $|1\rangle$ basis to obtain a shared secret key.

Figure 12.14. A QKD protocol which is secure, by virtue of use of perfect quantum computers, error-correction, and random testing of EPR pairs.

Exercise 12.33: Let $\{M_1, M_2, \ldots, M_n\}$ be a set of measurement observables which produce respective results X_i when an input state ρ is measured. Argue that the random variables X_i obey classical probability arguments if $[M_i, M_j] = 0$, that is, they commute with each other.

Exercise 12.34: (Entanglement distillation by error-correction) In Section 10.5.8, we saw that codewords of an $[n, m]$ qubit stabilizer code can be constructed by measuring its generators $g_1, \ldots, g_{n-m}$ on an *arbitrary* n qubit quantum state, then applying Pauli operations to change the result to be a simultaneous $+1$ eigenstate of the generators. Using that idea, show that if we start out with n EPR pairs in the state $|\beta_{00}\rangle^{\otimes n}$, and perform identical generator measurements on the two n qubit halves of the pairs, followed by Pauli operations to correct for *differences* in the measurement results between the pairs, then we obtain an encoded $|\beta_{00}\rangle^{\otimes m}$ state. Also show that if the stabilizer code corrects up to δn errors, then even if δn errors are suffered by an n qubit half, we still obtain $|\beta_{00}\rangle^{\otimes m}$.

A quantum error-correction protocol

The modified Lo–Chau protocol makes use of quantum error-correction to perform entanglement distillation, and is built essentially upon the EPR protocol. Entanglement is a frail resource, and quantum error-correction generally requires robust quantum computers, which are challenging to realize. Fortunately, however, this protocol can be systematically simplified in a series of steps, each of which provably does not compromise the security of the scheme. Let us begin by removing the need to distribute EPR pairs.

Note that the measurements which Alice perform at the end of the modified Lo–Chau protocol can be performed at the very start, with no change in any of the states held by the rest of the world. Alice's measurements of her halves of the check EPR pairs in step 8 collapse the pairs into n single qubits, so instead of sending entangled states, Alice can simply send single qubits. This gives us the modified steps

1′: Alice creates n random check bits, and n EPR pairs in the state $|\beta_{00}\rangle^{\otimes n}$. She also encodes n qubits as $|0\rangle$ or $|1\rangle$ according to the check bits.

2′: Alice randomly chooses n positions (out of $2n$) and puts the check qubits in these positions and half of each EPR pair in the remaining positions.

8′: Bob measures the n check qubits in the $|0\rangle$, $|1\rangle$ basis, and publicly shares the results with Alice. If more than t of these disagree, they abort the protocol.

Similarly, Alice's measurements in steps 9 and 10 collapse EPR pairs into *random* qubits *encoded in a random quantum code*. This can be seen in the following manner. A particularly convenient choice of code, which we shall employ for the remainder of this section, is an $[n, m]$ CSS code of C_1 over C_2, CSS(C_1, C_2), which encodes m qubits in n qubits and corrects up to t errors. Recall from Section 10.4.2 that for this code, H_1 and $H_2^\perp$ are the parity check matrices corresponding to the classical codes C_1 and $C_2^\perp$, in which each of the codeword states is

$$\frac{1}{\sqrt{|C_2|}} \sum_{w \in C_2} |v_k + w\rangle, \tag{12.201}$$

for v_k being a representative of one of the 2^m cosets of C_2 in C_1 (the notation v_k is chosen to suggest a vector v indexed by a key string k). Also recall that there exists a family of codes equivalent to this one, $\mathrm{CSS}_{z,x}(C_1, C_2)$, with codeword states

$$|\xi_{v_k,z,x}\rangle = \frac{1}{\sqrt{|C_2|}} \sum_{w \in C_2} (-1)^{z \cdot w} |v_k + w + x\rangle. \qquad (12.202)$$

These states form an orthonormal basis for a 2^n-dimensional Hilbert space (see Exercise 12.35), and thus we may write Alice's n EPR pair state as

$$|\beta_{00}\rangle^{\otimes n} = \sum_{j=0}^{2^n} |j\rangle|j\rangle = \sum_{v_k,z,x} |\xi_{v_k,z,x}\rangle|\xi_{v_k,z,x}\rangle. \qquad (12.203)$$

Note that in this expression, we have separated the labels into two kets, where the first denotes the qubits Alice keeps, and the second, the qubits which are sent to Bob. When Alice measures the stabilizer generators corresponding to H_1 and $H_2^\perp$ on her qubits in step **9**, she obtains *random* values for x and z, and similarly, her final measurement in step **10** gives her a *random* choice of v_k. The remaining n qubits are thus left in the state $|\xi_{v_k,z,x}\rangle$, which is the codeword for v_k in $\mathrm{CSS}_{z,x}(C_1, C_2)$. This is just the encoded counterpart of a 2^m qubit state $|k\rangle$. Therefore, as claimed above, Alice's measurements produce random qubits encoded in a random code.

Thus, instead of sending halves of EPR pairs, Alice can equivalently randomly choose x, z, and k, then encode $|k\rangle$ in the code $\mathrm{CSS}_{z,x}(C_1, C_2)$, and send Bob the encoded n qubits. This gives us the modified steps

1″: Alice creates n random check bits, a random m bit key k, and two random n bit strings x and z. She encodes $|k\rangle$ in the code $\mathrm{CSS}_{z,x}(C_1, C_2)$. She also encodes n qubits as $|0\rangle$ or $|1\rangle$ according to the check bits.

2″: Alice randomly chooses n positions (out of $2n$) and puts the check qubits in these positions and encoded qubits in the remaining positions.

6′: Alice announces b, x, z, and which n qubits are to provide check bits.

9′: Bob decodes the remaining n qubits from $\mathrm{CSS}_{z,x}(C_1, C_2)$.

10′: Bob measures his qubits to obtain the shared secret key k.

The resulting scheme, known as the CSS codes protocol, is shown in Figure 12.15.

Exercise 12.35: Show that the states $|\xi_{v_k,z,x}\rangle$ defined in (12.202) form an orthonormal basis for a 2^n-dimensional Hilbert space, that is,

$$\sum_{v_k,z,x} |\xi_{v_k,z,x}\rangle\langle\xi_{v_k,z,x}| = I. \qquad (12.204)$$

Hint: for C_1 an $[n, k_1]$ code, C_2 an $[n, k_2]$ code, and $m = k_1 - k_2$, note that there are 2^m distinct values of v_k, 2^{n-k_1} distinct x, and 2^{k_2} distinct z.

Exercise 12.36: Verify Equation (12.203).

Exercise 12.37: This is an alternative way to understand why Alice's measurements in steps **9** and **10** collapse EPR pairs into random qubits encoded in a random quantum code. Suppose Alice has an EPR pair $(|00\rangle + |11\rangle)/\sqrt{2}$. Show that if she measures the first qubit in the X basis, then the second qubit collapses into an eigenstate of X determined by the measurement result. Similarly, show that if she measures in the Z basis, then the second qubit is left in a Z eigenstate labeled by the measurement result. Using this observation and the results of Section 10.5.8, conclude that Alice's measurements of H_1, $H_2^\perp$, and $\bar{Z}$ on her EPR pair halves result in a random codeword of $\text{CSS}_{z,x}(C_1, C_2)$ determined by her measurement results.

QKD protocol: CSS codes

$1''$: Alice creates n random check bits, a random m bit key k, and two random n bit strings x and z. She encodes $|k\rangle$ in the code $\text{CSS}_{z,x}(C_1, C_2)$. She also encodes n qubits as $|0\rangle$ or $|1\rangle$ according to the check bits.

$2''$: Alice randomly chooses n positions (out of $2n$) and puts the check qubits in these positions and the encoded qubits in the remaining positions.

3: Alice selects a random $2n$-bit string b, and performs a Hadamard transform on each qubit for which b is 1.

4: Alice sends the resulting qubits to Bob.

5: Bob receives the qubits and publicly announces this fact.

$6'$: Alice announces b, x, z, and which n qubits are to provide check bits.

7: Bob performs Hadamards on the qubits where b is 1.

$8'$: Bob measures the n check qubits in the $|0\rangle$, $|1\rangle$ basis, and publicly shares the results with Alice. If more than t of these disagree, they abort the protocol.

$9'$: Bob decodes the remaining n qubits from $\text{CSS}_{z,x}(C_1, C_2)$.

$10'$: Bob measures his qubits to obtain the shared secret key k.

Figure 12.15. A QKD protocol which is secure, by virtue of simplification of the modified Lo–Chau protocol via CSS codes.

Reduction to BB84

The CSS codes QKD protocol is secure by virtue of direct reduction from the modified Lo–Chau protocol, and is much simpler because it does not make any evident use of EPR pairs. Unfortunately, it is still unsatisfactory, because it requires perfect quantum computation to perform the encoding and decoding (instead of just single qubit preparation and measurements), and Bob needs to temporarily store qubits in a quantum memory while waiting for communication from Alice. The use of CSS codes, however, enables these two requirements to be removed, essentially because they decouple phase flip correction from bit flip correction.

First, note that Bob measures his qubits in the Z basis immediately after decoding; thus, the phase correction information Alice sends as z is unnecessary. Thus, since C_1 and C_2 are classical codes, instead of decoding then measuring, he can immediately measure to obtain $v_k + w + x + \epsilon$ (where ϵ represents some possible error, due to the channel and to Eve), then decode *classically*: he subtracts Alice's announced value of x, then corrects the result to a codeword in C_1, which is definitely $v_k + w$, if the distance of the code is not exceeded. The final key k is the coset of $v_k + w + C_2$ in C_1 (see Appendix 2 for an explanation of cosets, and this notation). This gives us:

9″: Bob measures the remaining qubits to get $v_k + w + x + \epsilon$, and subtracts x from the result, correcting it with code C_1 to obtain $v_k + w$.

10″: Bob computes the coset of $v_k + w + C_2$ in C_1 to obtain the key k.

Second, since Alice need not reveal z, the state she effectively sends is a mixed state, averaged over random values of z,

$$\rho_{v_k,x} = \frac{1}{2^n} \sum_z |\xi_{v_k,z,x}\rangle\langle\xi_{v_k,z,x}| \tag{12.205}$$

$$= \frac{1}{2^n|C_2|} \sum_z \sum_{w_1,w_2 \in C_2} (-1)^{z \cdot (w_1 + w_2)} |v_k + w_1 + x\rangle\langle v_k + w_2 + x| \tag{12.206}$$

$$= \frac{1}{|C_2|} \sum_{w \in C_2} |v_k + w + x\rangle\langle v_k + w + x|. \tag{12.207}$$

This state is simple to create: Alice need only *classically* choose $w \in C_2$ at random, and construct $|v_k + w + x\rangle$, using her randomly determined x and k. We thus have:

1‴: Alice creates n random check bits, a random n bit string x, a random $v_k \in C_1/C_2$, and a random $w \in C_2$. She encodes n qubits in the state $|0\rangle$ or $|1\rangle$ according to $v_k + w + x$, and similarly, n qubits according to the check bits.

Steps $1‴$ and $9″$ can be simplified further, by changing step $6'$ slightly. Currently, Alice sends $|v_k + w + x\rangle$, Bob receives and measures to obtain $v_k + w + x + \epsilon$, then Alice sends x, which Bob subtracts to obtain $v_k + w + \epsilon$. But if Alice chooses $v_k \in C_1$ (as opposed to C_1/C_2), then w is unnecessary. Moreover, $v_k + x$ is then a completely random n bit string, and instead of the above protocol, it is equivalent if Alice chooses x at random, sends $|x\rangle$, Bob receives and measures to obtain $x + \epsilon$, then Alice sends $x - v_k$, which Bob subtracts to obtain $v_k + \epsilon$. Now, there is no difference between the random check bits and the code bits! This gives us:

1⁗: Alice chooses a random $v_k \in C_1$, and creates $2n$ qubits in the state $|0\rangle$ or $|1\rangle$ according to $2n$ random bits.

2‴: Alice randomly chooses n positions (out of $2n$) and designates these as the check qubits, and the remainder as $|x\rangle$.

6″: Alice announces b, $x - v_k$, and which n qubits are to provide check bits.

9‴: Bob measures the remaining qubits to get $x + \epsilon$, and subtracts $x - v_k$ from the result, correcting it with code C_1 to obtain v_k.

10″: Alice and Bob compute the coset of $v_k + C_2$ in C_1 to obtain the key k.

Next, note that Alice need not perform Hadamard operations (although in practice, single qubit operations are not so difficult to accomplish with photons). She can instead encode her qubits directly in either the $|0\rangle$, $|1\rangle$ (Z) basis or in the $|+\rangle$, $|-\rangle$ (X) basis, depending on the bits of b:

1⁗′: Alice creates $(4 + \delta)n$ random bits. For each bit, she creates a qubit in either the $|0\rangle$, $|1\rangle$ basis, or the $|+\rangle$, $|-\rangle$ basis, according to a random bit string b.

We are almost done: encoding and decoding are now performed classically. The remaining problem is to remove the need for a quantum memory. To solve this problem, suppose that Bob goes ahead and measures immediately after receiving qubits from Alice, choosing randomly to measure in either the X or Z bases. When Alice subsequently announces b, they can keep only those bits for which their bases happened to be the same. This allows Bob to do away completely with his quantum storage device. Note that with high probability they discard half their bits, so in order to obtain the same number of key bits as before, they should start with a little (say, δ) more than twice the number of original random bits. Of course, Alice now must delay her choice of which bits are to be check bits, until after the discarding step. This gives us the final protocol, shown in Figure 12.16. This protocol is exactly the same as BB84, with only minor cosmetic differences. Note how the use of the classical code C_1 performs information reconciliation, and computing the coset of $v_k + C_2$ in C_1 performs privacy amplification (see Section 12.6.2).

QKD protocol: Secure BB84

1: Alice creates $(4 + \delta)n$ random bits.
2: For each bit, she creates a qubit in either the Z basis, or the X basis, according to a random bit string b.
3: Alice sends the resulting qubits to Bob.
4: Alice chooses a random $v_k \in C_1$.
5: Bob receives the qubits, publicly announces this fact, and measures each in the Z or X bases at random.
6: Alice announces b.
7: Alice and Bob discard those bits Bob measured in a basis other than b. With high probability, there are at least $2n$ bits left; if not, abort the protocol. Alice decides randomly on a set of $2n$ bits to continue to use, randomly selects n of these to be check bits, and announces the selection.
8: Alice and Bob publicly compare their check bits. If more than t of these disagree, they abort the protocol. Alice is left with the n bit string x, and Bob with $x + \epsilon$.
9: Alice announces $x - v_k$. Bob subtracts this from his result, correcting it with code C_1 to obtain v_k.
10: Alice and Bob compute the coset of $v_k + C_2$ in C_1 to obtain the key k.

Figure 12.16. The final QKD protocol arrived at by systematic reduction of the CSS codes protocol, which is exactly the same as BB84 (up to minor cosmetic differences). For clarity, we have dropped the $'$ notation.

In summary, we have systematically proven the security of the BB84 quantum key distribution protocol, by starting with a manifestly secure scheme requiring perfect quantum computation and quantum memories, and systematically reducing it to BB84. By virtue of having only made modifications which manifestly leave Eve's quantum state (conditioned on all revealed classical information) unchanged, we conclude that BB84 is secure. Naturally, there are some caveats. This proof only applies to an ideal situation, where the states sent are those described. In practice, qubit sources are imperfect; for example, such sources are often lasers attenuated to approximately generate single photons, representing qubits (as described in Section 7.4.1). Furthermore, the proof does not place any bounds on the amount of effort Alice and Bob must go to in decoding; for practical key distribution, C_1 must be efficiently decodable. This proof also does not provide an upper bound on the tolerable eavesdropping; it utilizes CSS codes, which are not optimal. It is estimated that a rate of bit and phase errors up to 11% is acceptable using a protocol similar to BB84, but with the aid of quantum computers to encode and decode, higher error rates may be tolerable. The ultimate capability of quantum cryptography is an interesting open issue, and we expect such fundamental questions about the physical limits of computation and communication to continue to intrigue and challenge researchers in the future.

Exercise 12.38:   Show that if you had the ability to distinguish non-orthogonal states, then it would be possible to compromise the security of BB84, and indeed, all of the QKD protocols we have described.

Problem 12.1:   In this problem we will work through an alternate proof of the Holevo bound. Define the *Holevo chi quantity*,

$$\chi \equiv S(\rho) - \sum_x p_x S(\rho_x). \tag{12.208}$$

(1) Suppose the quantum system consists of two parts, A and B. Show that

$$\chi_A \leq \chi_{AB}. \tag{12.209}$$

(*Hint*: Introduce an extra system which is correlated with AB, and apply strong subadditivity.)

(2) Let $\mathcal{E}$ be a quantum operation. Use the previous result to show that

$$\chi' \equiv S(\mathcal{E}(\rho)) - \sum_x p_x S(\mathcal{E}(\rho_x)) \leq \chi \equiv S(\rho) - \sum_x p_x S(\rho_x). \tag{12.210}$$

That is, the Holevo chi quantity decreases under quantum operations. This is an important and useful fact in its own right.

(3) Let E_y be a set of POVM elements. Augment the quantum system under consideration with an 'apparatus' system, M, with an orthonormal basis $|y\rangle$. Define a quantum operation by

$$\mathcal{E}(\rho \otimes |0\rangle\langle 0|) \equiv \sum_y \sqrt{E_y} \rho \sqrt{E_y} \otimes |y\rangle\langle y|, \tag{12.211}$$

where $|0\rangle$ is some standard pure state of M. Prove that after the action of $\mathcal{E}$, $\chi_M = H(X:Y)$. Use this and the previous two results to show that

$$H(X:Y) \leq S(\rho) - \sum_x p_x S(\rho_x), \tag{12.212}$$

which is the Holevo bound.

Problem 12.2:   This result is an extension of the previous problem. Provide a proof of the no-cloning theorem by showing that a cloning process for non-orthogonal pure states would necessarily increase χ.

Problem 12.3: For a fixed quantum source and rate $R > S(\rho)$, design a quantum circuit implementing a rate R compression scheme.

Problem 12.4: (Linearity forbids cloning) Suppose we have a quantum machine with two slots, A and B. Slot A, the *data slot*, starts out in an unknown quantum state ρ. This is the state to be copied. Slot B, the *target slot*, starts out in some standard quantum state, σ. We will assume that any candidate copying procedure is *linear* in the initial state,

$$\rho \otimes \sigma \rightarrow \mathcal{E}(\rho \otimes \sigma) = \rho \otimes \rho, \tag{12.213}$$

where $\mathcal{E}$ is some linear function. Show that if $\rho_1 \neq \rho_2$ are density operators such that

$$\mathcal{E}(\rho_1 \otimes \sigma) = \rho_1 \otimes \rho_1 \tag{12.214}$$
$$\mathcal{E}(\rho_2 \otimes \sigma) = \rho_2 \otimes \rho_2, \tag{12.215}$$

then any mixture of ρ_1 and ρ_2 is not copied correctly by this procedure.

Problem 12.5: (Classical capacity of a quantum channel – Research) Is the product state capacity (12.71) the true capacity of a noisy quantum channel for classical information, that is, the capacity when entangled inputs to the channel are allowed?

Problem 12.6: (Methods for achieving capacity – Research) Find an efficient construction for codes achieving rates near the product state capacity (12.71) of a noisy quantum channel for classical information.

Problem 12.7: (Quantum channel capacity – Research) Find a method to evaluate the capacity of a given quantum channel $\mathcal{E}$ for the transmission of quantum information.

Summary of Chapter 12: Quantum information theory

- **No-cloning**: No quantum device can be constructed which outputs $|\psi\rangle|\psi\rangle$, given $|\psi\rangle$, for arbitrary $|\psi\rangle$.

- **Holevo's bound**: The maximum accessible classical information when trying to distinguish between quantum states ρ_x sent with probability distribution p_x is

$$H(X:Y) \leq \chi \equiv S\left(\sum_x p_x \rho_x\right) - \sum_x p_x S(\rho_x).$$

- **Schumacher's quantum noiseless channel coding theorem**: $S(\rho)$ can be interpreted as the number of qubits needed to faithfully represent a quantum source described by ρ.

- **The Holevo–Schumacher–Westmoreland theorem**: The capacity of a noisy quantum channel $\mathcal{E}$ for classical information is given by:

$$C(\mathcal{E}) = \max_{\{p_x, |\psi_x\rangle\}} S\left(\sum_x p_x \mathcal{E}(|\psi_x\rangle\langle\psi_x|)\right) - \sum_x p_x S(\mathcal{E}(|\psi_x\rangle\langle\psi_x|)). \quad (12.216)$$

- **Majorization condition for entanglement transformation**: Alice and Bob can transform $|\psi\rangle$ to $|\varphi\rangle$ by local operations and classical communication if and only if $\lambda_\psi \prec \lambda_\varphi$, where λ_ψ is a vector of the eigenvalues of the reduced density matrix of $|\psi\rangle$ (similarly for λ_φ).

- **Pure state entanglement distillation and dilution**: Alice and Bob can convert between n copies of a joint state $|\psi\rangle$ and $nS(\rho)$ Bell pairs, by local operations and classical communication alone, as $n \to \infty$, where ρ is the reduced density matrix.

- **Quantum cryptography**: Provably secure key distribution is possible by communicating using non-orthogonal quantum states, with a protocol such as BB84. Eavesdropping on the channel will cause a detectable increase in the error rate, because information gain implies disturbance.

History and further reading

The book of Cover and Thomas[CT91] is a terrific introduction to classical information theory. The reader looking for a more advanced yet still readable treatment of information theory should consult Csiszár and Körner[CK81]. Also well worth the read are Shannon's original papers, among the most influential in twentieth century science. These have been reprinted together in a single volume by Shannon and Weaver[SW49]. Bennett and Shor[BS98] and Bennett and DiVincenzo[BD00] have written excellent review articles on quantum information theory.

The no-cloning theorem is due to Dieks[Die82] and to Wootters and Zurek[WZ82]. An enormous amount of work has been done extending these results. By far the majority of the papers consider various schemes for cloners which are interesting in some particular way – they optimize some measure of cloning fidelity, or some other property one might wish a cloner to have. We will not attempt to give a full review of this work here, but note that many of these papers may be found on the internet at http://arXiv.org/ in the quant-ph archive. Some papers of especial interest include the work of Barnum, Caves, Fuchs, Jozsa and Schumacher[BCF+96] extending the range of application of the no-cloning theorem to mixed states and non-unitary cloning devices; by Mor[Mor98] on the cloning of states of composite systems; by Westmoreland and Schumacher[WS98] on a possible equivalence between cloning and faster-than-light communication; and a rebuttal by van Enk[van98b].

The Holevo bound was conjectured by Gordon in 1964[Gor64], and proved by Holevo in 1973[Hol73]. The conceptually simple proof we have given is based upon the difficult-to-prove strong subadditivity inequality, however Holevo used a more direct approach which has been simplified by Fuchs and Caves[FC94]. The approach via strong subadditivity is due to Yuen and Ozawa[YO93]; see also Schumacher, Westmoreland and Wootters[SWW96].

The classical noiseless channel coding theorem is due to Shannon[Sha48, SW49]. The quantum noiseless channel coding theorem is due to Schumacher[Sch95], and is described in a pioneering paper that introduced in an integrated fashion many of the fundamental notions of quantum information theory, including sources, fidelity measures, and the notion of quantum states as a resource that could be treated in information-theoretic terms. This last observation, simple but profound, was driven home by Schumacher's introduction in the paper of the now-ubiquitous term *qubit*, attributed to a conversation between Schumacher and Wootters. A paper by Jozsa and Schumacher[JS94] simplified Schumacher's original approach; this paper was published earlier than [Sch95], but was was written at a later time. These papers were based on the *ensemble-average* fidelity measure discussed in Exercise 12.8, rather than the entanglement fidelity based proof we have given here, which is based on the approach of Nielsen[Nie98]. A slight hole in the original papers by Schumacher, and Schumacher and Jozsa was filled by the work of Barnum, Fuchs, Jozsa and Schumacher[BFJS96]. M. Horodecki[Hor97] subsequently provided a more powerful proof of the same result that also points the way towards a theory of quantum data compression of ensembles of mixed states. The compression scheme described in Box 12.4, which is the quantum analogue of Cover's enumerative coding method[CT91], is originally attributed to Schumacher[Sch95], and quantum circuits for it are explicitly given by Cleve and DiVincenzo[CD96]. Braunstein, Fuchs, Gottesman, and Lo have generalized this to provide a quantum analogue to Huffman encoding[BFGL98], and Chuang and Modha to arithmetic coding[CM00].

The Holevo–Schumacher–Westmoreland (HSW) theorem has an interesting history. The problem it addresses was first discussed by Holevo[Hol79] in 1979, who made some partial progress on the problem. Unaware of this work, Hausladen, Jozsa, Schumacher, Westmoreland, and Wootters[HJS+96] solved a special case of the problem in 1996. Independently, and shortly thereafter, Holevo[Hol98] and Schumacher and Westmoreland[SW97] proved the HSW theorem giving the product state capacity of a noisy quantum channel for classical information. Fuchs[Fuc97] has described some interesting examples of the product state capacity, where the ensemble of states maximizing the expression (12.71) for the capacity contains non-orthogonal members. King and Ruskai[KR99] have made some promising headway on the problem of showing that the product state capacity is the same as the capacity unrestricted to product states, but the general problem remains open.

The entropy exchange was defined by Lindblad[Lin91], and rediscovered by Schumacher[Sch96b], who proved the quantum Fano inequality. The coherent information was introduced by Lloyd[Llo97] and by Schumacher and Nielsen[SN96] in the context of the capacity of a noisy quantum channel; [SN96] proves the quantum data processing inequality. A table containing the inequalities mentioned in Exercise 12.15 may be found in Nielsen's Ph.D. dissertation[Nie98]. The as yet unsolved problem of determining the quantum channel capacity (Problem 12.7) has an interesting history. Initial work on the subject came from several different perspectives, as may be seen from the papers of Barnum, Nielsen and Schumacher[BNS98], of Bennett, DiVincenzo, Smolin, and Wootters[BDSW96], of Lloyd[Llo97], of Schumacher[Sch96b] and of Schumacher and Nielsen[SN96]. The equivalence of several of these points of view has been understood through the work of Barnum, Knill and Nielsen[BKN98], and Barnum, Smolin and Terhal[BST98]. The capacity has been established for some specific channels (most notably the quantum erasure channel) by Bennett, DiVincenzo and Smolin[BDS97], and a lower bound on the capacity of the depolarizing channel making intriguing use of degenerate quantum codes has been obtained by Shor and Smolin[SS96], and refined by DiVincenzo, Shor and Smolin[DSS98]. Zurek[Zur89], Milburn[Mil96], and Lloyd[Llo96] analyzed examples of quantum Maxwell's demons, though not in the context of error-correction. The analysis here is based on the work of Nielsen, Caves, Schumacher and Barnum[NCSB98]. This point of view has also been pursued by Vedral[Ved99] to obtain limits on entanglement distillation procedures. The quantum Singleton bound is due to Knill and Laflamme[KL97]. The proof we give is due to Preskill[Pre98b].

The study of entanglement has blossomed into a major area of research, and there are far too many papers on the subject to even begin to give an account here. Once again, see the quant-ph archive at http://arXiv.org/. The conditions for entanglement transformation based on majorization (Theorem 12.15) are due to Nielsen[Nie99a]. Theorem 12.13 is due to Uhlmann[Uhl71, Uhl72, Uhl73]. Proposition 12.14 is due to Lo and Popescu[LP97]. Entanglement catalysis was discovered by Jonathan and Plenio[JP99]. Marshall and Olkin[MO79] is a comprehensive introduction to majorization, including the proof of Birkhoff's theorem. The limits for entanglement dilution and distillation are due to Bennett, Bernstein , Popescu and Schumacher[BBPS96]. Entanglement distillation protocols for mixed states were introduced by Bennett, Brassard, Popescu, Schumacher, Smolin and Wootters[BBP+96], and the connection to quantum error-correction developed in a ground-breaking paper by Bennett, DiVincenzo, Smolin and Wootters[BDSW96] that has stimulated a lot of subsequent research. The example illustrated in Figure 12.11 was noted by Gottesman and Nielsen (unpublished). We mention just a few more papers on entanglement of exceptional interest that may serve as an entry-point to the literature; unfortunately many papers of note are omitted as a result. A series of papers by members of the Horodecki family (Michal, Pawel and Ryszard) have investigated the properties of entanglement in depth; of especial note are [HHH96, HHH98, HHH99a, HHH99b, HHH99c]. Also of great interest are the papers by Vedral and Plenio[VP98] and by Vidal[Vid98].

For an excellent (early) overview of quantum cryptography at the lay level see the article by Bennett, Brassard, and Ekert in Scientific American[BBE92]. Quantum cryptographic ideas were first put forward by Wiesner in the late 1960s. Unfortunately, Wiesner's ideas were not accepted for publication, and it wasn't until the early 1980s that the ideas became known. Wiesner proposed that (entangled) quantum states, if they could be stored for long periods of time, could be used as a kind of counterfeit-proof money[Wie, Wie83]. Bennett developed several further protocols, one of which lead to the first experimental implementation, by Bennett, Bessette, Brassard, Salvail, and Smolin[BBB+92], which is of historical interest (in principle) since it transmitted its information less than a meter and, moreover, eavesdropping was facilitated by a loud buzzing sound which emanated from the power supply whenever a 'one' was sent! The concept of privacy amplification was first introduced by Bennett, Brassard, and Robert[BBR88]. For information reconciliation protocols see [BBB+92] and [BS94]. Theorem 12.16 is stated and proved in Bennett, Brassard, Crépeau, and Maurer[BBCM95], in a very readable general treatment of privacy amplification. Note that the information disclosed during reconciliation has an important impact on the threshold for privacy amplification, as bounded in Theorem 12.17, proven by Cachin and Maurer[CM97]. Privacy amplification has applications to classical key generation using distant correlated random sources such as starlight sensed by satellites[Mau93]. The four state protocol known as BB84 is named after the authors, Bennett and Brassard[BB84], and similarly, the two state B92 protocol is named after Bennett[Ben92]. The EPR protocol was devised by Ekert[Eke91]. The proof of the random sampling bound in Exercise 12.27 is due to Ambainis. The limitations and security of quantum cryptography have been discussed in depth in many publications. For a sample, see the works by Barnett and Phoenix[BP93]; Brassard[Bra93]; Ekert, Huttner, Palma, and Peres[EHPP94]; also [Phy92]. The connection between coherent information and privacy was recognized by Schumacher and Westmoreland[SW98]. Numerous papers have been published on experimental implementations of quantum cryptographic systems. For a good introduction, see Hughes, Alde, Dyer, Luther, Morgan, and Schauer[HAD+95]; the demonstration of quantum cryptography under Lake Geneva was by Muller, Zbinden, and Gisin[MZG96]. The experiment described in Box 12.7 was performed at IBM, by Bethune and Risk[BR98, BR00], and we thank them for the schematic of their apparatus. A large number of proofs of the security of various quantum key distribution protocols, under different circumstances, have been presented. Of particular note is a complete (and extensive, but somewhat complicated) proof of the security of QKD with BB84 given by Mayers[May98]. Biham, Boyer, Brassard, van de Graaf, and Mor have also considered attacks against BB84[BBB+98]. A simpler proof, which uses EPR states and requires perfect quantum computation has been given by Lo and Chau[LC99]; this is the protocol we began with in Section 12.6.5. Lo has simplified it to begin with a test which ascertains the error rate, before transmitting key material[Lo99]. The even simpler (and beautiful!) proof given in Section 12.6.5 is due to Shor and Preskill[SP00], who also give the 11% estimate mentioned in Section 12.6.5. Our presentation of this proof also benefited greatly from conversations with Gottesman.

Appendix 1: Notes on basic probability theory

This appendix collects some elementary definitions and results about probability theory. It is assumed that the reader already has some familiarity with this material. The reader who is not familiar with any of the results should take time out to prove them, as indicated in the exercises.

The basic notion of probability theory is that of a *random variable*. A random variable X may take one of a number of values, x, with probabilities $p(X = x)$. We use upper case to denote the random variable, and x to denote the values that random variable may take. We often use the notation $p(x)$ instead of $p(X = x)$, leaving the '$X =$' implicit. In this book we shall only be concerned with random variables which take their value from a finite set of values, and we always assume that this is the case. Occasionally it is convenient to consider random variables whose values are vectors taking values, for example, in the set $(i, j), i = 1, \dots, m_1, j = 1, \dots, m_2$.

The *conditional probability* that $Y = y$ given that $X = x$ is defined by

$$p(Y = y | X = x) \equiv \frac{p(X = x, Y = y)}{p(X = x)}, \tag{A1.1}$$

where $p(X = x, Y = y)$ is the probability that $X = x$ *and* $Y = y$. When $p(X = x) = 0$ we make the convention that $p(Y = y | X = x) = 0$. We often use the notation $p(y|x)$ leaving the '$Y =$' and '$X =$' implicit. Random variables X and Y are said to be *independent* if $p(X = x, Y = y) = p(X = x)p(Y = y)$ for all x and y. Note that when X and Y are independent, it follows that $p(y|x) = p(y)$ for all x and y.

Bayes' rule relates the conditional probabilities for Y given X to those for X given Y,

$$p(x|y) = p(y|x)\frac{p(x)}{p(y)}. \tag{A1.2}$$

The probability $p(y)$ appearing in this expression is often re-expressed using the law of total probability discussed below.

Exercise A1.1: Prove Bayes' rule.

One of the most important and frequently used results in probability theory is the *law of total probability*. It states that if X and Y are two random variables, then the probabilities for Y can be expressed in terms of the probabilities for X, and the conditional probabilities for Y given X,

$$p(y) = \sum_x p(y|x)p(x), \tag{A1.3}$$

where the sum is over all values x that X can take.

Exercise A1.2: Prove the law of total probability.

The *expectation, average,* or *mean* of a random variable X is defined by

$$\mathbf{E}(X) \equiv \sum_x p(x)x, \tag{A1.4}$$

where the sum is over all values x which the random variable X can take.

Exercise A1.3: Prove that there exists a value of $x \geq \mathbf{E}(X)$ such that $p(x) > 0$.

Exercise A1.4: Prove that $\mathbf{E}(X)$ is linear in X.

Exercise A1.5: Prove that for independent random variables X and Y,
$\mathbf{E}(XY) = \mathbf{E}(X)\mathbf{E}(Y)$.

The *variance* of a random variable X is defined by the expression

$$\mathrm{var}(X) \equiv \mathbf{E}[(X - \mathbf{E}(X))^2] = \mathbf{E}(X^2) - \mathbf{E}(X)^2. \tag{A1.5}$$

The *standard deviation,* $\Delta(X) \equiv \sqrt{\mathrm{var}(X)}$, is a measure of the spread of a random variable about the average. *Chebyshev's inequality* quantifies more precisely in what sense the standard deviation is a measure of the spread of values a random variable may take. It states that for any $\lambda > 0$ and random variable X with finite variance,

$$p(|X - \mathbf{E}(X)| \geq \lambda\Delta(X)) \leq \frac{1}{\lambda^2}. \tag{A1.6}$$

Thus, the probability of being more than λ standard deviations away from the mean gets small as λ goes to infinity.

Exercise A1.6: Prove Chebyshev's inequality.

The main text of the book contains many other results in probability theory, including the *Chernoff bound* on page 154, *Fano's inequality* on page 536, and the *law of large numbers* on page 541.

History and further reading

Probability theory enjoys a surfeit of superb texts. We especially recommend the text of Grimmett and Stirzaker[GS92] as an excellent introduction to many basic ideas in the theory of probability and stochastic processes. From a more purely mathematical point of view, Williams[Wil91] is a superb introduction to the modern theory of probability, with an emphasis on the beautiful theory of martingales. Finally, the classic two-volume text of Feller[Fel68a, Fel68b] is an excellent indepth introduction to the ideas of probability theory.

Appendix 2: Group theory

The mathematical theory of groups is useful at several points in the study of quantum computation and quantum information. The generalization of the order-finding, factoring, and period finding algorithms in Chapter 5 is based on the *hidden subgroup problem*; the stabilizer formalism for quantum error-correction in Chapter 10 is based on some elementary group theory. The number theory described in Appendix 4 uses the properties of the group $\mathbf{Z}_n^*$. And, implicitly, the quantum circuits used throughout the book are an example of the use of Lie groups. In this appendix we review some basic material on group theory. We summarize many of the fundamental concepts and provide important definitions, but do not attempt to explain very much, as group theory is a vast subject!

A2.1 Basic definitions

A group $(G, \cdot)$ is a non–empty set G with a binary group multiplication operation '$\cdot$', with the following properties: (*closure*) $g_1 \cdot g_2 \in G$ for all $g_1, g_2 \in G$; (*associativity*) $(g_1 \cdot g_2) \cdot g_3 = g_1 \cdot (g_2 \cdot g_3)$, for all $g_1, g_2, g_3 \in G$; (*identity*) there exists $e \in G$ such that $\forall g \in G, g \cdot e = e \cdot g = g$; (*inverses*) for all $g \in G$, there exists $g^{-1} \in G$ such that $g \cdot g^{-1} = e$ and $g^{-1} \cdot g = e$. We often leave out $\cdot$ and write $g_1 \cdot g_2$ as simply $g_1 g_2$. We also often refer to the group G without referring explicitly to its multiplication operation, but it must be defined.

A group G is *finite* if the number of elements in G is finite. The *order* of a finite group G is the number of elements it contains, denoted as $|G|$. A group G is said to be *Abelian* if $g_1 g_2 = g_2 g_1$ for all $g_1, g_2 \in G$. A simple example of a finite Abelian group is the additive group $\mathbf{Z}_n$ of integers modulo n, with 'multiplication' operation the operation of modular addition. It is easily checked that this operation satisfies the closure and associativity axioms; there is an identity element, 0, since $x + 0 = x \pmod{n}$ for all x, and every $x \in G$ has an inverse, $n - x$, since $x + (n - x) = 0 \pmod{n}$.

The *order* of an element $g \in G$ is the smallest positive integer r such that g^r (g multiplied with itself r times) equals the identity element e.

A *subgroup* H of G is a subset of G which forms a group under the same group multiplication operation as G.

Theorem A2.1: (**Lagrange's theorem**) If H is a subgroup of a finite group G then $|H|$ divides $|G|$.

Exercise A2.1: Prove that for any element g of a finite group, there always exists a positive integer r such that $g^r = e$. That is, every element of such a group has an order.

Exercise A2.2: Prove Lagrange's theorem.

Exercise A2.3: Show that the order of an element $g \in G$ divides $|G|$.

If g_1 and g_2 are elements of G, then the conjugate of g_2 with respect to g_1 is the element $g_1^{-1}g_2g_1$. If H is subgroup of G, then it is known as a normal subgroup if $g^{-1}Hg = H$ for all $g \in G$. The *conjugacy class* G_x of an element x in a group G is defined by $G_x \equiv \{g^{-1}xg | g \in G\}$.

Exercise A2.4: Show that if $y \in G_x$ then $G_y = G_x$.

Exercise A2.5: Show that if x is an element of an Abelian group G then $G_x = \{x\}$.

An interesting example of a group which is not Abelian is the *Pauli group* on n qubits. For a single qubit, the Pauli group is defined to consist of all the Pauli matrices, with multiplicative factors $\pm 1, \pm i$ allowed by the definition:

$$G_1 \equiv \{\pm I, \pm iI, \pm X, \pm iX, \pm Y, \pm iY, \pm Z, \pm iZ\}. \tag{A2.1}$$

This set of matrices forms a group under the operation of matrix multiplication. You might wonder why we don't omit the multiplicative factors ± 1 and $\pm i$; the reason these are included is to ensure that G_1 is closed under multiplication, and thus forms a legitimate group. The general Pauli group on n qubits is defined to consist of all n-fold tensor products of Pauli matrices, and again we allow multiplicative factors of $\pm 1, \pm i$.

A2.1.1 Generators

The study of groups is often greatly simplified by the use of a set of *group generators* for the group being studied. A set of elements $g_1, \ldots, g_l$ in a group G is said to *generate* the group G if every element of G can be written as a product of (possibly repeated) elements from the list $g_1, \ldots, g_l$, and we write $G = \langle g_1, \ldots, g_l \rangle$. For example, $G_1 = \langle X, Z, iI \rangle$, since every element in G can be written as a product of X, Z and iI. On the other hand, $\langle X \rangle = \{I, X\}$, a subgroup of G_1, since not every element of G_1 can be written as a power of X. The notation $\langle \cdots \rangle$ which we use for group generators may potentially be confused with the notation for observable averages introduced in Section 2.2.5 (page 87); however, in practice it is always clear from context how the notation is being used.

The great advantage of using generators to describe groups is that they provide a *compact* means of describing the group. Suppose G has size $|G|$. Then it is pretty easy to show that there is a set of $\log(|G|)$ generators generating G. To see this, suppose $g_1, \ldots, g_l$ is a set of elements in a group G, and g is not an element of $\langle g_1, \ldots, g_l \rangle$. Let $f \in \langle g_1, \ldots, g_l \rangle$. Then $fg \notin \langle g_1, \ldots, g_l \rangle$, since if it were then we would have $g = f^{-1}fg \in \langle g_1, \ldots, g_l \rangle$, which we know is false by assumption. Thus for each element $f \in \langle g_1, \ldots, g_l \rangle$ there is an element fg which is in $\langle g_1, \ldots, g_l, g \rangle$ but not in $\langle g_1, \ldots, g_l \rangle$. Thus adding the generator g to $\langle g_1, \ldots, g_l \rangle$ doubles (or more) the size of the group being generated, from which we conclude that G must have a set of generators containing at most $\log(|G|)$ elements.

A2.1.2 Cyclic groups

A *cyclic group* G possesses an element a such that any element $g \in G$ can be expressed as a^n for some integer n. a is known as a *generator* of G, and we write $G = \langle a \rangle$. A *cyclic subgroup* H generated by $g \in G$ is the group formed by $\{e, g, g^2, \ldots, g^{r-1}\}$, where r is the order of g. That is, $H = \langle g \rangle$.

Exercise A2.6:   Show that any group of prime order is cyclic.

Exercise A2.7:   Show that every subgroup of a cyclic group is cyclic.

Exercise A2.8:   Show that if $g \in G$ has finite order r, then $g^m = g^n$ if and only if $m = n(\mathrm{mod}\ r)$.

A2.1.3   Cosets

For H a subgroup of G, the *left coset* of H in G determined by $g \in G$ is the set $gH \equiv \{gh | h \in H\}$. The right coset is defined similarly. Often whether a coset is a 'left' or 'right' coset is implied by context. In the case of a group like $\mathbf{Z}_n$ where the group operation is addition it is conventional to write cosets of a subgroup H in the form $g + H$, for $g \in \mathbf{Z}_n$. Elements of a particular coset gH are known as *coset representatives* of that coset.

Exercise A2.9:   Cosets define an equivalence relation between elements. Show that $g_1, g_2 \in G$ are in the same coset of H in G if and only if there exists some $h \in H$ such that $g_2 = g_1 h$.

Exercise A2.10:   How many cosets of H are there in G?

A2.2   Representations

Let M_n be the set of $n \times n$ complex matrices. A matrix group is a set of matrices in M_n which satisfy the properties of a group under matrix multiplication. We shall denote the identity element in such groups as I. A *representation* ρ of a group G may be defined as a function which maps G to a matrix group, preserving group multiplication. Specifically, $g \in G$ is mapped to $\rho(g) \in M_n$, such that $g_1 g_2 = g_3$ implies $\rho(g_1)\rho(g_2) = \rho(g_3)$. If the map is many to one, it is known as a *homomorphism*; if it is one to one, then the map is an *isomorphism*. A representation ρ which maps into M_n has *dimension* $d_\rho = n$. The representations we have defined are also referred to as being *matrix representations*; there are more general ones, but these are sufficient for our purposes. For the remainder of this appendix all the groups G which we deal with should be taken to be finite groups.

A2.2.1   Equivalence and reducibility

Two important concepts about representations are *equivalence* and *reducibility*. The *character* of a matrix group $G \subset M_n$ is a function on the group defined by $\chi(g) = \mathrm{tr}(g)$, for $g \in G$, where $\mathrm{tr}(\cdot)$ is the usual trace function on matrices. It has the following properties: (1) $\chi(I) = n$, (2) $|\chi(g)| \leq n$, (3) $|\chi(g)| = n$ implies $g = e^{i\theta} I$, (4) χ is constant on any given conjugacy class of G, (5) $\chi(g^{-1}) = \chi^*(g)$, and (6) $\chi(g)$ is an algebraic number for all g. Two matrix groups are said to be *equivalent* if they are isomorphic, and corresponding elements under the isomorphism have the same character.

Exercise A2.11: (Characters)   Prove the properties of characters given above.

Exercise A2.12: (Unitary matrix groups)   A unitary matrix group is comprised solely of unitary matrices (those which satisfy $U^\dagger U = I$). Show that every matrix group is equivalent to a unitary matrix group. If a representation of a group consists entirely of unitary matrices, we may refer to it as being a *unitary representation*.

A matrix group G in M_n is said to be *completely reducible* if it is equivalent to another matrix group H which is of block diagonal form, that is all elements $m \in H$ are of the form $\text{diag}(m_1, m_2)$, for some $m_1 \in M_{n_1}$ and $m_2 \in M_{n_2}$. If no such equivalence exists, then the matrix group is irreducible. The following is a useful property of irreducible matrix groups:

Lemma A2.2: (**Schur's lemma**) Let $G \subset M_n$ and $H \subset M_k$ be two matrix groups of the same order, $|G| = |H|$. If there exists a k by n matrix S such that $Sg_i = h_i S$ for some ordering of all elements $g_i \in G$ and $h_i \in H$, then either S is the zero matrix, or $n = k$ and S is a square nonsingular matrix.

Exercise A2.13: Show that every irreducible Abelian matrix group is one dimensional.

Exercise A2.14: Prove that if ρ is an irreducible representation of G, then $|G|/d_\rho$ is an integer.

The following theorem connects irreducibility with characters:

Theorem A2.3: A matrix group G is irreducible if and only if

$$\frac{1}{|G|} \sum_{g \in G} |\chi(g)|^2 = 1 . \tag{A2.2}$$

A2.2.2 Orthogonality
The key theorem of representation theory is the following:

Theorem A2.4: (**Fundamental theorem**) Every group G has exactly r inequivalent irreducible representations, where r is the number of conjugacy classes of G. And if $\rho^p \in M_{d_p}$ and ρ^q are any two of these, then the matrix elements satisfy the orthogonality relations

$$\sum_{g \in G} \left[\rho^p(g) \right]^{-1}_{ij} \left[\rho^q(g) \right]_{kl} = \frac{|G|}{d_p} \delta_{il} \delta_{jk} \delta_{pq} , \tag{A2.3}$$

where $\delta_{pq} = 1$ if $\rho^p = \rho^q$ and is zero otherwise.

Exercise A2.15: Using the Fundamental Theorem, prove that characters are orthogonal, that is:

$$\sum_{i=1}^{r} r_i \left(\chi_i^p\right)^* \chi_i^q = |G| \delta_{pq} \quad \text{and} \quad \sum_{p=1}^{r} \left(\chi_i^p\right)^* \chi_j^p = \frac{|G|}{r_i} \delta_{ij} , \tag{A2.4}$$

where p, q, and δ_{pq} have the same meaning as in the theorem, and χ_i^p is the value the character of the pth irreducible representation takes on the ith conjugacy class of G, and r_i is the size of the ith conjugacy class.

Exercise A2.16: S_3 is the group of permutations of three elements. Suppose we order these as mapping 123 to: $123; 231; 312; 213; 132$, and 321, respectively. Show that there exist two one-dimensional irreducible representations of S_3, one of which is trivial, and the other of which is $1, 1, 1, -1, -1, -1$, corresponding in order to the six permutations given earlier. Also show that there exists a two-dimensional irreducible representation, with the matrices

$$\begin{bmatrix} 1 & 0 \\ 0 & 1 \end{bmatrix}, \quad \frac{1}{2}\begin{bmatrix} -1 & -\sqrt{3} \\ \sqrt{3} & -1 \end{bmatrix}, \quad \frac{1}{2}\begin{bmatrix} -1 & \sqrt{3} \\ -\sqrt{3} & -1 \end{bmatrix},$$

$$\begin{bmatrix} -1 & 0 \\ 0 & 1 \end{bmatrix}, \quad \frac{1}{2}\begin{bmatrix} 1 & \sqrt{3} \\ \sqrt{3} & -1 \end{bmatrix}, \quad \frac{1}{2}\begin{bmatrix} 1 & -\sqrt{3} \\ -\sqrt{3} & -1 \end{bmatrix}. \tag{A2.5}$$

Verify that the representations are orthogonal.

A2.2.3 The regular representation

The number 1 is a valid one-dimensional matrix representation of any group. However, it is trivial. A representation is *faithful* if the matrix group of the representation is isomorphic to the original group. The *regular representation* is a faithful representation which exists for any group, and is constructed in the following manner. Let $\vec{v} = [g_1, g_2, \ldots, g_{|G|}]^T$ be a column vector of elements from G. Multiplying each element of $\vec{v}$ by an element $g \in G$ permutes the entries of the vector; this permutation can be represented by a $|G| \times |G|$ matrix acting by matrix multiplication on $\vec{v}$. The $|G|$ such matrices corresponding to the possible different permutations form a faithful representation of G, under matrix multiplication.

Exercise A2.17: Prove that the regular representation is faithful.

Exercise A2.18: Show that the character of the regular representation is zero except on the representation of the identity element, for which $\chi(I) = |G|$.

The decomposition of arbitrary representations into tensor sums of irreducible representations obeys the following theorem:

Theorem A2.5: If ρ is an arbitrary representation of G with character χ, and ρ^p are the inequivalent irreducible representations of G with characters χ^p, then $\rho = \oplus_p c_p \rho^p$, where $\oplus$ denotes a direct sum, and c_p are the numbers determined by

$$c_p = \frac{1}{|G|} \sum_{i=1}^{r} r_i \, (\chi_i^p)^* \chi_i . \tag{A2.6}$$

Exercise A2.19: Use Theorem A2.5 to show that the regular representation contains d_{ρ^p} instances of each irreducible representation ρ^p. Thus, if R denotes the regular representation, and $\hat{G}$ denotes the set of all inequivalent irreducible representations, then

$$\chi_i^R = \sum_{\rho \in \hat{G}} d_\rho \chi_i^\rho . \tag{A2.7}$$

Exercise A2.20: The character of the regular representation is zero except for the conjugacy class i containing e, the identity element in G. Show, therefore, that

$$\sum_{\rho \in \hat{G}} d_\rho \chi^\rho(g) = N \delta_{ge} .$$
(A2.8)

Exercise A2.21: Show that $\sum_{\rho \in \hat{G}} d_\rho^2 = |G|$.

A2.3 Fourier transforms

Let G be a finite group of order N, and f be a function which maps group elements to complex numbers. For an irreducible representation ρ of G, of dimension d_ρ, we define the Fourier transform of f to be $\hat{f}$,

$$\hat{f}(\rho) \equiv \sqrt{\frac{d_\rho}{N}} \sum_{g \in G} f(g) \rho(g) .$$
(A2.9)

Note that for ρ a matrix representation, $\hat{f}(\rho)$ maps matrices to matrices. Let $\hat{G}$ be a complete set of inequivalent irreducible representations of G. We define the inverse Fourier transform of $\hat{f}$ to be

$$f(g) = \frac{1}{\sqrt{N}} \sum_{\rho \in \hat{G}} \sqrt{d_\rho} \, \text{tr}(\hat{f}(\rho)\rho(g^{-1})) .$$
(A2.10)

Because $\sum_\rho d_\rho^2 = N$, f and $\hat{f}$ can both be expressed as vectors of complex numbers of length N. Note that the coefficients in the above equations have been chosen such that if $\hat{G}$ consists of unitary representations, then the Fourier transformations are unitary.

The above definitions can be understood by substituting (A2.9) into (A2.10) to obtain

$$f(g) = \frac{1}{N} \sum_{\rho \in \hat{G}} \sum_{g' \in G} d_\rho f(g') \, \text{tr} \left(\rho(g')\rho(g^{-1}) \right)$$
(A2.11)

$$= \frac{1}{N} \sum_{\rho \in \hat{G}} \sum_{g' \in G} d_\rho f(g') \, \text{tr} \left(\rho(g' \, g^{-1}) \right)$$
(A2.12)

$$= \frac{1}{N} \sum_{g' \in G} f(g') \sum_{\rho \in \hat{G}} d_\rho \, \chi^\rho(g' \, g^{-1}) .$$
(A2.13)

Using (A2.8), we may simplify (A2.13) to

$$f(g) = \sum_{g' \in G} f(g') \delta_{g' g} ,$$
(A2.14)

as desired.

Exercise A2.22: Substitute (A2.10) into (A2.9) and prove that $\hat{f}(\rho)$ is obtained.

Exercise A2.23: Let us represent an Abelian group G by $g \in [0, N-1]$, with addition as the group operation, and define $\rho_h(g) = \exp[-2\pi igh/N]$ as the h representation of g. This representation is one-dimensional, so $d_\rho = 1$. Show that the Fourier transform relations for G are

$$\hat{f}(h) = \frac{1}{\sqrt{N}} \sum_{g=0}^{N-1} f(g) \, e^{-2\pi igh/N} \quad \text{and} \quad f(h) = \frac{1}{\sqrt{N}} \sum_{g=0}^{N-1} \hat{f}(g) \, e^{2\pi ihg/N} . \quad \text{(A2.15)}$$

Exercise A2.24: Using the results of Exercise A2.16, construct the Fourier transform over S_3 and express it as a 6×6 unitary matrix.

History and further reading

There are many outstanding texts on group theory, and virtually any book on algebra has a section devoted to the theory of groups. The discussion here borrowed much notation from the text on finite groups by Lomont[Lom87]. Hammermesh is a standard reference for group theory in physics[Ham89]. Discussions of Fourier transforms over groups are not so common. Diaconis and Rockmore have written a good article on the efficient computation of Fourier transforms over groups[DR90]; many of their results are also reviewed in Fässler and Stiefel[FS92]. Beth independently discovered the fast Fourier transform over groups[Bet84], as did Clausen[Cla89].

Appendix 3: The Solovay–Kitaev theorem

In Chapter 4 we showed that an arbitrary unitary operation U may be implemented on a quantum computer using a circuit consisting of single qubit and controlled-NOT gates. Such *universality* results are important because they ensure the equivalence of apparently different models of quantum computation. For example, the universality results ensure that a quantum computer programmer may design quantum circuits containing gates which have four input and output qubits, confident that such gates can be simulated by a constant number of controlled-NOT and single qubit unitary gates.

An unsatisfactory aspect of the universality of controlled-NOT and single qubit unitary gates is that the single qubit gates form a *continuum*, while the methods for fault-tolerant quantum computation described in Chapter 10 work only for a *discrete* set of gates. Fortunately, also in Chapter 4 we saw that *any* single qubit gate may be approximated to arbitrary accuracy using a finite set of gates, such as the controlled-NOT gate, Hadamard gate H, phase gate S, and $\pi/8$ gate. We also gave a heuristic argument that approximating the chosen single qubit gate to an accuracy ϵ required only $\Theta(1/\epsilon)$ gates chosen from the finite set. Furthermore, in Chapter 10 we showed that the controlled-NOT, Hadamard, phase and $\pi/8$ gates may be implemented in a fault-tolerant manner.

In this appendix we show that a much faster rate of convergence than $\Theta(1/\epsilon)$ may be achieved. The *Solovay–Kitaev theorem* shows that for any gate U on a single qubit, and given any $\epsilon > 0$, it is possible to approximate U to a precision ϵ using $\Theta(\log^c(1/\epsilon))$ gates from a fixed finite set, where c is a small constant approximately equal to 2. The best possible value for c isn't known yet, so we are going to explain the proof of the Solovay–Kitaev theorem for c approximately equal to 4, and then in the end of appendix problems outline a method that may be used to reduce c down closer to 2. We will also prove that c cannot be less than 1; determining the best possible value of c between 1 and 2 is an open problem!

To appreciate the importance of the Solovay–Kitaev theorem, imagine a quantum computer programmer designs an algorithm using $f(n)$ gates to solve a problem. Suppose the algorithm he or she comes up with uses many gates outside the usual fault-tolerant set of controlled-NOT, Hadamard, phase and $\pi/8$ gates. How many gates are required to implement the algorithm fault-tolerantly? If the tolerance to error for the entire algorithm is to be ϵ, then the individual gates must be accurate to a tolerance $\epsilon/f(n)$. By the heuristic argument of Chapter 4 it would take $\Theta(f(n)/\epsilon)$ gates in the fault-tolerant set to approximate each one of the gates used in the algorithm, for a total cost $\Theta(f(n)^2/\epsilon)$, a polynomial increase in the number of gates required by the algorithm. Using the Solovay–Kitaev theorem each gate in the algorithm can be simulated by $O(\log^c(f(n)/\epsilon))$ gates in the fault-tolerant set, for a total cost $O(f(n)\log^c(f(n)/\epsilon))$ for the fault-tolerant algorithm, which is only polylogarithmically more than for the original algorithm. For many problems, such a polylogarithmic cost is quite acceptable, whereas the polynomial cost provided by the heuristic argument of Chapter 4 may be much less desirable.

To state the Solovay–Kitaev theorem more precisely, we need to define some notation and nomenclature. Recall that $SU(2)$ is the set of all single qubit unitary matrices with determinant equal to one. We restrict our attention to $SU(2)$, since all single qubit unitary gates may be written as the product of an element of $SU(2)$ with an unimportant global phase factor. Suppose $\mathcal{G}$ is a finite set of elements of $SU(2)$; $\mathcal{G}$ plays the role of the finite set of elementary gates our quantum computer programmer is using to simulate all the other gates. For the sake of concreteness, think of $\mathcal{G}$ as containing the fault-tolerant set H, S, and T, with appropriate global phases added to ensure the determinants are all equal to one. We suppose for the sake of convenience that $\mathcal{G}$ *contains its own inverses*, that is, if $U \in \mathcal{G}$, then $U^\dagger \in \mathcal{G}$. In the case of the fault-tolerant set, that means adding $S^\dagger = S^3$ and $T^\dagger = T^7$ to the set, which fortunately can be expressed in terms of gates already in the set. A *word* of length l from $\mathcal{G}$ is a product $g_1 g_2 \ldots g_l \in SU(2)$, where $g_i \in \mathcal{G}$ for each i. We define $\mathcal{G}_l$ to be the set of all words of length at most l, and $\langle \mathcal{G} \rangle$ to be the set of all words of finite length.

We need some notion of *distance* to quantify what we mean by an approximation to a unitary matrix. The exact measure used is not all that important. It is convenient for our purposes to use the *trace distance* studied in Chapter 9, $D(U, V) \equiv \mathrm{tr}|U - V|$, where $|X| \equiv \sqrt{X^\dagger X}$ is the positive square root of $X^\dagger X$. Actually, this definition differs by a factor 2 from the definition used in Chapter 9; the reason for using a different normalization is that it makes geometric visualization of the proof of the Solovay–Kitaev theorem easier, as we shall see (it will also be helpful to think of elements in $SU(2)$ as being points in space). A subset S of $SU(2)$ is said to be *dense* in $SU(2)$ if for any element U of $SU(2)$ and $\epsilon > 0$ there is an element $s \in S$ such that $D(s, U) < \epsilon$. Suppose S and W are subsets of $SU(2)$. Then S is said to form an ϵ-net for W, where $\epsilon > 0$, if every point in W is within a distance ϵ of some point in S. Our interest is in how fast $\mathcal{G}_l$ 'fills up' $SU(2)$ as l is increased. That is, for how small an ϵ is $\mathcal{G}_l$ an ϵ-net for $SU(2)$? The Solovay–Kitaev theorem says that ϵ gets small very quickly indeed as l is increased.

Exercise A3.1: In Chapter 4 we made use of the distance measure
$E(U, V) = \max_{|\psi\rangle} \|(U - V)|\psi\rangle\|$, where the maximum is over all pure states $|\psi\rangle$.
Show that when U and V are single qubit rotations, $U = R_{\hat{m}}(\theta), V = R_{\hat{n}}(\varphi)$,
$D(U, V) = 2E(U, V)$, and thus it does not matter whether we use the trace
distance or the measure $E(\cdot, \cdot)$ for the Solovay–Kitaev theorem.

Theorem A3.1: (**Solovay–Kitaev theorem**) Let $\mathcal{G}$ be a finite set of elements in $SU(2)$
containing its own inverses, such that $\langle \mathcal{G} \rangle$ is dense in $SU(2)$. Let $\epsilon > 0$ be given.
Then $\mathcal{G}_l$ is an ϵ-net in $SU(2)$ for $l = O(\log^c(1/\epsilon))$, where $c \approx 4$.

As already noted, the best possible value of c is somewhat lower than 4, but it is convenient to give the proof for this particular case. In Problem 3.1 we explain how modifications of the proof can be used to lower c. The first part of the proof is to show that the points of $\mathcal{G}_l$ get very dense in a small neighbourhood of the identity matrix I as l is increased, a conclusion encapsulated in the following lemma. To state the lemma, we define S_ϵ to be the set of all points U in $SU(2)$ such that $D(U, I) \leq \epsilon$.

Lemma A3.2: Let $\mathcal{G}$ be a finite set of elements in $SU(2)$ containing its own inverses, and such that $\langle\mathcal{G}\rangle$ is dense in $SU(2)$. There exists a universal constant ϵ_0 independent of $\mathcal{G}$, such that for any $\epsilon \leq \epsilon_0$, if $\mathcal{G}_l$ is an ϵ^2-net for S_ϵ, then $\mathcal{G}_{5l}$ is a $C\epsilon^3$-net for $S_{\sqrt{C}\epsilon^{3/2}}$, for some constant C.

We prove Lemma A3.2 shortly, but first let's see how it implies the Solovay–Kitaev theorem. There are two steps to the proof. The first step is to apply Lemma A3.2 iteratively to show that the neighbourhood of the origin fills in very quickly as the word length l is increased. Since $\mathcal{G}$ is dense in $SU(2)$ we can find an l_0 such that $\mathcal{G}_{l_0}$ is an ϵ_0^2-net for $SU(2)$, and thus also for S_{ϵ_0}. Applying Lemma A3.2 with $\epsilon = \epsilon_0$ and $l = l_0$ implies that $\mathcal{G}_{5l_0}$ is a $C\epsilon_0^3$-net for $S_{\sqrt{C}\epsilon_0^{3/2}}$. Applying Lemma A3.2 again with $\epsilon = \sqrt{C}\epsilon_0^{3/2}$ and $l = 5l_0$ implies that $\mathcal{G}_{5^2 l_0}$ is a $C(\sqrt{C}\epsilon_0^{3/2})^3$-net for $S_{\sqrt{C}(\sqrt{C}\epsilon_0^{3/2})^{3/2}}$. Iterating this procedure k times, we find that $\mathcal{G}_{5^k l_0}$ is an $\epsilon(k)^2$-net for $S_{\epsilon(k)}$, where

$$\epsilon(k) = \frac{(C\epsilon_0)^{(3/2)^k}}{C}. \tag{A3.1}$$

Without loss of generality we may suppose ϵ_0 has been chosen such that $C\epsilon_0 < 1$, and therefore $\epsilon(k)$ gets small very fast as k increases. It will also be useful to note that, provided ϵ_0 is chosen small enough, $\epsilon(k)^2 < \epsilon(k+1)$.

The second step is to let U be any element of $SU(2)$ and use the translation idea illustrated in Figure A3.1 to approximate U using products of elements of $\mathcal{G}$. Let $U_0 \in \mathcal{G}_{l_0}$ be an $\epsilon(0)^2$-approximation to U. Now define V so that $VU_0 = U$, that is, $V \equiv UU_0^\dagger$. Thus $D(V, I) = \mathrm{tr}|V - I| = \mathrm{tr}|(U - U_0)U_0^\dagger| = \mathrm{tr}|U - U_0| < \epsilon(0)^2 < \epsilon(1)$. From the iterated application of Lemma A3.2 discussed above, we can find $U_1 \in \mathcal{G}_{5l_0}$ which is an $\epsilon(1)^2$-approximation to V. It follows that $U_1 U_0$ is an $\epsilon(1)^2$-approximation to U. Now define V' so that $V'U_1 U_0 = U$, that is, $V' \equiv UU_0^\dagger U_1^\dagger$. Thus $D(V', I) = \mathrm{tr}|V - I| = \mathrm{tr}|(U - U_1 U_0)U_0^\dagger U_1^\dagger| = \mathrm{tr}|U - U_1 U_0|/ < \epsilon(1)^2 < \epsilon(2)$. It follows from the iterated application of Lemma A3.2 that we can find $U_2 \in \mathcal{G}_{5^2 l_0}$ which is an $\epsilon(2)^2$-approximation to V', and thus $U_2 U_1 U_0$ is an $\epsilon(2)^2$-approximation to U. Continuing in this way we construct $U_k \in \mathcal{G}_{5^k l_0}$ so that $U_k U_{k-1} \ldots U_0$ is an $\epsilon(k)^2$-approximation to U.

Putting it all together, a sequence of $l_0 + 5l_0 + \cdots + 5^k l_0 < \frac{5}{4}5^k l_0$ gates can be used to approximate any unitary gate U to an accuracy $\epsilon(k)^2$. To approximate to some desired accuracy ϵ, we therefore must choose k such that

$$\epsilon(k)^2 < \epsilon. \tag{A3.2}$$

Substituting (A3.1) this can be restated as

$$\left(\frac{3}{2}\right)^k < \frac{\log(1/C^2\epsilon)}{2\log(1/C\epsilon_0)}. \tag{A3.3}$$

It follows that the number of gates required to approximate to within ϵ satisfies ($c = \log 5/\log(3/2) \approx 4$)

$$\text{number of gates} < \frac{5}{4}5^k l_0 = \frac{5}{4}\left(\frac{3}{2}\right)^{kc} l_0 < \frac{5}{4}\left(\frac{\log(1/C^2\epsilon)}{2\log(1/C\epsilon_0)}\right)^c l_0. \tag{A3.4}$$

That is, the number of gates required to approximate to within ϵ is $O(\log^c(1/\epsilon))$, completing the proof of the Solovay–Kitaev theorem.

The proof of Lemma A3.2 uses a few elementary facts about multiplication of elements

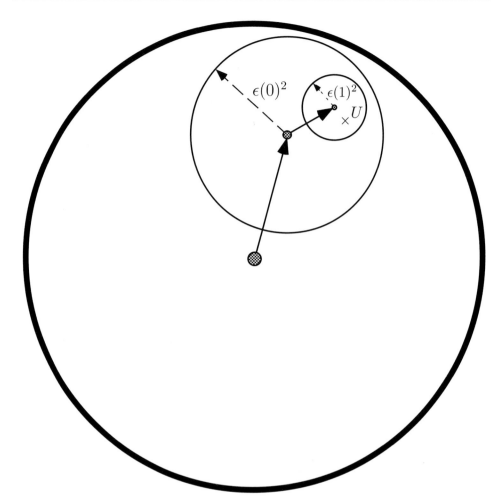

Figure A3.1. The translation step used in the proof of the Solovay–Kitaev theorem. To approximate an arbitrary single qubit gate we first approximate to within a distance $\epsilon(0)^2$ using l_0 gates from $\mathcal{G}$. Then we improve the approximation by adding $5l_0$ more gates, for a total accuracy better than $\epsilon(1)^2$, and continue on in this way, quickly converging to U.

of $SU(2)$, which we now recall. The key idea of the lemma is to work in the neighborhood of the identity, which greatly simplifies the rather complicated operation of multiplication in $SU(2)$. More precisely, suppose U and V are elements of $SU(2)$, and define the *group commutator* of U and V by

$$[U, V]_{\text{gp}} \equiv UVU^\dagger V^\dagger. \tag{A3.5}$$

Suppose U and V are both close to the identity, so that they may be written $U = e^{-iA}$ and $V = e^{-iB}$, where A and B are Hermitian matrices such that $\text{tr}|A|, \text{tr}|B| \leq \epsilon$ for some small ϵ. Expanding $e^{\pm iA}$ and $e^{\pm iB}$ to terms quadratic in A and B gives

$$D([U, V]_{\text{gp}}, e^{-[A,B]}) = O(\epsilon^3), \tag{A3.6}$$

where $[A, B] = AB - BA$ is the usual commutator for matrices (in fact, the commutator for the Lie algebra of $SU(2)$). Thus, in the neighborhood of the identity we can study the group commutator by studying instead the much simpler matrix commutator.

Indeed, for qubits the matrix commutator has an especially nice form. An arbitrary element of $SU(2)$ may be written $U = u(\vec{a}) \equiv \exp(-i\vec{a} \cdot \vec{\sigma}/2)$ for some real vector $\vec{a}$. Similarly $V = u(\vec{b}) = \exp(-i\vec{b} \cdot \vec{\sigma}/2)$ for some real vector $\vec{b}$. Recalling from Exercise 2.40 that

$$[\vec{a} \cdot \sigma, \vec{b} \cdot \sigma] = 2i \left(\vec{a} \times \vec{b} \right) \cdot \vec{\sigma}, \tag{A3.7}$$

we see from (A3.6) that

$$D \left([U, V]_{\mathrm{gp}}, u \left(\vec{a} \times \vec{b} \right) \right) = O(\epsilon^3). \tag{A3.8}$$

The basic idea of the proof of Lemma A3.2 is now easy to understand. The details, most of which relate to approximation issues, are filled in below for completeness; for now we just give the main idea, as illustrated in Figure A3.2. Suppose we wish to approximate some element $U = u(\vec{x})$ in S_{ϵ^2}. You will see in Exercise A3.4 that trace distances like $D(U, I)$ are equal (up to small corrections) to the Euclidean distance $\|\vec{x}\|$, so to a good approximation $\|\vec{x}\| \leq \epsilon^2$. We can always choose $\vec{y}$ and $\vec{z}$ of length at most ϵ, such that $\vec{x} = \vec{y} \times \vec{z}$. Pick $\vec{y}_0$ and $\vec{z}_0$ such that $u(\vec{y}_0)$ and $u(\vec{z}_0)$ are elements of $\mathcal{G}_l$ that ϵ^2-approximate $u(\vec{y})$ and $u(\vec{z})$, respectively. Applying (A3.6) to the commutator $[u(\vec{y}_0), u(\vec{z}_0)]_{\mathrm{gp}}$ we get an $O(\epsilon^3)$-approximation to U. This gives a $O(\epsilon^3)$-net for S_{ϵ^2}; to finish the proof of the lemma we apply a translation step like that in the main part of the proof of the Solovay–Kitaev theorem, obtaining in $5l$ gates an $O(\epsilon^3)$-approximation to any element of $S_{O(\epsilon^{3/2})}$.

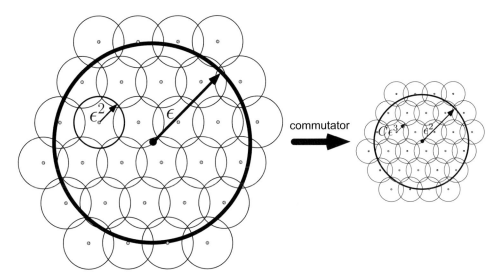

Figure A3.2. The main idea in the proof of Lemma A3.2. Taking group commutators of elements U_1 and U_2 in S_ϵ fills in S_{ϵ^2} much more densely. Note that the density of circles appearing on the right hand side ought to be much higher than is shown, as there should be one for each pair of circles on the left; the lower density is merely for clarity. The proof of the lemma is completed by applying a translation step (not shown) to get good approximations to any element of $S_{\sqrt{3}\epsilon^{3/2}}$.

Exercise A3.2: Suppose A and B are Hermitian matrices such that $\mathrm{tr}|A|, \mathrm{tr}|B| \leq \epsilon$. Prove that for all sufficiently small ϵ,

$$D \left(\left[e^{-iA}, e^{-iB} \right]_{\mathrm{gp}}, e^{-[A,B]} \right) \leq d\epsilon^3, \tag{A3.9}$$

for some constant d, establishing Equation (A3.6). (*Comment*: for practical purposes it may be interesting to obtain good bounds on d.)

Exercise A3.3: Let $\vec{x}$ and $\vec{y}$ be any two real vectors. Show that

$$D(u(\vec{x}), u(\vec{y})) = 2\sqrt{2}\sqrt{1 - \cos(x/2)\cos(y/2) - \sin(x/2)\sin(y/2)\hat{x} \cdot \hat{y}}, \quad \text{(A3.10)}$$

where $x \equiv \|\vec{x}\|, y \equiv \|\vec{y}\|$, and $\hat{x}$ and $\hat{y}$ are unit vectors in the $\vec{x}$ and $\vec{y}$ directions, respectively.

Exercise A3.4: Show that in the case $\vec{y} = 0$ the formula for $D(u(\vec{x}), u(\vec{y}))$ reduces to

$$D(u(\vec{x}), I) = 4\sin\left|\frac{x}{4}\right|. \quad \text{(A3.11)}$$

Exercise A3.5: Show that when $x, y \leq \epsilon$,

$$D(u(\vec{x}), u(\vec{y})) = \|\vec{x} - \vec{y}\| + O(\epsilon^3). \quad \text{(A3.12)}$$

Proof
(Lemma A3.2)

Suppose $\mathcal{G}_l$ is an ϵ^2-net in S_ϵ. The first step of the proof is to show that $[\mathcal{G}_l, \mathcal{G}_l]_{\text{gp}}$ is a $C\epsilon^3$-net for S_{ϵ^2} and some constant C.

Let $U \in S_{\epsilon^2}$ and pick $\vec{x}$ such that $U = u(\vec{x})$. By Exercise A3.4 it follows that $x \leq \epsilon^2 + O(\epsilon^6)$. Choose any pair of vectors $\vec{y}$ and $\vec{z}$ of length at most $\epsilon + O(\epsilon^3)$ such that $\vec{x} = \vec{y} \times \vec{z}$. $\mathcal{G}_l$ is an ϵ^2-net for S_ϵ, so choose U_1 and U_2 in $\mathcal{G}_l \cap S_\epsilon$ such that

$$D(U_1, u(\vec{y})) < \epsilon^2 + O(\epsilon^5) \quad \text{(A3.13)}$$
$$D(U_2, u(\vec{z})) < \epsilon^2 + O(\epsilon^5), \quad \text{(A3.14)}$$

and let $\vec{y}_0$ and $\vec{z}_0$ be chosen such that $U_1 = u(\vec{y}_0)$ and $U_2 = u(\vec{z}_0)$. By Exercise A3.4 it follows that $y_0, z_0 \leq \epsilon + O(\epsilon^3)$. Our goal is to show that $D(U, [U_1, U_2]_{\text{gp}})$ is smaller than $C\epsilon^3$. To do this, we use the triangle inequality,

$$D(U, [U_1, U_2]_{\text{gp}}) \leq D(U, u(\vec{y}_0 \times \vec{z}_0)) + D(u(\vec{y}_0 \times \vec{z}_0), [U_1, U_2]_{\text{gp}}). \quad \text{(A3.15)}$$

The second term is at most $d'\epsilon^3$, by Exercise A3.2, where d' is a constant slightly larger than d because of the possible contribution due to the fact that $y_0, z_0 < \epsilon + O(\epsilon^3)$, rather than $y_0, z_0 < \epsilon$. Substituting $U = u(\vec{x})$, making use of Exercise A3.5, introducing an appropriate constant d'' and doing some elementary algebra gives

$$D(U, [U_1, U_2]_{\text{gp}}) \leq D(u(\vec{x}), u(\vec{y}_0 \times \vec{z}_0) + d'\epsilon^3 \quad \text{(A3.16)}$$
$$= \|\vec{x} - \vec{y}_0 \times \vec{z}_0\| + d''\epsilon^3 \quad \text{(A3.17)}$$
$$= \|\vec{y} \times \vec{z} - \vec{y}_0 \times \vec{z}_0\| + d''\epsilon^3. \quad \text{(A3.18)}$$
$$= \|[(\vec{y} - \vec{y}_0) + \vec{y}_0] \times [(\vec{z} - \vec{z}_0) + \vec{z}_0] - \vec{y}_0 \times \vec{z}_0\| + d''\epsilon^3 \quad \text{(A3.19)}$$
$$\leq (d'' + 2)\epsilon^3 + O(\epsilon^4) \quad \text{(A3.20)}$$
$$\leq C\epsilon^3, \quad \text{(A3.21)}$$

where C is an appropriately chosen constant.

The second step in the proof of the lemma is to apply a translation step like that used in the main part of the proof of the Solovay–Kitaev theorem. Specifically, given

$U \in S_{\sqrt{C\epsilon^3}}$, we can find V in $\mathcal{G}_l$ such that $D(U,V) \leq \epsilon^2$, and thus $UV^\dagger \in S_{\epsilon^2}$. Then find W_1 and W_2 in $\mathcal{G}_l$ such that $D([W_1,W_2]_{\text{gp}}, UV^\dagger) \leq C\epsilon^3$, and therefore

$$D([W_1, W_2]_{\text{gp}} V, U) \leq C\epsilon^3, \tag{A3.22}$$

which completes the proof. □

Exercise A3.6: Fixing the set $\mathcal{G}$ of elementary gates, describe an algorithm which, given a description of a single qubit unitary gate U and a desired accuracy $\epsilon > 0$, efficiently computes a sequence of gates from $\mathcal{G}$ that ϵ-approximates U.

The analysis in this appendix is rather crude, and a much tighter analysis can be made. One issue of especial interest is the best possible value of the exponent c in the $O(\log^c(1/\epsilon))$ bound. It is not difficult to show that c can be no less than 1. To see this, imagine we have a collection of N little balls, all of radius ϵ, in $SU(2)$. The volume of these balls scales like ϵ^d, for some unimportant constant d. Therefore, if the balls are to cover $SU(2)$, N must be of size $\Omega(1/\epsilon^d)$. Suppose we consider all possible sequences $U_1 U_2 \ldots U_g$ consisting of g gates chosen from $\mathcal{G}$. Clearly such sequences can generate at most $|\mathcal{G}|^g$ distinct unitary operations. Thus we must have $|\mathcal{G}|^g = \Omega(1/\epsilon^d)$, which implies the desired lower bound on the number of gates,

$$g = \Omega\left(\log\left(\frac{1}{\epsilon}\right)\right). \tag{A3.23}$$

Problem 3.1: The following problem outlines a more elaborate construction that achieves an $O(\log^2(1/\epsilon) \log^c(\log(1/\epsilon)))$ bound on the number of gates required to approximate to within ϵ of a desired target, for any $c > 2$.

(1) Suppose $\mathcal{N}$ is a δ-net in S_ϵ, for $0 < \delta < \epsilon \leq \epsilon_0$, ϵ_0 sufficiently small. Show that $[\mathcal{N}, \mathcal{N}]_{\text{gp}}$ is a $d\delta\epsilon$-net in S_{ϵ^2}, for some constant d.

(2) Suppose $\mathcal{G}_l$ is a δ-net in S_ϵ, for $0 < \delta < \epsilon \leq \epsilon_0$. Show that $\mathcal{G}_{4^k l}$ is a $d^k \delta \epsilon^{2^k - 1}$-net in $S_{\epsilon^{2^k}}$.

(3) Suppose we define k by

$$k \equiv \left\lceil \log\left(\frac{\log(1/\epsilon)}{\log(1/\epsilon_0)}\right) \right\rceil, \tag{A3.24}$$

and suppose we can find l such that $\mathcal{G}_l$ is a δ_0-net for S_{ϵ_0}, where

$$d^k \delta_0 = \epsilon_0. \tag{A3.25}$$

Show that $\mathcal{G}_{4^k l}$ is an ϵ-net for $S_{\epsilon_0^{2^k}}$.

(4) Use the already-proved version of the Solovay–Kitaev theorem to show that choosing $l = O(k^c)$ suffices in the previous part of this problem, where $c = \log(5)/\log(3/2)$ is the constant appearing in the exponent in the already-proved version of the Solovay–Kitaev theorem.

(5) Combine the previous results to prove that $O(\log^2(1/\epsilon) \log^c(\log(1/\epsilon)))$ gates can be used to ϵ-approximate an arbitrary gate in $SU(2)$.

(6) Show that any $c > 2$ can appear in the conclusion of the previous result.

Problem 3.2: (**Research**) If it exists, find an approximation procedure asympoticially faster than the result found in the previous problem. Ideally, a procedure would (a) saturate the $\Omega(\log(1/\epsilon))$ lower bound on the number of gates required to perform the approximation, and (b) provide an efficient algorithm for constructing such approximating sequences of gates.

Problem 3.3: (**Research**) Fix a finite set of single qubit gates $\mathcal{G}$ which can be performed fault-tolerantly and which generate a set dense in the single qubit gates; say the $\pi/8$ gate and the Hadamard gate. Develop an elegant, efficient and reasonably tight method which, given an arbitrary single qubit gate U and some $\epsilon > 0$, produces a sequence of gates from the fault-tolerant set giving an ϵ-approximation to U, up to global phase.

History and further reading

The results in this appendix were proved by Solovay in 1995 (unpublished manuscript), and independently by Kitaev, who gave an outline of the proof in [Kit97b]. In the same paper Kitaev observed that the result can be generalized to many Lie groups other than $SU(2)$; roughly speaking, the key fact about $SU(2)$ used in the proof was that $[S_\epsilon, S_\epsilon]_{\text{gp}} \supseteq S_{\Omega(\epsilon^2)}$, and other Lie groups for which this fact holds also obey some version of the Solovay–Kitaev theorem. The Solovay–Kitaev theorem is true, for example, for the Lie group $SU(d)$ of d by d unitary matrices with unit determinant. After hearing of this result, Solovay subsequently generalized his proof in a similar fashion. Our presentation has benefited substantially from a 1999 lecture of Freedman, and by discussions with Freedman, Kitaev and Solovay.

Appendix 4: Number theory

Understanding some elementary number theory is necessary if we are to understand cryptosystems and how quantum computers can be used to break them. In this appendix we review some basic facts about number theory.

A4.1 Fundamentals

Let's start off by agreeing about a few conventions for nomenclature and notation. The set of *integers* is the set $\{\ldots, -2, -1, 0, 1, 2, \ldots\}$, denoted $\mathbf{Z}$. We may occasionally refer to the *natural numbers*, meaning non-negative integers, but more often we'll say *non-negative integer* or *positive integer*, in order to make the distinction between the case when zero is included, and when zero is not included.

Suppose n is an integer. An integer d *divides* n (written $d|n$) if there exists an integer k such that $n = dk$. We say in this case that d is a *factor* or *divisor* of n. Notice that 1 and n are always factors of n. When d does not divide (is not a factor of) n we write $d \nmid n$. For example, $3|6$ and $3|18$, but $3 \nmid 5$ and $3 \nmid 7$.

Exercise A4.1: (Transitivity) Show that if $a|b$ and $b|c$ then $a|c$.

Exercise A4.2: Show that if $d|a$ and $d|b$ then d also divides linear combinations of a and b, $ax + by$, where x and y are integers.

Exercise A4.3: Suppose a and b are positive integers. Show that if $a|b$ then $a \leq b$. Conclude that if $a|b$ and $b|a$ then $a = b$.

A *prime number* is an integer greater than 1 which has only itself and 1 as factors. The first few prime numbers are $2, 3, 5, 7, 11, 13, 17, \ldots$. Perhaps the most important single fact about the positive integers is that they may be represented uniquely as a product of factors which are prime numbers. This result is given an appropriately impressive name, the *fundamental theorem of arithmetic*:

Theorem A4.1: (**Fundamental theorem of arithmetic**) Let a be any integer greater than 1. Then a has a *prime factorization* of the form

$$a = p_1^{a_1} p_2^{a_2} \ldots p_n^{a_n}, \tag{A4.1}$$

where $p_1, \ldots, p_n$ are distinct prime numbers, and $a_1, \ldots, a_n$ are positive integers. Moreover, this prime factorization is unique, up to the order of the factors.

Proof

The reader who has never seen a proof of the fundamental theorem of arithmetic is strongly encouraged to attempt to supply it for themselves. Failing that, a proof may be found in any elementary number theory text; see the end of appendix 'History and further reading' for references.   □

For small numbers it is easy to find the prime factorization by trial and error, for example $20 = 2^2 \cdot 5^1$. For large numbers no efficient algorithm is known to find the prime factorization on a classical computer, despite immense effort aimed at finding such an algorithm.

Exercise A4.4:   Find the prime factorizations of 697 and 36 300.

A4.2   Modular arithmetic and Euclid's algorithm

We're all thoroughly familiar with the techniques of ordinary arithmetic. Another type of arithmetic, *modular arithmetic*, is extremely useful in understanding the properties of numbers. We assume that you are familiar with the elementary ideas of modular arithmetic, and so will quickly breeze through the basic ideas and notation, before coming to more advanced theory.

Modular arithmetic can be thought of as the arithmetic of *remainders*. If we divide 18 by 7 we get the answer 2, with a remainder of 4. More formally, given any positive integers x and n, x can be written (uniquely) in the form

$$x = kn + r, \qquad\qquad\qquad (A4.2)$$

where k is a non-negative integer, the result of dividing x by n, and the *remainder* r lies in the range 0 to $n - 1$, inclusive. Modular arithmetic is simply ordinary arithmetic in which we only pay attention to remainders. We use the notation (mod n) to indicate that we are working in modular arithmetic. For instance, we write $2 = 5 = 8 = 11 (\text{mod } 3)$, because 2, 5, 8 and 11 all have the same remainder (2) when divided by 3. The appellation '(mod n)' reminds us that we are working in modular arithmetic, with respect to the number n.

Addition, multiplication, and subtraction operations for modular arithmetic may all be defined in the obvious ways, but it is perhaps not so obvious how to define a division operation. To understand how this may be done we introduce another key concept from number theory, that of the *greatest common divisor* of two integers. The greatest common divisor of integers a and b is the largest integer which is a divisor of both a and b. We write this number as $\gcd(a, b)$. For example, the greatest common divisor of 18 and 12 is 6. An easy way of seeing this is to enumerate the positive divisors of 18 (1, 2, 3, 6, 9, 18) and 12 (1, 2, 3, 4, 6, 12), and then pick out the largest common element in the two lists. This method is quite inefficient and impractical for large numbers. Fortunately, there is a much more efficient way of working out the greatest common divisor, a method known as *Euclid's algorithm*, whose explication occupies us for the next few pages.

Theorem A4.2: **(Representation theorem for the gcd)** The greatest common divisor of two integers a and b is the least positive integer that can be written in the form $ax + by$, where x and y are integers.

Proof

Let $s = ax + by$ be the smallest positive integer that can be written in this form. Since $\gcd(a, b)$ is a divisor of both a and b it is also a divisor of s. It follows that $\gcd(a, b) \leq s$. To complete the proof we demonstrate that $s \leq \gcd(a, b)$ by showing that s is a divisor of both a and b. The proof is by contradiction. Suppose s is not a divisor of a. Then $a = ks + r$, where the remainder r is in the range 1 to $s - 1$. Rearranging this equation and using $s = ax + by$ we see that $r = a(1 - kx) + b(-ky)$ is a positive integer that can be written as a linear combination of a and b, and which is smaller than s. But this contradicts the definition of s as the smallest positive integer that can be written as a linear combination of a and b. We conclude that s must divide a. By symmetry s must also be a divisor of b, which completes the proof. □

Corollary A4.3: Suppose c divides both a and b. Then c divides $\gcd(a, b)$.

Proof

By Theorem A4.2, $\gcd(a, b) = ax + by$ for some integers x and y. Since c divides a and b it must also divide $ax + by$. □

When does a number, a, have a multiplicative inverse in modular arithmetic? That is, given a and n, when does there exist a b such that $ab = 1 (\text{mod } n)$? For example, note that $2 \cdot 3 = 1 (\text{mod } 5)$, so the number 2 has multiplicative inverse 3 in arithmetic modulo 5. On the other hand, trial and error shows that 2 has no multiplicative inverse modulo 4. Finding multiplicative inverses in modular arithmetic turns out to be related to the gcd by the notion of *co-primality*: integers a and b are said to be *co-prime* if their greatest common divisor is 1. For example, 14 and 9 are co-prime, since the positive divisors of 14 are 1, 2, 7 and 14, while 1, 3 and 9 are the positive divisors of 9. The following corollary characterizes the existence of multiplicative inverses in modular arithmetic using co-primality.

Corollary A4.4: Let n be an integer greater than 1. An integer a has a multiplicative inverse modulo n if and only if $\gcd(a, n) = 1$, that is, a and n are co-prime.

Proof

Suppose a has a multiplicative inverse, which we denote a^{-1}, modulo n. Then $aa^{-1} = 1 + kn$ for some integer k, and thus $aa^{-1} + (-k)n = 1$. From Theorem A4.2 we conclude that $\gcd(a, n) = 1$. Conversely, if $\gcd(a, n) = 1$ then there must exist integers a^{-1} and b such that $aa^{-1} + bn = 1$, and therefore $aa^{-1} = 1 (\text{mod } n)$. □

Exercise A4.5: For p a prime prove that all integers in the range 1 to $p - 1$ have multiplicative inverses modulo p. Which integers in the range 1 to $p^2 - 1$ do not have multiplicative inverses modulo p^2?

Exercise A4.6: Find the multiplicative inverse of 17 modulo 24.

Exercise A4.7: Find the multiplicative inverse of $n + 1$ modulo n^2, where n is any integer greater than 1.

Exercise A4.8: (Uniqueness of the inverse) Suppose b and b' are multiplicative inverses of a, modulo n. Prove that $b = b' (\text{mod } n)$.

The next theorem is the key to Euclid's efficient algorithm for finding the greatest common divisor of two positive integers.

Theorem A4.5: Let a and b be integers, and let r be the remainder when a is divided by b. Then provided $r \neq 0$,

$$\gcd(a, b) = \gcd(b, r). \tag{A4.3}$$

Proof

We prove the equality by showing that each side divides the other. To prove that the left hand side divides the right note that $r = a - kb$ for some integer k. Since $\gcd(a, b)$ divides a, b and linear combinations of these it follows that $\gcd(a, b)$ divides r. By Corollary A4.3, $\gcd(a, b)$ divides $\gcd(b, r)$. To prove that the right hand side divides the left note that $\gcd(b, r)$ divides b, and since $a = r + kb$ is a linear combination of b and r it follows that $\gcd(b, r)$ also divides a. By Corollary A4.3, $\gcd(b, r)$ divides $\gcd(a, b)$. □

Exercise A4.9:   Explain how to find $\gcd(a, b)$ if the prime factorizations of a and b are known. Find the prime factorizations of 6825 and 1430, and use them to compute $\gcd(6825, 1430)$

Euclid's algorithm for finding the greatest common divisor of positive integers a and b works as follows. First, order a and b so that $a > b$. Divide b into a, with result k_1 and remainder r_1: $a = k_1 b + r_1$. By Theorem A4.5 $\gcd(a, b) = \gcd(b, r_1)$. Next, we perform a second division with b playing the role of a, and r_1 playing the role of b: $b = k_2 r_1 + r_2$. By Theorem A4.5 $\gcd(a, b) = \gcd(b, r_1) = \gcd(r_1, r_2)$. Next, we perform a third division with r_1 playing the role of a and r_2 the role of b: $r_1 = k_3 r_2 + r_3$. By Theorem A4.5 $\gcd(a, b) = \gcd(b, r_1) = \gcd(r_1, r_2) = \gcd(r_2, r_3)$. We continue in this manner, each time dividing the most recent remainder by the second most recent remainder, obtaining a new result and remainder. The algorithm halts when we obtain a remainder that is zero, that is, $r_m = k_{m+1} r_{m+1}$ for some m. We have $\gcd(a, b) = \gcd(r_m, r_{m+1}) = r_{m+1}$, so the algorithm returns r_{m+1}.

As an example of the use of Euclid's algorithm we find $\gcd(6825, 1430)$:

$$6825 = 4 \times 1430 + 1105 \tag{A4.4}$$
$$1430 = 1 \times 1105 + 325 \tag{A4.5}$$
$$1105 = 3 \times 325 + 130 \tag{A4.6}$$
$$325 = 2 \times 130 + 65 \tag{A4.7}$$
$$130 = 2 \times 65. \tag{A4.8}$$

From this we see that $\gcd(6825, 1430) = 65$.

An adaptation of Euclid's algorithm may be used to efficiently find integers x and y such that $ax + by = \gcd(a, b)$. The first stage is to run through the steps of Euclid's algorithm, as before. The second stage begins at the second last line of the running of Euclid's algorithm, and involves successive substitution of the lines higher up in the algorithm as illustrated by the following example:

$$65 = 325 - 2 \times 130 \tag{A4.9}$$

$$= 325 - 2 \times (1105 - 3 \times 325) = -2 \times 1105 + 7 \times 325 \qquad \text{(A4.10)}$$

$$= -2 \times 1105 + 7 \times (1430 - 1 \times 1105) = 7 \times 1430 - 9 \times 1105 \qquad \text{(A4.11)}$$

$$= 7 \times 1430 - 9 \times (6825 - 4 \times 1430) = -9 \times 6825 + 37 \times 1430. \qquad \text{(A4.12)}$$

That is, $65 = 6825 \times (-9) + 1430 \times 37$, which is the desired representation.

What resources are consumed by Euclid's algorithm? Suppose a and b may be represented as bit strings of at most L bits each. It is clear that none of the divisors k_i or remainders r_i can be more than L bits long, so we may assume that all computations are done in L bit arithmetic. The key observation to make in a resource analysis is that $r_{i+2} \leq r_i/2$. To prove this we consider two cases:

- $r_{i+1} \leq r_i/2$. It is clear that $r_{i+2} \leq r_{i+1}$ so we are done.
- $r_{i+1} > r_i/2$. In this case $r_i = 1 \times r_{i+1} + r_{i+2}$, so $r_{i+2} = r_i - r_{i+1} \leq r_i/2$.

Since $r_{i+2} \leq r_i/2$, it follows that the divide-and-remainder operation at the heart of Euclid's algorithm need be performed at most $2\lceil \log a \rceil = O(L)$ times. Each divide-and-remainder operation requires $O(L^2)$ operations, so the total cost of Euclid's algorithm is $O(L^3)$. Finding x and y such that $ax + by = \gcd(a, b)$ incurs a minor additional cost: $O(L)$ substitutions are performed, at a cost of $O(L^2)$ per substitution to do the arithmetic involved, for a total resource cost of $O(L^3)$.

Euclid's algorithm may also be used to efficiently find multiplicative inverses in modular arithmetic. This is implicit in the proof of Corollary A4.4; we now make it explicit. Suppose a is co-prime to n, and we wish to find a^{-1}, modulo n. To do so, use Euclid's algorithm and the co-primality of a and n to find integers x and y such that

$$ax + ny = 1. \qquad \text{(A4.13)}$$

Note then that $ax = (1 - ny) = 1(\text{mod } n)$, that is, x is the multiplicative inverse of a, modulo n. Furthermore, this algorithm is computationally efficient, taking only $O(L^3)$ steps, where L is the length in bits of n.

Now that we know how to efficiently find inverses in modular arithmetic, it is only a short step to solve simple linear equations, such as

$$ax + b = c(\text{mod } n). \qquad \text{(A4.14)}$$

Suppose a and n are co-prime. Then using Euclid's algorithm we may efficiently find the multiplicative inverse a^{-1} of a, modulo n, and thus the solution to the previous equation,

$$x = a^{-1}(c - b)(\text{mod } n). \qquad \text{(A4.15)}$$

An important result known as the *Chinese remainder theorem* extends the range of equations we may solve much further, allowing us to efficiently solve systems of equations in modular arithmetic.

Theorem A4.6: (**Chinese remainder theorem**) Suppose $m_1, \ldots, m_n$ are positive integers such that any pair m_i and m_j $(i \neq j)$ are co-prime. Then the system of equations

$$x = a_1(\text{mod } m_1) \qquad \text{(A4.16)}$$

$$x = a_2(\text{mod } m_2) \qquad \text{(A4.17)}$$

$$\cdots\cdots\cdots$$

$$x = a_n (\text{mod } m_n) \qquad (A4.18)$$

has a solution. Moreover, any two solutions to this system of equations are equal modulo $M \equiv m_1 m_2 \ldots m_n$.

Proof

The proof is to explicitly construct a solution to the system of equations. Define $M_i \equiv M/m_i$ and observe that m_i and M_i are co-prime. It follows that M_i has an inverse modulo m_i, which we denote N_i. Define $x \equiv \sum_i a_i M_i N_i$. To see that x is a solution to the system of equations, note that $M_i N_i = 1(\text{mod } m_i)$ and $M_i N_i = 0(\text{mod } m_j)$ when $i \neq j$, so $x = a_i(\text{mod } m_i)$, which demonstrates the existence of a solution.

Suppose x and x' are both solutions to the system of equations. It follows that $x - x' = 0(\text{mod } m_i)$ for each i, and thus m_i divides $x - x'$ for each i. Since the m_i are co-prime, it follows that the product $M = m_1 \ldots m_n$ also divides $x - x'$, so $x = x'(\text{mod } M)$, as we set out to show.

$\square$

Euclid's algorithm and the Chinese remainder theorem are two of the signal triumphs of algorithmic number theory. How ironic then that they should play a role in the sequence of ideas leading up to the RSA cryptosystem, whose presumed security is based on the *difficulty* of performing certain algorithmic tasks in number theory. Nevertheless, this is indeed the case! We turn now to the number-theoretic background necessary to understand the RSA cryptosystem. The key ideas are a famous result of classical number theory, *Fermat's little theorem* – not to be confused with Fermat's last theorem – and a generalization of Fermat's little theorem due to Euler. The proof of Fermat's little theorem relies on the following elegant lemma.

Lemma A4.7: Suppose p is prime and k is an integer in the range 1 to $p - 1$. Then p divides $\binom{p}{k}$.

Proof
Consider the identity

$$p(p - 1) \cdots (p - k + 1) = \binom{p}{k} k(k - 1) \cdots 1. \qquad (A4.19)$$

Since $k \geq 1$ the left hand side (and thus the right) is divisible by p. Since $k \leq p - 1$ the term $k(k - 1) \cdots 1$ is not divisible by p. It follows that $\binom{p}{k}$ must be divisible by p. $\square$

Theorem A4.8: (**Fermat's little theorem**) Suppose p is a prime, and a is any integer. Then $a^p = a(\text{mod } p)$. If a is not divisible by p then $a^{p-1} = 1(\text{mod } p)$.

Proof
The second part of the theorem follows from the first, since if a is not divisible by p then a has an inverse modulo p, so $a^{p-1} = a^{-1}a^p = a^{-1}a = 1(\text{mod } p)$. We prove the first part of the theorem for positive a (the case of non-positive a follows easily) by induction on a. When $a = 1$ we have $a^p = 1 = a(\text{mod } p)$, as required. Suppose the result holds true

for a, that is, $a^p = a(\bmod p)$ and consider the case of $a + 1$. By the binomial expansion,

$$(1 + a)^p = \sum_{k=0}^{p} \binom{p}{k} a^k. \tag{A4.20}$$

By Lemma A4.7, p divides $\binom{p}{k}$ whenever $1 \le k \le p - 1$, so all terms except the first and last vanish from the sum modulo p, $(1 + a)^p = (1 + a^p)(\bmod p)$. Applying the inductive hypothesis $a^p = a(\bmod p)$ we see that $(1 + a)^p = (1 + a)(\bmod p)$, as required. □

There is a remarkable generalization of Fermat's little theorem due to Euler, based on the *Euler φ function*. $\varphi(n)$ is defined to be the number of positive integers less than n which are co-prime to n. As an example, note that all positive integers less than a prime p are co-prime to p, and thus $\varphi(p) = p - 1$. The only integers less than p^α which are not co-prime to p^α are the multiples of p: $p, 2p, 3p, \ldots, (p^{\alpha-1} - 1)p$, from which we deduce

$$\varphi(p^\alpha) = (p^\alpha - 1) - (p^{\alpha-1} - 1) = p^{\alpha-1}(p - 1). \tag{A4.21}$$

Furthermore, if a and b are co-prime, then the Chinese remainder theorem can be used to show that

$$\varphi(ab) = \varphi(a)\varphi(b). \tag{A4.22}$$

To see this, consider the system of equations $x = x_a(\bmod a), x = x_b(\bmod b)$. Applying the Chinese remainder theorem to this set of equations we see that there is a one-to-one correspondence between pairs (x_a, x_b) such that $1 \le x_a < a, 1 \le x_b < b, \gcd(x_a, a) = 1, \gcd(x_b, b) = 1$, and integers x such that $1 \le x < ab, \gcd(x, ab) = 1$. There are $\varphi(a)\varphi(b)$ such pairs (x_a, x_b) and $\varphi(ab)$ such x, from which we deduce (A4.22).

Equations (A4.21) and (A4.22) together imply a formula for $\varphi(n)$ based on the prime factorization of n, $n = p_1^{\alpha_1} \cdots p_k^{\alpha_k}$:

$$\varphi(n) = \prod_{j=1}^{k} p_j^{\alpha_j - 1}(p_j - 1). \tag{A4.23}$$

Exercise A4.10: What is $\varphi(187)$?

Exercise A4.11: Prove that

$$n = \sum_{d|n} \varphi(d), \tag{A4.24}$$

where the sum is over all positive divisors d of n, including 1 and n. (*Hint:* Prove the result for $n = p^\alpha$ first, then use the multiplicative property (A4.22) of φ to complete the proof.)

Fermat's little theorem has the following beautiful generalization, due to Euler:

Theorem A4.9: Suppose a is co-prime to n. Then $a^{\varphi(n)} = 1(\bmod n)$.

Proof
We first show by induction on α that $a^{\varphi(p^\alpha)} = 1(\bmod p^\alpha)$. For $\alpha = 1$ the result is just Fermat's little theorem. Assume the result is true for $\alpha \ge 1$, so

$$a^{\varphi(p^\alpha)} = 1 + kp^\alpha, \tag{A4.25}$$

for some integer k. Then by (A4.21),

$$a^{\varphi(p^{\alpha+1})} = a^{p^\alpha(p-1)} \tag{A4.26}$$

$$= a^{p\varphi(p^\alpha)} \tag{A4.27}$$

$$= (1 + kp^\alpha)^p \tag{A4.28}$$

$$= 1 + \sum_{j=1}^{p} \binom{p}{j} k^j p^{j\alpha}. \tag{A4.29}$$

Using Lemma A4.7 it is easy to see that $p^{\alpha+1}$ divides every term in the sum, so

$$a^{\varphi(p^{\alpha+1})} = 1(\mathrm{mod}\ p^{\alpha+1}), \tag{A4.30}$$

which completes the induction. The proof of the theorem is completed by noting that for arbitrary $n = p_1^{\alpha_1} \cdots p_m^{\alpha_m}$, $a^{\varphi(n)} = 1(\mathrm{mod}\ p_j^{\alpha_j})$ for each j, as $\varphi(n)$ is a multiple of $\varphi(p_j^{\alpha_j})$. Applying the construction in the proof of the Chinese remainder theorem we find that any solution to the set of equations $x = 1(\mathrm{mod}\ p_j^{\alpha_j})$ must satisfy $x = 1(\mathrm{mod}\ n)$, and thus $a^{\varphi(n)} = 1(\mathrm{mod}\ n)$. $\qquad\square$

Define $\mathbf{Z}_n^*$ to be the set of all elements in $\mathbf{Z}_n$ which have inverses modulo n, that is, the set of all elements in $\mathbf{Z}_n$ which are co-prime to n. $\mathbf{Z}_n^*$ is easily seen to form a group of size $\varphi(n)$ under multiplication, that is, it contains the multiplicative identity, products of elements in $\mathbf{Z}_n^*$ are in $\mathbf{Z}_n^*$, and $\mathbf{Z}_n^*$ is closed under the multiplicative inverse operation. (For an overview of elementary group theory, see Appendix 2.) What is not so obvious is the remarkable structure $\mathbf{Z}_n^*$ has when n is a power of an odd prime p, $n = p^\alpha$. It turns out that $\mathbf{Z}_{p^\alpha}^*$ is a *cyclic* group, that is, there is an element g in $\mathbf{Z}_{p^\alpha}^*$ which *generates* $\mathbf{Z}_{p^\alpha}^*$ in the sense that any other element x may be written $x = g^k(\mathrm{mod}\ n)$ for some non-negative integer k.

Theorem A4.10: Let p be an odd prime, α a positive integer. Then $\mathbf{Z}_{p^\alpha}^*$ is cyclic.

Proof
The proof of this fact is a little beyond the scope of this book. It can be found in many texts containing any considerable amount of number theory. See for example Section 3.2 of Knuth[Knu98a], especially pages 16 through 23.

$\qquad\square$

Exercise A4.12: Verify that $\mathbf{Z}_n^*$ forms a group of size $\varphi(n)$ under the operation of multiplication modulo n.

Exercise A4.13: Let a be an arbitrary element of $\mathbf{Z}_n^*$. Show that $S \equiv \{1, a, a^2, \ldots\}$ forms a subgroup of $\mathbf{Z}_n^*$, and that the size of S is the least value of r such that $a^r = 1(\mathrm{mod}\ n)$.

Exercise A4.14: Suppose g is a generator for $\mathbf{Z}_n^*$. Show that g must have order $\varphi(n)$.

Exercise A4.15: *Lagrange's theorem* (Theorem A2.1 on page 610) is an elementary result of group theory stating that the size of a subgroup must divide the order of the group. Use Lagrange's theorem to provide an alternate proof of Theorem A4.9, that is, show that $a^{\varphi(n)} = 1(\mathrm{mod}\ n)$ for any $a \in \mathbf{Z}_n^*$.

A4.3   Reduction of factoring to order-finding

The problem of factoring numbers on a classical computer turns out to be equivalent to another problem, the *order-finding* problem. This equivalence is important as it turns out that quantum computers are able to quickly solve the order-finding problem, and thus can factor quickly. In this section we explain the equivalence between these two problems, focusing on the reduction of factoring to order-finding.

Suppose N is a positive integer, and x is co-prime to N, $1 \leq x < N$. The *order* of x modulo N is defined to be the least positive integer r such that $x^r = 1 (\mathrm{mod}\ N)$. The *order-finding problem* is to determine r, given x and N.

Exercise A4.16: Use Theorem A4.9 to show that the order of x modulo N must divide $\varphi(N)$.

The reduction of factoring to order-finding proceeds in two basic steps. The first step is to show that we can compute a factor of n if we can find a non-trivial solution $x \neq \pm 1 (\mathrm{mod}\ N)$ to the equation $x^2 = 1 (\mathrm{mod}\ N)$. The second step is to show that a randomly chosen y co-prime to N is quite likely to have an order r which is even, and such that $y^{r/2} \neq \pm 1 (\mathrm{mod}\ N)$, and thus $x \equiv y^{r/2} (\mathrm{mod}\ N)$ is a solution to $x^2 = 1 (\mathrm{mod}\ N)$.

Theorem A4.11: Suppose N is a composite number L bits long, and x is a non-trivial solution to the equation $x^2 = 1 (\mathrm{mod}\ N)$ in the range $1 \leq x \leq N$, that is, neither $x = 1 (\mathrm{mod}\ N)$ nor $x = N - 1 = -1 (\mathrm{mod}\ N)$. Then at least one of $\gcd(x - 1, N)$ and $\gcd(x + 1, N)$ is a non-trivial factor of N that can be computed using $O(L^3)$ operations.

Proof
Since $x^2 = 1 (\mathrm{mod}\ N)$, it must be that N divides $x^2 - 1 = (x+1)(x-1)$, and thus N must have a common factor with one or the other of $(x+1)$ and $(x-1)$. But $1 < x < N-1$ by assumption, so $x - 1 < x + 1 < N$, from which we see that the common factor can not be N itself. Using Euclid's algorithm we may compute $\gcd(x - 1, N)$ and $\gcd(x + 1, N)$ and thus obtain a non-trivial factor of N, using $O(L^3)$ operations.   □

Lemma A4.12: Let p be an odd prime. Let 2^d be the largest power of 2 dividing $\varphi(p^\alpha)$. Then with probability exactly one-half 2^d divides the order modulo p^α of a randomly chosen element of $\mathbf{Z}_{p^\alpha}^*$.

Proof
Note that $\varphi(p^\alpha) = p^{\alpha-1}(p-1)$ is even, since p is odd, and thus $d \geq 1$. By Theorem A4.10 there exists a generator g for $\mathbf{Z}_{p^\alpha}^*$, so an arbitrary element may be written in the form $g^k (\mathrm{mod}\ p^\alpha)$ for some k in the range 1 through $\varphi(p^\alpha)$. Let r be the order of g^k modulo p^α and consider two cases. The first case is when k is odd. From $g^{kr} = 1 (\mathrm{mod}\ p^\alpha)$ we deduce that $\varphi(p^\alpha) | kr$, and thus $2^d | r$, since k is odd. The second case is when k is even. Then

$$g^{k\varphi(p^\alpha)/2} = \left(g^{\varphi(p^\alpha)} \right)^{k/2} = 1^{k/2} = 1 (\mathrm{mod}\ p^\alpha). \tag{A4.31}$$

Thus $r | \varphi(p^\alpha)/2$ from which we deduce that 2^d does not divide r.

Summarizing, $\mathbf{Z}_{p^\alpha}^*$ may be partitioned into two sets of equal size: those which may be written g^k with k odd, for which $2^d|r$, where r is the order of g^k, and those which may be written g^k with k even, for which $2^d \nmid r$. Thus with probability $1/2$ the integer 2^d divides the order r of a randomly chosen element of $\mathbf{Z}_{p^\alpha}^*$, and with probability $1/2$ it does not.   □

Theorem A4.13: Suppose $N = p_1^{\alpha_1} \cdots p_m^{\alpha_m}$ is the prime factorization of an odd composite positive integer. Let x be chosen uniformly at random from $\mathbf{Z}_N^*$, and let r be the order of x, modulo N. Then

$$p(r \text{ is even and } x^{r/2} \neq \; - 1(\text{mod } N)) \geq 1 - \frac{1}{2^m}. \tag{A4.32}$$

Proof
We show that

$$p(r \text{ is odd or } x^{r/2} = -1(\text{mod } N)) \leq \frac{1}{2^m}. \tag{A4.33}$$

By the Chinese remainder theorem, choosing x uniformly at random from $\mathbf{Z}_N^*$ is equivalent to choosing x_j independently and uniformly at random from $\mathbf{Z}_{p_j^{\alpha_j}}^*$, and requiring that $x = x_j(\text{mod } p_j^{\alpha_j})$ for each j. Let r_j be the order of x_j modulo $p_j^{\alpha_j}$. Let 2^{d_j} be the largest power of 2 that divides r_j and 2^d be the largest power of 2 that divides r. We will show that to have r odd or $x^{r/2} = -1(\text{mod } N)$ it is necessary that d_j takes the same value for all values of j. The result then follows, as from Lemma A4.12 the probability of this occurring is at most $1/2^m$.

The first case we consider is when r is odd. It is easy to see that $r_j|r$ for each j, and therefore r_j is odd, so $d_j = 0$ for all $i = 1, \ldots, k$. The second and final case is when r is even and $x^{r/2} = -1(\text{mod } N)$. Then $x^{r/2} = -1(\text{mod } p_j^{\alpha_j})$, so $r_j \nmid (r/2)$. Since $r_j|r$ we must have $d_j = d$ for all j.   □

Theorems A4.11 and A4.13 can be combined to give an algorithm which, with high probability, returns a non-trivial factor of any composite N. All the steps in the algorithm can be performed efficiently on a classical computer except (as far as is known today) an order-finding 'subroutine' which is used by the algorithm. By repeating the algorithm we may find a complete prime factorization of N. The algorithm is summarized below.

(1) If N is even, return the factor 2.
(2) Use the algorithm of Exercise 5.17 to determine whether $N = a^b$ for integers $a \geq 1$ and $b \geq 2$, and if so return the factor a.
(3) Randomly choose x in the range 1 to $N - 1$. If $\gcd(x, N) > 1$ then return the factor $\gcd(x, N)$.
(4) Use the order-finding subroutine to find the order r of x, modulo N.
(5) If r is even and $x^{r/2} \neq \; - 1(\text{mod } N)$ then compute $\gcd(x^{r/2} - 1, N)$ and $\gcd(x^{r/2} + 1, N)$, and test to see which is a non-trivial factor, returning that factor. Otherwise, the algorithm fails.

Steps 1 and 2 of the algorithm either return a factor, or else ensure that N is an odd integer with more than one prime factor. These steps may be performed using $O(1)$ and $O(L^3)$ operations, respectively. Step 3 either returns a factor, or produces a randomly chosen element x of $\mathbf{Z}_N^*$. Step 4 calls the order-finding subroutine, computing the order r of x, modulo N. Step 5 completes the algorithm, since Theorem A4.13 guarantees that with probability at least one-half r will be even and $x^{r/2} \neq -1 \pmod{N}$, and then Theorem A4.11 guarantees that either $\gcd(x^{r/2} - 1, N)$ or $\gcd(x^{r/2} + 1, N)$ is a non-trivial factor of N.

Exercise A4.17: (Reduction of order-finding to factoring)   We have seen that an efficient order-finding algorithm allows us to factor efficiently. Show that an efficient factoring algorithm would allow us to efficiently find the order modulo N of any x co-prime to N.

A4.4 Continued fractions

There are many remarkable connections between the continuum of real numbers and the integers. One such connection is the beautiful theory of *continued fractions*. In this section we develop a few elements of the theory of continued fractions, elements crucial to the application of the fast quantum algorithms for order-finding and factoring detailed in Chapter 5.

As an example of a continued fraction, consider the number s defined by the expression

$$s \equiv \cfrac{1}{2 + \cfrac{1}{2 + \cfrac{1}{2 + \cdots}}}. \tag{A4.34}$$

Informally, note that $s = 1/(2+s)$, from which it is easy to satisfy oneself that $s = \sqrt{2}-1$. The idea of the continued fractions method is to describe real numbers in terms of integers alone, using expressions such as (A4.34). A *finite simple continued fraction* is defined by a finite collection $a_0, \ldots, a_N$ of positive integers,

$$[a_0, \ldots, a_N] \equiv a_0 + \cfrac{1}{a_1 + \cfrac{1}{a_2 + \cfrac{1}{\cdots + \cfrac{1}{a_N}}}}. \tag{A4.35}$$

We define the nth convergent ($0 \leq n \leq N$) to this continued fraction to be $[a_0, \ldots, a_n]$.

Theorem A4.14: Suppose x is a rational number greater than or equal to one. Then x has a representation as a continued fraction, $x = [a_0, \ldots, a_N]$, which may be found by the *continued fractions algorithm*.

Proof
The continued fractions algorithm is best understood by example. Suppose we are trying to decompose $31/13$ as a continued fraction. The first step of the continued fractions algorithm is to split $31/13$ into its integer and fractional part,

$$\frac{31}{13} = 2 + \frac{5}{13}. \tag{A4.36}$$

Next we invert the fractional part, obtaining

$$\frac{31}{13} = 2 + \frac{1}{\frac{13}{5}}. \tag{A4.37}$$

These steps – split then invert – are now applied to $13/5$, giving

$$\frac{31}{13} = 2 + \frac{1}{2 + \frac{3}{5}} = 2 + \frac{1}{2 + \frac{1}{\frac{5}{3}}}. \tag{A4.38}$$

Next we split and invert $5/3$:

$$\frac{31}{13} = 2 + \frac{1}{2 + \frac{1}{1 + \frac{2}{3}}} = 2 + \frac{1}{2 + \frac{1}{1 + \frac{1}{\frac{3}{2}}}}. \tag{A4.39}$$

The decomposition into a continued fraction now terminates, since $3/2 = 1 + 1/2$ may be written with a 1 in the numerator without any need to invert, giving a final continued fraction representation of $31/13$ as

$$\frac{31}{13} = 2 + \frac{1}{2 + \frac{1}{1 + \frac{1}{1 + \frac{1}{2}}}}. \tag{A4.40}$$

It's clear that the continued fractions algorithm terminates after a finite number of 'split and invert' steps for any rational number, since the numerators which appear ($31, 3, 2, 1$ in the example) are strictly decreasing. How quickly does this termination occur? We'll come back to that question shortly. □

The theorem above has been stated for $x \geq 1$; however, in practice it is convenient to relax the requirement that a_0 to be positive and allow it to be any integer, which results in the restriction $x \geq 1$ becoming superfluous. In particular, if x is in the range 0 through 1 as occurs in applications to quantum algorithms, then the continued fraction expansion has $a_0 = 0$.

The continued fractions algorithm provides an unambiguous method for obtaining a continued fraction expansion of a given rational number. The only possible ambiguity comes at the final stage, because it is possible to split an integer in two ways, either $a_n = a_n$, or as $a_n = (a_n - 1) + 1/1$, giving two alternate continued fraction expansions. This ambiguity is actually useful, since it allows us to assume without loss of generality that the continued fraction expansion of a given rational number has either an odd or even number of convergents, as desired.

Exercise A4.18: Find the continued fraction expansion for $x = 19/17$ and $x = 77/65$.

Theorem A4.15: Let $a_0, \ldots, a_N$ be a sequence of positive numbers. Then

$$[a_0, \ldots, a_n] = \frac{p_n}{q_n}, \tag{A4.41}$$

where p_n and q_n are real numbers defined inductively by $p_0 \equiv a_0, q_0 \equiv 1$ and $p_1 \equiv 1 + a_0 a_1, q_1 \equiv a_1$, and for $2 \leq n \leq N$,

$$p_n \equiv a_n p_{n-1} + p_{n-2} \tag{A4.42}$$
$$q_n \equiv a_n q_{n-1} + q_{n-2}. \tag{A4.43}$$

In the case where a_j are positive integers, so too are the p_j and q_j.

Proof

We induct on n. The result is easily checked directly for the cases $n = 0, n = 1, 2$. By definition, for $n \geq 3$,

$$[a_0, \ldots, a_n] = [a_0, \ldots, a_{n-2}, a_{n-1} + 1/a_n]. \tag{A4.44}$$

Applying the inductive hypothesis, let $\tilde{p}_j/\tilde{q}_j$ be the sequence of convergents associated with the continued fraction on the right hand side:

$$[a_0, \ldots, a_{n-2}, a_{n-1} + 1/a_n] = \frac{\tilde{p}_{n-1}}{\tilde{q}_{n-1}}. \tag{A4.45}$$

It is clear that $\tilde{p}_{n-3} = p_{n-3}, \tilde{p}_{n-2} = p_{n-2}$ and $\tilde{q}_{n-3} = q_{n-3}, \tilde{q}_{n-2} = q_{n-2}$, so

$$\frac{\tilde{p}_{n-1}}{\tilde{q}_{n-1}} = \frac{(a_{n-1} + 1/a_n)p_{n-2} + p_{n-3}}{(a_{n-1} + 1/a_n)q_{n-2} + q_{n-3}} \tag{A4.46}$$

$$= \frac{p_{n-1} + p_{n-2}/a_n}{q_{n-1} + q_{n-2}/a_n}. \tag{A4.47}$$

Multiplying top and bottom of the right hand side by a_n we see that

$$\frac{\tilde{p}_{n-1}}{\tilde{q}_{n-1}} = \frac{p_n}{q_n}. \tag{A4.48}$$

Combining Equations (A4.48), (A4.45) and (A4.44) gives

$$[a_0, \ldots, a_n] = \frac{p_n}{q_n}, \tag{A4.49}$$

as required. $\qquad\square$

Exercise A4.19: Show that $q_n p_{n-1} - p_n q_{n-1} = (-1)^n$ for $n \geq 1$. Use this fact to conclude that $\gcd(p_n, q_n) = 1$. (*Hint*: Induct on n.)

How many values of a_n must be determined to obtain a continued fraction expansion for a rational number $x = p/q > 1$, where p and q are co-prime? Suppose $a_0, \ldots, a_N$ are positive integers. From the definition of p_n and q_n it follows that p_n and q_n are increasing sequences. Therefore $p_n = a_n p_{n-1} + p_{n-2} \geq 2p_{n-2}$ and similarly $q_n \geq 2q_{n-2}$, from which it follows that $p_n, q_n \geq 2^{\lfloor n/2 \rfloor}$. Thus the $2^{\lfloor N/2 \rfloor} \leq q \leq p$, and so $N = O(\log(p))$. It follows that if $x = p/q$ is a rational number, p and q are L bit integers, then the continued fraction expansion for x can be computed using $O(L^3)$ operations – $O(L)$ 'split and invert' steps, each using $O(L^2)$ gates for elementary arithmetic.

Theorem A4.16: Let x be a rational number and suppose p/q is a rational number such that

$$\left|\frac{p}{q} - x\right| \leq \frac{1}{2q^2}. \tag{A4.50}$$

Then p/q is a convergent of the continued fraction for x.

Proof

Let $p/q = [a_0, \ldots, a_n]$ be the continued fraction expansion for p/q, and define p_j, q_j as in Theorem A4.15, so that $p_n/q_n = p/q$. Define δ by the equation

$$x \equiv \frac{p_n}{q_n} + \frac{\delta}{2q_n^2}, \tag{A4.51}$$

so $|\delta| < 1$. Define λ by the equation

$$\lambda \equiv 2 \left(\frac{q_n p_{n-1} - p_n q_{n-1}}{\delta} \right) - \frac{q_{n-1}}{q_n}. \tag{A4.52}$$

The reason we define λ this way is because with a little algebra we can see that it satisfies the equation

$$x = \frac{\lambda p_n + p_{n-1}}{\lambda q_n + q_{n-1}}, \tag{A4.53}$$

and therefore $x = [a_0, \ldots, a_n, \lambda]$. Choosing n even we see from Exercise A4.19 that

$$\lambda = \frac{2}{\delta} - \frac{q_{n-1}}{q_n}. \tag{A4.54}$$

By the increasing property of q_n it follows that

$$\lambda = \frac{2}{\delta} - \frac{q_{n-1}}{q_n} > 2 - 1 > 1. \tag{A4.55}$$

Therefore λ is a rational number greater than 1, and so has a simple finite continued fraction, $\lambda = [b_0, \ldots, b_m]$, and so $x = [a_0, \ldots, a_n, b_0, \ldots, b_m]$ is a simple finite continued fraction for x with p/q as a convergent. □

Problem 4.1: (Prime number estimate) Let $\pi(n)$ be the number of prime numbers which are less than n. A difficult-to-prove result known as the *prime number theorem* asserts that $\lim_{n \to \infty} \pi(n) \log(n)/n = 1$ and thus $\pi(n) \approx n/\log(n)$. This problem gives a poor man's version of the prime number theorem which gives a pretty good lower bound on the distribution of prime numbers.

(1) Prove that $n \leq \log \binom{2n}{n}$.

(2) Show that

$$\log \binom{2n}{n} \leq \sum_{p \leq 2n} \left\lfloor \frac{\log(2n)}{\log p} \right\rfloor \log p, \tag{A4.56}$$

where the sum is over all primes p less than or equal to $2n$.

(3) Use the previous two results to show that

$$\pi(2n) \geq \frac{n}{\log(2n)}. \tag{A4.57}$$

History and further reading

There are many excellent books on number theory. We have made considerable use of the excellent book by Koblitz[Kob94], which combines much introductory material about number theory, algorithms, and cryptography all in one location. A similar combination forming a small part of a much more comprehensive presentation oriented towards algorithms may be found in Chapter 33 of Cormen, Leiserson and Rivest[CLR90]. Our discussion of continued fractions is based upon Chapter 10 of the classic text on number theory by Hardy and Wright[HW60]. Problem 4.1 is adapted from Papadimitriou[Pap94].

Appendix 5: Public key cryptography and the RSA cryptosystem

Cryptography is the art of enabling two parties to communicate in private. For example, a consumer wishing to make a purchase on the internet wants to transmit their credit card number over the internet in such a way that only the company they are purchasing from gains access to the number. Rather more ominously, in wartime each of the warring parties wants the means to carry on private communication. To achieve privacy a *cryptographic protocol* or *cryptosystem* is used. Effective cryptosystems make it easy for parties who wish to communicate to do so, but make it very difficult for third parties to 'eavesdrop' on the contents of the conversation.

A particularly important class of cryptosystems are the *public key cryptosystems*. The basic idea of public key cryptography is illustrated by the analogy depicted in Figure A5.1. Alice sets up a mailbox with the property that *anybody* can send her mail, by putting it into the mailbox, but only she can retrieve mail out of the mailbox. To achieve this she gives the mailbox *two* doors. On top of the mailbox is a locked trap door. Any person able to open the trap door can drop mail into the box. However, the chute from the trap door into the box is one way, so they can't reach into the box and fish mail out. Alice makes the key to the trapdoor *freely available* to the public – it is a *public key* – so that she can receive mail from absolutely *anybody*. On the front of the mail box is a second door, from which mail already inside the box can be retrieved. Alice is in possession of the sole key for that door; it is her own *secret key*. This arrangement – involving two keys, one secret and one public – allows anybody in the world to communicate with Alice while maintaining privacy.

Public key cryptosystems operate according to similar principles. Suppose Alice wishes to receive messages using a public key cryptosystem. She must first generate two *cryptographic keys*, one a *public key*, P, the other a *secret key*, S. The exact nature of these keys depends on the details of the cryptosystem being used. Some cryptosystems use simple objects like numbers as keys, while other cryptosystems use much more complicated mathematical objects, like elliptic curves, as keys. Once Alice has generated her keys, she publishes the public key so that anybody can obtain access to the key.

Now suppose Bob wishes to send Alice a private message. He first obtains a copy of Alice's public key P, and then *encrypts* the message he wishes to send Alice, using Alice's public key to perform the encryption. Exactly how the encryption transformation is performed depends on the details of the cryptosystem in use. The key point is that in order to be secure against eavesdropping the encryption stage needs to be very difficult to reverse, even making use of the public key used to encrypt the message in the first place! It's like the trap door for mail – what you can put in you can't take back out, even if you have the key to the trap door. Since the public key and the encoded message is the only information available to an eavesdropper it won't be possible for the eavesdropper to recover the message. Alice, however, has an additional piece of information not available

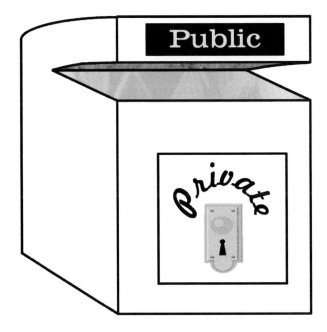

Figure A5.1. The key ideas of public key cryptography, illustrated in more familiar terms. Essentially the same scheme is implemented by the Post Office in many countries.

to an eavesdropper, the secret key, S. The secret key determines a second transformation, this time on the encrypted message. This transformation is known as *decryption*, and is inverse to encryption, allowing Alice to recover the original message.

In an ideal world that is how public key cryptography would work. Unfortunately, at the time of writing it is not known whether there are any such secure schemes for doing public key cryptography. There do exist several schemes which are *widely believed* to be secure, and which are in common use for applications such as internet commerce, but wide belief is not equivalent to a proof of security. The reason these schemes are believed to be secure is because so much effort has been devoted to finding a means for breaking these schemes (without success!), a sort of proof by attrition. The most widely used of these public key cryptosystems is the RSA cryptosystem, named RSA for the initials of its creators, Rivest, Shamir, and Adleman. The presumed security of the RSA cryptosystem is based, as we shall now see, on the apparent difficulty of factoring on a classical computer. Understanding RSA requires a little background in number theory, which is covered in Appendix 4, notably Sections A4.1 and A4.2.

Suppose Alice wishes to create public and private keys for use with the RSA cryptosystem. She uses the following procedure:

(1) Select two large prime numbers, p and q.
(2) Compute the product $n \equiv pq$.
(3) Select at random a small odd integer, e, that is relatively prime to
     $\varphi(n) = (p - 1)(q - 1)$.
(4) Compute d, the multiplicative inverse of e, modulo $\varphi(n)$.
(5) The *RSA public key* is the pair $P = (e, n)$. The *RSA secret key* is the pair
     $S = (d, n)$.

Suppose a second party, Bob, wishes to encrypt a message M to send to Alice, using the public key (e, n). We assume the message M has only $\lfloor \log n \rfloor$ bits, as longer messages may be encrypted by breaking M up into blocks of at most $\lfloor \log n \rfloor$ bits and then encrypting the blocks separately. The encryption procedure for a single block is to compute:

$$E(M) = M^e (\text{mod } n). \tag{A5.1}$$

$E(M)$ is the encrypted version of the message M, which Bob transmits to Alice. Alice can quickly decrypt the message using her secret key $S = (d, n)$, simply by raising the encrypted message to the dth power:

$$E(M) \rightarrow D(E(M)) = E(M)^d (\text{mod } n). \tag{A5.2}$$

For the decryption to be successful we need $D(E(M)) = M(\text{mod } n)$. To see that this is the case, note that by construction $ed = 1(\text{mod } \varphi(n))$ and thus $ed = 1 + k\varphi(n)$ for some integer k. The proof now proceeds by considering two different cases. In the first case, M is co-prime to n. By Euler's generalization of Fermat's little theorem, Theorem A4.9, it follows that $M^{k\varphi(n)} = 1(\text{mod } n)$ and thus,

$$D(E(M)) = E(M)^d (\text{mod } n) \tag{A5.3}$$
$$= M^{ed} (\text{mod } n) \tag{A5.4}$$
$$= M^{1+k\varphi(n)} (\text{mod } n) \tag{A5.5}$$
$$= M \cdot M^{k\varphi(n)} (\text{mod } n) \tag{A5.6}$$
$$= M(\text{mod } n), \tag{A5.7}$$

which establishes that the decryption is successful when M is co-prime to n. Suppose next that M is not co-prime to n, so that one or both of p and q divide M. To be specific, we consider the case where p divides M and q does not divide M; the other possible cases requre only minor modifications. Because p divides M we have $M = 0(\text{mod } p)$ and thus $M^{ed} = 0 = M(\text{mod } p)$. Because q does not divide M we have $M^{q-1} = 1(\text{mod } q)$ by Fermat's little theorem, and thus $M^{\varphi(n)} = 1(\text{mod } q)$, since $\varphi(n) = (p-1)(q-1)$. Using $ed = 1 + k\varphi(n)$ we see that $M^{ed} = M(\text{mod } q)$. By the Chinese remainder theorem it follows that we must have $M^{ed} = M(\text{mod } n)$, and thus the decryption is also successful when M is not co-prime to n.

Exercise A5.1: Written examples of the application of RSA tend to be rather opaque. It's better to work through an example yourself. Encode the word 'QUANTUM' (or at least the first few letters!), one letter at a time, using $p = 3$ and $q = 11$. Choose appropriate values for e and d, and use a representation of English text involving 5 bits per letter.

How efficiently can RSA be implemented? There are two implementation issues to be considered. First is the generation of public and private keys for the cryptosystem. If this can't be done quickly then RSA won't be good for much. The main bottleneck is the generation of the prime numbers p and q. The way this is attacked is to randomly select a number of the desired length, and then to apply a *primality* test to determine if the number is, in fact, prime. Fast primality tests such as the Miller–Rabin test can be used to determine whether a number is prime using roughly $O(L^3)$ operations, where L is the desired size of the cryptographic key. If the number is found to be composite then we simply repeat the procedure until a prime is found. The prime number theorem (see Problem 4.1) implies that the probability of any given number being prime is about $1/\log(2^L) = 1/L$, so with high probability $O(L)$ trials are required to obtain a prime number, for a total cost of $O(L^4)$ operations to do key generation.

The second issue in the implementation of RSA is the efficiency of the encryption and decryption transformations. These are accomplished by modular exponentiation, which we know can be done efficiently using $O(L^3)$ operations – see Box 5.2 on page 228. Thus all the operations required to use the RSA cryptosystem can be done quite quickly on a classical computer, and in practice modest computing power can quite easily cope with keys up to a few thousand bits in length.

How can RSA be broken? We describe two methods by which one might hope to break RSA, one based on order-finding, the other based on factoring. Suppose Eve receives an encrypted message $M^e(\text{mod } n)$, and knows the public key (e, n) used to encrypt the message. Suppose she can find the order of the encrypted message, that is, she can find the smallest positive integer r such that $(M^e)^r = 1(\text{mod } n)$. (Without loss of generality, we may suppose such an order exists, that is, M^e is co-prime to n. If this is not the case, then $M^e(\text{mod } n)$ and n have a common factor that may be extracted by Euclid's algorithm, which would allow us to break RSA, as in the second method described below.) Then Exercise A4.16 implies that r divides $\varphi(n)$. Since e is co-prime to $\varphi(n)$ it must also be co-prime to r, and thus has a multiplicative inverse modulo r. Let d' be such a multiplicative inverse, so $ed' = 1 + kr$ for some integer k. Then Eve can recover the original message M by raising the encrypted message to the d'th power:

$$(M^e)^{d'}(\text{mod } n) = M^{1+kr}(\text{mod } n) \tag{A5.8}$$

$$= M \cdot M^{kr}(\text{mod } n) \tag{A5.9}$$

$$= M(\text{mod } n). \tag{A5.10}$$

It is interesting that Eve never actually learns the secret key (d, n); she only learns (d', n). Of course, d' is closely related to d, since d' is the inverse of e modulo r, d is the inverse of e modulo $\varphi(n)$, and r divides $\varphi(n)$. Nevertheless, this example shows that it is possible to break RSA without necessarily determining the exact value of the secret key. Of course, this method only works if Eve has an efficient method for order-finding, and no such method is currently known for a classical computer. On a quantum computer, however, order-finding can be accomplished efficiently, as described in Section 5.3.1, and thus RSA can be broken.

Exercise A5.2: Show that d is also an inverse of e modulo r, and thus $d = d'(\text{mod } r)$.

A second method for breaking RSA allows one to determine the secret key completely. Suppose Eve could factor $n = pq$, extracting p and q, and thus giving a means for efficiently computing $\varphi(n) = (p-1)(q-1)$. It is then an easy matter for Eve to compute d, the inverse of e modulo $\varphi(n)$, and thus completely determine the secret key (d, n). So, if factoring large numbers were easy then it would be easy to break RSA.

The presumed security of RSA rests on the fact that these attacks rely on having algorithms to solve problems which are believed (but not known) to be intractable on a classical computer, the order-finding and factoring problems. Unfortunately, it's not even *known* to be the case that RSA is secure if these problems are hard. It could be that these problems really are difficult, yet there is some other way of breaking RSA. Despite these caveats, more than two decades of attempts to break RSA have resulted in failure, and it is widely believed that RSA is secure against attacks by classical computers.

Problem 5.1:   Write a computer program for performing encryption and decryption using the RSA algorithm. Find a pair of 20 bit prime numbers and use them to encrypt a 40 bit message.

History and further reading

Public key cryptosystems were invented by Diffie and Hellman in 1976[DH76], and independently by Merkle at about the same time, although his work was not published until 1978[Mer78]. The RSA cryptosystem was invented shortly after by Rivest, Shamir, and Adleman[RSA78]. In 1997 it was disclosed that these ideas – public key cryptography, the Diffie-Hellman and RSA cryptosystems – were actually invented in the late 1960s and early 1970s by researchers working at the British intelligence agency GCHQ. An account of this work may be found at 'http://www.cesg.gov.uk/about/nsecret/'. Primality tests such as the Miller–Rabin and Solovay–Strassen tests are described in Koblitz's excellent book[Kob94] on number theory and cryptography, which contains a wealth of additional material on public key cryptography. These primality tests were two of the earliest indicators that randomized algorithms may be more efficient for some purposes than deterministic algorithms. The Solovay–Strassen algorithm is due to Solovay and Strassen[SS76], and the Miller–Rabin test is due jointly to Miller[Mil76] and Rabin[Rab80].

Appendix 6: Proof of Lieb's theorem

One of the most important and useful results in quantum information theory is the *strong subadditivity inequality* for von Neumann entropies. This states that for a trio of quantum systems, A, B, C,

$$S(A, B, C) + S(B) \leq S(A, B) + S(B, C). \tag{A6.1}$$

Unfortunately, no transparent proof of strong subadditivity is known. Chapter 11 presents a relatively simple proof, based upon a deep mathematical result known as *Lieb's theorem*. In this appendix we prove Lieb's theorem. We begin with a few simple notations and definitions.

Suppose $f(A, B)$ is a real-valued function of two matrices, A and B. Then f is said to be *jointly concave* in A and B if for all $0 \leq \lambda \leq 1$,

$$f(\lambda A_1 + (1 - \lambda)A_2, \lambda B_1 + (1 - \lambda)B_2) \geq \lambda f(A_1, B_1) + (1 - \lambda)f(A_2, B_2). \tag{A6.2}$$

For matrices A and B, we say $A \leq B$ if $B - A$ is a positive matrix. We say $A \geq B$ if $B \leq A$. Let A be an arbitrary matrix. We define the *norm* of A by

$$\|A\| \equiv \max_{\langle u|u\rangle=1} |\langle u|A|u\rangle|. \tag{A6.3}$$

In our proof of Lieb's theorem we will have occasion to use the following easily verified observations:

Exercise A6.1: ($\leq$ **is preserved under conjugation**) If $A \leq B$, show that $XAX^{\dagger} \leq XBX^{\dagger}$ for all matrices X.

Exercise A6.2: Prove that $A \geq 0$ if and only if A is a positive operator.

Exercise A6.3: ($\leq$ **is a partial order**) Show that the relation $\leq$ is a partial order on operators – that is, it is transitive ($A \leq B$ and $B \leq C$ implies $A \leq C$), asymmetric ($A \leq B$ and $B \leq A$ implies $A = B$), and reflexive ($A \leq A$).

Exercise A6.4: Suppose A has eigenvalues λ_i. Define λ to be the maximum of the set $|\lambda_i|$. Prove that:

(1) $\|A\| \geq \lambda$.
(2) When A is Hermitian, $\|A\| = \lambda$.
(3) When

$$A = \begin{bmatrix} 1 & 0 \\ 1 & 1 \end{bmatrix}, \tag{A6.4}$$

$\|A\| = 3/2 > 1 = \lambda$.

Exercise A6.5: (AB **and** BA **have the same eigenvalues**) Prove that AB and
BA have the same eigenvalues. (*Hint*: For invertible A, show that
$\det(xI - AB) = \det(xI - BA)$, and thus the eigenvalues of AB and BA are the
same. By continuity this holds even when A is not invertible.)

Exercise A6.6: Suppose A and B are such that AB is Hermitian. Using the previous
two observations show that $\|AB\| \le \|BA\|$.

Exercise A6.7: Suppose A is positive. Show that $\|A\| \le 1$ if and only if $A \le I$.

Exercise A6.8: Let A be a positive matrix. Define a superoperator (linear operator on
matrices) by the equation $\mathcal{A}(X) \equiv AX$. Show that $\mathcal{A}$ is positive with respect to
the Hilbert–Schmidt inner product. That is, for all X, $\mathrm{tr}(X^\dagger \mathcal{A}(X)) \ge 0$.
Similarly, show that the superoperator defined by $\mathcal{A}(X) \equiv XA$ is positive with
respect to the Hilbert–Schmidt inner product on matrices.

With these results in hand, we are now in a position to state and prove Lieb's theorem.

Theorem A6.1: (**Lieb's theorem**) Let X be a matrix, and $0 \le t \le 1$. Then the
function

$$f(A, B) \equiv \mathrm{tr}(X^\dagger A^t X B^{1-t}) \tag{A6.5}$$

is jointly concave in positive matrices A and B.

Lieb's theorem is an easy corollary of the following lemma:

Lemma A6.2: Let $R_1, R_2, S_1, S_2, T_1, T_2$ be positive operators such that
$0 = [R_1, R_2] = [S_1, S_2] = [T_1, T_2]$, and

$$R_1 \ge S_1 + T_1 \tag{A6.6}$$
$$R_2 \ge S_2 + T_2. \tag{A6.7}$$

Then for all $0 \le t \le 1$,

$$R_1^t R_2^{1-t} \ge S_1^t S_2^{1-t} + T_1^t T_2^{1-t} \tag{A6.8}$$

is true as a matrix inequality.

Proof
We begin by proving the result for $t = 1/2$, and then use this to establish the result for
general t. It will be convenient to assume that R_1 and R_2 are invertible, and it is left as
an exercise to make the minor technical modifications to the proof necessary to establish
the result when this is not the case.

Let $|x\rangle$ and $|y\rangle$ be any two vectors. Applying the Cauchy–Schwarz inequality twice
and performing some straightforward manipulations, we have

$$|\langle x|(S_1^{1/2} S_2^{1/2} + T_1^{1/2} T_2^{1/2})|y\rangle|$$
$$\le |\langle x|S_1^{1/2} S_2^{1/2}|y\rangle| + |\langle x|T_1^{1/2} T_2^{1/2}|y\rangle| \tag{A6.9}$$
$$\le \|S_1^{1/2}|x\rangle\| \, \|S_2^{1/2}|y\rangle\| + \|T_1^{1/2}|x\rangle\| \, \|T_2^{1/2}|y\rangle\| \tag{A6.10}$$
$$\le \sqrt{\left(\|S_1^{1/2}|x\rangle\|^2 + \|T_1^{1/2}|x\rangle\|^2\right)\left(\|S_2^{1/2}|y\rangle\|^2 + \|T_2^{1/2}|y\rangle\|^2\right)} \tag{A6.11}$$

$$= \sqrt{\langle x|(S_1 + T_1)|x\rangle\langle y|(S_2 + T_2)|y\rangle}. \tag{A6.12}$$

By hypothesis, $S_1 + T_1 \leq R_1$ and $S_2 + T_2 \leq R_2$, so

$$|\langle x|(S_1^{1/2}S_2^{1/2} + T_1^{1/2}T_2^{1/2})|y\rangle| \leq \sqrt{\langle x|R_1|x\rangle\langle y|R_2|y\rangle}. \tag{A6.13}$$

Let $|u\rangle$ be any unit vector. Then applying (A6.13) with $|x\rangle \equiv R_1^{-1/2}|u\rangle$ and $|y\rangle \equiv R_2^{-1/2}|u\rangle$ gives

$$\langle u|R_1^{-1/2}(S_1^{1/2}S_2^{1/2} + T_1^{1/2}T_2^{1/2})R_2^{-1/2}|u\rangle$$

$$\leq \sqrt{\langle u|R_1^{-1/2}R_1R_1^{-1/2}|u\rangle\langle u|R_2^{-1/2}R_2R_2^{-1/2}|u\rangle} \tag{A6.14}$$

$$= \sqrt{\langle u|u\rangle\langle u|u\rangle} = 1. \tag{A6.15}$$

Thus

$$\|R_1^{-1/2}(S_1^{1/2}S_2^{1/2} + T_1^{1/2}T_2^{1/2})R_2^{-1/2}\| \leq 1. \tag{A6.16}$$

Define

$$A \equiv R_1^{-1/4}R_2^{-1/4}(S_1^{1/2}S_2^{1/2} + T_1^{1/2}T_2^{1/2})R_2^{-1/2} \tag{A6.17}$$

$$B \equiv R_2^{1/4}R_1^{-1/4}. \tag{A6.18}$$

Note that AB is Hermitian, so by Exercise A6.6 on page 646,

$$\|R_1^{-1/4}R_2^{-1/4}(S_1^{1/2}S_2^{1/2} + T_1^{1/2}T_2^{1/2})R_2^{-1/4}R_1^{-1/4}\|$$

$$= \|AB\| \leq \|BA\| \tag{A6.19}$$

$$= \|R_1^{-1/2}(S_1^{1/2}S_2^{1/2} + T_1^{1/2}T_2^{1/2})R_2^{-1/2}\| \tag{A6.20}$$

$$\leq 1, \tag{A6.21}$$

where the last inequality is just (A6.16). AB is a positive operator, so by Exercise A6.7 on page 646 and the previous inequality,

$$R_1^{-1/4}R_2^{-1/4}(S_1^{1/2}S_2^{1/2} + T_1^{1/2}T_2^{1/2})R_2^{-1/4}R_1^{-1/4} \leq I. \tag{A6.22}$$

Finally, by Exercise A6.1 on page 645, and the commutativity of R_1 and R_2,

$$S_1^{1/2}S_2^{1/2} + T_1^{1/2}T_2^{1/2} \leq R_1^{1/2}R_2^{1/2}, \tag{A6.23}$$

which establishes that (A6.8) holds for $t = 1/2$.

Let I be the set of all t such that (A6.8) holds. By inspection, we see that 0 and 1 are elements of I, and we have just shown that $1/2$ is an element of I. We now use the $t = 1/2$ case to prove the result for any t such that $0 \leq t \leq 1$. Suppose μ and η are any two elements of I, so that

$$R_1^\mu R_2^{1-\mu} \geq S_1^\mu S_2^{1-\mu} + T_1^\mu T_2^{1-\mu} \tag{A6.24}$$

$$R_1^\eta R_2^{1-\eta} \geq S_1^\eta S_2^{1-\eta} + T_1^\eta T_2^{1-\eta}. \tag{A6.25}$$

These inequalities are of the form (A6.6) and (A6.7) for which the $t = 1/2$ case has already been proved. Using the $t = 1/2$ result we see that

$$\left(R_1^\mu R_2^{1-\mu}\right)^{1/2}\left(R_1^\eta R_2^{1-\eta}\right)^{1/2} \geq \left(S_1^\mu S_2^{1-\mu}\right)^{1/2}\left(S_1^\eta S_2^{1-\eta}\right)^{1/2}$$

$$+ \left(T_1^\mu T_2^{1-\mu}\right)^{1/2}\left(T_1^\eta T_2^{1-\eta}\right)^{1/2}. \tag{A6.26}$$

Using the commutativity assumptions $0 = [R_1, R_2] = [S_1, S_2] = [T_1, T_2]$, we see that for $\nu \equiv (\mu + \eta)/2$,

$$R_1^\nu R_2^{1-\nu} \geq S_1^\nu S_2^{1-\nu} + T_1^\nu T_2^{1-\nu}. \tag{A6.27}$$

Thus whenever μ and η are in I, so is $(\mu + \eta)/2$. Since 0 and 1 are in I, it is easy to see that any number t between 0 and 1 with a finite binary expansion must be in I. Thus I is dense in $[0, 1]$. The result now follows from the continuity in t of the conclusion, (A6.8).

$\square$

The proof of Lieb's theorem is a simple application of Lemma A6.2. The clever idea that makes this possible is to choose the operators in Lemma A6.2 to be *superoperators* – linear maps on operators. These will be chosen in such a way as to positive with respect to the Hilbert–Schmidt inner product $(A, B) \equiv \mathrm{tr}(A^\dagger B)$.

Proof

(Lieb's theorem)

Let $0 \leq \lambda \leq 1$ and define superoperators $\mathcal{S}_1, \mathcal{S}_2, \mathcal{T}_1, \mathcal{T}_2, \mathcal{R}_1, \mathcal{R}_2$ as follows:

$$\mathcal{S}_1(X) \equiv \lambda A_1 X \tag{A6.28}$$
$$\mathcal{S}_2(X) \equiv \lambda X B_1 \tag{A6.29}$$
$$\mathcal{T}_1(X) \equiv (1 - \lambda) A_2 X \tag{A6.30}$$
$$\mathcal{T}_2(X) \equiv (1 - \lambda) X B_2 \tag{A6.31}$$
$$\mathcal{R}_1 \equiv \mathcal{S}_1 + \mathcal{T}_1 \tag{A6.32}$$
$$\mathcal{R}_2 \equiv \mathcal{S}_2 + \mathcal{T}_2. \tag{A6.33}$$

Observe that $\mathcal{S}_1$ and $\mathcal{S}_2$ commute, as do $\mathcal{T}_1$ and $\mathcal{T}_2$, and $\mathcal{R}_1$ and $\mathcal{R}_2$. Recall Exercise A6.8 on page 646, that all these operators are positive with respect to the Hilbert–Schmidt inner product. By Lemma A6.2,

$$\mathcal{R}_1^t \mathcal{R}_2^{1-t} \geq \mathcal{S}_1^t \mathcal{S}_2^{1-t} + \mathcal{T}_1^t \mathcal{T}_2^{1-t}. \tag{A6.34}$$

Using the Hilbert–Schmidt inner product to take the $X \cdot X$ matrix element of the previous inequality gives

$$\mathrm{tr}\left[X^\dagger (\lambda A_1 + (1 - \lambda) A_2)^t X (\lambda B_1 + (1 - \lambda) B_2)^{1-t} \right]$$
$$\geq \mathrm{tr}\left[X^\dagger (\lambda A_1)^t X (\lambda B_1)^{1-t} \right] + \mathrm{tr}\left[X^\dagger ((1 - \lambda) A_2)^t X ((1 - \lambda) B_2)^{1-t} \right] \tag{A6.35}$$
$$= \lambda \mathrm{tr}(X^\dagger A_1^t X B_1^{1-t}) + (1 - \lambda) \mathrm{tr}(X^\dagger A_2^t X B_2^{1-t}) \quad , \tag{A6.36}$$

which is the desired statement of joint concavity.

$\square$

History and further reading

The history of Lieb's theorem is tied up with the proof of the strong subadditivity inequality for quantum entropies, and may be found together with the history of the proof of that inequality in the 'History and further reading' for Chapter 11.

Bibliography

Citations with '*arXive e-print quant-ph/xxxxxxx*' designations are available on the internet at http://www.arXiv.org

[ABO97]   D. Aharonov and M. Ben-Or. Fault tolerant computation with constant error. In *Proceedings of the Twenty-Ninth Annual ACM Symposium on the Theory of Computing*, pages 176–188, 1997.

[ABO99]   D. Aharonov and M. Ben-Or. Fault-tolerant quantum computation with constant error rate. *SIAM J. Comp.*, page to appear, 1999. *arXive e-print quant-ph/9906129*.

[ABOIN96] D. Aharonov, M. Ben-Or, R. Impagliazzo, and N. Nisan. Limitations of noisy reversible computation. *arXive e-print quant-ph/9611028*, 1996.

[ADH97]   L. Adleman, J. Demarrais, and M. A. Huang. Quantum computability. *SIAM J. Comp.*, 26(5):1524–1540, 1997.

[Adl94]   L. M. Adleman. Molecular computation of solutions to combinatorial problems. *Science*, 266:1021, 1994.

[Adl98]   L. M. Adleman. Computing with DNA. *Sci. Am.*, 279:54–61, Aug. 1998.

[AE75]   L. Allen and J. H. Eberly. *Optical Resonance and Two-level Atoms*. Dover, New York, 1975.

[Aha99a]   D. Aharonov. *Noisy Quantum Computation*. Ph.D. thesis, The Hebrew Univesity, Jerusalem, 1999.

[Aha99b]   D. Aharonov. Quantum computation. In D. Stauffer, editor, *Annual Reviews of Computational Physics VI*. World Scientific, Singapore, 1999.

[AKN98]   D. Aharonov, A. Kitaev, and N. Nisan. Quantum circuits with mixed states. *STOC 1997*, 1998. *arXive e-print quant-ph/9806029*.

[AL70]   H. Araki and E. H. Lieb. Entropy inequalities. *Comm. Math. Phys.*, 18:160–170, 1970.

[AL97]   D. S. Abrams and S. Lloyd. Simulation of many-body Fermi systems on a quantum computer. *Phys. Rev. Lett.*, 79(13):2586–2589, 1997. *arXive e-print quant-ph/9703054*.

[AL99]   A. Ashikhmin and S. Lytsin. Upper bounds on the size of quantum codes. *IEEE Trans. Inf. Theory*, 45(4):1206–1215, 1999.

[Alb83]   P. M. Alberti. A note on the transition-probability over c-* algebras. *Lett. in Math. Phys.*, 7(1):25–32, 1983.

[Amb00]   A. Ambainis. Quantum lower bounds by quantum arguments. *arXive e-print quant-ph/0002066*, 2000.

[And79]   T. Ando. Concavity of certain maps on positive definite matrices and applications to Hadamard products. *Linear Algebra Appl.*, 26:203–241, 1979.

[Ash97]   A. Ashikhmin. Remarks on bounds for quantum codes. *arXive e-print quant-ph/9705037*, 1997.

[Bar78]   E. Barton. A reversible computer using conservative logic. Unpublished MIT 6.895 term paper, 1978.

[BB84]   C. H. Bennett and G. Brassard. Quantum cryptography: Public key distribution and coin tossing. In *Proceedings of IEEE International Conference on Computers, Systems and Signal Processing*, pages 175–179, IEEE, New York, 1984. Bangalore, India, December 1984.

[BBB+92]   C. H. Bennett, F. Bessette, G. Brassard, L. Salvail, and J. Smolin. Experimental quantum cryptography. *J. Cryptology*, 5:3–28, 1992.

[BBB+98]   E. Biham, M. Boyer, G. Brassard, J. van de Graaf, and T. Mor. Security of quantum key distribution against all collective attacks. *arXive e-print quant-ph/9801022*, 1998.

[BBBV97]   C. H. Bennett, E. Bernstein, G. Brassard, and U. Vazirani. Strengths and weaknesses of quantum computing. *SIAM J. Comput.*, 26(5):1510–1523, 1997. *arXive e-print quant-ph/9701001*.

[BBC+93]   C. H. Bennett, G. Brassard, C. Crépeau, R. Jozsa, A. Peres, and W. Wootters. Teleporting an unknown quantum state via dual classical and EPR channels. *Phys. Rev. Lett.*, 70:1895–1899, 1993.

[BBC+95]   A. Barenco, C. H. Bennett, R. Cleve, D. P. DiVincenzo, N. Margolus, P. Shor, T. Sleator, J. Smolin, and H. Weinfurter. Elementary gates for quantum computation. *Phys. Rev. A*, 52:3457–3467, 1995. *arXive e-print quant-ph/9503016*.

[BBC+98]   R. Beals, H. Buhrman, R. Cleve, M. Mosca, and R. de Wolf. Quantum lower bounds by polynomials. In *Proceedings of the 39th Annual Symposium on Foundations of Computer Science (FOCS'98)*, pages 352–361, IEEE, Los Alamitos, California, November 1998. *arXive e-print quant-ph/9802049*.

[BBCM95]   C. H. Bennett, G. Brassard, C. Crépeau, and U. M. Maurer. Generalized privacy amplification. *IEEE Trans. Inf. Theory*, 41:1915–1923, 1995.

[BBE92]   C. H. Bennett, G. Brassard, and A. K. Ekert. Quantum cryptography. *Sci. Am.*, 267(4):50, Oct. 1992.

[BBHT98]   M. Boyer, G. Brassard, P. Høyer, and A. Tapp. Tight bounds on quantum searching. *Fortsch. Phys. – Prog. Phys.*, 46(4–5):493–505, 1998.

[BBM+98]   D. Boschi, S. Branca, F. D. Martini, L. Hardy, and S. Popescu. Experimental realization of teleporting an unknown pure quantum state via dual classical and Einstein-Podolski-Rosen channels. *Phys. Rev. Lett.*, 80:1121–1125, 1998. *arXive e-print quant-ph/9710013*.

[BBP+96]   C. H. Bennett, G. Brassard, S. Popescu, B. Schumacher, J. A. Smolin, and W. K. Wootters. Purification of noisy entanglement and faithful teleportation via noisy channels. *Phys. Rev. Lett.*, 76:722, 1996. *arXive e-print quant-ph/9511027*.

[BBPS96]   C. H. Bennett, H. J. Bernstein, S. Popescu, and B. Schumacher. Concentrating partial entanglement by local operations. *Phys. Rev. A*, 53(4):2046–2052, 1996. *arXive e-print quant-ph/9511030*.

[BBR88]   C. H. Bennett, G. Brassard, and J. M. Robert. Privacy amplification by public discussion. *SIAM J. Comp.*, 17:210–229, 1988.

[BCDP96]   D. Beckman, A. N. Chari, S. Devabhaktuni, and J. Preskill. Efficient networks for quantum factoring. *Phys. Rev. A*, 54(2):1034, 1996. *arXive e-print quant-ph/9602016*.

[BCF+96]   H. Barnum, C. M. Caves, C. A. Fuchs, R. Jozsa, and B. Schumacher. Noncommuting mixed states cannot be broadcast. *Phys. Rev. Lett.*, 76(15):2828–2821, 1996. *arXive e-print quant-ph/9511010*.

[BCJ+99]   S. L. Braunstein, C. M. Caves, R. Jozsa, N. Linden, S. Popescu, and R. Schack. Separability of very noisy mixed states and implications for NMR quantum computing. *Phys. Rev. Lett.*, 83(5):1054–1057, 1999.

[BCJD99]   G. K. Brennen, C. M. Caves, P. S. Jessen, and I. H. Deutsch. Quantum logic gates in optical lattices. *Physical Review Letters*, 82:1060–1063, 1999.

[BD00]   C. H. Bennett and D. P. DiVincenzo. Quantum information and computation. *Nature*, 404:247–55, 2000.

[BDG88a]   J. L. Balcázar, J. Diaz, and J. Gabarró. *Structural Complexity*, Volume I. Springer-Verlag, Berlin, 1988.

[BDG88b]   J. L. Balcázar, J. Diaz, and J. Gabarró. *Structural Complexity*, Volume II. Springer-Verlag, Berlin, 1988.

[BDK92]   R. G. Brewer, R. G. DeVoe, and R. Kallenbach. Planar ion microtraps. *Phys. Rev. A*, 46(11):R6781–4, 1992.

[BDS97]   C. H. Bennett, D. P. DiVincenzo, and J. A. Smolin. Capacities of quantum erasure channels. *Phys. Rev. Lett.*, 78(16):3217–3220, 1997. *arXive e-print quant-ph/9701015*.

[BDSW96]   C. H. Bennett, D. P. DiVincenzo, J. A. Smolin, and W. K. Wootters. Mixed state entanglement and quantum error correction. *Phys. Rev. A*, 54:3824, 1996. *arXive e-print quant-ph/9604024*.

[Bel64]   J. S. Bell. On the Einstein-Podolsy-Rosen paradox. *Physics*, 1:195–200, 1964. Reprinted in J. S. Bell, *Speakable and Unspeakable in Quantum Mechanics*, Cambridge University Press, Cambridge, 1987.

[Ben73]   C. H. Bennett. Logical reversibility

of computation. *IBM J. Res. Dev.*, 17(6):525–32, 1973.

[Ben80] P. Benioff. The computer as a physical system: A microscopic quantum mechanical Hamiltonian model of computers as represented by Turing machines. *J. Stat. Phys.*, 22(5):563–591, 1980.

[Ben82] C. H. Bennett. The thermodynamics of computation - a review. *Int. J. Theor. Phys.*, 21:905–40, 1982.

[Ben87] C. H. Bennett. Demons, engines and the second law. *Sci. Am.*, 295(5):108, 1987.

[Ben89] C. H. Bennett. Time-space trade-offs for reversible computation. *SIAM J. Comput.*, 18:766–776, 1989.

[Ben92] C. H. Bennett. Quantum cryptography using any two nonorthogonal states. *Phys. Rev. Lett.*, 68(21):3121–3124, 1992.

[Bet84] T. Beth. *Methoden der Schnellen Fouriertransformation.* Teubner, Leipzig, 1984.

[BFGL98] S. L. Braunstein, C. A. Fuchs, D. Gottesman, and H. Lo. A quantum analog of Huffman coding. *arXive e-print quant-ph/9805080*, 1998.

[BFJS96] H. Barnum, C. A. Fuchs, R. Jozsa, and B. Schumacher. General fidelity limit for quantum channels. *Phys. Rev. A*, 54:4707, 1996. *arXive e-print quant-ph/9603014.*

[Bha97] R. Bhatia. *Matrix Analysis.* Springer-Verlag, New York, 1997.

[BHT98] G. Brassard, P. Høyer, and A. Tapp. Quantum counting. *arXive e-print quant-ph/9805082*, 1998.

[BK92] V. B. Braginsky and F. Y. Khahili. *Quantum Measurement.* Cambridge University Press, Cambridge, 1992.

[BK98a] S. L. Braunstein and H. J. Kimble. Teleportation of continuous quantum variables. *Phys. Rev. Lett.*, 80:869–72, 1998.

[BK98b] S. B. Bravyi and A. Y. Kitaev. Quantum codes on a lattice with boundary. *arXive e-print quant-ph/9811052*, 1998.

[BK99] S. L. Braunstein and H. J. Kimble. Dense coding for continuous variables. *arXive e-print quant-ph/9910010*, 1999.

[BKLW99] D. Bacon, J. Kempe, D. A. Lidar, and K. B. Whaley. Universal fault-tolerant computation on decoherence-free subspaces. *arXive e-print quant-ph/9909058*, 1999.

[BKN98] H. Barnum, E. Knill, and M. A. Nielsen. On quantum fidelities and channel capacities. *arXive e-print quant-ph/9809010*, 1998.

[BL95] D. Boneh and R. J. Lipton. Quantum cryptoanalysis of hidden linear functions (extended abstract). In Don Coppersmith, editor, *Lecture notes in computer science — Advances in Cryptology — CRYPTO'95*, pages 424–437, Springer-Verlag, Berlin, 1995.

[BMP+99] P. O. Boykin, T. Mor, M. Pulver, V. Roychowdhury, and F. Vatan. On universal and fault-tolerant quantum computing. *arXive e-print quant-ph/9906054*, 1999.

[BNS98] H. Barnum, M. A. Nielsen, and B. W. Schumacher. Information transmission through a noisy quantum channel. *Phys. Rev. A*, 57:4153, 1998.

[Boh51] D. Bohm. *Quantum Theory.* Prentice-Hall, Englewood Cliffs, New Jersey, 1951.

[BP93] S. M. Barnett and S. J. D. Phoenix. Information-theoretic limits to quantum cryptography. *Phys. Rev. A*, 48(1):R5–R8, 1993.

[BPM+97] D. Bouwmeester, J. W. Pan, K. Mattle, M. Eibl, H. Weinfurter, and A. Zeilinger. Experimental quantum teleportation. *Nature*, 390(6660):575–579, 1997.

[BR98] D. S. Bethune and W. P. Risk. An autocompensating quantum key distribution system using polarization splitting of light. In *IQEC '98 Digest of Postdeadline Papers*, pages QPD12–2, Optical Society of America, Washington, DC, 1998.

[BR00] D. S. Bethune and W. P. Risk. An autocompensating fiber-optic quantum cryptography system based on polarization splitting of light. *J. Quantum Electronics*, 36(3):100, 2000.

[Bra93] G. Brassard. A bibliography of quantum cryptography. *Université de Montréal preprint*, pages 1–10, 3 December 1993. A preliminary version of this appeared in *Sigact News*, vol. 24(3), 1993, pages 16-20.

[Bra98] S. L. Braunstein. Error correction for continuous quantum variables. *Phys. Rev. Lett.*, 80:4084–4087, 1998. *arXive e-print quant-ph/9711049.*

[BS94] G. Brassard and L. Salvail. Secret-key reconciliation by public discussion. In T. Helleseth, editor, *Lecture Notes in Computer Science: Advances in Cryptology – EUROCRYPT'93*, Volume

765, pages 410–423, Springer-Verlag, New York, 1994.

[BS98]   C. H. Bennett and P. W. Shor. Quantum information theory. *IEEE Trans. Inf. Theory*, 44(6):2724–42, 1998.

[BST98]   H. Barnum, J. A. Smolin, and B. Terhal. Quantum capacity is properly defined without encodings. *Phys. Rev. A*, 58(5):3496–3501, 1998.

[BT97]   B. M. Boghosian and W. Taylor. Simulating quantum mechanics on a quantum computer. *arXive e-print quant-ph/9701019*, 1997.

[BV97]   E. Bernstein and U. Vazirani. Quantum complexity theory. *SIAM J. Comput.*, 26(5):1411–1473, 1997. *arXive e-print quant-ph/9701001*.

[BW92]   C. H. Bennett and S. J. Wiesner. Communication via one- and two-particle operators on Einstein-Podolsky-Rosen states. *Phys. Rev. Lett.*, 69(20):2881–2884, 1992.

[CAK98]   N. J. Cerf, C. Adami, and P. Kwiat. Optical simulation of quantum logic. *Phys. Rev. A*, 57:R1477, 1998.

[Cav99]   C. M. Caves. Quantum error correction and reversible operations. *Journal of Superconductivity*, 12(6):707–718, 1999.

[CD96]   R. Cleve and D. P. DiVincenzo. Schumacher's quantum data compression as a quantum computation. *Phys. Rev. A*, 54:2636, 1996. *arXive e-print quant-ph/9603009*.

[CEMM98]   R. Cleve, A. Ekert, C. Macchiavello, and M. Mosca. Quantum algorithms revisited. *Proc. R. Soc. London A*, 454(1969):339–354, 1998.

[CFH97]   D. G. Cory, A. F. Fahmy, and T. F. Havel. Ensemble quantum computing by NMR spectroscopy. *Proc. Nat. Acad. Sci. USA*, 94:1634–1639, 1997.

[CGK98]   I. L. Chuang, N. Gershenfeld, and M. Kubinec. Experimental implementation of fast quantum searching. *Phys. Rev. Lett.*, 18(15):3408–3411, 1998.

[CGKL98]   I. L. Chuang, N. Gershenfeld, M. G. Kubinec, and D. W. Leung. Bulk quantum computation with nuclear-magnetic-resonance: theory and experiment. *Proc. R. Soc. London A*, 454(1969):447–467, 1998.

[Che68]   P. R. Chernoff. Note on product formulas for operator semigroups. *J. Functional Analysis*, 2:238–242, 1968.

[Cho75]   M.-D. Choi. Completely positive linear maps on complex matrices. *Linear Algebra and Its Applications*, 10:285–290, 1975.

[CHSH69]   J. F. Clauser, M. A. Horne, A. Shimony, and R. A. Holt. Proposed experiment to test local hidden-variable theories. *Phys. Rev. Lett.*, 49:1804–1807, 1969.

[Chu36]   A. Church. An unsolvable problem of elementary number theory. *Am. J. Math. (reprinted in [Dav65])*, 58:345, 1936.

[CK81]   I. Csiszár and J. Körner. *Information Theory: Coding Theorems for Discrete Memoryless Systems*. Academic Press, New York, 1981.

[CL83]   A. O. Caldeira and A. J. Leggett. Quantum tunnelling in a dissipative system. *Ann. Phys.*, 149(2):374–456, 1983.

[Cla89]   M. Clausen. Fast generalized Fourier transforms. *Theor. Comput. Sci.*, 67:55–63, 1989.

[Cle99]   R. Cleve. The query complexity of order-finding. *arXive e-print quant-ph/9911124*, 1999.

[CLR90]   T. H. Cormen, C. E. Leiserson, and R. L. Rivest. *Introduction to Algorithms*. MIT Press, Cambridge, Mass., 1990.

[CM97]   C. Cachin and U. M. Maurer. Linking information reconciliation and privacy amplification. *J. Cryptology*, 10:97–110, 1997.

[CM00]   I. L. Chuang and D. Modha. Reversible arithmetic coding for quantum data compression. *IEEE Trans. Inf. Theory*, 46(3):1104, May 2000.

[CMP+98]   D. G. Cory, W. Mass, M. Price, E. Knill, R. Laflamme, W. H. Zurek, T. F. Havel, and S. S. Somaroo. Experimental quantum error correction. *arXive e-print quant-ph/9802018*, 1998.

[CN97]   I. L. Chuang and M. A. Nielsen. Prescription for experimental determination of the dynamics of a quantum black box. *J. Mod. Opt.*, 44(11-12):2455–2467, 1997. *arXive e-print quant-ph/9610001*.

[Con72]   J. H. Conway. Unpredictable iterations. In *Proceedings of the Number Theory Conference*, pages 49–52, Boulder, Colorado, 1972.

[Con86]   J. H. Conway. Fractran: a simple universal programming language. In T. M. Cover and B. Gopinath, editors, *Open Problems in Communication and Computation*, pages 4–26, Springer-Verlag, New York, 1986.

[Coo71] S. A. Cook. The complexity of theorem-proving procedures. In *Proc. 3rd Ann. ACM Symp. on Theory of Computing*, pages 151–158, Association for Computing Machinery, New York, 1971.

[Cop94] D. Coppersmith. An approximate Fourier transform useful in quantum factoring. *IBM Research Report RC 19642*, 1994.

[CPZ96] J. I. Cirac, T. Pellizzari, and P. Zoller. Enforcing coherent evolution in dissipative quantum dynamics. *Science*, 273:1207, 1996.

[CRSS97] A. R. Calderbank, E. M. Rains, P. W. Shor, and N. J. A. Sloane. Quantum error correction and orthogonal geometry. *Phys. Rev. Lett.*, 78:405–8, 1997.

[CRSS98] A. R. Calderbank, E. M. Rains, P. W. Shor, and N. J. A. Sloane. Quantum error correction via codes over GF(4). *IEEE Trans. Inf. Theory*, 44(4):1369–1387, 1998.

[CS96] A. R. Calderbank and P. W. Shor. Good quantum error-correcting codes exist. *Phys. Rev. A*, 54:1098, 1996. *arXive e-print quant-ph/9512032*.

[CST89] R. A. Campos, B. E. A. Saleh, and M. C. Tiech. Quantum-mechanical lossless beamsplitters: SU(2) symmetry and photon statistics. *Phys. Rev. A*, 40:1371, 1989.

[CT91] T. M. Cover and J. A. Thomas. *Elements of Information Theory*. John Wiley and Sons, New York, 1991.

[CTDL77a] C. Cohen-Tannoudji, B. Diu, and F. Laloë. *Quantum Mechanics, Vol. I*. John Wiley and Sons, New York, 1977.

[CTDL77b] C. Cohen-Tannoudji, B. Diu, and F. Laloë. *Quantum Mechanics, Vol. II*. John Wiley and Sons, New York, 1977.

[CVZ+98] I. L. Chuang, L. M. K. Vandersypen, X. L. Zhou, D. W. Leung, and S. Lloyd. Experimental realization of a quantum algorithm. *Nature*, 393(6681):143–146, 1998.

[CW95] H. F. Chau and F. Wilczek. Simple realization of the Fredkin gate using a series of two-body operators. *Phys. Rev. Lett.*, 75(4):748–50, 1995. *arXive e-print quant-ph/9503005*.

[CY95] I. L. Chuang and Y. Yamamoto. Simple quantum computer. *Phys. Rev. A*, 52:3489–3496, 1995. *arXive e-print quant-ph/9505011*.

[CZ95] J. I. Cirac and P. Zoller. Quantum compu-tations with cold trapped ions. *Phys. Rev. Lett.*, 74:4091, 1995.

[Dav65] M. D. Davis. *The Undecidable*. Raven Press, Hewlett, New York, 1965.

[Dav76] E. B. Davies. *Quantum Theory of Open Systems*. Academic Press, London, 1976.

[DBE95] D. Deutsch, A. Barenco, and A. Ekert. Universality in quantum computation. *Proc. R. Soc. London A*, 449(1937):669–677, 1995.

[Deu83] D. Deutsch. Uncertainty in quantum measurements. *Phys. Rev. Lett.*, 50(9):631–633, 1983.

[Deu85] D. Deutsch. Quantum theory, the Church-Turing Principle and the universal quantum computer. *Proc. R. Soc. Lond. A*, 400:97, 1985.

[Deu89] D. Deutsch. Quantum computational networks. *Proc. R. Soc. London A*, 425:73, 1989.

[DG98] L.-M. Duan and G.-C. Guo. Probabilistic cloning and identification of linearly independent quantum states. *Phys. Rev. Lett.*, 80:4999–5002, 1998. *arXive e-print quant-ph/9804064*.

[DH76] W. Diffie and M. Hellman. New directions in cryptography. *IEEE Trans. Inf. Theory*, IT-22(6):644–54, 1976.

[DH96] C. Dürr and P. Høyer. A quantum algorithm for finding the minimum. *arXive e-print quant-ph/9607014*, 1996.

[Die82] D. Dieks. Communication by EPR devices. *Phys. Lett. A*, 92(6):271–272, 1982.

[DiV95a] D. P. DiVincenzo. Quantum computation. *Science*, 270:255, 1995.

[DiV95b] D. P. DiVincenzo. Two-bit gates are universal for quantum computation. *Phys. Rev. A*, 51(2):1015–1022, 1995.

[DiV98] D. P. DiVincenzo. Quantum gates and circuits. *Proc. R. Soc. London A*, 454:261–276, 1998.

[DJ92] D. Deutsch and R. Jozsa. Rapid solution of problems by quantum computation. *Proc. R. Soc. London A*, 439:553, 1992.

[DL98] W. Diffie and S. Landau. *Privacy on the Line: the Politics of Wiretapping and Encryption*. MIT Press, Cambridge Massachusetts, 1998.

[DMB+93] L. Davidovich, A. Maali, M. Brune, J. M. Raimond, and S. Haroche. *Phys. Rev. Lett.*, 71:2360, 1993.

[DR90] P. Diaconis and D. Rockmore. Efficient computation of the Fourier transform on finite groups. *J. Amer. Math. Soc.*, 3(2):297–332, 1990.

[DRBH87] L. Davidovich, J. M. Raimond, M. Brune, and S. Haroche. *Phys. Rev. A*, 36:3771, 1987.

[DRBH95] P. Domokos, J. M. Raimond, M. Brune, and S. Haroche. Simple cavity-QED two-bit universal quantum logic gate: The principle and expected performances. *Phys. Rev. Lett.*, 52:3554, 1995.

[DS96] D. P. DiVincenzo and P. W. Shor. Fault-tolerant error correction with efficient quantum codes. *Phys. Rev. Lett.*, 77:3260, 1996.

[DSS98] D. P. DiVincenzo, P. W. Shor, and J. Smolin. Quantum-channel capacities of very noisy channels. *Phys. Rev. A*, 57(2):830–839, 1998.

[Ear42] S. Earnshaw. On the nature of the molecular forces which regulate the constitution of the luminiferous ether. *Trans. Camb. Phil. Soc.*, 7:97–112, 1842.

[EBW87] R. R. Ernst, G. Bodenhausen, and A. Wokaun. *Principles of Nuclear Magnetic Resonance in One and Two Dimensions.* Oxford University Press, Oxford, 1987.

[EH99] M. Ettinger and P. Høyer. On quantum algorithms for noncommutative hidden subgroups. In *Symposium on Theoretical Aspects in Computer Science.* University of Trier, 1999. *arXive e-print quant-ph/9807029.*

[EHK99] M. Ettinger, P. Høyer, and E. Knill. Hidden subgroup states are almost orthogonal. *arXive e-print quant-ph/9901034*, 1999.

[EHPP94] A. K. Ekert, B. Huttner, G. M. Palma, and A. Peres. Eavesdropping on quantum-cryptographical systems. *Phys. Rev. A*, 50(2):1047–1056, 1994.

[EJ96] A. Ekert and R. Jozsa. Quantum computation and Shor's factoring algorithm. *Rev. Mod. Phys.*, 68:733, 1996.

[EJ98] A. Ekert and R. Jozsa. Quantum algorithms: Entanglement enhanced information processing. *Proc. R. Soc. London A*, 356(1743):1769–82, Aug. 1998. *arXive e-print quant-ph/9803072.*

[Eke91] A. K. Ekert. Quantum cryptography based on Bell's theorem. *Phys. Rev. Lett.*, 67(6):661–663, 1991.

[EM96] A. Ekert and C. Macchiavello. Error correction in quantum communication. *Phys.*

Rev. Lett., 77:2585, 1996. *arXive e-print quant-ph/9602022.*

[EPR35] A. Einstein, B. Podolsky, and N. Rosen. Can quantum-mechanical description of physical reality be considered complete? *Phys. Rev.*, 47:777–780, 1935.

[Eps73] H. Epstein. *Commun. Math. Phys.*, 31:317–325, 1973.

[Fan73] M. Fannes. A continuity property of the entropy density for spin lattice systems. *Commun. Math. Phys.*, 31:291–294, 1973.

[FC94] C. A. Fuchs and C. M. Caves. Ensemble-dependent bounds for accessible information in quantum mechanics. *Phys. Rev. Lett.*, 73(23):3047–3050, 1994.

[Fel68a] W. Feller. *An Introduction to Probability Theory and its Applications*, Volume 1. Wiley, New York, 1968.

[Fel68b] W. Feller. *An Introduction to Probability Theory and its Applications*, Volume 2. Wiley, New York, 1968.

[Fey82] R. P. Feynman. Simulating physics with computers. *Int. J. Theor. Phys.*, 21:467, 1982.

[FG98] E. Farhi and S. Gutmann. An analog analogue of a digital quantum computation. *Phys. Rev. A*, 57(4):2403–2406, 1998. *arXive e-print quant-ph/9612026.*

[FLS65a] R. P. Feynman, R. B. Leighton, and M. Sands. Volume III of *The Feynman Lectures on Physics.* Addison-Wesley, Reading, Mass., 1965.

[FLS65b] R. P. Feynman, R. B. Leighton, and M. Sands. Volume I of *The Feynman Lectures on Physics.* Addison-Wesley, Reading, Mass., 1965.

[FM98] M. H. Freedman and D. A. Meyer. Projective plane and planar quantum codes. *arXive e-print quant-ph/9810055*, 1998.

[FS92] A. Fässler and E. Stiefel. *Group Theoretical Methods and Their Applications.* Birkhaüser, Boston, 1992.

[FSB+98] A. Furusawa, J. L. Sørensen, S. L. Braunstein, C. A. Fuchs, H. J. Kimble, and E. S. Polzik. Unconditional quantum teleportation. *Science*, 282:706–709, 1998.

[FT82] E. Fredkin and T. Toffoli. Conservative logic. *Int. J. Theor. Phys.*, 21(3/4):219–253, 1982.

[Fuc96] C. A. Fuchs. *Distinguishability and Accessible Information in Quantum Theory.* Ph.D. thesis, The University of New

Mexico, Albuquerque, NM, 1996. *arXive e-print quant-ph/9601020*.

[Fuc97]   C. A. Fuchs. Nonorthogonal quantum states maximize classical information capacity. *Phys. Rev. Lett.*, 79(6):1162–1165, 1997.

[FvdG99]   C. A. Fuchs and J. van de Graaf. Cryptographic distinguishability measures for quantum-mechanical states. *IEEE Trans. Inf. Theory*, 45(4):1216–1227, 1999.

[Gar91]   C. W. Gardiner. *Quantum Noise*. Springer-Verlag, Berlin, 1991.

[GC97]   N. Gershenfeld and I. L. Chuang. Bulk spin resonance quantum computation. *Science*, 275:350, 1997.

[GC99]   D. Gottesman and I. L. Chuang. Quantum teleportation is a universal computational primitive. *Nature*, 402:390–392, 1999. *arXive e-print quant-ph/9908010*.

[GJ79]   M. R. Garey and D. S. Johnson. *Computers and Intractability*. W. H. Freeman and Company, New York, 1979.

[GN96]   R. B. Griffiths and C.-S. Niu. Semiclassical Fourier transform for quantum computation. *Phys. Rev. Lett.*, 76(17):3228–3231, 1996. *arXive e-print quant-ph/9511007*.

[Gor64]   J. P. Gordon. Noise at optical frequencies; information theory. In P. A. Miles, editor, *Quantum Electronics and Coherent Light*, Proceedings of the International School of Physics 'Enrico Fermi' XXXI, Academic Press, New York, 1964.

[Got96]   D. Gottesman. Class of quantum errorcorrecting codes saturating the quantum Hamming bound. *Phys. Rev. A*, 54:1862, 1996.

[Got97]   D. Gottesman. *Stabilizer Codes and Quantum Error Correction*. Ph.D. thesis, California Institute of Technology, Pasadena, CA, 1997.

[Got98a]   D. Gottesman. Fault-tolerant quantum computation with higher-dimensional systems. *arXive e-print quant-ph/9802007*, 1998.

[Got98b]   D. Gottesman. Theory of fault-tolerant quantum computation. *Phys. Rev. A*, 57(1):127–137, 1998. *arXive e-print quant-ph/9702029*.

[GP10]   D. Gottesman and J. Preskill. The Hitchiker's guide to the threshold theorem. *Eternally in preparation*, 1:1–9120, 2010.

[Gro96]   L. Grover. In *Proc. 28th Annual ACM Symposium on the Theory of Computation*, pages 212–219, ACM Press, New York, 1996.

[Gro97]   L. K. Grover. Quantum mechanics helps in searching for a needle in a haystack. *Phys. Rev. Lett.*, 79(2):325, 1997. *arXive e-print quant-ph/9706033*.

[Gru99]   J. Gruska. *Quantum Computing*. McGraw-Hill, London, 1999.

[GS92]   G. R. Grimmett and D. R. Stirzaker. *Probability and Random Processes*. Clarendon Press, Oxford, 1992.

[HAD+95]   R. J. Hughes, D. M. Alde, P. Dyer, G. G. Luther, G. L. Morgan, and M. Schauer. Quantum cryptography. *Contemp. Phys.*, 36(3):149–163, 1995. *arXive e-print quant-ph/9504002*.

[Hal58]   P. R. Halmos. *Finite-dimensional Vector Spaces*. Van Nostrand, Princeton, N.J., 1958.

[Ham89]   M. Hammermesh. *Group Theory and its Application to Physical Problems*. Dover, New York, 1989.

[HGP96]   J. L. Hennessey, D. Goldberg, and D. A. Patterson. *Computer Architecture: A Quantitative Approach*. Academic Press, New York, 1996.

[HHH96]   M. Horodecki, P. Horodecki, and R. Horodecki. Separability of mixed states: necessary and sufficient conditions. *Phys. Lett. A*, 223(1-2):1–8, 1996.

[HHH98]   M. Horodecki, P. Horodecki, and R. Horodecki. Mixed-state entanglement and distillation: is there a 'bound' entanglement in nature? *Phys. Rev. Lett.*, 80(24):5239–5242, 1998.

[HHH99a]   M. Horodecki, P. Horodecki, and R. Horodecki. General teleportation channel, singlet fraction, and quasidistillation. *Phys. Rev. A*, 60(3):1888–1898, 1999.

[HHH99b]   M. Horodecki, P. Horodecki, and R. Horodecki. Limits for entanglement measures. *arXive e-print quant-ph/9908065*, 1999.

[HHH99c]   P. Horodecki, M. Horodecki, and R. Horodecki. Bound entanglement can be activated. *Phys. Rev. Lett.*, 82(5):1056–1059, 1999.

[HJ85]   R. A. Horn and C. R. Johnson. *Matrix Analysis*. Cambridge University Press, Cambridge, 1985.

[HJ91]   R. A. Horn and C. R. Johnson. *Topics in Matrix Analysis*. Cambridge University Press, Cambridge, 1991.

[HJS+96] P. Hausladen, R. Jozsa, B. Schumacher, M. Westmoreland, and W. K. Wootters. Classical information capacity of a quantum channel. *Phys. Rev. A*, 54:1869, 1996.

[HJW93] L. P. Hughston, R. Jozsa, and W. K. Wootters. A complete classification of quantum ensembles having a given density matrix. *Phys. Lett. A*, 183:14–18, 1993.

[HK69] K.-E. Hellwig and K. Kraus. Pure operations and measurements. *Commun. Math. Phys.*, 11:214–220, 1969.

[HK70] K.-E. Hellwig and K. Kraus. Operations and measurements. II. *Commun. Math. Phys.*, 16:142–147, 1970.

[Hof79] D. R. Hofstadter. *Gödel, Escher, Bach: an Eternal Golden Braid*. Basic Books, New York, 1979.

[Hol73] A. S. Holevo. Statistical problems in quantum physics. In Gisiro Maruyama and Jurii V. Prokhorov, editors, *Proceedings of the Second Japan–USSR Symposium on Probability Theory*, pages 104–119, Springer-Verlag, Berlin, 1973. Lecture Notes in Mathematics, vol. 330.

[Hol79] A. S. Holevo. Capacity of a quantum communications channel. *Problems of Inf. Transm.*, 5(4):247–253, 1979.

[Hol98] A. S. Holevo. The capacity of the quantum channel with general signal states. *IEEE Trans. Inf. Theory*, 44(1):269–273, 1998.

[Hor97] M. Horodecki. Limits for compression of quantum information carried by ensembles of mixed states. *Phys. Rev. A*, 57:3364–3369, 1997.

[HSM+98] A. G. Huibers, M. Switkes, C. M. Marcus, K. Campman, and A. C. Gossard. Dephasing in open quantum dots. *Phys. Rev. Lett.*, 82:200, 1998.

[HW60] G. H. Hardy and E. M. Wright. *An Introduction to the Theory of Numbers, Fourth Edition*. Oxford University Press, London, 1960.

[IAB+99] A. Imamoglu, D. D. Awschalom, G. Burkard, D. P. DiVincenzo, D. Loss, M. Sherwin, and A. Small. Quantum information processing using quantum dot spins and cavity qed. *Phys. Rev. Lett.*, 83(20):4204–7, 1999.

[IY94] A. Imamoglu and Y. Yamamoto. Turnstile device for heralded single photons: Coulomb blockade of electron and hole tunneling in quantum confined p-i-n heterojunctions. *Phys. Rev. Lett.*, 72(2):210–13, 1994.

[Jam98] D. James. The theory of heating of the quantum ground state of trapped ions. *arXive e-print quant-ph/9804048*, 1998.

[Jay57] E. T. Jaynes. Information theory and statistical mechanics. ii. *Phys. Rev.*, 108(2):171–190, 1957.

[JM98] J. A. Jones and M. Mosca. Implementation of a quantum algorithm to solve Deutsch's problem on a nuclear magnetic resonance quantum computer. *arXive e-print quant-ph/9801027*, 1998.

[JMH98] J. A. Jones, M. Mosca, and R. H. Hansen. Implementation of a quantum search algorithm on a nuclear magnetic resonance quantum computer. *Nature*, 393(6683):344, 1998. *arXive e-print quant-ph/9805069*.

[Jon94] K. R. W. Jones. Fundamental limits upon the measurement of state vectors. *Phys. Rev. A*, 50:3682–3699, 1994.

[Joz94] R. Jozsa. Fidelity for mixed quantum states. *J. Mod. Opt.*, 41:2315–2323, 1994.

[Joz97] R. Jozsa. Quantum algorithms and the Fourier transform. *arXive e-print quant-ph/9707033*, 1997.

[JP99] D. Jonathan and M. B. Plenio. Entanglement-assisted local manipulation of pure states. *Phys. Rev. Lett.*, 83:3566–3569, 1999.

[JS94] R. Jozsa and B. Schumacher. A new proof of the quantum noiseless coding theorem. *J. Mod. Opt.*, 41:2343–2349, 1994.

[Kah96] D. Kahn. *Codebreakers: the Story of Secret Writing*. Scribner, New York, 1996.

[Kan98] B. Kane. A silicon-based nuclear spin quantum computer. *Nature*, 393:133–137, 1998.

[Kar72] R. M. Karp. Reducibility among combinatorial problems. In *Complexity of Computer Computations*, pages 85–103, Plenum Press, New York, 1972.

[KCL98] E. Knill, I. Chuang, and R. Laflamme. Effective pure states for bulk quantum computation. *Phys. Rev. A*, 57(5):3348–3363, 1998. *arXive e-print quant-ph/9706053*.

[Kit95] A. Y. Kitaev. Quantum measurements and the Abelian stabilizer problem. *arXive e-print quant-ph/9511026*,, 1995.

[Kit97a] A. Y. Kitaev. Fault-tolerant quantum computation by anyons. *arXive e-print quant-ph/9707021*, 1997.

[Kit97b]   A. Y. Kitaev. Quantum computations: algorithms and error correction. *Russ. Math. Surv.*, 52(6):1191–1249, 1997.

[Kit97c]   A. Y. Kitaev. Quantum error correction with imperfect gates. In A. S. Holevo O. Hirota and C. M. Caves, editors, *Quantum Communication, Computing, and Measurement*, pages 181–188, Plenum Press, New York, 1997.

[KL51]   S. Kullback and R. A. Leibler. On information and sufficiency. *Ann. Math. Stat.*, 22:79–86, 1951.

[KL97]   E. Knill and R. Laflamme. A theory of quantum error-correcting codes. *Phys. Rev. A*, 55:900, 1997. *arXive e-print quant-ph/9604034*.

[KL99]   E. Knill and R. Laflamme. Quantum computation and quadratically signed weight enumerators. *arXive e-print quant-ph/9909094*, 1999.

[Kle31]   O. Klein. *Z. Phys.*, 72:767–775, 1931.

[KLV99]   E. Knill, R. Laflamme, and L. Viola. Theory of quantum error correction for general noise. *arXive e-print quant-ph/9908066*, 1999.

[KLZ98a]   E. Knill, R. Laflamme, and W. H. Zurek. Resilient quantum computation. *Science*, 279(5349):342–345, 1998. *arXive e-print quant-ph/9702058*.

[KLZ98b]   E. Knill, R. Laflamme, and W. H. Zurek. Resilient quantum computation: error models and thresholds. *Proc. R. Soc. London A*, 454(1969):365–384, 1998. *arXive e-print quant-ph/9702058*.

[KMSW99]   P. G. Kwiat, J. R. Mitchell, P. D. D. Schwindt, and A. G. White. Grover's search algorithm: An optical approach. *arXive e-print quant-ph/9905086*, 1999.

[Kni95]   E. Knill. Approximating quantum circuits. *arXive e-print quant-ph/9508006*, 1995.

[Knu97]   D. E. Knuth. *Fundamental Algorithms 3rd Edition*, Volume 1 of *The Art of Computer Programming*. Addison-Wesley, Reading, Massachusetts, 1997.

[Knu98a]   D. E. Knuth. *Seminumerical Algorithms 3rd Edition*, Volume 2 of *The Art of Computer Programming*. Addison-Wesley, Reading, Massachusetts, 1998.

[Knu98b]   D. E. Knuth. *Sorting and Searching 2nd Edition*, Volume 3 of *The Art of Computer Programming*. Addison-Wesley, Reading, Massachusetts, 1998.

[Kob94]   N. Koblitz. *A Course in Number Theory and Cryptography*. Springer-Verlag, New York, 1994.

[KR99]   C. King and M. B. Ruskai. Minimal entropy of states emerging from noisy quantum channels. *arXive e-print quant-ph/9911079*, 1999.

[Kra83]   K. Kraus. *States, Effects, and Operations: Fundamental Notions of Quantum Theory*. Lecture Notes in Physics, Vol. 190. Springer-Verlag, Berlin, 1983.

[Kra87]   K. Kraus. Complementary observables and uncertainty relations. *Phys. Rev. D*, 35(10):3070–3075, 1987.

[KSC+94]   P. G. Kwiat, A. M. Steinberg, R. Y. Chiao, P. H. Eberhard, and M. D. Petroff. Absolute efficiency and time-response measurement of single-photon detectors. *Appl. Opt.*, 33(10):1844–1853, 1994.

[KU91]   M. Kitagawa and M. Ueda. Nonlinear-interferometric generation of number-phase correlated Fermion states. *Phys. Rev. Lett.*, 67(14):1852, 1991.

[Lan27]   L. Landau. Das dämpfungsproblem in der wellenmechanik. *Z. Phys.*, 45:430–441, 1927.

[Lan61]   R. Landauer. Irreversibility and heat generation in the computing process. *IBM J. Res. Dev.*, 5:183, 1961.

[LB99]   S. Lloyd and S. Braunstein. Quantum computation over continuous variables. *Phys. Rev. Lett.*, 82:1784–1787, 1999. *arXive e-print quant-ph/9810082*.

[LBW99]   D. A. Lidar, D. A. Bacon, and K. B. Whaley. Concatenating decoherence free subspaces with quantum error correcting codes. *Phys. Rev. Lett.*, 82(22):4556–4559, 1999.

[LC99]   H. Lo and H. F. Chau. Unconditional security of quantum key distribution over arbitrarily long distances. *Science*, 283:2050–2056, 1999. *arXive e-print quant-ph/9803006*.

[LCW98]   D. A. Lidar, I. L. Chuang, and K. B. Whaley. Decoherence-free subspaces for quantum computation. *Phys. Rev. Lett.*, 81(12):2594–2597, 1998.

[LD98]   D. Loss and D. P. DiVincenzo. Quantum computation with quantum dots. *Phys. Rev. A*, 57:120–126, 1998.

[Lec63]   Y. Lecerf. Machines de Turing réversibles. *Comptes Rendus*, 257:2597–2600, 1963.

[Leo97]   U. Leonhardt. *Measuring the Quantum*

State of Light. Cambridge University Press, New York, 1997.

[Lev73]   L. Levin. Universal sorting problems. *Probl. Peredaci Inf.*, 9:115–116, 1973. Original in Russian. English translation in *Probl. Inf. Transm. USSR* 9:265–266 (1973).

[Lie73]   E. H. Lieb. Convex trace functions and the Wigner-Yanase-Dyson conjecture. *Ad. Math.*, 11:267–288, 1973.

[Lie75]   E. H. Lieb. *Bull. AMS*, 81:1–13, 1975.

[Lin75]   G. Lindblad. Completely positive maps and entropy inequalities. *Commun. Math. Phys.*, 40:147–151, 1975.

[Lin76]   G. Lindblad. On the generators of quantum dynamical semigroups. *Commun. Math. Phys.*, 48:199, 1976.

[Lin91]   G. Lindblad. Quantum entropy and quantum measurements. In C. Bendjaballah, O. Hirota, and S. Reynaud, editors, *Quantum Aspects of Optical Communications*, Lecture Notes in Physics, vol. 378, pages 71–80, Springer-Verlag, Berlin, 1991.

[Lip95]   R. Lipton. DNA solution of hard computational problems. *Science*, 268:542–525, 1995.

[LKF99]   N. Linden, E. Kupce, and R. Freeman. NMR quantum logic gates for homonuclear spin systems. *arXive e-print quant-ph/9907003*, 1999.

[LL93]   A. K. Lenstra and H. W. Lenstra Jr., editors. *The Development of the Number Field Sieve*. Springer-Verlag, New York, 1993.

[Llo93]   S. Lloyd. A potentially realizable quantum computer. *Science*, 261:1569, 1993.

[Llo94]   S. Lloyd. Necessary and sufficient conditions for quantum computation. *J. Mod. Opt.*, 41(12):2503, 1994.

[Llo95]   S. Lloyd. Almost any quantum logic gate is universal. *Phys. Rev. Lett.*, 75(2):346, 1995.

[Llo96]   S. Lloyd. Universal quantum simulators. *Science*, 273:1073, 1996.

[Llo97]   S. Lloyd. The capacity of the noisy quantum channel. *Phys. Rev. A*, 56:1613, 1997.

[LLS75]   R. E. Ladner, N. A. Lynch, and A. L. Selman. A comparison of polynomial-time reducibilities. *Theor. Comp. Sci.*, 1:103–124, 1975.

[LMPZ96]   R. Laflamme, C. Miquel, J.-P. Paz, and W. H. Zurek. Perfect quantum error correction code. *Phys. Rev. Lett.*,

77:198, 1996. *arXive e-print quant-ph/9602019*.

[LNCY97]   D. W. Leung, M. A. Nielsen, I. L. Chuang, and Y. Yamamoto. Approximate quantum error correction can lead to better codes. *Phys. Rev. A*, 56:2567–2573, 1997. *arXive e-print quant-ph/9704002*.

[Lo99]   H. Lo. A simple proof of the unconditional security of quantum key distribution. *arXive e-print quant-ph/9904091*, 1999.

[Lom87]   J. S. Lomont. *Applications of Finite Groups*. Dover, New York, 1987.

[Lou73]   W. H. Louisell. *Quantum Statistical Properties of Radiation*. Wiley, New York, 1973.

[LP97]   H.-K. Lo and S. Popescu. Concentrating local entanglement by local actions – beyond mean values. *arXive e-print quant-ph/9707038*, 1997.

[LP99]   N. Linden and S. Popescu. Good dynamics versus bad kinematics. Is entanglement needed for quantum computation? *arXive e-print quant-ph/9906008*, 1999.

[LR68]   O. E. Lanford and D. Robinson. Mean entropy of states in quantum-statistical mechanics. *J. Math. Phys.*, 9(7):1120–1125, 1968.

[LR73a]   E. H. Lieb and M. B. Ruskai. A fundamental property of quantum-mechanical entropy. *Phys. Rev. Lett.*, 30(10):434–436, 1973.

[LR73b]   E. H. Lieb and M. B. Ruskai. Proof of the strong subadditivity of quantum mechanical entropy. *J. Math. Phys.*, 14:1938–1941, 1973.

[LR90]   H. Leff and R. Rex. *Maxwell's Demon: Entropy, Information, Computing*. Princeton University Press, Princeton, NJ, 1990.

[LS93]   L. J. Landau and R. F. Streater. On Birkhoff theorem for doubly stochastic completely positive maps of matrix algebras. *Linear Algebra Appl.*, 193:107–127, 1993.

[LS98]   S. Lloyd and J. E. Slotine. Analog quantum error correction. *Phys. Rev. Lett.*, 80:4088–4091, 1998.

[LSP98]   H.-K. Lo, T. Spiller, and S. Popescu. *Quantum information and computation*. World Scientific, Singapore, 1998.

[LTV98]   M. Li, J. Tromp, and P. Vitanyi. Reversible simulation of irreversible computation by pebble games. *Physica D*, 120:168–176, 1998.

[LV96]   M. Li and P. Vitanyi. Reversibility and adiabatic computation: trading time and space for energy. *Proc. R. Soc. London A*, 452:769–789, 1996. *arXive e-print quant-ph/9703022*.

[LVZ⁺99]   D. W. Leung, L. M. K. Vandersypen, X. Zhou, M. Sherwood, C. Yannoni, M. Kubinec, and I. L. Chuang. Experimental realization of a two-bit phase damping quantum code. *Phys. Rev. A*, 60:1924, 1999.

[Man80]   Y. Manin. *Computable and Uncomputable (in Russian)*. Sovetskoye Radio, Moscow, 1980.

[Man99]   Y. I. Manin. Classical computing, quantum computing, and Shor's factoring algorithm. *arXive e-print quant-ph/9903008*, 1999.

[Mau93]   U. M. Maurer. Secret key agreement by public discussion from common information. *IEEE Trans. Inf. Theory*, 39:733–742, 1993.

[Max71]   J. C. Maxwell. *Theory of Heat*. Longmans, Green, and Co., London, 1871.

[May98]   D. Mayers. Unconditional security in quantum cryptography. *arXive e-print quant-ph/9802025*, 1998.

[ME99]   M. Mosca and A. Ekert. The hidden subgroup problem and eigenvalue estimation on a quantum computer. *arXive e-print quant-ph/9903071*, 1999.

[Mer78]   R. Merkle. Secure communications over insecure channels. *Comm. of the ACM*, 21:294–299, 1978.

[Mil76]   G. L. Miller. Riemann's hypothesis and tests for primality. *J. Comput. Syst. Sci.*, 13(3):300–317, 1976.

[Mil89a]   G. J. Milburn. Quantum optical Fredkin gate. *Phys. Rev. Lett.*, 62(18):2124, 1989.

[Mil89b]   D. A. B. Miller. Optics for low energy communications inside digital processors: quantum detectors, sources, and modulators as efficient impedance converters. *Opt. Lett.*, 14:146, 1989.

[Mil96]   G. J. Milburn. A quantum mechanical Maxwell's demon. Unpublished, 1996.

[Mil97]   G. J. Milburn. *Scrödinger's Machines: the Quantum Technology Reshaping Everyday Life*. W. H. Freeman, New York, 1997.

[Mil98]   G. J. Milburn. *The Feynman Processor: Quantum Entanglement and the Computing Revolution*. Perseus Books, Reading, Mass., 1998.

[Min67]   M. L. Minsky. *Computation: finite and infinite machines*. Prentice-Hall, Englewood Cliffs, N.J., 1967.

[MM92]   M. Marcus and H. Minc. *A Survey of Matrix Theory and Matrix Inequalities*. Dover, New York, 1992.

[MMK⁺95]   C. Monroe, D. M. Meekhof, B. E. King, W. M. Itano, and D. J. Wineland. Demonstration of a fundamental quantum logic gate. *Phys. Rev. Lett.*, 75:4714, 1995.

[MO79]   A. W. Marshall and I. Olkin. *Inequalities: Theory of Majorization and its Applications*. Academic Press, New York, 1979.

[MOL⁺99]   J. E. Mooij, T. P. Orlando, L. Levitov, L. Tian, C. H. van der Waal, and S. Lloyd. Josephson persistent-current qubit. *Science*, 285:1036–1039, 1999.

[Mor98]   T. Mor. No-cloning of orthogonal states in composite systems. *Phys. Rev. Lett.*, 80:3137–3140, 1998.

[Mos98]   M. Mosca. Quantum searching, counting and amplitude amplification by eigenvector analysis. In R. Freivalds, editor, *Proceedings of International Workshop on Randomized Algorithms*, pages 90–100, 1998.

[Mos99]   M. Mosca. *Quantum Computer Algorithms*. Ph.D. thesis, University of Oxford, 1999.

[MR95]   R. Motwani and P. Raghavan. *Randomized Algorithms*. Cambridge University Press, Cambridge, 1995.

[MS77]   F. J. MacWilliams and N. J. A. Sloane. *The Theory of Error-correcting Codes*. North-Holland, Amsterdam, 1977.

[MU88]   H. Maassen and J. H. B. Uffink. Generalized entropic uncertainty relations. *Phys. Rev. Lett.*, 60(12):1103–1106, 1988.

[MvOV96]   A. J. Menezes, P. C. van Oorschot, and S. A. Vanstone. *Handbook of Applied Cryptography*. CRC Press, 1996.

[MWKZ96]   K. Mattle, H. Weinfurter, P. G. Kwiat, and A. Zeilinger. Dense coding in experimental quantum communication. *Phys. Rev. Lett.*, 76(25):4656–4659, 1996.

[MZG96]   A. Muller, H. Zbinden, and N. Gisin. Quantum cryptography over 23 km in installed under-lake telecom fibre. *Europhys. Lett.*, 33:334–339, 1996.

[NC97]   M. A. Nielsen and C. M. Caves. Reversible quantum operations and their application to teleportation. *Phys. Rev. A*, 55(4):2547–2556, 1997.

[NCSB98] M. A. Nielsen, C. M. Caves, B. Schumacher, and H. Barnum. Information-theoretic approach to quantum error correction and reversible measurement. *Proc. R. Soc. London A*, 454(1969):277–304, 1998.

[Nie98] M. A. Nielsen. *Quantum Information Theory*. Ph.D. thesis, University of New Mexico, 1998.

[Nie99a] M. A. Nielsen. Conditions for a class of entanglement transformations. *Phys. Rev. Lett.*, 83(2):436–439, 1999.

[Nie99b] M. A. Nielsen. Probability distributions consistent with a mixed state. *arXive e-print quant-ph/9909020*, 1999.

[NKL98] M. A. Nielsen, E. Knill, and R. Laflamme. Complete quantum teleportation using nuclear magnetic resonance. *Nature*, 396(6706):52–55, 1998.

[NPT99] Y. Nakamura, Y. A. Pashkin, and J. S. Tsai. Coherent control of macroscopic quantum states in a single-cooper-pair box. *Nature*, 398:786–788, 1999.

[OP93] M. Ohya and D. Petz. *Quantum Entropy and Its Use*. Springer-Verlag, Berlin, 1993.

[Pai82] A. Pais. *Subtle is the Lord: The Science and the Life of Albert Einstein*. Oxford University Press, Oxford, 1982.

[Pai86] A. Pais. *Inward Bound: Of Matter and Forces in the Physical World*. Oxford University Press, Oxford, 1986.

[Pai91] A. Pais. *Niels Bohr's Times: In Physics, Philosophy, and Polity*. Oxford University Press, Oxford, 1991.

[Pap94] C. M. Papadimitriou. *Computational Complexity*. Addison-Wesley, Reading, Massachusetts, 1994.

[Pat92] R. Paturi. On the degree of polynomials that approximate symmetric Boolean functions (preliminary version). *Proc. 24th Ann. ACM Symp. on Theory of Computing (STOC '92)*, pages 468–474, 1992.

[PCZ97] J. F. Poyatos, J. I. Cirac, and P. Zoller. Complete characterization of a quantum process: the two-bit quantum gate. *Phys. Rev. Lett.*, 78(2):390–393, 1997.

[PD99] P. M. Platzman and M. I. Dykman. Quantum computing with electrons floating on liquid helium. *Science*, 284:1967, 1999.

[Pen89] R. Penrose. *The Emperor's New Mind*. Oxford University Press, Oxford, 1989.

[Per52] S. Perlis. *Theory of Matrices*. Addison-Wesley, Reading, Mass., 1952.

[Per88] A. Peres. How to differentiate between non-orthogonal states. *Phys. Lett. A*, 128:19, 1988.

[Per93] A. Peres. *Quantum Theory: Concepts and Methods*. Kluwer Academic, Dordrecht, 1993.

[Per95] A. Peres. Higher order schmidt decompositions. *Phys. Lett. A*, 202:16–17, 1995.

[Pet86] D. Petz. Quasi-entropies for finite quantum systems. *Rep. Math. Phys.*, 23(1):57–65, 1986.

[Phy92] Physics Today Editor. Quantum cryptography defies eavesdropping. *Physics Today*, page 21, November 1992.

[PK96] M. B. Plenio and P. L. Knight. Realistic lower bounds for the factorization time of large numbers on a quantum computer. *Phys. Rev. A*, 53:2986–2990, 1996.

[PK98] M. B. Plenio and P. L. Knight. The quantum-jump approach to dissipative dynamics in quantum optics. *Rev. Mod. Phys.*, 70(1):101–144, 1998.

[Pop75] R. P. Poplavskii. Thermodynamical models of information processing (in Russian). *Usp. Fiz. Nauk*, 115(3):465–501, 1975.

[PRB98] M. Pueschel, M. Roetteler, and T. Beth. Fast quantum Fourier transforms for a class of non-abelian groups. *arXive e-print quant-ph/9807064*, 1998.

[Pre97] J. Preskill. Fault-tolerant quantum computation. *arXive e-print quant-ph/9712048*, 1997.

[Pre98a] J. Preskill. Fault-tolerant quantum computation. In H.-K. Lo, T. Spiller, and S. Popescu, editors, *Quantum information and computation*. World Scientific, Singapore, 1998.

[Pre98b] J. Preskill. *Physics 229: Advanced Mathematical Methods of Physics — Quantum Computation and Information*. California Institute of Technology, 1998. *URL: http://www.theory.caltech.edu/people/preskill/ph229/*

[Pre98c] J. Preskill. Reliable quantum computers. *Proc. R. Soc. London A*, 454(1969):385–410, 1998.

[Rab80] M. O. Rabin. Probabilistic algorithm for testing primality. *J. Number Theory*, 12:128–138, 1980.

[Rah99] H. Z. Rahim. Richard Feynman and Bill Gates: an imaginary encounter. 1999. *URL: http://www.trnsoft.com/features/1rfbg.htm*

[Rai98] E. M. Rains. Quantum weight enumerators. *IEEE Trans. Inf. Theory*, 44(4):1388–1394, 1998.

[Rai99a]   E. M. Rains. Monotonicity of the quantum linear programming bound. *IEEE Trans. Inf. Theory*, 45(7):2489–2492, 1999.

[Rai99b]   E. M. Rains. Nonbinary quantum codes. *IEEE Trans. Inf. Theory*, 45(6):1827–1832, 1999.

[Rai99c]   E. M. Rains. Quantum shadow enumerators. *IEEE Trans. Inf. Theory*, 45(7):2361–2366, 1999.

[RB98]   M. Roetteler and T. Beth. Polynomial-time solution to the hidden subgroup problem for a class of non-abelian groups. *arXive e-print quant-ph/9812070*, 1998.

[Res81]   A. Ressler. *The Design of a Conservative Logic Computer and A Graphical Editor Simulator*. Master's thesis, Massachusetts Institute of Technology, 1981.

[RHSS97]   E. M. Rains, R. H. Hardin, P. W. Shor, and N. J. A. Sloane. Nonadditive quantum code. *Phys. Rev. Lett.*, 79(5):953–954, 1997.

[Roy96]   A. Royer. Reduced dynamics with initial correlations, and time-dependent environment and Hamiltonians. *Phys. Rev. Lett.*, 77(16):3272–3275, 1996.

[RR67]   D. W. Robinson and D. Ruelle. *Commun. Math. Phys.*, 5:288, 1967.

[RSA78]   R. L. Rivest, A. Shamir, and L. M. Adleman. A method of obtaining digital signatures and public-key cryptosystems. *Comm. ACM*, 21(2):120–126, 1978.

[Rus94]   M. B. Ruskai. Beyond strong subadditivity: improved bounds on the contraction of generalized relative entropy. *Rev. Math. Phys.*, 6(5A):1147–1161, 1994.

[RWvD84]   S. Ramo, J. R. Whinnery, and T. van Duzer. *Fields and waves in communication electronics*. Wiley, New York, 1984.

[RZBB94]   M. Reck, A. Zeilinger, H. J. Bernstein, and P. Bertani. Experimental realization of any discrete unitary operator. *Phys. Rev. Lett.*, 73(1):58–61, 1994.

[Sak95]   J. J. Sakurai. *Modern Quantum Mechanics*. Addison-Wesley, Reading, Mass., 1995.

[SC99]   R. Schack and C. M. Caves. Classical model for bulk-ensemble NMR quantum computation. *Phys. Rev. A*, 60(6):4354–4362, 1999.

[Sch06]   E. Schmidt. Zur theorie der linearen und nichtlineuren integralgleichungen. *Math. Annalen.*, 63:433–476, 1906.

[Sch36]   E. Schrödinger. Probability relations between separated systems. *Proc. Cambridge Philos. Soc.*, 32:446–452, 1936.

[Sch95]   B. Schumacher. Quantum coding. *Phys. Rev. A*, 51:2738–2747, 1995.

[Sch96a]   B. Schneier. *Applied Cryptography*. John Wiley and Sons, New York, 1996.

[Sch96b]   B. W. Schumacher. Sending entanglement through noisy quantum channels. *Phys. Rev. A*, 54:2614, 1996.

[Sha48]   C. E. Shannon. A mathematical theory of communication. *Bell System Tech. J.*, 27:379–423, 623–656, 1948.

[Sho94]   P. W. Shor. Algorithms for quantum computation: discrete logarithms and factoring. In *Proceedings, 35th Annual Symposium on Foundations of Computer Science*, IEEE Press, Los Alamitos, CA, 1994.

[Sho95]   P. Shor. Scheme for reducing decoherence in quantum computer memory. *Phys. Rev. A*, 52:2493, 1995.

[Sho96]   P. W. Shor. Fault-tolerant quantum computation. In *Proceedings, 37th Annual Symposium on Fundamentals of Computer Science*, pages 56–65, IEEE Press, Los Alamitos, CA, 1996.

[Sho97]   P. W. Shor. Polynomial-time algorithms for prime factorization and discrete logarithms on a quantum computer. *SIAM J. Comp.*, 26(5):1484–1509, 1997.

[Sim79]   B. Simon. *Trace Ideals and Their Applications*. Cambridge University Press, Cambridge, 1979.

[Sim94]   D. Simon. On the power of quantum computation. In *Proceedings, 35th Annual Symposium on Foundations of Computer Science*, pages 116–123, IEEE Press, Los Alamitos, CA, 1994.

[Sim97]   D. R. Simon. On the power of quantum computation. *SIAM J. Comput.*, 26(5):1474–1483, 1997.

[SL97]   P. W. Shor and R. Laflamme. Quantum analog of the MacWilliams identities for classical coding theory. *Phys. Rev. Lett.*, 78(8):1600–1602, 1997.

[SL98]   D. Shasha and C. Lazere. *Out of Their Minds: The Lives and Discoveries of 15 Great Computer Scientists*. Springer-Verlag, New York, 1998.

[Sle74]   D. Slepian, editor. *Keys Papers in the Development of Information Theory*. IEEE Press, New York, 1974.

[Sli96]   C. P. Slichter. *Principles of Magnetic Resonance*. Springer, Berlin, 1996.

[SN96]   B. W. Schumacher and M. A. Nielsen. Quantum data processing and error correction. *Phys. Rev. A*, 54(4):2629, 1996. *arXive e-print quant-ph/9604022*.

[SP00]   P. W. Shor and J. Preskill. Simple proof of security of the BB84 quantum key distribution protocol. *arXive e-print quant-ph/0003004*, 2000.

[SS76]   R. Solovay and V. Strassen. A fast Monte-Carlo test for primality. *SIAM J. Comput.*, 6:84–85, 1976.

[SS96]   P. W. Shor and J. A. Smolin. Quantum error-correcting codes need not completely reveal the error syndrome. *arXive e-print quant-ph/9604006*, 1996.

[SS99]   A. T. Sornborger and E. D. Stewart. Higher order methods for simulations on quantum computers. *Phys. Rev. A*, 60(3):1956–1965, 1999. *arXive e-print quant-ph/9903055*.

[ST91]   B. E. A. Saleh and M. C. Teich. *Fundamentals of Photonics*. Wiley, NY, 1991.

[Ste96a]   A. M. Steane. Error correcting codes in quantum theory. *Phys. Rev. Lett.*, 77:793, 1996.

[Ste96b]   A. M. Steane. Multiple particle interference and quantum error correction. *Proc. R. Soc. London A*, 452:2551–76, 1996.

[Ste97]   A. Steane. The ion-trap quantum information processor. *Appl. Phys. B – Lasers and Optics*, 64(6):623–642, 1997.

[Ste99]   A. M. Steane. Efficient fault-tolerant quantum computing. *Nature*, 399:124–126, May 1999.

[STH⁺99]   S. Somaroo, C. H. Tseng, T. F. Havel, R. Laflamme, and D. G. Cory. Quantum simulations on a quantum computer. *Phys. Rev. Lett.*, 82:5381–5384, 1999.

[Str76]   G. Strang. *Linear Algebra and Its Applications*. Academic Press, New York, 1976.

[SV99]   L. J. Schulman and U. Vazirani. Molecular scale heat engines and scalable quantum computation. *Proc. 31st Ann. ACM Symp. on Theory of Computing (STOC '99)*, pages 322–329, 1999.

[SW49]   C. E. Shannon and W. Weaver. *The Mathematical Theory of Communication*. University of Illinois Press, Urbana, 1949.

[SW93]   N. J. A. Sloane and A. D. Wyner, editors. *Claude Elwood Shannon: Collected Papers*. IEEE Press, New York, 1993.

[SW97]   B. Schumacher and M. D. Westmoreland. Sending classical information via noisy quantum channels. *Phys. Rev. A*, 56(1):131–138, 1997.

[SW98]   B. Schumacher and M. D. Westmoreland. Quantum privacy and quantum coherence. *Phys. Rev. Lett.*, 80(25):5695–5697, 1998.

[SWW96]   B. W. Schumacher, M. Westmoreland, and W. K. Wootters. Limitation on the amount of accessible information in a quantum channel. *Phys. Rev. Lett.*, 76:3453, 1996.

[Szi29]   L. Szilard. Uber die entropieverminderung in einen thermodynamischen system bei eingriffen intelligenter wesen. *Z. Phys.*, 53:840–856, 1929.

[TD98]   B. M. Terhal and D. P. DiVincenzo. The problem of equilibration and the computation of correlation functions on a quantum computer. *arXive e-print quant-ph/9810063*, 1998.

[THL⁺95]   Q. A. Turchette, C. J. Hood, W. Lange, H. Mabuchi, and H. J. Kimble. Measurement of conditional phase shifts for quantum logic. *Phys. Rev. Lett.*, 75:4710, 1995.

[Tro59]   H. F. Trotter. On the product of semigroups of operators. *Proc. Am. Math. Soc.*, 10:545–551, 1959.

[Tsi80]   B. S. Tsirelson. Quantum generalizations of Bell's inequality. *Lett. Math. Phys.*, 4:93, 1980.

[Tur36]   A. M. Turing. On computable numbers, with an application to the Entscheidungsproblem. *Proc. Lond. Math. Soc. 2 (reprinted in [Dav65])*, 42:230, 1936.

[Tur97]   Q. A. Turchette. *Quantum optics with single atoms and single photons*. Ph.D. thesis, California Institute of Technology, Pasadena, California, 1997.

[Uhl70]   A. Uhlmann. On the Shannon entropy and related functionals on convex sets. *Rep. Math. Phys.*, 1(2):147–159, 1970.

[Uhl71]   A. Uhlmann. Sätze über dichtematrizen. *Wiss. Z. Karl-Marx-Univ. Leipzig*, 20:633–637, 1971.

[Uhl72]   A. Uhlmann. Endlich-dimensionale dichtematrizen I. *Wiss. Z. Karl-Marx-Univ. Leipzig*, 21:421–452, 1972.

[Uhl73]   A. Uhlmann. Endlich-dimensionale dichtematrizen II. *Wiss. Z. Karl-Marx-Univ. Leipzig*, 22:139–177, 1973.

[Uhl76]   A. Uhlmann. The 'transition probability' in the state space of a *-algebra. *Rep. Math. Phys.*, 9:273–279, 1976.

[Uhl77]   A. Uhlmann. Relative entropy and the

Wigner-Yanase-Dyson-Lieb concavity in an interpolation theory. *Commun. Math. Phys.*, 54:21–32, 1977.

[Ume62] H. Umegaki. *Kōdai Math. Sem. Rep.*, 14:59–85, 1962.

[Vai94] L. Vaidman. Teleportation of quantum states. *Phys. Rev. A*, 49(2):1473–6, 1994.

[van98a] W. van Dam. Quantum oracle interrogation: getting all information for half the price. In *Proceedings of the 39th FOCS*, 1998. *arXive e-print quant-ph/9805006*.

[van98b] S. J. van Enk. No-cloning and superluminal signaling. *arXive e-print quant-ph/9803030*, 1998.

[Ved99] V. Vedral. Landauer's erasure, error correction and entanglement. *arXive e-print quant-ph/9903049*, 1999.

[Vid98] G. Vidal. Entanglement monotones. *arXive e-print quant-ph/9807077*, 1998.

[Vid99] G. Vidal. Entanglement of pure states for a single copy. *Phys. Rev. Lett.*, 83(5):1046–1049, 1999.

[von27] J. von Neumann. *Göttinger Nachrichten*, page 245, 1927.

[von56] J. von Neumann. Probabilistic logics and the synthesis of reliable organisms from unreliable components. In *Automata Studies*, pages 329–378, Princeton University Press, Princeton, NJ, 1956.

[von66] J. von Neumann. Fourth University of Illinois lecture. In A. W. Burks, editor, *Theory of Self-Reproducing Automata*, page 66, University of Illinois Press, Urbana, 1966.

[VP98] V. Vedral and M. B. Plenio. Entanglement measures and purification procedures. *Phys. Rev. A*, 57(3):1619–1633, 1998.

[VR89] K. Vogel and H. Risken. Determination of quasiprobability distributions in terms of probability distributions for the rotated quadrature phase. *Phys. Rev. A*, 40(5):2847–2849, 1989.

[VYSC99] L. M. K. Vandersypen, C. S. Yannoni, M. H. Sherwood, and I. L. Chuang. Realization of effective pure states for bulk quantum computation. *Phys. Rev. Lett.*, 83:3085–3088, 1999.

[VYW+99] R. Vrijen, E. Yablonovitch, K. Wang, H. W. Jiang, A. Balandin, V. Roychowdhury, T. Mor, and D. DiVincenzo. Electron spin resonance transistors for quantum computing in silicon-germanium het-

erostructures. *arXive e-print quant-ph/9905096*, 1999.

[War97] W. Warren. The usefulness of NMR quantum computing. *Science*, 277(5332):1688, 1997.

[Wat99] J. Watrous. **PSPACE** has 2-round quantum interactive proof systems. *arXive e-print cs/9901015*, 1999.

[WC67] S. Winograd and J. D. Cowan. *Reliable Computation in the Presence of Noise*. MIT Press, Cambridge, MA, 1967.

[Weh78] A. Wehrl. General properties of entropy. *Rev. Mod. Phys.*, 50:221, 1978.

[Wel88] D. J. A. Welsh. *Codes and Cryptography*. Oxford University Press, New York, 1988.

[Wie] S. Wiesner. Unpublished manuscript, circa 1969, appeared as [Wie83].

[Wie83] S. Wiesner. Conjugate coding. *SIGACT News*, 15:77, 1983.

[Wie96] S. Wiesner. Simulations of many-body quantum systems by a quantum computer. *arXive e-print quant-ph/9603028*, 1996.

[Wil91] D. Williams. *Probability with Martingales*. Cambridge University Press, Cambridge, 1991.

[Win98] E. Winfree. *Algorithmic Self-Assembly of DNA*. Ph.D. thesis, California Institute of Technology, Pasadena, California, 1998.

[WMI+98] D. J. Wineland, C. Monroe, W. M. Itano, D. Leibfried, B. E. King, and D. M. Meekhof. Experimental issues in coherent quantum-state manipulation of trapped atomic ions. *J. Res. Natl. Inst. Stand. Tech.*, 103:259, 1998.

[WS98] M. D. Westmoreland and B. Schumacher. Quantum entanglement and the nonexistence of superluminal signals. *arXive e-print quant-ph/9801014*, 1998.

[WY63] E. P. Wigner and M. M. Yanase. *Proc. Natl. Acad. Sci. (U.S.A.)*, 49:910–918, 1963.

[WY90] K. Watanabe and Y. Yamamoto. Limits on tradeoffs between third-order optical nonlinearity, absorption loss, and pulse duration in self-induced transparency and real excitation. *Phys. Rev. A*, 42(3):1699–702, 1990.

[WZ82] W. K. Wootters and W. H. Zurek. A single quantum cannot be cloned. *Nature*, 299:802–803, 1982.

[Yao93] A. C. Yao. Quantum circuit complexity. *Proc. of the 34th Ann. IEEE Symp. on*

Foundations of Computer Science, pages 352–361, 1993.

[YK95]   S. Younis and T. Knight. Non dissipative rail drivers for adiabatic circuits. In *Proceedings, Sixteenth Conference on Advanced Research in VLSI 1995*, pages 404–14, IEEE Computer Society Press, Los Alamitos, CA, 1995.

[YKI88]   Y. Yamamoto, M. Kitagawa, and K. Igeta. In *Proc. 3rd Asia-Pacific Phys. Conf.*, World Scientific, Singapore, 1988.

[YO93]   H. P. Yuen and M. Ozawa. Ultimate information carrying limit of quantum systems. *Physical Review Letters*, 70:363–366, 1993.

[YY99]   F. Yamaguchi and Y. Yamamoto. Crystal lattice quantum computer. *Appl. Phys. A*, pages 1–8, 1999.

[Zal98]   C. Zalka. Simulating quantum systems on a quantum computer. *Proc. R. Soc. London A*, 454(1969):313–322, 1998.

[Zal99]   C. Zalka. Grover's quantum searching algorithm is optimal. *Phys. Rev. A*, 60(4):2746–2751, 1999.

[Zan99]   P. Zanardi. Stabilizing quantum information. *arXive e-print quant-ph/9910016*, 1999.

[ZG97]   P. Zoller and C. W. Gardiner. Quantum noise in quantum optics: the stochastic Schrödinger equation. In S. Reynaud, E. Giacobino, and J. Zinn-Justin, editors, *Quantum Fluctuations: Les Houches Summer School LXIII*, Elsevier, Amsterdam, 1997.

[ZHSL99]   K. Zyczkowski, P. Horodecki, A. Sanpera, and M. Lewenstein. Volume of the set of separable states. *Phys. Rev. A*, 58(2):883–892, 1999.

[ZL96]   W. H. Zurek and R. Laflamme. Quantum logical operations on encoded qubits. *Phys. Rev. Lett.*, 77(22):4683–4686, 1996.

[ZLC00]   X. Zhou, D. W. Leung, and I. L. Chuang. Quantum logic gate constructions with one-bit "teleportation". *arXive e-print quant-ph/0002039*, 2000.

[ZR98]   P. Zanardi and M. Rasetti. Noiseless quantum codes. *Phys. Rev. Lett.*, 79(17):3306–3309, 1998.

[Zur89]   W. H. Zurek. Thermodynamic cost of computation, algorithmic complexity and the information metric. *Nature*, 341:119, 1989.

[Zur91]   W. H. Zurek. Decoherence and the transition from quantum to classical. *Phys. Today*, October 1991.

Index

Bold page numbers indicate the place where the concept is introduced, explained, or defined. Major theorems are listed together under 'theorem', and end-of-chapter problems are listed together under 'problem'.